# SELECTED TOPICS ON OPTICAL FIBER TECHNOLOGY

Edited by **Moh. Yasin,**
**Sulaiman W. Harun**
and **Hamzah Arof**

**Selected Topics on Optical Fiber Technology**
Edited by Moh. Yasin, Sulaiman W. Harun and Hamzah Arof

**Published by InTech**
Janeza Trdine 9, 51000 Rijeka, Croatia



**Publishing Process Manager** Martina Blecic

**Technical Editor** Teodora Smiljanic

**Cover Designer** InTech Design Team

**Edition** 2014

Additional hard copies can be obtained from orders@intechopen.com

Selected Topics on Optical Fiber Technology,
Edited by Moh. Yasin, Sulaiman W. Harun and Hamzah Arof
p. cm.
ISBN 978-953-51-0091-1

For EU product safety concerns:
IN TECH d.o.o., Prolaz Marije Krucifikse Kozulić 3, 51000 Rijeka, Croatia,
info@intechopen.com or visit our website at intechopen.com.

# Meet the editors

Dr Moh. Yasin received his BSc in 1990 from Airlangga University. He went on to receive MSc and PhD degrees from Gadjah Mada University in 1999 and 2010. His research interests include fiber optic communications and sensors. Currently, he is an associate professor at the Faculty of Science and Technology, Airlangga University, Indonesia.

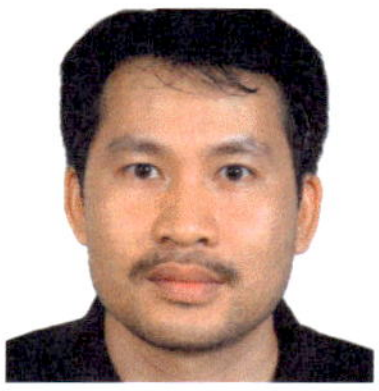

Dr Hamzah Arof received his BSc from Michigan State University, and PhD from the University of Wales. Both degrees were in electrical engineering. His current research interests include signal processing and photonics. Currently he is affiliated with the Department of Electrical Engineering, University of Malaya, Malaysia.

Dr Sulaiman Wadi Harun received the BE degree in Electrical and Electronics System Engineering from Nagaoka University of Technology, Japan in 1996. He received his MSc and PhD degrees in photonic technology from the University of Malaya in 2001 and 2004. He is actively working on optical amplifiers, fiber optic sensors and fiber lasers. Sulaiman Wadi Harun is also an Associate Fellow Member for both Academy of Sciences Malaysia and Malaysia Science Association. He is currently working as a full Professor in the Department of Electrical Engineering, University of Malaya, Malaysia.

# Contents

# Preface

Optical fibers, an important and promising material, have been the subject of intensive research and development due to their many scientific and practical applications. They are designed to guide light along its length by confining as much light as possible in its core. The interaction between the propagating light with the fiber material is the foundation of the development of various applications such as optical amplifiers, fiber lasers, sensors etc. The use and demand for optical fibers have grown in tandem with numerous new applications that have been continuously introduced by researchers and engineers. The development of optical fiber technology for communication networks, medical applications and other areas represents a unique confluence of the physics, electronics and mechanical engineering disciplines. This new book presents the latest researches in the field optical fiber technology, which consists of four sections.

Many current research efforts are focused on comprehending the theories, operating characteristics and technology of fiber laser and amplifier devices, which are mainly based on rare-earth-doped silica, as newfound technologies are expected to have profound impacts on a broad variety of communication and industrial applications. Section 1 presents the recent advances on fiber laser researches. The role of rare-earth optical fibers in fiber laser development is highlighted in Chapter 1. Chapters 2 and 3 describe the fabrication technique of rare-earth doped fibers using a modified chemical vapor deposition (MCVD) in conjunction with solution doping processes. Chapter 4 reviews on various efforts to comprehend and improve the spectroscopic properties of some rare-earth and thulium ions doped into silica. Chapter 5 demonstrates tunable fiber laser systems based on multimode interference effect. Chapter 6 describes microlens based fiber grating external cavity laser modules with low cost and good performance. Chapter 7 presents a review on the generation of few-cycle fs light pulses using gas-filled hollow-core fiber. Chapter 8 demonstrates a novel whispering–gallery–mode (WGM) fiber laser. The properties of the fiber laser, including energy threshold, produced length and polarization of lasing emission are discussed. Two important applications of the fiber lasers on optoelectronics, linearly polarized three-color lasing emission and single WGM lasing emission, are also demonstrated in this chapter.

Section 2 reviews the applications of optical fibers to medical, imaging, spectroscopy and measurement. Chapter 9 discusses the development of near infrared spectroscopy system based on optical fiber for skin moisture measurement. Chapter 10 investigates the presence of cellular optical fibers in the retina. Chapter 11 gives a technical review of near-infrared light and systems, which are applicable to optical brain imaging. Besides for medical purpose, optical fiber imaging can also be used in spectroscopy and other applications. Chapter 12 introduces a novel optical fiber application in the form of conductive and transparent optical fiber probe in multifunctional scanning tunneling microscropy. This kind of probe can be utilized for high-quality scanning tunneling microscope (STM) imaging, near-field excitation and detection of high-intensity STM-induced electroluminescence (STML). Chapter 13 describes the applications of coupling optical fibers to spectroscopic instrumentation for applications in chemical and biological threats, and explosives detection. Chapter 14 presents a research work on point diffraction interferometry (PDI) and phase shifting (PS) method. The development of PS/PDI having two point sources of optical fibers for absolute surface figure measurement of large apertured optics is demonstrated. In this work, attempts are made to eliminate the inevitable wavefront distortion via numerical reconstruction of the wave based on inverse problem. Highly precise measurements of spherical and spherical mirrors are realized.

Recently, several types of new fibers are developed for various applications. Section 3 reports on research advances on these fibers. Chapter 15 describes a newly developed "crystalline" plastic optical fiber with excellent heat resistance and dimensional stability. In this chapter, changes in the fine structure and lamella arrangement of the fibers formed by tetrafluoroethylene copolymers upon drawing are investigated by using wide-angle X-ray diffraction (WAXD) and small-angle X-ray scattering (SAXS) methods. This study is valuable as a fundamental research in the field of polymer physics. Chapter 16 demonstrates a single mode microstructured optical fiber (MOF) with improved bend performance. This chapter describes on research efforts on the finding and implementation of a few novel MOF designs that could effectively combine the large core dimensions and the expanded spectral operation range as compared to classical MOFs. It is obvious that new structures should be actualized by applying principles different from the basic concepts of the standard MOF technology. Chapter 17 thoroughly describes the fabrication of microfibers and its structures such as microfiber loop resonator (MLR), microfiber coil resonator (MCR) and microfiber knot resonator (MKR). A variety of applications of these structures will also be presented in this chapter. Chapter 18 discusses and compares how the shape of the modulated signal (e.g, exponential-wave, sine-wave, Gaussian, etc.) can improve the system performance when using both positive and negative dispersion fibers. With this method it is possible to improve each of the WDM system channels individually, offering a low-cost solution since it only involves changes in the transmitters and avoids replacing the fiber. This chapter also presents analytical and simulation results pertaining to the transmission of chirped optical signals in a dispersive fiber.

Current progress of optical fiber sensors is reviewed in section 4. Chapter 19 describes the mechanical properties of commercial optical fibers such as the elastic constant, the

Young modulus and the mean strain limit. The understanding of these properties can be useful for the design and modeling of optical sensors, and on the determination of aging performance of optical fiber deployed in telecommunication networks. Chapter 20 briefly summarizes recent studies of macroscopic fiber fuse propagation behaviour in silica fibers. The strong heat-induced absorption of silica glass and the highly confined supply of laser energy cause captured plasma to shift to the light source along the fiber leaving catastrophic damage behind it. Chapter 21 discusses the phenomenon of light production by a particle passing near the fiber, which is referred to as particle induced guided light (PIGL). Chapter 22 presents simulation results of nonlinear parametric studies of photon in a fiber Bragg grating (FBG). It is shown that it is plausible to use soliton for FBG writing and the soliton can be controlled by manipulating the parametric effects. Chapter 23 opens with some general considerations on the radiation–matter interaction, and continues with a review of irradiation effects on different types of optical fibers (silica optical fibers, plastic optical fibers, special optical fibers), effects which should be considered when developing radiation sensors. The bulk of this chapter is dedicated to the designs of promising intrinsic and extrinsic optical fiber sensors for radiation measurements. Chapter 24 presents nano-optic sensors, which uses optical fibers as a pointed probe to exploit optical and mechanical effects on the nano-scale, i.e., of very close proximity, even less than 5 nm.

**Dr Moh. Yasin**,
Department of Physics, Faculty of Science,
Airlangga Univ. Surabaya,
Indonesia

**Professor Sulaiman W. Harun**,
Deptartment of Electrical Engineering,
Faculty of Engineering, Univ. of Malaya,
Malaysia

**Dr Hamzah Arof**,
Deptartment of Electrical Engineering,
Faculty of Engineering, Univ. of Malaya,
Malaysia

# Part 1

## Fiber Lasers

# 1

# Rare-Earth Doped Optical Fibers

Efraín Mejía-Beltrán
*Centro de Investigaciones en Óptica*
*México*

## 1. Introduction

An optical fiber becomes active by doping its core with one or more atomic elements, usually (but not restricted to) rare-earths (RE's), more specifically, the *lanthanides* that occupy the atomic numbers 57 to 71 of the periodic table. As it will be mentioned later in more detail, they use three electrons in bonding to the condensed materials such as crystals and glasses to become triply ionized ions. Because they present absorption and emission bands from UV to NIR, the materials doped with these become very active in converting the properties of optical signals. Most optical fibers are made of crystal quartz ($SiO_2$) that is melted and cooled down such that stays "frozen" in its vitreous state. This disordered pattern of the constituents, Silicon and Oxygen, produce randomly distorted unit cells of the crystal (quartz) to become *silica*. Other important fiber materials with special properties have been discovered (and studied) during the last decades; among the vast variety, the zirconium-fluorides (also named fluorozirconates) have been of special importance because the RE's notably change their spectral properties. Among those changes are the broadening of the absorption and emission bands and much longer excited state lifetimes of up to some orders of magnitude compared to silica. In addition, their operation region covers and further exceeds the silica transparency band (~200 to 2100 nm) to be ~200-6000 nm (Lucas, 1989). Broader absorption bands allow the use of non wavelength-stable or even multi-line delivering pumps, usually provided by the cheapest semiconductor lasers; whereas with broader emission bands, it is possible to cover a wide range of emitting wavelengths. For illumination or broadband sources this characteristic becomes important (Digonnet, 2001). Also the optical fiber amplifiers (FA's) that amplify weak signals such as the channels of telecommunication systems increase their capacity thanks to this characteristic. Because a laser is an amplifier with a resonant cavity, it is possible to take advantage of this broad emission spectrum to generate several laser lines by designing the appropriate resonator or even, it is possible to insert a wavelength selecting device within the cavity to select the desired wavelength to be emitted. This topic is described in more detail in another chapter of this book. Longer lifetimes benefits efficiency in some fiber lasers (FL's) and FA's and also increase the probability for already excited ions to absorb another photon that re-excites them to a more energetic state from which, if the lifetime is also long a third photon may be absorbed, and so on. This re-excitation is called *excited state absorption* (ESA) and when two or more photons are absorbed to excite a higher energy level capable of producing or amplifying signals of shorter wavelengths, it is said that *upconversion* occurs. This phenomenon will be sufficiently discussed later in this chapter. For now it is enough to say that some ions inside some glasses may absorb IR signals and produce laser emissions and amplification in the VIS-UV region.

Apart from the first flash-lamp pumped FL's and FA's in which a spectral portion of the incoherent light converts into monochromatic-coherent light, FL's are strictly speaking *fiber laser converters* because they convert the coherent wavelength(s) of the laser pump into different (also coherent) ones. Because their optical-to-optical conversion efficiencies range from 5 to >95%, FL's are among the most efficient lasers.

In order to illustrate the advantages of typical fiber-based devices compared to their bulk counterparts, let us depart from the following example. In typical solid-state laser-pumped lasers such as the $Nd^{3+}$:YAG or Ti:Sapphire, the active length of the laser material is at most few centimeters. Just to give an idea of this, let us suppose that we have a single- mode (SM) 2mm-diameter collimated gaussian beam from a 808-nm laser to pump a $Nd^{3+}$-doped glass laser. If we want to produce a 6-μm beam waist ($2w_0$) in the middle of the glass, an 11.7 mm focal length lens would be necessary as illustrated in Figure 1.

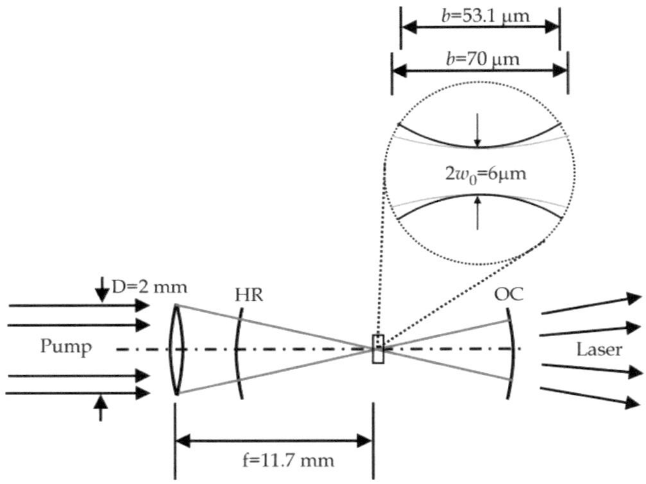


Fig. 1. Typical scheme of a bulk laser.

The region in which the beam is practically collimated is called the confocal parameter that gives $b$~70 μm (Siegman, 1986). An active cavity for the laser signal desired, 1064 nm for example, may be formed by using two mirrors; a high reflector (HR) and an output coupler (OC). The curvatures of the mirrors, the distance between them and the position of the active material have to be such that the lowest order gaussian mode of the 1064-nm signal is perfectly confined inside the cavity. The confocal parameter for this signal with the same beam waist is different than the one for the pump and gives $b$~53.1 μm meaning that the spatial overlap between both beams is partial. Larger active volumes are achieved by increasing the beam waists; however, this implies to use very large focal lengths for the lenses. For example for a ~100-μm waist, f≈19.4 cm and $b$≈19.4 mm; with these values the system is reaching the limit of being compact. The focused pumping beam in larger systems might be "bended" by using mirrors but a new issue appears because the larger active volume produces more heat that needs to be dissipated and a cooling system is required. Apart from these differences, larger beam waists imply higher pump power to reach laser operation because energy density within the active volume decreases. Then, larger systems are appropriated for non-compact high-power lasers. In our 6-μm waist example, low operating threshold in the order of tens of milliwatts, no cooling requirements for power delivery in the order of ~100 mW and low power delivery are possible. Even supposing that the concentration of dopants is the right one, pump absorption by such a short sample would be very inefficient because absorption also depends on length. In order to assure the oscillation of the cavity with such a tiny active region, the reflectivities of both mirrors would be close to 100 % because lower reflectivities means high cavity loss to overcome by the cavity gain. Let us say that the high reflector (HR) is 100 % and the output coupler (OC) has 99%. In this way, it will only deliver 1% of the signal generated inside the cavity. In addition to that, the glass surfaces may need anti-reflection coating to reduce loss from the air-glass interfaces. This would be an inefficient system delivering <1% of the total signal generated by the cavity, i.e. less than 1mW for 100 mW pump power. Any attempt to extract more power, let's say replacing the OC to 95 % to produce ≤5 mW, would most possibly lead to: a) no oscillation at all, b) oscillation with very unstable behavior or c) higher pump power threshold for oscillation. Let us now introduce some changes to our system. Suppose the replacement of the glass bulk by a 6-μm core $Nd^{3+}$-doped optical fiber as in Figure 2, the mirrors attached (or deposited) at the fiber flat ends, and the pump signal focused at one end (the HR transparent for this signal). Now we have a sufficiently large material that more efficiently absorbs the pump and has a larger gain length (usually from tens of cm to hundreds of meters) with a total overlap of the beams. In such high-gain cavity, it is possible to change the OC to extract more than 50 % of the optical power. These changes briefly describe the upgrade from conventional bulk lasers to RE-doped fiber lasers (REDFL's). Coiled optical fibers mean much larger fiber-cavities occupying a modest space. Apart from being very stable, FL's that deliver up to some watts, easily connect to a fiber-link and most have an excellent, usually circular, beam quality. The enlarged active volume (the fiber core) is in the center of a fine glass strand (the optical fiber) and because the ratio of the fiber surface to the active volume is immense the heath produced in the active volume easily dissipates through the large surface. This makes these systems not to require cooling systems when delivering up to some watts. These unique characteristics make them among the best candidates for the development of new laser sources. In fact, the limitations of laser diodes, such as low power and poor beam quality were overcome by the invention of the double-cladd optical fiber (Po et al., 1989). On that guiding structure, the low-quality pump-beam provided by an array of laser diodes couples into a very large core (up to 125 microns and sometimes non-circular) and as it propagates, it pumps the single-mode RE-doped core (usually at the center). The single-mode signals emitted by such multimode-excited core preferentially amplify along its axis. In this way, the low quality beam transforms into a high-quality one.

Most recent applications of RE-doped fibers (REDF's) include temperature sensors and optically-controlled fiber attenuators (OCFA's). Concerning temperature sensors, a REDF is optically pumped to excite certain transition whose upper energy level has a close lower-energy one. At low temperatures, the atomic vibrations of the glass weakly (non-radiatively) de-excite ions from the upper to the lower-energy one. At room temperature the ions from both levels directly decay to a much lower energy level, the ground state for example; in doing so they release the corresponding photons that form two bands whose peak wavelengths are very close. A temperature increase increases the non-radiative vibration-induced transitions that change the shape of the optical spectrum emitted; the emitted higher-energy photons decrease while the lower-energy ones increase. The intensity ratio of these bands becomes proportional to temperature values (Castrellón-Uribe & García-Torales, 2010). Advantages of this type of sensor include non-electrical and remote operation because a large optical fiber guides the pump signal to the transducer (the doped fiber) and also collects the signals to be processed. OCFA's consist on guiding a non ground-state resonant-signal. This probe signal becomes absorbed by co-guiding a control signal that excites the ions from the ground state to a excited state that activates the absorption of the probe (Mejia & Pinto, 2009). This device will be detailed later in the chapter.

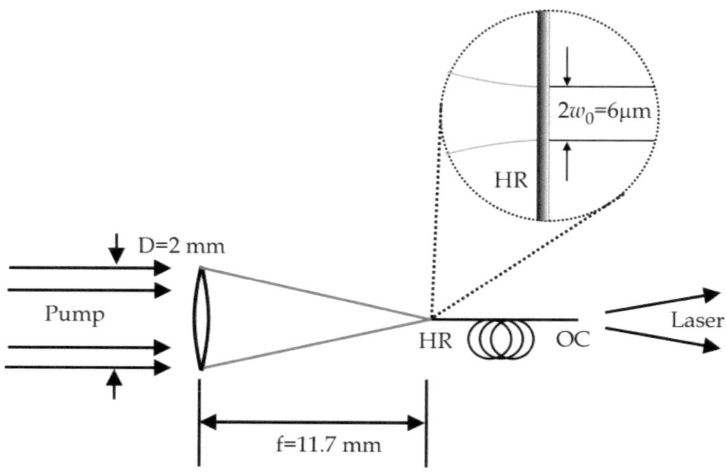


Fig. 2. Fiber laser.

## 2. Principles of operation

In this section, the electronic configuration of the rare earths becomes the natural departing point that allows understanding the light-matter interactions. Because the optical fibers are usually made of glass, these interactions change slightly depending of the type of glass they are immersed in. The basic pumping schemes that exist to produce light amplification are described and qualitatively compared, being described more quantitatively in the next section. Also an important phenomenon called upconversion in which the production of high energy photons are produced by the absorption of two or more lower–energy ones is described and its special qualities highlighted. The most general and basic formulas that are common to all the fiber devices covered in the chapter are obtained here.

### 2.1 Triply ionized rare-earths as active centers

As mentioned before, the lanthanides have atomic numbers from 57(La) to 71(Lu). When immersed in glass, three of their electrons are used in bonding to the glass-molecules, two from the s-type orbital of the outer (n=6) shell, and one from an f-type orbital of an inner shell (n=4). The absence of those negative charges, unbalances the atomic charge to 3+ because the number of protons in the nucleus do not vary. Thus, it is said that the RE atoms become triply ionized when immersed in glass. This ionization changes their electron configuration to be equivalent to the elements in the periodic table that are three places below concerning atomic number (Z). For example, Lanthanum (La) that has 57 electrons (Z=57) in its neutral state, becomes $La^{3+}$ with the electron configuration of Xe (Z=54): $1s^22s^22p^63s^23p^64s^23d^{10}4p^65s^24d^{10}5p^6$. In words, this electron configuration may be described as: "two electrons are in the first shell occupying the simplest type of orbital ($1s^2$); eight in the second shell, two occupying the s-type orbital or sub-shell ($2s^2$) and the rest in the other sub-shell that is composed of three, more complex in shape and hence more energetic p-type orbitals that are orthogonally-oriented and allocate two electrons each ($2p^6$); and so on..." From all, only $La^{3+}$ and $Lu^{3+}$ have all the sub-shells completely filled (or closed); their electron configurations are (Xe)$4f^{0}$ and (Xe)$4f^{14}$ respectively; this makes them optically inactive because there are no vacant sites (quantum states) in the orbitals within the *f* sub-shell that might temporarily allocate optically-excited electrons from other orbitals i.e. it is not possible to temporarily change the electron arrangement. In contrast, all others from $Ce^{3+}$[(Xe)$f^{1}$] to $Yb^{3+}$[(Xe)$f^{13}$] have quantum states "available" within the non-closed *f* sub-shell. On these, the non-occupied *f*-type orbitals may be temporarily occupied by optically-excited electrons coming from lower energy orbitals. In short, energy changes of the ions are produced by electronic re-arrangements within the 4*f* sub-shell. These transitions are relatively unaffected by external perturbations such as electro-magnetic fields and vibrations from the host material because the most external 5s and 5p closed sub-shells produce a *shielding effect*. According to Pauling (Pauling, 1988) the atoms with higher Z have their inner shells closer to the nucleus because of the increase of charge (more protons and electrons). Since the 4*f* sub-shell is physically closer to the nucleus than the 5s and 5p, they become more compact and thus more protected as Z increases. Then, the 4*f* electronic transitions of $Ce^{3+}$ are more influenced by the glass perturbations than those of the one with the highest atomic number, $Yb^{3+}$. Without excitation, a RE ion stays in its more stable electron arrangement called the ground state (GS) that has the energy that corresponds to the vector superposition of all the quantum states of the electrons (Miniscalco, 2001). Changes in the electron arrangement means changes of energy of the ions, then, the number of energy levels that one ion can take when excited depends on the number of electron sites available. For example, $Yb^{3+}$ that has only one quantum state available can be in one of two possible energies, the GS or the excited state (ES). Before describing more complicated sets of levels that correspond to the rest of RE's let us briefly describe the origin of the energy intrinsic to an atom.

As established by quantum mechanics, each electron has a unique set of quantum numbers *n*, *l*, *m* and *s* that quantify its energy and position within an atom. The first one describes the radial region (number of shell) in which it stays most of the time regardless of its orbital movement. The second quantum number indicates the number of units of angular momentum that is proportional to the magnetic field produced by its movement and as such, it depends on the type of orbital (also called sub-shell); the s-type corresponds to $l$=0 because its shape is simply spherical, the *p*-type corresponds to the lowest angular momentum ($l$=1) because it has a slightly more complex shape; with even more "exotic" shapes are *d* and *f* with $l$=2 and $l$=3 respectively. The *s*-type sub-shell has a pair of electrons occupying a single orbital; the *p*-type has three pairs occupying three orbitals; the *d* five, and the *f* seven. Within a sub-shell, the orbitals have in general the same shape ($l$) but differ in spatial orientation ($m$). Then the vector of angular momentum composed of magnitude ($l$) and direction ($m$) of the translational movement of the electron is complete. The fourth quantum number ($s$) refers to the vector that describes the magnetic field produced by the

electron's rotation on its axis and it is called the *spin* number. Then, the two electrons that occupy a specific orbital rotate in opposite directions (Pauling, 1988).
As mentioned before, without excitation a $RE^{3+}$ keeps its most stable energy, the GS which is given by the natural arrangement of the electrons in the 4*f* orbitals. Any re-arrangements of these electrons by excitation cause discrete energy changes. The orbital angular momentum vectors of all the individual 4*f*-electrons may be added to form a resultant vector (**L**) and in the same manner, the vector addition of their spins gives **S**, being the total angular momentum **J=L+S**; this is called the Russell-Saunders or spin-orbital coupling. Because every electron arrangement has its own **JLS** set of values, the possible energy values are labeled accordingly as $^{2S+1}L_J$ (Miniscalco, 2001). The possible values of **L** are given by the letters S, P, D, F, G, H, I, K..., that correspond respectively to 0, 1, 2, 3, 4, 5, 6, 7.... In this way, the labeling of the energy levels commonly used in the literature make sense because the superscript on the left gives information of the spins, the letter that corresponds to **L** gives information of the orbitals that are occupied and **J** is a combination of both.

### 2.2 Optical properties of RE-doped glasses

In spite of the shielding of the 4*f* transitions, when the RE ions are introduced as dopants in condensed materials such as crystals or glasses, weak interactions with the electrostatic field of the atomic arrangement take place and as a consequence each **JLS** level split into discrete sub-levels (called multiplet) because of the weak electrostatic interactions with the atoms of the material. This is called the Stark effect and it is so weak that the sub-levels are spaced between 10 and 100 $cm^{-1}$. The strength of the effect depends on the type of host material, and in most materials their broadening produces overlap due to the material vibration (temperature). Then, except for very low (close to cryogenic) temperatures, the net effect is a band creation whose width depends on the host material. Because in crystals all the atoms of the network are perfectly ordered, all the RE ions are affected by identical fields and it is said that they are *homogeneously broadened*. Glasses by contrary have site-to-site field variations because their atoms are not as ordered as in crystals and hence each ion has its own multiplet. This is called *inhomogeneous broadening* and even in a small sample the overlap of all the multiplets creates bands whose widths depend on the type of glass. The optical properties of the RE's immersed into two very different type of glasses are described right away.
In general, most optical fiber glasses used as hosts for RE's have optical properties between the silica and fluorozirconates from which the most common is the ZBLAN that owe its name to its constituents $ZrF_4$, $BaF_2$, $LaF_3$, $AlF_3$ and NaF. Each glass responds different to temperature that manifests as vibrations because of its molecular composition. For example, the superposition of all the possible vibrational energies (phonons) for silica form a continuum that covers a band of 40 THz ($\lambda_{phonon} \geq \sim 7.5$ μm, $\Delta E \leq 1300$ $cm^{-1}$) with the strongest mode overlapping at ~7.5-15.5 THz that corresponds to 19.35 and 40 μm ($250 \leq \Delta E \leq 520$ $cm^{-1}$), respectively (Agrawal, 1989). By contrast, a fluoride-based glass, ZBLAN for example, presents a much narrower band with mode overlapping at the edge 15 THz that corresponds to $\lambda_{phonon}$ ~20 μm ($\Delta E \approx 580$ $cm^{-1}$) (Luu-Gen & Chen-Ke, 1996; Quin et al. 1997). Then, all the energy levels separated less than this GAP, 1300 $cm^{-1}$ for silica and 580 $cm^{-1}$ for ZBLAN, are thermally connected and the higher instantaneously feeds population into the lower. Because the phonon spectrum of ZBLAN is narrower, two levels thermally connected in Silica, may not be alike in ZBLAN; then, more radiative transitions are possible in ZBLAN.

## 2.3 Three and four-level pumping schemes for light amplification

The most important pumping schemes for fiber amplifiers and lasers depend on the energy level arrangement of the ions in the glass. In general, they can be classified as four or three-level systems (Fig. 3). In real systems there may be many more levels involved in, but they can be simplified to either of these as follows. Fig. 3(a) shows the four-level scheme. Without any excitation, all the ions are in the ground state $E_1$ with a total density $N_1$. When exciting with the wavelength that corresponds to the energy difference between $E_1$ and $E_4$, part of the ions populate $E_4$ from where they decay very rapidly down to $E_3$ by releasing phonons because $E_4$-$E_3$ lies within the energy vibrations of the glass. Hence, the level $E_4$ may be considered as practically empty because it does not retain ions, they just pass through it to populate $E_3$ (density $N_3$) where they tend to accumulate because this level is not vibration-connected to $E_2$ and hence the only way to decay is by an emission of the corresponding $E_3$-$E_2$ photons. The ions stay for a short time in $E_3$ which produces energy accumulation. This level is called the *metastable level* and typical lifetime values are from tens of microseconds to some milliseconds and depend on the type of glass.

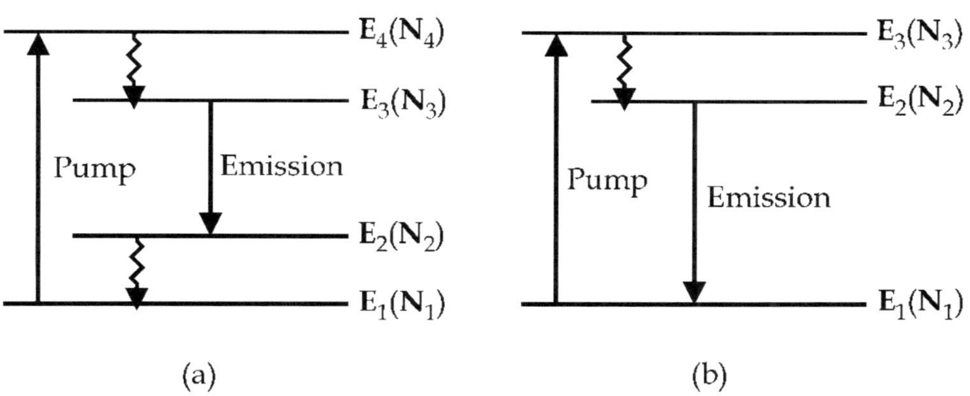


Fig. 3. (a) Four-level pumping scheme (b) Three-level pumping scheme.

Certain level of a RE may exhibit a lifetime more than one order of magnitude in ZBLAN (or other low-energy phonon glass) than its lifetime in silica for example. As the ions make the transition to $E_2$, spontaneously emit the corresponding photons and from this level that is also thermally connected to the ground state they instantaneously decay to the ground state by providing energy vibrations to the glass (phonons). Thus, $E_2$ does not accumulate population ($N_2 \approx 0$) similar to $E_4$ ($N_4 \approx 0$) and as a consequence it is possible to obtain population inversion ($N_3 >> N_2$). Under this condition, the spontaneously-emitted photons of energy $E_3$-$E_2$ propagate mostly through excited ions on $E_3$ and because of resonance with $E_3$-$E_2$, they become "negatively absorbed" by stimulating ions from $E_3$ to make transitions to the practically empty $E_2$. Negative absorption means stimulated emission in which one photon generates an identical one and hence optical amplification takes place because these two produce four, and so on. The signal produced in this way and that amplifies through a long material such as an optical fiber is called amplified spontaneous emission (ASE). Also an external signal of energy $E_3$-$E_2$ becomes amplified as it propagates through the fiber, this is the principle of the optical amplifiers. Also, two cavity mirrors may be placed such that they "see" each other through the fiber and reflect the $E_3$-$E_2$ signal that as it goes and returns becomes amplified, the signal that could be extracted from such cavity is the laser signal.

Apart from being easy to achieve population inversion, if part of active material (doped fiber in this case) is not excited, no signal re-absorption occurs because the $E_2$-$E_3$ absorbing transition is inactive. Then, unlike three level systems, the optical fiber can be longer than necessary without inducing losses other than those produced by the glass (usually negligible).

Fig. 3(b) shows the three-level scheme. Here, the pumping transition is $E_1$-$E_3$ and fast nonradiative relaxations accumulate the population on $E_2$. Observe that the active transition here is $E_2$-$E_1$ and then, if part of fiber is not excited, signal re-absorption occurs because the high density of ions in the ground state ($E_1$) excite to $E_2$. Then, unlike four level systems, three level systems present higher pump threshold because not only the entire fiber should be excited but all the fiber should be sufficiently excited to present population inversion; otherwise, considerable loss occurs and reaching the pump threshold for lasing becomes harder. Then, in these systems, there exists an optimal fiber length for each pump power level (as studied in the next section) whereas in the four-level case, the fiber may exceed the required length. One important advantage of three-level is: less fiber heating because there is only one step of non-radiative relaxation that implies an improvement of the conversion efficiency.

### 2.4 Population dynamics of a three-level system

Several amplifiers and fiber lasers that have been developed are modeled as three-level systems. Among these are the transition $^2F_{5/2}$-$^2F_{7/2}$ of $Yb^{3+}$, the $^4I_{13/2}$-$^4I_{15/2}$ of $Er^{3+}$, the $^3H_4$-$^3H_6$ of $Tm^{3+}$ and the $^5I_6$-$^5I_8$ of $Ho^{3+}$. All these pumping schemes have in common that the pump level is composed of at least two very close bands as shown in Figure 4(a) for $Ho^{3+}$ in ZBLAN glass. Figure 4(b) shows an equivalent energy level diagram that represents such schemes.

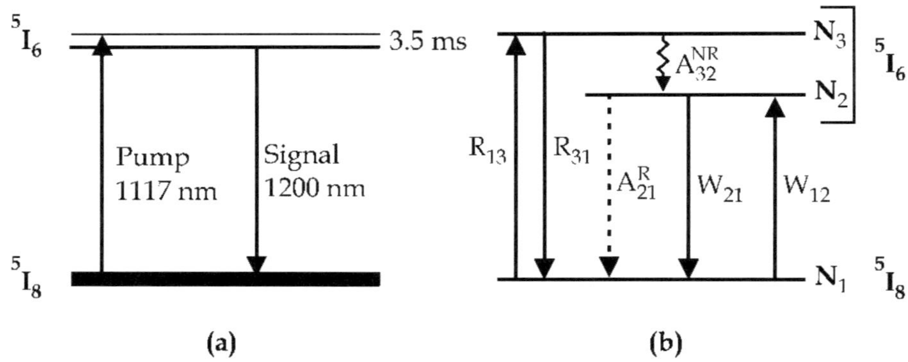


Fig. 4. (a) Partial energy level diagram of $Ho^{3+}$ in ZBLAN glass (b) Three-level equivalent scheme.

Based on the equivalent scheme, $R_{13}$ and $R_{31}$ are respectively the rates of excitation and de-excitation of the pump level, their formulae are

$$R_{13} = \frac{I_p \sigma_a(\lambda_p)}{h\nu_p} \tag{1}$$

$$R_{31} = \frac{I_p \sigma_e(\lambda_p)}{h\nu_p} \tag{2}$$

$W_{12}$ and $W_{21}$ are the ones corresponding to the signal

$$W_{12} = \frac{I_s \sigma_a(\lambda_s)}{h\nu_s} \tag{3}$$

$$W_{21} = \frac{I_s \sigma_e(\lambda_s)}{h\nu_s} \tag{4}$$

$$A_{21} = \frac{1}{\tau_{21}} \tag{5}$$

And the latter refers to the radiative ratio where $\tau_{21}$ is the $^5I_6$ level lifetime.
With pump and generated signal varying, the ions dynamically distribute on the energy levels as [From Fig. 4(a)]

$$\frac{dN_1}{dt} = -N_1(W_{12} + R_{13}) + N_2(A_{21}^R + W_{21}) + N_3R_{31} \tag{6}$$

$$\frac{dN_2}{dt} = -N_2(A_{21}^R + W_{21}) + N_1W_{12} + N_3A_{32}^{NR} \tag{7}$$

$$\frac{dN_3}{dt} = -N_3A_{32}^{NR} + N_1R_{13} \tag{8}$$

$$N_t = N_1 + N_2 + N_3 \tag{9}$$

The last one refers to the energy conservation law, *NR* to non-radiative and $N_t$ is the RE concentration. Solving this system of equations by supposing CW signals (i.e. the *dN's/dt=0*) one may obtain the population densities at each level. Another equation is added for a four-level system, and so on. However, in the system treated here, simplifications may be realized. With the arguments of section 2.3 and the fact that in the RE's mentioned the absorption and emission bands overlap as shown for $Ho^{3+}$ in Figure 5(a), the system may be reduced to a two-level one with $N_3 \approx 0$ (very high $A_{32}^{NR}$) and including the *R's* within the *W's* with the right changes [see Fig. 5(b)]. Then Equations (3) and (4) change to

$$W_{12} = \frac{I_p \sigma_a(\lambda_p)}{h\nu_p} + \frac{I_s \sigma_a(\lambda_s)}{h\nu_s} \tag{10}$$

$$W_{21} = \frac{I_s \sigma_e(\lambda_s)}{h\nu_s} + \frac{I_p \sigma_e(\lambda_p)}{h\nu_p} \tag{11}$$

And hence the ions dynamically distribute on the energy levels as

$$\frac{dN_1}{dt} = -N_1W_{12} + N_2(A_{21}^R + W_{21}) \tag{12}$$

$$\frac{dN_2}{dt} = -N_2(A_{21}^R + W_{21}) + N_1W_{12} \tag{13}$$

That under CW operation become

$$N_1W_{12} = N_2(A_{21}^R + W_{21}) \tag{14}$$

$$N_2(A_{21}^R + W_{21}) = N_1W_{12} \tag{15}$$

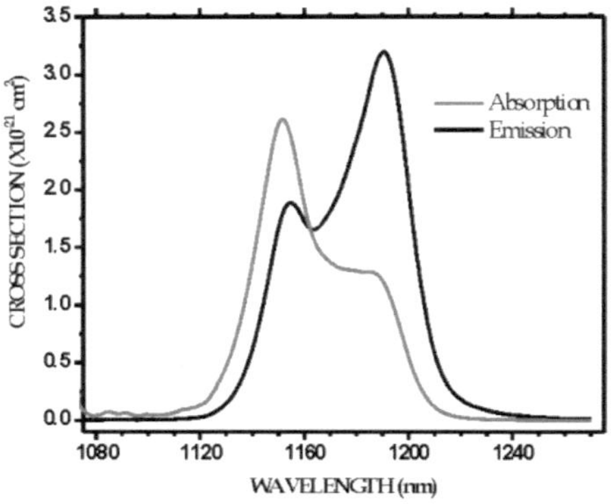


(a)

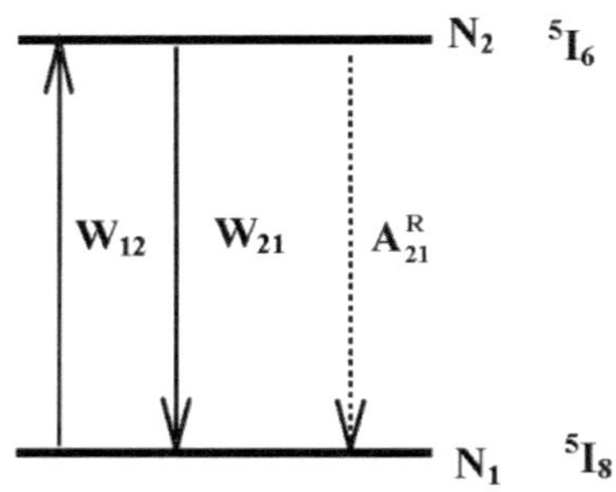


(b)

Fig. 5. (a) Measured spectra of $Ho^{3+}$ in ZBLAN glass (b) Two-level simplified scheme.

With the energy conservation law

$$N_t = N_1 + N_2 \tag{16}$$

From these equations and equation (5) one obtains

$$N_1 = N_t \frac{1+W_{21}\tau}{1+W_{21}\tau+W_{12}\tau} \tag{17}$$

$$N_2 = N_t \frac{W_{12}\tau}{1+W_{21}\tau+W_{12}\tau} \tag{18}$$

These are the population densities at each energy level of our simplified system.

### 2.5 Propagation equations for pump and signal

Let us suppose that a REDF operating as a quasi two-level system as mentioned in section 2.4 guides the pump beam with power $P_p$. The pump variation when traveling from point to point along the fiber is

$$\frac{dP_p}{dz} = \left[-N_1\sigma_a(\lambda_p) + N_2\sigma_e(\lambda_p)\right]P_p \tag{19}$$

Where

$$g_p = -N_1\sigma_a(\lambda_p) + N_2\sigma_e(\lambda_p) \tag{20}$$

Is the *gain coefficient* produced by the dopants that in most conditions is dominated by the first term being a negative gain. In the same manner, it is possible to establish that for the propagating signal

$$\frac{dP_s}{dz} = [N_2\sigma_e(\lambda_s) - N_1\sigma_a(\lambda_s)]P_s \tag{21}$$

Where

$$g_s = N_2\sigma_e(\lambda_s) - N_1\sigma_a(\lambda_s) \tag{22}$$

Refers also to the gain coefficient.

When the fiber is not pumped and a weak signal propagates, $N_2 \approx 0$ and hence all the population is at the ground state ($N_1 = N_t$), $g_p \approx -N_1\sigma_a(\lambda_p)$ and then the "gain coefficient" becomes the *small signal loss coefficient* as

$$\alpha_0 = N_t\sigma_a(\lambda) \tag{23}$$

The loss coefficient is a function of the pump power (or intensity) as follows (Siegman, 1986)

$$\alpha(P_p) = \frac{\alpha_0}{1+\frac{P_p}{P_p^{sat}}} \tag{24}$$

And for transitions in which there is not spectra overlap $P_p^{sat} = \frac{h\nu_p a}{\sigma_a \tau}$ is the pump power that makes possible to have $\alpha = \frac{\alpha_0}{2}$ and obviously it is called the *saturation pump power*; here $a$ is the cross section area of the fiber core. Hence, a definition for the *saturation intensity* is $I^{sat} = h\nu/\sigma\tau$.

Establishing a *small signal gain coefficient* is more complicated because it depends on the population distribution. It is better to say that under population inversion conditions a weak signal that do not notably redistributes the population experiences a small signal gain; whereas more powerful signals that redistribute population tend to experience less gain. Then

$$g(P_s) = \frac{g_0}{1+\frac{P_s}{P_s^{sat}}} \tag{25}$$

With $P_s^{sat} = \frac{h\nu_s a}{\sigma_e \tau}$ as the *saturation signal power*. For quasi two-level systems (Desurvire, 1994) these saturation signal equations are

$$P_p^{sat} = \frac{h\nu_p a}{[\sigma_a(\lambda_p)+\sigma_e(\lambda_p)]\tau} \tag{26}$$

$$P_s^{sat} = \frac{h\nu_s a}{[\sigma_a(\lambda_s)+\sigma_e(\lambda_s)]\tau} \tag{27}$$

The ratio emission-to-absorption cross sections at $\lambda_s$ or $\lambda_p$ may be defined as

$$\eta_i = \frac{\sigma_e(\lambda_i)}{\sigma_a(\lambda_i)} \qquad i = s, p \tag{28}$$

That introduced in (26) and (27) makes them

$$P_p^{sat} = \frac{h\nu_p a}{[1+\eta_p]\sigma_a(\lambda_p)\tau} = \frac{\eta_p h\nu_p a}{[1+\eta_p]\sigma_e(\lambda_p)\tau} \tag{29}$$

$$P_s^{sat} = \frac{h\nu_s a}{[1+\eta_s]\sigma_a(\lambda_s)\tau} = \frac{\eta_s h\nu_s a}{[1+\eta_s]\sigma_e(\lambda_s)\tau} \tag{30}$$

Because $P = Ia$ and multiplying (10) and (11) by $\tau$, they become

$$W_{12}\tau = \frac{P_p\sigma_a(\lambda_p)\tau}{h\nu_p a} + \frac{P_s\sigma_a(\lambda_s)\tau}{h\nu_s a} \tag{31}$$

$$W_{21}\tau = \frac{P_s\sigma_e(\lambda_s)\tau}{h\nu_s a} + \frac{P_p\sigma_e(\lambda_p)\tau}{h\nu_p a} \tag{32}$$

Solving (29) and (30) for $\sigma\tau/h\nu a$ and normalizing signal powers as

$$p = \frac{P_s}{P_s^{sat}} \text{ and } q = \frac{P_p}{P_p^{sat}} \tag{33}$$

(31) and (32) in a normalized version are

$$W_{12}\tau = \frac{1}{[1+\eta_s]}p + \frac{1}{[1+\eta_p]}q \tag{34}$$

$$W_{21}\tau = \frac{\eta_s}{[1+\eta_s]}p + \frac{\eta_p}{[1+\eta_p]}q \tag{35}$$

Also using (28), the propagation equations (19) and (21) may be expressed as

$$\frac{dP_p}{dz} = \sigma_a(\lambda_p)[\eta_p N_2 - N_1]P_p \tag{36}$$

$$\frac{dP_s}{dz} = \sigma_a(\lambda_s)[\eta_s N_2 - N_1]P_s \tag{37}$$

Where the populations $N_1$ and $N_2$ are now expressed as [introducing (34) and (35) in (17) and (18)]:

$$N_1 = N_t\frac{1+\frac{1}{1+\eta_s}p+\frac{1}{1+\eta_p}q}{1+p+q} \tag{38}$$

$$N_2 = N_t\frac{\frac{1}{1+\eta_s}p+\frac{1}{1+\eta_p}q}{1+p+q} \tag{39}$$

That introduced in the propagation equations give

$$\frac{dP_p}{dz} = -\sigma_a(\lambda_p)N_t\frac{1+\frac{\eta_s-\eta_p}{1+\eta_s}p}{1+p+q}P_p \tag{40}$$

$$\frac{dP_s}{dz} = \sigma_a(\lambda_s)N_t\frac{\frac{\eta_s-\eta_p}{1+\eta_p}q-1}{1+p+q}P_s \tag{41}$$

Using the definition for the small signal loss [equation (23)], normalizing to the saturation powers, defining $U' = \frac{\eta_s-\eta_p}{1+\eta_s}$ and $U = \frac{\eta_s-\eta_p}{1+\eta_p}$; the propagation equations reduce to

$$\frac{dq}{dz} = -\alpha_p\frac{U'p+1}{1+p+q}q \tag{42}$$

$$\frac{dp}{dz} = \alpha_s\frac{Uq-1}{1+p+q}p \tag{43}$$

## 3. Fiber lasers

In 1960, while working for the Huges Research Laboratory in California, T. H. Maiman reported on August the first laser by irradiating (pumping) a crystal (ruby)-cavity with a powerful flash lamp.[1] Soon after, on December 1961, while working for the American

Optical Company in Massachusetts, E. Snitzer reported laser oscillation in an optical fiber based cavity.[2] It consisted of a $Nd^{3+}$-doped optical fiber cavity pumped by a flash lamp that for obvious reasons was termed *fiber laser* (FL). The fact that absorption losses in optical fibers were gradually decreasing (at present for example the best telecommunication fibers have a loss of less than 0.2 dB/km), together with the development of semiconductor lasers that were introduced as pumps for this type of lasers strongly motivated its investigation that boomed in the 1980's. Laser diodes are among the most efficient with typical overall electrical-to-optical conversion efficiencies superior to 50 %.

The pumping signal excites the atoms of the medium into a higher energy level to create population inversion that means amplification and therefore lasing. The pump is usually provided by another laser. In the work described here, the pump source was a diode-pumped fiber laser system operating at 1064-nm wavelength and the active material was an $Ho^{3+}$-doped optical fiber.

The optical cavity is created by two mirrors arranged such that the light amplifies as it travels back and forth through the gain medium. Regularly one of the two mirrors (the output coupler) is partially transparent with the purpose that part of the signal is emitted through it. These mirrors can be dichroic filters, Bragg gratings or simply perpendicular cleaved facets of fiber-ends. In the later, highly efficient lasers only require the ~4% of the amplified signal to travel back into the cavity to be re-amplified and the rest (96%) is delivered as useful laser light.

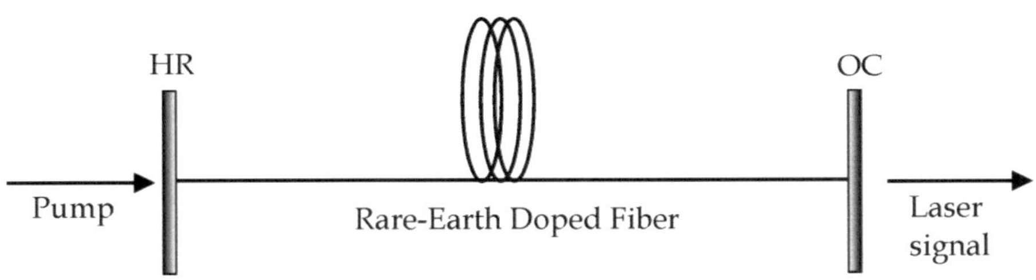


Fig. 6. Typical configuration of a fiber laser. The mirror **HC** has high transmission for the pump wavelength ($\lambda_p$) and high reflection for the laser signal ($\lambda_s$); the mirror **OC** partially reflects $\lambda_s$ and the transmitted part is the laser signal.

### 3.1 Modeling of a quasi two-level fiber laser

As introduced on section 2, three-level pumping amplifying systems may be studied as quasi two-level when absorption and emission spectra overlap. Because these systems need to present population inversion all along the optical fiber, a serious level of pump power is not absorbed and hence emitted as residual. This is a serious limitation of these systems concerning optical conversion efficiency. However, in order to assure that most of the absorbed power transfers to the amplifying signal, an optimal length needs to be estimated. This is our goal in this section.

Along a fiber laser there are two stimulated-emission generated signals that propagate in opposite directions. Figure 6 depicts these signals together with the pump from zero to the optimal fiber length ($L_{opt}$). The total (normalized) power at any point ($z$) is

$$p(z) = \beta^+(z) + \beta^-(z) \tag{44}$$

Where

$$\beta^{+}(z) = \frac{P_s^{+}(z)}{P_s^{sat}} \tag{45}$$

Propagates from the pump end ($z$=0) to the delivery end ($z$= $L_{opt}$) and

$$\beta^{-}(z) = \frac{P_s^{-}(z)}{P_s^{sat}} \tag{46}$$

Using equation (43) for the forward propagating signal we may write

$$\frac{1}{\beta^{+}}\frac{d\beta^{+}}{dz} = \alpha_s \frac{Uq-1}{1+p+q} \tag{47}$$

In forward pumping, the propagation equation for the pump signal is [from equation (42)]

$$\frac{1}{q}\frac{dq}{dz} = -\alpha_p \frac{U'p+1}{1+p+q} \tag{48}$$

The ratio of the propagation equations gives

$$(\alpha_p U'p + \alpha_p)\frac{d\beta^{+}}{\beta^{+}} = (-\alpha_s Uq + \alpha_s)\frac{dq}{q} \tag{49}$$

Before integrating this equation let us establish the limits. For now the left part is from $\beta^{+}(0)$ to $\beta^{+}(L_{opt})$ whereas the ones for the right part may be obtained under the criterion (see Figure 7): at $z$=0 the power is $q_0$ but at $z$=$L_{op}$ the gain saturates and hence $\beta^{+}$ stops growing up. This condition makes ${d\beta^{+}}/{dz} = 0$ in equation (47) for $q = {1}/{U}$ and is the key criterion for optimal performance. Integrating (49) we obtain

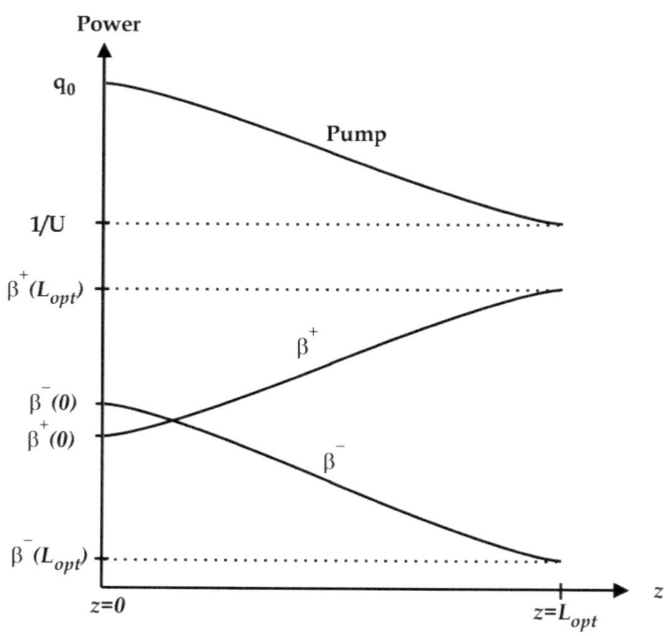


Fig. 7. Normalized powers of propagating signals along a quasi two-level fiber laser.

$$\alpha_p U' p \ ln \frac{\beta^+(L_{opt})}{\beta^+(0)} + \alpha_p \ ln \frac{\beta^+(L_{opt})}{\beta^+(0)} = -\alpha_s + \alpha_s U q_0 + \alpha_s \ ln \frac{1}{U q_0} \quad (50)$$

In order to simplify this equation such that known parameters are involved (mirror reflectivities), at z=0 and $z= L_{opt}$ the next boundary conditions are satisfied

$$\beta^+(0) = r_1 \beta^-(0)$$

(51)

$$\beta^-(L_{opt}) = r_2 \beta^+(L_{opt})$$

Where $r_1$ and $r_2$ are (respectively) the HR and OC reflectivities. Because both signals experience the same gain, the product of $\beta^+(L_{opt}) = \beta^+(0)e^{gL_{opt}}$ and $\beta^-(L_{opt}) = \beta^-(0)e^{-gL_{opt}}$ gives $\beta^+(L_{opt})\beta^-(L_{opt}) = \beta^+(0)\beta^-(0)$; the next constant may be defined for any length (See Fig. 7)

$$\beta^2 = \beta^+(z)\beta^-(z) \quad (52)$$

Then, from $\beta^+(0)\beta^-(0) = \beta^+(L_{opt})\beta^-(L_{opt})$ and using the boundary conditions the single-pass gain becomes

$$\frac{\beta^+(L_{opt})}{\beta^+(0)} = \frac{1}{\sqrt{r_1 r_2}} \quad (53)$$

That introduced in equation (50) and solving for the intra-cavity power signal, we obtain a numerical value as

$$p = -\frac{\alpha_p \ ln\sqrt{r_1 r_2} - \alpha_s(1 - U q_0 + ln U q_0)}{\alpha_p U' ln\sqrt{r_1 r_2}} \quad (54)$$

And the optimal length may be obtained by re-arranging equation (48) as

$$\frac{1+p+q}{q} dq = -\alpha_p (U'p + 1) dz \quad (55)$$

That integrated over the same limits gives the value of the optimal length for maximum efficiency as

$$L_{opt} = \frac{ln U q_0 (p+1) + q_0 - {}^1/_U}{\alpha_p(\ U'p + 1)} \quad (56)$$

For example, two estimations of optimal length as function of pump power are given in Figure 9 for a $Ho^{3+}$-doped ZBLAN fiber laser oscillating at $\lambda_s \approx 1194$ nm (see the emitting spectrum in Fig. 8) with 5000-ppmwt concentration, 7.5-micron core diameter and $\tau$=3.5 ms. In Fig. 9(a) the pump was $\lambda_p$=1117 nm and the stars correspond to the cavity lengths of the experiments; then, the lengths used were optimum for very low powers between 100 and 200 mW.

Fig. 9(b) corresponds to an estimation when pumping at $\lambda_p$=1175 nm; here the 84-cm cavities, one with HR=0.04 ($r_1$) and OC=0.04 ($r_2$) and the other with HR=1 and OC=0.04, are good for very low powers; but the 1.5-m cavity is optimum for powers above 1W.

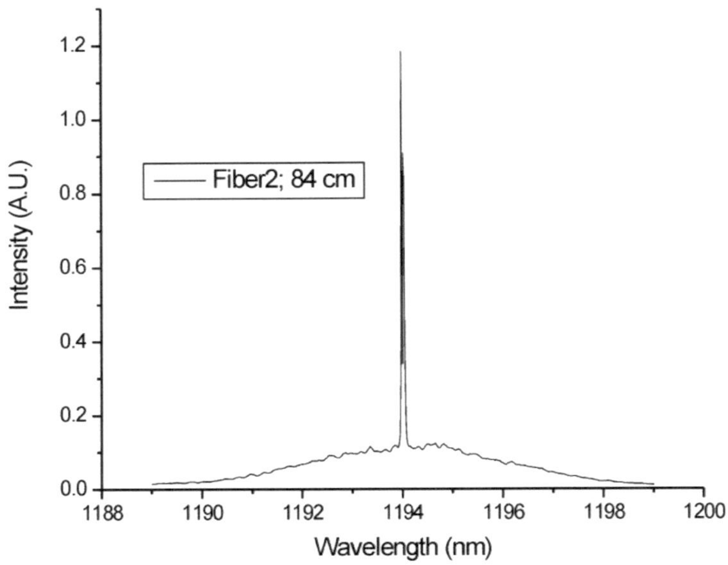


Fig. 8. Typical emission spectrum of a $Ho^{3+}$-doped fiber laser when pumped at 1117 or 1175 nm.

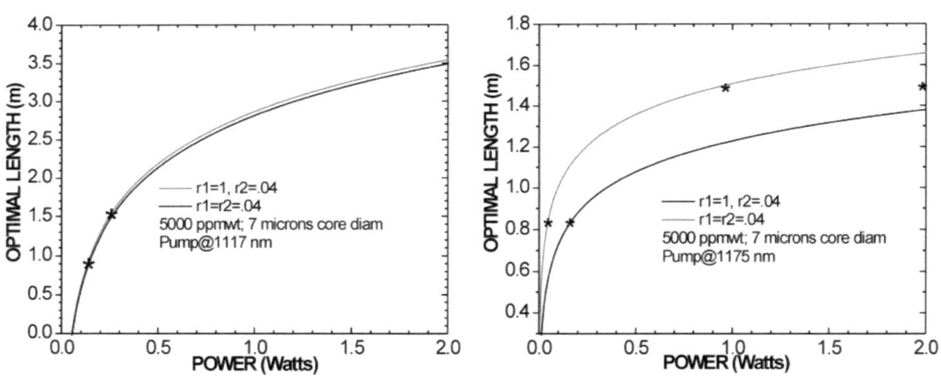


Fig. 9. Optimal fiber laser length (a) 1117-nm pump (b) 1175-nm pump.

The model has been proven with the fiber lasers mentioned at the beginning of this section and presents good agreement with the experimental results.

## 4. Other rare-earth fiber devices

In this section, two relatively modern applications of REDF's are described, both based on the upconversion phenomenon that is responsible of multi-photon absorption. These devices depend on the development of special types of glasses that extended the possibilities for light converters.

### 4.1 Upconversion fiber lasers and amplifiers

After the parallel development of $Nd^{3+}$-doped systems emitting at 1064 nm and laser diodes (LD's) at ~809 nm, most FL research has been devoted to 980-nm pumped $Er^{3+}$-doped fiber lasers (EDFL's) and amplifiers (EDFA's) operating in the highest transparency of optical fibers, around 1550 nm, most specifically the C-Band at 1530-1565 nm (Digonnet, 2001). At present, the necessity of more optical channels has attracted interest for the $Er^{3+}$ 1565- 1625 nm (L-Band) and the $Tm^{3+}$ 1460-1530 nm (S-Band). Although somehow mentioned, there are in general, two main facts that limit operation of silica-based fiber lasers to operate from ~800 nm to ~2200 nm. The main one is pump availability. The most mature semiconductor technology that is in the market includes laser diodes emitting in the ~800-850 nm, ~900-980 nm and ~1500-1600 nm regions. Besides, there is a quantum rule that establishes that, in general, the lowest energy excited states of $RE^{3+}$ are the most stable. In other words, one ion that is excited to the highest energy levels will make quick transitions down to the low energy levels. During their multi-step decay, they will stay shorter times at the superior levels and, in general, the lifetime will increase as it approaches the lowest energy levels. Then, in high phonon-energy materials such as silica, the metastable levels are the low-energy ones that are resonant with shorter wavelengths. Because of this, the VIS-UV regions started being explored until the development of low-energy phonon materials such as the fluoride-zirconium based or the tellurites. On these, two adjacent energy levels are less *thermally-connected* and as a consequence the multi-phonon decay rate is much lower. Then, the accumulation of population in all the levels (especially the highest) is more probable. In addition, the lifetimes increase and even the highest energy levels are metastable. Optical glasses such as ZBLAN or tellurites are efficient in subsequently absorbing two or more low-energy photons to produce another with higher energy. This up-conversion phenomenon permits the conversion of two or more IR photons into a UV-VIS one. At present, VIS-UV laser diodes deliver quite modest powers. Although this is the ideal source for any laser wavelength they present several challenges because material damage occurs at modest powers. A general rule for laser diodes is that good beam shape (single-mode and quasi-circular) is associated to low power of some milliwatts whereas high power LD's usually consist of an arrangement of low power ones or an arrangement of highly-rectangular end-emitting LD's. An improvement from the oldest technology of electrically exciting gases like Argon, Xenon or Neon is represented by frequency doubling or tripling solid state lasers such as the Nd:YAG or Nd:$YVO_4$. In general, producing low-energy photons from high-energy ones is easier because the conservation law of energy tell us that *the energy delivered by a system is equal to the energy absorbed minus the energy lost during the conversion*. This is not the case when producing high-energy photons departing from low energy ones; in this case, more than one photon is necessary.

Several upconversion fiber lasers in the UV-VIS region have been demonstrated; $Pr^{3+}$:ZBLAN for example covers RGB regions that are important for laser displays; $Nd^{3+}$ has produced 380 and 410 nm; $Ho^{3+}$ and $Er^{3+}$ green; $Tm^{3+}$ blue (450 and 480 nm), UV (284 nm) and red (650 nm) (Funk & Eden, 2001; El-Agmy, 2008). Other important applications include those in which small spot size represents high-density storage, high resolution printing and fine lithography; although strong light-mater interactions makes them important for surgery being the most common eye-surgery; other applications extent to UV-curing of polymers and epoxies, sterilization of medical instruments, etc. Blue fiber lasers in particular are important for undersea optical communications because cold water from the sea is transparent in the 470-500 nm region.

### 4.2 Light controlling light fiber attenuators

Although the activation of the absorption in the upper states has been reported before most authors have not exploited the use of this phenomenon for photonic devices other than optical sources. Recently, we have demonstrated that it is possible to attenuate or modulate a guided beam inside an optical fiber by another beam (Mejia & Pinto, 2009). Attenuation in optical fibers is usually realized by using bulk attenuators (or modulators) between two fibers which, in general, implies extracting light from one fiber, attenuate (or modulate) and then coupling back to the fiber link. This is basically a bulky approach that as such has the disadvantage of presenting high insertion losses. One of the simplest all-fiber approaches consists on physically deforming an optical fiber to induce the losses. It implies fiber fatigue and hence limitations in the life of the device. Purely photonic approaches in which one beam of light controls another one have been recently demonstrated and have still several limitations such as non-transparency recovery.

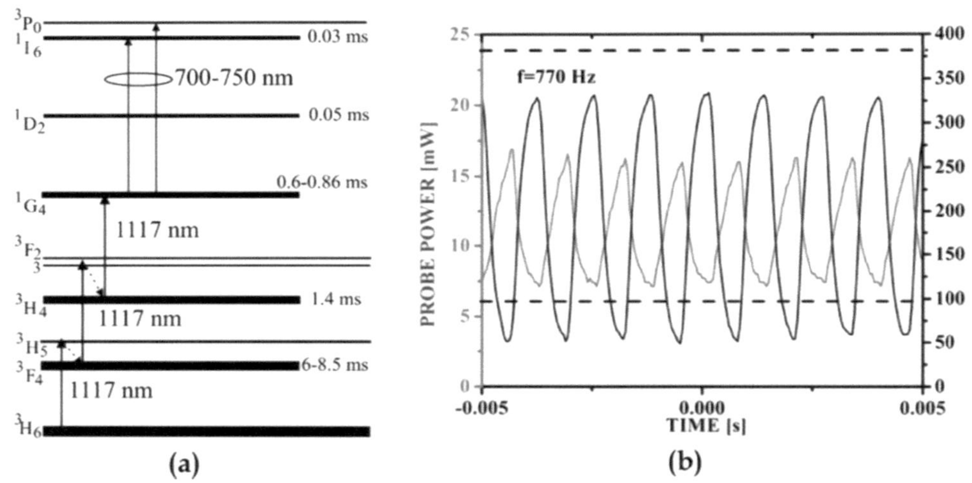


Fig. 10. (a) Energy levels involved in 700-750 nm attenuation by a control 1117-nm signal (b) Traces of probe signal in gray color (use left scale) and control (black with scale on the right).

An energy-level diagram that displays the transitions involved in our experiments is depicted in Figure 10 (a). An initial 1117-nm photon is absorbed by a Tm ion in the transition $^3H_6 \rightarrow ^3H_5$ and from $^3H_5$ the ion rapidly decays non-radiatively to the high laying $^3F_4$ state (~6-8.5 ms); a second photon populates the $^3F_{3,2}$ short lived states, from where the ions suffer a fast relaxation to the $^3H_4$ level (~1.4 ms), and *third-step* photons populate $^1G_4$ that is also metastable (~0.6-0.86 ms). In this way, the transitions $^1G_4 \rightarrow ^3P_0$ and $^1G_4 \rightarrow ^1I_6$ become active and their band overlapping absorbs in the 700-750 nm interval. In this way, a 725-nm probe signal was controlled as guided through an optical fiber. Figure 10 (b) shows both signals; note that the operation was superior to 700 Hz. The fiber was a ~45-cm ZBLAN fiber, doped with 2000 ppmwt of $Tm^{3+}$, 3-μm core diameter. The system was capable of 79-%

attenuating 700 mW of 711.2-nm probe signal by co-propagating 800 mW of control signal. The system perfectly operated as an optical inverter up to 200 Hz. The theoretical limit for working as an optical inverter was ≈1100 Hz. The dashed horizontal lines in the figure are the maximum and minimum reached by the probe at low frequencies. Because the phenomenon responsible for the attenuation (upconversion) depends on the time taken to absorb three photons, the induced attenuation is practically instantaneous. Then, the response of the system is imposed by $^1G_4$-lifetime. This system may be important in those applications requiring uniform attenuation of all the cross section of a beam because it attenuates the whole signal coupled in the fiber core at once. Other opportunities are optically-controlled Q-switching of lasers because loss-modulations within 5-10% are typical. Because the system modulated at least 37% above 700 Hz, smaller modulations imply an increase of operating frequency. As the control signal in a commercial device of this type would be produced from a laser diode, the non-mechanical and purely photonic nature of the system (driven by low-voltage electronics) makes it robust. An additional advantage of the scheme is that it is polarization independent.

## 5. Conclusion

Rare-earth doped optical fibers had played a prominent role in laser development. Their geometry, that usually includes a circular core, has proven to be among the main reasons to choose them as laser converters. The devices based on these fibers are very compatible with the optical fiber infrastructure that covers the globe. Laser efficiencies have over-passed the dreams of first laser researchers; their powers have scaled up in such a way that also the dreams of first fiber-laser researchers has been over-passed; quite modest cooling systems for REDFL's have made possible the production of kilowatts of optical power. In telecommunications the optical amplifiers made possible the high speed regeneration of optical channels within the optical fiber networks and the amplification windows have expanded. Other devices like sensors, broadband sources and optical attenuators are still to be developed because new types of optical fibers (the photonic crystal type, for example) improve their performance. Because these devices depend on diode laser development, every time a new diode laser appears, their possibilities increase.

## 6. Acknowledgment

This work was supported by CONCYTEG and CONACYT-México.

## 7. References

Agrawal, GP (1989). *Nonlinear Fiber Optics* (Second Edition), Academic Press, ISBN 0-12-045142-5, California, USA

Castrellón-Uribe, J. & García-Torales, G. Remote Temperature Sensor Based on the Up-Conversion Fluorescence Power Ratio of an Erbium-Doped Silica Fiber Pumped at 975 nm. *Fiber and Integrated Optics*, Vol. 19, No. 4, (July 2010), pp. 272-283, ISSN 1096-4681

Desurvire, E. (1994). *Erbium-Doped Fiber Amplifiers: principles and applications* (First Edition), Wiley-Interscience, ISBN 0-471-58977-2 , New York, USA

Digonnet, M. (2001). *Rare-Earth-Doped Fiber Lasers and Amplifiers* (Second Edition), Marcel Dekker, ISBN 0-8247-0458-4, New York, USA

El-Agmi, RM. Upconversion CW Laser at 284 nm in a Nd:YAG-Pumped Double-Cladding Thulium-Doped ZBLAN Fiber Laser, *Laser Physics*, Vol. 18, No. 6, (June 2008), pp. 803-806

Funk, D. & Eden, G. (2001). Visible Fluoride Fiber Lasers, In: *Rare-Earth-Doped Fiber Lasers and Amplifiers*, Michael J.F. Digonnet, (Ed.), 171-242, Marcel Dekker, ISBN 0-8247-0458-4, New York, USA

Lucas, J. (1989). Fluoride glasses. *Journal of Materials science*, Vol. 24, No.1, (January 1989), pp. 1-13, ISSN 0022-2461

Luu-Gen, H. & Chen-Ke, S. The Structural Investigation of a ZBLAN Glass by Vibrational Spectroscopy, *Chinese Journal of Physics*, Vol. 34, No. 5, (October 1996), pp. 1270-1275

Mejía, E. & Pinto, V. Optically controlled loss in an optical fiber, *Optics Letters*, Vol. 34, No. 18, (September 2009), pp. 2796-2798

Miniscalco, W. (2001). Optical and Electronic Properties of Rare Earth Ions in Glasses, In: *Rare-Earth-Doped Fiber Lasers and Amplifiers*, Michael J.F. Digonnet, (Ed.), 17-105, Marcel Dekker, ISBN 0-8247-0458-4, New York, USA

Pauling, L. (1988). *General chemistry* (First Dover Edition), Dover Publications, ISBN 0-486-65622-5, New York, USA

Po, H.; Snitzer, E.; Tumminelli, R.; Zenteno, L.; Hakimi, F.; Cho, N. & Haw, T. (1989). Double Cladd High Brightness Nd Fiber Laser Pumped by GaAlAs Phased Array, *proceedings of Optical Fiber Communication Conference*, ISBN 978-9999904467, Houston, Texas, USA, February 1989

Quin, L., Shen, ZX., Low, BL., Lee, HK., Lu, TJ., Dai, DS., Tang, SH., & Kuok, MH. Crystallization Study of Heavy Metal Fluoride Glasses ZBLAN by Raman Spectroscopy, *Journal of Raman Spectroscopy*, Vol. 28, No. 5, (July 1997), pp. 495-499

Siegman, A. (1986). *Lasers* (First Edition), University Science Books, ISBN 0-935702-11-5, Mill Valley, California, USA

# 2

# Fabrication of Large Core $Yb_2O_3$ Doped Phase Separated Yttria-Alumino Silicate Nano-Particles Based Optical Fiber for Use as Fiber Laser

M. C. Paul[1], A. V. Kir'yanov[2], S. Bysakh[3], S. Das[1], M. Pal[1], S. K. Bhadra[1], M. S. Yoo[4], A. J. Boyland[4] and J. K. Sahu[4]

*[1]Fiber Optics and Photonics Division, Central Glass & Ceramic Research Institute (CGCRI), CSIR, Jadavpur,*
*[2]Centro de Investigaciones en Optica, Guanajuato,*
*[3]SEM-ESCA Laboratory, Central Glass & Ceramic Research Institute (CGCRI), CSIR, Jadavpur,*
*[4]Optoelectronics Research Center (ORC), University of Southampton, Southampton,*
*[1,3]India*
*[2]Mexico*
*[4]UK*

## 1. Introduction

Development of new rare-earth (RE)-doped optical fibers for high power amplifiers and lasers require continuous improvements in the fiber spectroscopic properties like gain and efficiency characteristics, resistance to spectral hole burning and photodarkening (PD) phenomena besides reduction in device size and economical efficiency. Silica glass as a host material for fibers has proved to be very attractive though some potential applications of RE-doped fibers suffer from limitations in terms of spectroscopic properties resulting from clustering or inappropriate local environment when doped into silica. Several mixed-oxide and non-oxide alternatives to silica have been successfully proposed to obtain spectroscopically improved amplifying single mode fibers in spite of fabrication difficulties and high cost.

The route of interest here consists of using silica as a mechanical host and support of the fiber optical waveguide, and of embedding RE-ions within oxide nanoparticles of composition and structure different from those of silica. When nanoparticles have a crystalline structure the glass becomes transparent glass ceramics (Gonçalves et al., 2002). However, the nanoparticle may also be amorphous, such as those obtained by phase separation (Zarzycki, 1991). Some reports on RE-doped transparent glass ceramic based single mode fibers use low melting mixed oxides prepared by a rod-in-tube technique (Samson et al., 2002), or mixed oxyfluorides using a double-crucible technique (Samson et al., 2001). However, the low melting point of these materials causes low compatibility with silica components. Transition metal-doped silica-based transparent ceramic fibers were prepared by MCVD (modified chemical vapor deposition) process and using a slurry method (Yoo et al., 2003), where the particles were synthesized before insertion into the silica tube substrate.

Over the past several decades considerable work have been carried out on incorporation of RE oxide nano-crystallites into different glass hosts. Different processes have been developed such as the co-sputtering technique (Fujii et al., 1998), pyrolysis (John et al., 1999), ion implantation (Chryssou et al., 1999), laser ablation (Nichols et al., 2001) and sol-gel processes (Yeatman et al., 2000). Another process which has recently been developed by a Finnish company, Liekki, is the direct deposition of nano-particles (Rajala et al., 2003). In addition, Leikki Company (Koponen et al., 2006) proposed an amplification fiber doped with RE element nano-particles. Compared to conventional Er doped active fiber; it has many advantages in optical gain, amplification bandwidth, photo-darkening, efficiency, quenching phenomena, etc. All these processes are related to the outside vapour deposition technique except for the sol-gel, which involves longer preform fabrication time. Many applications of nano-fiber have been found in telecommunication and sensor such as chemical sensor (Shi et al., 2007) and fiber ring laser (Jiang et al., 2007). An alternative technology is to dope nano-materials into optical fiber. Cho et al. (Cho et al., 2001) doped PbTe nano-particles into optical fiber core and demonstrated its nonlinear optical features. Dove et al. (Dove et al., 2001) fabricated a glass optical fiber doped with $Cd_3P_2$ nano-material, obtaining a gain of 7.1dB in a 4 mm-length special fiber. Kawanishi et al. (Kawanishi et al., 2006) injected semiconductor quantum dots solution into a holey optical fiber.

On the other hand solution doping technique (Bandyopadhyay et al., 2004; Bhadra et al., 2003; Sen et al., 2005; Townsend et al., 1987) in the modified chemical vapour deposition (MCVD) method (Li , 1985) is the most common way to incorporate the RE ions into the core of silica optical fiber preform. However, the incorporation of REs into a suitable nano-crystalline host that are dispersed within the silica rich matrix of optical fiber preform, through MCVD and solution doping process is challenging compared to the fabrication of such type of bulk material by normal crucible melting process. In earlier work, we have reported the synthesis of $Er_2O_3$ doped phase-separated amorphorous nano-particles into calcium-germano silicate core glass host by applying the basic principle of phase-separation phenomena (Blanc et al., 2009). This is to improve the spectroscopic properties of Er-doped fiber, mainly the spectral broadening of fluorescence band.

RE ions into nano-crystalline hosts becomes very important as it experience very dissimilar site and different crystalline fields which give rise to broadening of the individual stark levels. When the RE ions are confined in crystalline environments of low phonon energy, they yield large excited state lifetime and absorption cross-section compared to vitreous surroundings. Generally, $Yb^{3+}$ in $Y_2O_3$ or YAG ($Y_3Al_5O_{12}$) nano-crystalline low silica host exhibits a promising material for high power, high brightness, and high efficiency laser systems because of its small quantum defect between the pump and lasing transitions (Shirakawa et al., 2004). Furthermore, the glass host matrix in which the nano-crystals are immersed possesses the chemical durability and mechanical property of oxide glass. To develop more efficient fiber laser sources based on rare-earths doped materials, hosts with low phonon energies are required. This lower phonon energy reduces significantly non-radiative decay due to multi-phonon relaxation, allowing increased lifetime of some excited levels that can relax radiatively or can store energy for further up-conversion, cross-relaxation, or energy transfer processes. Considering such importances, the incorporation of $Yb_2O_3$ into yttrium-alumino silicate phase-separated nano-crystallites was reported within the core region of silica preform through chemical impregnation of porous phospho-silicate or pure silica layer deposited via MCVD process followed by post-thermal treatment of the perform (Paul et al., 2010).

$Yb^{3+}$ in $Y_2O_3$ or YAG nano-crystalline host is suitable for making up-conversion and high power lasers (De et al., 2006; Lu et al., 2008a, 2008b; Mun et al., 2005; Patra et al., 2005; Shirakawa et al., 2003; Vetrone et al., 2003). Various $Yb_2O_3$ doped host materials have been progressively investigated earlier for fiber lasers, and the Yb:YAG laser is scaled up to an average power of 60 W with an 810-fs duration in a laser with thin-disk geometry (Innerhofer et al., 2003). Yb-doped sesquioxides ($RE_2O_3$, RE = Y, Sc, Lu) serve as potential alternatives to Yb:YAG for power scaling because of their desirable thermal properties. In addition, the strong electron–phonon interaction causes characteristic spectral broadening, especially in the case of $Y_2O_3$. Due to these characteristics, Yb-doped sesquioxides are expected to be a promising laser material for high-power and ultrashort pulse lasers. In this work $Y_2O_3$ was selected as an attractive host material for laser applications as it is a refractory oxide with a melting point of 2380°C, a very high thermal conductivity, $k_{Y2O3}$= 27 W/mK, two times YAG's one, $k_{YAG}$= 13 W/mK. Another interesting property allowing radiative transitions between electronic levels is that the dominant phonon energy is 377cm$^{-1}$ which is one of the smallest phonon cutoff among oxides (Ubaldini & Carnasciali, 2008).

Laser operation has been also demonstrated with sesquioxide crystals fabricated by melt-growth methods (Petermann et al., 2002) and a mode-locked $Yb^{3+}$:$Sc_2O_3$ crystalline laser has also been reported (Klopp et al., 2003). The laser ceramics based on rare-earth-doped $Y_2O_3$, $(Y_{0.5}Gd_{0.5})_2O_3$, $Sc_2O_3$, and $Lu_2O_3$ with neodymium (Lu et al., 2001) and ytterbium (Takaichi et al., 2004) have been demonstrated. The passive mode locking of a diode-pumped $Yb^{3+}$:$Y_2O_3$ ceramic laser was demonstrated (Shirakawa et al., 2003, 2004). The lasing of the 1 at.% Yb:YAG ceramic laser was also demonstrated with the maximum output power of 1.02 W and a slope efficiency of 25% (Yusong et al., 2007). All such type of glass ceramic based laser containing $Y_2O_3$ or YAG crystals possesses low lasing efficiency. Here we have made Yttria alumino rich $Yb_2O_3$ doped silica glass based phase-separated nano-particles containing optical fibers to demonstrate good lasing efficiency where the maximum vibrational energy in YAS ($Y_2O_3$–$Al_2O_3$–$SiO_2$) glass (Jander and Brocklesby, 2004) is about 950 cm$^{-1}$ which is less than the Maximum vibrational energy value of 1100 cm$^{-1}$ in silica glass (Tomozawa & Doremus, 1978).

Considering the importance of Yb:YAG nano-crystals as a lasing host, we have reviewed here the formation of such type of nano-crystals within the silica based core glass matrix of optical fiber preform by solution doping technique under suitable thermal annealing conditions. In this paper, we have discussed about the formation of nanostructure in optical fiber samples made from the annealed nano-crystalline host based preform. The role of phosphorous (P) and fluorine (F) was studied on the formation of $Yb_2O_3$ doped yttrium-alumino silicate phase-separated crystalline nano-particles. Study of the nature of the particles within the doping host of optical fibers was also done. The change in the local environments of $Yb^{3+}$ ion was elucidated from the high-resolution transmission electron microscopy imaging, electron diffraction, X-ray diffraction analyses. The average dopant levels within the core region were evaluated by electron probe micro-analyses (EPMA). We also report the critical fabrication parameters, the material characterization results, spectroscopic properties, PD phenomena along with their lasing characteristics of such kind of optical fibers.

The purpose of this work is to develop nano-engineering glass based large core optical fibers having diameter around 20-35 micron containing $Yb_2O_3$ doped phase separated nano particles which may improve the photo-darkening phenomenon, lasing property of the fibers mainly the lasing efficiency as well as spectral broadening of the lasing spectrum compared to the normal $Yb_2O_3$ doped YAG crystal based ceramic laser (Yusong et al., 2007) as well as normal alumino-silica based optical fibers.

## 2. Fabrication of nano-engineering glass based optical preforms and fibers

### 2.1 Benefit of the choice of $Yb_2O_3$ into nano-crystalline host

- Rare-earth ions into nano-crystalline hosts experience very dissimilar side and experience different crystalline fields, which give rise to broadening of the individual stark levels. When the rare-earth-doping ions are confined in crystalline environments of low phonon energy, it yields large excited state lifetime and optical absorption cross-section compared to vitreous surroundings (Shirakawa et al., 2004).
- Generally $Yb^{3+}$ in $Y_2O_3$ or $Y_3Al_5O_{12}$ nano-crystalline low silica host exhibit a promising material for high power, high brightness, and high efficiency laser systems because of its small quantum defect between pump and lasing transitions (Shirakawa et al., 2004).
- Another interesting property allowing radiative transitions between electronic levels is that the dominant phonon energy is 377cm$^{-1}$ which is one of the smallest phonon cutoff among oxides (Ubaldini & Carnasciali, 2008).
- Such nano-crystalline structures will be obtained after a thermally controlled growth of the crystal phase directly in the bulk glass through suitable thermal treatment after making of optical preform.
- Purposes of the work is to develop the glass preforms for drawing into Yb-doped optical fibers where nano-structuration of the host should result in improvement of the characteristics of Yb-doped fiber lasers.

### 2.2 Mechanism of the formation of phase-separated nano-engineering glass based optical preform

Incorporation of glass formers and modifiers occurs through solution doping process followed by the MCVD technique. Under appropriate perturbation, such as a thermal treatment, the glass forming the core will be separate into two phases of low and rich silica content, respectively. As the low-silica phase constitutes a small portion of the total core volume, microparticles or even nanoparticles may be expected.

Doping of Yb ion into lithium-aluminosilicate based glass containing $Y_2O_3$ was done through solution doping process followed by phase separation technique in which addition of $P_2O_5$ serve as a nucleating agent to increase phase separation with generation of $Yb_2O_3$ doped micro or nano-crystallites into the core matrix of optical preform. The glass formers incorporated by the vapour phase deposition process involves $SiO_2$, $P_2O_5$ along with glass modifiers $Al_2O_3$, BaO, $Li_2O$, $Yb_2O_3$ and $Y_2O_3$ incorporated by solution doping technique.

Incorporation occurs through viscous sintering phenomena. At sintering temperature the core glass will be in a metastable immisicibility under condition of $T_C$ (crystallization temperature) < $T_m$ (melting temperature) where phase separation kinetics are faster than crystallization kinetics. More and more negative value of the free energy change of the system for mixing of the oxide components greater will be the phase separation. The core composition of the doping host was modified to minimize the larger phase-separation along with crystallization through optimization of the doping levels of $P_2O_5$, $Y_2O_3$ and $Al_2O_3$ along with $Li_2O$ content with incorporation of the other dopants such as BaO so that the following conditions are satisfied:-

- The core glass should be in high transparency to obtain low optical scattering
- Closely matched indices of refraction
- Low birefringency crystals
- Crystal size much smaller than wavelength of light
- Particle size < 15 nm
- Interparticle spacing 15 nm
- Narrow particle size distribution
- No crystal clustering

Liquid-liquid phase separation is a common and well-known phenomenon in silicate liquids, and can be observed in the high-silica regions of many binary and more complex systems, at temperatures either above or below the liquidus (Vogel, 1994). As the field strength (charge/radius) of the modifier oxide cation increases, the resulting consulate (critical) temperatures Tc become systematically higher (Hageman and Oonk, 1986; Hudon & Baker, 2002). Thermodynamically, it results from strongly positive heats of mixing between $SiO_2$ and modifier oxide components in silica-rich liquids (Hudon & Baker, 2002), microscopically, the clustering of nonbridging oxygens around high field strength modifier cations lowers energy by facilitating local charge balance and, if extensive enough, stabilizes two coexisting liquids. Here phase-separation has been induced in the clear glasses through suitable thermal annealing process with the appearance of either a crystalline or amorphous phase separation.

One of the reasons may be that yttria-alumino silicate glass undergoes phase-separation under suitable doping levels of $Al_2O_3$ and $Y_2O_3$ where the glass enters within the immiscible region of yttria-alumino silicate (YAS) glass. A ternary diagram of YAS glass system was shown in Fig. 1 derived from FactSage (Facility for the analysis of chemical thermodynamics) 5.5 thermo chemical software and database. The composition of such kind of glass having silica content around 90 mol% form both two liquid and clear glass zones. The glass transition temperatures for fluorine doped yttria-alumino-silicate glass based optical preform are found to be between 985 and 1115°C which is explained in Section 3.3. The glass transition temperature of oxide glass is related to a combination of several factors such as the density of covalent cross-linking, the number and strength of the coordinate links formed between oxygen and the cation, and the oxygen density of the network (Ray, 1974). With increasing Y content, more coordinate links are formed between oxygen and yttrium, which is opposed by the lower oxygen density of the network from the more open structure needed to accommodate larger yttrium ions and depolymerization in the network with decreasing silica content or increasing Y/Al. Such type of nano-structuration retain within the core glass matrix of optical fiber.

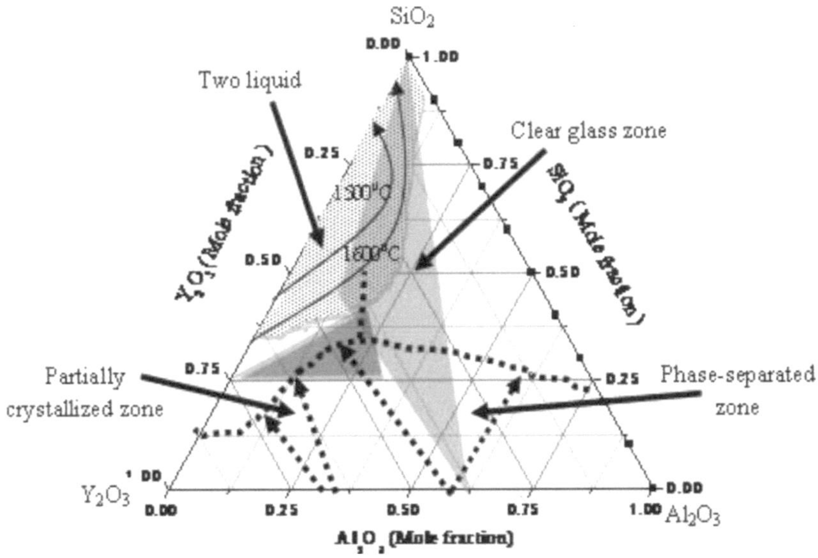


Fig. 1. Phase-diagram of $SiO_2$-$Al_2O_3$-$Y_2O_3$ system derived from Fact Sage software indicating the partially crystallized, glass forming and phase-separated zones.

### 2.3 Role of different co-dopants for making of nano-engineering glass based optical fibers

- The composition of $Yb_2O_3$ doped nano-engineering optical fiber was selected as $SiO_2$-$P_2O_5$-$Y_2O_3$-$Al_2O_3$-$Li_2O$-BaO.
- $P_2O_5$ serve as a nucleating agent for promotion of phase-separation phenomena along with crystallization.
- $Y_2O_3$ and $Al_2O_3$ serve as a formation of crystalline host of composition of $Y_3Al_5O_{12}$ under suitable thermal annealing process.
- $Li_2O$ serve for formation of glass-ceramic based material of composition (lithium alumino-silica) LAS glass to increase the optical transparency of the doping host.
- BaO serve as an agent which increase the glass formation region of the matrix as well as reduce the viscosity of the glass host.
- In some cases fluorine was also incorporated for enhancement of the phase-separation followed by reducing the phonon-energy of the glass host.

### 2.4 Modified chemical vapour deposition (MCVD) process with solution doping technique

Incorporation of Yb ions into nano-enginnering glass based on yttria-alumino-silica host was done through solution doping process followed by suitable thermal treatment of the preform (Paul et al, 2010a). The inner diameter of the tube is typically 17.0-18.0 mm. $P_2O_5$ was added into the deposited porous layer where $P_2O_5$ serve as a nucleating agent to increase the phase separation with generation of $Yb_2O_3$ doped micro or nano-crystallites into the core matrix of optical fiber preform. $SiO_2$ and $P_2O_5$ which serve as glass formers were incorporated through the vapour phase deposition process. The glass modifiers such as $Al_2O_3$, BaO, $Li_2O$, $Yb_2O_3$ and $Y_2O_3$ are incorporated by the solution doping technique using an alcoholic-water (1:5) mixture of appropriate strength of $YbCl_3.6H_2O$, $AlCl_3.6H_2O$, $YCl_3.6H_2O$, $LiNO_3$ and $BaCl_2.2H_2O$. All the preforms fabricated by the conventional MCVD technique, where the gaseous chlorides vapour of $SiCl_4$ and $POCl_3$ including $O_2$ were passed through a rotating silica tube and heated by an external burner which moves along the tube. Due to high temperature, the chloride precursors oxidise, forming particles which deposit on the inner wall of the tube. At this stage depending upon the radial temperature gradient occurring across the hot zone, a homogeneous gas phase reaction takes place and solid particles nucleate from the reaction products. Particle growth occurs primarily as a result of coagulation of colliding particles as the finely dispersed solids suspended in the gas pass through the radial-gradient of hot zone. Deposition of porous layers occurs through collisions induced by Brownian motion which result in aggregates of individual glass particles. Such kind of deposition is caused by thermophoresis, a phenomenon in which suspended particles experiences a net force in the direction of decreasing temperature due to a greater rate of collisions with gas molecules on the hot side of each particle which sinter together by viscous flow if the glass is sufficiently fluid (Miller & Chynoweth, 1979; Shelby, 1997; Izawa & Sudo, 1987). The deposited porous layer (Fig. 2) after solution doping turned into a glassy layer when the burner passes over it at temperature around 1600°C. In the final stage, the tube was collapsed into a solid preform at a temperature higher than 2000°C. The collapsing phenomenon occurs due to inward viscous flow of mass driven by surface tension and pressure difference between inside and outside of the tube. Controls of flow of $O_2$, tube temperature and pressure difference between inside and outside of the tube are very important during collapsing process to maintain circular geometry of core. The diameter of the doped region was increased from 20 to 35 micron through modification of several fabrication parameters such as number of deposited porous layers, composition of deposited porous layers and CSA of the starting deposited tube. The uniform doping levels within the large core fiber preforms was maintained through optimization of the deposition temperature with respect to composition as well as numbers of the deposited porous layers based on enlargement of the silica tube prior to deposition. Large core (1.5-2.85 mm) $Yb_2O_3$ doped preforms (Paul et al., 2010b) based on yttria-alumino-silica glass hosts were fabricated through modification of several fabrication parameters such as number of deposited porous layers, composition of deposited porous layers and CSA of the starting deposited tube. To make 19.0-20.0 micron diameter fiber from the preform, we have used 82-85 $mm^2$ CSA tube having inner diameter of 17.0 mm with deposition of 6 numbers of porous layers at optimum deposition temperature of 1350±10°C. The fiber diameter of 22.0±2.5 μm was drawn from the preform made through deposition of 6 numbers of porous layers at optimum deposition temperature of 1275±10°C using silica tube of CSA of 58-60 $mm^2$ having inner diameter of 18.0 mm. On the other hand, the fiber diameter of 32±2.5 μm was made from the preform fabricated through deposition of 10 numbers of porous layers at optimum deposition temperature of 1240±10°C within inner surface of silica tube of CSA of 58-60 $mm^2$ having inner diameter of 22.0 mm enlarged from 18.0 mm inner diameter under suitable pressurization prior to deposition of multiple porous phospho-silica layers. The same composition of deposited porous phospho-silica layer was used at each case.

The uniform doping levels within the large core of each fiber preform was maintained through optimization of the deposition temperature with respect to composition and numbers of the deposited porous layers as well as varying the inner diameter of the tube under suitable pressurization prior to deposition. Nano-engineering fibers varying NA between 0.07-0.12 and core diameter ranging from 20 to 35 micron were drawn from the

annealed perform under suitable heating conditions made through modification of several fabrication parameters such as number of deposited porous layers, composition of deposited porous layers and CSA of the starting silica tube with varying the inner diameter of the silica tube under suitable pressuriazation prior to deposition. The details about the making of large core $Yb_2O_3$ doped nano-engineering glass based optical preform through MCVD process followed by solution doping technique is described below:
The following important parameters are involved in this process.

- Composition of the gas phase mixture within the silica tube
- Number of deposited porous layers with pre-sintering
- Deposition temperature of the porous layers
- Composition of the solution of dopant precursors
- Soaking , drying and heat treatment of solution impregnated porous layers
- Dehydration
- Sintering
- Collapsing process

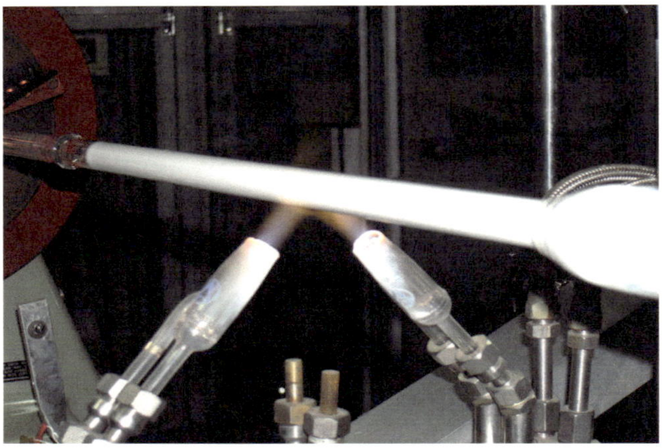

Fig. 2. Deposition of multiple porous layers of composition of $SiO_2$-$P_2O_5$ along forward direction by the MCVD process

### 2.5 Composition of the gas phase mixture within the silica tube

The composition of a gas mixture of 13-15 mol% of $SiCl_4$, 1.0-2.0 mol% of $POCl_3$, 2.5-4.0 mol% of He and 79-83.5 mol% of $O_2$ was found to be the best for smooth deposition to prevent coagulation of the particles. If coagulation arises, the formation of bubbles starts during sintering stage.

### 2.6 Number of deposited porous layers with pre-sintering

The fiber diameter around 19-20 micron was obtained from the preform made through deposition of maximum 6 numbers of porous silica layers having thickness around 10 µm containing 0.15 to 0.20 mol% of $P_2O_5$ under optimum deposition temperature of 1350 ±10°C followed by pre-sintering within 1300-1350°C which possesses uniform porosity along both axial and radial directions in order to obtain uniform distribution of Al, Yb, Y and Ba ions along the whole diameter (Paul et al., 2010a, 2010b). In this case we have used silica tube of

CSA around 82-85 mm$^2$ and inner diameter around 17.0 mm prior to deposition of porous layers. Under such condition multiple porous layers form uniform networking of the deposited oxide particles between each layer and not disturbed before going to solution impregnation process. From fabrication of a series of large core $Yb_2O_3$ doped nano-engineering preforms shown in Table 1 under different conditions, it was observed that uniform porosity in terms of uniform doping levels of Al, Yb, Y and P ions along the entire core region can be obtained through the way of solution soaking process maintaining uniform thickness (10 μm) of each layer up to deposition of 6 numbers of porous layers under deposition temperature of 1350±10°C followed by pre-sintering at 1300-1350°C (Paul et al., 2010a). The radial view of the microscopic picture (Fig. 3) of such six numbers of multiple porous layers shows a uniform networking of the deposited oxide particles between each layers having uniform thickness of 10 micron as well as uniform porosity varying slightly within ± 2%.

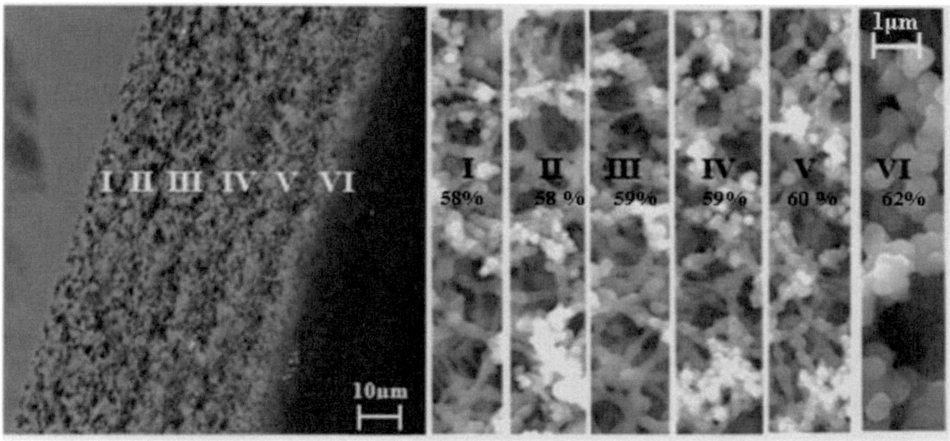


Fig. 3. Radial view of deposited six numbers of phospho-silica layers with the enlarged surface morphology of each layer.

The porosity of the layers varies from 58 to 62% between innermost and outermost layer. The axial view of such thick porous layer shown in Fig. 4 clearly indicates that a uniform networking of the deposited oxide particles between each layer is formed well as observed clearly in the enlarged microscopic view of the particular portion of the deposited layers.
The fiber diameter of 25.0 μm was drawn from the preform made through deposition of 6 numbers of porous layers at optimum deposition temperature of 1275±10°C using silica tube of CSA of 58-60 mm$^2$ having inner diameter around 18.0mm. The pre-sintering was done within 1250-1275°C for 3 passes. The purpose was to deposit thin (8.0–10.0 μm) individual layers as shown in Fig. 5a for each deposition pass in order to reduce the penetration length for uniform soaking of the dissolved ions along the radial direction. On the other hand, fiber diameter of 32±2.5 μm was made from the preform fabricated through deposition of 10 numbers of porous layers at optimum deposition temperature of 1240±10°C within inner surface of silica tube of CSA of 58-60 mm$^2$ having inner diameter around 22.0 mm enlarged from the initial inner diameter of 18.0mm under suitable pressurization. In this case tube was enlarged through suitable pressurization prior to deposition of multiple porous phospho-silica layers. The pre-sintering was done within 1275

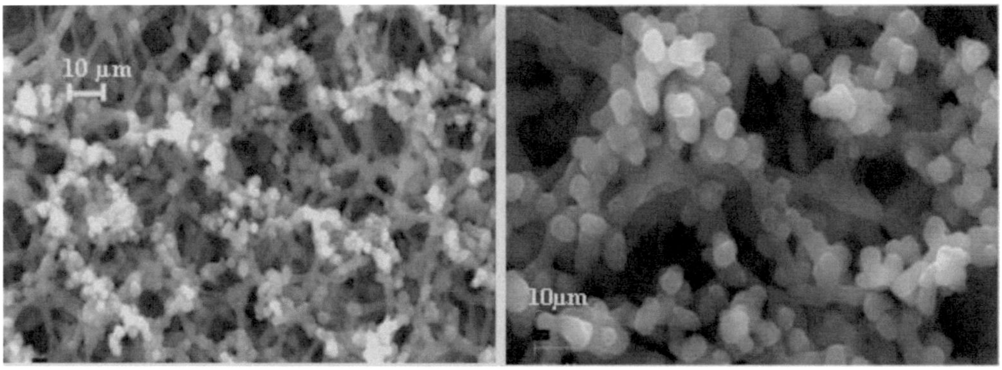


Fig. 4. The microscopic axial view of six multiple porous layers deposited at around 1350 ±10°C along with the enlarged view of the particular portion of the deposited.

to 1300°C for 3 passes. The radial view of deposited 10 numbers of porous layers having thickness around 5-6 micron was shown in Fig. 5b. To make LMA fibers with core diameter around 30–35 microns, we used thin wall silica tube of cross sectional area (CSA) around 58-60 mm$^2$ having inner diameter around 18.0 mm of thickness 1.0 mm. The outer surface of such type thin-wall tube was further enlarged to make inner diameter around 22.0 mm under optimum pressurization prior to deposition of multiple porous phospho-silicate layers for getting large core diameter. The purpose was to deposit thin individual layer (5.0-6.0 µm) for each deposition pass in order to reduce the penetration length for uniform soaking of the dissolved ions along the radial direction.

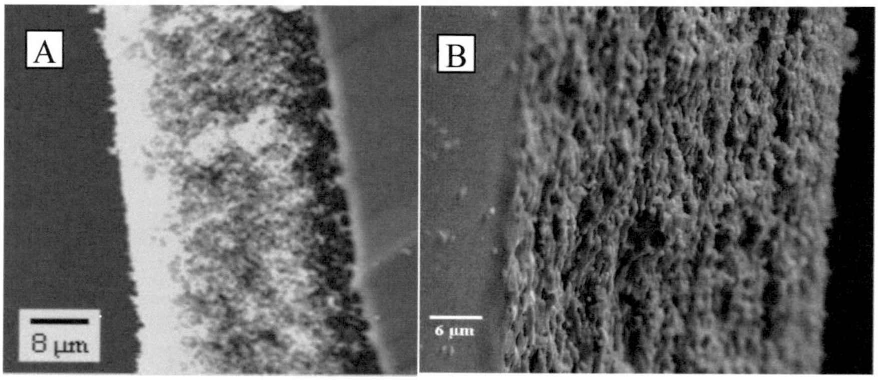


Fig. 5. The microscopic radial view of six multiple porous layers (A) and ten multiple porous layers (B) deposited at around 1275 ± 10°C and 1240 ± 10°C, respectively.

### 2.7 Composition of the solution of dopant precursors

Several preforms were made using various strength of $AlCl_3.6H_2O$, $YCl_3.6H_2O$, $LiNO_3$ and $YbCl_3.6H_2O$ which maintain the suitable viscosity around 1.5 cP (Paul et al., 2010). The viscosity of the solution of the dopant precursors plays an important role to prevent the disturbances of thick porous layer. Such disturbance of thick highly porous medium occurs if the viscosity of the soaking solution is high. The viscosity of 1.5 cP of the solution is found to be most suitable for soaking of six as well as ten numbers of porous layers deposited at

1350±10°C, 1275±10°C and 1240±10°C, respectively. The effect of high viscosity of the doping solution gives rise to an accumulation of the deposited soot particles after solution soaking process due to generation of high viscous force within the pores along with loosing out of the soot layers from the inner surface during drying with inert gas as well as heating process.

### 2.8 Soaking, drying and heat treatment of solution impregnated porous layers

Solution soaking process was done for a period of 30-60 minutes (depending on the numbers of deposited porous layers) followed by drying with flow of $N_2$ gas for a period of 45-60 minutes under very low pressure (~0.5-1.0 Psig) as well as drying thermally by heating around 600-700°C for a period of 10 minutes with flow of $O_2$ at the rate of 500cc/min along with He flow at the rate of 75cc/min for elimination of the problem of any disturbances of the thick porous layer (50-75μm).

### 2.9 Dehydration

Dehydration step is very important to reduce OH content of the fabricated fiber. Dehydration of core layer is carried out at lower temperature range (600-700°C) in presence of $Cl_2$, $O_2$ and He due to evaporation of LiCl from $Li_2O$ during dehydration as the equilibrium constant value of the chlorination of lithium oxide is very high (Fig. 6) derived from FactSage (Facility for the analysis of chemical thermodynamics) 5.5 thermo chemical software and database. $Li_2O$ plays an important role to make transparent glass-ceramic (Morimoto, 2004; Morimoto & Emem, 2004). The optical transparency of yttria-alumino silica based core glass has been lost; if more amount of lithium evaporate.

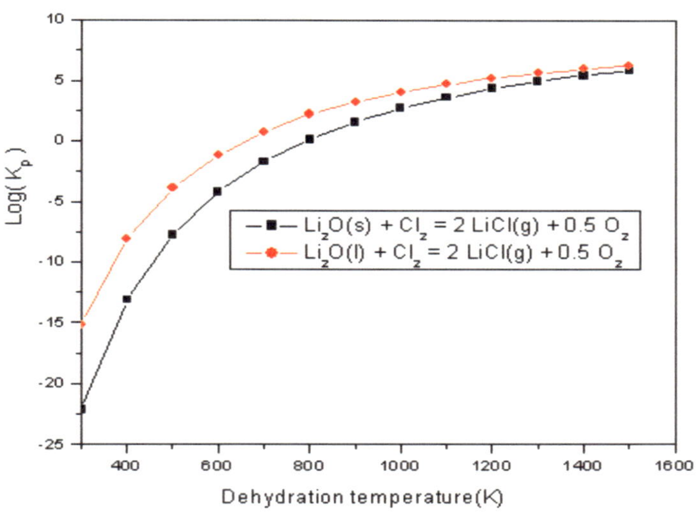


Fig. 6. Dehydration was done at low temperature around 700°C due to very high quilibrium constant of the evaporation of LiCl from $Li_2O$

### 2.10 Sintering

At sintering stage, the heating temperature was increased gradually from 1300 to 1900°C by a step-wise increment of 100°C for elimination of the problem of formation of small bubbles either at the core clad boundary or within the whole core region in presence of a mixture of $O_2$ and He having the ratio of He:$O_2$ (1:4 to 1:1). The disturbances of thick soot layer through the formation of bubbles either at the core-clad boundary or within the whole core region arises if the porous layers are not deposited at optimum deposition temperature. The formation of small bubbles also generate at the core-clad boundary during sintering of the thick soaked porous layers if the deposition temperature varies much from one layer to another layer specifically if the first layer deposited at higher temperature compared to the other layers or all the porous layers are deposited at higher temperature around 1500°C. The main reason arises due to lack of suitable porosity of the innermost layer compared to the other layers in terms of soaking ability of the dopant precursors for smooth sintering. In this case, the bubble formation occurs due to high viscosity mismatch between the inner and uppermost layers. During viscous sintering process, the upper layers get sintered at lower temperature compared to the innermost layer. As a result of it, small bubbles may generate during sintering as well as collapsing stage from the innermost un-sintered layer.

### 2.11 Collapsing process

The collapsing was done in 3 steps. In the first step, partial collapsing occurs under high flow of $O_2$ around 250cc/min and He of 65cc/min at 15.0 psig to maintain circulartity of the tube with burner speed of 40mm/min. In the second step the flow of $O_2$ was reduced to 75cc/min and He of 25cc/min at 5-6 psig with burner speed of 30mm/min. In the last pass, the flow of $O_2$ was increased to 125cc/min and He of 50 cc/min at 10 psig with burner speed of 15-20mm/min to maintain the perfect circularity of the large core after opening of the by-pass line at the moment of final collapsing of one end.

The solution doping technique followed by MCVD process using the above fabrication parameters provides homogeneous distribution of Al, Y, P and Yb ions as well as very uniform refractive index within the whole large core region of nano-engineering preform or fiber. The uniform doping levels within such large core optical fiber preform and refractive index profiles of corresponding optical fibers are also mentioned in this chapter.

### 2.12 Incorporation mechanishm of fluorine into porous silica layer

Fluorine was incorporated into porous layer through solution soaking process using $H_2SiF_6$ (fluorosilicic acid) for 10-15 minutes. To prevent etching of porous layer in presence of F, the optimized 10-15% diluted $H_2SiF_6$ was used. Furthermore, to avoid the formation of gelatinous precipitation with direct use of fluorosilicic acid into the alcoholic solution of chloride precursors, the double solution soaking process was performed for the incorporation of fluorine. Incorporation of fluorine into deposited porous silica surface was done based on the following mechanism.

The dilution of supersaturated fluorosilicic acid solution with water controls the incorporation of fluorine into the resulting porous silica matrix. An intermediate, hydrolyzed species, $SiF_n(OH)_{4-n}$ (n < 4), will be formed by the reaction shown in Eq. (1) (Chou & Lee, 1994; Homma et al., 1993). These species present in the solution react with the porous silica substrate surface. The fluorine-containing siloxanes are subsequently formed, which adsorb onto the surface where condensation and bonding occurs between the oligomers and surface hydroxyl groups (Homma et al., 1993).

$$H_2SiF_6 + (4-n)H_2O \leftrightarrow SiF_n(OH)_{4-n} + (6-n)HF \tag{1}$$

$$5H_2SiF_6 + SiO_2 \rightarrow 3[SiF_6SiF_4]^{2-} + 2H_2O + 6H^+ \quad (2)$$

On the other hand, if we have used concentrated fluorosilicic acid, porous silica can be dissolved through formation of fluoro silicon complexes such as $[SiF_6.SiF_4]^{2-}$ shown in Eq. (2). The bridged fluoro silicon complex has electron deficient silicon because of the high electronegativity of the bonded fluorine, which create weak Si-F bonds. Such kind of bonds is very much prone to nucleophilic attack by water where fluorine ion ($F^-$) combines with the proton to form hydrofluoric acid (HF). pH of the solution was adjusted properly where the high acidity of the solution allows protons to react with $[SiF_4(OH)_2]^{2-}$ to form $SiF_4$ and water through Eq. (3). Hydrolysis of $SiF_4$ will then yield hexa fluorosilicate anion, protons and silicic acid through Eq. (4) for incorporation of fluorine into porous silica layers.

$$[SiF_4(OH)_2]^{2-} + 2H^+ \rightarrow SiF_4 + 2H_2O \quad (3)$$

$$3SiF_4 + 4H_2O \rightarrow 2SiF_6{}^{2-} + Si(OH)_2 + 4H^+ \quad (4)$$

A list of preforms having different doping levels along with fiber parameters are given in the following two tables.

| Nano-preforms | Value of Al/Y (molar ratio) | $Yb_2O_3$ (mol%) | $P_2O_5$ (mol%) | BaO (mol%) | Fluorine (mol%) | Average particle size (nm) | Background loss at 1285nm (dB/Km) |
|---|---|---|---|---|---|---|---|
| NYb-1 | 20 :1 | 0.07 | ---- | 0.03 | ----- | ---- | ---- |
| NYb-2 | 13 :1 | 0.12 | ---- | 0.05 | ----- | ---- | ---- |
| NYb-3 | 6 :1 | 0.12 | ---- | 0.05 | ----- | ---- | 30 |
| NYb-4 | 1.75 :1 | 0.14 | ---- | 0.07 | ----- | 20-30 | 75 |
| NYb-5 | 1.75 :1 | 0.14 | 0.40 | 0.08 | ----- | 10-25 | 125 |
| NYb-6 | 1.75 :1 | 0.14 | 0.50 | 0.07 | ----- | 25-50 | 450 |
| NYb-7 | 1.75:1 | 0.14 | 0.60 | 0.08 | 0.25 | 25-60<br>3-5 | 2200 |
| NYb-8 | 1.75:1 | 0.14 | 0.60 | 0.08 | 0.30 | 25-65<br>3-5 | 2400 |
| NYb-9 | 1.75:1 | 0.14 | ----- | 0.08 | 0.25 | 6-10 | 40-50 |
| NYb-10 | 1.7:1 | 0.10 | 0.20 | ---- | ----- | 10-15 | 45-60 |
| IMNYb3 | 1.7:1 | 0.15 | 0.20 | ---- | 0.25 | 5-10 | 40-55 |
| IMNYb4 | 1.7:1 | 0.15 | 0.20 | ---- | 0.25 | 5-10 | 45-60 |
| IMNYb-12D | 1.7:1 | 0.15 | .20 | --- | 0.25 | 5-10 | 40-65 |
| IMNYb-12P | 1.7:1 | 0.15 | .20 | --- | 0.25 | 5-10 | 45-60 |

Table 1. The doping levels of different nano-preforms evaluated by EPMA along with size of the particles and background loss of the fabricated nano-particles fibers.

| Doping host | Fiber ID | No. of porous layers | Deposition temperature (±10°C) | CSA of the deposited tube ($mm^2$) | NA (±0.02) | Cladding absorption at 976 nm (dB/m) | Core dia. (μm) |
|---|---|---|---|---|---|---|---|
| $SiO_2$-$Y_2O_3$-$Al_2O_3$-$P_2O_5$-$Yb_2O_3$ | NYb 4-NYb10 | 6 | 1350 | 82-85 | 0.09 | 5-10 | 15.0-20.0 |
| $SiO_2$-$Y_2O_3$-$Al_2O_3$-$P_2O_5$-$Yb_2O_3$ | IMNYb-3, IMNYb-4 | 6 | 1275 | 58-60 | 0.09 | 5-10 | 22.0±2.5 |
| $SiO_2$-$Y_2O_3$-$Al_2O_3$-$P_2O_5$-$Yb_2O_3$ | IMNYb-12D & IMNYb-12P | 10 under expanded condition | 1240 | 58-60 | 0.09 | 5-10 | 32±2. 5 |

Table 2. Fabrication and fiber parameters of different large core nano-engineering fibers.

### 2.13 Thermal annealing process

The preform samples were annealed at different temperatures for appropriate time to find out the optimum temperature for the formation of nano-crystallites inside the silica glass matrix. After fabrication of preforms, we have made five circular sections of thickness around 1-1.5 mm having 10 mm diameter from each preform. Each section was heated in a temperature-controlled furnace of 1000, 1200, 1300, 1450 and 1650°C at the rate of 20°C/min and left for 3 hours once the temperature has been reached, to achieve the structural equilibrium, followed by the cooling to room temperature at a rate of 20°C/min. Depending on the glass composition the annealing step was optimized to obtain a highly transparent glass with the nano-sized crystallites distributed homogeneously in that, as well as a complete segregation of the RE ions into the nano-crystal site.

The diagram of annealing sequences of fiber preform samples heated at different temperatures in a closed furnance of Ar gas environment was shown in Fig. 7. The microscopic picture of the core region of fiber preform (NYb-7) with and without annealing was given in Fig. 8. The core glass of preform annealed at 1650°C and without annealing as shown in Fig. 8 becomes almost transparent. Whereas the preform annealed at 1450°C becomes highly opaque. This phenomena indicates that the nature of $Yb_2O_3$ doped yttria-alumino silica host under thermal annealing at 1450°C becomes different than that of core glass matrix of preform samples without annealing and with annealing under heat treatment at 1650°C. The nature of such kind of glass was evaluated from XRD and TEM along with EDX as well as electron diffraction pattern analyses. The observation of the thermal annealing of preform samples at different temperatures is given in Table 3.

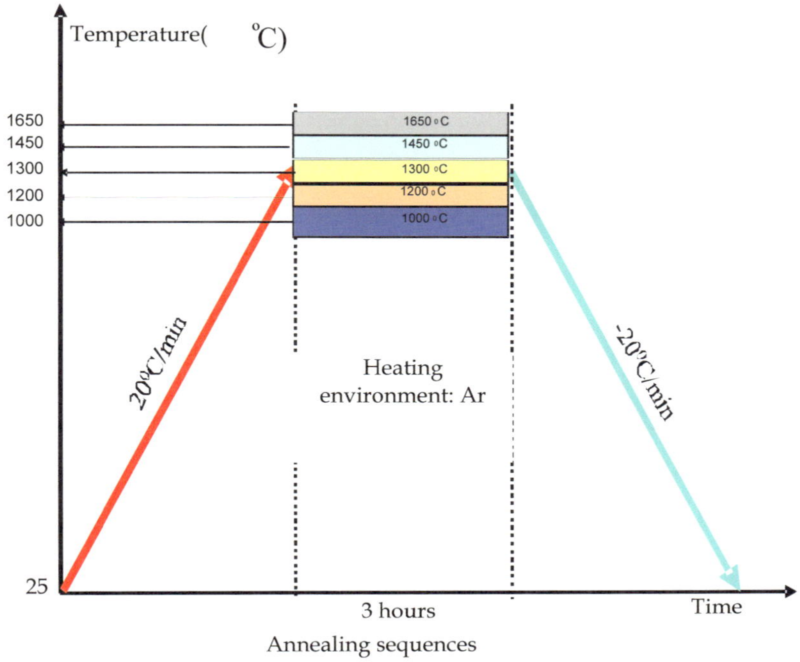


Fig. 7. The diagram of annealing sequences fiber preform samples heated at different temperatures in a closed furnance.

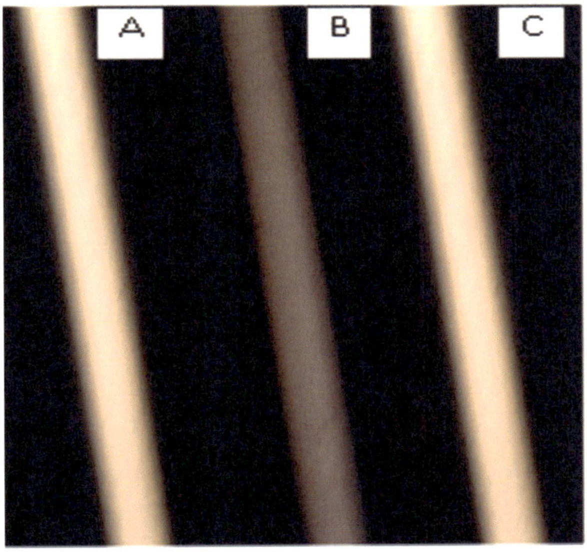


Fig. 8. Microscopic picture of the core region of preforms (A) Preform without annealing (B) Annealed at 1450°C for 3 hours (C) Annealed at 1650°C for 3 hours.

| Sample numbers | Heating rates (°C/min) | Soaking period (hr) | Observation against thermal soaking temperature | | | | |
|---|---|---|---|---|---|---|---|
| NYb-7 Preform | 20 | 3 | 1000°C Transparent | 1200°C DNP (slightly milky) | 1300°C DNP (slightly milky) | 1450°C CNP (strongmilky) | 1650°C (slightly milky) (No particles) |
| NYb-8 Preform (Fluorine doped) | 20 | 3 | 1000°C Transparent | 1200°C DNP (slightly milky) | 1300°C DNP (slightly milky) | 1450°C CNP (strong milky) | 1650°C (slightly milky) (No Particles) |

Table 3. Conditions and effect of thermal annealing of preform samples at different temperatures.

After that milling and polishing of the annealed preforms are done. Milling was done for removal of the silica mass equivalent to 1mm depth from one side of the preform. Preform was milled using ultrasonic machine. After milling, D-shaped fiber with low index polymer coated was drawn using fibre draw tower.

## 3. Material characterizations of the nano-engineering doping host

The material characterizations of nano-engineering doping host have been done at different stages from deposition of multiple porous layers to making of optical preforms through the following various techniques.

### 3.1 Scanning electron microscopy (SEM) analyses of porous layers

Different soot parameters namely density, pore sizes, porosity, surface area and soot layer thickness have been evaluated for optimization of deposition temperature of multiple porous layers against the numbers of deposited layers to get uniform distribution of different co-dopants along the diameter of fabricated preforms. The soot density was found to be around 0.258 gm/cc at optimum deposition temperature regions namely 1350±10°C. The porosity of six numbers of porous layers deposited at 1350±10°C along radial direction was found to be 58% to 62% from innermost layer to outermost layer as described earlier in Fig 3. The reason behind the slight decrease in porosity is a consequence of partial sintering. Both scanning electron microscopy (SEM LEO, S43Oi, LEO, UK) and field emission scanning electron microscopy Supra 35VP (Carl Zeiss, Germany) images of soot layer samples were captured before and after solution doping shown in Fig. 9 to investigate the average pore sizes, their distribution and soot layer thickness. The average pore size was evaluated from SEM micrographs using Image analyzer and found to be 1.25-1.5 micron.

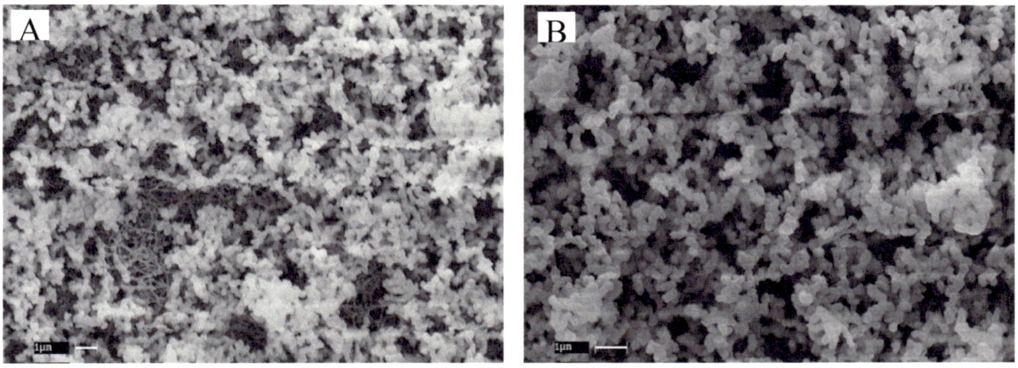


Fig. 9. SEM pictures of axial view of the porous layers (6 numbers) deposited at temperature around 1350±10°C (A) before (B) after solution soaking.

### 3.2 Atomic force microscopy (AFM) study

The technique which used for the analysis of porous silica layers is the Atomic Force Microscopy (AFM) Model: Multiview 3000 (Nanonics Imaging Ltd., Israel). AFM images are obtained by measurement of the force on a sharp tip (insulating or not) created by the proximity to the surface of the sample. This force is kept small and at constant level with a feedback mechanism. When the tip is moved sideways it will follow the surface contours. By this way, three dimensional images of the surface of the porous samples are obtained. A three-dimensional view and a corresponding cross section graph of the deposited porous surface shows a hillock profile with a dome height of 3 ± 0.25 μm and distance between dome tops 4.0±1.0 μm shown in Fig. 10 and Fig. 11. Both circular and irregular shaped pores were observed, with diameters of 3 ± 1 μm and a pore-to pore distance of 4.5 ± 0.5 μm. The porosity of the deposited thick layer was determined as 60 ± 2.0%.

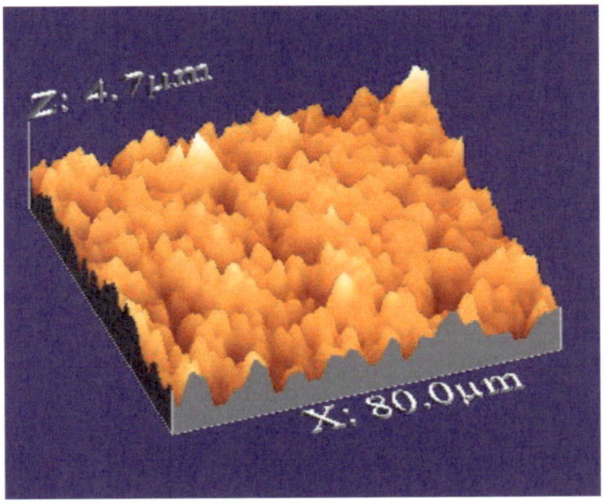


Fig. 10. A three dimensional AFM image of the surface of porous phospho-silica multi-layers.

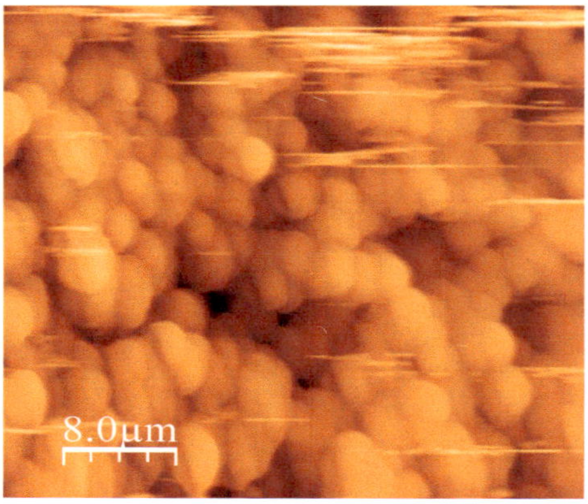


Fig. 11. A cross-sectional AFM image of the surface of porous phospho-silica multi-layers.

### 3.3 Differential thermal analyses (DTA) of the porous material

The determination of glass transition and crystallisation temperatures as well as melting temperature of such nano-engineering core glass material was also important for better understanding of the phenomena of phase separation as well as crystallisation mechanism with respect to the doping solution composition. Differential thermal analyser (DTA) Model: STA 409C (Netzsch, Germany) on samples ground to an average particle size of less than 25 µm was prepared. In such case, unsintered oxide porous layer of composition $SiO_2$-$P_2O_5$-$Y_2O_3$-$Yb_2O_3$-$Li_2O$-BaO formed by MCVD process followed by solution doping technique (soaked with an alcoholic solution of $YbCl_3$, $6H_2O$, $YCl_3$ $6H_2O$, $AlCl_3$ $6H_2O$, $LiNO_3$, $BaCl_2$ $2H_2O$) and after oxidation of the soot was removed from the inner surface of the silica tube and then ground to a powdered material for DTA. The DTA measurements were carried out using minimum about 30 mg of sample in a Pt crucible. Data for each run was automatically collected from the DTA apparatus. The glass transition and crystallization temperature of such type of glass was found to be around 985 and 1115°C respectively shown in Fig. 12. The exact melting temperature was not detected due to the temperature limitation of DTA upto 1450°C.

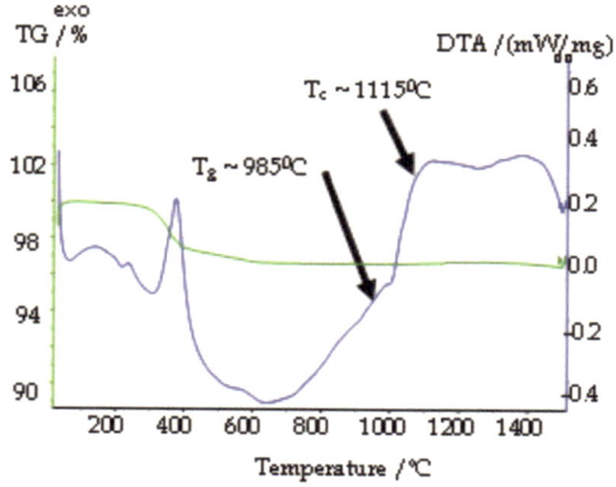


Fig. 12. Differential thermal analysis (DTA) curve of fluorinated $Yb_2O_3$ doped silica soot.

### 3.4 Electron probe micro-analyses (EPMA)

The average dopant percentages in samples were measured by an electron probe microanalyser (EPMA) at University of Minnesota, Electron Microprobe Laboratory, USA. For this, around 3-4 mm of the preform sample was grounded and polished both sides. Finally EPMA of the polished preform samples of thickness 1-1.5 mm was done with the maximum spatial resolution of 1μm after applying a thin graphite coating layer. Data of EPMA of preform samples were taken with a spatial gap of 30 micron within the whole core region shown in Fig. 13. The corresponding different dopant distribution of large core fiber preforms IM-NYb3 and IMNYb-12D are described in Fig. 14 and Fig. 15 respectively.

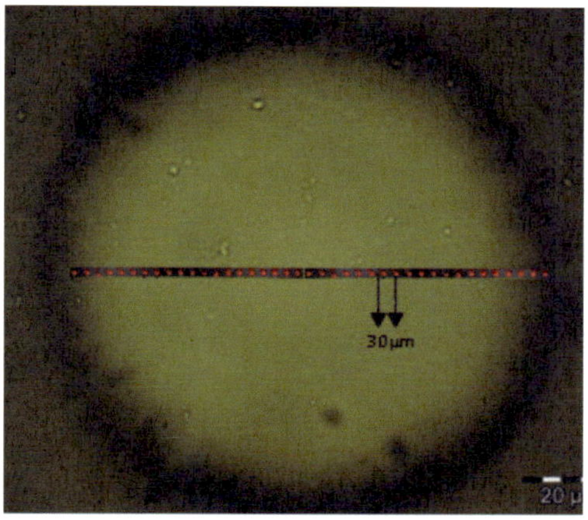


Fig. 13. Data of EPMA of preform samples taken with a spatial gap of 30μm within the core region

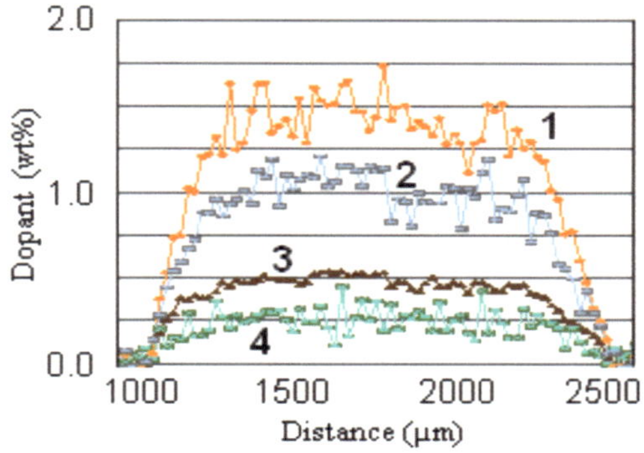


Fig. 14. Distributions of doping levels of $Al_2O_3$ (1), $Y_2O_3$ (2), $Yb_2O_3$ (3) and F (4) within the core area of fiber preform (IM-NYb-3).

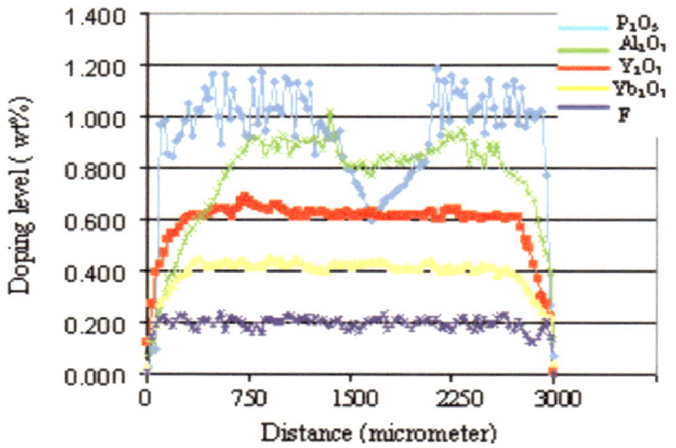


Fig. 15. Distributions of doping levels of $Al_2O_3$ (1), $Y_2O_3$ (2), $Yb_2O_3$(3) and F(4) within the core area of fiber preform (IMNYb-12D).

### 3.5 X-ray diffraction (XRD) study

To detect the exact crystalline phases formed during the heat treatment, X-ray diffraction (XRD) was performed on finely ground core glass specimens. The X-ray Patterns was collected using a powder diffractometer. X-ray diffraction spectra were extracted from the raw 2-dimensional data using the Fit2D program (Hammersley et al., 1996). Spectra were collected on nano-engineering preform sample (NYb-8) annealed at different temperatures shown in Fig. 16. The calibration of the distance between the sample and the detector was carried out by using a $LaB_6$ sample (NIST SRM660). Only the samples annealed at 1450ºC present crystalline peaks. The preform annealed at other temperatures does not show any characteristics diffraction peak of Yb:YAG crystals. The exact diffraction peaks of Yb:YAG crystals embedded in silica matrix is found to be 33.43(a), 46.606(b), 56.34(c) when annealed at 1450°C in comparison to the exact characteristics peaks at 33.064, 46.183 and 55.732 given in Joint Committee on Powder Diffraction Standards (JCPDS-38-0222) (Joint Committee on Powder Diffraction Standards, data file. 38-0222) and also observed by W. Yusong et al. (Yusong et al., 2007)

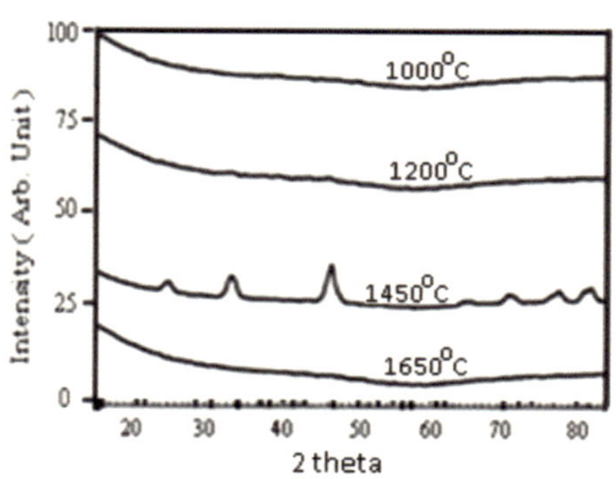


Fig. 16. X-ray diffraction spectra of the doped region of preform NYb-8 annealed at four different temperatures for 3 hours and different crystalline peaks developed for sample annealed at 1450ºC.

### 3.6 X-ray absorption spectroscopy

Distribution of Yb ions in the glass preform was also determined by the fluorescence yield of the Yb Lα line given in Fig. 17. The vertical line marks the point where the XAS measurement has been carried out. The following figures shows a typical distribution profile of Yb ions in the preform measured by X-ray fluorescence: these permits to determine the size of the Yb-doped zone that resulted to be of about 1.5 mm. The vertical bars indicate the point where the EXAFS spectra will be carried out that was chosed to be in the middle of the fibre core.

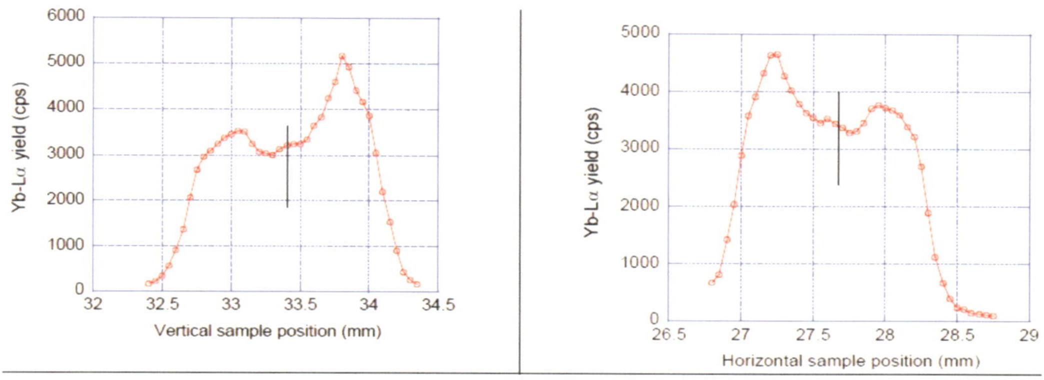


Fig. 17. The fluorescence yield of the Yb Lα line of nano-engineering preform (NYb-5). The vertical line marks the point where the XAS measurement has been carried out.

### 3.7 High resolution transmission electron microscopy (HRTEM) with electron diffraction pattern and EDX analyses

High resolution transmission electron microscope (HTEM) Model:Tecnai G2 30ST (FEI Company, USA) images of preform samples were used to study the core glass morphology using $LaB_6$ emitter. The TEM specimens were mechanically polished and dimpled to thickness of about 10 μm. The final thinning of the samples to electron transparency was carried out using an Ar-ion mill. To evaluate the composition of phase-separated particles the electron beam was focused on the particles and then focused in an area outside of the particles, when the energy dispersive X-ray analyses (EDX) data were taken. The nature of the particles was evaluated from their electron diffraction pattern. TEM picture along with electron diffraction pattern of $Yb_2O_3$ doped nano-crystalline glass based optical fiber preform (NYb-4) after thermal annealing at 1450°C temperature was shown in Fig. 18.

## 4. Effect of suitable dopant levels and thermal annealing on formation of crystalline nano-particles

Study on the effect of suitable dopant levels specifically the ratio of Al:Y and thermal annealing on formation of crystalline nano-particles within the core region of optical fiber preform samples were done through TEM along with electron diffraction pattern, EDX and XRD analyses. For this purpose several $Yb_2O_3$ doped large core preforms (listed in table 1) were fabricated with varying Al:Y ratio from 20:1 to 1.75:1 and other co-dopants, in order to study the dependence of glass composition on the formation of nano-crystalline particles. All co-dopants, except for P are incorporated in the structure through solution doping process.

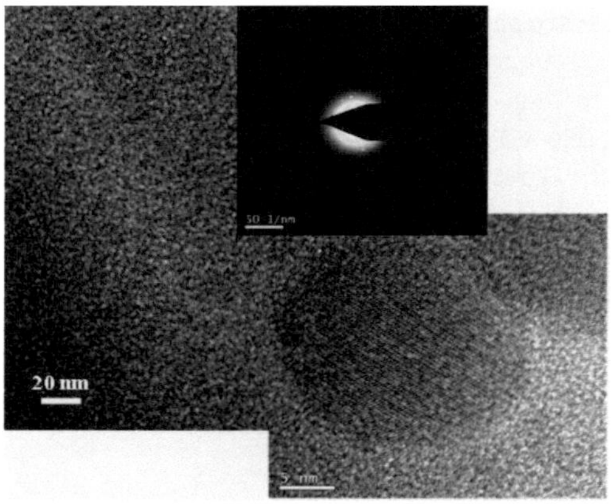


Fig. 18. TEM picture along with electron diffraction pattern as well as $Yb_2O_3$ doped crystalline nano-particles developed within yttria-alumino silicate glass of optical fiber preform (NYb-4) after thermal annealing at 1450°C

Two important parameters were involved for formation of $Yb_2O_3$ doped phase-separated yttria alumino-silicate nano-crystalline host. First, the doping ratio of Al:Y and second, suitable thermal treatment of the fabricated preform. The optimised thermal annealing conditions of preforms were identified to get $Yb_2O_3$ doped phase-separated nano-crystalline particles of composition Yb:YAG. The strength of dopant precursors of Y and Al within the doping solution was varied gradually from preform NYb-1 to NYb-4 to get the initial condition for formation of the phase-separated nano-particles within the preform samples. The ratio of Al:Y within the fiber preform samples were calculated from EPMA results. No-phase-separated particles were developed within the preforms of NYb-1, NYb-2 and NYb-3 under any thermal annealing conditions when the ratio of Al:Y varied from 20:1 to 6:1. Fig. 19 shows the TEM image of preform NYb-2 having Al:Y ratio of 13:1 and annealed at temperature of 1300°C with no evidence of nano-particle formation. However, the preform NYb-4 having Al:Y ratio of 1.75:1 shows phase-separated nano-particles under thermal annealing at 1300°C for 3 hours (Fig. 20). The formula of YAG crystal is $Y_3Al_5O_{12}$ where the ratio of Al:Y is around 1.7. Here the role of Al and Y at that ratio favours the formation of YAG crystals stoichiometrically. Higher than this ratio, excess Al prevents the formation of phase-separation which does not enter within the immisicibilty region of yttria alumino silicate glass.

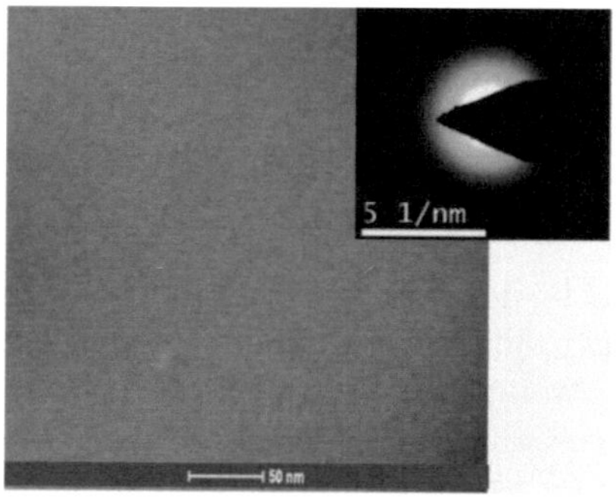


Fig. 19. TEM image of the doped area of preform NYb-2 annealed thermally at around 1300°C for 3 hours.

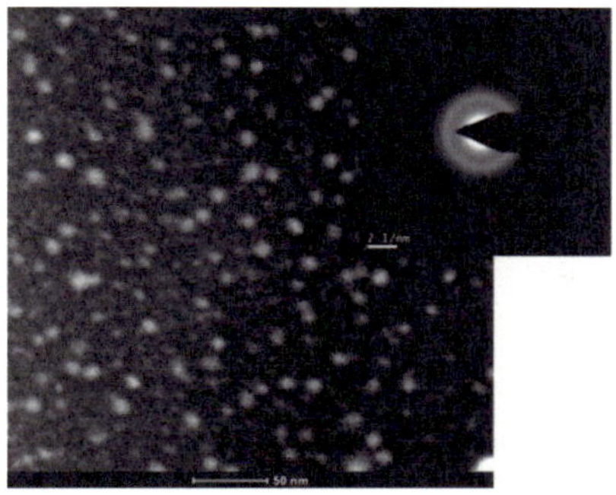

Fig. 20. TEM image of the doped area of preform NYb-4 annealed thermally at around 1300°C for 3 hours along with electron diffraction pattern.

On the other hand preform NYb-4 when annealed at 1000°C for 3 hours does not form any phase-separated particles as shown in Fig. 21. The formation of phase-separated amorphous particles started when the preform samples NYb-5 and NYb-6 annealed at 1200°C, as shown in Fig. 22 and Fig. 23.

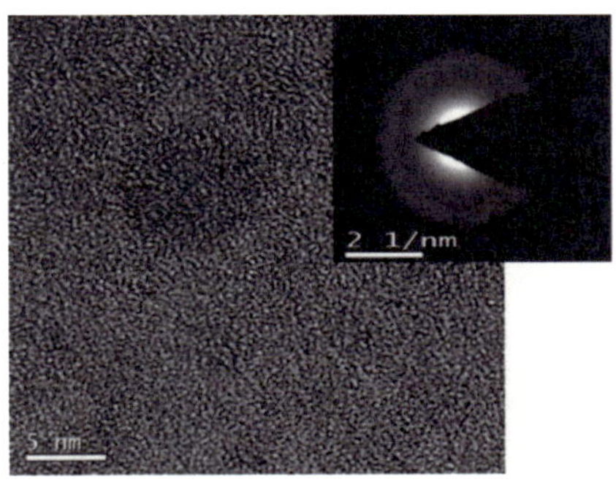


Fig. 21. TEM image of doped area of preform NYb-4 annealed thermally at around 1000°C.

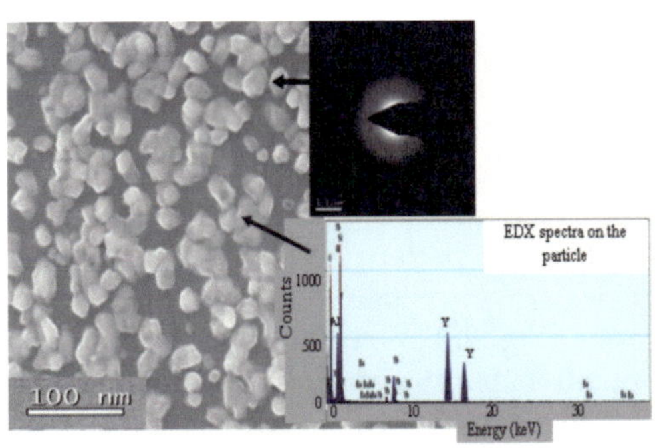


Fig. 22 . TEM image of doped area of preform NYb-5 annealed at 1200°C containing 0.40 mol% $P_2O_5$ along with electron diffraction pattern and EDX spectra.

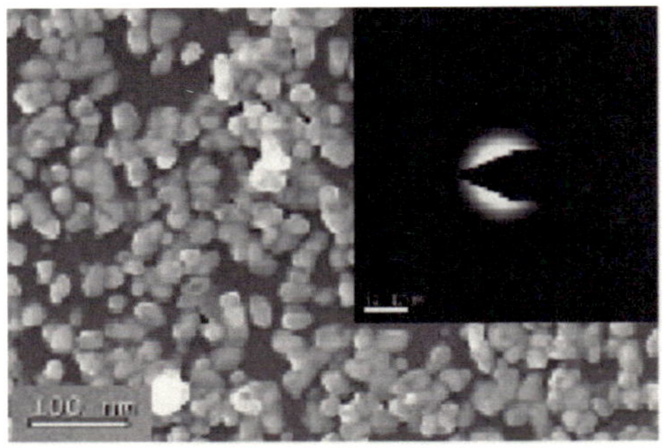


Fig. 23. TEM image of doped area of preform NYb-6 annealed at 1200°C containing 0.50 mol% of $P_2O_5$ along with electron diffraction pattern.

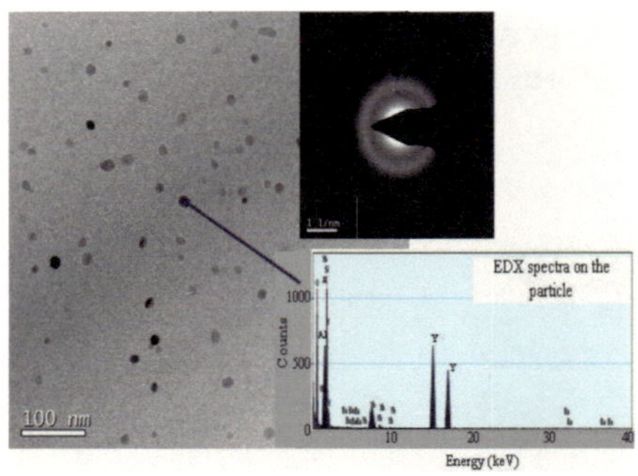


Fig. 24. TEM image of the doped area of preform NP-5 annealed at 1300°C containing 0.40 mol% of $P_2O_5$ along with electron diffraction pattern and EDX spectra.

At this annealing temperature the nano-particles which have grown together resulted in an interconnected nano-structure. The electron diffraction pattern shows the formation of amorphous phase (inset Fig. 22 and Fig. 23). When the same preform samples heated at around 1300°C, the nano-particles are separated from each other and mostly single isolated particles distributed uniformly in the core glass matrix were observed, as shown in Fig. 24 and Fig. 25. The TEM pictures shown in Fig. 22 to Fig. 25 for heat treated samples under 1200 and 1300°C are to study the effect of annealing temperature on the formation of phase-separated particles along with its nature. The isolated particles are not YAG nano-crystallites. The particles are $Yb_2O_3$ doped phase-separated yttria-rich alumino-silicate nano-particles evaluated from their corresponding EDX data.

Further annealing of the preform at around 1450°C, all the phase-separated particles became crystalline. The electron diffraction pattern as shown in the inset of Fig. 26 and Fig. 27 demonstrates that the particles are crystalline in nature.

High resolution micrographs (Fig. 28) were extracted from the previous TEM pictures of preforms NYb-5 and NYb-6 after thermal annealing at 1450°C for 3 hours, which show the crystalline nano-structure of less than 20 nm in size and surrounded by the amorphous $SiO_2$ matrix. At the crystalline orientation the phase contrast allows to differentiate lattice plans of ~ 0.3nm separation. This matches with the value reported by Jian-Cheng Chen (Chen, 2006). The lattice planes show that the original nano-crystals have combined to form a continuous crystalline lattice. Absence of grain boundaries within the crystalline phase

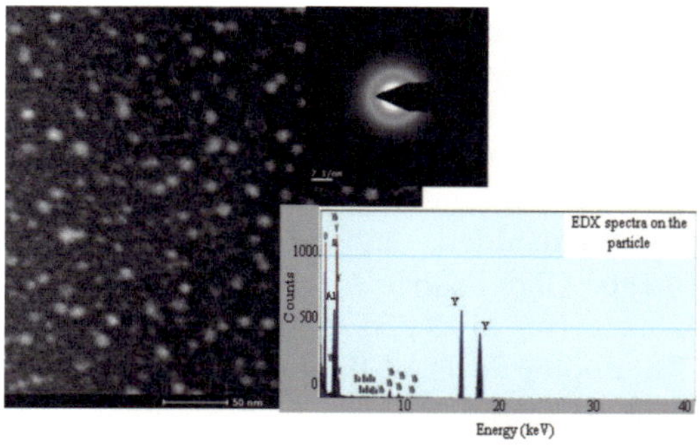


Fig. 25. TEM image of the doped area of preform NYb-6 annealed at 1300°C containing 0.50 mol% of $P_2O_5$ along with electron diffraction pattern and EDX spectra.

might be due to the agglomeration of independent nano-crystals. Here, two nano-crystals are separated by a layer of amorphous phase. The interconnection of the phase-separated particles developed during thermal annealing at 1450°C leads to the growth of a nano-crystalline structure shown in Fig. 28. All the particles were observed to be near round shape. The driving force behind the roundish morphology and interconnection of individual nano-crystals is due to the lowering of surface energy of the connected crystal network over individual nano-crystals.

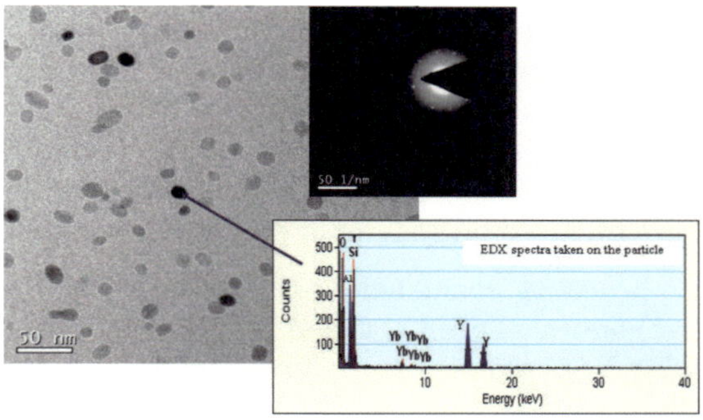


Fig. 26. TEM image of the doped area of preform NYb-5 annealed at 1450°C containing 0.40 mol% of $P_2O_5$ along with electron diffraction pattern and EDX spectra.

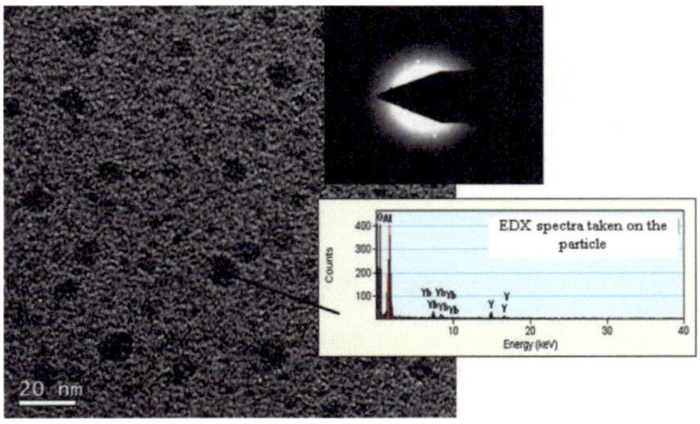


Fig. 27. TEM image of the doped area of preform NP-6 annealed at 1450°C containing 0.50 mol% of $P_2O_5$ along with electron diffraction pattern and EDX spectra.

The composition of such phase-separated crystalline particles was evaluated by comparing their energy dispersive X-ray analyses (EDX) data, taken from directly on the crystal particles and from the areas of outside the particles shown in Fig. 28. The EDX spectra taken from the crystal particles detected as Si, Al, O, Yb and Y signals, whereas the EDX data taken from the areas outside the particles detected Si, Al and O without Yb and Y. Si signal appear as the crystals are embedded with $SiO_2$. The dopant concentration of nano-crystallites with respect to the silica matrix was evaluated from EDX curve and shown in Fig. 28. The evaluated data consists of 13.0 atom% of Al, 0.02 atom% of Yb, 8.0 atom % of Y and 34.0 atom% of O. The doping level of the formation of Yb:YAG crystals within fiber preform sample (NYb-5) is 0.02 atom%. Such compositional analyses indicates that the phase-separated crystalline nano-particles were Yb:YAG crystals dispersed into the silica glass matrix.

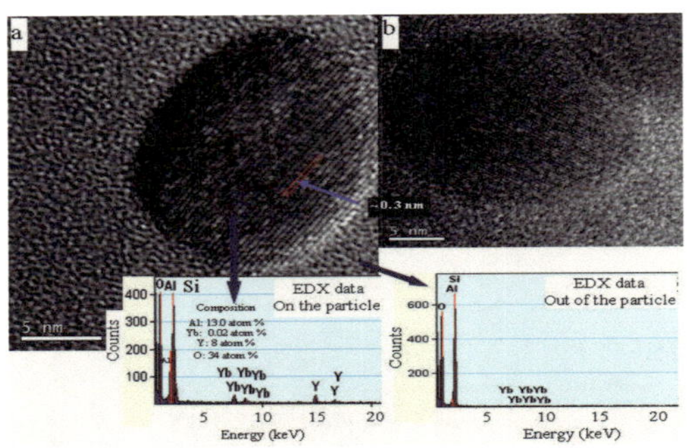


Fig. 28. HRTEM image showing the lattice plan of the crystalline nano-particles developed within the doped region of preforms (a) NYb-5 and (b) NYb-6 annealed at 1450°C for 3 hours along with EDX spectra taken on and out of the particles.

The nature of the glass host was evaluated from X-ray diffraction spectra of the doped region of preform (NYb-5) annealed at 1300°C and 1450°C for 3 hours shown in Fig. 29. The vertical lines correspond to YAG phase. The preform annealed at 1300°C does not show any characteristics diffraction peak of Yb:YAG crystals. The exact diffraction peaks corresponding to the lattice plan of 420, 611 and 633 of Yb:YAG crystals embedded in silica matrix is found to be 33.43, 46.606, 56.34 respectively when annealed at 1450°C in comparison to the exact characteristics peaks at 33.064, 46.183 and 55.732 respectively described in Joint Committee on Powder Diffraction Standards (JCPDS-38-0222) (Joint Committee on Powder Diffraction Standards, data file. 38-0222) and also observed by W. Yusong et. Al (Yusong et al., 2007). Such a small variation occurs as Yb:YAG crystals are embedded in silica glass based matrix and not in a perfectly Yb:YAG ceramic state.
Generally the formation of such type of Yb:YAG crystals in bulk glass samples occurs by the following way :

$$5Al_2O_3 + 3(1-x)\ Y_2O_3 + 3xYb_2O_3 = 2Y_{3(1-x)}Yb_{3x}Al_5O_{12} \text{ where } x = 0.1$$

Similar type of reaction may takes place within yttria-alumino-silicate core glass matrix of optical fiber preform where the substitution of $Y^{3+}$ by $Yb^{3+}$ occurs due to their same valence state as well as similar ionic radius. On the other hand, $Si^{4+}$ may substitute for $Al^{3+}$ or it may sit at an interstitial position since the ionic radii of $Si^{4+}$ are 0.040 and 0.054 nm, and the ionic radii of $Al^{3+}$ are 0.053 and 0.068 nm, if their coordination numbers are 4 or 6, respectively[59].

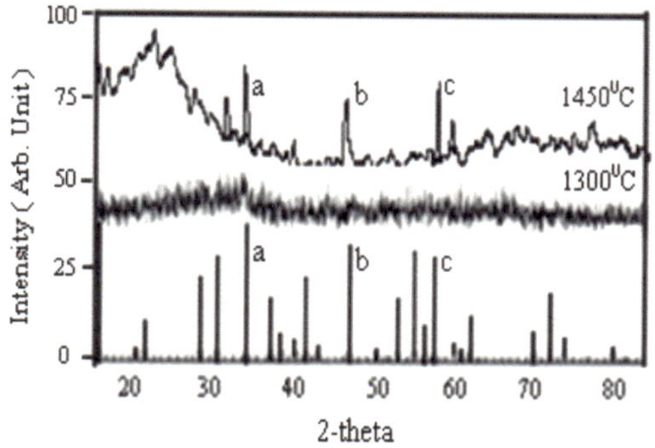


Fig. 29. X-ray diffraction spectra of the doped region of preform NYb-5 annealed at 1300°C and 1450°C for 3 hours. The vertical lines corresponds to YAG phase.

The possibility of sitting $Si^{4+}$ to an interstitial site can be neglected as the large strain and higher number of vacancies per Si in an interstitial location will make the energies required for the incorporation very high. We have not observed formation of $Yb_2O_3$ doped mullite crystals within the preform samples due to the large difference in ionic radii of $Al^{3+}$ and $Yb^{3+}$ (~0.05 nm) which results in lattice strain to restricts the formation of a solid solution comparable to mullite.

When the preform samples annealed above 1600°C, the crystalline behaviour of the particles were destroyed and most of the particles disappear, as shown in Fig. 30. This may be due to reaching of the melting temperature of the glass. The glass transition and crystallization temperature of such type of glass was found to be around 985 and 1115°C respectively as mentioned earlier in Fig. 12. The exact melting temperature was not detected due to the maximum limit of 1400°C temperature of our DTA apparatus. However from the fabrication point of view, the sintering of deposited layer starts around 1600°C which correspond to the melting temperature of the doped porous layer.

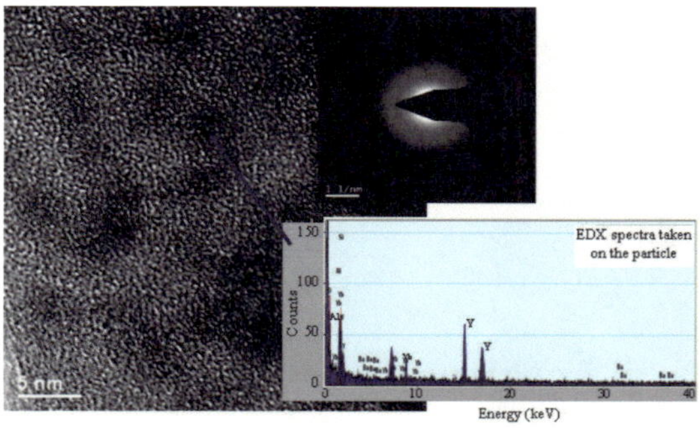


Fig. 30. TEM image of the doped region of preform NYb-5 annealed at 1600°C along with electron diffraction pattern and EDX spectra.

## 5. Effect of $P_2O_5$ and fluorine on formation of nano-particles within the core of preforms and fibers

The effect of phosphorous and fluorine doping on formation of $Yb_2O_3$ doped nano-crystalline particles within silica glass matrix was studied through high-resolution TEM analyses. The core glass remains transparent after annealing at 1000 and 1600°C. On the other hand, when annealed at 1450°C, the core region becomes slightly opaque which indicates the formation of phase-separated nano-crystalline particles in the preform sample. Fig. 26 and Fig. 27 shows the TEM pictures of two preform sections ( NYb-5 and NYb-6 ) having different content of $P_2O_5$ for investigation of the effect of $P_2O_5$ on the size and nature of the nano-particles. The maximum particle size in high $P_2O_5$ content (0.50 mol%) core glass preform, NP-6, was found to be larger compared to the preform NYb-5 containing less $P_2O_5$ content (0.40 mole%) in the core.

Increasing the $P_2O_5$ content in glass accelerate the growth of formation of phase separated particles upon heating through thermal perturbatation, where $P_2O_5$ serves as a nucleating agent owning to the higher field strength difference (>0.31) between $Si^{4+}$ and $P^{5+}$ (Tomazawa & Doremus, 1979). We have taken several TEM pictures from different sections of the core region of samples NYb-5 and NYb-6 to find out the variation of particle sizes. A wide variation from 10-25 nm and 25-50 nm in particle sizes were observed from the preform samples NYb-5 and NYb-6 respectively. The size of the nano-particles increased with increasing of the $P_2O_5$ content. At the ratio of 20:1, 13:1 and 6:1 within the glass matrix formation of any nano-particles within the doped region of optical fiber preform under annealing upto the temperature of 1600°C does not occur.

The effect of fluorine on the formation of $Yb_2O_3$ doped crystalline nano-particles within the preform samples NYb-7 and NYb-8 after thermal annealing at 1450°C are shown in Fig. 31 and Fig. 32 respectively. Here fluorine was incorporated using 10-15% fluorosilicic acid. Both preforms contain the same doping levels of $P_2O_5$ (0.60 mol%). The doping level of fluorine content was measured by electron probe microanalyses (EPMA). The distribution profile of fluorine content along the diameter of core region of two fiber preform NYb-7 and NYb-8 was shown in Fig. 33. The average doping level of fluorine in preforms NYb-7 and NYb-8 was found to be 0.25 and 0.3 mole%. The spherical particles in Fig. 31 and 32 show bimodal size distribution. The large particles are of the order of 60 nm and the smaller ones are of the order of 3-5 nm. In addition, a very dense distribution of nano-particles (~2-10 nm size) was detected which are rich in high atomic number elements of Yb and Y. This was confirmed from the corresponding EDX spectra taken in the region of densely distributed nano-particles of preform NYb-7 which shows strong signals of Yb and Y.

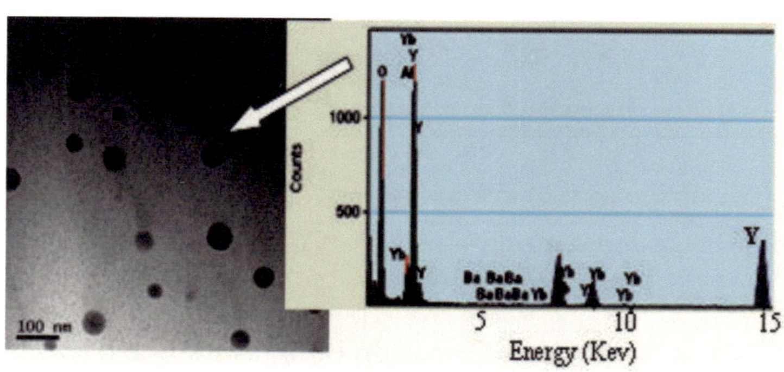


Fig. 31. TEM image and EDX spectra of the area of a dense distribution of nano-particles (2-10 nm) of preform NYb-7 annealed at 1450°C containing 0.6 mol% of $P_2O_5$ and 0.25 mol% of fluorine.

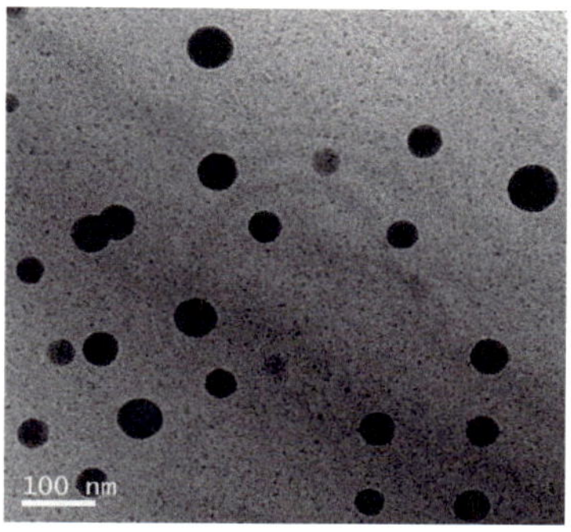


Fig. 32. TEM image of the area of preform NYb-8 annealed at 1450°C containing 0.6 mol% of $P_2O_5$ and 0.30 mol% of fluorine

The doping of fluorine in yttria-alumino silicate glass composition may have a dramatic effect on the nucleation and crystallisation behaviour of the glass contributing to the bimodal size distribution. This effect was not only a result of the stoichiometric considerations of crystal formation but also of the network disrupting role of fluorine within a glass network. Here fluorine may act as a nucleating agent, promoting crystallization of $Yb_2O_3$ doped YAG crystals within the silica glass matrix and also serves as a facilitator of the kinetics of crystallization through rearrangement of the glass network. The non-bridging oxygens preferentially bond to the silicon atoms present, thereby preventing the formation of silicon–fluorine bonds and the formation of volatile $SiF_4$. In such glass, P and F may play a different kind of nucleation as well as crystallization phenomenon for the formation of observed bimodal size distribution.

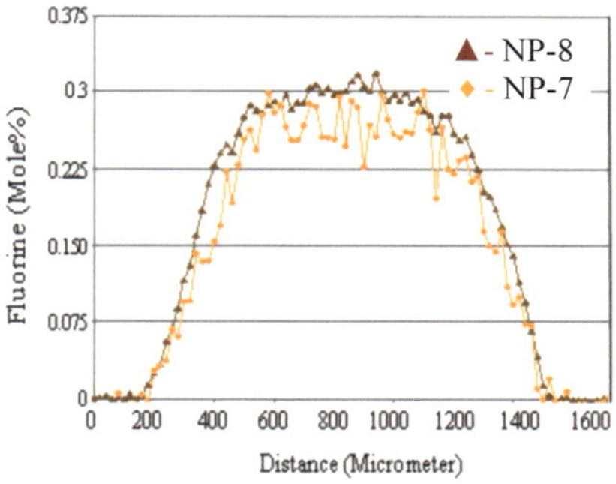


Fig. 33. Distribution profile of fluorine along diameter of two preforms NYb-7 and NYb-8.

The preform sample NYb-9, which contains 0.25 mol% F and no $P_2O_5$, after thermal annealing at 1450°C shows uniform distribution of crystalline nano-particles (Fig. 34) having sizes around 6-10 nm. HRTEM image of the black dots of doped area of preform (NYb-9) shows the formation of the nano-crystalline particles of Yb:YAG. The corresponding electron diffraction pattern shows the crystalline behaviour of the particles. The HRTEM picture clearly shows the crystalline plan of Yb:YAG crystals having 0.3 nm separation of lattice plans. The exact composition of the crystalline nano-particles was evaluated from the corresponding EDX spectra. The evaluated data shows 15.0 atom% of Al, 0.02 atom% of Yb, 9.0 atom% of Y and 36.0 atom% of O.

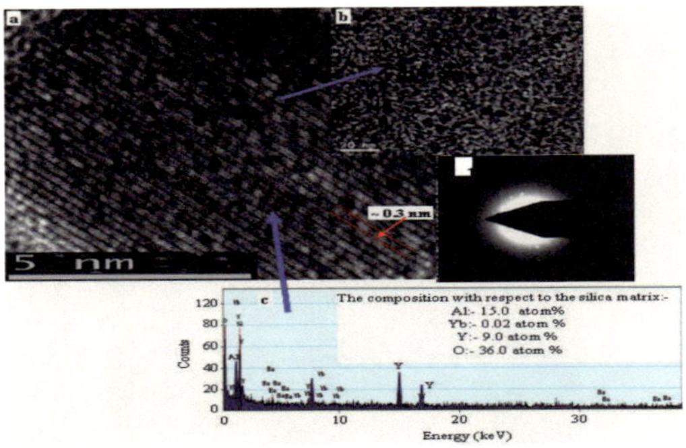

Fig. 34. HTEM image of the doped area of preform NYb-9 annealed at 1450°C containing 0.25 mol% of fluorine showing (a) the nano-crystalline particles of Yb:YAG along with (c) EDX curve and (d) electron diffraction spectra.

## 6. Nature and sizes of the particles of $Yb_2O_3$ doped nano-engineering fibers

Nature and sizes of the particles of nano-engineering fiber was evaluated from HRTEM analyses along with EDX and electron diffraction pattern.To make the TEM analyses of optical fiber samples first, fiber coating was removed. After that fiber was crashed in two different marble mortars and ground to fine powder. The glass powder was dispersed in acetone liquid. Cu saver was rinsed into the liquid and took out. After drying, the glass powder was sticked on the surface of the saver. The Cu saver was put under TEM to check the individual small particles by EDX. Nature and sizes of the particles of nano-engineering fiber was evaluated from HRTEM analyses along with EDX and electron diffraction pattern. The small phase-separated nano-particles within the doped core region of optical fiber NYb-9 was observed under TEM analyses shown in Fig. 35.

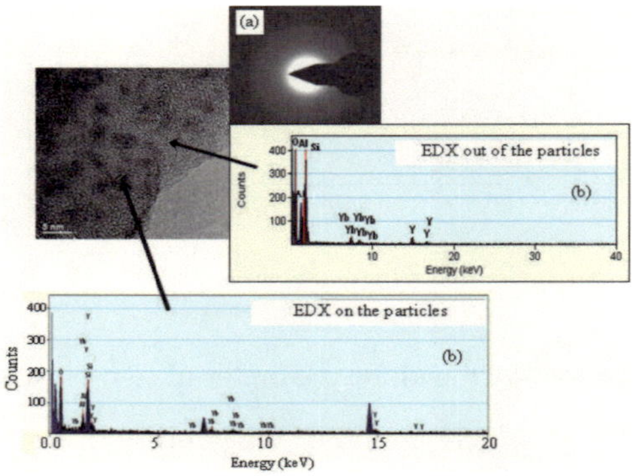


Fig. 35. TEM analyses of Yb-doped nano-particle silica optical fiber sample NYb-9 (a) Electron diffraction pattern and (b) EDX spectra on and out of the particles.

The size of the particles of the fiber core region occurs within 5.0-8.0 nm region. The EDX spectra shows that both phase-separated and non-phase separated regions contain Al, Y and Yb. However the intensity related to Y and Yb in phase-separated particles becomes very much larger than the non-phase-separated particles. On the other hand, the intensity related to Al in phase-separated particles becomes lower than that of non-phase-separated particles.

Such analysis indicates that most of ytterbium is located in the phase-separated yttria-rich region compared to Al dominated non-phase separated region. The nature of the phase-separated particles is found to be non-crystalline confirmed from their electron diffraction pattern. Such TEM analyses confirm the presence of $Yb_2O_3$ doped phase-separated yttria-rich alumino-silicate amorphous particles. The corresponding EDX spectra indicates the presence of two different kind of Yb ions showing different surrounding environments. Fig. 35 confirm that the crystalline nano-particles initially formed in the preform (NYb-9) annealed at 1450°C was maintained within the fiber samples after drawing above 1900°C but nature of the particles changed from crystalline to amorphous state.

Here both $P_2O_5$ and F serve as a nucleating agent to induce phase-separation in bulk silica glass matrix. We have studied the effect of $P_2O_5$, F and both on formation of YAG crystals within the fiber preform sample systematically. Both $P_2O_5$ and F are important from view point of phase-separation phenomena to induce the formation of Yb:YAG crystals within silica glass matrix. On the other hand, incorporation of $P_2O_5$ reduces the PD phenomena of the $Yb_2O_3$ doped fiber laser. Fluorine incorporation into the doping host is also important as it reduces the refractive index of silica glass. As a result, incorporation of fluorine helps to maintain the required refractive index difference around 0.002-0.0025 with incorporation of larger content of Y and Al in order to achieve the ideal parameter of large mode area $Yb_2O_3$ doped fiber for use as high power fiber laser with good lasing efficiency and beam quality. In this research work fluorine is found to be the most important as we are able to maintain the whole particles within 5-8 nm range in order to achieve the low background loss of the fiber within 40-50 dB/Km at wavelength 1285 nm. The details spectroscopic and lasing characteristics of Yb-doped nano-particles fibers drawn from the nano-crystalline preforms were mentioned earlier.

The formation of these nano-particles can be explained in terms of the phase separation phenomena and crystal growth mechanism, which normally occur in bulk silica based glass. Growth of $Yb_2O_3$ doped nano-crystallites was based on the metastable nature of yttria-alumino silicate glass and will transform to the stable crystalline state, if enough thermal energy is available which occurs by a two step nucleation and crystal growth process (Varshneya, 1994; Rawson, 1967). As RE doped yttria-alumino silicate glass is heated, the viscosity decreases which infact increases the tendency for structural rearrangement and crystallization. When the temperature was increased high enough (above 1350°C), crystal nuclei begin to form. The nuclei are the tiny regions in the glass structure. As the temperature was further increased, the rate of nuclei formation, or nucleation, increased and reached to a maximum at around 1450-1500°C, depending on the composition. At higher temperatures, additional thermal energy causes the nuclei to grow by crystal growth mechanisms. The rate of crystal growth increases with increasing temperature from 1450°C to a maximum of 1575°C and then the rate decreases to zero at the liquidous temperature around 1600°C. At temperature above 1600°C most of the particles disappears with destruction of their crystalline behaviour. Here, $P_2O_5$ serves as a nucleating agent for promoting the phase-separation phenomena along with the crystallization. Both $Y_2O_3$ and $Al_2O_3$ serve as a formation of crystalline host of composition $Y_3Al_5O_{12}$. The $Li_2O$ helps formation of glass-ceramic based material increasing the transparency of glass host. BaO acts as an agent for increases the glass formation region of the matrix as well as to reduce the viscosity of the glass host. A uniform distribution of $Yb_2O_3$ doped YAG nano-crystals having particle size of 6-10 nm was obtained for the first time to our knowledge into a 0.25

mol% fluorine co-doped silica optical fiber preform made by the MCVD and solution doping technique, and under suitable thermal treatment.

## 7. Geometrical and optical characterizations of nano-engineering fiber

The refractive index profile of nano-engineering fibers was measured by fiber analyser. Using the transmission microscope (Model: Nikon Eclipse LV 100) connected to a high resolution digital colour camera controlled by imaging software, the cross-sectional views of all the fiber sections are taken along with the values of core-clad dimensions. The cross-sectional view along with refractive index profile of $Yb_2O_3$ doped nano-engineering fibers (NYb-6 and NYb-10) having diameter around 17-20 micron are given in Fig. 36 and Fig. 37, respectively.

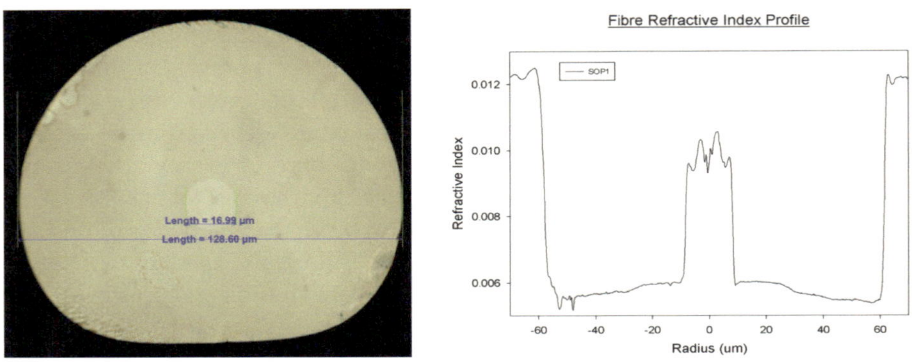


Fig. 36. Cross-sectional view and refractive-index profile of $Yb_2O_3$ doped nano-enginnering fiber (NYb-6).

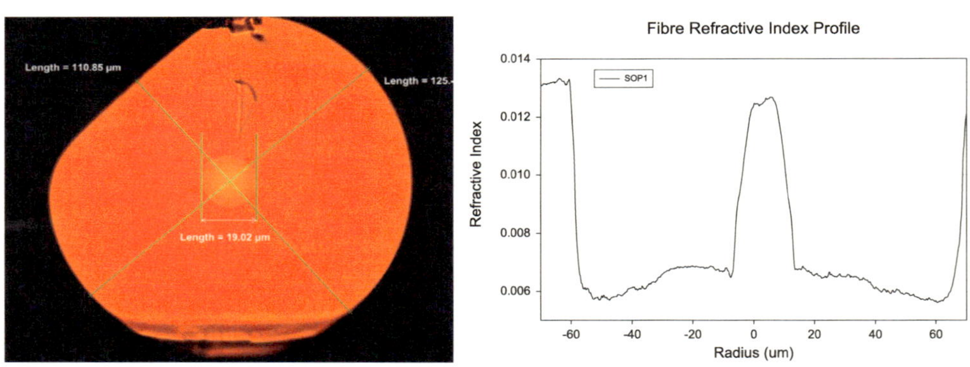


Fig. 37. Cross-sectional view and refractive-index profile of $Yb_2O_3$ doped nano-enginnering fiber (NYb-10).

The cross-sectional view of large core nano-engineering fibers (IMNYb-3, IMNYb-12D & IMNYb-12P ) having diameter around 25.0 - 35.0 micron along with their refractive index profiles are given in Fig. 38 and Fig. 39 respectively.

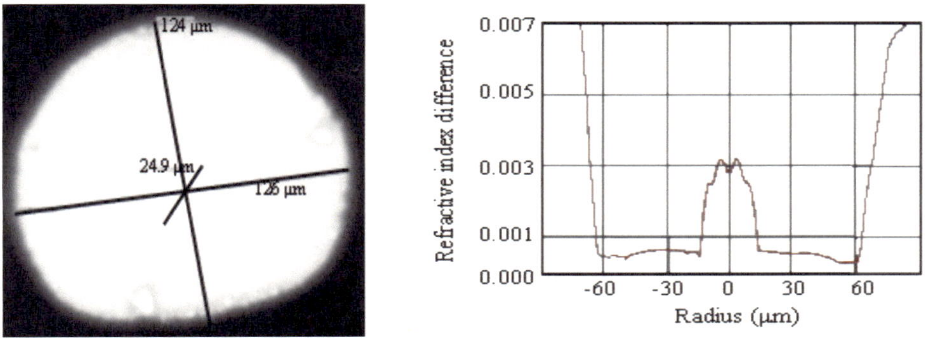


Fig. 38. Cross-sectional view and refractive-index profile of $Yb_2O_3$ doped large core nano-enginnering fiber (IMNYb-3) having core diameter around 25.0 micron.

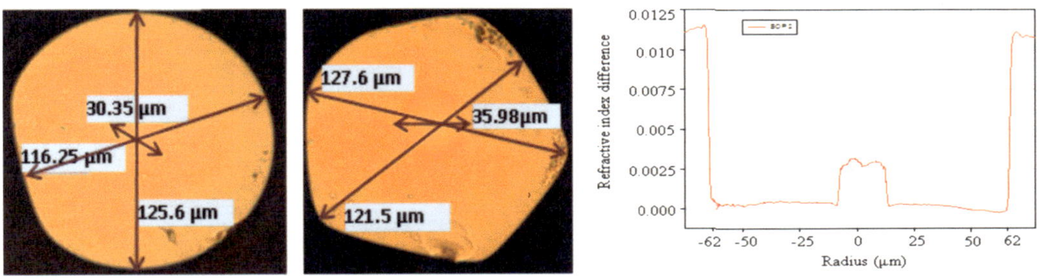


Fig. 39. Cross-sectional view and refractive-index profile of $Yb_2O_3$ doped large core nano-enginnering fiber (IMNYb-12D & IMNYb-12P) having core diameter of 30.0 - 35.0 µm.

## 8. Spectroscopic properties of $Yb_2O_3$ doped nano-engineering preforms and fibers

The UV-VIS spectra of three preform samples NYb-6, NYb-8 and NYb-9 annealed at 1450°C was shown in Fig. 40. The instrument used for this purpose was UV-VIS-NIR-3101PC scanning spectrophotometer. The peaks at 220 nm and 380 nm are attributed to the presence of ytterbium oxygen defect center (YbODC) (Yoo et al., 2007) and $Yb^{2+}$ ions (Yusong et al., 2007).

The absorbance value of fiber preform NYb-9 becomes much lower than that of NYb-6 and NYb-8. This phenomena indicates that the size of the particles generated within the preform sample NYb-9 becomes lower than that of preforms NYb-6 and NYb-8. The optical absorption spectra of preform NYb-5 annealed at 1300°C and 1450°C was shown in Fig. 41. The fiber preform sample NYb-5 annealed at 1300°C shows two normal absorption peaks centered at 915 and 975nm wavelengths which confirm that Yb ions are present in normal glass matrix. The same preform annealed at 1450°C had two additional absorption peaks centered at 935 and 960nm which becomes comparable to the absorption peaks centered at 941 and 968 observed in Yb:YAG ceramic samples[41]. Such a small variations in absorption peak positions arise as the crystals are dispersed in silica glass matrix not in a Yb:YAG pure state. Thus optical absorption confirms that YAG crystals are developed within silica glass matrix based fiber preform sample under annealing at 1450°C.

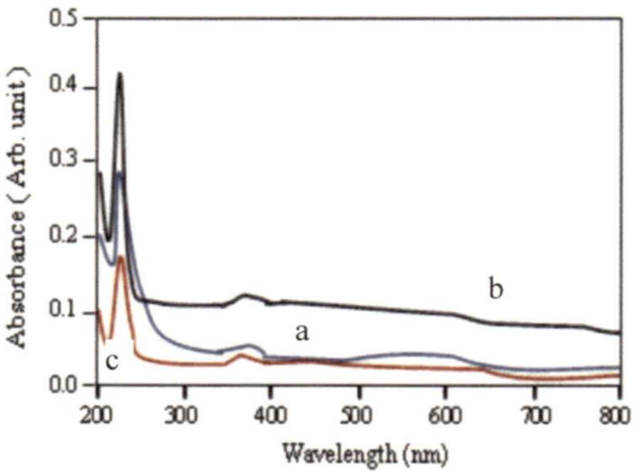


Fig. 40. UV-VIS spectra of three preform samples (a) NYb-6, (b) NYb-8 and (c) NYb-9 annealed at 1450°C.

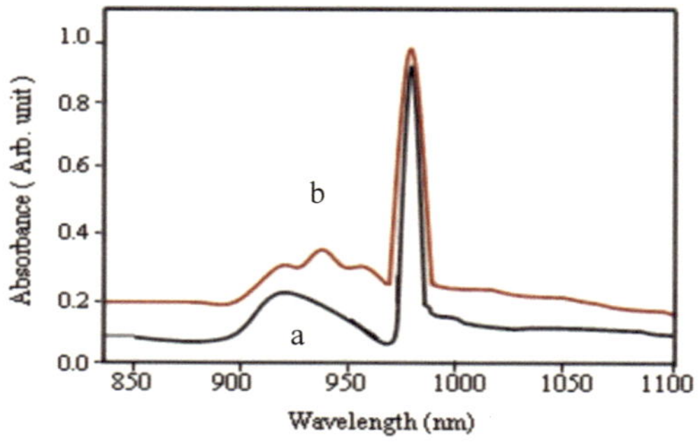


Fig. 41. Optical absorption spectra of preform NYb-5 (a) annealed at 1300°C and (b) 1450°C in NIR region.

The spectroscopic properties of the prepared fibers were investigated and compared to a Yb-doped aluminosilicate (Yb:Al) fiber fabricated in-house. The background loss of the fibers at 1285 nm wavelength was measured by high resolution optical time-domain reflectrometer (OTDR) and presented in Table-1. It was observed that background loss of the fibers very much related to the sizes of the $Yb_2O_3$ doped nano-particles. The nano-particles silica fiber NYb-9 having sizes 5-8 nm shows the minimum background loss around 40-50 dB/Km at 1285 nm wavelength. Whereas the nano-particles silica fibers NYb-6, NYb-7 and NYb-8 containing the larger sized particles shows the high background loss. The background loss at 1285 nm was 40-2400 dB/km, depending on the core composition and size of the nano-particles as shown in Table-1.The optical absorption spectra of the fabricated nano-particles fibers NYb-3, NYb-5, NYb-6, NYb-7 and NYb-9 are measured from 350-1750 nm wavelength using white light source and shown in Fig. 42.

The peaks corresponding to 915 and 975 nm are characteristics absorption peaks of $Yb^{3+}$ ions (Langnera et al., 2008).The normal fiber NYb-3 and nano-particles fiber NYb-9 shows the minimum absorption loss compared to the fibers NYb-5, NYb-6 and NYb-7 containing larger sized particles. The nano-particle fiber NYb-9 is found to be the best from view point of the formation of very small sized nano-particles within 5-8 nm showing low background loss of 40-50 dB/Km at 1285 nm wavelength.

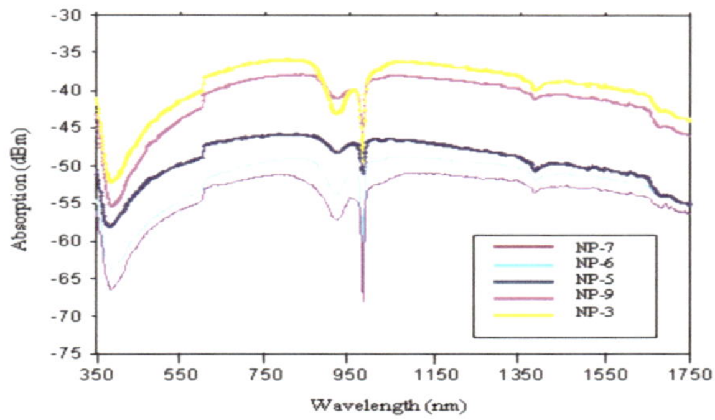


Fig. 42. Optical absorption of different $Yb_2O_3$ doped nano-particles silica fibers measured using white light source.

The fluorescence and lifetime of the fibre samples are measured under pumping at 915 and 976 nm wavelengths (Yoo et al., 2010). A fiber-coupled single mode laser diode at 915 nm was used as a pump source. The pump fiber was spliced to a fiber under test and the other end of the test fiber was angle cleaved to suppress any feedback of signal light. Index matching fluid was applied to the output end to further suppress the undesired feedback. The fluorescence was captured by placing one end of a multi-mode fiber to the side of the test fibers. An InGaAs photodetector was connected to the other end of the multi-mode fiber to record fluorescence decay time. The 915 nm laser diode was modulated externally by an acousto-optic modulator in the course of the lifetime measurement. We used less than 5 cm long fibers to avoid amplified spontaneous emission and reabsorption.

Fig. 43 shows the measured decay time with the nanoparticle fiber and the Yb:Al fiber (Yoo et al., 2010). The decay time was determined at the point where the intensity drops to 1/e of its original value. Both decay curves were well fitted with single exponential form. The decay time of the nanoparticle fiber was recorded as ~ 860 μsec which is close to that of Al:Yb fiber. The fitting goodness was better than 0.9996.

Fig. 44 shows measured absorption and fluorescence spectra for the nanoparticle fiber compared to those of Yb:Al fiber. Each peak was scaled to unity to make a spectral shape comparison. We see the spectral shapes of Yb absorption in the nanoparticle fiber are modified against the Yb:Al counterpart, which again indicates the different environments for the Yb ions. However, the line shapes do not follow the Yb-doped $Y_2O_3$ ceramic (Bourova et al., 2008) due to the glassy nature of the nanoparitcles as confirmed in the TEM analysis (Ballato et al., 2009).

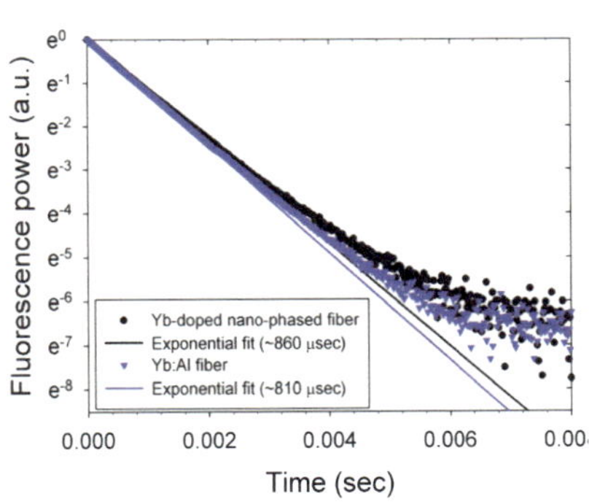


Fig. 43. Fluorescence decay time of Yb-doped nanoparticle silica fiber and Yb:Al fiber.

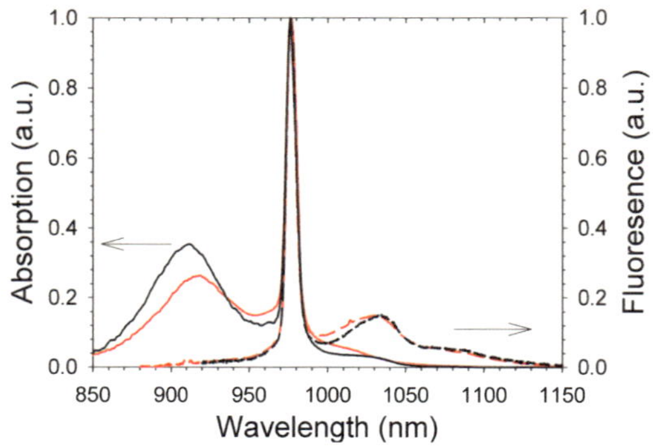


Fig. 44. Spectral shapes of absorption (solid line) and fluorescence (dashed line) of Yb-doped nanoparticle silica fiber (red) and Yb:Al fiber (black).

## 9. Lasing performances of $Yb_2O_3$ doped nano-engineering fibers

The Yb-doped nanoparticle fibers were tested under laser configuration (Yoo et al, 2010; Sahu et al, 2009). A D-shaped fiber (NYb-7) in double clad structure was pulled in 400 µm inner cladding diameter. The large inner cladding diameter was chosen to enable an efficient pump launch from the high-power pump diodes. The experimental arrangements are schematically shown in Fig. 45. The fiber was end-pumped by a 975 nm laser diode. Pump launch end of the fiber was cleaved perpendicularly to the fiber axis to provide 4% Fresnel reflection for the laser cavity. At the other end, a high reflective mirror (100%) at signal band was used to close the laser cavity. Most of the launched pump beam was absorbed through the 5 m long fiber. The output power linearly increased with the launched pump power. The output reached 85 W for a launched pump power of 120 W, representing good slope efficiency of 76% with respect to the launched pump power (Sahu et al., 2009) as shown in Fig. 46.

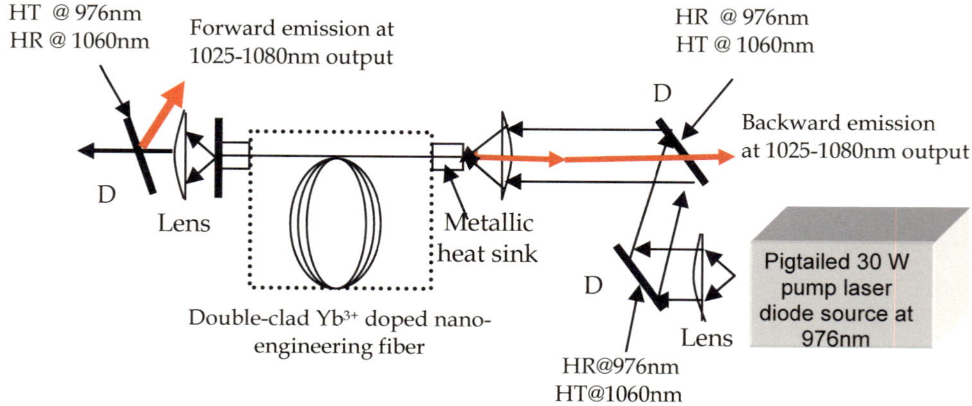


Fig. 45. Experimental arrangement for the laser performance characterization of nano-engineering fibers.

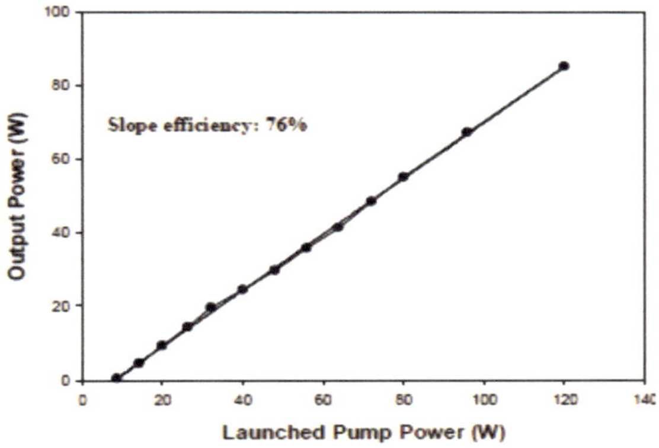


Fig. 46. Laser performance of 4 meter length of nano-engineering fiber (NYb-7).

We further investigated the lasing performance of another nano-engineering fiber (NYb-8) having 125 μm inner cladding diameter along with lasing bands. Fibers with 125 μm inner cladding diameter were placed in a free-running linear 4%-4% laser cavity test bed and the output spectrum was recorded by an OSA. We found that both lasing wavelength and 3dB bandwidth are pump power dependent. Fig. 47(a) represents the pump power dependent of laser spectra in 0.2 nm of OSA resolution. In a 4 m long fiber, the oscillation started at 1057 nm with total 13 dB of pump power absorption. As we increased the pump power, another band appeared at longer wavelength, ~ 1070 nm. The laser operated at two wavelength bands at ~ 1050 and 1070 nm with a gap between.

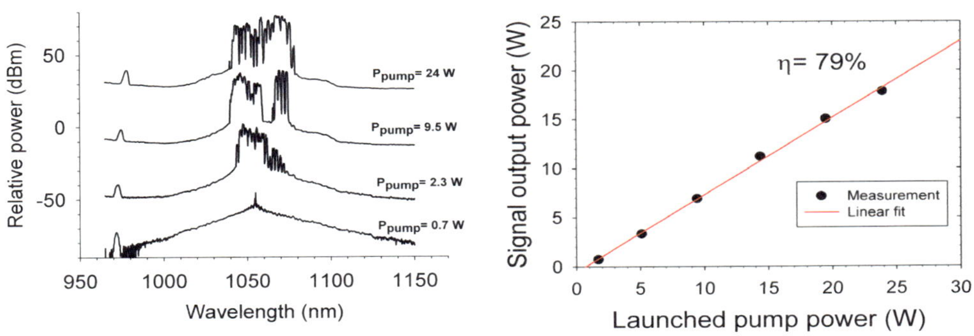


(a) (b)

Fig. 47. (a) Pump power dependent lasing band in Yb-doped nanoparticle silica fiber in 4 m length (b) Laser performance of 4 m long Yb-doped nanoparticle silica fiber (NYb-8).

More intense pump power filled the gap and made up a broad band oscillation from 1040-1075 nm. The observed pump power dependent of laser operation band is different as compared to Yb-doped silica fibers in conventional hosts. However, it did not compromise the laser efficiency. The fiber provided laser efficiency of 79% with respect to the launched pump power as shown in Fig. 47 (b). Identification of the cause of this behaviour is under progress.

The lasing characteristic of other nano-partcles silica fiber NYb-9 containing fluorine was shown in Fig. 48. The lasing efficiency of such nano-particles silica fiber was observed around 83.0 % with respect to the absorbed pump power (Paul et al., 2010). The maximum output power was well above 10 W, and limited by available pump power. The fiber containing $Yb_2O_3$ doped yttria rich alumino silica based nano-particles sized within 5-8 nm shows good lasing efficiency around 83.0% with respect to the absorbed pump power compared to the Yb:YAG crystal based ceramic laser (Yusong et al., 2007). The output spectrum of the laser for both forward and backward signal was shown in Fig. 49. Such nano-particles silica fiber shows a broad band oscillation from 1040-1075 nm in their output spectrum.

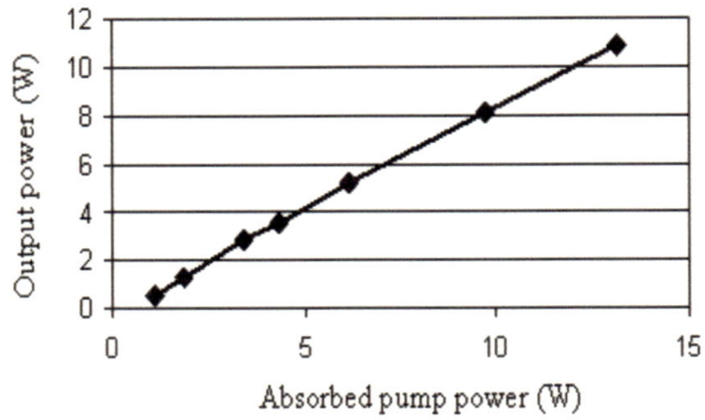


Fig. 48. Laser performance of 4 meter long Yb-doped nano-particle silica fiber NYb-9 under pumping at 975 nm wavelength.

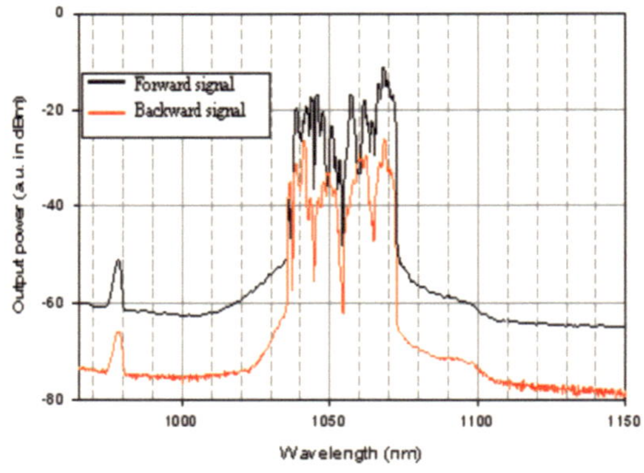


Fig. 49. Output spectrum of the lasing band of forward and backward signal of Yb doped nano particle silica fiber NYb-9 under launched pump power of 16 W by 975 nm pump laser diode.

The lasing characteristics of the fabricated Yb-doped nano-engineering large core fibers (IMNYb-3, IMNYb-12D & IMNYb-12P ) pumping at 976 nm by a multi-mode laser diode in the cladding-pump configuration through pump combiner was measured as shown in Fig. 50, Fig. 51 and Fig. 52 respectively. The laser cavity was formed by splicing two fiber bragg gratings (FBG) with reflectivity 99% (HR) and 10% (LR) at 1080 nm placed at each end of the Yb fiber. The pump wavelength was again 976 nm which provided the excitation within the zero-phonon line of $Yb^{3+}$ ion.

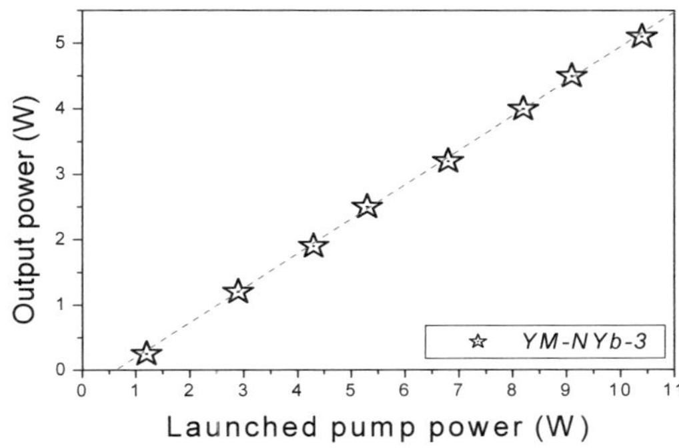


Fig. 50. Laser performance of 4.5 meter long Yb-doped nano-particle silica fiber IMNYb-3 under pumping at 976 nm wavelength.

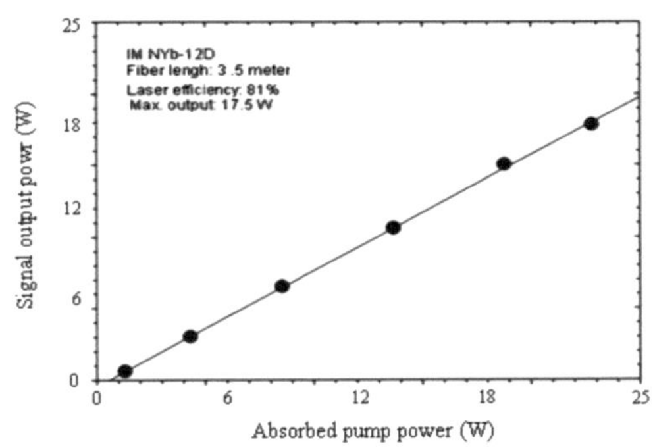


Fig. 51. Laser performance of 3.5 meter long Yb-doped nano-particle silica fiber IMNYb-12D under pumping at 976 nm wavelength.

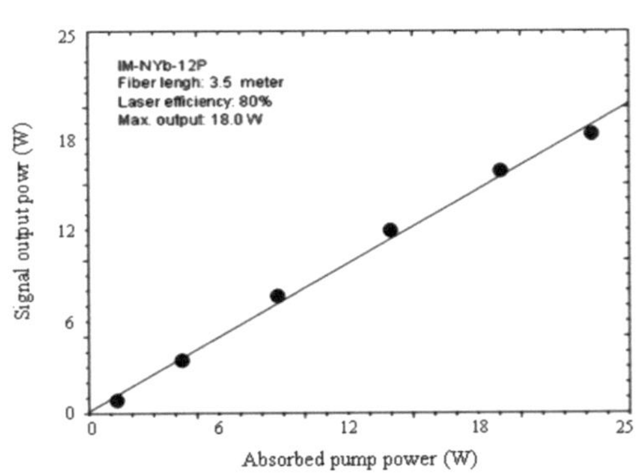


Fig. 52. Laser performance of 3.5 meter long Yb-doped nano-particle silica fiber IMNYb-12P under pumping at 976 nm wavelength.

## 10. Photo-darkening phenomena of nano-engineering fibers

A lot of study was done about PD in silica fibers doped with different RE ions (Broer et al., 1993; Atkins & Carter, 1994; Koponen et al., 2006). PD deteriorates the performance of fiber lasers or amplifiers (Hönninger et al., 2007; Morasse et al., 2007). Such phenomenon is initiated by the high Yb inversion (Koponen et al., 2006; Hönninger et al., 2007). The generation of color centers are responsible for the induced excess loss in the UV/VIS/IR wavelength range and was attributed to a photo-ionization process powered by the energy of excited Yb ions. PD in Yb doped fibers can also be induced by 488 nm irradiation (Yoo et al., 2007). Several measurement techniques to investigate the temporal and spectral characteristics of the loss evolution were developed (Koponen et al., 2006; Hönninger et al., 2007; Morasse et al., 2007; Yoo et al., 2007; Jasapara et al., 2006; Shubin et al., 2006; Koponen et al., 2007). Cladding pumping is advantageous, because a homogeneous Yb inversion can be achieved, independent of the core diameter and adjustable in a wide range with moderate pump powers. The pump induced PD in Yb-doped fibers has been recognized as a bottleneck for power scaling in many applications. It was found that the induced loss is proportional to the inversion level of the Yb3+ ions. Host material dependence was also reported and phosphosilicate can suppress the PD in a significant amount compared to the aluminosilicate counterpart (Shubin et al., 2007; Jetschke et al., 2008).

Here we have investigated the PD phenomena of Yb doped nanoparticles in a silica rich matrix based optical fibers as an alternative to the Yb in a 'standard', such as aluminium or phosphorous co-doped, silica host for use in high power fiber lasers. It is expected that the $Y_2O_3$ nanoparticles within a silica host will modify the Yb environments, which influences host material dependent processes such as PD and rare-earth solubility.

As the spectroscopic properties of the nanoparticle fiber described earlier indicate modification in the surrounding environments of the Yb ions, we expect different behavior of PD in the fibers. The PD was evaluated by monitoring the transmitted probe power at 633 nm through the fiber under test. The PD measurement setup is presented in Fig. 53. We used fiber-coupled single mode 977 nm laser diode as a pump source. The output end of the pump fiber was spliced to one port of wavelength-division multiplexing (WDM) coupler and the pump beam was delivered to the test fiber by splicing the output end of the WDM coupler and the test fiber. A He–Ne laser at 633 nm was used as a probe beam which coupled to the test fiber through the WDM coupler. The probe beam propagated the same direction as the pump beam. The output end of the test fiber was spliced to another WDM coupler to separate the pump and probe beam. The probe beam was chopped by mechanical chopper and the output power was detected by photodiode and lock-in amplifier after passing through the monochromator. We used ~ 1 cm of the test fiber to create uniform universion level of Yb ions and to suppress unwanted amplified spontaneous emission. The pump input power was maintained to provide >35% of population inversion of $Yb^{3+}$ throughout the fibers. We carried the PD measurement with the Yb-doped nanoparticle fiber and the Yb:Al fiber. Both fibers were in 125 μm diameter. The small signal absorption in both fibers was around 3 dB/m at 976 nm. The temporal characteristics of the transmitted probe power are represented in Fig. 54. The PD induced loss is significantly reduced in the Yb-doped nanoparticle fiber. When fitted the measured results with stretched exponential form (Shubin et al., 2007), we found that the saturated induced loss in Yb-doped nanocrystalline fiber reduced by 20 times compared to the aluminosilicate counterpart.

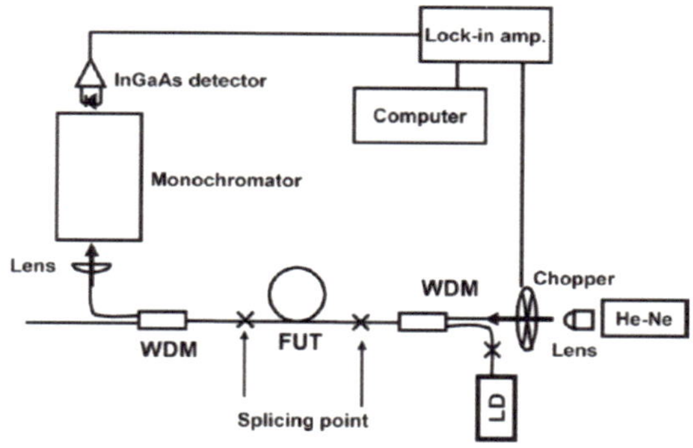


Fig. 53. Photodarkening induced loss measurement setup. FUT: Fiber under test, LD: Laser diode, WLS: White light source.

The PD process is associated with non-binding oxygen near surfaces of ytterbium/aluminium clusters which forms in silica material co-doped with ytterbium and aluminium. The non-binding oxygen originates from $Yb^{3+}$ in substitute of $Si^{4+}$ sites. When subjecting the ytterbium to 976 nm pump radiation, excess energy is radiated as phonons which cause a lone electron of one non-binding oxygen atom to shift to a nearest neighbour non-binding oxygen atom with creation of a hole and a pair of lone electrons that results in a coulomb field between the oxygen atoms to form an unstable colour centre. The conversion of such unstable colour centre to a semi-stable centre requires the shifting of one electron of the lone electron pair to a nearest neighbour site (Mattsson et al., 2008). In yttria alumino silicate glass matrix the yttrium ($Y^{3+}$) is chosen as codoping ion since their ionic radius is similar (0.0892 nm for $Y^{3+}$ and 0.0985 nm for $Yb^{3+}$). $Yb^{3+}$ and $Y^{3+}$ have also the same valence; therefore they can easily substitute its-self and can increase the distance between two $Yb^{3+}$ ions. In the case of codoping with $Yb^{3+}$ and $Y^{3+}$, the formation of Yb-O-Y-O-Yb may be take place just like to the formation of Er-O-Y-O-Er as reported by Trinh Ngoc Ha et. al (Ha et al., 2004). As a result of it, the formation of Yb related ODC (Yb-Yb) will be prevented. Hence the phenomenon of PD effect may be reduced.

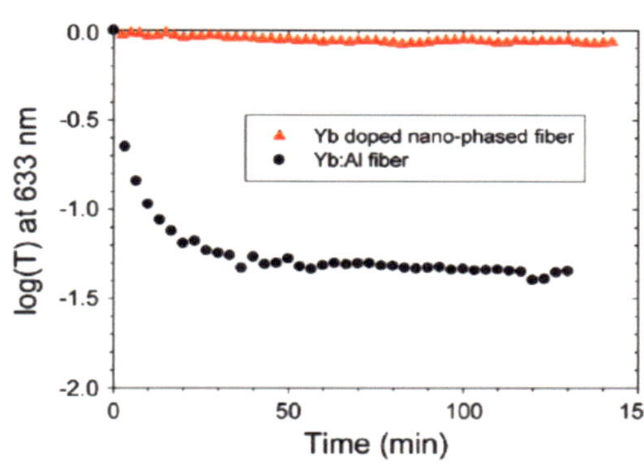


Fig. 54. Temporal characteristics of transmitted power at 633 nm for Yb-doped nanoparticle fiber and Yb:Al fiber.

On the other hand, the UV absorption spectra of $Yb_2O_3$ doped alumino-silicate glass based optical preform and $Yb_2O_3$ doped yttria-alumino silicate glass based optical preform samples having almost the same doping levels of $Yb_2O_3$ ( 0.13 mol%) and $Al_2O_3$ (0.9 mol%) was shown in Fig. 55.

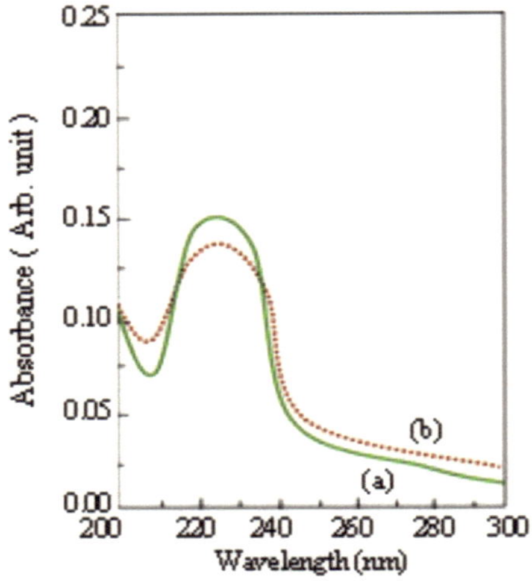


Fig. 55. Absorption spectra of (a) $Yb_2O_3$ doped alumino silicate preform and (b) $Yb_2O_3$ doped yttria alumino silicate perform (NYb-7).

In $Yb_2O_3$ doped alumino silicate glass the peak related to 220-225 nm is associated with the formation of YbODC (Yoo et al., 2007). The peak intensity related to YbODC becomes lower in yttria alumino silicate glass based optical preform compared to alumino-silicate glass based optical preform. This phenomena signifies that the presence of yttria reduces the formation of ytterbium related oxygen defect centers. The PD in the Yb-doped aluminosilicate preform takes place through the breaking of ODCs under two-photon absorption which gives rise to release of free electrons. The released electrons may be trapped at Al or Yb sites and form a color center resulting in PD. The choice of Al and Y as codopants is of due to the same valence of $Al^{3+}$, $Y^{3+}$, and $Yb^{3+}$ as well as similar lattice structure of $Al_2O_3$, $Y_2O_3$ and $Yb_2O_3$. As the field strength of the modifier cation increases, either through decreasing ionic radius or increased charge, it is expected to begin to perturb the aluminosilicate network more strongly, because of the energetic stabilization provided by closer association of negatively charged species, in particular non-bridging oxygen (NBO). In yttria-alumino silicate glass with increasing Y/Si ratio, the $^{29}Si$ peak shifts to a higher frequency, indicating a smaller shielding of the Si nucleus because of the substitution of Al by Y, which may be related to the conversion of bridging oxygen (BO) to non-bridging oxygen (NBO) (Schaller & Stebbins, 1998). $^{27}Al$ MAS NMR study (Kohli et al., 1992) also demonstrate that one major effect of $Y^{3+}$ is to promote the conversion of $AlO_4$ groups to higher-coordinate species which provides more oxygen surrounding spaces, thus reducing the number of ODCs, the hypothesized precursor to PD.

## 11. Conclusions

A detail review was done about the development of large core nano-engineered $Yb_2O_3$ doped fiber having diameter around 20-35 micron drawn from nano-crystalline YAG based silica glass preforms made by MCVD and solution doping technique followed by suitable thermal annealing conditions of the preform. Such nano-partciles are developed during drawing of optical fiber from the modified preform which annealed at 1450-1500°C for 3 hours under heating and cooling rates of 20°C/min. The diameter of the doped region was increased from 20 to 35 micron through modification of several fabrication parameters such as number of deposited porous layers, composition of deposited porous layers and CSA of the starting deposited tube. The uniform doping levels within the large core fiber preforms was maintained through optimization of the deposition temperature with respect to composition and numbers of the deposited porous layers as well as thickness of the silica tube with varying inner diameter of the tube prior to deposition. The effect of phosphorous and fluorine doping on nano-structuration of $Yb_2O_3$ doped fiber was discussed in details. The formation of nano-particles occurs through the phase separation phenomena and crystal growth mechanisms which normally occur in bulk silica-based glass. The size and shape of the nano-particles were evaluated from the TEM results. The crystalline nature of nano-particles was observed after annealing at 1450°C. In presence of P the average nano-crystallites size of 25-50 nm were homogeneously dispersed within the core region. The crystal size increases as the $P_2O_5$ doping level increases due to the strong field strength difference of $P^{5+}$ and $Si^{4+}$ ( >0.3) where $P_2O_5$ serves as a nucleating agent to accelerate the growth of formation of large crystallites. A bimodal size distribution with larger particles of the order of 60 nm and the smaller ones of 3-5 nm in diameter was observed in presence of P where most likely P and F together play a different kind of nucleation and crystallization phenomenon for such kind of bimodal distribution. By adding fluorine to the silica based composition, nano particles of 6-10 nm in diameter are obtained. The composition of phase-separated crystalline particles was evaluated by comparing their EDX data, taken directly from the crystal particles and from an area outside of the particles. The EDX spectra taken from the crystal particles detected Si, Al, O, Yb and Y signals, whereas the areas outside of the particles detected Si, Al and O and no Yb and Y. The compositional analyses of the phase-separated nano-particles consists of 13.0 atom % of Al, 0.02 atom% of Yb, 9.0 atom % of Y and 36.0 atom% of O. The doping level of the formation of Yb:YAG crystals (confirmed from the XRD spectra) within fiber preform sample is found to be 0.02 atom%. The phase-separated crystalline nano-particles were developed within the central core glass matrix under post-heat treatment within 1450-1500°C for a period of 3 hours. Such annealed preform shows two additional absorption peaks centered at 935 and 960 nm indicating the formation of Yb:YAG crystals. The main feature of such process involves the direct synthesis of Yb:YAG nano-crystallites within the core of optical fiber preform through the conventional MCVD and solution doping technique. However nature of the particles was changed from crystalline to amorphous at fiber drawing stage when fiber drawn from such annealed preform after heating at 2000°C within fiber drawing furnace. The TEM spectra reveal the size of the particles within 6-10 nm. The energy dispersive X-ray (EDX) spectra signifies that yttrium and ytterbium are dominant in phase-separated particles whilst they are sparse in the non-phase-separated region.The background loss of such nano-particles based fibers is maintained within 40-50 dB/Km by keeping the particle sizes within 6-10 nm range. The fiber containing 6-10 nm sizes of $Yb_2O_3$ doped yttria alumino rich silica based nano-particles shows good lasing efficiency around 80% with respect to the absorbed pump power compared to the Yb:YAG crystal based ceramic laser. Such nano-engineering fiber shows a broad band lasing oscillation from 1040-1075 nm.

More generally, this concept has great potentials as possible solutions to now a days issues in lasing fibers specifically PD phenomenon under high power application as well as fiber amplifiers for intrinsic gain flattening and spectral hole burning.

## 12. Future direction

The future work involves the fabrication of optical fiber based on suitable host of advanced nano-engineered rare-earth doped modified silica glass of 85-90% $SiO_2$ containing heavy metals and semi-conductors through solution doping process followed by MCVD technique. The aspect of such kind of future work will reveal the suppression of PD phenomena to stabilize the lasing output power through modification of the surrounding environments of RE ions which embedded into different nano-engineering hosts. The future work will also focus the making of large core area upto 35.0 μm diameter nano-engineering optical fibers with uniform distribution of phase-separated nano-particles from view point of good lasing efficiency at high power applications.

Extended X-ray absorption fine structure (EXAFS) analyses will be carried out for determination of the structural details of the site occupied by the $Yb^{3+}$ ion into such nano-engineered glass. X-ray Absorption Spectroscopy at the Yb-$L_{III}$ edge (E=8944 eV) and X-ray Powder Diffraction measurements will be done at the GILDA-CRG beamline at the European Synchrotron Radiation Facility (ESRF), France.

Although nano-material doping technology has opened a new way in developing novel special optical fiber, however, as far as we know, there are few work about optical fiber amplification based on silica fiber doped with nano-semiconductor materials. In future work, we will demonstrate a novel special silica fiber doped with InP and ZnS semiconductor nano-particles into the core. Due to the nano size, semiconductor nano-particles will show remarkable quantum confinement effect and size tunable effect, which may provide excellent amplification features. Such nanostructuration of doped fiber will be proposed as a new route to 'engineer' the local dopant environment. All these results will benefit to optical fiber components such as lasers, amplifiers and sensors, which can now be realized with silica glass.

## 13. Acknowledgements

The authors wish to acknowledge CSIR and the Royal Society, UK for provision of financial support under CSIR-Royal Society Joint Collaborative Research Work (2007/R2). The authors also wish to acknowledge DST (Government of India) and CONACYT (Government of Mexico) for providing financial support through the Program of cooperation in science and technology between India and Mexico. The authors would like to thank to the Electron Microprobe Laboratory (University of Minnesota, USA) for performing the EPMA. The authors (1 and 3) are thankful to Prof. Indranil Manna, Director, CGCRI for his continuous encouragement, guidance and support in this work.

## 14. References

Atkins, G. R. & Carter, A. L. G. (1994). Photodarkening in $Tb^{3+}$-doped phosphosilicate and germanosilicate optical fibers. *Optics Letter*, Vol. 19, No. 12, pp. 874-876, ISSN 0146-9592.

Ballato, J., Hawkins, T., Foy, P., Kokuoz, B., Stolen, R., McMillen, C., Daw, M., Su, Z., Tritt, T. M., Dubinskii, M., Zhang, J., Sanamyan, T. & Matthewson, M. J. (2009). On the fabrication of all-glass optical fibers from crystals. *Journal of Applied Physics*, Vol. 105, No. 5, pp. 053110-053118, ISSN 0021-8979.

Bandyopadhyay, T., Sen, R., Dasgupta, K., Bhadra, S. K. & Paul, M. C. (2004). Process for making rare earth doped optical fiber. *US patent 6751990*.

Bhadra, S. K., Sen, R., Pal, M., Paul, M. C. Naskar, M. K., Chatterjee, S., Chatterjee, M. & Dasgupta, K. (2003). Development of rare-earth doped fibres for amplifiers in WDM system. *IEE Proceedings of Circuits, Devices & Systems*, Vol. 150, No. 6, pp. 480-485, ISSN 1350-2409.

Blanc, W., Dussardier, B. & Paul, M. C. (2009). Er-doped oxide nanoparticles in silica-based optical-fibers. *GlassTechnology: European Journal of Glass Science and Technology*, Vol. A50, pp. 79-81, ISSN 0017-1050.

Bourova, E., Billancourt, B., Blanchandin, S., Leplingard, F. & Lasas, J. (April 8, 2008). Active optical fiber for Raman amplification. *US patent 7355788B2*.

Broer, M. M., Krol, D. M. & DiGiovanni, D. J., (1993). Highly nonlinear near-resonant photodarkening in a thulium-doped aluminosilicate glass fiber. Optics Letter, Vol. 18, No. 10, pp. 799-801, ISSN 0146-9592.

Chryssou, C. E., Kenyon, A. J., Iwayama, T. S., Pitt, C. W. & Hole, D. E. (1999). *Journal of Applied Physics Letter,* Vol. 75, No. 14, pp. 2011-2013, ISSN 0003-6951.

Cho, J. S., Paek, U. C., Han, W. T. & Heo, J. (2001). Fabrication and heat treatment effects on absorption characteristics of glass fibers doped with PbTe semiconductor quantum dots. *Optical Fiber Communication Conference*, OSA Technical Digest Series, Optical Society of America, ISSN 1-55752-655-9, California, March 2001.

Chou, J. S. & Lee, S. C. (1994). The initial growth mechanism of silicon oxide by liquid phase deposition. *Journal of Electrochemical Society,* Vol. 141, pp. 3214-3218, ISSN 0013-4651.

Chen, J. C. (2006). Spectroscopic study on the fluorescence of Cr ions in double-clad Cr:YAG crystal fiber. *Ph. D Thesis*, Institute of Electro-Optical Engineering, National Sun Yat-Sen University, Taiwan.

Dove, J. F., Russell, H., Kim, J.-S., Nivartvong, N., Flattery, J., Keller, D., & Kornreich, P. G. (2001). Light amplification by a $Cd_3P_2$ cylinder fiber. *Proceeding* of SPIE, No. 4216, pp. 62–66, ISBN 9780819438812, February 2001.

De, G., Qin, W., Zhang, J., Wang, Y., Cao, C. & Cui, Y. (2006). Upconversion luminescence properties of $Y_2O_3$:$Yb^{3+}$,$Er^{3+}$ nanostructures. *Journal of Luminescence,* Vol. 119, pp. 258-263, ISSN 0022-2313.

Fujii, M., Yoshida, M., Hayashi, S. & Yamamoto, K. (1998). Photoluminescence from $SiO_2$ films containing Si nanocrystals and Er: Effects of nanocrystalline size on the photoluminescence efficiency of $Er^{3+}$. *Journal Applied Physics.* Vol. 84, pp. 4525-4531, ISSN 0021-8979.

Gonçalves, M. C., Santos, L. F. & Almeida, R. M. (2002). Rare-earth-doped transparent glass Ceramics. *Comptes Rendus Chimie*, Vol. 5, pp. 845-854, ISSN 1631-0748.

Ha, T. N., Thang, P. N., Chung, L. N., Nga, P. T., Hoi, P. V., Luat, L.V., Cham, T. T., Lien, V. T. K., Barthou L. C. & Benalloul, P. (2004). *International Workshop on Photonics and Applications*. Hanoi, Vietnam, April 2004.

Hageman, V. B. M. & Oonk, H. A. J. (1986). Liquid Immiscibility in the $SiO_2$–MgO, $SiO_2$–SrO, $SiO_2$–$La_2O_3$, and $SiO_2$–$Y_2O_3$ Systems. *Journal of Physics Chemistry of Glasses,* Vol. 27, No. 5, pp. 194-198, ISSN 0031-9090.

Hudon, H. & Baker, D. R. (2002). The Nature of Phase Separation in Binary Oxide Melts and Glasses: I. Silicate Systems. *Journal of Non-Crystalline Solids,* Vol. 303, pp. 299-345, ISSN 0022-3093.

Homma, T., Katoh, T., Yamada, Y. & Murao, Y. (1993). A Selective $SiO_2$ Film-Formation Technology Using Liquid-Phase Deposition for Fully Planarized Multilevel Interconnections. *Journal of the Electrochemical Society,* Vol. 140, No. 8, pp. 2410-2414, ISSN 0013-4651.

Hammersley, A. P., Svensson, S. O., Hanfland, M., Fitch, A. N. & Häusermann, D. (1996). Two-dimensional detector software: From real detector to idealised image or two-theta scan. *High Pressure Research,* Vol. 14, No. 4, pp. 235-248, ISSN 0895-7959.

Hönninger, I. M., Boullet, J., Cardinal, T., Guillen, F., Podgorski, M., Doua, R. B. & Salin, F. (2007). Photodarkening and photobleaching of an ytterbium-doped silica double-clad LMA fiber. *Optics Express, Vol.* 15, No. 4, pp. 1606-1611, ISBN 1094-4087.

Izawa, T. & Sudo, S. (1987). *Optical Fibers: Materials and Fabrication,* KTK Scientific Publishers, ISBN 9027723788, Tokyo.

Jander, P. & Brocklesby, W. S. (2004). Spectroscopy of yttria-alumina-silica glass doped with thulium and erbium. *IEEE Journal of Quantum Electron,* Vol. 40, No. 5, pp. 509-512, ISSN 0018-9197.

Jasapara, J., Andrejco, M., DiGiovanni, D. & Windeler, R. (2006). Effect of heat and $H_2$ gas on the photodarkening of $Yb^{3+}$ fibers. *Proceeding of Lasers and Electro-Optics/Quantum Electronics and Laser Science Conference and Photonic Applications Systems Technologies, Technical Digest,* ISBN 978-1-55752-813-1, Optical Society of America, Washington, DC, May 2006.

Jetschke, S., Unger, S., Schwuchow, A., Leich, M. & Kirchhof, J. (2008). Efficient Yb laser fibers with low photodarkening by optimization of the core composition, *Optics Express,* Vol. 16, No. 20, pp. 15540-15545, ISSN 1094-4087.

Jiang, X., Song, Q., Xu, L., Fu, J. & Tong, L. (2007). Microfiber knot dye laser based on the evanescent-wave-coupled gain. *Journal of Applied Physics Letters,* Vol. 90, No. 23, pp. 233501-233503 (2007), ISSN 0003-6951.

John, J. S., Coffer, J. L., Chen, Y. D. & Pinizzotto, R. F. (1999). Synthesis and characterization of discrete luminescent erbium-doped silicon nanocrystals. *Journal of American Chemical Society.* Vol. 121, No. 9, pp. 1888-1892, ISSN 0002-7863.

Joint Committee on Powder Diffraction Standards, data file. 38-0222.

Kawanishi, S., Ohmori, M., Tanakad, M. & Sakaki, H. (2006). Observation of photoluminescence of semiconductor nanocrystal quantum dotsin thecore of photonic band gap fiber. *Tech. Digest of IEEE LEOS meeting,* ISBN 978-1-55752-856-8, San Diego, CA February 2008.

Klopp, P., Griebner, U., Petrov, V., Griebner, U., Petermann, K., Peters, V. & Erbert, G. (2004). Highly efficient mode locked Yb:Sc2O3 laser. *Optics Letter,* Vol. 29, No. 4, pp. 391-393, ISSN 0146-9592.

Kohli, J. T., Shelby, J. E. & Frye, J. S. (1992). A structural investigation of yttrium aluminosilicate glasses using Si-29 and Al-27 magic-angle spinning nuclear magnetic resonance *Journal Physics Chemistry of Glasses*, Vol. 33, pp. 73-78, ISSN 0031-9090.

Koponen, J. J., Söderlund, M. J. & Hoffman, H. J. (2006). Measuring photodarkening from single-mode ytterbium doped silica fibers. *Optics Express*, Vol. 14, pp. 11539-11544, ISSN 1094-4087.

Koponen, J., Söderlund, M., Hoffman, H. J., Kliner, D. & Koplow, J. (2007). Photodarkening Measurements in Large-Mode-Area Fibers. *Proceedings of the SPIE 6453*, ISBN 9780819465665, March 2007.

Langnera, A., Schotza, G., Suchb, M., Kayserb, T., Reichelc, V., Grimmc, S., Kirchhofc, J., Kraused, V. & Rehmannd, G. (2008). A new material for high-power laser fibers. *Proceedings of the SPIE*, 6873, ISBN 9780819470485, April 2008.

Lu, J., Lv, J., Murai, T., Takaichi, T., Uematsu, T., Ueda, K., Yagi, H., Yanagitani, T. & Kaminskii, A. (2001). $Nd^{3+}$:$Y_2O_3$ ceramic laser. *Japanese Journal of Applied Physics*, Vol. 40, L1277-1279, ISSN 0021-4922.

Li, T. (1985). *Optical Fiber Communications*, Academic press, ISBN 0 12 447301 6, New York.

Lu, Q., Li, A., Guo, F. U., Sun L. & Zhao, L. C. (2008). The two-photon excitation of $SiO_2$-coated $Y_2O_3$:$Eu^{3+}$ nanoparticles by a near-infrared femtosecond laser. *Nanotechnology*, Vol. 19, No. 20, pp. 205704-205711, ISSN 1687-9503.

Lu, Q., Li, A., Guo, F. U., Sun L. & Zhao, L. C. (2008). Experimental study on the surface modification of $Y_2O_3$:$Tm^{3+}$/$Yb^{3+}$ nanoparticles to enhance upconversion fluorescence and weaken aggregation. *Nanotechnology*, Vol. 19, No. 14, pp. 145701-205709, ISSN 1687-9503.

Mattsson, K. E., Knudsen, S. N., Cadier, B. & Robin, T. (2008). Photo darkening in ytterbium co-doped silica material. *Proceedings of the SPIE*, 6873, ISBN 9780819470485, May 2008.

Miller, S. E. & Chynoweth, A. G. (1979). *Optical Fiber Telecommunications*, Academic Press, ISBN 0124973507, 9780124973503, New York.

Morasse, B., Chatigny, S., Gagnon, E., Hovington, C., Martin, J. P. & De Sandro, J. P. (2007). Low photodarkening single cladding ytterbium fibre amplifier. in Fiber Lasers IV: Technology, Systems, and Applications, D. J. Harter,A. Tünnermann, J. Broeng, and C. Headley, eds., *Proceedings of the SPIE* 6453, 64530H-1-9.

Morimoto, S. (2004). Optical Absorption of Cr-containing $Li_2O$-$SiO_2$ system transparent glass-ceramics. *Journal of the Ceramic Society of Japan*, Vol. 112, pp. 130-132, ISSN 1882-0743.

Morimoto, S., & Emem, W. (2005). Strength of Li2O-SiO2 System Transparent Glass-Ceramics. *Journal of the Ceramic Society of Japan*, 112, 259-262, ISSN 1882-0743.

Mun, J. H., Novoselov, A., Yoshikawa, A., Boulon, G. & Fukuda, T. (2005). Growth of $Yb^{3+}$-doped $Y_2O_3$ single crystal rods by the micro-pulling-down method. *Materials Research Bulletin*, Vol. 40, No. 8, pp. 1235-1243, ISSN 0025-5408.

Nichols, W.T., Keto, J.W., Henneke, D. E., Brock, J., Malyavanatham, R. G. & Becker, M. F. (2001). Large-scale production of nanocrystals by laser ablation of microparticles in a flowing aerosol *Journal of Applied Physics Letter*, Vol. 78, pp. 1128-1130, ISSN 0003-6951.

Patra, A., Saha, S., Alencar, M. A. R. C., Rakov, N. & Maciel, G. S. (2005). Blue upconversion emission of $Tm^{3+}$- $Yb^{3+}$ in $ZrO_2$ nanocrystals: Role of $Yb^{3+}$ ions. *Chemical Physics Letters*, Vol. 407, No. 4-6, pp. 477-481, ISSN 0009-2614.

Paul, M. C., Bysakh, S., Das, S., Bhadra, S. K., Pal, M., Yoo, S., Kalita, M. P., Boyland, A. & Sahu, J. K. (2010). $Yb_2O_3$-doped YAG nano-crystallites in silica-based core glass matrix of optical fiber preform. *Materials Science and Engineering B,* Vol. 175, No. 2, pp. 108-119, ISSN 0921-5107.

Paul, M. C., Upadhyaya, B. N., Das, S., Dhar, A., Pal, M., Kher, S., Dasgupta, K., Bhadra, S. K. & Sen, R. (2010). Study of the fabrication parameters of large core $Yb_2O_3$ doped optical fibre through solution doping technique. *Optics Communications,* Vol. 283, pp. 1039-1046, ISBN 0030-4018.

Petermann, K., Fornasiero, L., Mix, E. & Peters,V . (2002). High melting sesquioxides: crystal growth, spectroscopy and laser experiments. *Optical Materials,* Vol. 19, pp. 67-71, ISSN 0925-3467.

Rajala, M., Janka, K. & Kykkänen, P. (2003). An industrial method for nanoparticle synthesis with a wide range of compositions. *Review on Advanced Materials Science,* Vol. 5, pp. 493-497, ISSN 1605-8127.

Rawson, H. (1967). Inorganic Glass-Forming Systems, *Academic Press, ISBN 012583750X, London.*

Ray, N. H. (1974). Composition-property relationships in inorganic oxide glasses. *Non-Crystalline Solids,* Vol. 15, pp. 423, ISSN 0022-3093.

Sahu, J. K., Paul, M. C., Kalita, M. P., Boyland, A., Codemard, C., Yoo, S., Webb, A., Standish, R. J., Nilsson, J., Das, S., Bhadra, S. K., Pal, M., Dhar, A. & Sen, R. (2009). Ytterbium doped nanostructured optical fibers for high power fiber lasers. *Conference on Lasers and Electro-Optics, CLEO-2009,* Baltimore, MD, ISBN 978-1-55752-869-8.

Samson, B.N., Pinckney, L. R., Wang, J., Beall, G. H. & Borrelli, N. F. (2002). Nickel-doped nanocrystalline glass- ceramic fiber. *Optics Letters,* Vol. 27, No. 15, pp. 1309-1311, ISSN 0146-9592.

Samson, B. N., Tick, P. A. & Borrelli, N. F. (2001). Efficient neodymium-doped glass-ceramic fiber laser and amplifier. *Optics Letters,* Vol. 26, No. 3, pp. 145-147, ISSN 0146-9592.

Schaller, T. & Stebbins, J. F. (1998). The structural role of lanthanum and yttrium in aluminosilicate glasses: A $^{27}Al$ and $^{17}O$ MAS NMR study. *Journal of Physical Chemistry B,* Vol. 102, No. 52, pp. 10690-10697, ISSN 1089-5647.

Sen, R., Pal, M., Paul, M. C., Bhadra, S. K., Chatterjee, S. & Dasgupta, K. (2005). *Method of fabricating rare earth doped optical fiber.* US patent 6851281.

Shannon, R.D. (1976). *Revised effective ionic radii and systematic studies of interatomic distances in halides and chalcogenides.* Acta Crystallographica Section A, Vol. 32, No. 5, pp. 751-767, ISSN 0108-7673.

Shelby, J. E. (1997). *Introduction to glass science and technology,* Royal Society of Chemistry, ISBN 0854045333, UK.

Shi, L., Xu, Y., Tan, W. & Chen, X. (2007). Simulation of optical microfiber loop resonators for ambient refractive index sensing. *Sensors* Vol. 7, No. 5, pp. 689-696, ISSN 1424-8220.

Shirakawa, A., Takaichi, K., Yagi, H., Tanisho, M., Bisson, J.-F., Lu, J., Ueda, K., Yanagitani T. & Kaminskii, A. A. (2004). First mode-locked ceramic laser: Femtosecond $Yb_3+:Y_2O3$ ceramic laser, *Journal of Laser Physics,* Vol. 14, pp. 1375-1381, ISSN 1054-660X.

Shirakawa, A., Takaichi, K., Yagi, H., Bisson, J. F., Lu, J., Musha, M., Ueda, K., Yanagitani, T., Petroy, T. S. & Kaminskii, A. A. (2003). Diode-pumped mode-locked Yb3+:Y2O3 ceramic laser. *Optics Express*, Vol. 11, No. 22, pp. 2911-2916, ISSN 1094-4087.

Shirakawa, A., Takaichi, K., Yagi, H., Bisson, J., Lu, J., Musha, M., Ueda, K., Yanagitani, T., Petrov, T. & Kaminskii, A. (2003). Diode-pumped mode-locked Yb3+:Y2O3 ceramic laser. Optics Express, Vol. 11, No. 22, pp. 2911-2916, ISSN 1094-4087.

Shubin, A. V., Yashkov, M. V., Melkumov, M. A., Smirnow, S. A., Bufetov, I. A. & Dianov, E. M. (2007). Photodarkening of aluminosilicate and phosphosilicate Yb-doped fibers. *Lasers and Electro-Optics 2007 and International Quantum Electronics Conference, CLEOE-IQEC 2007*, ISBN 978-1-4244-0931-0, June 2007.

Tomozawa, M., Doremus, R. H., Tomozawa, M. & Doremus, R. H. (1978) *Glass I: Interaction with Electromagnetic Radiation*, ISBN Academic, NewYork, Vol. 12, pp. 5-89

Takaichi, K., Yagi, H., Lu, J., Bisson, J. F., Shirakawa, A., Ueda, K., Yanagitani, K. & Kaminskii, A. A. (2004). Highly efficient continuous-wave operation at 1030 and 1075 nm wavelengths of LD pumped $Yb^{3+}$ : $Yb_2O_3$ ceramic lasers. *Applied Physics Letter*, Vol. 84, pp. 317-319, ISSN 0003-6951.

Tomazawa, M. & Doremus, R. H. (1979). *Treatise on Materials Science and Technology*, Academic Press, New York.

Townsend, J. E., Poole, S. B. & Payne, D. N. (1987). Solution doping technique for fabrication of rare earth doped optical fibers. *Electronics Letter*, Vol. 23, pp. 329-331, ISSN 0013-5194.

Ubaldini, A. & Carnasciali, M. (2008). Raman characterisation of powder of cubic $RE_2O_3$ (RE = Nd, Gd, Dy, Tm, and Lu), $Sc_2O_3$ and $Y_2O_3$. *Journal of Alloys Compounds*, Vol. 454, No. 1-2, pp. 374-378, ISSN 0925-8388.

Varshneya, A. K. (1994). *Fundamentals of inorganic glasses*, Academic Press. ISBN 978-0-12-714970-7, San Diego, CA.

Vetrone, F., Boyer, J. C., Capobianco, J. A., Speghini, A. & Bettinelli, M. (2003). Effect of $Yb^{3+}$ codoping on the upconversion emission in nanocrystalline $Y_2O_3$:$Er^{3+}$. *Journal of Physical Chemistry B*. Vol. 107, No. 5, pp. 1107-1112, 1089-5647.

Vogel, W. (1994). *Glass Chemistry*.,ISBN 3–540-57572–3, Springer-Verlag, Berlin.

Yeatman, E. M., Ahmad, M. M., McCarthy, O., Martucci, A., & Guglielmi, M. (2000). Sol-gel fabrication of rare-earth doped photonic components. *Journal of sol-gel science technology, Vol.*19, pp. 231-236, ISSN 0928-0707.

Yoo, S., Paek, U. C. & Han, W.T. (2003). (2003). Development of a glass optical fiber containing $ZnO$–$Al_2O_3$–$SiO_2$ glass-ceramics doped with $Co^{2+}$ and its optical absorption characteristics. *Journal of Non-Crystalline Solids. Vol. 315, No. 1-2, pp. 180-186, ISSN 0022-3093.*

Yoo, S., Basu, C., Boyland, A. J., Sones, C., Nilsson, J., Sahu, J. K., Payne, D. (2007). Photodarkening in Yb-doped aluminosilicate fibers induced by 488 nm irradiation. *Optics Letter*, Vol. 32, No. 12 pp. 1626-1628, ISSN 0146-9592.

Yoo, S., Kalita, M. P., Boyland, A. J., Webb, A. S., Standish, R. J., Sahu, J. K., Paul, M. C., Das, S., Bhadra, S. K. & Pal, M. (2010). Ytterbium-doped Y2O3 nanoparticle silica optical fibers for high power fiber lasers with suppressed photodarkening. *Optics Communications* Vol. 283, No. 18, pp. 3423-3427, ISBN 0030-4018.

Yusong, W., Jiang, L., Yubai, P., Jingkun, G., Benxue, J., Yi, X. & Jun, X. (2007). Diode-pumped Yb:YAG ceramic laser. *Journal of American Ceramic Society,* Vol. 90, No. 10, pp. 3334-3337, ISSN 1551-2916.

*Zarzycki, J. (1991).* Glasses and the vitreous state *(First edition), Cambridge University Press, ISBN 0 521 35582 6, Press Syndicate of the University of Cambridge.*

# 3

# An Improved Method of Fabricating Rare Earth Doped Optical Fiber

Ranjan Sen and Anirban Dhar
*CSIR-Central Glass & Ceramic Research Institute, Kolkata, India*

## 1. Introduction

Optical fiber, a technology well-known to have revolutionized the telecommunication industry, is now becoming the key component behind the success of a number of enabling technologies such as sensing, biomedicine, defence, security and novel light sources. Along with standard telecommunication fibers, rare earth (RE) doped specialty optical fibers are the backbones of the photonics industry. Elements such as erbium (Er), ytterbium (Yb), yttrium (Y), neodymium (Nd), thulium (Tm) and europium (Eu) are vital optically active ingredients at the heart of many lasers, optical amplifiers and phosphors. Erbium-doped fiber amplifier (EDFA) — to overcome losses in long-distance optical communication links, fiber lasers — a technology having great potential for medical, defence and manufacturing sectors, Light Emitting Diode (LED) - across the visible, ultraviolet and infrared wavelengths, potential for vast range of applications like traffic signals, instrumentation, communication, televisions, display etc. are just a few obvious examples. A range of optical fiber designs and materials are now being developed to meet the needs of both established and growing industries.

Several research groups and industries around the world are developing the technology required to fabricate RE doped fibers with enhanced performance. In contrast to the standard telecommunication fibers, active fibers demand a greater variety of materials and structures in order to realize superior characteristics of amplification and lasing at a variety of operating wavelengths. Thus an optimized fabrication process needs to be established.

Out of the various processes used for manufacturing optical fibers, the Modified Chemical Vapor Deposition (MCVD) process [Nagel et al. 1982, MacChesney, 2000] has been widely accepted for making specialty fibers because of better control over refractive index (RI) profile and superior process flexibility. In this process, the reactant gases ($SiCl_4$, $O_2$ and various dopant precursors) passing down a rotating silica tube is heated by an external oxy-hydrogen burner that slowly traverses in the direction of input gas flow. The proportion of $SiCl_4$ and the dopants is precisely controlled by controlling the amount of carrier gas (mostly oxygen) through the bubblers containing liquid halide precursors having reasonably high vapor pressure (~$10^2$ mm of Hg) over the temperature range of 20-30°C. The reactant gases undergo high-temperature oxidation to form submicrometer range soot particles which get deposited downstream of the hot zone according to thermophoretic mechanism [Li, 1985]. The particulate layer is consolidated into thin pore free glassy film as the burner traverses over the deposited region. The desired RI profile is built up by repeating the above step with variation of vapor phase composition. Subsequent to completion of deposition, the tube is collapsed to a solid rod, known as preform through a few passes of the burner at temperatures above 2000°C.

The above technique is however not suitable for doping RE inside the preform core due to low vapor pressure of the RE precursors [Digonnet, 1993] at temperatures of 20-30°C. Thus RE doped fibers are prepared by modifying standard fabrication processes and adopting special techniques. The MCVD coupled with RE-doping process viz. solution doping method [Stones & Burrus, 1973; Townsend et.al 1987], sol-gel process [Matjec et al. 1997; Chatterjee et al. 2003] and direct nano-particle deposition (DND) [Tammela et al. 2002] are some of the well known methods that have been exploited successfully. On the contrary, the vapor phase delivery techniques namely heated frit source delivery [Ainslie et al. 1988], heated source delivery [Poole, 1985, 1986], heated source injector delivery [MacChesney & Simpson, 1985], aerosol delivery [Laoulacine et al. 1988; Morse et al. 1989], chelate delivery method [Thompson et al. 1984; Tumminelli et al. 1990], which require relatively complex set-up and have not yet been standardized for commercial production.

Since its inception, solution doping process has been well accepted due to its process simplicity and low implementation cost for fabricating RE-doped preforms among the various known methods. Although, MCVD coupled with solution doping is now well established and regularly employed even for commercial fiber production, the process still suffers from poor control over RE incorporation and reproducibility. The non-uniform thickness, porosity and pore size distribution in the deposited core layer leads to variation in RE concentration and inhomogeneity along the length of the preform/fiber. On the other hand solution strength, Al/RE proportion, dipping period etc influence the RE incorporation and fiber properties during solution doping stage. Thus judicious adjustment of the different process parameters based on their interdependence is vital for any further improvement in the process and the fiber performance. Till date no detailed exercise has been done to achieve such process optimization. Unless one understands the effect and interrelation of different process steps involved in the solution-doping technique it is not feasible to get control over the entire process.

This Chapter describes a systematic investigation on the process steps and associated parameters related to the above concerns. The experimental methodologies for optimization of soot deposition conditions and solution doping parameters are reported with important results. The correlation of soot morphology and solution parameters with ultimate fiber properties is presented along with optimized fabrication conditions which led to about 80% process reproducibility.

## 2. MCVD-solution doping process

MCVD process coupled with solution doping technique involves two major steps, viz. deposition of porous core layer within a silica tube using MCVD process at appropriate deposition temperature and impregnation of the porous deposit with a solution containing salts of RE (or combination of REs) and a codopant, mostly Al. The soaked layer is gently heated in presence of oxygen for conversion of the RE- and Al-salts into oxides. This is done to avoid evaporation of the salts during subsequent processing stages at high temperature and reduction of the RE content in the core in an uncontrolled manner. However, control of temperature at this stage is critical to ensure complete oxidation without evaporation which is strongly dependent on the characteristics of the salt used. The porous layer is subsequently dehydrated in presence of chlorine to eliminate OH adsorbed in the soot during solution doping. A temperature between 800-1200°C is usually maintained at this stage. However the duration needs to be optimized based on the thickness of the porous layer. The final step is sintering of the layer to obtain the RE doped core. A slow sintering process comprising several passes of the burner and gradual increase of temperature is helpful [Bandyopadhyay et al. 2001] to ensure smooth sintering without formation of imperfections within the sintered layer. The tube containing the RE-doped core is finally collapsed to produce the preform. A soft collapsing technique like the method adopted during sintering is useful to avoid loss of RE and Al-oxides from the core, particularly from the innermost region causing a central dip. The technique also provides preforms with good geometry. Fibers of desired dimension with resin coating are drawn under optimum conditions using a fiber drawing tower to achieve the desired optical and mechanical properties. A schematic of the solution doping method is shown in Fig. 1.

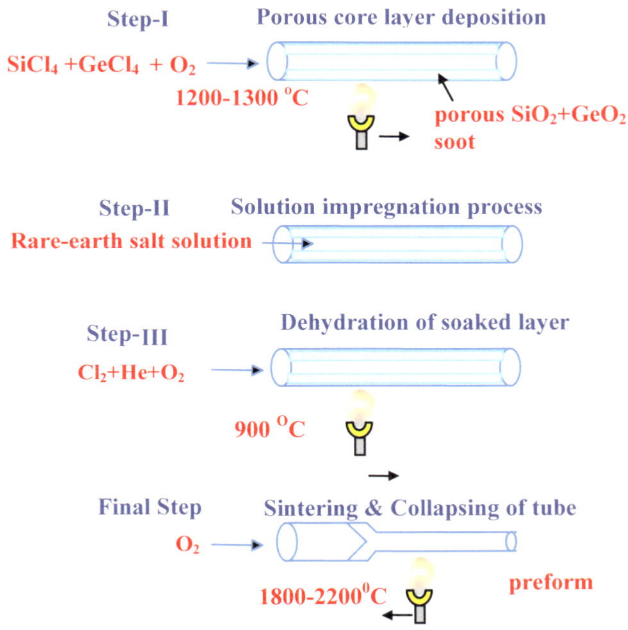


Fig. 1. Schematic of MCVD-solution doping process.

The critical step in this process is the deposition of uniform soot layer of suitable composition and porosity which serves as the precursor for solution impregnation. The variation in porosity as well as the pore size distribution leads to poor control over RE incorporation and inhomogeneity along the length of the preform/fiber. As a result, identical lengths of fiber from different sections of the preform do not provide the same performance. Unless the soot layer morphology is completely known and can be controlled by suitably adjusting the deposition conditions, the fiber performance cannot be improved. On the other hand the composition of the soaking solution, the Al/RE ratio, the nature of solvent, etc. are the controlling factors during the solution doping step to achieve the desired properties in the fiber.

Keeping in view the above factors, experiments have been performed on two broad aspects: a) control of soot deposition following MCVD process and b) optimization of solution doping parameters subsequent to deposition. This is elaborated in the following two sections. Each section consists of experimental methodology, characterization procedures adopted and significant observations with optimized conditions achieved. The results of the two parts are ultimately combined to realize the optimized fabrication conditions that enable to achieve enhanced repeatability with good control over RE incorporation.

## 3. Optimization of soot deposition parameters

In this section the influence of parameters controlling soot layer morphology is discussed. Two important parameters namely vapor phase composition and soot deposition temperature have been identified for an in-depth study to correlate their influence on the porous deposit microstructure and consequent effect on final preform/fiber characteristics. The soot layer characteristics have been critically analyzed and results obtained are correlated with the experimental parameters to achieve optimized deposition conditions.

### 3.1 Experimental methodologies

#### 3.1.1 Soot composition

The MCVD process was adopted to deposit porous silica layers of various compositions at selected temperatures inside a 20 mm outer diameter silica tube (F-300 grade from Heraeus) with tube thickness of 1.25 mm. In order to investigate the effect of vapor phase composition on deposited soot layer morphology, three different vapor phase compositions, specifically pure $SiO_2$, $SiO_2$+$GeO_2$ ($GeCl_4/SiCl_4$=0.86) and $SiO_2$+$P_2O_5$ ($POCl_3/SiCl_4$=0.48) were selected. A temperature in the range of 1200-1300°C was selected for depositing pure $SiO_2$ and $SiO_2$+$GeO_2$ layers. Since at this temperature $SiO_2$+$P_2O_5$ soot starts sintering, a temperature close to 1100°C was selected for depositing $SiO_2$-$P_2O_5$ composition which enabled to obtain soot layer with appreciable porosity for the selected vapor phase composition. Except a lower deposition temperature in case of $SiO_2$-$P_2O_5$, the other experimental conditions remained same for all the compositions.

Following the soot deposition, solution impregnation was carried out for fixed time span of 1 hour using an ethanolic solution containing 0.3 M $AlCl_3$ + 0.01 M $ErCl_3$. The soaked soot deposit was then oxidized and dehydrated. The dehydration was carried out in presence of $Cl_2$ at a temperature of 900°C. The oxidized layer was sintered in presence of $O_2$ and helium at temperatures in the range of 1400°C to 1700°C. Finally collapsing was performed at a temperature above 2200°C to obtain the preform. Fibers of 125±0.2 μm diameter with dual resin coating were drawn from the preforms in a conventional tower (Heathway, UK make).

#### 3.1.2 Soot deposition temperature

In order to examine the influence of deposition temperature on the porous deposit characteristics as well as the ultimate fiber properties in terms of RE incorporation, another series of experiment was carried out. In this series $GeO_2$ doped porous core layer was deposited at temperatures of 1220, 1255 & 1295°C maintaining the same vapor phase composition ($GeCl_4/SiCl_4$=0.86). The deposited soot was then processed following similar experimental conditions mentioned above (Sec 3.1.1) to obtain the final preform/fiber and to characterize their properties.

### 3.2 Characterization procedure

Scanning electron microscope (SEM Model:LEO-S430i) has been extensively used to analyze deposited soot morphology with variation of composition and deposition temperature. The

samples of deposit were collected from different sections of the overall length of tube both before and after solution doping in a careful manner to preserve the soot structure and avoid generation of defects. The selected specimens were dried under infra-red (IR) lamp for about 30 minutes to remove adsorbed moisture from the soot surface and coated with silver (Ag) using EMSCOPE (UK made) DC sputtering instrument prior to SEM investigation. Both secondary electron (SE) and back scattered electron (BSE) images were captured based on the required information. An image analysis software (LEICA Q500Mc) was utilized to calculate size and shape of the pores within the soot deposited from captured SEM images.
In order to correlate the results obtained from the SEM micrograph analysis, surface area measurement was performed for different soot samples following BET (Brunaur-Emmett-Teller) method to get an idea about the sizes of pores formed within the soot deposit. Accordingly, samples of soot deposit were collected from the substrate tube by taking utmost care to avoid any damage in soot structure. The samples were dried by heating around 100°C and precise weight was taken before putting inside the instrument (Quantachrome). Liquid Nitrogen gas was used as an adsorbate. The sample weight was varied from 0.05 gm to 0.15 gm in different measurement to check the repeatability of the result obtained.
The composition of collected soot deposit before solution doping had been determined using chemical analysis technique. This was specially aimed for Ge and P doped soot in order to compare the composition of deposited soot layer with that of input vapor mixture and sintered core in the final preform. Usually Inductively Coupled Plasma (ICP) method was used to determine the soot composition, but for Ge doped soot particles, besides ICP, a special analytical method [Mukhopadhyay & Kundu, 2006] using mannitol as complexing agent has been developed. This alternative technique provides very good result for the soot samples and matches closely with result obtained from ICP method.
The Electron Probe Micro Analysis (EPMA) of the polished preform samples was carried out using JEOL 8900 Electron Probe Microanalyzer to determine the RE and Al distribution across the core of the preform sample. For this investigation preform samples of 1.5 mm thickness polished on both the faces were prepared and measurements were carried out across the preform core at an interval of 5 μm to acquire the data on elemental distribution.
Longitudinal and radial homogeneity of the fabricated preforms in terms of RI were evaluated using preform analyzer (PK2600). Similarly employing fiber analyzer (NR20, EXFO), RI profile and numerical aperture (NA) of the drawn fiber were analyzed. The attenuation of the fabricated fibers was measured using "cutback technique" employing spectral attenuation measurement set-up (Bentham, UK-made). The measurement was carried out in a spectral range of 800 nm to 1600 nm. Fibers drawn from different parts of the preform were measured and compared for verifying compositional uniformity and repeatability of measurement.

### 3.3 Results & discussion

#### 3.3.1 Pore size vs soot composition

The SEM investigation of porous soot deposit reveals interdependence of soot layer morphology and layer thickness with input vapor phase composition. The addition of Ge or P-oxide is found to influence the particle growth dynamics and the viscosity of the silicate, which control the formation of the deposited soot network. With the lowering of viscosity, the porous structure collapses and unites together more easily, forming larger pores even at lower temperatures besides reducing deposited layer thickness. The SEM micrographs presented in Fig. 2 show the change in porous layer microstructure with change in soot composition. Pure $SiO_2$ deposit is homogeneous with finer network structure while the $SiO_2$+$P_2O_5$ network is composed of pores with wide variation in shape and size. Such variation is a consequence of pore collapsing and combining together as P facilitates sintering at lower temperature. The $GeO_2$ doped deposit has an intermediate structure between $SiO_2$ and $SiO_2$+$P_2O_5$ composition. Image analysis result shown in Fig. 2 represents the variation in pore size distribution in relation to soot composition. Accordingly the mean pore size is found to be 0.5 μm for $SiO_2$, 1.5 μm for $SiO_2$+$GeO_2$ and 5.5 μm for $SiO_2$+$P_2O_5$ soot samples. The observations indicate that pure $SiO_2$ deposit is most suitable for using as a precursor during solution doping as the network homogeneity and pore size uniformity are superior compared to other compositions. However, experiments reveal that pure $SiO_2$ deposit has the disadvantage of disengaging during soaking due to poor adhesion with the silica tube surface. So the amount of dopant in the soot deposit and the deposition temperature needs to be judiciously selected to achieve good compositional homogeneity.

The gradual addition of dopants such as $GeO_2$ and $P_2O_5$ led to decrease in the thickness compared to that of pure $SiO_2$ layer due to lowering of viscosity and sintering temperature, which results in densification of the layer. The thickness is least (mostly below 5 μm) for P-doped soot layer as P facilitates the sintering appreciably and highest for pure $SiO_2$ soot layer. In the case of $GeO_2$ doped soot layer, the thickness varied from 3.5 μm to 7.5 μm corresponding to change in temperature from 1295°C to 1255°C.

A new term 'pore area fraction' proposed in this regard and defined by the ratio of total pore area to the area of the deposit under consideration, showed a variation of 50% to 8% for the different compositions. The analytical data are presented in Table 1.

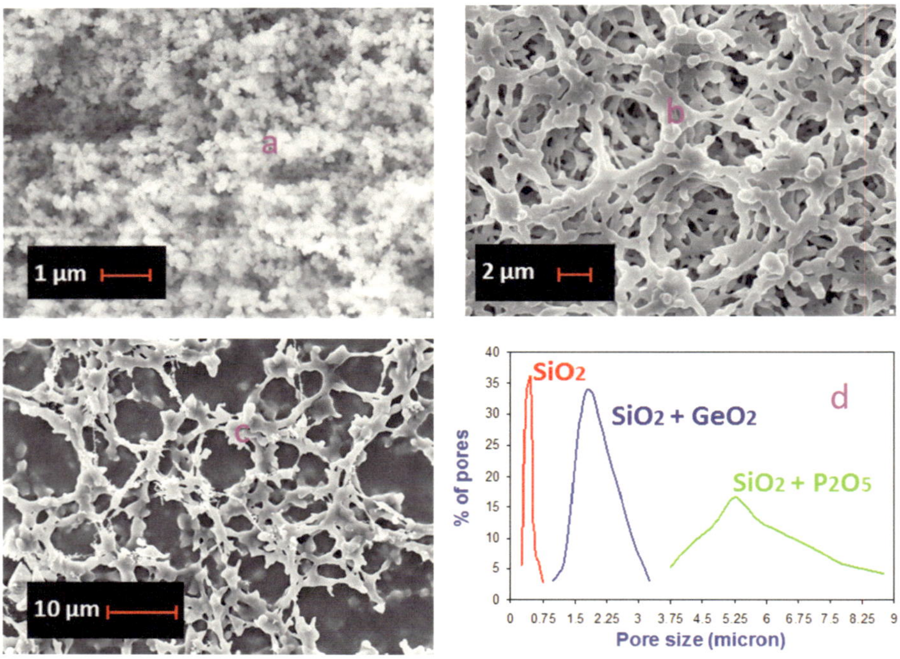


Fig. 2. SEM micrograph of (a) pure $SiO_2$, (b) $GeO_2$ doped, (c) $P_2O_5$ doped soot and (d) pore size distribution with change in soot composition.

| Deposition temperature (°C) | Vapor phase composition | Average pore size in µm | Pore area fraction |
|---|---|---|---|
| 1250 | Pure $SiO_2$ | 0.3-0.5 | 8% |
| 1250 | $SiO_2$+$GeO_2$ | 1.5-2.0 | 27% |
| 1100 | $SiO_2$+$P_2O_5$ | 5.0-5.5 | 50% |

Table 1. Variation of pore sizes with change in composition.

### 3.3.2 Pore size vs. soot deposition temperature

The SEM micrograph of $GeO_2$ doped soot layers deposited at temperatures of 1220° and 1295°C is presented in figs. 3 (a) & (b). Comparison of the said micrographs clearly reveals strong influence of deposition temperature over the network formation of the soot particles and the pore size distribution. The pore size distribution evaluated from image analysis of the micrographs provides a quantitative comparison as shown in Fig. 3 (c). The curves in the Fig.3 (c) represent pore size analysis of the soot deposit at temperatures of 1255°and 1295°C respectively. It is observed that the pore size has a wider distribution at higher temperature although mean pore size remains in the range of 1.5 µm. Thus at higher temperature the pores seem to collapse and fuse together resulting in greater pore size variation and decrease in porosity due to partial consolidation. If the number of pores in the range of 1-2 µm is taken as an indicator of uniformity, it is observed that the uniformity of pore size distribution reduces from 70% to 59% corresponding to the above temperature increment of 75°C. The pore area fraction showed a variation of 32% to 24% for the said increase in temperature from 1255° to 1295°C.

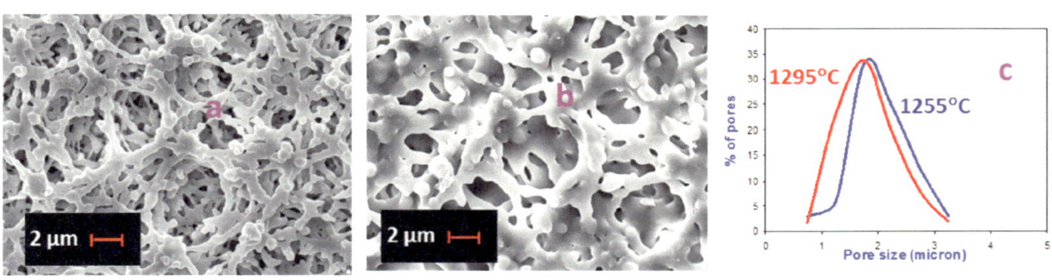


Fig. 3. SEM micrograph (a) $SiO_2$-$GeO_2$ soot deposited at 1220°C, (b) $SiO_2$-$GeO_2$ soot deposited at 1295°C and (c) Pore size distribution against deposition temperature for $SiO_2$-$GeO_2$ doped soot at 1255°C and at 1295°C.

### 3.3.3 Effect of composition and deposition temperature on surface area

The analysis of surface area of the soot particles is found to be in good agreement with interpretation drawn from SEM analysis as already discussed. The surface area of pure $SiO_2$ soot is about 195 $m^2/g$ at 1200°C and decreases significantly with addition of dopant like Ge or P. About 2 to 3 folds enhancement in surface area is observed irrespective of soot composition for a 55°C reduction in deposition temperature as evidenced by result given in Table 2. This reduction is a result of partial consolidation of the soot layer at higher temperatures. On the other hand, the reduction in soot layer viscosity with addition of dopant like Ge or P is the reason behind lower surface area of doped silicate soot layer compared to pure silica.

| Soot composition | Deposition temperature (°C) | Surface area, $m^2/g$ |
|---|---|---|
| $SiO_2$ | 1200 | 195.01 |
| | 1255 | 66.79 |
| $SiO_2$+$GeO_2$ ($GeCl_4$/$SiCl_4$=0.86) | 1200 | 18.07 |
| | 1255 | 10.32 |
| $SiO_2$+$P_2O_5$ ($POCl_3$/$SiCl_4$=0.48) | 1100 | 8.8 |

Table 2. Surface area of soot samples of different compositions.

### 3.3.4 Composition of soot and its influence

Chemical analysis of soot deposit when compared to input vapor composition and final core composition provides important information related to process mechanism involved during processing of soot deposit subsequent to solution soaking. $GeO_2$ proportion in the soot corresponds to about 73% and 81% completion of oxidation reaction of $GeCl_4$ at core deposition temperatures of 1250°C and 1295°C respectively. The values are found to be in close proximity to the theoretically calculated conversion of 82% and 87% at the said temperatures. The theoretical analysis has been done by considering the reaction as first order with respect to the chloride reactant and evaluating the rate constant values at specific temperatures [Pal et al. 2005]. The similarity of $GeO_2$ proportion in the input vapor mixture with that in the unsintered soot layer is significant as it is directly related to the viscosity of the deposit and the extent of sintering at the deposition temperature. The NA of the ultimate fiber corresponding to this composition is near to 0.20 which corresponds to 10.5 mol% $GeO_2$ in the core. Thus the $GeO_2$ content in soaked soot substantially decreases in the ultimate core layer indicating that the major amount of $GeO_2$ is lost during sintering and collapsing stages. The result reveals that the network formation and pore collapsing take place in soot with much higher $GeO_2$ concentration than that present in the final fiber. Similar result was obtained while analyzing the soot containing $P_2O_5$ where the problem of evaporation is more pronounced.

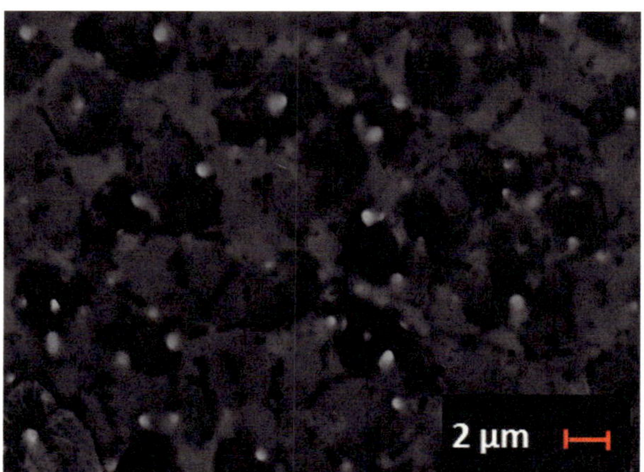


Fig. 4. SEM micrograph of $SiO_2$-$GeO_2$ soot deposited at 1295°C after solution doping.

### 3.3.5 RE incorporation dependence on soot composition and deposition temperature

The BSE micrograph of the $SiO_2$+$GeO_2$ layer subsequent to soaking with a solution of 0.3 M $AlCl_3$ + 0.01 M $ErCl_3$ in ethanol for one hour is presented in Fig. 4. The solution impregnation through the pores is prominent with white region in the image showing the presence of higher molecular weight substances like Er (white) and Al (grey). The point that is evident is that the uniformity of dopant impregnation through solution is dependent on the morphology of the soot layer deposit. Larger pores draw greater amount of solution leading to increase in both RE and Al concentration and consequently act as precursor for development of RE and Al clusters. It is known that RE clustering leads to pair induced quenching and degradation in final fiber performance. Similarly the Al clusters are also probable sources for Al rich phase separation and defect generation in the core. Thus a uniform pore size distribution is a prerequisite to obtain uniform RE distribution in the ultimate preform/fiber.

The analysis of SEM images and surface area results indicates that pure silica soot network consists of smaller pores with higher surface to volume ratio compared to doped silica soot deposit. Accordingly based on the soot porosity and pore size uniformity factor, pure silica stands as the preferred soot composition to obtain RE-doped preforms with uniform distribution along the length compared to Ge or P doped core layer. However, weak adhesion of pure $SiO_2$ deposit with the silica tube surface is associated with the risk of peeling off during soaking (particularly for solution of higher concentration) besides poor RE solubility. All these restrict the use of pure silica as the core layer composition. Considering essential requirement of RE doped fibers and analyzing the above results, $GeO_2$-doped silica soot is found to be the most suitable for formation of core layer and a deposition temperature of 1260°C is found to be optimum to obtain acceptable pore size uniformity for the selected vapor phase composition.

A correlation has been worked out with regard to final Er incorporation in three different soot layers deposited with variation in temperature and composition. The absorption at 980 nm due to Er content differs by about 0.90 dB/m for two $GeO_2$ doped fibers where the deposition temperatures had a difference of 40°C [Dhar et al. 2006]. This corresponds to an Er concentration difference of 200 ppm. The fiber F1 prepared using lower soot deposition temperature (1255°C) has higher absorption compared to fiber F2 prepared from soot deposited at 1295°C as shown in Fig.5. Thus a concentration difference of 100 ppm is observed for a temperature variation of 20°C for the selected composition. Partial sintering with collapsing of the pores at higher temperature leads to reduced surface to volume ratio and consequently less solution retention resulting in lesser incorporation of rare earth ion. The observation is important to optimize the solution doping parameters and control RE concentration in the fiber. However, $P_2O_5$ doped soot demonstrates reverse trend and the fabricated fibers contain higher Er concentration instead of soot layer densification due to viscosity lowering and reduced intake of solution. This reverse trend can be explained on the basis of the soot chemistry of Si-P [Digiovanni et al. 1989] in comparison to that of Si-Ge. Substitution of Si by P helps to avoid charge imbalance, which results in additional Al absorption and consequent increases in Er level in the layer. Additionally investigation of the soot network structure indicates that P doped layer contains larger pores compared to the Ge doped deposit and has the capacity of retaining more amount of solution. This is because in case of P addition, the pores have a tendency to combine together with the disappearance of intermediate wall during partial consolidation instead of mere size reduction.

An interesting empirical relation has been established analyzing the results on soot morphology in terms of pore area fraction and Er ion incorporation level. The absorption of 3.12, 2.27 and 4.99 dB/m (Table 3) is observed against pore area fraction of 32%, 24% and 50% in the porous layer microstructures of fibers F1, F2 and F3 respectively. This corresponds to a relation of B=c.Q, where B represents Er ion absorption at 980 nm in dB/m, Q the percent pore area fraction and c is a constant. The value of c is found to be 0.1 for the solution used in the present case. The relation is found to be valid for several fibers fabricated for this purpose. The point that is obvious from all the above results is that an analysis of the porous layer microstructure provides an indication about the rare earth concentration, distribution and the homogeneity in the final preform/fiber.

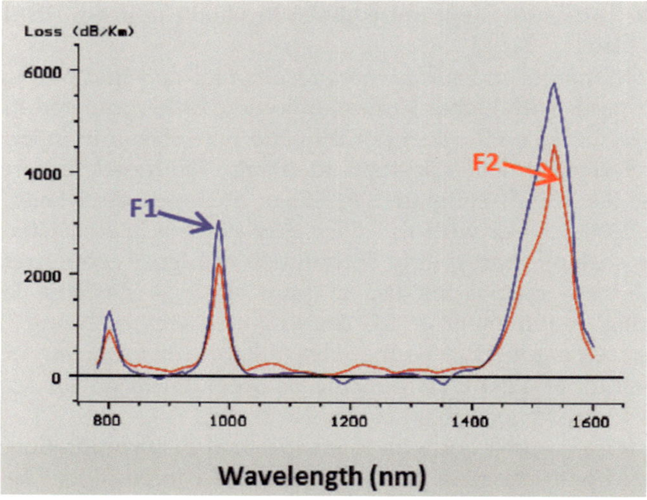


Fig. 5. Spectral attenuation curves of F1 and F2 showing $Er^{3+}$concentration variation.

| Fiber | Input vapor mixture | Deposition temperature (°C) | Er ion absorption peak at 980 nm, dB/m | Fiber Characterization | |
|---|---|---|---|---|---|
| | | | | NA | Core-diameter in μm |
| F1 | $GeCl_4/SiCl_4$=0.86 | 1255 | 3.120 | 0.22 | 7.6 |
| F2 | $GeCl_4/SiCl_4$=0.86 | 1295 | 2.265 | 0.20 | 7.5 |
| F3 | $POCl_3/SiCl_4$=0.48 | 1100 | 4.990 | 0.13 | 6.5 |

Table 3. Fiber properties and parameters.

## 4. Solution doping parameters

During the solution doping phase, there are a number of parameters which play a significant role. In this section we have identified the parameters which are important for controlling solution impregnation and consequently RE/Al incorporation into the core which determine the final fiber performance. Emphasis is given on optimization of solution doping parameters and correlation with soot characteristics to achieve enhanced process repeatability.

### 4.1 Solution doping parameters

The SEM images of soot deposit clearly show that the porous layer comprises of open or closed pores interconnected to form 3-dimensional structures. Although the nature of network and extent of open and closed pores depend on soot-composition and chosen deposition temperature, the soot structure can be seen as series of capillaries interconnected to each other with irregular arrangement. Thus during solution doping the flow through this porous layer is equivalent to flow of soaking solution through capillaries. It is well known that the flow of solution through porous medium is inversely proportional to solution viscosity as demonstrated by the following equation:

$$V = \frac{r}{4\mu l}\gamma\cos\theta \qquad (1)$$

where $V$ = velocity of solution flowing through porous medium
r = radius of capillary tube (can be taken as pore size), $l$ = length of porous frit,
$\mu$ = viscosity of solution, $\gamma$ = surface tension of liquid, $\theta$ = contact angle
Thus the solution viscosity and surface tension are critical factors to achieve desired dopant incorporation into the preform core. Both the parameters are dependent on the selected solution composition. According to equation (1), higher viscosity of the solution leads to reduction in flow through the pores. Therefore, the dipping period assumes significant importance according to solution properties for desired RE doping in the core. The other important parameters are RE-salt concentration and proportion of Al/RE in the solution. It is well known that Al codoping helps in solvation of RE ions in silica network [Arai et al. 1986] and reduces the loss of RE from the core during collapsing stages [Ainslie et al. 1988]. Although it is established that increased dopant concentration in the solution leads to higher dopant intake into the core, the proportion of Al/RE needs to be optimized to avoid unwanted clustering effect which degrades final fiber performance. The core-clad interface also shows defect generation when the preform core is rich in $Al_2O_3$ which increases scattering loss of the fiber. Accordingly a systematic investigation was carried out to achieve optimized soaking solution composition and process conditions during solution impregnation stage. Experimental methodologies used, characterization techniques employed and results obtained are discussed in the following sub-sections.

### 4.2 Experimental methodologies

#### 4.2.1 Solution properties

The viscosities of solutions with varying dopant concentrations prepared using different solvents viz. water, methanol, ethanol and n-propanol were measured for comparison and establishing a database which can be utilized for selection of appropriate solvent and concentration during preparation of the soaking solution. Al ion concentration was varied from 0.15 M to 0.98 M keeping RE ion concentration fixed at 0.01 M. Either $AlCl_3$ or $Al(NO_3)_3$ was used as a source of Al-salt while $ErCl_3$ was the RE ion source in these experiments. The density and viscosities of different soaking solutions was measured using pycnometer and Ostwald Viscometer at 25°C while surface tension value was taken using "du Nouy Tensiometer" at 25°C.

#### 4.2.2 Selection of solvent

The solvent used for preparation of soaking solution, should be polar in nature with appreciable dielectric constant in order to dissolve Al and RE salt easily, it must be low boiling liquid so that evaporates easily on passing inert gas like $N_2$ at room temperature after solution doping and should not chemically react with deposited soot layer. These requirements restrict the choice of solvent to water and alcohol. In order to assess the effect of solvent on final fiber performance, we fabricated a series of preforms where two parts were soaked with soaking solution of same strength but dissolved using either water or ethanol.

For this investigation a 60 cm long wave guide tube (Suprasil F-300) of dimension 20/17.5 mm was selected and porous core layer of optimized $GeO_2$-doped silica soot (already discussed under section 3) was deposited at a temperature region of 1255°C subsequent to deposition of F-doped cladding layers. The deposited tube was then cut into two equal parts and separately dipped for one hour into two different soaking solutions containing 0.3 M $AlCl_3$+0.01 M $ErCl_3$ prepared using water and ethanol as solvent. Subsequent to solution doping, two parts were remounted on the glass working lathe, joined and processed further to obtain the final preform. Thus the preform fabricated consists of two parts soaked with ethanolic and aqueous solutions. Fibers were drawn from two ends of the preform and Er-incorporation was compared to investigate the influence of solvent effect.

#### 4.2.3 Influence of dipping period

In order to investigate the effect of dipping period, we carried out the experiment using optimized $GeO_2$–doped vapor phase composition using a 60 cm long waveguide tube of dimension 20/17.5 mm where different parts of the soot deposit were soaked for different time span ranging from 15 to 60 minutes. The solution containing 0.3 M $AlCl_3$+0.01 M $ErCl_3$ made using ethanol was used for this experiment. The tube with the deposit was marked into different parts of equal length and during solution doping; the solution was partly drained out (at a fixed draining rate of 2.5 cm/min) after fixed time intervals from the deposited tube so that different portions of the deposit were soaked for different time periods. The input end from where the deposition took place was soaked for minimum time period while the other end of the tube was soaked for maximum time period. The experiments were performed using two different deposition temperature regimes namely at 1250±5°C and at 1300±5°C. Subsequent to solution soaking the soaked tube was processed further to obtain the final preform. Fibers were drawn from all the different parts and the incorporated Er-ion in different parts soaked for different time spans was evaluated to investigate the effect of dipping period on the final fiber composition.

#### 4.2.4 Influence of Al/RE proportion

It has already been discussed under section 4.1 that codoping with $Al_2O_3$ helps to enhance the RE ion solubility in silica and restricts the evaporation of $RE_2O_3$ resulting in a uniform radial distribution of RE and Al in the preform core. Further Al doping improves the flat-gain characteristics of an Er doped fiber (EDF). However, Al-doping beyond a certain limit gives rise to phase-separation and generation of defect centres at the core-clad boundary (star-like pattern) leading to enhancement of scattering loss. So it is essential to find out suitable Al/RE proportion to fabricate preform/fibers of improved performance. Accordingly a series of preform runs were carried out by maintaining optimized vapor phase composition (to deposit $GeO_2$-doped silica soot) and other processing parameters by changing the solution composition from run to run. The porous core layer deposition temperature was fixed at 1250±5°C and 1300±5°C in two different series. In these experiments, preform runs were carried out by changing the solution composition where for a fixed Er concentration of 0.01 M, the Al/Er proportion was varied from 0 to 100 with several intermediate values. Preform runs with the soaking solution comprising only 0.01 M $ErCl_3$ in ethanol corresponding to Al/Er ratio of 0 was carried out to investigate the incorporation of Er ion in absence of Al ion. Subsequent processing of the soaked porous deposit produced preforms/fibers with variation in Al/Er ratio.

### 4.2.5 Core-clad interface problem

Formation of defects ('star-like' pattern) along core-clad interface is a universal phenomenon that occurs in $Al_2O_3$ rich preform/fiber core. The effect becomes more adverse in presence of $GeO_2$ in the core region. The presence of such defects results in unwanted scattering loss in the ultimate fiber. Although many investigations have been carried out by different groups to explain the origin of this defect and possible route to overcome the problem [An et al. 2004], no one has been able to suggest a reliable method to eliminate it. It is believed that viscosity mismatch between core-clad material is important factor responsible for this phenomena.

A novel approach that has been adopted in the present investigation in getting rid of this problem is employing a soaking solution containing dispersed fumed silica instead of conventional soaking solution. The amount of dispersed fumed silica (Alfa Aesar, surface area 175-200 $m^2/gm$) was varied in different preform runs from 0.1 to 0.6 gm based on Al-ion content of 0.3 to 1.75 M in the soaking solution. Both aqueous and alcoholic solutions were used to verify the effectiveness of the proposed process.

## 4.3 Characterization procedure

The characterization process used here is alike that discussed under section 3.2. The RI profiles of the fabricated preforms and ultimate fibers were evaluated using preform analyzer (PK2600) and fiber analyzer (NR-20). The concentration of Er was determined from the absorption peak at 980 nm in the spectral attenuation curve. The distribution of different dopants inside the preform/fiber core was evaluated using Electron Microprobe analysis. In addition, the core-clad interface boundary was observed under high resolution optical microscope to examine the defects formed, if any, due to addition of Al.

## 4.4 Results & discussion

### 4.4.1 Viscosity of solution and its influence

The measured value of viscosity reveals that solution viscosity increases with enhancement of solution strength as presented in Fig. 6. About 10 fold increase in viscosity value (1.98 to 19.5 cP) was observed for a variation in the concentration of $AlCl_3$ from 0.15 to 1.0 M in an ethanolic solution. Since the final RE concentration is proportional to the amount of solution retained by the porous structure, solution viscosity influences RE incorporation level in the final fiber. The solution retention capacity of the porous medium is dependent on viscosity of the soaking solution [Kim et al. 2003] and equation (1) predicts that highly viscous solution requires longer infiltration time through porous medium. The solution retention is higher for high viscous solution. Thus based on the deposited soot layer morphology and its adhesion strength with glass surface, solution viscosity needs to be adjusted. For $GeO_2$ doped ($GeCl_4/SiCl_4$=0.86) silica soot layer deposited at a temperature in the region of 1200-1230°C porous structure shows good uniformity but its adhesion to silica glass surface is weak showing tendency of disengagement or crack development during solution doping stage if a solution of viscosity greater than 3.0 cP is used. In comparison, a soot layer of above composition deposited at a temperature region of 1260-1280°C shows good adhesion with no sign of imperfection for solutions up to a viscosity value of 7.0 cP. For solutions of higher viscosity it is essential to deposit the soot at temperatures above 1290°C. But the porosity decreases appreciably leading to poor rare earth incorporation. This is also accompanied by formation of larger pores having enhanced probability of local composition variation. Considering the above factors, a temperature of 1260°C is found to be optimum for deposition of $SiO_2$-$GeO_2$ (10.5 mol% $GeO_2$) core layer. The deposit can be used for solution impregnation without problem up to a solution strength of 0.60 M $AlCl_3$ in ethanol and leads to a NA of 0.20 or greater.

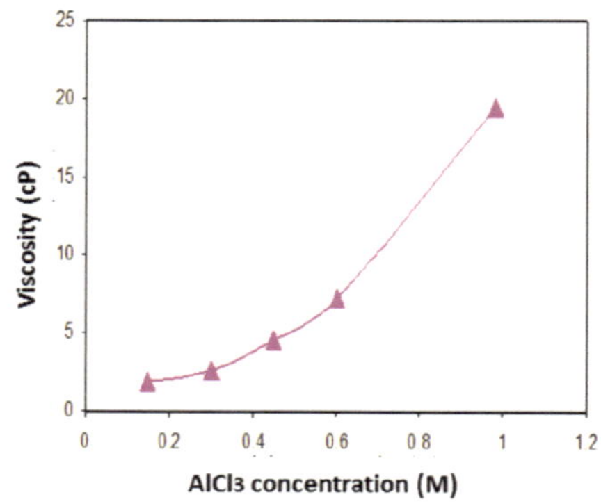


Fig. 6. Variation of Viscosity with changes in $AlCl_3$ concentration in ethanol.

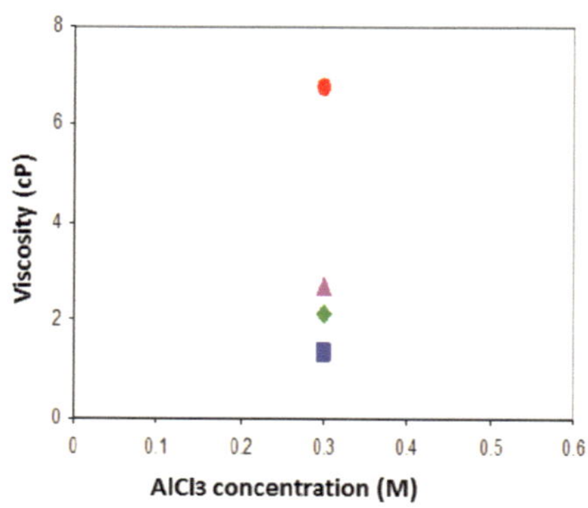


Fig. 7. Variation of Viscosity with change in solvent ● (propanol) ▲ (ethanol) ◆ (methanol) ■ (water).

### 4.4.2 Selected solvent vs. RE incorporation level

The nature of solvent is found to play a significant role on solution characteristics especially on viscosity as evident from Fig. 7 which shows the viscosity of a 0.3 M $AlCl_3$ solution increases from 1.3 cP to 6.7 cP by changing the solvent from water to propanol. It has been observed that for fibers obtained using a solution of 0.3 M $AlCl_3$+0.01 M $ErCl_3$ soaked for 60 minute, Er concentration difference equal to 120 ppm occurs between an ethanolic solution and an aqueous solution as evident from the spectral attenuation curve presented in Fig.8.

EPMA result [Fig.9] indicated that additional Al ion was adsorbed by alcohol soaked preform part compared to aqueous counterpart (preform part soaked with aqueous solution). This can be explained on the basis of surface tension differences of alcoholic and aqueous solution. The surface tension of fixed strength of aqueous solution selected for the study is found to be 2 to 3 times higher than that of the ethanolic solution. It has been reported [Khopin et al. 2005] that higher infiltration time is required for aqueous solution compared to alcoholic solution as a consequence of higher surface tension of aqueous solution.

Since Al and RE salts formed different complexes when added in solvents such as water or ethanol, size of the complex formed also governs the rate of dopant ion adsorption during solution impregnation. Due to the smaller size of the Al complex [Al $(OC_2H_5)x.Cl_{3-x}$] in alcohol compared to complex formed in aqueous solution $[Al(OH)(H_2O)_5]^{2+}$, porous deposit will soak more dopant ions during soaking in alcoholic solution compared to the aqueous solution.

Thus lower surface tension as well as smaller complex size of dopant ions are responsible for faster soaking in case of ethanolic solution compared to same strength of aqueous solution when soaked for fixed time span. The investigation reveals that solvent plays an important role in controlling incorporation of RE ion in fiber. Using water as solvent, longer soaking period is required to achieve saturation of pores and the pore size influence becomes greater compared to ethanolic solution of same concentration.

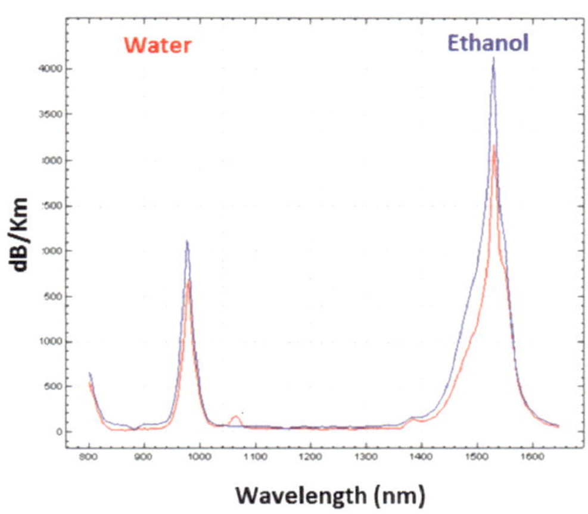


Fig. 8. Attenuation curve of fibers fabricated by using aqueous and ethanolic solutions of 0.3 M $AlCl_3$+0.01 M $ErCl_3$.

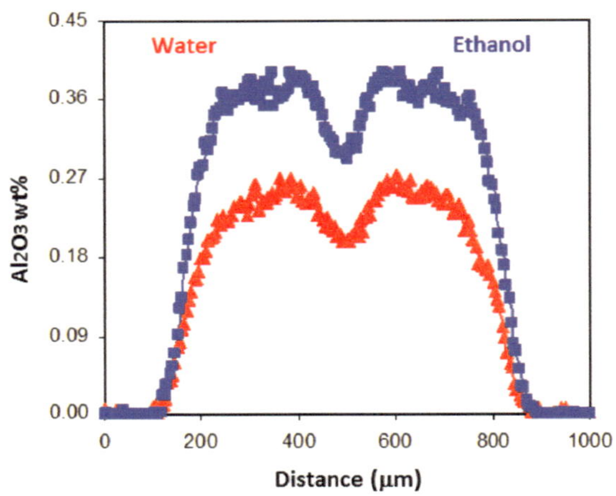


Fig. 9. EPMA result showing Al distribution across the preform core using aqueous and ethanolic solutions of 0.3 M $AlCl_3$+0.01 M $ErCl_3$.

#### 4.4.3 RE incorporation level against dipping period

Conventionally dipping period of 1 hour is used in the solution doping process to achieve the saturation level during soaking stage but our investigation points out an interesting trend. From the above set of experiments, it is realized that the optimum dipping period is dependent on soot layer thickness, composition, porosity and also on solution characteristics like nature of solvent and Al/RE concentration. Accordingly, the dipping period can be suitably adjusted for getting a specific rare earth ion concentration in the sintered core layer. It is observed that during soaking, initially the rare earth incorporation rate is high, which decreases with time before getting saturated, depending upon soot morphology and solution nature. The variation in final rare earth ion concentration with dipping period when plotted produces Fig. 10 where effect of deposition temperature is clearly revealed. At 1250°C, the Er incorporation is nearly 300 ppm for a dipping period of 15 minutes which increases about 2 fold (~ 640 ppm) after a time period of 45 minutes. The maximum concentration after 60 minutes of dipping is about 690 ppm. The RE incorporation at various dipping periods is much lower at deposition temperature of 1300°C because of lower porosity of the porous deposit. Thus the investigation helps to obtain an optimized dipping period of 45 minute for a porous germanosilicate core layer having about 7.5 μm thickness, deposited at 1250-1260°C. This would correspond to an Er ion concentration of about 650 ppm when 0.3 M $AlCl_3$ + 0.01 M $ErCl_3$ solution in ethanol is used for soaking purpose. This observation essentially indicates presence of equilibrium between bulk solution and adsorbed dopant ion.

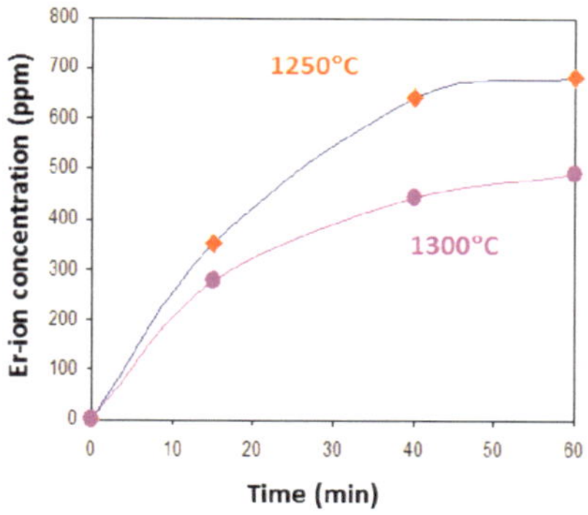


Fig. 10. Variation of Er ion concentration in the core with change in dipping period at deposition temperatures of 1250°C and 1300°C.

#### 4.4.4 Influence of Al/RE proportion

Another important parameter under present investigation is the judicious selection of Al/RE proportion. It was observed that adjustment in Al proportion in the solution not only alters the Al incorporation but also $RE^{3+}$ concentration in the fiber. Thus for deposited layers of fixed composition and porosity, it is possible to vary the final $RE^{3+}$ concentration in the glass by maintaining the same RE ion concentration in the soaking solution with only adjustment in the Al ion proportion. This step is also responsible for controlling $Al^{3+}$ concentration in the core. The observation is completely new and important. Usually, in order to change the RE concentration in the core, either the porosity is varied by changing soot composition or by altering the deposition temperature, otherwise the RE concentration in the soaking solution is adjusted proportionately. Fig. 11 represents the variation of Er concentration in final fiber using a fixed concentration of 0.01 M $ErCl_3$ but different $AlCl_3$ concentrations in the soaking solution for soot layer deposited at temperatures of 1250°C and 1300°C. The result indicates that the increase in $AlCl_3$ concentration in the solution from 0.1 M to 0.6 M leads to about 60% increase in Er ion incorporation. The RE incorporation for a fixed Al concentration can also be increased further by increasing the amount of RE salt in the soaking solution. The method provides much better control over RE incorporation and its uniformity along the preform/fiber length compared to the known techniques. As a result, the reproducibility is also improved compared to the conventional methods. The chemistry behind this can be explained as follows. During soaking, the RE absorption efficiency is increased with increase of Al concentration because of a 'cooperative phenomenon'. The increase in Al ion concentration along with the RE ions in the porous deposit facilitates formation of $SiO_2$-$Al_2O_3$-$RE_2O_3$ network during sintering. The RE ions thus become embedded in silica network, and their evaporation/ diffusion during sintering and collapsing at high temperature decreases. The overall effect is an effective increase in RE ion concentration. The effect is linear up to an Al ion concentration of 0.65 M in the soaking solution and then follows a decreasing trend.

The cooperative phenomenon described above has also been explained by a theoretical model [Dhar et al. 2008]. The model is based on the assumption that apart from direct adsorption from the solution phase, whose rate will be proportional to the concentration in the solution and also to the fraction of uncovered surface, there is also a parallel cooperative adsorption, in which an adsorbed Al particle helps additional adsorption of dopant ions. Rate of this cooperative adsorption will be proportional to the following factors, namely concentration of Al ions in solution, fraction of uncovered surface and also to the fraction of surface covered with Al species. Desorption rate will be proportional to the fraction of surface covered by Al species. However, when Al concentration in soaking solution exceeds a certain limit, the number of vacant surface sites decreases significantly resulting in decrease of Er incorporation. This observation was also experimentally verified.

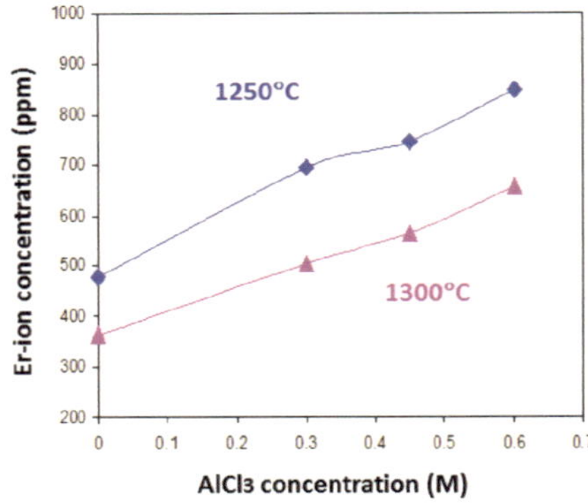


Fig. 11. Variation of Er ion concentration in the fiber with change in $AlCl_3$ concentration for a fixed $ErCl_3$ concentration in solution ▲ (1300°C) ◆ (1250°C).

#### 4.4.5 Removal of core-clad interface defects

Use of dispersed fumed silica is found to be an excellent route to overcome the problem of defect generation (star-like pattern development) along the core-clad interface. The basis of the idea was to avoid segregation of Al-rich phase during sintering of soaked soot layer by providing free silica which helps to create alumino-silicate glass phase. Employing this technique about 4.6 wt% of $Al_2O_3$ containing preform in presence of $GeO_2$ (10.5 mol%) has been successfully fabricated without any core-clad interface disturbance. This preform was prepared using a solution containing about 1.75 M concentration of $Al^{3+}$ ion where only 0.3 gm fumed silica was dispersed using ordinary ultra-sonication process [Dhar et al. 2010]. Figure 12 presents high resolution optical microscopic view of core-clad boundaries obtained employing (a) normal solution containing RE and Al salts without addition of fumed silica and (b) fumed silica dispersed soaking solution. It clearly shows how the addition of fumed silica has been effective in eliminating the defects formed due to the presence of high concentration of $Al_2O_3$ in the core.

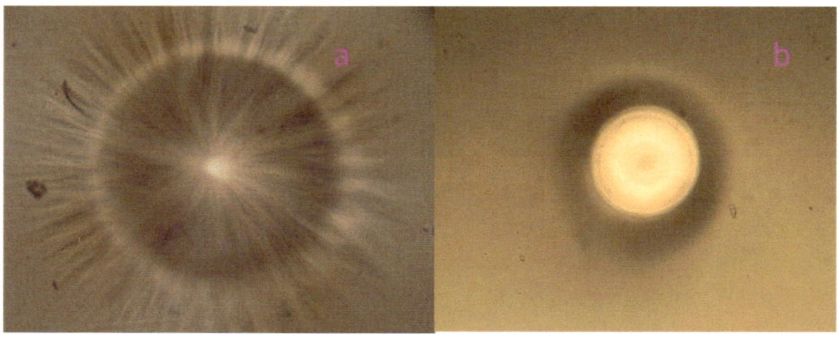


Fig. 12. Optical microscopic view of core-clad interface in preforms using (a) normal RE and Al containing solution and (b) fumed silica dispersed solution.

## 5. Correlation of different parameters

The overall investigation reveals the interdependence of soot layer characteristics with the solution properties and the dipping parameters. The soot composition and deposition temperature are found to significantly influence the porous layer microstructure which ultimately controls the amount of RE incorporation. Analysis of soot morphology and incorporated Er amount in final preform/fiber reveals existence of an empirical relation presented by a simple equation B=c.Q. Here "B" is Er ion absorption at 980 nm in dB/m, "c" a constant having value of 0.1 for the solution used in this experiment and "Q" pore area fraction (ratio of total pore area to the area of the deposit under SEM investigation). This analogy can be extended for other REs to predict the RE ion incorporation based on known soot parameters irrespective of soot composition. The viscosities of the solution are found to vary appreciably with change in solution strength and nature of solvent. About 10 fold increase in the viscosity is observed for a variation in the $AlCl_3$ concentration from 0.15 to 1.0 M in ethanolic solution. Thus the selection of solution of appropriate viscosity becomes an important criteria depending upon the porous soot layer morphology and its adhesion to the glass surface in order to avoid defects in the core. For $GeO_2$ doped ($GeCl_4/SiCl_4$=0.86) silica soot layer deposited at temperature in the region of 1260-1280°C, the porous structure shows good adhesion with little sign of imperfection for solution up to a viscosity of 7.0 cP. An important effect first time observed is the interdependence of Al ion concentration with the RE incorporation into the core because of a cooperative phenomenon. Thus for deposited layers of fixed composition and porosity, it is possible to vary the final $RE^{3+}$ concentration in the glass by maintaining the same RE ion concentration in the soaking solution, with only adjustment in the Al ion proportion. It is also possible to adjust the average RE concentration to the desired levels by appropriate selection of dipping period. A soaking period of 45 minute is observed to be optimum for a porous core layer of 7.5 μm thickness, as the incorporation rate slows down appreciably after this period. Thus for fabricating a germanosilicate core fiber with numerical aperture of 0.20 or above, a process condition, where the porous core is deposited maintaining 1250-1260°C temperature, the soaking solution contains 0.3 M $AlCl_3$+ 0.01 M $ErCl_3$ made in ethanol and dipping time is selected as 45 minute, is observed to be optimum and the fiber core is found to contain around 650 ppm of $Er^{3+}$ ion with appreciably good doping uniformity along the length. Several preforms and fibers with good reproducibility have been fabricated by the above process.

## 6. Conclusion

The influence of parameters associated with different phases of the solution doping process viz. soot deposition by MCVD technique and RE impregnation via solution doping method has been revealed through a systematic investigation. This has led to optimization of fabrication conditions to achieve better control over RE incorporation and enhanced repeatability. Experimental results indicate strong interdependence of soot morphology and solution doping parameters with ultimate preform/fiber properties. For the first time, the influence of the parameters has been exposed in a quantitative manner. The cooperative phenomenon not observed earlier between Al and RE offers a new approach of controlling RE concentration in the core. The method provides better control over RE incorporation and its uniformity along the preform/fiber length compared to the known technique. A process reproducibility of about 80% has been achieved by maintaining the optimized fabrication conditions.

## 7. References

Ainslie, B. J., Armitage, R., Craig, S. P. & Wakefield, B. (1988), Fabrication and optimization of the erbium distribution in silica based doped fibres, *Proceedings of ECOC*, p. 62, Brighton, UK, September 1988

An, H., Tang, Y., McNamara, P. & Fleming, S. (2004), Viewing structural inhomogeneities at the core-cladding interface of re-heated MCVD optical fibre preforms with optical microscopy, *Optics Express*, Vol. 12, No. 25, (December 2004), pp. 6153-6158, ISSN 1094-4087

Arai, K., Namikawa, H., Kumata, K. & Honda, T. (1986), Aluminum or phosphorous co-doping effects on the fluorescence and structural properties of neodymium-doped silica glass, *Journal of Applied Physics*, Vol. 59, (January 1986), pp. 3430-3436, ISSN 0021-8979

Becker, P. C., Olsson, N. A., & Simpson, J. R. (1999), *Erbium doped fiber amplifiers-Fundamentals and technology* (1st Edition), Academic Press, ISBN 978-0-12-084590-3, San Diego, CA USA

Bandyopadhyay, T., Sen, R., Bhadra, S.K., Dasgupta, K. & Paul M. (2001), Process for making rare earth doped optical fibre, US Patent 6,751,990, 6.22.2004, Available from http://www.freepatentsonline.com/6751990.html

Chatterjee, M., Sen, R., Pal, M., Naskar, M., Paul, M., Bhadra, S., Dasgupta, K., Ganguli, D., Bandyopadhyay, T., Gedanken, A. & Reisfeld, R. (2003), Rare-earth doped optical fibre from oxide nano-particles, *Acta Optica Sinica*, Vol. 23, (October 2003), pp. 35-56, ISSN 0253-2239

Dhar, A., Paul, M. C., Pal, M., Mondal, A. Kr., Sen, S., Maiti, H. S. & Sen, R. (2006), Characterization of porous core layer for controlling rare earth incorporation in optical fiber, *Optics Express*, Vol. 14, No. 20, (October 2006), pp. 9006-9015, ISSN 1094-4087

Dhar, A., Paul, M. C., Pal, M., Bhadra, S. K., Maiti, H. S. & Sen, R. (2007), An improved method of controlling rare earth incorporation in optical fiber, *Optics Communication*, Vol. 277, (September 2007), pp. 329-334, ISSN 0030-4018

Dhar, A., Pal, A., Paul, M. Ch, Ray, P., Maiti, H. S. & Sen, R. (2008), The mechanism of rare earth incorporation in solution doping process, *Optics Express*, Vol. 16, No. 17, (August 2008), pp. 12835-12846, ISSN 1094-4087

Dhar, A., Das, S., Maiti, H.S. & Sen, R. (2010), Fabrication of high aluminium containing rare-earth doped fiber without core-clad interface defects, *Optics Communication*, Vol. 283. pp. 2344-2349, ISSN 0030-4018

Digonnet, M. J. F. (1993), *Rare Earth Doped Fiber Lasers and Amplifiers* (2nd Edition), Marcel Dekker Inc., ISBN 0824704584, New York, USA

Digiovanni, D. J., Macchesney, J. B. & Kometani, T. Y. (1989), Structure and properties of silica containing aluminum and phosphorus near the $AlPO_4$ join, *Journal of Non-Crystalline Solids*, Vol. 113, (November 1989), pp. 58-64, ISSN 0022-3093

Khopin, V. F., Umnikov, A. A., Gur'yanov, A. N., Bubnov, M. M., Senatorov, A. K. & Dianov, E. M. (2005), Doping of optical fiber preforms via porous silica layer infiltration with salt solution, *Inorganic Materials,* (Engl. Transl), Vol. 41, No. 3, (March 2005), pp. 303-307, ISSN 1608-3172

Kim, Y. H., Paek, U. C. & Han, W. T. (2001), Effect of soaking temperature on concentrations of rare-earth ions in optical fibre core in solution doping process, *Proceedings of Rare-earth-doped materials and DevicesV, SPIE,* Vol. 4282, ISBN 9780819439604, San Jose, CA, USA, January 2001

Laoulacine, R., Morse, T. F., Charilaou, P. & Cipolla, J. W. (1988), Aerosol delivery of non-volatile dopants in the MCVD System, Extended Abstracts of the AIChE Annual Meeting, Washington, DC, USA, December, 1988

MacChesney, J. B. & Simpson, J. R. (1985), Optical waveguides with novel compositions, *Proceedings of Optical Fibre Communication Conference,* Paper WH5, p. 100, Technical Digest, San Diego, CA, USA, February, 1985

MacChesney, J. B. (2000), MCVD: Its origin and subsequent developments, *IEEE Journal of Quantum Electronics,* Vol. 6, (December 2000), pp. 1305-1306, ISSN 0018-9197

Matejec, V., Hayer, M., Pospisilova, M. & Kasik, I. (1997), Preparation of optical cores of silica optical fiber by the sol-gel method, *Journal of Sol-Gel Science Technology,* Vol.8, (February 1997), pp. 889-893, ISSN 1573-4846

Mears R. J., Reekie, L., Jauncey, I. M., & Payne, D. N. (1987), Low-noise Erbium-doped fiber amplifier at 1.54μm, *Electronics Letter,* Vol. 23, pp. 1026-1028, ISSN 0013-5194

Morse, T. F., Reinhart, L., Kilian, A., Risen, W. & Cipolla, J. W (1989), Aerosol doping technique for MCVD and OVD, *Proceedings of Fibre Laser Sources and Amplifiers, SPIE,* Vol. 1171, pp. 72-79, ISBN 9780819402073

Mukhopadhyay, S. S. & Kundu, D. (2006), Chemical composition analysis of germinate glass by alkalimetric titration of Germinate Mannitol complex, *Journal of Indian Chemical Society,* Vol. 83, No. 3, (March 2006), pp. 255-258. ISSN 0019-4522

Nagel, S. R., MacChesney, J. B. & Walker, K. L. (1982), Overview of the Modified Chemical vapour Deposition (MCVD) process and performance, *IEEE Journal of Quantum Electronics,* Vol. QE-18, No. 4, (April 1982), pp. 459-476, ISSN 0018-9197

Pal, M., Sen, R., Paul, M. C., Bhadra, S. K., Chatterjee, S., Ghosal, D. & Dasgupta, K. (2005), Investigation of the deposition of porous layers by the MCVD method for the preparation of rare-earth doped cores of optical fibers, *Optics Communications,* Vol. 254, (October 2005), pp. 88-95, ISSN 0030-4018

Poole, S. B., Payne, D. N. & Fermann, M. E. (1985), Fabrication of low-loss optical fibres containing rare earth ions, *Electronics Letter,* Vol. 21, No. 17, (August 1985), pp. 737-738, ISSN 0013-5194

Poole, S. B., Payne, D. N., Mears, R. J., Fermann, M. E. & Laming, R. I. (1986), Fabrication and characterization of low-loss optical fibres containing rare-earth ions, *Journal of Lightwave Technology,* Vol. LT-4, No. 7, (July 1986), pp. 870-876, ISSN 0733-8724

Stone, J. & Burrus, C. A. (1973), $Nd^{3+}$ doped $SiO_2$ lasers in end pumped fiber geometry, *Applied Physics Letters,* Vol. 23, (October 1973), pp. 388-389, ISSN 1077-3118

Tammela, S., Kiiveri, P., Sarkilahti, S., Hotoleanu, M., Vaikonen, H., Rajala, M., Kurki, J. & Janka, K. (2002), Direct Nanoparticle Deposition process for manufacturing very short high gain Er-doped silica glass fibers, *Proceedings of ECOC 2002,* ISBN 8790974638, Copenhagen, September, 2002

Thompson, D. A., Bocko, P. L. & Gannon, J. R. (1984), New source compounds for fabrication of doped optical waveguide fibres, *Proceedings of Fibre Optics adverse Environment II*, SPIE, Vol. 506, ISBN 9780892525416

Tingye, L. (1985), *Optical Fiber Communications Volume 1 Fiber Fabrication,* Academic Press, Inc., ISBN 0124473016, Orlando, Florida.

Townsend, J. E., Poole, S. B. & Payne, D. N. (1987), Solution-doping technique for fabrication of rare-earth-doped optical fibers, *Electronics Letters*, Vol. 23, No. 7, (March 1987), pp. 329-331, ISSN 0013-5194

Tumminelli, R. P., McCollum, B. C. & Snitzer, E. (1990), Fabrication of high-concentration rare-earth doped optical fibres using chelates, *Journal of Lightwave Technology*, Vol. LT-8, No. 11, (November 1990), pp. 1680-1683, ISSN 0733-8724

4

# Tailoring of the Local Environment of Active Ions in Rare-Earth- and Transition-Metal-Doped Optical Fibres, and Potential Applications

Bernard Dussardier[1], Wilfried Blanc[1] and Pavel Peterka[2]

[1]*Laboratoire de Physique de la Matière Condensée,*
*Université de Nice Sophia-Antipolis - CNRS, Nice*
[2]*Institute of Photonics and Electronics AS CR, Prague*
[1]*France*
[2]*Czech Republic*

## 1. Introduction

During the last two decades, the development of optical fibre-based sophisticated devices have benefited from the development of very performant optical fibre components. In particular, rare-earth (RE)-doped optical fibres have allowed the extremely fast development of fibre amplifiers for optical telecommunications (Desurvire, 1994, 2002), lasers (Digonnet, 2001) and temperature sensors (Grattan & Sun, 2000). The most frequently used RE ions are $Nd^{3+}$, $Yb^{3+}$, $Er^{3+}$ and $Tm^{3+}$ for their optical transitions in the near infrared (NIR) around 1, 1.5 and 2 µm. A great variety of RE-doped fibres design have been proposed for specific applications: depending of the RE concentration and nature of the fibre glass, various schemes (in terms of electronic transition within the RE populations) have been implemented. For example, $Er^{3+}$-doped fibre amplifiers (EDFA) for long haul telecommunications use the very efficient 1.55 µm optical transition, whereas high concentrations of $Yb^{3+}$ and $Er^{3+}$ codoped in the same fibre allowed efficient non-radiative energy transfers from the 'sensitizer' ($Yb^{3+}$) to the 'acceptor' ($Er^{3+}$) in order to increase the power yield of the system, applied in power amplifiers and lasers. All the developed applications of amplifying optical fibres are the result of time consuming and careful optimization of the material properties, particularly in terms of dopant incorporation in the glass matrix, transparency and quantum efficiency.

RE-doped fibres are made of a choice of glasses: silica is the most widely used, sometimes as the result of some compromises. Alternative glasses, including low maximum phonon energy (MPE) ones, are also used because they provide better quantum efficiency or emission bandwidth to some optical transitions of particular RE ions. The icon example is the $Tm^{3+}$-doped fibre amplifier (TDFA) for telecommunications in the S-band (1.48-1.53 µm) (Komukai et al., 1995), for which low MPE glasses have been developed: oxides (Minelly & Ellison, 2002; Lin et al., 2007), fluorides (Durteste et al., 1991), chalcogenides (Hewak et al., 1993), etc. Although some of these glasses have a better transparency than silica in the infrared spectral region (above 2.3 µm), for applications in the NIR these glasses have some drawbacks not acceptable at a commercial point of view: high fabrication cost, low reliability, difficult connection to silica components and, in the case of fibre lasers, low optical damage threshold and resistance to heat. To our knowledge, silica glass is the only material able to meet most of applications requirements, and therefore the choice of vitreous silica for the active fibre material is of critical importance. However a pure silica TDFA would suffer strong non-radiative de-excitation (NRD) caused by multiphonon coupling from $Tm^{3+}$ ions to the matrix. Successful insulation of $Tm^{3+}$-ions from matrix vibrations by appropriate ion-site 'engineering' would allow the development of a practical silica-based TDFA. However, note that $Tm^{3+}$-doped silica fibre are very good for 2 μm laser sources (Jackson & King, 1999). This shows that the choice of optimal host glass is also dictated by the seeked application.

Other dopants have recently been proposed to explore amplification over new wavelength ranges. Bi-doped glasses with optical gain (Murata et al., 1999) and fibre lasers operating around 1100-1200 nm have been developed (Dianov et al., 2005; Razdobreev et al., 2007), although the identification of the emitting center is still not clear, and optimization of the efficiency is not yet achieved. Transition metal (TM) ions of the Ti-Cu series would also have interesting applications as broad band amplifiers, super-fluorescent or tunable laser sources, because they have in principle ten-fold spectrally larger and stronger emission cross-sections than RE ions. However, important NRD strongly reduces the emission quantum efficiency of TM ions in silica. Bi- and TM-doped fibres optical properties are extremely sensitive to the glass composition and/or structure to a very local scale. As we have shown, Cr ions in silica using standard fabrication methods provide strong and ultrabroad NIR luminescent, but only at low temperature (Dussardier et al., 2002). Interestingly, other applications of TM-doped fibre have been proposed in lasers.

As for $Tm^{3+}$ ions, practical applications based on silica doped with alternative dopants along 'low efficiency' optical transitions would be possible when the 'ion site engineering' will be performed in a systematic approach. This approach is proposed via 'encapsulation' of dopants inside glassy or crystalline nanoparticles (NP) embedded in the fibre glass, like reported for oxyfluoride fibres (Samson et al., 2001) and multicomponent silicate fibres (Samson et al., 2002). In NP-doped-silica fibres, silica would act as support giving optical and mechanical properties to the fibre, whereas the dopant spectroscopic properties would be controlled by the NP nature. The NP density, mean diameter and diameter distribution must be optimized for transparency (Tick et al., 1995). In this context, our group has made contributions in various aspects introduced above. Our motivations are usually application oriented, but we address fundamental issues. First, the selected dopants act as probes of the local matrix environment, via their spectroscopic variations versus ligand field intensity, site structure, phonon energy, statistical proximity to other dopants, etc. The studies are always dedicated to problems or limitation in applications, such as for EDFA and TDFA, or high temperature sensors. It is also important to use a commercially derived fabrication technique, here the Modified Chemical Vapor Deposition (MCVD), to assess the potential of active fibre components for further development.

The aim of this chapter is reviewing of our contributions towards the comprehension and improvement of the spectroscopic properties of some RE and TM ions doped into silica. Its outline is as follow : Section 2 will describe the MCVD fabrication method of preform and fibre samples, and the characterization techniques applied to all samples. In Section 3, we report on the spectroscopic investigations of $Tm^{3+}$-doped fibres versus the material composition, including phonon interactions and non-radiative relaxations. It includes the proposal of potential applications by numerical simulations. Section 4 summarizes our original investigations on transition metals, focusing on chromium ions ($Cr^{3+}$ and $Cr^{4+}$) in silica-based fibres. Applications of $Cr^{4+}$-doped optical fibres as integrated saturable absorbers for passively Q-switched lasers are investigated. In section 5 are reported our recent discoveries in RE-doped dielectric nanoparticles, grown by phase separation. Finally, perspectives and conclusions are drawn in section 6.

## 2. Experimental

### 2.1 Preforms and fibres fabrication

All fibres investigated in this article were drawn from preforms prepared by the MCVD technique (Nagel et al., 1985) at the Laboratoire de Physique de la Matière Condensée (Nice, France). In this process, chemicals (such as $O_2$, $SiCl_4$) are mixed inside a silica tube that is rotating on a lathe. The flame from a burner translating along the lathe axis heats the tube and locally produces a chemical oxidizing reaction that transforms $SiCl_4$ and $O_2$ into $SiO_2$ molecules and gaseous chlorine. Extremely fine silica particles are deposited as soot on the inner side of the tube. This soot is transformed into a glass layer when the burner is passing over. The cladding layers are synthesized first, followed by the core layer(s). Germanium and phosphorus can also be incorporated directly through the MCVD process. They are added to raise the refractive index. Moreover, the latter serves also as a melting agent, decreasing the melting temperature of the glass. All the other elements like RE, TM, Al (a common glass modifier) are incorporated through the so-called solution doping technique (Townsend et al., 1987). The core layer is deposited at lower temperature than the preceding cladding layers, so that they are not fully sintered and left porous. Then the substrate tube is filled with an alcoholic ionic solution and allowed to impregnate the porous layers. After soaking the solution is removed, the porous layer is dried and sintered. The tube is then collapsed at 2000 °C into a cylindrical preform. The preform is drawn into a fibre using a vertical tube furnace on a drawing tower. The preform tip is heated above 2000°C and as the glass softens, a thin drop falls by gravity and pulls a thin glass fibre. The diameter of the fibre is adjusted by varying the pulling capstan speed, under controlled tension. The fibre is then coated with a UV-curable polymer and is finally taken up on a rotating drum.

### 2.2 Material characterizations

Refractive index profiles (RIP) of the preforms and fibres were measured using dedicated commercial refractive index profilers. The oxide core compositions of the samples were deduced from RIP measurements on preforms, knowing the correspondence between index rising and $AlO_{3/2}$, $GeO_2$, $PO_{5/2}$ concentration in silica glass from the literature. The composition was also directly measured on some preforms using electron probe microanalysis technique in order to compare results. The concentration of these elements is generally around few mol%. Luminescent ions concentrations are too low to be measured through the RIP. They were measured through absorption spectra. For example, the $Tm^{3+}$ ion concentration has been deduced from the 785 nm ($^3H_6 \rightarrow {}^3H_4$) absorption peak measured in fibres and using absorption cross-section reported in the literature (Jackson & King, 1999) : $\sigma_{abs}(785\ nm) = 8.7x10^{-25}\ m^2$.

## 3. Thulium-doped fibres

### 3.1 Improvement of the 800 and 1470 nm $Tm^{3+}$ emission efficiencies

Thulium-doped fibres have been widely studied in the past few years. Because of $Tm^{3+}$ ion rich energy diagram (Fig. 1), lasing action and amplification at multiple infrared and visible wavelengths are allowed. In this paragraph, we will focus on the $^3H_4$ manifold. Thanks to the possible stimulated emission peaking at 1.47 μm ($^3H_6 \rightarrow {}^3F_4$, see Fig. 1), discovered by (Antipenko, 1983), one of the most exciting possibilities of $Tm^{3+}$ ion is amplifying optical signal in the S-band (1.47–1.52 μm), in order to increase the available bandwidth for future optical communications. The $^3H_4$ level can also decay radiatively by emitting 800 nm radiation which is of primary importance for high power laser and medical applications. Unfortunately, the upper $^3H_4$ level of this transition is very close to the next lower $^3H_5$ level so NRD are likely to happen in high phonon energy glass hosts, causing detrimental gain quenching. Improvement of the $Tm^{3+}$ spectroscopy was proposed through Tm-Tm (Simpson et al., 2006) and Yb-Tm (Simpson et al., 2008) energy transfers mechanims. In this paragraph, we discuss on the modification of the local phonon energy around $Tm^{3+}$ ions to reduce NRD. Then, we present the resulting improvement in emission efficiencies for specific applications through numerical modeling.

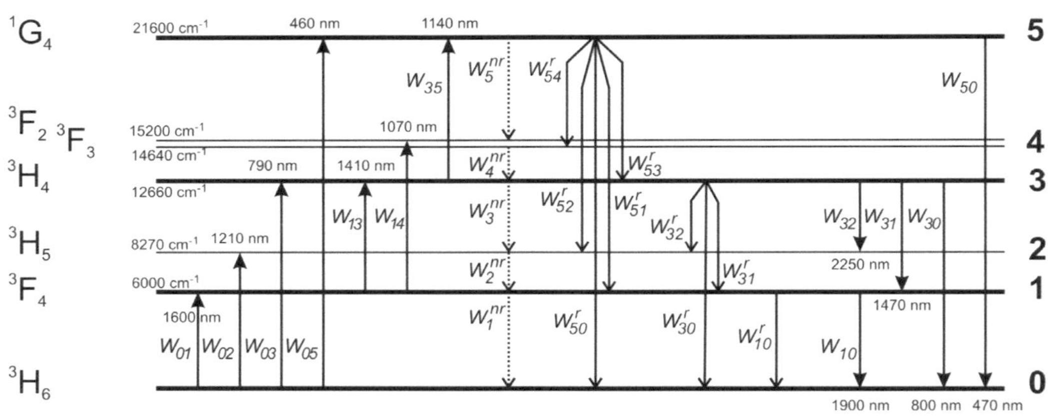


Fig. 1. Energy diagram of $Tm^{3+}$ ion. Parameters are described in the text.

#### 3.1.1 Local phonon interactions

The relevant NRD transition of interest here is from the $^3H_4$ level downto the $^3H_5$ level. The NRD rate is expressed as (Van Dijk, 1983):

$$W_3^{nr} = W_0 \times \exp\left[-\alpha_{VD}\left(\Delta E - 2E_p\right)\right] \tag{1}$$

where $W_0$ and $\alpha_{VD}$ are constants depending on the material, $\Delta E$ is the energy difference between the both levels and $E_p$ is the phonon energy of the glass. To improve the quantum efficiency (which depends on the fluorescence lifetime) of the 800 and 1470 nm emission bands, modifications of $Tm^{3+}$ ion local environment were investigated by co-doping with selected modifying oxides (Faure et al., 2007). $GeO_2$ and $AlO_{3/2}$ have a lower maximum phonon energy than silica. As opposite demonstration, high phonon energy $PO_{5/2}$ was also used. To investigate the modification of the local environment, decay curves of the 810 nm fluorescence from the $^3H_4$ level were recorded. All the measured decay curves are non-exponential. This can be attributed to several phenomena and will be discussed in the next paragraph. Here, we study the variations of 1/e lifetimes ($\tau$) versus concentration of oxides of network modifiers (Al or P) and formers (Ge). The lifetime strongly changes with the composition of the glass host. The most striking results are observed within the *Tm(Al)* sample series: $\tau$ linearly increases with increasing $AlO_{3/2}$ content, from 14 μs in pure silica to 50 μs in sample *Tm(Al)* containing 17.4 mol% of $AlO_{3/2}$. The lifetime was increased about 3.6 times. The lifetime of the 20 mol% $GeO_2$ doped fibre *Tm(Ge)* was increased up to 28 μs whereas that of the 8 mol% $PO_{5/2}$ doped fibre *Tm(P)* was reduced down to 9 μs. Aluminum codoping seems the most interesting route among the three tested codopants.

### 3.1.2 Non-exponential shape of the 810-nm emission decay curves

In aluminium-doped fibres, fluorescence decay curves from the $^3H_4$ level were found to be non-exponential. It is thought that $Tm^{3+}$ ions are inserted in a glass which is characterized by a multitude of different sites available for the RE ion, leading to a multitude of decay constants. This phenomenological model was first proposed by Grinberg et al. and applied to $Cr^{3+}$ in glasses (Grinberg et al., 1998). This model was applied for the first time to $Tm^{3+}$-doped glass fibres (Blanc et al., 2008). In this method, the luminescence decay is given by :

$$I(t) \approx \sum_i A_i \exp[-t/\tau_i] \tag{2}$$

where $A_i$ and $\tau_i$ are discrete distributions of amplitudes and decay constants (lifetimes), respectively. For the fitting procedure on the $A_i$ series, 125 fixed values for $\tau_i$ were considered, logarithmically spaced from 1 to 1000 μs. For a given composition (Fig. 2a), we can notice two main distributions of the decay constant. With the aluminium concentration, they increase from 6 to 15 μs and from 20 to 100 μs, respectively (Fig. 2b). From the histograms of the amplitude distributions obtained from the fittings, characteristic lifetimes were correlated with those expected for thulium located either in a pure silica or pure $Al_2O_3$ environment. The $^3H_4$ lifetime is calculating by using this equation:

$$1/\tau = 1/\tau_{rad} + W_3^{nr} \tag{3}$$

where $\tau_{rad}$ corresponds to the radiative lifetime which is given to be 670 μs in silica (Walsh & Barnes, 2004). $W_0$ and $\alpha$ were estimated for different oxide glasses (Van Dijk & Schuurmans, 1983; Layne et al., 1977). The energy difference $\Delta E$ was estimated by measuring the absorption spectrum of the fibres. When Al concentration varies, this value is almost constant around 3700 $cm^{-1}$ (Faure et al., 2007). With these considerations, the $^3H_4$ expected lifetime can be calculated. In the case of silica glass, $\tau_{silica}$ = 6 μs and for a pure $Al_2O_3$ environment, $\tau_{alumina}$ = 110 μs. Both values agree with those obtained from the fitting procedure. The distribution of decay constant around 10 μs corresponds to $Tm^{3+}$ ions located in almost pure silica environment whereas the second distribution is attributed to $Tm^{3+}$ located in $Al_2O_3$ -rich sites. This result shows that the global efficiency is increased by increasing the Al concentation in TDFAs. However concentrations above 20 at% would cause excess loss and glass stability problems in silica.

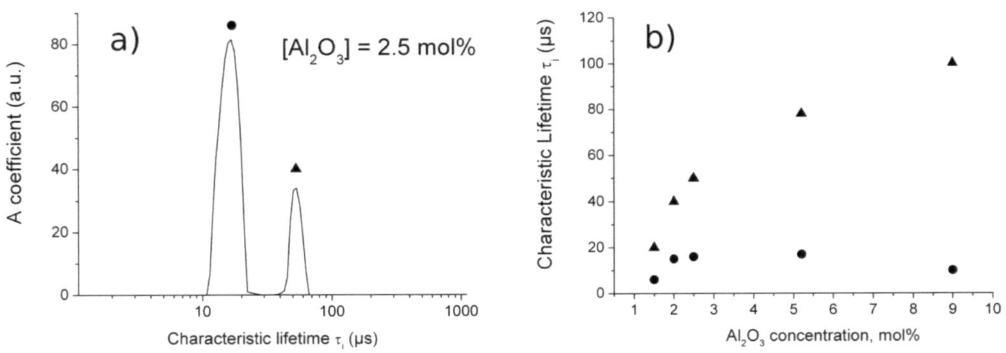


Fig. 2. a) Histogram of the recovered amplitude distributions obtained for silica-based $Tm^{3+}$-doped fibres. b) variations of the short (circle) and long (triangle) characteristic lifetimes vs $Al_2O_3$ concentration.

### 3.2 Modeling of Tm-doped silica-fibre devices

Numerical models are useful tools for investigating thulium-doped fibre devices, predicting their performance and optimization of their parameters. We have developed comprehensive, spectrally and spatially resolved numerical model that is based on simultaneous solution of the laser rate equations and propagation equations that describe evolution of the optical power along the fibre. The rate equations for the relevant energy levels can be written according to the energy level diagram in Fig. 1 as follows:

$$\frac{dn_1}{dt} = n_0(W_{01} + W_{02}) - n_1(W_{10} + W_{13} + W_{14} + W_1^{nr} + W_{10}^r) + \\ + n_3(W_{31} + W_3^{nr} + W_{32}^r + W_{31}^r) + n_5(W_{51}^r + W_{52}^r), \tag{4}$$

$$\frac{dn_3}{dt} = n_0(W_{03} + W_{04}) + n_1(W_{13} + W_{14}) + n_5(W_5^{nr} + W_{54}^r + W_{53}^r) - \\ - n_3\left(W_{35} + W_{31} + W_{30} + W_3^{nr} + \sum_{j=0}^{2} W_{3j}^r\right), \tag{5}$$

$$\frac{dn_5}{dt} = n_0 W_{05} + n_3 W_{35} - n_5(W_{50} + W_5^{nr} + \sum_{j=0}^{4} W_{5j}^r), \tag{6}$$

while it holds that the sum of population $n_i$ on each respective level is equal to the total thulium ion concentration $n_t$ in the core. It is assumed that the thulium ions are homogenously distributed and excited in the doped section of radius $b$ in the fibre core of radius $a$ and it holds that $b \leq a$. The transition rates $W_{ij}$ accounts for the stimulated absorption and emission between the respective levels. Spontaneous decay processes are described by $W_{ij}^r$ and $W_i^{nr}$, the radiative and nonradiative decay rates, respectively. The formulae for the transition rates can be found in (Peterka et al., 2004). In steady state the rate equations become a set of four linear algebraic equations. The propagation of optical power at respective wavelength is governed by the following propagation equation:

$$\frac{dP^{\pm}(\lambda)}{dz} = \pm\Gamma(\lambda)P^{\pm}(\lambda)\sum_{ij}^{\{10,30,31,50\}}\left(n_i\sigma_{ij}(\lambda) - n_j\sigma_{ji}(\lambda)\right)$$
$$\mp\Gamma(\lambda)P^{\pm}(\lambda)\left(n_0\sigma_{02}(\lambda) + n_0\sigma_{04}(\lambda) + n_1\sigma_{14}(\lambda) + n_3\sigma_{35}(\lambda)\right) \quad (7)$$
$$\pm\Gamma(\lambda)\sum_{ij}^{\{10,30,31,50\}} 2h\nu_{ij}\Delta\nu n_i\sigma_{ij}(\lambda) \mp \alpha(\lambda)P^{\pm}(\lambda).$$

where $h$ is the Planck constant, $P^{\pm}$ are the spectral power densities of the radiation propagating in both directions along the fibre axis and $\sigma_{ij}$ is the respective transition cross-section. The cross-sections are shown in Fig. 3. The spectral dependence of the cross-sections are approximated by a set of Gaussian functions, the respective coefficients can be found in (Peterka et al., 2011). The overlap factor $\Gamma$ represents the fraction of the transversal field distribution that interacts with the RE ions. The first term in equation (7) describes the amplification and reabsorption of optical signals, the second term represents ground-state absorption and excited state absorption (ESA) in spectral bands with no significant emission, the third term accounts for spontaneous emission and the fourth term stands for background loss $\alpha$ of the fibre. Evolution of optical power in each ASE spectral slot around a wavelength $\lambda$ (typical 1 nm wide slot is used) is governed by its respective propagation equation. The propagation equations together with the set of the rate equations under steady state conditions are solved simultaneously along the fibre using Runge-Kutta-Gill method of the fourth order. Since the boundary conditions for the counterpropagating partial waves $P^-$ are not known at the beginning of the fibre, an iterative solution is applied. The above described numerical model was verified by experimental comparisons (Blanc et al., 2006; Lüthi et al., 2007).

### 3.3 Applications

Thulium-doped fibres are renowned for their applications in high power fibre lasers at around 2 μm. It is despite the low quantum conversion efficiency of $^3F_4$ level in thulium-doped silica fibres, which is about 10% compared to ~100% quantum conversion efficiency of the ytterbium- and erbium-doped fibres, at around 1 μm and 1.5 μm, respectively. The lower quantum conversion efficiency increases the 2 μm laser threshold but has almost no effect on the laser slope efficiency. Indeed, kW-class thulium-doped fibre lasers have been demonstrated recently (Moulton, 2011). Quantum conversion efficiency of the $^3H_4$ level in non-modified silica fibres is much lower than that of $^3F_4$ level, only about 2% and therefore most of the applications of laser transitions originating from $^3H_4$ level are hindered by the lack of reliable low-phonon fibre host as discussed above. In the following two paragraphs we will show potential of the developed thulium-doped fibres with enhanced $^3H_4$ level lifetime for applications in fibre amplifiers for communication S-band and for fibre lasers around 810 nm.

#### 3.3.1 Thulium-doped fibre amplifier at 1.47 μm

The low-loss and single-mode spectral range of the most common single-mode telecommunication fibre spans from about 1260 nm to 1675 nm. However, this tremendous transmission bandwidth is nowadays used for long-haul transmission only in a restricted portion in the C- (1530-1565 nm) and L-bands (1565-1625 nm), where reliable EDFA are available. The next logical frontier is the S-band (1460-1530 nm). One of the most promising

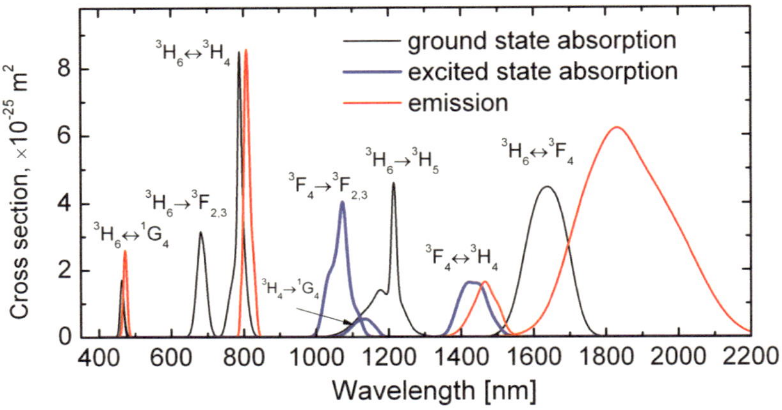


Fig. 3. Absorption and emission cross section spectra of thulium.

candidates for amplification in the S-band is the TDFA. Together with EDFA it can substantially increase the bandwidths. An example of a TDFA structure is shown in Fig. 4.

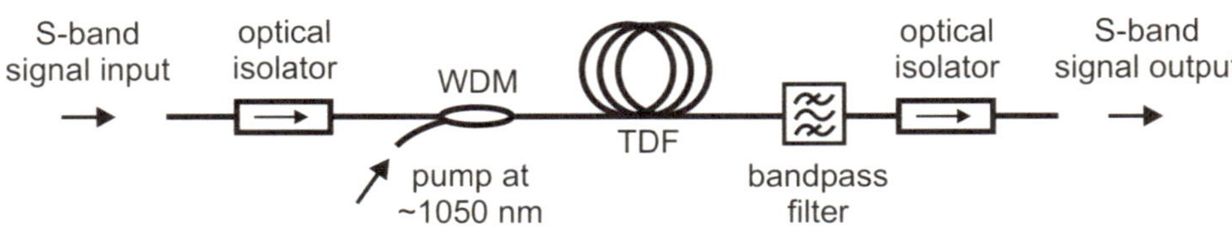


Fig. 4. Typical layout of the TDFA.

Two main obstacles have to be solved in order to use the $^3H_4 \rightarrow {}^3F_4$ transition in the TDFA. Firstly, the quantum conversion efficiency of the $^3H_4$ thulium level should be increased, or in other words, the $^3H_4$ fluorescence lifetime should be made longer. One possible solution of the first problem is described above. Secondly, the lifetime of the upper laser level $^3H_4$ is shorter than that of the lower level $^3F_4$. Therefore, direct pumping of the $^3H_4$ level will hardly constitute the necessary population inversion. The second problem can be solved by using more complex pumping scheme employing gradual upconversion of the thulium ion to the upper laser level. It can be done by a single laser source as shown in Fig. 5a. The disadvantage is the possibility of the ESA $^3H_4 \rightarrow {}^1G_4$ that results in loosing pump photons. Many dual-wavelength pumping scheme have been also reported as reviewed in (Peterka et al., 2004). Several results of TDFA performance simulation and of optimization of the TDFA parameters are shown in Fig. 5 and Fig. 6. The TDFA spectral gain is shown in Fig. 5b for different silica fibre hosts represented by the fluorescence lifetime of the $^3H_4$ level: 14 μs of the typical non-modified silica fibre and 45 μs of the alumina-doped fibre developed by the authors and 55 μs of the highly germanium-oxide-doped fibre (Cole & Dennis, 2001). Similar value of 58 μs was also measured in highly alumina-doped fibre (Peterka et al., 2007). Codirectional pump of 1 W at 1064 nm is assumed. Unless otherwise stated, we consider the thulium concentration $n_t$=1.56×10$^{25}$ m$^{-3}$, the core diameter 2.6 μm, numerical aperture $NA$=0.3 and the radiative lifetimes of the levels $^3F_4$, $^3H_4$ and $^1G_4$ are 3500, 650 and 860 μs, respectively. The fluorescence lifetimes of the levels $^3F_4$, $^3H_4$ and $^1G_4$ are 430, 45 and 784 μs, respectively. For sake of comparisons we set the cross sections the same for all fibre host types, although in real fibres they also depend on the host material. We have checked that their values have little effect on the calculated S-band gain compared to the effect of $^3H_4$ lifetime. The relevant branching ratios were estimated using Judd-Ofelt theory and were evaluated as follows: $\beta_{54}$=0.03, $\beta_{53}$=0.11, $\beta_{52}$=0.30, $\beta_{51}$=0.06, $\beta_{50}$=0.50, $\beta_{32}$=0.03, $\beta_{31}$=0.09, and $\beta_{30}$=0.88. Zero background loss is assumed. Signal power evolution along the thulium-doped fibre is shown in Fig. 5c for three input pump powers at 1064 nm. The available gain is significantly reduced by the presence of ASE, mainly around 800 nm. Suppression of the ASE, e.g., by cascaded inscription of long-period fibre gratings into the doped fibre or by using thulium-doped photonic crystal fibre with tailored band-gaps, would ameliorate the gain as shown also in Fig. 5.

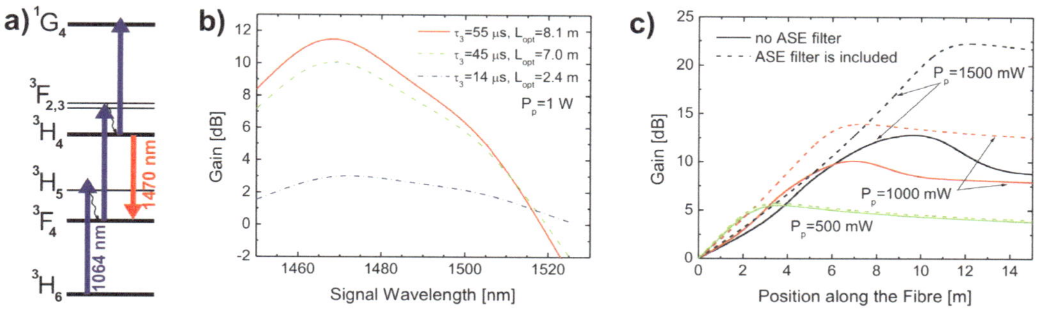


Fig. 5. Upconversion pumping of TDFA at around 1050 nm (a). The effect of glass host composition, mainly manifested by the $^3H_4$ level lifetime, on the gain in the S-band (b) and optimization of the TDFA length (c).

Optimization of the fibre waveguide parameters is shown in Fig. 6a. Similarly as in the case of the EDFA, the gain gets higher with increasing numerical aperture and narrower core diameter. The effect of distributed filtering out of the ASE around 800 nm is also shown in the graph. Optimization of the pump wavelength is calculated in Fig. 6b. The optimum pump wavelength shifts towards shorter wavelength with increasing power levels. Shorter wavelength's pump sees lower GSA but at the same time higher first step ESA and consequently the integral value of the population inversion along the fibre is higher.

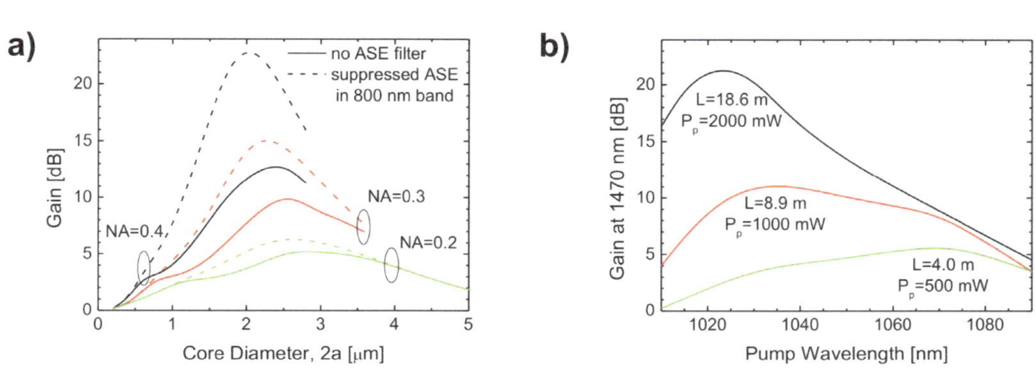


Fig. 6. Optimization of the waveguide parameters, namely the NA and core radius (a) and optimiziation of the pump wavelength (b).

### 3.3.2 Thulium-doped fibre laser at 800 nm

The silica-based fibre lasers around 800 nm would extend the spectral range covered by high-power fibre lasers. The single-transversal mode, high-power laser source in the 800 nm spectral band is of interest for a variety of applications. The laser can be used for fibre sensors, instrument testing and for pumping of special types of lasers and amplifiers, e.g. the bismuth-doped lasers. Bismuth-doped fibres pumped around 800 nm may shift their gain to 1300 nm telecommunication band, where highly reliable silica-based fibre amplifiers are still unavailable. An efficient fibre laser in 800 nm spectral region could potentially be used as a replacement for titanium sapphire laser in some applications. A high-power amplifier in the 800 nm band would be useful in the optical fibre communications and short-haul free-space communications. Although the laser diodes at this wavelength have been available for a long time, to our knowledge, commercially available single-mode laser diodes are limited to about 200 mW of output power in diffraction limited beam. It should not be confused with, e.g., laser diode stacks of ~kW output power, that are highly multimode with very high $M^2$ >1000 factor. The amplification and lasing at 800 nm band has been already investigated using fluoride-based TDFs and output power of up to 2 W and 37% slope efficiency was achieved (Dennis et al., 1994). The output power was limited by the pump damage threshold of the fluoride fibre. Fibre-host reliability problems might be solved by using of silica-based fibres with enhanced $^3H_4$ lifetime. We proposed compact all-silica-fibre setup, example of which is shown in Fig. 7a. The proposed laser utilizes upconversion pumping scheme according to Fig. 7b.

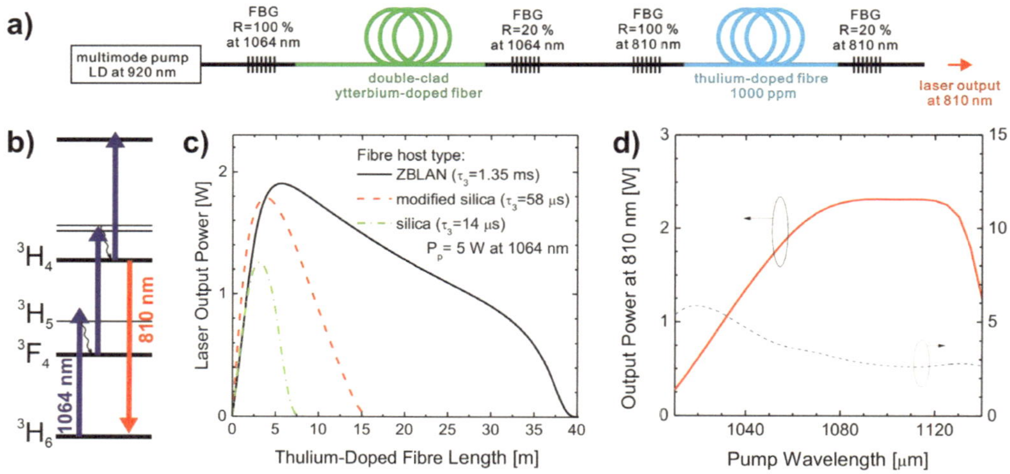


Fig. 7. Fibre laser setup in compact all-fibre arrangement (a), single wavelength up-conversion pumping scheme (b), effect of the host material (c) and pump wavelength (d). The waveguide parameters used for the parts (c) and (d) are: the core diameter of 3.4 μm and numerical aperture of 0.2.

We have performed optimization of the waveguide parameters, namely the core radius and *NA* (Peterka et al., 2011). The optimum length of the TDF can be determined from the calculated dependence of the laser output on the TDF length as it is shown in Fig. 7c for the three laser hosts. The laser output power vs. pump wavelength is shown in Fig. 7d. In comparison with the TDFA for the S-band telecommunications with optimal pump wavelength around 1020 nm (Fig. 6), the region of optimal pump wavelengths is shifted towards longer wavelength because in this case high inversion between $^3H_4$ and $^3H_6$ levels and high pump absorption from $^3H_6$ level is desirable. The background loss of 0.1 dB/m was considered. It can be seen in Fig. 7c that lasing at 810 nm is hard to achieve with silica-based Tm-doped fibre in contrast to the fluoride host materials. However, the lasing might be possible even for silica-based fibre for specific short range of the fibre lengths. Especially when the $^3H_4$ lifetime is enhanced, the laser output is very close to the one of ZBLAN host.

### 3.4 Conclusions

Using a comprehensive numerical model we have shown the potential of the developed thulium-doped fibres for applications in fibre amplifiers for S-band telecommunication and for fibre lasers around 810 nm. The gain exceeding 20 dB of the S-band TDFA could be obtained with optimized fibre waveguiding parameters and optimized pump wavelength. Although the required pump power levels are relatively high, of the orders of Watts, thanks to the progress in ytterbium-doped fibre lasers even these pump powers can provide a cost effective solution. We have shown that efficient lasing at 810 nm can be achieved using silica-based Tm-doped fibre with enhanced $^3H_4$ lifetime for specific ranges of the fibre and laser cavity parameters. Such a fibre laser would broaden the spectral range currently covered by silica-based fibre lasers and it may replace the conventional laser systems in numerous applications.

## 4. Transition metal ions in silica-based optical fibres

### 4.1 Introduction

Many applications need ultra-broad band gain optical fibre materials. The well established tunability of RE-doped fibre devices is limited by shielding of the optically active electronic orbitals of RE ions, whereas unshielded orbitals are found in transition metal (TM) ions. One of the most promising is $Cr^{4+}$ because of its NIR ultra-broad band transitions. However incorporating of $Cr^{4+}$ ion in optical materials is difficult because it is less stable than the $Cr^{3+}$ ion. Some cristals such as $Cr^{4+}$:YAG are very good broad-band gain media (Sennaroglu et al., 1995), but little literature deals with $Cr^{4+}$-doped bulk glasses (Cerqua-Richardson et al., 1992; Hömmerich et al., 1994), and even fewer on Cr-doped silica fibres (Schultz, 1974; Chen et al.; 2007). Note that $Ni^{2+}$ in vitroceramic fibres is also a promising amplifying dopant (Samson et al., 2002). However, the proposed fibres lack of high transparency, reliability and cost-effectiveness of silica-based fibres, as produced by MCVD for instance. We have explored this field through basic studies on the optical properties of TM ions in silica optical fibres. In particular, the final TM oxidation states in the fibre core are strongly process-dependent. Also, the optical properties of one particular TM oxidation state (say $Cr^{4+}$) is difficult to interprete because they depend (1) on the host composition and final structure, due to crystal-field (or ligand field, in glass) fluctuations (Henderson & Imbush, 1989) and (2) on the experimental conditions (temperature, pressure, excitation wavelength,…). We describe the specific preparation details used for the Cr-doped fibres, we sum up spectroscopic results and interpretations, and finally we summarize our original studies on Cr-doped fibre saturable absorbers for all-fibre passively Q-switched (PQS) fibre lasers.

### 4.2 Fabrication and characterization of Chromium-doped samples

The fibres were prepared as in section 2.1, using $Cr^{3+}$ alcoholic doping solutions and oxidizing or neutral atmosphere for the drying-to-collapse stages. Samples containing Ge or/and Al were prepared, referred to as *Cr(Ge)*, *Cr(Ge-Al)* and *Cr(Al)*, respectively. We determined the absolute Cr content and relative concentrations of oxidation states using plasma emission and electron paramagnetic resonance spectroscopies, respectively. We assigned the optical transition of $Cr^{3+}$ and $Cr^{4+}$ using the Tanabe-Sugano (T.-S.) formalism (Sugano et al., 1970) and we qualitatively determined their strength, energy and bandwidth. From composition and sharing out of valencies, we have determined the absorption cross-sections. Absorption, emission and decay measurements were performed at room (RT) and low temperatures (LT, 12 or 77 K), using various pump wavelengths. Full details of the experimental procedures and analysis are given in (Felice et al., 2000, 2001; Dussardier et al., 2002).

### 4.3 Local structure, valency states and spectroscopy of transition metal ions

The main finding is that only $Cr^{3+}$ and $Cr^{4+}$ oxidation states were found. $Cr^{3+}$ is favoured by Ge co-doping in *O*-symmetry as in other oxide glasses (Henderson & Imbush, 1989 Rasheed et al., 1991), whereas $Cr^{4+}$ in distorded tetrahedral site symmetry ($C_s$) (Anino et al., 1997) is present in all samples, and is promoted by Al or when *[Cr]* is high (Fig. 8). The most interesting samples are the low-doped *Cr(Al)* ones containing only $Cr^{4+}$. The $Cr^{3+}$ absorption cross section $\sigma_{abs}{}^{3+}$(670 nm)= 43 x $10^{-24}$ $m^2$ is consistent with reports in other materials like ruby (Cronemeyer, 1966) and silica glass (Schultz, 1974), while $\sigma_{abs}{}^{4+}$(1000 nm)~3.5 x $10^{-24}$ $m^2$ is lower than in reference crystals for lasers (Sennaroglu et al., 2006) or saturable absorbers (Lipavsky et al., 1999), but consistent with estimated values in alumino-silicate glass (Hömmerich et al., 1994).

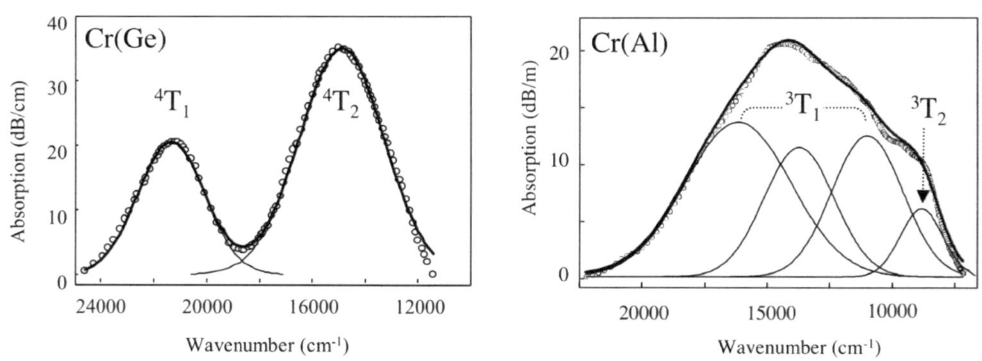


Fig. 8. Absorption from a *Cr(Ge)* preform (*[Cr]* = 1400 mol ppm) and a *Cr(Al)* fibre (*[Cr]* = 40 mol ppm). Solid lines are gaussian fittings to $Cr^{3+}$ (left) and $Cr^{4+}$ (right) transitions, respectively. Assignments are from the ground level $Cr^{3+}$:$^4A_2$ or $Cr^{4+}$:$^3A_2$ to the indicated level, respectively. The $Cr^{4+}$ energy states are referred to by their irreductible representation in the $T_d$ symmetry coordination. The $Cr^{4+}$:$^3T_2$ level three-fold splitting is due to distorsion from perfect $T_d$ symmetry.

The main normalized parameter from the T.-S. formalism (*Dq/B=1.43*) was lower than the so-called 'crossing value' (*Dq/B* = 1.6), and lower than those reported for $Cr^{4+}$ in laser materials like YAG and forsterite (Anino et al., 1997). As expected consequences, a broad featureless NIR emission band along the $^3T_2 \rightarrow {}^3A_2$ transition is observed and no narrow emission line from the $^1E$ state is seen (Fig. 9, left). The LT fluorescence from $Cr^{4+}$ spreads

from 850 to 1700 nm, and strongly varies depending on core chemical composition, *[Cr]* and $\lambda_p$ (pump wavelength). The observed bands were all attributed to $Cr^{4+}$ ions, in various sites. Fig. 9 shows examples of fluorescence spectra of $Cr^{4+}$ in different samples and experimental conditions. Possible emission from $Cr^{3+}$, $Cr^{5+}$ or $Cr^{6+}$ centers was rejected (Dussardier et al., 2002). The fluorescence sensitivity to *[Cr]* and $\lambda_p$ suggests that Cr-ions are located in various host sites, and that several sites are simultaneously selected by scanning of $\lambda_p$. It is also suggested that although Al promotes $Cr^{4+}$ over $Cr^{3+}$ when *[Cr]* is low, $Cr^{4+}$ is also promoted at high *[Cr]* in Ge-modified fibres. The strong decrease of fluorescence from LT to RT (not shown) is attributed to temperature quenching caused by NRD through multiphonon relaxations, like in crystalline materials where the emission drops by typically an order of magnitude from 77 K to 293 K (Sennaroglu et al., 1995).
The non-exponential LT fluorescence decays (Fig. 9, right) depend on *[Cr]* and $\lambda_s$. The fast decay part is assigned to Cr clusters or $Cr^{4+}$-rich phases within the glass. The 1/e-lifetimes ($\tau$) at $\lambda_s$ = 1100 nm are all within the 15-35 µs range in Al-containing samples, whereas $\tau$ ~ 3-11 µs in *Cr(Ge)* samples, depending on *[Cr]*. The lifetime of isolated ions ($\tau_{iso}$), measured on the exponential tail decay curves (not shown) reach high values: $\tau_{iso}$~200 to 300 µs at $\lambda_s$~1100 nm, $\tau_{iso}$~70 µs at $\lambda_s$~1400 nm. In heavily-doped *Cr(Ge)* samples, $\tau_{iso}$ is an order of magnitude less. Hence, $Cr^{4+}$ ions are hosted in various sites: the lowest energy ones suffer more NRD than the higher energy ones. Also presence of Al improves the lifetime, even at high *[Cr]*. It is estimated that at RT, lifetime $\tau$ would be much less than 1 µs.
As a summary of these spectroscopic studies, the main results are: (i) the observed LT fluorescence of $Cr^{4+}$ is extremely sensitive to glass composition, total *[Cr]* and excitation wavelength; (ii) using Al as a glass network modifier has advantages: longer excited state lifetime and broader fluorescence bandwidth than in Ge-modified silica; (3) the RT emission is strongly limited by NRD in silica. Broadband light sources would need a better control of the local environment around TM ions. However this study has allowed to propose an alternative application to Cr-doped fibre for self pulsing lasers.

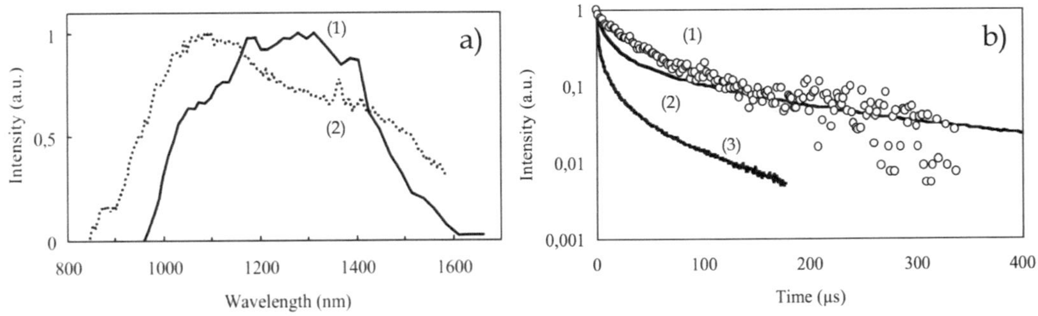


Fig. 9. Fluorescence spectra (a) : (1) fibre *Cr(Al)*, $\lambda_p$ = 900 nm, *T*=77 K ; (2) preform *Cr(Ge-Al)*, $\lambda_p$ = 673 nm, *T*=12K. Fluorescence decays (b) from *Cr(Al)* samples, $\lambda_p$ = 673 nm, *T*=12 K: (1) $\lambda_s$~1100 nm and *[Cr]*=40 ppm, (2) $\lambda_s$~1100 nm, *[Cr]*=4000 ppm, and (3) $\lambda_s$~1400 nm, *[Cr]*=4000 ppm.

### 4.4 Passively Q-switched fibre lasers using a $Cr^{4+}$–doped fibre saturable absorber

The development of Q-switched fibre lasers is an alternative to bulk pulsed solid-state lasers operating in the ns- to µs-range for applications such as material processing and metrology. Actively Q-switched (AQS) fibre lasers have been extensively studied (Wang & Xu, 2007), as well as passively Q-switched (PQS) lasers (Siegman, 1986). Most AQS and PQS fibre lasers use externally driven bulk modulators or bulk passive components (Hideur et al., 2001; Paschotta et al., 1999) or intracavity TM-doped crystals (Laroche et al., 2002), located in a free space section of the cavity, causing alignment and reliability critical problems. To avoid them, we are interested into all-fibre systems comprising a saturable absorber (SA) fibre. We had experimentally demonstrated the first all-fibre PQS laser using a Cr-doped optical fibre saturable absorber (SA) (Tordella et al., 2003). Before, only one numerical study had mentioned this possibility (Luo & Chu, 1999), and 'auto' self-pulsing had been observed in heavily $Er^{3+}$-doped fibres (Sanchez et al., 1993). However, to obtain stable PQS behaviour, the criterium of 'second laser threshold' must be met (Siegman, 1986). It imposes that the SA absorption cross-section be larger than the gain medium emission cross section. Also, the SA absorption must spectrally overlap the emission gain curve and the SA relaxation time must be short to allow for high repetition rates. RE absorption transitions have limited spectral width and sometimes very long decay times. For these reasons, TM ions are in principle better SA elements: they offer broad absorption and fast decay. For example, $\sigma_{abs}^{4+}$(1 µm) is at least 10 times larger than $\sigma_{em}^{Nd3+}$ (Felice et al., 2000) and $\tau_{Cr4+}$ <1 µs is much less than of $\tau_{Nd3+}$(500 µs) or $\tau_{Yb3+}$ (800 µs) (Dussardier et al., 2002).
Our first all-fibre PQS laser, operating at 1084 nm, used a core-pumped Nd-doped fibre spliced to a Cr-doped fibre saturable absorber (CrSA) (Tordella et al., 2003). Although the pulse energy was low (~15 nJ), the characteristic behaviour agreed with theory, and the principle was demonstrated. Since, few PQS fibre lasers have implemented alternative SA ions in fibres, like RE ($Ho^{3+}$, $Tm^{3+}$, $Sm^{3+}$) or Bi, using sometimes rather complex cavity configurations, or heavily RE doped fibre SA (Kurkov, 2011). More recently, double-clad ytterbium-doped fibre (DCYF) lasers have raised a great interest. In some cases, DCYF lasers self-pulse in a random manner (Upadhyaya et al., 2010), along two main operating regimes: the sustained self-pulsing (SSP: period longer than the cavity round-trip time), and the self mode-locking (SML: period equal to the cavity round-trip time) (Brunet et al., 2005). The setup of our second system is schematically shown in Fig. 10.

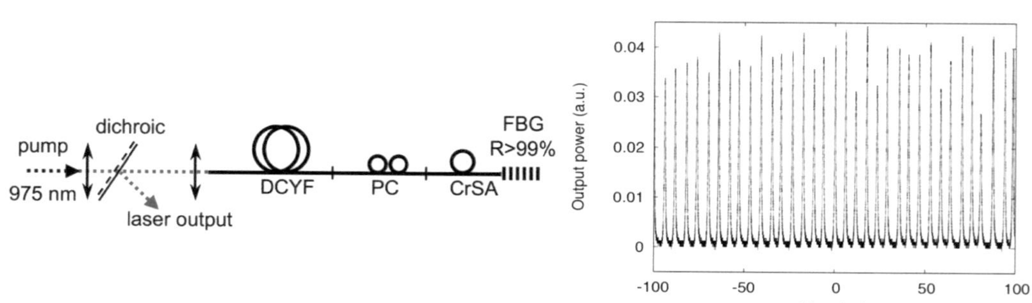


Fig. 10. $Yb^{3+}$:$Cr^{4+}$ PQS laser setup, and typical pulse train. DCYF: double-clad Yb-doped optical fibre, PC: polarization controler, CrSA: Cr-doped fibre saturable absorber, FBG: high reflectivity fibre Bragg grating mirror.

It is based on a home-made D-shaped DCYF. All experimental details are found in (Dussardier et al., 2011). The cavity parameters were chosen so that it would provide favorable conditions for SSP. As a consequence, the laser slope efficiency (SE) was not optimized. That also increased the threshold of appearance of stimulated Brillouin scattering (SBS) far above the achieved output power, therefore SBS had not effect on the reported results, and all the observed dynamic could be attributed to PQS mode of operation.
The DCYF laser without the CrSA had a chaotic behaviour, randomly switching between continuous-wave (CW) and Q-switched operation. The pulse envelopes (few μs long) were strongly modulated with a period of one cavity return-trip time (Fig. 11(a)), a sign of a combination of SSP and SML. The same was observed using a commercial $Yb^{3+}$-heavily doped fibre. The CrSA fibre drastically changed the laser characteristics. The μs-long pulse trains were stabilized over the whole available pump range (peak RMS fluctuations < 10 % at 9.4 times the threshold) (Fig. 10). The PQS mode was present at any pump power. The smooth Q-switched pulses did not show the sub-modulation at the return trip period that was visible in the 'DCYF only' laser (Fig. 11(b)). The reasons of the stabilization are not fully understood yet, and are still under study. The PQS laser threshold and SE were 1 W and 6 %, respectively. Although the SE has not been optimized it compares well with other fibre lasers using RE- or Bi-doped fibre SA (up to ~10%) (Kurkov, 2011). At 16 W of incident pump power the minimum pulse duration, highest repetition rate and peak power were 480 ns, 350 kHz and 20 W, respectively. Such a PQS laser source could be optimized to increase the intracavity optical density, and hence to saturate the CrSA more and faster, and produce more energetic pulses. It could also be a seed in a master-oscillator power amplifier configuration to be amplified by power Yb-doped fibre amplifiers.

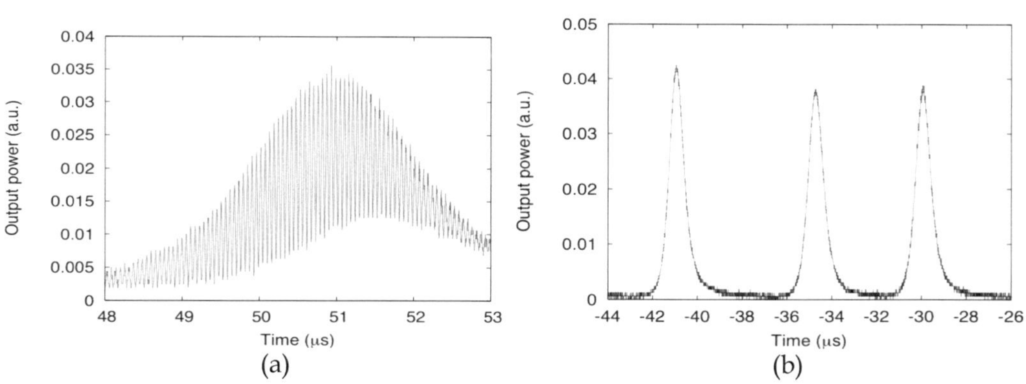


Fig. 11. (a) zoom on a typical pulse envelope from the 'DCYF only' chaotic laser output. (b) zoom on few typical pulses from the stabilized 'DCYF+CrSA' laser.

### 4.5 Conclusion on the study of Cr-doped optical fibres

Some glass modifiers like Al in silica-based fibres induce major spectroscopic changes, even at low concentrations (~1-2 mol%). This would help engineering the Cr optical properties in silica-based fibres, using possibly alternative modifiers. An interesting application for PQS all-fibre lasers was found. However, increasing of the quantum efficiency would be necessary for amplifiers and light sources. One needs to control the TM local environment by some material engineering. Preliminary work was reported (Dvoyrin et al., 2003) on post-heat treatment of $Cr^{3+}$-doped Ga-modified silica fibres. Other possible material engineering would be the nanoscale spontaneous phase separation.

## 5. Rare-earth-doped dielectric nanoparticles

Optical fibres are mostly based on silica glass for its interesting properties, as mentioned in section 1. However, some of its characteristics (high phonon energy, low solubility of luminescent ions, ...) induce drawbacks in spectroscopic properties of rare-earth (RE)-doped fibre lasers and amplifiers. Here we investigate the growing of RE-doped dielectric oxide nanoparticles (DNP) in silica-based optical fibres. Through this route, we keep the optical and mechanical advantages of silica glass, and we investigate the potential of engineering of RE spectroscopic properties through the choice of the DNP chemical composition.

Scarce reports on RE-doped transparent glass ceramics (TGC) singlemode fibres use low melting mixed oxides prepared by a rod-in-tube technique (Samson et al., 2002), or mixed oxyfluorides using a double-crucible technique (Samson et al., 2001), both with a subsequent ceramming stage. However the low melting point of these materials causes low compatibility with silica components. Transition metal-doped silica-based TGC fibres were prepared by MCVD and using a slurry method (Yoo et al., 2003), i.e. the particles were synthesized before insertion into the silica tube-substrate.

In this section, we discuss a more straightforward and original technique to embed RE ions within *in-situ* grown oxide nanoparticles in silica-based preforms (Blanc et al., 2009a). The implemented principle is the spontaneous phase separation process (Zarzycki, 1991). Silicate systems can exhibit strong and stable immiscibility when they contain divalent metals oxides (MO, where M= Mg, Ca or Sr) (Fig. 12). For example, if a silicate glass containing few mol% MO is heated it will decompose into two phases : one silica-rich and one MO-rich in shape of spherical particles. Two key advantages of this process are that (i) nanoparticles are grown *in-situ* during the course of the fabrication process and (ii) there is no need (and associated risks) of nanoparticles manipulation by an operator. Further, the process takes advantage of the high compositional control and purity typical of the MCVD technique. The so-called 'solution doping technique' was applied to incorporate alkaline-earth and erbium ions. To raise the core refractive index and ease the fabrication, germanium (~2 mol%) and small amounts of phosphorus (~1 mol%) were added. In the next paragraphs, we discuss the influence of the alkaline-earth ions on the nanoparticles formation. We also present the attenuation measurements as well as the nanoparticles composition. Finally, modifications of the erbium spectroscopy are described.

### 5.1 Effect of Mg, Ca and Sr on the formation of the nanoparticles

When alkaline-earth (Mg, Ca and Sr) ions are incorporated, nanoparticles are observed both in preforms and fibres. Typical SEM pictures from the exposed core section of cleaved fibres are shown on Fig. 13 (Mg, Ca and Sr concentrations in the doping solution are 0.1 mol/l). The gray disk corresponds to the fibre core (~8 μm). The dark central part of the core is caused by the evaporation of germanium element; this is a common artifact of the MCVD technique that can be corrected through process optimization. Nanoparticles are visible as bright spots. They show an important compositional contrast compared to the silica background. The size distribution of the NP depends strongly on the nature of the alkaline earth ions. It is about 50 nm for Mg-doped fibre while it reaches 100 nm or more for Sr- and Ca-doped samples. The size of the nanoparticles is also dependent on the concentration of

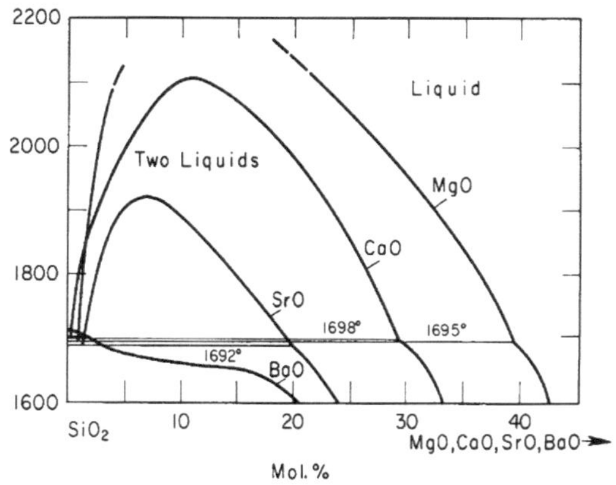


Fig. 12. Immiscibility-gap in the phase-diagram of binary $SiO_2$-MO glass (M = Mg, Ca, Sr) (Ol'shankii, 1951).

the alkaline-earth ions (Blanc et al., 2011). For the fibre doped with the solution with 0.1 mol/l of $MgCl_2$, the mean particle size is 48 nm and no particle bigger than 100 nm was observed in these MgO-doped fibres, unlike the case of CaO-doped fibres. Analysis of the SEM images also revealed that the inter-particle distance is in the 100 - 500 nm range. When the solution concentration increases from 0.1 to 1 mol/l of $MgCl_2$, the inter-particle distance remains nearly the same but the mean particle diameter almost doubles to reach 76 nm. With this concentration, NPs up to 160 nm were observed.

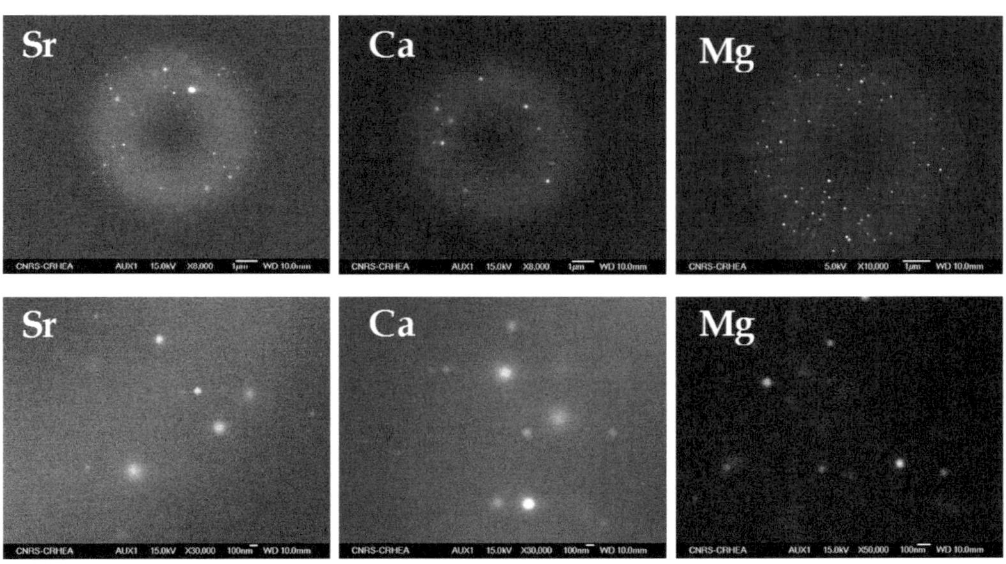


Fig. 13. SEM pictures of a MO-doped fibre (M = Sr, Ca, Mg). For each fibres, alkaline-earth concentration of the doping solution is 0.1 mol/l.

### 5.2 Transparency

The scattering loss in Mg-doped fibre was measured through the cut-back method. The fibre bending radius was kept > 20 cm to minimize bend loss. The 0.1 mol/l Mg-doped fibre attenuation spectrum is displayed in Fig. 14. At the wavelength of 1350 nm, losses were measured to be 0.4 dB/m only. This value is comparable with the attenuation measured in low-melting temperature transparent glass ceramics fibres (Samson et al., 2002) and is compatible with amplifier applications. The loss variation *vs* wavelength is attributable to light scattering induced by the nanoparticles. Due to the small particle size we assume Rayleigh scattering to be the major source of scatter loss, which was estimated according to the formula: $\alpha$ (dB/m)=4.34.$C_{Rayleigh}$.$N$.$\Gamma$, where $N$ is the DNP density ($m^{-3}$), $\Gamma$ is the overlap factor between the field and the core containing the DNP ($\Gamma = 0.3$ in fibre A) and $C_{Rayleigh}$ ($m^2$) is the Rayleigh scattering coefficient (Bohren & Huffman, 2004):

$$C_{Rayleigh} = \frac{(2\pi)^5}{48} \times \frac{d^6}{\lambda^4} \times n_m^4 \times \left( \frac{n_n^2 - n_m^2}{n_n^2 + 2n_m^2} \right)^2 \quad (8)$$

where $d$ is the nanoparticle diameter, $n_m$ and $n_n$ the host material and particles refractive indices, respectively. The actual particle composition is not known but a high content of MgO is expected (Stebbins et al., 2009). The nanoparticle refractive index is estimated at ~ 1.65, like that of Mg-based oxide such as $Mg_2SiO_4$ (Burshtein et al., 2003). Under these considerations, fitting of the experimental data with Eq. 8 yields a particle density $N$~ 0.4 x $10^{20}$ $m^{-3}$, or equivalently a mean inter-particle distance ~300 nm. These values agree well with the results collected from the SEM pictures taken from cleaved fibres.

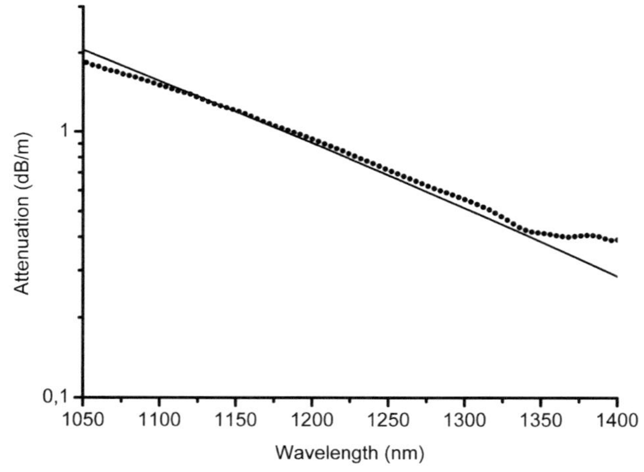


Fig. 14. Transmission spectrum of a Mg-doped fibre (0.1 mol/l solution). Dots: experimental data, full line: calculated Rayleigh scattering curve.

At 1350 nm, the normalized frequency V was lower than 1.3 ($LP_{11}$ mode cut-off wavelength is 700 nm). The difference between experimental data and Rayleigh scattering curve above 1350 nm is attributed to bending loss only. However, for practical application, the presented fabrication technique allows any necessary waveguide optimization without preventing the TGC growth. The attenuation of other fibres (1 mol/l of $MgCl_2$ or Ca- and Sr-doped) was extremely high (several 100s dB/m), in agreement with the Rayleigh formula (120 dB/m) assuming large monodisperse 100 nm particles. This shows that the particle mean size must be less than 50 nm for potential applications, and that the technique presented here is able to produce fibres with acceptable scattering loss.

### 5.3 Composition of the nanoparticles

The composition was investigated by EDX analyses. When nanoparticles are analyzed, Ca or Mg, P and Si are found while only Si is detected outside of the particles. Germanium seems to be homogeneously distributed over the entire glass. When erbium is added to the composition, it is found to be inserted into the particles as it is presented on Fig. 15. Such conclusion was also drawn when emission intensity of $Er^{3+}$ was mapped on the end-face of a cleaved fibre under a confocal microscope (Blanc, 2009b). No erbium fluorescence is detected outside the DNP. This result was expected due to the low solubility of RE ions in silica (or even in germano-silicate), whereas strongly modified and amorphous silicates, such as in the DNP, have a high solubility for these ions. These results indicate that erbium ions are located inside or very close to the DNP.

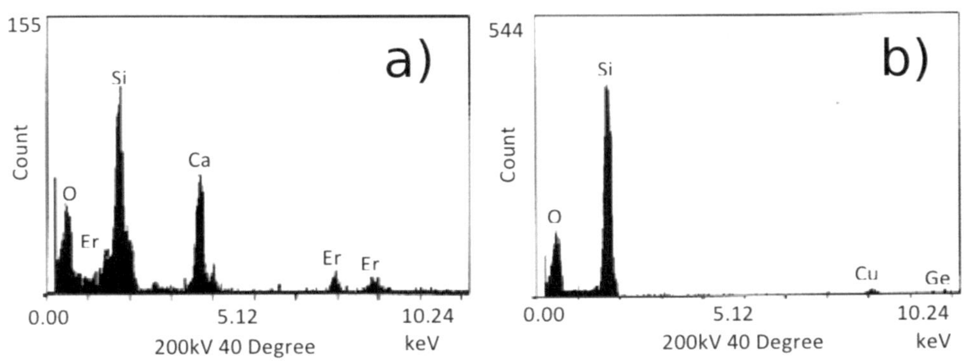


Fig. 15. Energy dispersive x-ray analyses spectra of the preform sample doped with Ca and Er. The area analysed corresponds to the nanoparticle (a) and outside (b).

### 5.4 Erbium spectroscopy

The emission spectra and lifetime from fibres doped with 0.1 (Fibre A) and 1 mol/l (Fibre B) were measured at room temperature around 1.55 µm under 980-nm pump excitation. The emission spectrum from Fibre A (Fig. 16) is similar to that in a silicate environment. Fibre B shows a distinct broadening of its spectrum (FWHM is 44 nm) by as much as ~ 50 % compared to Fibre A. In comparison, more than 10 at% of aluminum, as network modifier, would have been necessary to obtain the same FWHM in 'type III' Er-doped fibres in optical amplifiers for telecommunications (Desurvire, 1994), because Al and Er are evenly distributed across the core volume. Moreover, the shape of the fluorescence spectrum from Fibre B is quite unusual: it decreases monotonically between the peaks at 1.53 and 1.55 µm and the commonly observed dip at 1.54 µm is absent. These features would be attractive for realizing intrinsically gain flattened fibre amplifiers, provided sufficient minimization of the scattering loss is ensured through process optimization. Both fibres A and B produced single-exponential decaying fluorescence with 11.7 and 6.7 ms lifetimes, respectively.

Modifications of the $Er^{3+}$ spectroscopic properties in the TGC optical fibres are clearly evident from the above results when the Mg concentration increases. Such observations were previously reported in Ca-doped preforms. The broadening of the fluorescence curve and the lower lifetime obtained from Fibre B is tentatively interpreted as an effect of the modification of the erbium ions averaged local field. In other words, erbium ions in Fibre B are, on average, located in a medium with stronger local field compared to Fibre A. This induces stronger spontaneous transition probability, and hence a shorter lifetime (Henderson, 1989). Although the exact composition of the NPs is not yet known, emission spectrum and fluorescence lifetime of erbium ions in Fibre B are closely related to the results reported in the literature in phosphate glasses (Liu et al., 2003 ; Jiang et al., 1998). When Mg concentration increases, erbium ions environment changes from silicate to phosphate.

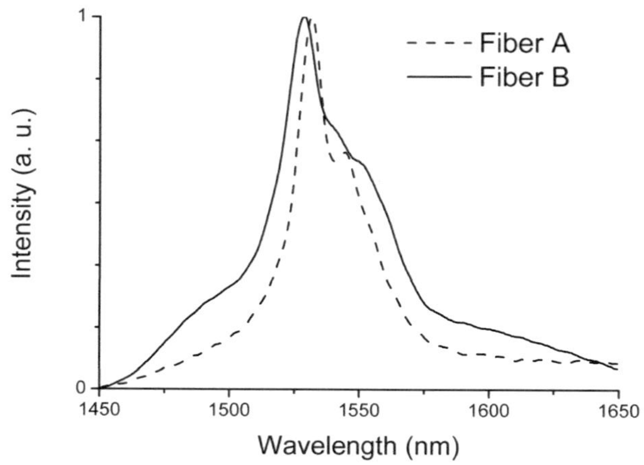


Fig. 16. Room temperature emission spectra from Mg-doped fibres. The Mg concentration in the doping solution is 0.1 (A) and 1.0 (B) mol/l, respectively. Excitation wavelength is 980 nm.

### 5.5 Conclusion

A method to fabricate $Er^{3+}$-doped TGC fibers entirely through the MCVD process is demonstrated. By adding magnesium to the silica-based composition, a low-loss fiber is obtained through *in situ* growth without requiring a separate process to realize NP (such as postprocess ceramming). An important result of this study is that the type of obtained nanoparticle by this technique could be as large as ~50 nm for applications such as in fiber amplifiers and lasers. Moreover, a broadening of the emission spectrum by as much as ~50% is observed with attractive features to realize gain flattened fiber amplifiers. More generally, this concept might have great potential as a possible solution to address various current issues in amplifying fibers, including (but not limited to) realization of intrinsically gain flattened amplifiers, etc.

## 6. General conclusion

The choice of a glass to develop new optical fibre component is most of the time a result of compromises. Silica glass is the most widely used for its many advantages (reliability, low cost fabrication, ...). The overall composition suffers from different drawbacks, such as high phonon energy or low luminescent ions solubility, which affect quantum efficiency or emission bandwidth of luminescent ions, for example. However, spectroscopic properties of active ions are not directly related to the average properties, but to their local environment. Then, new luminescent properties of dopants can be obtained by slightly modifying the silica composition.

In this chapter, we have shown that thulium emission efficiency can be improved by reducing the local phonon energy. Using a comprehensive numerical model we have shown that efficient S-band TDFA and lasing at 810 nm can be obtained. The chromium valence state was controlled by adding some glass modifiers like Al. This open the way to PQS all-fibre lasers. Then, nanostructuration of doped fibre is proposed as a new promising route to 'engineer' the local dopant environment. All these results will benefit to optical fibre components such as lasers, amplifiers and sensors, which can now be realized with silica glass.

## 7. Acknowledgments

The work was supported by CNRS Office of International Relations project No. PICS 5304, Czech Science Foundation project No. P205/11/1840, by the Ministry of Education, Youth and Sports of the Czech Rep., project No. ME10119 "FILA", by the INSIS-CNRS. LPMC is with GIS "GRIFON" (CNRS, http://grifon.xlim.fr/).

## 8. References

Anino, C. ; Théry J. & Vivien, D. (1997). New $Cr^{4+}$ activated compounds in tetrahedral sites for tunable laser applications. *Optical Materials*, Vol. 8, No. 1-2, pp 121-128.

Antipenko, B.M. ; Mak, A.A. ; Raba, O.B. ; Seiranyan, K.B. & Uvarova, T.V. (1983). New lasing transition in the $Tm^{3+}$ ion, *Soviet Journal of Quantum Electronics*, Vol. 13, No 4, pp. 558-560.

Blanc, W.; Peterka, P.; Faure, B.; Dussardier, B. ; Monnom, G. ; Kasik, I. ; Kanka, J. ; Simpson, D. & Baxter, G. (2006). Characterization of thulium-doped silica-based optical fibre for S-band amplifier. *SPIE 6180 Proceedings of Photonics, Devices, and Systems III*. ISBN: 80-86742-08-3, Prague (Czech Republic), June 2005.

Blanc, W. ; Sebastian, T.L. ; Dussardier, B. ; Michel, C. ; Faure, B. ; Ude, M. & Monnom, G. (2008). Thulium environment in a silica doped optical fibre. *Journal of non-crystalline solids*. Vol. 354, No 2-9, pp. 435-439.

Blanc, W. ; B. Dusssardier & M.C. Paul (2009a). Er-doped oxide nanoparticles in silica-based optical fibres, *Glass Technology: European Journal of Glass Science and Technology*. Vol. 50, No 1, pp. 79-81.

Blanc, W. ; Dussardier, B. ; Monnom, G. ; Peretti, R. ; Jurdyc, A.-M. ; Jacquier, B. ; Foret, M. & Roberts, A. (2009b). Erbium emission properties in nanostructured fibers. *Applied Optics* Vol. 48, No 31, pp. G119-124.

Blanc, W. ; Mauroy, V. ; Nguyen, L. ; Bhaktha, S.N.B. ; Sebbah, P. ; Pal, B.P. & Dussardier, B. (2011). Fabrication of rare-earth doped transparent glass ceramic optical fibers by modified chemical vapor deposition, *Journal of the American Ceramic Society*. DOI: 10.1111/j.1551-2916.2011.04672.x, published on-line.

Bohren, C.F. & Huffman, D.R..(2004). *Absorption and scattering of light by small particles,* Wiley Science, ISBN 0471293407, New-York.

Brunet, F. ; Taillon, Y. ; Galarneau, P. & LaRochelle, S. (2005). A simple model describing both self-mode locking and sustained self-pulsing in ytterbium-doped ring fiber lasers. IEEE *Journal of Lightwave Technology,* Vol.23, pp. 2131-2138.

Burshtein, Z. & Shimony, Y. (2002). Refractive index dispersion and anisotropy in $Cr^{4+}$:$Mg_2SiO_4$, *Optical materials,* Vol. 20, No. 2, pp 87-96.

Cerqua-Richardson, K. ; Peng, B. & Izumitani, T. (1992). Spectroscopic Investigation of $Cr^{4+}$-doped glasses. *OSA Proceedings of Advanced Solid-State Lasers,* ISBN 1557522235, Santa Fe (NM, USA, February 1992), Vol. 13. pp 52-55.

Chen, J.-C. ; Lin, Y.-S. ; Tsai, C.-N. ; Huang, K.-Y. ; Lai, C.-C. & Su, W.-Z. (2007). 400-nm-bandwidth emission from a Cr-doped glass fiber. *IEEE Photonics Technology Letters,* Vol. 19, No. 8, pp 595 - 597.

Cole, B. & Dennis, M. L. (2001). S-band amplification in a thulium doped silicate fiber. *Proceedings of Conference on Optical Fiber Communication,* ISBN 1557526559, Anaheim (CA, USA), March 2001.

Cronemeyer, D.C. (1966). Optical absorption characteristics of pink ruby. *Journal of the Optical Society of America,* Vol. 56, No. 12, pp 1703-1705.

Dennis, M. L.; Dixon, J. W. & Aggarwal, I. (1994). High power upconversion lasing at 810 nm in Tm:ZBLAN fibre, *Electronics Letters,* Vol. 30, No. 2, pp. 136-13

Desurvire, E. (1994). *Erbium Doped Fiber Amplifiers: principles and applications,* Wiley Interscience, ISBN 0-471-58977-2, New-York.

Desurvire, E. ; Bayart, D. ; Desthieux, B. & Bigo, S. (2002). *Erbium Doped Fiber Amplifiers: device and system developments,* Wiley Interscience, ISBN 0-471-41903-6, New-York.

Dianov, E.M. ; Dvoyrin, V.V. ; Mashinsky, V.M. ; Umnikov, A.A. ; Yashkov, M.V. & Gur'yanov, A.N. (2005). CW bismuth fiber laser. *Quantum Electronics.* Vol. 35, No. 12, pp 1083-1084.

Digonnet, M. (2001). *Rare-Earth-Doped Fiber Lasers and Amplifiers* (2nd ed.) Marcel Dekker. ISBN 0-203-90465-6, New-York.

Durteste, Y. ; Monerie, M. ; Allain, J.-Y. & Poignant, H. (1991). Amplification and lasing at 1.3 μm in praseodymium-doped fluoridezirconate fibers. *Electronics Letters,* Vol. 27, pp. 626-628.

Dussardier, B. ; Guyot, Y. ; Felice, V. ; Monnom, G. & Boulon, G. (2002). $Cr^{4+}$-doped silica-based optical fibers fluorescence from 0.8 μm to 1.7 μm. *Proceedings of Advanced Solid State Lasers,* in Trends in Optics and Photonics Series (OSA), ISBN: 1-55752-697-4, Vol. 68, pp. 104-108.

Dussardier, B; Maria, J. & Peterka, P. (2011). Passively Q-switched ytterbium and chromium doped fibre laser. *Applied Optics,* Vol. 50, No. 25, pp. E20–E23.

Dvoyrin, V.V. ; Mashinsky, V. ; Neustruev, B. ; Dianov, E.M. ; Guryanov, A.N. & Umnikov, A.A. (2003). Effective room-temperature luminescence in annealed chromium-doped silicate optical fibers. *Journal of the Optical Society of America B,* Vol. 20, No. 2, pp. 280-283.

Faure, B. ; Blanc, W. ; Dussardier, B. & Monnom, G. (2007). Improvement of the $Tm^{3+}$:$^3H_4$ level lifetime in silica optical fibers by lowering the local phonon energy. *Journal of Non-Crystalline Solids*. Vol. 353, No 29, pp. 2767-2773.

Felice, V.; Dussardier, B. ; Jones, J.K. ; Monnom, G. & Ostrowsky, D.B. (2000). $Cr^{4+}$-doped silica optical fibres : absorption and fluorescence properties. *The European Physical Journal-Applied Physics*, Vol. 11, pp. 107-110.

Felice, V.; Dussardier, B. ; Jones, J.K. ; Monnom, G. & Ostrowsky, D.B. (2001). Chromium-doped silica optical fibres : influence of the core composition on the Cr oxidation states and crystal field. *Optical Materials*. Vol. 16, No. 1-2, pp. 269-277

Grattan, K.T.V. & Sun, T. (2000). Fiber optic sensor technology: an overview. *Sensors and Actuators A*: Physical, Vol. 82, No. 1-3, pp. 40-61.

Grinberg, M. ; Russell, D.L. ; Holliday, K. ; Wisniewski, K. & Koepke, Cz. (1998). Continuous function decay analysis of a multisite impurity activated solid. *Optics Communications*, Vol. 156, pp. 409–418.

Henderson, B. and Imbush, G.F. (1989). *Optical Spectroscopy of Inorganic Solids*, Clarendon Press, ISBN 0-19-851372-0, Oxford.

Hewak, D.W. ; Deol, R.S. ; Wang, J. ; Wylangowski, G. ; Mederios Neto, J.A. ; Samson, B.N. ; Laming, R.I. ; Brocklesby, W.S. ; Payne, D.N. ; Jha, A. ; Poulain, M. ; Otero, S. ; Surinach, S. & Baro., M.D. (1993). Low phonon-energy glasses for efficient 1.3 μm optical fiber amplifiers. *Electronics Letters*, Vol. 29, No. 2, pp. 237-239.

Hideur, A. ; Chartier, T. ; Brunel, M. ; Salhi, M. ; Özkul, C. & Sanchez, F. (2001). Mode-lock, Q-switch and CW operation of an Yb-doped double-clad fiber ring laser. *Optics Communications*, Vol. 198, No. 1-3, pp. 141–146.

Hömmerich, U. ; Eilers, H. ; Yen, W.M. ; Hayden, J.S. & Aston, M.K. (1994). Near infrared emission at 1.35 μm in Cr doped glass. *Journal of Luminescence*, Vol. 60&61, pp. 119-122.

Jackson, S.D. & King, T.A. (1999). Theoretical modelling of Tm-doped silica fiber lasers. *Journal of Lightwave Technology*. Vol. 17, No. 5, pp. 948-956.

Jiang, S. ; Myers, M. & Peyghambarian, N. (1998). $Er^{3+}$ Doped Phosphate Glasses and Lasers. *Journal of Non-Crystalline Solids*. Vol. 239, No 1-3, pp. 143-148.

Komukai, T. ; Yamamoto, T. ; Sugawa, T. & Miyajima, Y. (1995). Upconversion pumped Thulium-doped fluoride fiber amplifier and laser operating at 1.47 μm. *IEEE Journal of Quantum Electronics*. Vol. 31, No. 11, pp. 1880-1889.

Kurkov, A. S. (2011). Q-switched all-fiber lasers with saturable absorbers. *Laser Physics Letters*, Vol. 8, No. 5, pp. 335-342 + references therein.

Laroche, M. ; Chardon, A.M. ; Nilsson, J. ; Shepherd, D.P. ; Clarkson, W.A. ; Girard, S. & Moncorgé, R. (2002). Compact diode-pumped passively Q-switched tunable Er-Yb double-clad fiber laser. *Optics Letters*, Vol. 27, No. 22, pp. 1980-1982.

Layne, C.B. ; Lowdermilk W.H. & Weber, M.J. (1977). Multiphonon relaxation of rare-earth ions in oxide glasses. *Physical Review B*. Vol. 16, No 1, pp.10-21.

Lin, H. ; Tanabe, S. ; Lin, L. ; Hou, Y.Y. ; Liu, K. ; Yang, D.L. ; Ma, T.C. ; Yu, J.Y. & Pun, E.Y.B. (2007). Near-infrared emissions with widely different widths in two kinds of Er3+-doped oxide glasses with high refractive indices and low phonon energies. *Journal of Luminescence*, Vol. 124, No. 1, pp. 167-172.

Lipavsky, B. ; Kalisky, Y. ; Burshtein, Z. ; Shimony, Y. & Rotman, S. (1999). Some optical properties of $Cr^{4+}$-doped crystals. *Optical Materials*, Vol. 13, No. 1, pp. 117-127.

Liu, Z. ; Qi, C. ; Dai, S. ; Jiang, Y. & Huet, L. (2003). Spectra and Laser Properties of $Er^{3+}$, $Yb^{3+}$:Phosphate Glasses. *Optical Materials.* Vol. 21, No 4, pp. 789-794.

Luo, L. & Chu, P.L. (1999) Passive Q-switched erbium-doped fibre laser with saturable absorber, *Optics Communications*, Vol. 161, No. 4-6, pp. 257 - 263.

Lüthi, S. R.; Gomes, A. S. L.; Sundheimer, M. L.; Dussardier, B; Blanc, W. & Peterka, P. (2007) Distributed gain in a Tm-doped silica fiber: experiment and modeling. Proceedings of CLEO-Europe, ISBN 1424409306, Munchen (Germany), June 2007.

Minelly, J. & Ellison, A. (2002). Applications of antimony-silicate glasses for fiber optic amplifiers. *Optical Fiber Technology*, Vol. 8, No. 2, pp. 123-138.

Moulton, P. M. (2011). High power Tm:silica fiber lasers: current status, prospects and challenges, Proceedings of CLEO-Europe, ISBN : 978-1-4577-0532-8, Munchen (Germany), May 2011

Murata, K. ; Fujimoto, Y. ; Kanabe, T. ; Fujita, H. & Nakatsuka, M. (1999). Bi-doped SiO2 as a new laser material for an intense laser. *Fusion Engineering and Design.* Vol. 44, No. 1-4, pp. 437-439.

Nagel, S.R. ; MacChesney, J.B. & Walker, K.L. (1985). *Modified chemical vapor deposition. in : Optical Fiber Communications: Vol 1 'Fiber Fabrication'*, ed. T. Li, pp 1-64, Academic Press, ISBN 0-12-447301-6, Orlando.

Ol'shankii, Ya. I. (1951). Equilibria of two immiscible-liquids in systems of alkaline-earth silicates. *Dokl. Akad. Nauk S.S.S.R.* Vol. 76, No. 1, pp 93-96.

Paschotta, R. ; Häring, R. ; Gini, E. ; Melchior, H. ; Keller, U. ; Offerhaus, H. L. & Richardson, D. J. (1999). Passively Q-switched 0.1-mJ fiber laser system at 1.53 μm. *Optics Letters*, Vol. 24, pp. 388–390.

Peterka, P. ; Faure, B.; Blanc W.; Karasek, M. & Dussardier, B. (2004). Theoretical modelling of S-band thulium-doped silica fibre amplifiers. *Optical and Quantum Electronics*, Vol. 36, No. 1-3, pp. 201-212.

Peterka, P. ; Kasik, I.; Dhar, A.; Dussardier, B. & Blanc, W. (2011). Theoretical modeling of fiber laser at 810 nm based on thulium-doped silica fibers with enhanced $^3H_4$ level lifetime. *Optics Express*, Vol. 19, No. 3, pp. 2773-2781.

Peterka, P. ; Kasik, I.; Matejec V.; Blanc W.; Faure, B.; Dussardier, B.; Monnom, G. & Kubecek, V. (2007). Thulium-doped silica-based optical fibers for cladding-pumped fiber amplifiers, *Optical Materials*, Vol. 30, No. 1, pp. 174-176.

Peterka, P.; Blanc, W.; Dussardier, B.; Monnom, G.; Simpson, D. A. & Baxter, G. W. (2008). Estimation of energy transfer parameters in thulium- and ytterbium-doped silica fibers. SPIE 7138 Proceedings of Photonics, Devices, and Systems IV, ISBN: 978-081947379-0, Prague (Czech Republic), august 2008.

Rasheed, F. ; O'Donnell, K.P. ; Henderson, B. & Hollis, D.B. (1991). Disorder and the optical spectroscopy of $Cr^{3+}$-doped glasses: I. Silicate glasses. *Journal of Physics: Condensed Materials.* Vol. 3, No. 12, pp. 1915-1930.

Razdobreev, I. ; Bigot, L. ; Pureur, V. ; Favre, A. ; Bouwmans, G. & Douay, M. (2007). Efficient all-fiber bismuth-doped laser. *Applied Physics Letters.* Vol. 90, pp. 031103 (3 pages).

Samson, B.N. ; Pinckney, L.R. ; Wang, J. ; Beall, G.H. & Borrelli, N.F. (2002). Nickel-doped nanocrystalline glass-ceramic fiber. *Optics Letters.* Vol. 27, No. 15, pp 1309-1311.

Samson, B.N. ; Tick, P.A. & Borrelli, N.F. (2001). Efficient neodymium-doped glass-ceramic fiber laser and amplifier. *Optics Letters.* Vol. 26, No. 3, pp. 145-147.

Sanchez, F. ; Le Boudec, P. ; François, P.-L. ; Stephan, G. (1993). Effects of ion pairs on the dynamics of erbium-doped fiber lasers. *Physical Review A.* Vol. 48, No. 3, pp. 2220-2229.

Schultz, P.C. (1974). Optical Absorption of the Transition Elements in Vitreous Silica. *Journal of the American Ceramics Society.* Vol. 57, No. 7 pp. 309-313.

Sennaroglu, A. ; Demirbas, U. ; Ozharar, S. & Yaman, F. (2006). Accurate determination of saturation parameters for $Cr^{4+}$-doped solid-state saturable absorbers. *Journal of the Optical Society of America B.* Vol. 23, No. 2, pp. 241-249.

Sennaroglu, A. ; Pollock, C.R. & Nathel, H. (1995). Efficient continuous-wave chromium-doped YAG laser. *Journal of the Optical Society of America B.* Vol. 12, No. 5, pp. 930-937.

Siegman, A.E. (1986) *Lasers*, Univ. Science Books, ISBN 0-935702-11-5, Mill Valley.

Simpson, D.A.; Baxter, G.W.; Collins, S.F.; Gibbs, W.E.K.; Blanc, W.; Dussardier, B. & Monnom, G. (2006). Energy transfer up-conversion in $Tm^{3+}$-doped silica fiber. *Journal of Non-Crystalline Solids.* Vol. 352, No 2, pp. 136-141.

Simpson, D.A. ; Gibbs, W.E.K. ; Collins, S. F. ; Blanc, W. ; Dussardier, B. ; Monnom, G. ; Peterka, P. & Baxter, G. W. (2008). Visible and near infra-red up-conversion in $Tm^{3+}/Yb^{3+}$ co-doped silica fibers under 980 nm excitation. *Optics Express.* Vol. 16, No 18, pp. 13781-13799.

Stebbins, J.F. ; Kim, N. ; Andrejcak, M.J. ; Boymel, P.M. & Zoitos B.K. (2009). Characterization of phase separation and thermal history effects in magnesium silicate glass fibers by nuclear magnetic resonance spectroscopy. *Journal of the American Ceramic Society.* Vol. 92, No 1, pp. 68-74.

Sugano, S. ; Tanabe, Y. & Kamimura, H. (1970) *Multiplets of Transition-Metal Ions in Crystals*, Academic Press, New York.

Tick, P.A. ; Borrelli, N.F. ; Cornelius, L.K. & Newhouse, M.A. (1995). Transparent glass ceramics for 1300 nm amplifier applications. *Journal of Applied Physics.* Vol. 78, No. ll, pp 6367-6374.

Townsend, J.E. ; Poole, S.B. & Payne, D.N. (1987). Solution doping technique for fabrication of rare earth doped optical fibers. *Electronics Letters.* Vol. 23, No. 7, pp. 329-331.

Upadhyaya, B. N. ; Chakravarty, U. ; Kuruvilla, A. ; Oak, S.M. ; Shenoy, M.R. & Thyagarajan, K. (2010). Self-pulsing characteristics of a high-power single transverse mode Yb-doped CW fiber laser. *Optics Communications.* Vol. 283, No. 10, pp. 2206-2213.

Van Dijk, J.M.F. & Schuurmans, M.F.H. (1983). On the nonradiative and radiative decay rates and a modified exponential energy gap law for 4f–4f transitions in rare-earth ions. *The journal of Chemical Physics.* Vol. 78, No 9, pp. 5317-5313.

Walsh, B.M. & Barnes, N.P. (2004). Comparison of Tm:ZBLAN and Tm:silica fiber lasers: Spectroscopy and tunable pulsed laser operation around 1.9 μm. *Applied Physics B.* Vol. 78, No 3-4, pp. 325-333.

Wang, Y. & Xu, Ch.-Q. (2007). Actively Q-switched fiber lasers: Switching dynamics and nonlinear processes (A review). *Progress in Quantum Electronics.* Vol. 31, No. 3-5, pp. 131–216.

Yoo, S. ; Paek, U.-C. & Han, W.-T. (2003). Development of a glass optical fiber containing $ZnO–Al_2O_3–SiO_2$ glass-ceramics doped with $Co^{2+}$ and its optical absorption characteristics. *Journal of Non-Crystalline Solids.* Vol. 315, No 1-2, pp. 180–186.

Zarzycki, J (1991). *Glasses and the vitreous state.* Cambridge university press, ISBN 0521355826, Cambridge.

# 5

# Tunable Rare-Earth Doped Fiber Laser

Arturo A. Castillo-Guzman and Romeo Selvas-Aguilar
*Physical and Mathematical Sciences Research Center (CICFIM) UANL-FCFM*
*San Nicolás de los Garza, Nuevo León,*
*México*

## 1. Introduction

Recently, researches worldwide have been focused in building up systems capable to increase the output optical powers as well as to improve the beam quality of the delivering system of fiber lasers. Notwithstanding, the demanding of sources in the telecom industry together with the increasing in required bandwidth, need sources with multi-wavelength emissions or tunable sources. The main advantage of tunable sources is due to the fact that by using a single source is possible to tune a laser gain medium at different wavelengths. For example, this feature allows having high quality sources for the telecom application.
Many techniques have been then developed in order to obtain a wide tuning range. Using bulk grating (Chen et al., 2003), fiber Bragg grating (Xia et al., 2006; Goh et al., 2003; Mokhtar et al., 2003), SOA fiber laser based on Sagnac loop mirror via polarization tuning (Zhang et al., 2009), and Fabry-Perot cavity (Chawki et al., 1993) are some example of them. However these techniques are relative expensive or might required very sophisticated setups.
Recently, the multimode interference (MMI) effect has demonstrated to be applied in a wide variety of applications. Because of its wavelength dependence, it works as a wider band-pass filter which is therefore well useful to our tuning mechanism.
On the first part of the chapter, it is analyzed the basic feature of the MMI effect as well as the research carried out by applying this idea in tunable fiber laser systems. These two aspects are well detailed reviewed and subsequently established as the bases for an application of a high power tunable rare-earth optical fiber laser. Parameters like total tunable range, signal to noise ratio, laser line-width are presented.
The second part describes an erbium-doped fiber optical amplifier numerical simulation in a MOPA configuration with a seed signal obtained from a tunable erbium-doped fiber laser (EDFL) based on the MMI effect. The MOPA amplifier was designed using a double-cladding erbium-doped fiber (DC EDF), pumped with a 10 Watts high-power 980nm diode laser. Based on the MMI effect, the tunable filter source of the tunable EDFL is composed by a no-core optical fiber and a ferrule of 127 microns of diameter in a ring laser cavity. The tunable range achieved was 24 nm. This tunablility is from 1576-1600nm which corresponds to the C and L band of the telecomm. The output power was roughly 2.7 Watts with a line-width of 0.1nm. This setup demonstrates to be a portable, no expensive optical fiber laser.

## 2. Multimode interference effect

The self-imaging feature of a homogeneous multimode planar optical waveguide has been applied in the design of passive planar monomodal optical couplers based on multimode interference (MMI) (Soldano et al., 1992). This characteristic is described as an input field profile which is reproduced in single or multimode images by mode interference at periodic intervals along the propagation direction of the guide. The self-imaging was firstly suggested by (Bryhgdahl, 1973) and by (Ulrich, 1975a, 1975b).

On one hand, the theory of multimode interference has been also introduced by (Bachmann et al., 1994, 1995) and (Soldano & Pennings, 1995). A high performance multimode interference filter was also proposed by (Li et al., 2002) for 1.3 and 1.5μm regions. These works do not described the self-imaging for optical fibers. On the other hand and using this effect a wavelength tuning fiber laser was reported by (Selvas et al., 2005). Later on, an enhancement of this work was reported by increasing the tunable range from 8nm to 12nm (Anzueto-Sanchez et al., 2006).

Taking into account Fig. 1 which is a schematic diagram of the minimum length required to this effect takes place, the mathematical expressions behind this effect are as followed.

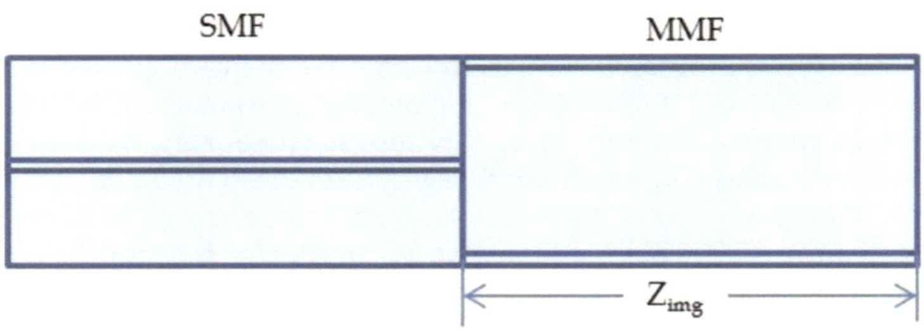


Fig. 1. A schematic diagram in which the multimode interference effect takes place in optical fibers.

The field distribution at the end of the MMF segment of length $z$, can be written as (Mohammed et al., 2004)

$$E_{MMF}(r,\theta)=\sum_{j=0}^{M}a_j\psi_j(r,\theta)\exp\left(i\beta_j z\right), \tag{1}$$

where $\psi_j$ and $a_j$ are the $j$th mode electric field amplitude and the field expansion coefficient, respectively. The constant $M$ is the total number of guided modes inside the MMF. For on-axis coupling, the number of modes is reduced to the number of radially symmetric modes. The field expansion coefficients in Eq. (1) can be calculated from the following cross-correlation formula:

$$\alpha_j=\frac{\int_{\theta=0}^{2\pi}\int_{r=0}^{\infty}E_s(r,\theta)\times\psi_j(r,0)^{*}\,rdrd\theta}{P_j},$$

$$P_j=\int_{\theta=0}^{2\pi}\int_{r=0}^{\infty}\left|\psi_j(r,\theta)\right|^2 rdrd\theta. \tag{2}$$

In Eq. (2), $E_s$ is the field distribution of the SMF fundamental mode. The power coupling efficiency to the second SMF is calculated through the overlap integral between $E_{MMF}$ and the second SMF fundamental mode.

Now, considering the single reimaging of the input field, it occurs at specific longitudinal locations within the MMF where $|\psi(r,z)|=|\psi(r,0)|$. This occurs at periodic longitudinal $z$ locations (Mehta et al., 2003)

$$z_{img}(\lambda)=\frac{\frac{m\pi}{2}}{\beta_{inside,0}(\lambda)-\beta_{inside,1}(\lambda)}, \qquad m=1,2,3,... \tag{3}$$

where $\beta_{inside,0}$ and $\beta_{inside,1}$ represent the longitudinal propagation constants within the MMF. These terms are determined using the weakly guiding approximation and applying physical dimensions and material properties of the MMF into the dispersion relations that characterize propagation within the optical fiber.

Continuing with the analysis of this effect, the reimage that comes out of the MMF has been used in order to get an optical device (Soldano et al., 1992; Li et al., 2002; Selvas et al., 2005; Anzueto-Sanchez et al., 2006; Mehta et al., 2003). All these works used a high reflector mirror to get the light back into the MMF again as is showed in Fig 2. The analysis goes further because of the free space condition. Eq. (4) gives the relation between the mirror distance expression and the maximum coupling of light back into the SMF (Mohammed et al., 2004)

$$z_{out}(\lambda)=\frac{\frac{\pi}{2}-\{(L_{MMF})(\beta_{inside,0}(\lambda)-\beta_{inside,1}(\lambda))\}}{\beta_{out,0}(\lambda)-\beta_{out,1}(\lambda)} \tag{4}$$

where $L_{MMF}$ is the length of the multimode optical fiber.

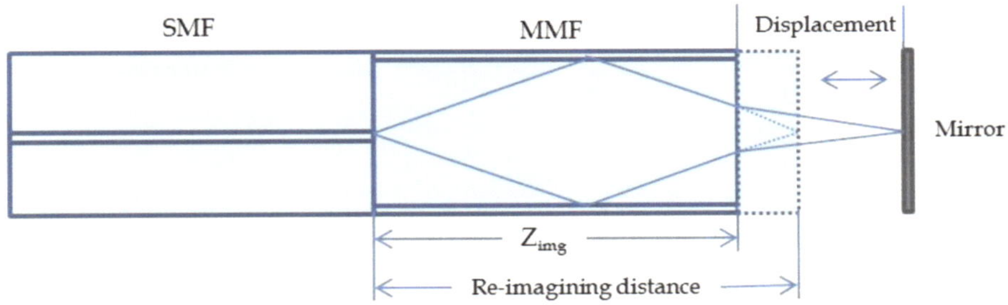


Fig. 2. SMF/MMF/mirror in free space setup.

In other words, the length at which single images are formed can be obtained from the restricted symmetric interference condition which is given by (Selvas et al., 2005 or Mehta et al., 2003)

$$L=p\left(\frac{3L_{\pi}}{4}\right) \quad \text{with p} = 0,1,2,..., \tag{5}$$

where $L_{\pi}$ is the beat length

$$L_{\pi}\cong\frac{4n_{MMF}W_{MMF}^{2}}{3\lambda_{0}}. \tag{6}$$

where $n_{MMF}$ and $W_{MMF}$ correspond respectively to the refractive index and diameter of the MMF core, with $\lambda_0$ as the free space wavelength.

Once established the basic principle and the mathematical equations of the MMI effect, in the next section, it will present three different tunable optical fiber lasers in which their tunable systems are based on the MMF theory, and each of them have modified aspects of the MMI effect in order to improve the optical results.

## 3. Tunable optical doped fiber laser based on MMI effect

Recently, the multimode interference effect has showed so much interest to build up different optical device. A lot of groups all around the world have introduced this effect into their own research. The multimode interference effect has been applied on several optical devices, especially in optical fiber based source, such as: multi-wavelength fiber laser (Poustie et al., 1994), all-fiber refractometer sensor (Wang & Farrell, 2006), high-temperature sensor (Li et al,. 2005) and wavelength tunable laser (Selvas et al., 2005). The latter research area is mainly the focused of this chapter.

### 3.1 Tunable double-clad ytterbium-doped fiber (DCYDF)

This laser consists of a double-clad ytterbium-doped fiber (DCYDF) with 6/125 μm of core/cladding diameter and an 0.14/0.45 numerical apertures (Martinez-Rios et al., 2003). The length of the DCYDF was 16 m, which corresponds to 7.2 dB of pump absorption (just the 80% of the pump is absorbed). The setup employed is depicted in Fig. 3.

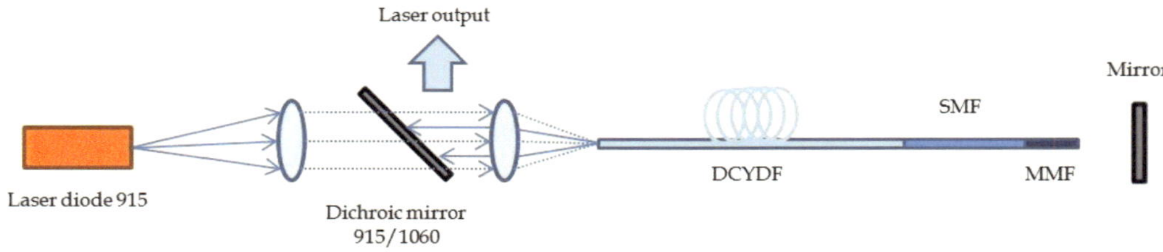


Fig. 3. Set up of the tunable double-clad ytterbium-doped fiber laser.

This tunable laser consists in a laser diode at 915 nm and 5 watts as output power. The light was launched into the DCYDF by two lenses which were AR coated. The DCYDF was fusion spliced to the MMI filter (SMF and MMF), after that, it was set a mirror in order to get the light feedback into the MMF. It was also added a dichroic mirror in order to obtain the output laser power as well as protection to the laser diode.

It is reported by the authors that the broadband mirror was placed at this position, and wavelength tuning was then demonstrated by varying the distance between the broadband mirror and the output facet of the MMF. The separation distance between the mirror and the MMF were achieved by steps of 25 micros where the output power and the optical spectrum of the tunable laser were measured. Fig. 4 shows the relation between the separation distance in steps of the MMF from the mirror and the wavelength tuned.

It can be notice from the Fig. 4. that the linearity tendency are coming from the steps moving away the MMF and the mirror in relation with the wavelength range tuned. It is also demonstrated that the total tuning range was only 8 nm (1088nm to 1097nm). The measured

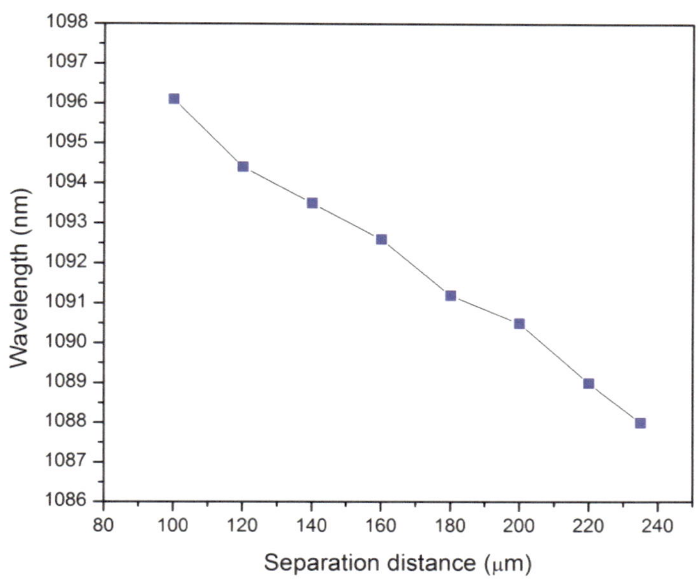


Fig. 4. Tuning wavelength range of the YDFL vs. separation distance of the mirror.

line-width was roughly 0.5 nm and the signal-to-noise ratio better than 30 dB. This optical spectrum is showed in Fig.5.

Additionally, it is reported that an average of 500 mW was obtained with the minimum and maximum power being 460 and 550 mW, respectively. For separation distances shorter than 100 microns and longer than 250 microns, the lasing emission vanished and it is notice that does not correlate to the emission cross section spectrum of the Yb ions. If the goal is to extend the tuning range, it is needed to use a mirror with a flat spectral response within the Yb-gain spectra, and consequently to take advantage of the whole gain range of the ytterbium-doped fibers. The reflectivity of the mirror at 1100-1320nm is about 95% and this reflectivity drops drastically to about 73% from 1100-1070 nm. It is also mentioned that the alignment process have been converted in a tedious work.

### 3.2 Enhanced tuning mechanism in fiber laser based on multimode interference effect

Taking into account the problems faced on the previous work, a new proposal was then analyzed. Now, an enhanced design for a tuning mechanism was developed (Anzueto-Sanchez et al., 2006). In this occasion, an SU-8 fiber gripper structure was fabricated. This device would align much easier the MMF along with a gold-coated fiber (golden facet) as a fiber mirror. It was also implemented a micrometric stepping motor which was attached to the gold-coated fiber to control the separation distance between the two fibers. Fig. 6. shows the experimental set up of this tunable optical fiber laser.

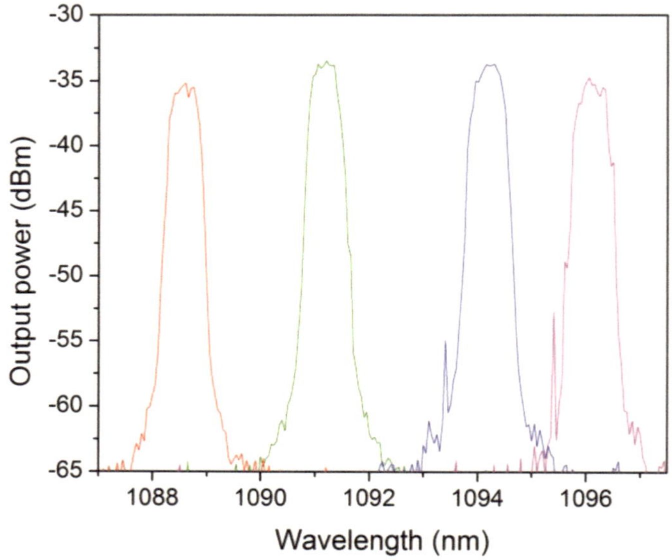


Fig. 5. Total optical tuning range of a DCYFL.

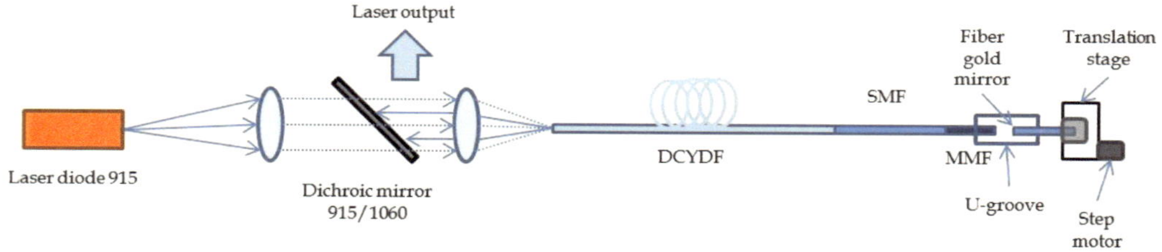


Fig. 6. Schematic diagram of the experimental setup.

The experimental setup was design similarly to the previous one. The important difference lay out in the tunable mechanism. The DCYDF is pumped by a fiber pigtailed multimode laser diode launching the light through two aspheric lenses to one end of the Ytterbium fiber with a maximum power of 1.8 W at 915 nm. To separate the pumping wavelength and the laser emission, it was set a dichroic mirror. The other end of the Ytterbium fiber was fusion spliced to the filter. Fig. 7 shows the tuning device proposed.

The tuning device consists of a SU-8 fiber gripper which is a channel with negative slope walls. The structure was able to grip both fibers once these were inserted in the groove, let them moving along their longitudinal direction giving them an automatic x-y alignment. The fabrication process of the SU-8 structures is well explained in (Raymond et al., 2004). In this case, one fiber is the MMF while the other is the mirror-fiber. It was used a standard thermal evaporator in order to coating a gold film on one facet of a single mode fiber. The MMF was fixed to the SU-8 and the fiber mirror was attached to a computer controlled micrometric stepping motor getting a better displacement control of the fiber, as is showed in Fig. 6.

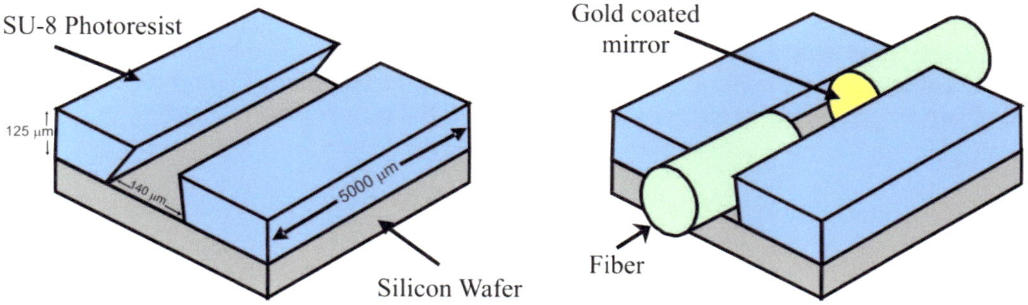


Fig. 7. Wavelength tuning SU-8 device.

Fig. 8. Shows the relation between the separation distance between the MMF and the mirror.

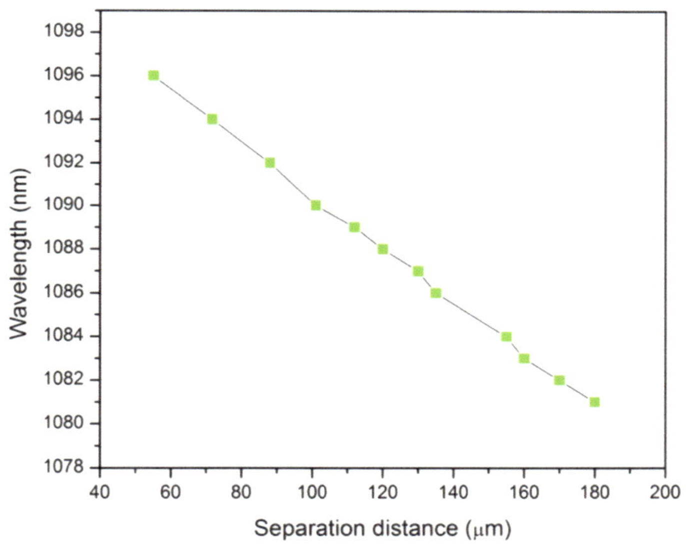


Fig. 8. Experimental results of displacement of facet mirror fiber versus emitted laser wavelength.

It is understood from Fig. 8 that a quasi-linear wavelength tuning against the separation between the MMF and the fiber mirror is obtained. A total tunable range of 12.24 nm was measured and the slope efficiency was around 30% at each wavelength.

The alignment problems were reduced considerably; and an improvement in the total tunable range was therefore achieved, increasing it from 8 nm to 12.24 nm when comparing it with the last section work.

### 3.3 Tunable multimode interference filter

The maximum tuning wavelength range achieved with the previous set up has been limited to only 12 nm. This limitation is not due to the MMI filter itself, but rather from other effects arising from the way of its implementation. Considering that the peak wavelength response of the filter exhibits a linear dependence when the length of the MMF is modified, a capillary tube was filled with a refractive index matching liquid as a new proposal. This implementation was figure out in order to increase the length of the MMF as well as the total tuning range (Antonio-Lopez et al., 2010). From the equations (5) and (6) it is possible to obtain an expression in terms of other parameter, the emission wavelength,

$$\lambda_0 = p\left(\frac{n_{MMF} D_{MMF}^2}{L}\right) \quad \text{with p = 0,1,2,....} \qquad (7)$$

This equation is indicating us that if we want to tune in wavelength, the parameters to modify are: the refractive index, the length of the MMF, or the diameter of the MMF. This work proposed the modification of the MMF length as a first option. It was then used a fused silica (n=1.444) ferrule with an inner diameter of 127 μm, and an outer diameter of 5 mm. The process consists in fill the ferrule with a high refractive index liquid with n=1.62 (Cargille Index Matching Liquid), so a liquid multimode waveguide (MMW) would take place inside of the ferrule. As it is explained in[14], the SMF and MMF were inserted in the ferrule, then, at every step the two fibers were moving apart to each other, the space created between them would fill with the refractive index liquid; so the effective length of the MMF will be the sum of the real MMF length plus the liquid MMW segment. As a consequence, if the effective length of the MMF is increased, the wavelength response should be tuned. The set up used is showed in Fig. 9. An important factor was the MMF used. It is also demonstrated that having a bigger refractive index difference between core and cladding provides a MMI filter with narrower linewidth and better contrast (Mohammed et al., 2006). The MMF used is known as no-core, which it is a 125 μm of diameter with air as the cladding.

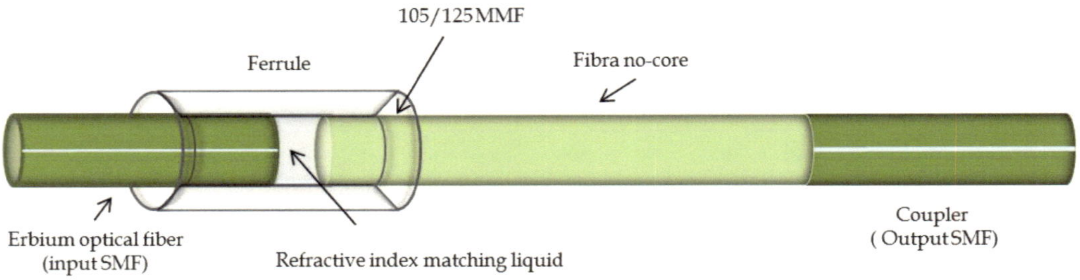


Fig. 9. Schematic of the tuning mechanism with a liquid MMW segment.

There was a deep study in order to obtain the best multimode fiber for the filter. The finite Difference Beam Propagation Method (FD-BPM) was used to numerically model the MMI effect that occurs in the MMF. All the fibers considered for the simulations were those which are commercial. Considering Eq. (5) the autoimages must be formed periodically along the MMF. Nevertheless, due to the nature of the MMI effect, the real output field images are given each 4$^{th}$ image. Fig. 10 to Fig. 13 shows these simulations.

Fig. 10 to Fig. 13 were obtained from a simulation on the BeamPro program. The parameters used in Fig. 13 were the no-core´s. This multimode fiber shows to be the better option for the filter due to its own optical properties along with the possibility to do the filter in the practice. As can be seen in Fig. 13 there are different reimages with different intensities, these formed images at several distances are known as pseudoimages. Most of them are similar to the input field but they show some losses, therefore, the filter operates with the 4th image.

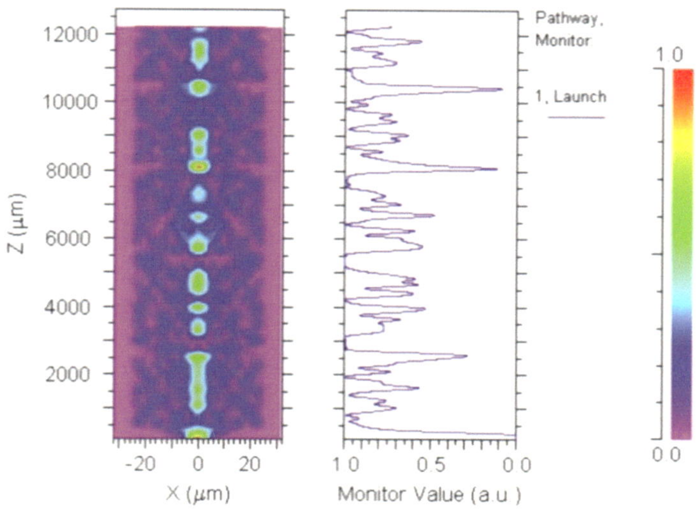


Fig. 10. Beam propagation of a 50/125 µm MMF for 1575 nm.

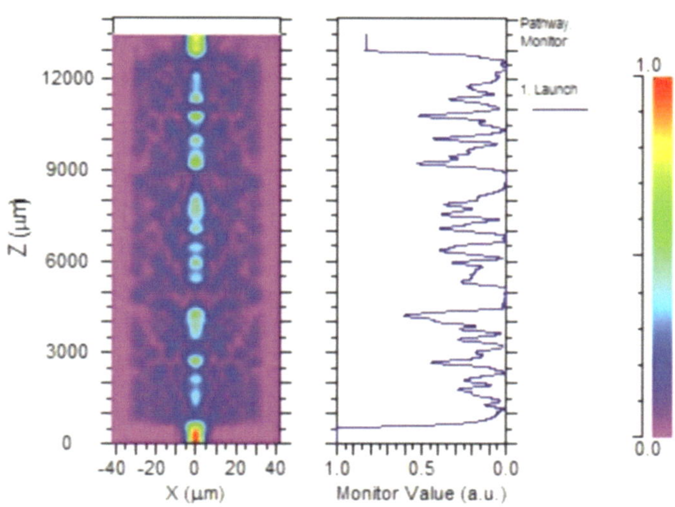


Fig. 11. Beam propagation of a 62.5/125 µm MMF for 1575 nm.

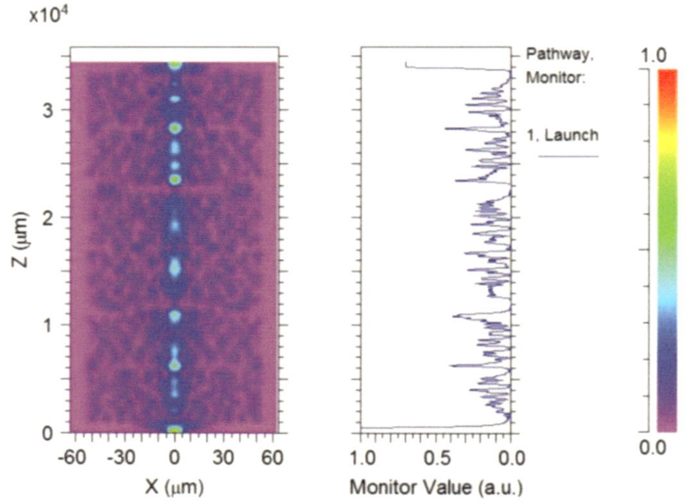


Fig. 12. Beam propagation characteristics of a 105/125 μm MMF for 1575 nm.

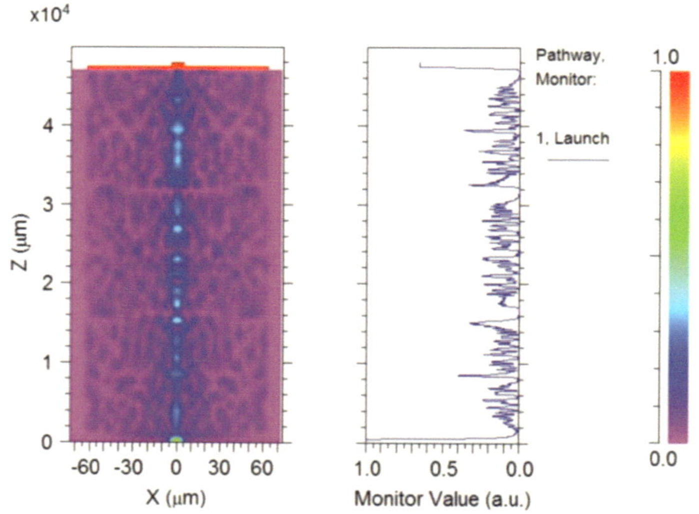


Fig. 13. Beam propagation characteristics of a 125/125 μm MMF for 1575 nm.

Now, since it was used a refractive index liquid which trends to accumulate in the end of the ferrule, it was decided to split the numbers of images in two different multimode fibers. The first re-image was obtained on the 105/125 MMF, the next third re-images were obtained on the no-core fiber. The length of both fibers would be calculated from Eq. (7). By this way, losses due to the contact of the liquid refractive index with the no-core fiber were avoided.

This tunable multimode interference bandpass filter is added to an erbium doped fiber ring-laser. The set up used is showed in Fig. 14.

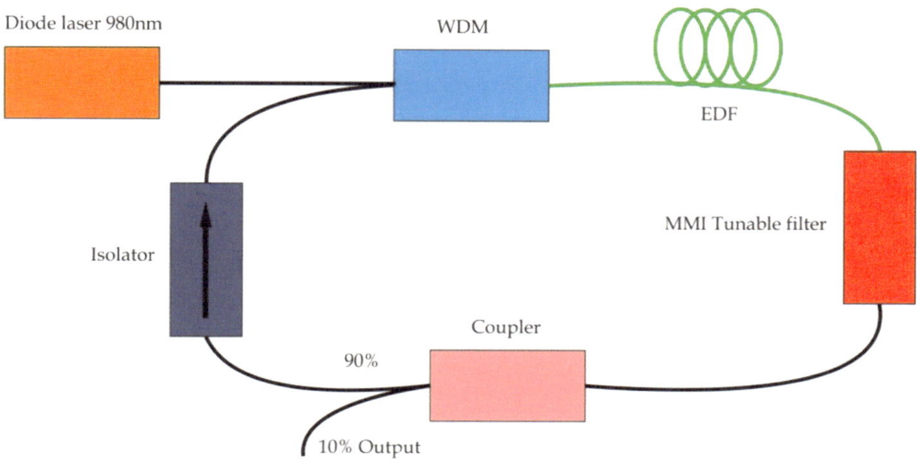


Fig. 14. Schematic set up of a tunable erbium doped fiber laser based on MMI effect.

The experimental set up consists on a laser diode at 980 nm with an output power of 150 mW, a WDM coupler of 980 nm/1550 nm, 5 meters long of Erbium doped fiber (EDF), a C-band optical isolator to keep the laser unidirectional , a 10/90 coupler to monitor the tunability of the laser, and the MMI tunable filter.

Fig. 15 shows the relation between the separation distance between the SMF and the 105/125 MMF. The response of the filter was at every 200 μm.

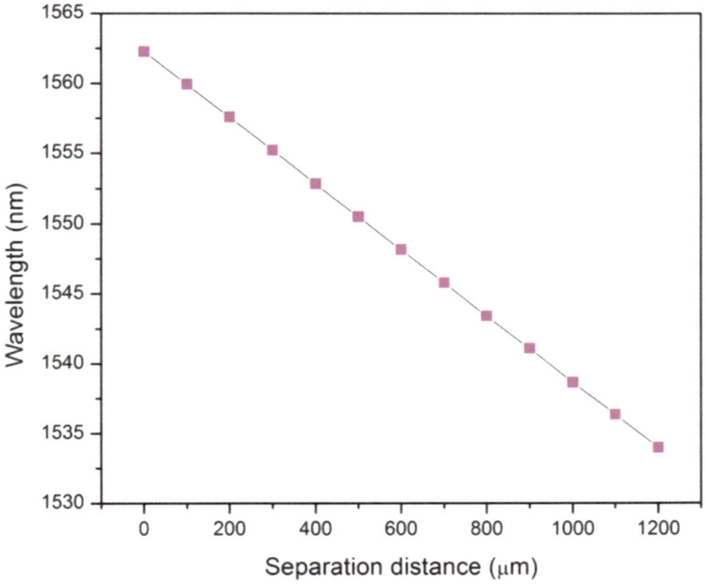


Fig. 15. Wavelength versus separation distance between SMF and 105/125 MMF.

Finally with the implementation of a liquid index matching fluid (n=1.64) inside of a ferrule a tuning range of almost 30 nm was achieved with an insertion losses of 0.4 dB with a 3 dB bandwidth of 0.35 nm as it is showed in Fig. 16.

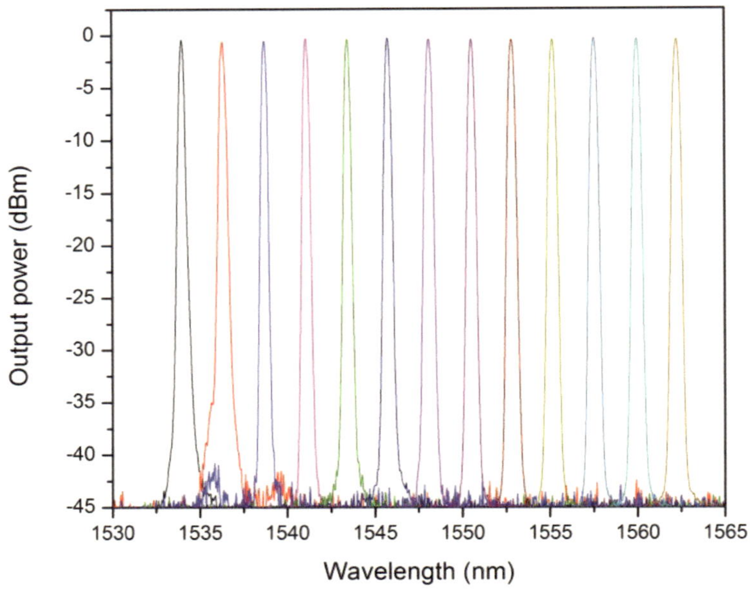


Fig. 16. Wavelength versus Output power with a multimode interference filter.

### 3.4 A liquid multimode waveguide segment means more tunable range

After the work done so far with the implementation of a liquid index matching liquid (n=1.64), several questions were on the table, for example, how far can be gone with the tunable range?, depending on what?. So, on a try for answering these questions, the spectral response of a group of liquid index matching fluids were studied (Castillo-Guzman et al., 2010). Taking into account that as it is demonstrated in Eq. (7), the peak wavelength response of the MMI filter can be selected by simply changing the length of the MMF, and it was necessary to design a mechanism to change the length of the MMI in real time, creating a liquid waveguide. Previously, it was analysed a liquid index matching fluid of 1.64 as refractive index. Now, the proposal was a deep study of several index matching fluids in order to enlarge the waveguide which gives a larger tunable range as a consequence. Fig. 17 shows the relation between the refractive index of the index matching fluids versus the FWHM.

The index matching liquids were provided by Cargille Labs. It is important to remember that the refractive index of the liquid has to be higher than that of the ferrule and lower than the no-core fiber in order to prevent losses. Fig. 17 shows that the bandwidth increases along with the refractive index. Therefore, since the ferrule refractive index is 1.444 and the no-core fiber is 1.463, it was selected the 1.45. Another change done in this work was that the MMF length was design using only no-core fiber instead of two different MMF.

The set up used was similarly of the Fig. 9, with the exceptions of the variations of the index matching fluids and the use of the no-core fiber as the unique MMF. Fig. 18 shows the set up

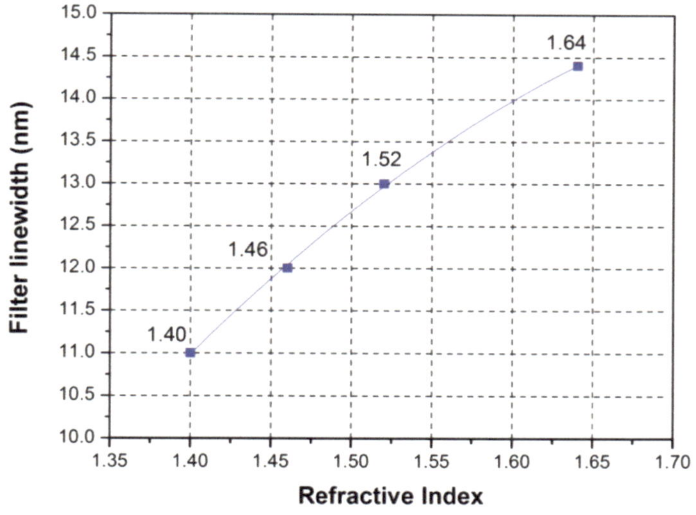


Fig. 17. Refractive index of different index matching fluids vs. MMI filters bandwidth at FWHM.

used in this occasion. About of the elements used in the set up, it was reported a L-band Erbium-doped fiber (EDF) with 2.85 m long, 0.25 NA and 3000 ppm of Er-concentration. A 980 nm laser diode with 150 mW of maximum output power from Lucent Technologies as the pump source. The rest of the elements remained the same.

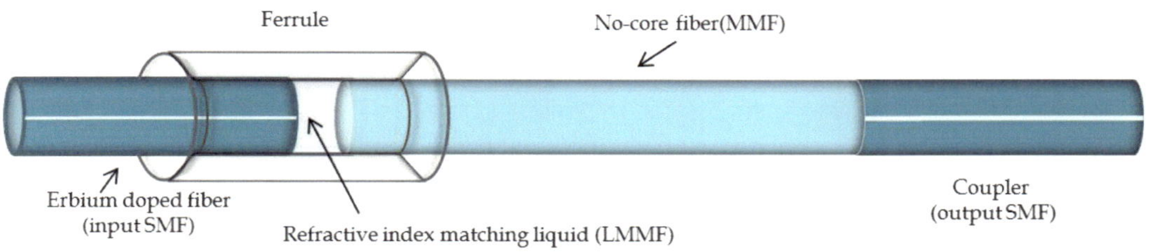


Fig. 18. Set up of a tunable erbium doped fiber laser with a liquid waveguide as a filter.

In order to obtain a tunable range, the EDF and the no-core fiber were separated. The first results obtained are shown in Fig. 19. Here, is shown the relation between the separation distance and the wavelength tuned. The wavelength tuning was achieved varying this separation in 50 μm steps, within a total of 1.6 mm. The index matching fluid used was the 1.64.

The total tuning range was of 39nm. It was from 1561.98nm to 1600.76nm. The linewidth measured was of 0.4nm and the signal to noise ratio was of 40dBm. These experimental results are showed in Fig. 20. The MMI filter has proved to be confident for an erbium optical fiber laser. The tuning mechanism was applied back and forth maintaining the output power always stable.

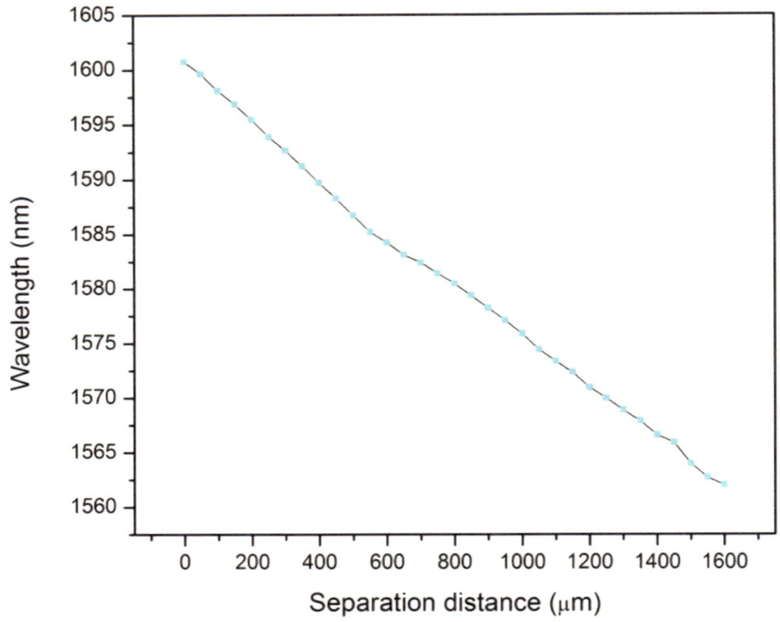


Fig. 19. Peak wavelength response of the MMI filter as a function of the separation distance between the fibers in the ferrule.

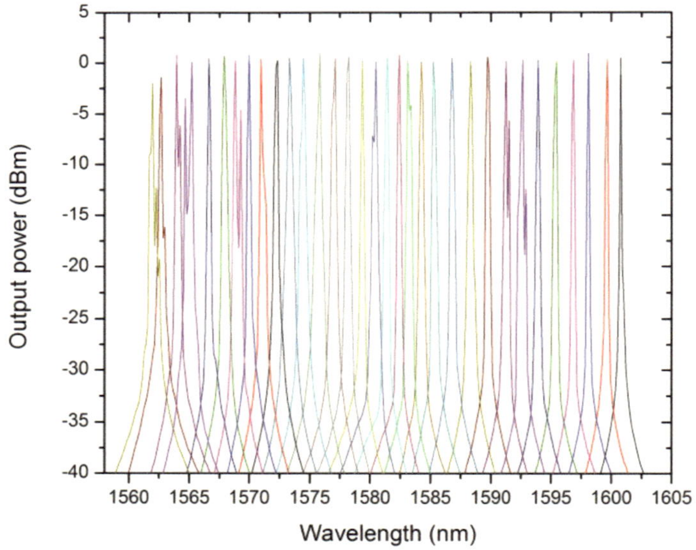


Fig. 20. Tuning range of 39nm, output power versus wavelength.

The second results obtained are shown in Fig. 21. Here, is shown the relation between the separation distance and the wavelength tuned. The wavelength tuning was achieved varying this separation in 1000 μm steps, within a total of 1.7 mm. the index matching fluid used was 1.45.

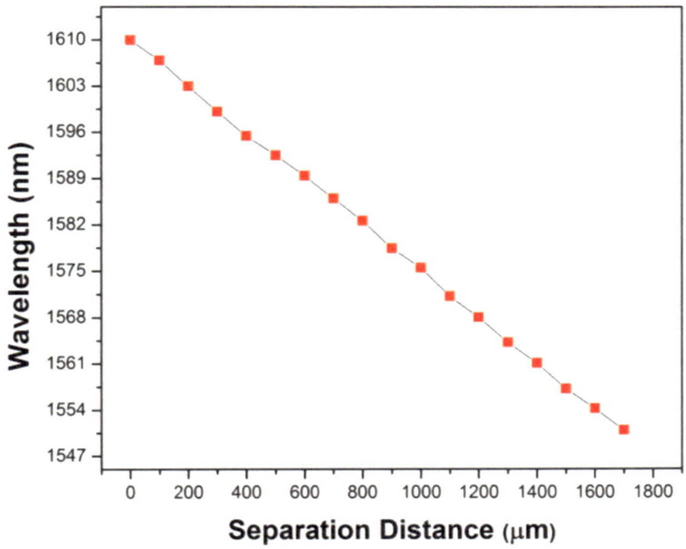


Fig. 21. Separation distance between fibers (single mode fiber and no-core) versus wavelength tuned.

As it can be seen, the total tunable range was limited by 60 nm. It was reported that a possible reason for this is the small mismatch between the diameter of the silica ferrule (127 μm) and that of the no-core fiber (125 μm) and also for the inner ferrule imperfections. So, every time the fibers were moving apart to each other, some kind of tilting between the facets end-fibers could deteriorate the coupling of the self-image to the output SMF. It is understood that a more uniform capillary with a diameter closer to the diameter of the MMF would increase the tuning range.

The complete tunable range is showed in Fig. 22. It is cleared that the peak wavelengths achieved in both tunable set ups differs. It is assumed that the difference on the index matching fluids affects directly in the optical spectrum wavelength as well as on the total tunable range because of the reduction of the loss inside of the liquid waveguide.

The total insertion loss was of 0.64 dB. The tunable range covered was of 60 nm which was from 1549 nm to 1609 nm. The measured laser linewidth was of 0.4 nm with a signal-to-noise ratio (SNR) of about 40 dB.

Since the wavelength peak response is a function of the MMF length, it could be possible to apply this system to other kind of rare-earth doped fiber laser in order to develop different wavelength ranges. Also, it is possible to apply this filter to high power application since the index matching liquids have boiling points of at least 100°C or implement it together with a master oscillator power amplifier (MOPA) configuration.

### 3.5 Numerical model for a high power erbium doped fiber laser

It was compute the evolution of a signal, pump, backward and forward ASE (Amplified Spontaneous emission) in a fiber laser in a model that takes into account the modal shape of the radiation fields, the spectral shape of cross sections and of forward and backward propagation ASE, pump power depletion, and saturations due to ASE and or signal. The

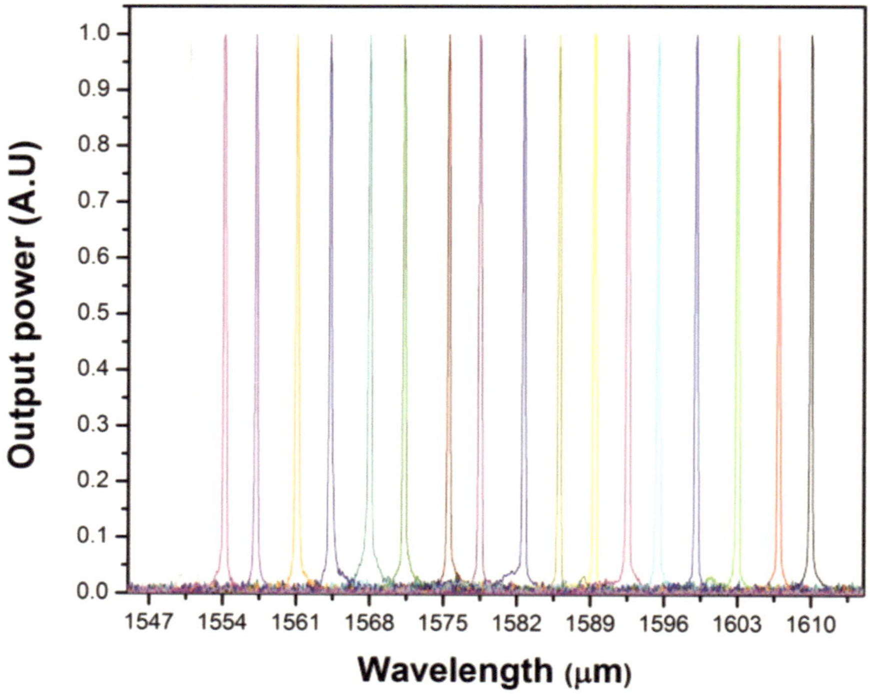


Fig. 22. Total tunable range of an EDFL based on the MMI effect.

model can be easily extended to different rare-earth fiber laser configuration subject to some constrains, since the physics remains the same in many cases. It is used the spectroscopy data of an erbium doped fiber from a published paper.

Rare-earth doped fibers offer many advantages for example: they are lightweight and flexible and can be coiled so that compactly packaged and robust devices can be carried out; they are well-controlled beam shape, owing to the confinement of light in the core of the fiber; the thermal management is simple thanks to the extended length, large surface area, and small transverse dimension of the fiber (Ueda et al., 2002). It can also find the rare-earth fibers good devices offering a high gain, and the use of a glass host leads to a wide bandwidth, which makes them suitable for amplifiers as well as for widely tunable lasers. Other attractive features of fibers lasers include narrow linewidth (with wavelength-selection), efficiency, stability, reliability, and temporal characteristics ranging from CW to femtosecond pulses. All these attractions make fiber lasers very useful in areas such as thermal printing, drilling, welding, cutting, material processing, nonlinear frequency conversion, remote pumping of EDFAs (Erbium Doped fiber Amplifiers), range finding, defense, aerospace, and medicine (Anthon et al., 2001; Jackson & Lauto, 2002). For those reasons, the rare-earth modeling is acquiring so much importance now in our days. This work is focus in the modeling of an erbium doped fibers seeded with a tunable fiber source based on the multimode interference effect.

### 3.5.1 The model

The equations do need to be solved numerically. Assuming again that the pump and the signal light are co-propagating, the numerical solution proceeds as follows: First, the (discretized) forward-travelling ASE spectrum and the pump and signal powers are propagated forward according to Eqs. (1) - (6) with $S-ASE = 0$ in Eq. (5).

$$\frac{dP(z,v)}{dz} = g_m(z,v)P(z,v) \tag{8}$$

$$g_m(v) = \int_0^{\infty} \Psi(r,v)\left[N_2(r,z)\left(\sigma_e(v) - \sigma_{ESA}(v)\right) - N_1(r,z)\sigma_a(v)\right]2\pi r dr \tag{9}$$

$$\gamma_e(z,v) = \sigma_e(v)\int \Psi(r,v)N_2(r,z)2\pi r dr \tag{10}$$

$$\frac{dS_{ASE}^{+}(z,v)}{d(\pm z)} = g_m(z,v)S_{ASE}^{\pm}(z,v) + 2hv\gamma_e(z,v) \tag{11}$$

$$\begin{aligned} R(r,z) &= \frac{P_p(z)\Psi(r,v_p)\sigma_a(v)}{hv_p} + \frac{P_p(z)\Psi(r,v_s)\sigma_a(v_s)}{hv_s} \\ &+ \int_0^{\infty} \Psi(r,v)\frac{S_{ASE}^{+}(z,v) + S_{ASE}^{-}(z,v)}{hv}\sigma_a(v)dv \end{aligned} \tag{12}$$

$$n_2 = \frac{R}{(R+W+A)} \tag{13}$$

The equations are integrated numerically, e.g. with a Runge-Kutta method. At the same time, the population inversion throughout the fiber is calculated. The resulting forward-travelling light powers are then used in Eq. (12) when Eq. (11) is integrated backward to yield the backward-travelling ASE spectrum. In the process, the population inversion is recalculated. Appropriate boundary conditions (determined by the endreflectors) are used to couple forward and backward-propagating light fields to each other. Then, Eq. (8) and Eq. (11) are integrated forward again, now including the *S- ASE(z,v)* just found in Eq. (12). This process is repeated until convergence is reach. Other methods for solving Eqs. (8 and 12) exist, too. In the case of a laser, no signal is injected; instead, the lasing field builds up from spontaneous emission.

### 3.5.2 Results and discussion

Erbium-doped fiber laser is modeled as a three level system. It was pumped at 980nm, 10 W diode laser, to amplified light at 1567, that is, it was launched into it a seed signal at 1567nm in order to configure our experiment that would consists of a tunable erbium fiber laser feeding a double-clad erbium-doped fiber amplifier in a MOPA configuration and thus realizes a high power system together with our previous results of a MMI-effect-widely-tunable-fiber laser. Double-cladding-pumping has revolutionized fiber lasers over the last decades. The perspective of that a single-mode diodes are limited in power to a few hundred miliwatts changed with the breakthrough idea, referred to as cladding-pumping (using double-clad fibers), patented by James Kafka at Spectra Physics ( Kafka, 1989). A double-clad fiber, indeed, enables a good match with the output beam from broad-stripe diode. These are multi-mode devices that can generate much higher powers than single-mode ones. Thus, with cladding-pumping, a double-clad fiber laser can produce much higher output power, in a beam that nevertheless can be diffraction-limited. Cladding-pumped fiber lasers are currently considered for many applications, where high laser powers are required.

In Fig. 23 is plotted the values of the laser signal calculated in this fashion versus absorbed pump power. The system achieved a total output power of 2.64 Watts and with a threshold absorbed power of 0.3W. We obtained a slope efficiency of around 30%. The dopant concentration for our power regime was 300ppm wt% of erbium with a total fiber length of 10 m. Some parameters of the double cladding optical fiber were an inner cladding diameter of 30 μm, and out cladding diameter of 125 μm, a NA of 0.1 and a radius of $1.8 \times 10^{-6}$ m. The model utilized can be easily extended and complete the all widely tunable seed signals (1549 nm to 1609 nm) for our double-clad erbium fiber laser simulated configuration, since the optics remains the same.

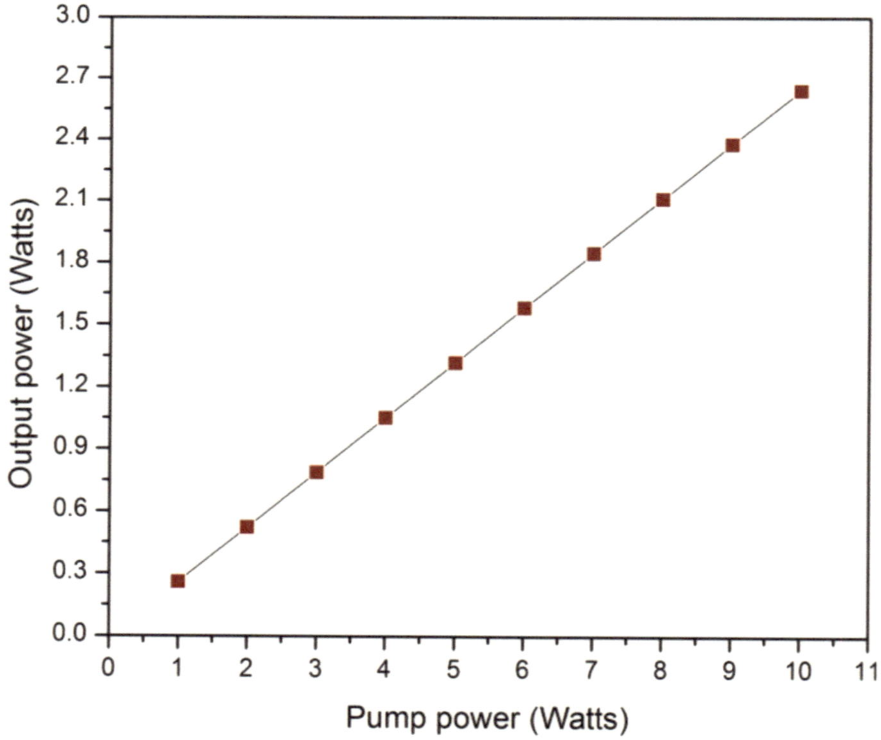


Fig. 23. Erbium results, output power versus pump power.

The results obtained in this part, were compared with other rare-earth doped-fiber simulations and it was realized that this work is highly reliable. It was utilized some spectroscopy data of important fiber fabrication companies as a base like the diameters of the fiber, indexes of refraction, pump wavelength, etc; that guarantee even more our work.

## 4. Conclusion

It has been demonstrated tunable rare-earth doped fiber lasers which based their tunable system in the multimode interference effect that occurs on the optical fibers. Different arrangements were used in order to apply this effect. Firstly, it was used a mirror and a facet fiber with a gold film as a reflecting device to induce the light back into the system and selecting the wavelength to tune by moving away the reflecting device from the final facet of the multimode fiber. The tunable range achieved with these optical lasers was as many as 8 nm and 12 nm respectively.

The following setup presented shows to be the most promising device, and it consisted in a device that works as a liquid waveguide inside of a ferrule filled it with an index matching fluid for a longer tunable range and for conserving the light propagation in one axis. The highlighted results showed a total tunable range of 60 nm which goes from 1549 nm to 1609 nm. The measured laser linewidth was of 0.4 nm with a signal-to-noise ratio (SNR) of about 40 dB.

In general this device is inexpensive and portable due to the simplicity of its elements. Also, these types of lasers are well applied to a variety of technologies such as optical communications and medicine.

It is also concluded by a simulation that a high power fiber laser that uses our optical filter as tunable device is possible to build up. An output power of 2.64 Watts was achieved within the optical range tuned of 60 nm. This device is very likely to scale even to higher powers along with different rare earth doped fiber lasers.

## 5. Acknowledgments

Special thanks to the National Mexican Council of Science and Technology (CONACYT), the Publical and Educational Secretary (SEP) and Physical and Mathematical Sciences Research Center (CICFIM) UANL-FCFM for all the support in the last years.

## 6. Nomenclature

SOA Semiconductor Optical Amplifier

MMI Multimode Interference

MOPA Master Oscillator Power Amplifier

EDFL Erbium Doped fiber Laser

DCEDF Double Cladding Erbium Doped Fiber

MMF Multimode Fiber

| | |
|---|---|
| SMF | Singlemode fiber |
| DCYDF | Double Cladding Ytterbium Doped Fiber |
| AR | Anti-Reflection Coating |
| YDFL | Yterbium Doped Fiber Laser |
| DCYFL | Double Cladding Yterbium Fiber Laser |
| SU-8 | SU-8 photoresist |
| TEDFL | Tunable Erbium Doped Fiber Laser |
| MMW | Multimode Waveguide |
| FWHM | Full Width at Half-Maximum |
| SNR | Signal to Noise |
| ASE | Amplified Spontaneous Emission |
| EDFA | Erbium Doped Fiber Amplifier |
| NA | Numerical Aperture |

## 7. References

Chen, H., Babin, F., Leblanc, M., He, G. & Schinn G. W. (2003). 70-Nm Tunable Single-Longitudinal Mode Erbium-Doped Fiber Laser, *Proc. SPIE,* Vol. 4833, No. 956.

Xia, L., Shum, P., Wang, Y. X. & Cheng, T. H. (2006). Stable Triple-Wavelength Fiber Ring Laser With Ultranarrow Wavelength Spacing Using A Triple-Transmission-Band Fiber Bragg Grating Filter, *IEEE Photon. Technol. Lett.,* Vol. 18. No. 20, pp. 2162-2164.

Goh, C. S., Mokhtar, M.R., Butler, S.A., Set, S.Y., Kikuchi, K., & Ibsen, M. (2003). Wavelength Tuning Of Fiber Bragg Grating Over 90nm Using A Simple Tuning Package, *IEEE Photon. Technol. Lett.*, Vol. 15, No. 4, pp. 557-559.

Mokhtar, M. R., Goh, C.S., Butler, S.A., Set, S.Y., Kikuchi, K., Richardson, D.J & Ibser, M. (2003). Fiber Bragg Grating Compression-Tuned Over 110nm, *Electron. Lett.,* Vol. 39, No. 6, pp. 509-511.

Zhang, Z., Wu, J., Xiu, K., Hong, X. & J. Lin. (2009). Tunable Multiwavelength SOA Fiber Laser With Ultra-Narrow Wavelength Spacing Based On Nonlinear Polarization Rotation, *Opt. Express*, Vol. 17, No. 17, pp. 200-205.

Chawki, M. J., Valiente, I., Auffret, R. & Tholey, V. (1993). All Fibre, 1.5 Mu M Widely Tunable Single Frequency And Narrow Linewidth Semiconductor Ring Laser With Fibre Fabry Perot Filter, *Electron. Lett.,* Vol. 42, No. 23, pp. 2034 - 2035.

Soldano, L., Veerman, F., Smit, M., Verbeek, B., Dubost, A. & Pennings, E. (1992). Planar Monomode Optical Couplers Based on Multimode Interference Effects, *Journal of Ligthwave Technology*, Vol. 10, No.12, pp. 1843-1850.

Mohammed, W., Mehta, A. & Johnson, E. (2004). Wavelength Tunable Fiber Lens Based on Multimode Interference, *Journal of Ligthwave Technology*, Vol. 22, No.2, pp. 469-477.

Soldano, L. & Pennings E. (1995). Optical Multi-Mode Interference Devices Based on Self-Imaging: Principles and Applications, *Journal of Ligthwave Technology*, Vol. 13, No.4, pp. 615-627.

Li, B., Chua, S., Leitz, C. & Fitzgerald, E. (2002). 1x2 Optical Waveguide Filters Based on Multimode Interference for 1.3- and 1.55-μm Operation, *Optical Engineering*, Vol. 41, pp. 723.

Selvas, R., Torres-Gomez, I., Martinez-Rios, A., Alvarez-Chavez, J.; May-Arrioja, D., LiKamWa, P., Mehta, A. & Johnson, E. (2005). Wavelength Tuning of Fiber Lasers Using Multimode Interference Effects, *Optics Express*, Vol. 13, No. 23, pp. 9439-9445.

Anzueto-Sánchez, G., Martínez-Ríos, A., May-Arrioja, D., Torres-Gómez, I., Selvas-Aguilar, R. & Alvárez-Chávez, J. (2006). Enhanced Tuning Mechanism in Fiber Laser Based on Multimode Interference Effects, *Electronic Letters*, Vol. 42, No. 23, pp. 1337-1338.

Mehta, A., Mohammed, W. & Johnson, E. (2003). Multimode Interference-Based Fiber-Optic Displacement Sensor, *Photonic Technology Letters*, Vol. 15, No. 8, pp. 1129-1131.

Poustie, J., Finlayson, N. & P. Harper. (1994). Multiwavelength Fiber Laser Using A Spatial Mode Beating Filter, *Opt. Lett.*, Vol. 19, pp. 716-718.

Wang, Q. & Farrell, G. (2006). All-Fiber Multimode-Interference-Based Refractometer Sensor: Proposal And Design, *Opt. Lett.* Vol. 31, pp. 317-319.

Li, Q., Lin, C., Tseng, P. & Lee, H. P. (2005). Demonstration Of High Extinction Ratio Modal Interference In A Two-Mode Fiber And Its Application For All-Fiber Comb Filter And High-Temprerature Sensor, *Opt. commun.*, Vol. 250, pp. 280-285.

Martinez-Rios, A., Starodumov, A., Po, H., Wang, Y. & Demidov, A. (2003). Efficient Operation of Double-Clad Yb-Doped Fiber Lasers with a Novel Circular Cladding Geometry, *Optics Letters*, Vol. 28, pp. 1642-1644.

Raymond, C. & Johnson, E. (2004). Micro-Photonic Systems Utilizing SU-8. *Proceedings of SPIE 2004 MOEMS and Miniaturized Systems IV*, pp. 64–69.

Antonio-Lopez J., Castillo-Guzman A., May-Arrioja D., Selvas-Aguilar R. & LiKamWa P. (2010). Tunable Multimode-Interference Bandpass Fiber Filter, *Opt. Lett.* Vol. 35, pp. 324-326.

Mohammed W., Smith P. & Gu X. (2006). All-Fiber Multimode Interference Bandpass Filter, *Opt. Lett.*, Vol. 31, pp. 2547-2549.

Castillo-Guzman, A., Antonio-Lopez, J., Selvas-Aguilar, R., May-Arrioja, D., Estudillo-Ayala, J. & LiKamWa P. (2010). Widely Tunable Erbium-Doped Fiber Laser Based On Multimode Interference Effect, *Opt. Express*, Vol. 18, pp. 591-597.

Ueda, K.I., Sekiguchi, H. & Kan H. (2002). 1 Kw Cw Output From Fiber Embedded Lasers, *in Proc. Conference on Lasers and Electro-Optics*, Vol. CPDC4.

Anthon, D., Fisher, J., Keur, M., Sweeney, K., Ott, D., Matton, P. & Emslie, C. (2001). High Power Optical Amplifiers For CATV Applications, *Optical Fiber Communication Conference*, Vol. 2.

Jackson, S. D & Lauto, A. (2002). Diode-Pumped Fiber Lasers: A New Clinical Tool, *Lasers in surgery and Medicine*, Vol. 30, pp. 184-190.

Kafka J. (1989). Laser Diode Pumped Fiber Laser With Pump Cavity. US Patent 4, 829.

# 6

# Design of High Performance and Low-Cost Single Longitudinal Mode Laser Module for DWDM Application

Huei-Min Yang
*I-Shou University/Communication Engineering Department*
*Taiwan*

## 1. Introduction

Over the past years, the internet core rapidly increases, transparent protocols become more available for DWDM system in the fiber to the home, fiber to the institutions service. In densely spaced wavelength-routed and large-capacity optical network systems, single-mode lasers with narrow linewidth and low temperature sensitivity are of particular importance as the light sources in transmitter modules. The efficient coupling of a laser diode to an optical fiber has been a problem of general concern since the advent of fiber-optic communication systems. This chapter will introduce to fabricate a fiber grating external cavity laser (FGECL) module with a low cost while still maintaining a good single longitudinal mode performance by using a low-cost AR(Anti-reflection)-coated($5\times10^{-3}$) laser and a tapered hyperbolic-end fiber (THEF) microlens for strong coupling the fiber grating external cavity. Previously, high-performance FGECL modules have only been available by using a complicated AR-coated($1\times10^{-5}$) laser process that leads to a high packaging cost. The FGECL consisted of a HR/AR-coated diode laser, an uncoated THEF microlens, and a fiber grating. The results showed that the FGECL module exhibits a side-mode suppression ratio (SMSR) higher than 44 dB, a higher output power of more than 2 mW, and a larger operation current range of over 50 mA. In addition, excellent wavelength stability and a low-penalty directly modulated 2.488 Gbit/s were also obtained for FGECL modules. The THEF microlens demonstrated up to 86% coupling efficiency for a laser with an aspect ratio of 1:1.5. A fiber microlens which provides an efficient coupling mechanism to match the spot size of the diode laser to the fiber is commonly used for optical alignment in FGECL modules. Low-cost FGECL modules with good performance were achieved primarily owing to the THEF microlens having a high coupling efficiency (typically 75%) which enhanced the feedback power from the fiber grating external cavity to the HR/AR-coated laser compared with the currently available hemispherical microlens which has low coupling efficiency (typically 50%). Therefore, the packaged FGECL module is suitable for use in low cost 2.5 and 10 Gbit/s lightwave transmission systems, such as gigabit passive optical networks (GPONs), metropolitan area networks (MANs), and fiber-to-the-home (FTTH) applications.

This chapter is organized as follows:

Part I: A tapered hyperbolic-shaped microlens has been improved for efficient the coupling of the output from a laser diode into an optical fiber. A tapered hemispherical-end fibermicrolens is also evaluated for comparison purposes. The Fresnel diffraction theory is used to evaluate laser to fiber coupling using microlenses. Experimental results demonstrate that, for an elliptical Gaussian laser diode with an aspect ratio of 1:1.5, the coupling efficiencies for the hyperbolic-shaped microlenses and the hemispherical-end fiber microlenses are 87% and 62%, respectively. Calculation results show excellent agreement with the measurements. The research results illustrate that a hyperbolic-shaped microlens has a much higher coupling efficiency than a hemispherical-end microlens due to efficient mode matching and phase matching. An empirical model and a comparison of coupling efficiency between hyperbolic-end and hemispheric-end microlenses are presented.

Part II describes a new scheme of the tapered hyperbolic-end fiber (THEF) and experimental setup. A new lensed-fiber scheme for producing a THEF microlens with an efficient coupling between a laser diode and a single-mode fiber. The high-coupling performance of microlenses with a hyperbolic profile was due to the improved mode matching between the laser and the fiber when compared to currently available hemispherical microlenses, The THEF microlens is ideally suited to collect all the available radiation emitting from a laser source. The diffraction theory predicted that near 100% coupling efficiency could be achieved by a THEF with a circular laser beam pattern, based on both the horizontal and vertical far-field angles of 20 degrees.

Part III describes a low-cost FGECL modules with good performance which were achieved primarily owing to the THEF microlenses having a high coupling efficiency (typically 75%). In general, a high-performance FGECL requires a low AR-coated ($1\times10^{-5}$) laser. However, such packages require complicated processes, which make their cost too high. Using a temperature-insensitive and low-chirp fiber Bragg grating reflector as the external cavity for the semiconductor gain element. The possibility fabricating FGECL modules at low cost while still maintaining a good performance by employing a low-cost AR-coated ($5\times10^{-3}$) laser and a THEF microlens can be demonstrated in the part III. The fiber grating external cavity lasers (FGECLs) have been developed for 2.5 Gbit/s WDM with a low bit error rate penalty. It has been shown that the side-mode suppression ratio (SMSR) of the non-AR-coated FGECL exhibits a current-dependent oscillation.

## 2. Theoretical fundamentals between a tapered hyperbolic-shaped microlens and a hemispherical-end fiber microlens

A hyperbolic-shaped microlens, the shape of a perfect microlens was modeled in the shape of a hyperbolic curve, which was different from the previous studies. The hyperbolic-shaped microlens of this study transforms accurately the incident Gaussian pattern wave into a plane wave according to geometric optics. The novel hyperbolic-shaped microlenses were fabricated by symmetrically tapering the fibers during the unique etching process and hyperbolically lensing the tips during the fusing process. In comparison with other techniques of making hyperbolic-shaped microlenses, the fabrication process of our tapered hyperbolic-shaped fiber microlens is reproducible and suitable for high-volume production.

### 2.1 A model of the hemispherical-end microlens

Fig. 1(a) shows the hemispherical-end microlens with a radius of curvature of 9.4 μm. The equation for the hemispherical-end microlens can be expressed as Eq. (1)

$$(R_\ell - z)^2 + r^2 = R_\ell{}^2 \tag{1}$$

where $z$ is the axial coordinate along the optical axis of the lens (the direction of traveling wave), r is perpendicular to the optical axis, $R_\ell$ is one radius of curvature of the hemispherical-end microlens, as shown in Fig. 1(b). The phase transformations of the hemispherical-end microlens can be expressed as Eq. (2)

$$t_\ell(r) = \exp\left[-i \cdot \kappa \cdot (n-1) \cdot \left(R_\ell - \sqrt{R_\ell^{\,2} - r^2}\right)\right] \tag{2}$$

where $n$ is the refractive index of the fiber core and $\kappa$ represents the wave number of a propagating wave, given by $2\pi/\lambda$. The Gaussian laser beam entering into the mode field of the hemispherical-end microlens can be stated as Eq. (3)

$$U_n(x,y,z) = U_{sx}(x,z) \cdot U_{sy}(y,z) \cdot t_\ell(r) \tag{3}$$

with Eqs. (1), (2) and (3), we can calculate the coupling efficiency of an elliptical Gaussian mode laser to the hemispherical-end microlens.

### 2.2 A model of the hyperbolic-shape microlens

The lens shape of a perfect the hyperbolic-shape microlens was modeled by a hyperbolic curve. Fig. 2(a) shows the hyperbolic-shape microlens with a radius of curvature of 9.4 μm and the simulated hyperbolic curve from the enlarged tip. The feature of the hyperpolic-shape microlens has a smaller taper angle than the hemispherical-end microlens and a larger tilt taper angle of the lens tip of the hyperpolic-shape. The hyperbolic equation can be expressed as

$$\frac{z^2}{a^2} - \frac{r^2}{b^2} = 1 \tag{4}$$

This formula was derived from the fitted lens radius, $R_\ell$, within the core region and the tilt angle, $\theta_a$, of the asymptotic line with respect to the fiber axis, where z is the axial coordinate along the optical axis of the lens (i.e. the direction of traveling wave), r is perpendicular to the optical axis and is equal to the square root of $x^2 + y^2$, as shown in Fig. 2(b).

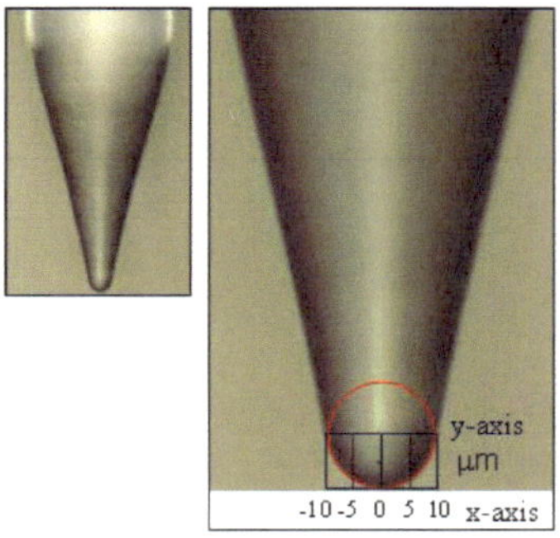


(a)

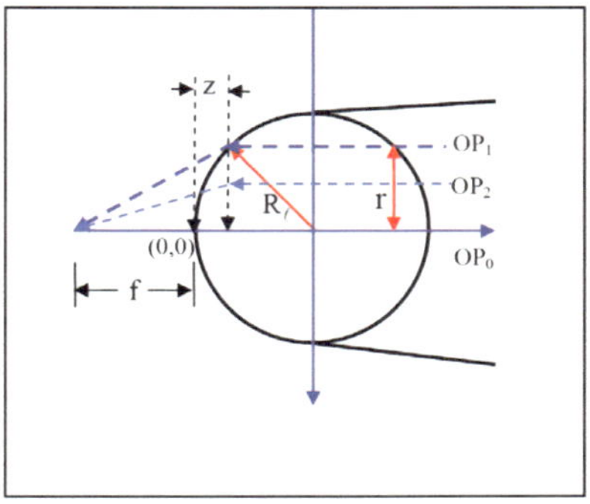


(b)

Fig. 1. (a) Hemispherical-end microlens with a radius of curvature of 9.4 μm, right figure is a simulated hemispherical curve from the enlarged tip. (b) $z$ is the axial coordinate along the optical axis of the lens (the direction of traveling wave), and $r$ is perpendicular to the optical axis.

We assume that the circle was fitted perfectly within the fiber core region. The repeatability of the measurement is within 1μm. Hence, the center point, (h, 0), of the circle can be calculated with the following equation,

$$h(a) = r_f^2 \cdot \frac{\left[1/\tan(\theta_a)^2\right] + 1}{2 \cdot \left[\sqrt{\left[r_f^2/\tan(\theta_a)\right]^2 + a^2} - a\right]} \tag{5}$$

where $r_f$ is the fiber core radius. Since the following equation needs to be satisfied,

$$R_\ell + a - h(a) = 0 \tag{6}$$

$R_\ell$ is one radius of curvature of the hyperbolic-shape microlens, the value of $a$ can be solved in terms of the measured parameters. The value of $b$ is calculated from the asymptotic relationship,

$$b = a \cdot \tan(\theta_a) \tag{7}$$

The phase transformations for the hyperbolic lens can be represented by the following equations (8) & (9)

$$t_{\ell x}(x) = \exp\left[-i \cdot \kappa \cdot (n-1) \cdot \left((a/b) \cdot \sqrt{b^2 + (x + d_x)^2} - a\right)\right] \tag{8}$$

$$t_{\ell y}(y) = \exp\left[-i \cdot \kappa \cdot (n-1) \cdot \left((a/b) \cdot \sqrt{b^2 + (y + d_y)^2} - a\right)\right] \tag{9}$$

where $n$ is the refractive index of the fiber core, and $d_x$ and $d_y$ represent the deviation (offset) of the lens from the fiber center axis in the horizontal and vertical directions, respectively. The mode field after the hyperbolic-shape microlens can be expressed as

$$U_n(x,y,z) = U_{sx}(x,z) \cdot U_{sy}(y,z) \cdot t_{\ell x}(x) \cdot t_{\ell y}(y) \tag{10}$$

### 2.3 Experiments and analysis of transmission properties

The laser-to-fiber coupling was modeled based on the Fresnel diffraction theory. From the elliptical Gaussian laser source to the fiber microlens, which used Fresnel diffraction for beam propagation through the free space. At the lens tip, a phase delay caused by the fiber microlens was added to the electric field. Finally, an overlap integral between the transformed field and the fiber mode field was calculated to obtain a coupling efficiency.

The mode field patterns for both the laser and fiber were assumed to be Gaussian and the 1550 nm laser diode had an elliptical emission pattern. The measured full-width-half-maximum (FWHM) far-field angles along the horizontal and vertical directions, $\theta_x$ and $\theta_y$, were used to calculate the laser mode-field radii at the front facet, $\omega_{ox}$ and $\omega_{oy}$, respectively. The relationship between the far-field angle and the mode-field radius at the waist can be derived as

$$\omega_{ox} = \left(\sqrt{\ln(2)/2}\right)\lambda/\left(\pi \cdot \tan(\theta_x/2)\right) \tag{11}$$

$$\omega_{oy} = \left(\sqrt{\ln(2)/2}\right)\lambda\Big/\left(\pi \cdot \tan\left(\theta_y/2\right)\right) \tag{12}$$

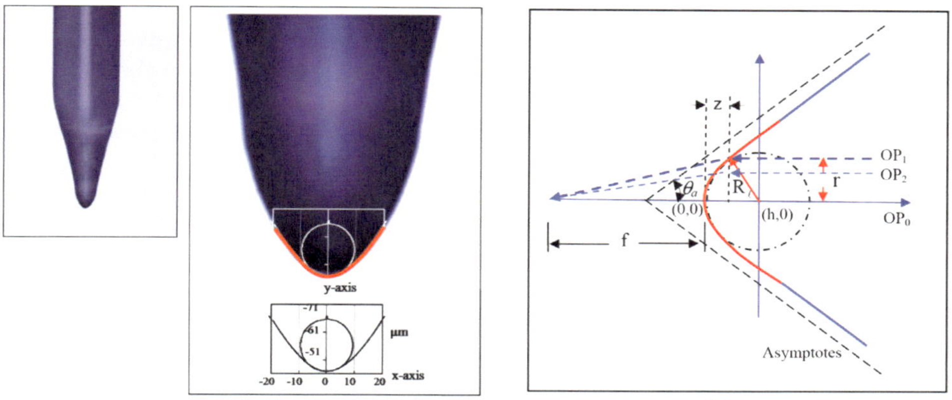


(a) (b)

Fig. 2. (a) Enlarged tip of hyperbolic-shape microlens with a radius of curvature of 9.4 μm. (b) The circle was fitted perfectly within the fiber core region, z is the axial coordinate along the optical axis of the lens, r is perpendicular to the optical axis, the tilt angle of the asymptotic line with respect to the fiber axis with $\theta_a$.

The laser mode fields at a distance Z from the facet were thus represented by the Fresnel diffraction theory,

$$U_{sx}(x,z) = \int_{-\infty}^{\infty} \exp\left[-\left(x_1/\omega_{ox}\right)^2 + i\cdot(\kappa/2\cdot z)\cdot(x-x_1)^2\right]dx_1 \cdot \sqrt{\exp(i\cdot\kappa\cdot z)/(i\cdot\lambda\cdot z)} \tag{13}$$

$$U_{sy}(y,z) = \int_{-\infty}^{\infty} \exp\left[-\left(y_1/\omega_{oy}\right)^2 + i\cdot(\kappa/2\cdot z)\cdot(y-y_1)^2\right]dy_1 \cdot \sqrt{\exp(i\cdot\kappa\cdot z)/(i\cdot\lambda\cdot z)} \tag{14}$$

where $\lambda$ is the wavelength, and $\kappa$ is the wave number, given by $2\pi/\lambda$. Equations (13) and (14) do not include the astigmatism for the laser diode. This is the case for the well-designed index-guided lasers. Finally, the coupling efficiency is calculated from the overlap integral of this new mode field with the fiber fundamental mode field, $U_f(x, y)$,

$$\eta(z) = \frac{\left|\int_{-\infty}^{\infty}\int_{-\infty}^{\infty} U_n(x,y,z)\cdot U_f^*(x,y)\,dxdy\right|^2}{\int_{-\infty}^{\infty}\int_{-\infty}^{\infty} U_n(x,y,z)\cdot U_n^*(x,y,z)\,dxdy \cdot \int_{-\infty}^{\infty}\int_{-\infty}^{\infty} U_f(x,y)\cdot U_f^*(x,y)\,dxdy} \tag{15}$$

For a fixed fiber lens radius, the coupling efficiency between the laser diode and the fiber was calculated. The hyperbolic-shape microlenses had a much higher coupling efficiency than the hemispherical-end microlenses for an elliptical Gaussian laser with an aspect ratio of 1:1.5. An experimental comparison of the hyperbolic-shape microlenses has demonstrated a coupling efficiency of up to 86.8% while the hemispherical-end fiber microlenses with taper asymmetry have demonstrated imperfect coupling of at best

61.5%. An elliptical Gaussian laser diode with an aspect ratio of 1:1.5, a widely adopted commercial product, has a full-width-half-maximum (FWHM) far-field divergence of 20°×30°(lateral × vertical) at 20°C and 60 mA, both horizontal angle and vertical angle as shown in Figs. 3(a) and (b), respectively. Theoretically, a high coupling efficiency can be obtained under specific conditions, for example, the coupling loss is relative to the radii of curvature of microlenses and laser diodes aspect ratio. The diffraction theory predicted that near 100 % coupling efficiency could be achieved by a hyperbolic-shape microlens with a circular laser beam pattern, based on both the horizontal and vertical far-field angles of 20 degrees.

Fig. 4 shows the coupling efficiency as a function of radius of curvature for a laser diode with an aspect ratio of 1:1.5; the hyperbolic-shape microlenses can reach a 95% coupling efficiency and the hemispherical-end fiber microlenses with a typical coupling efficiency of only 70% for calculation results. The theory does not take into account any astigmatism in the endface of the fiber microlens. There is a difference in optical field distributions between an elliptical Gaussian laser diode and the radius of curvature, not only in the size, but also in the profile. The coupling loss of a spot-size transformation between the tapered hyperbolic-shape fiber lens and an elliptical Gaussian laser diode is extremely reduced by controlling the shape of the hyperbolic lens tapered portion. The hyperbolic-shape microlenses have a higher coupling efficiency due to larger acceptance angles which was the major reason to bring about excellent mode matching, as shown in Fig. 5.

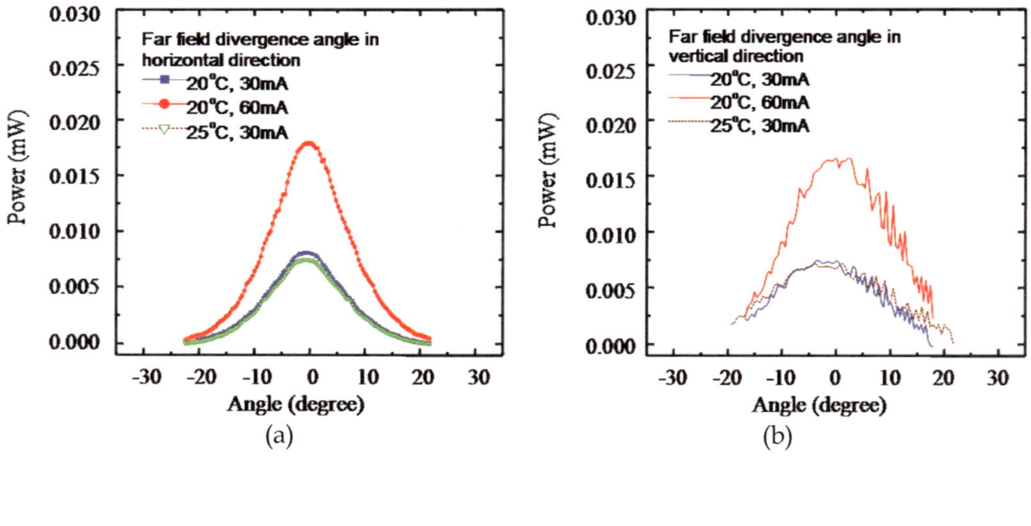


Fig. 3. An elliptical Gaussian laser diode with an aspect ratio of 1:1.5 has a full-width-half-maximum (FWHM) far-field divergence of 20°×30° (lateral × vertical) at 20°C and 60 mA (a) horizontal direction and (b) vertical direction.

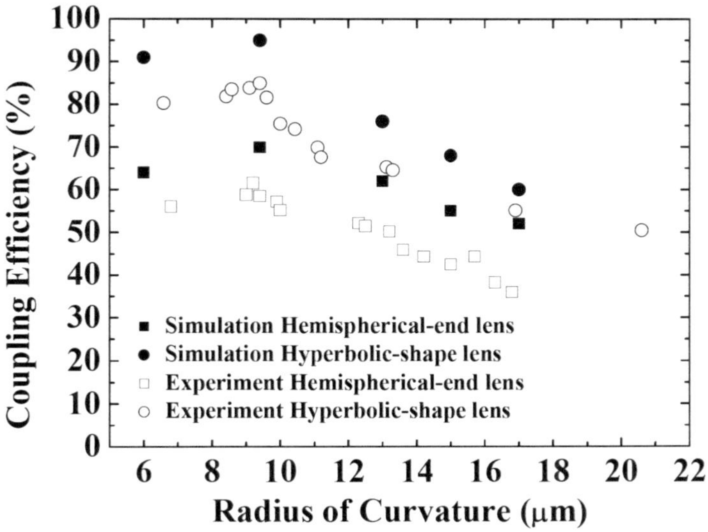


Fig. 4. The coupling efficiency as a function of radius of curvature for a laser diode with an aspect ratio of 1:1.5, the oupling loss is relative to the radii of curvature of microlenses and laser diodes aspect ratio. The hyperbolic-shape microlenses have a much higher coupling efficiency than the hemispherical-end microlenses for experimental and calculated results.

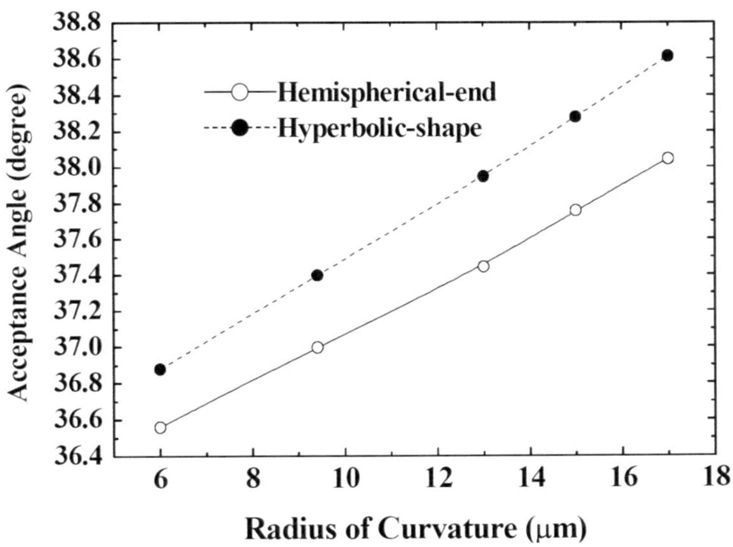


Fig. 5. Hyperbolic-shape microlenses have a higher coupling efficiency due to larger acceptance angles than the hemispherical-end microlenses.

For calculating the intensity and wavefront inside the optical waveguide, we have designed two models of the hyperbolic-shape and the hemispherical-end microlens with Solidworks software, as shown in Fig. 6. Figs. 6(a-1) and (a-2) show the pictures of the hemispherical-end microlens and an enlarged tip, respectively. Figs. 6(b-1) and (b-2) show the hyperbolic-shape microlens and an enlarged tip, respectively. Figs. 7(a) and (b) show the optical intensity and the two-dimensional contour plots at 4.7 μm from the tip of microlens in the optical waveguide of the hemispherical-end microlenses and the hyperbolic-shape microlenses, respectively; both microlenses were of a radius curvature of 9.4 μm and the beam was propagating along the z-direction (the fiber axis). Figs. 8(a) and (b) show the optical intensity and the two-dimensional contour plots at 18.8 μm from the tip of microlens in the optical waveguide of the hemispherical-end microlenses and the hyperbolic-shape microlenses, respectively. Obviously, .the hyperbolic-shape microlens shows a strengthened intensity along the waveguide path after transformation by the microlens. The optical intensity of the hyperbolic-shape microlenses with radii of curvature were of 9.4 μm and 15 μm at 4.7 μm from the tip of microlens in the optical waveguide, the strengths were 0.933 a.u. and 0.789 a.u., respectively, as shown in Figs. 9(a) and (b), respectively. Therefore, a radius of curvature of 9.4 μm is an optimal lens of the hyperbolic-shape microlenses of standard single-mode fibers owing to higher coupling efficiencies after lenses' transformation. The calculation of the effect of the optical coupling of the hyperbolic-shape microlenses based on an empirical model is in agreement with the measured results.

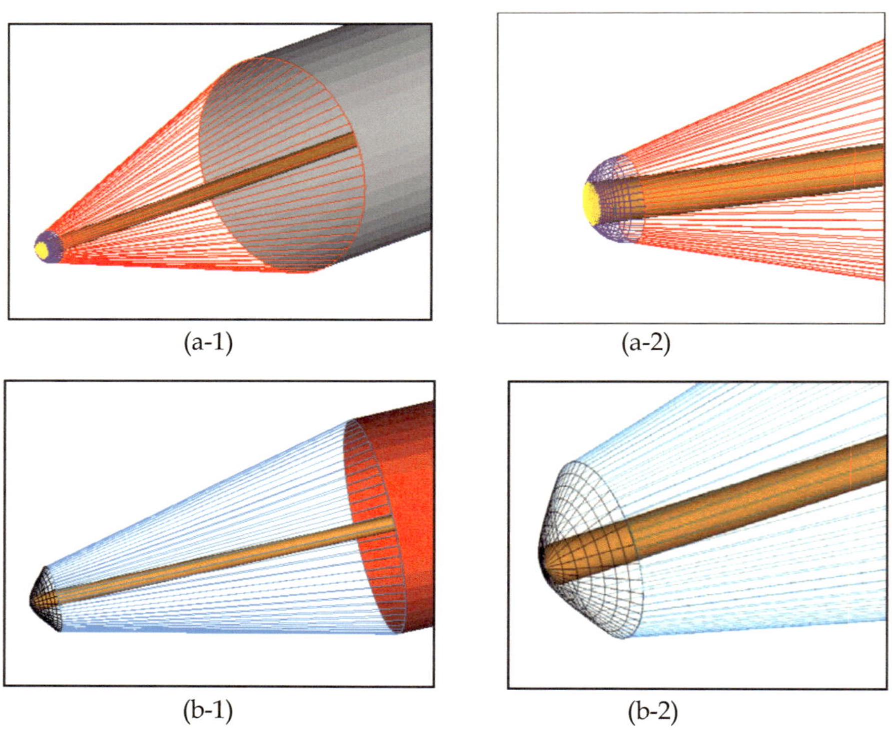


Fig. 6. Two models with Solidworks software, (a-1) hemispherical-end microlens, (a-2) enlarged tip of the hemispherical-end microlens , and (b-1) hyperbolic-shape microlens, (b-2) enlarged tip of the hyperbolic-shape microlens.

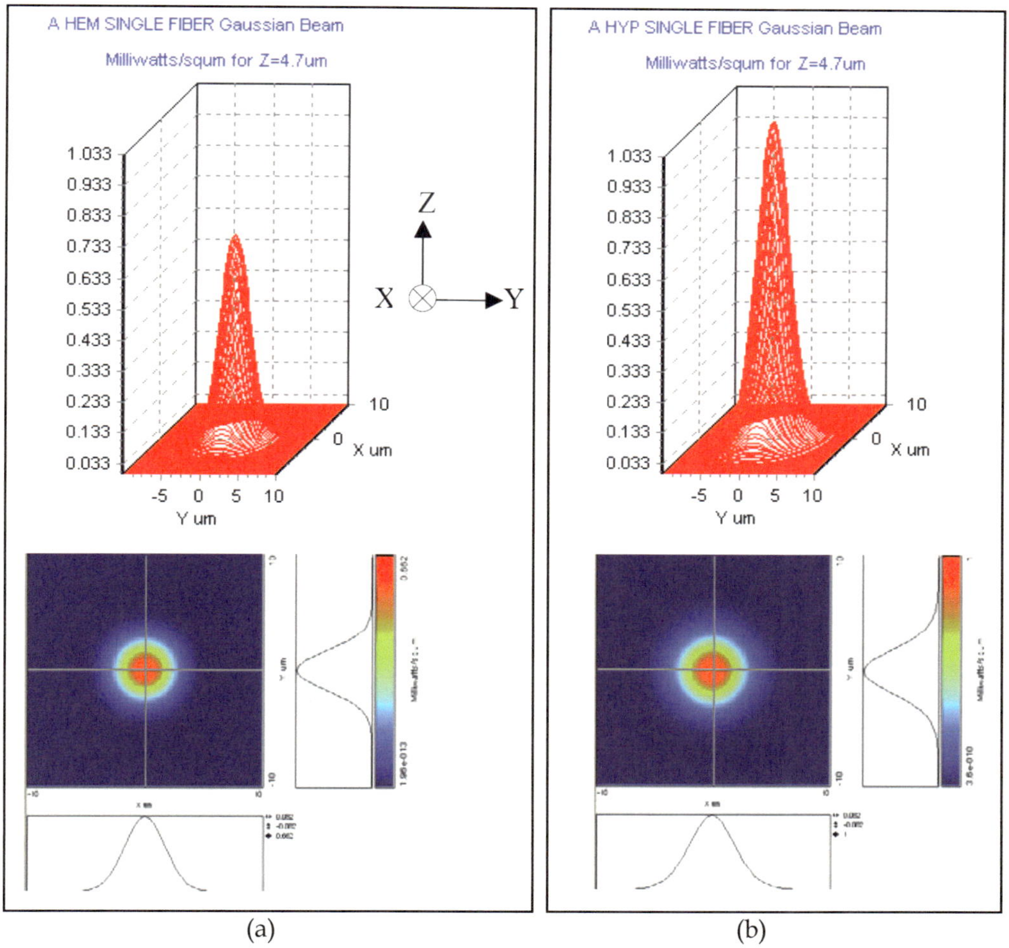


Fig. 7. The optical intensity and the two-dimensional contour plots at 4.7μm from the tip of microlens in the optical waveguide. Both the hyperbolic-shape and the hemispherical-end microlens were of a radius curvature of 9.4 μm, the beam was propagating along the z-direction (the fiber axis). (a) hemispherical-end microlens, and (b) hyperbolic-shape microlens.

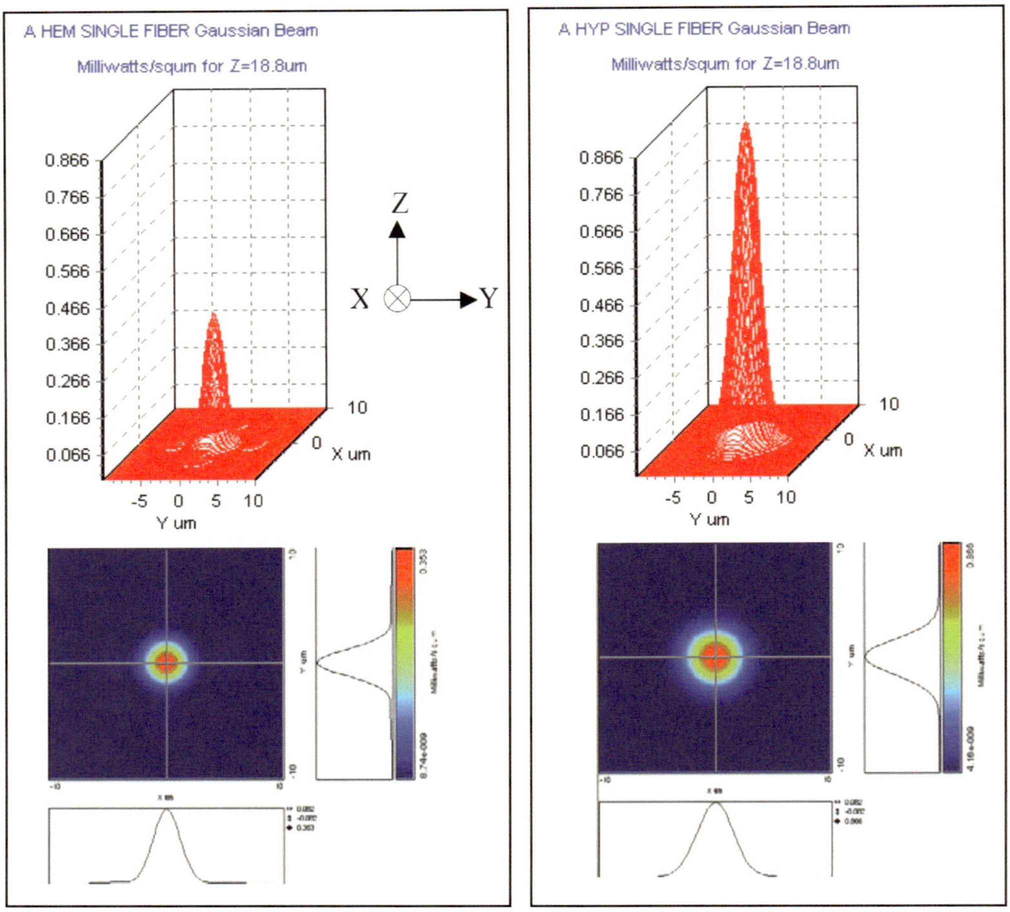


(a) (b)

Fig. 8. The optical intensity and the two-dimensional contour plots at 18.8 μm from the tip of microlens in the optical waveguide. Both the hyperbolic-shape and the hemispherical-end microlens were of a radius curvature of 9.4 μm, the beam was propagating along z-direction (the fiber axis). (a) hemispherical-end microlens, and (b) hyperbolic-shape microlens.

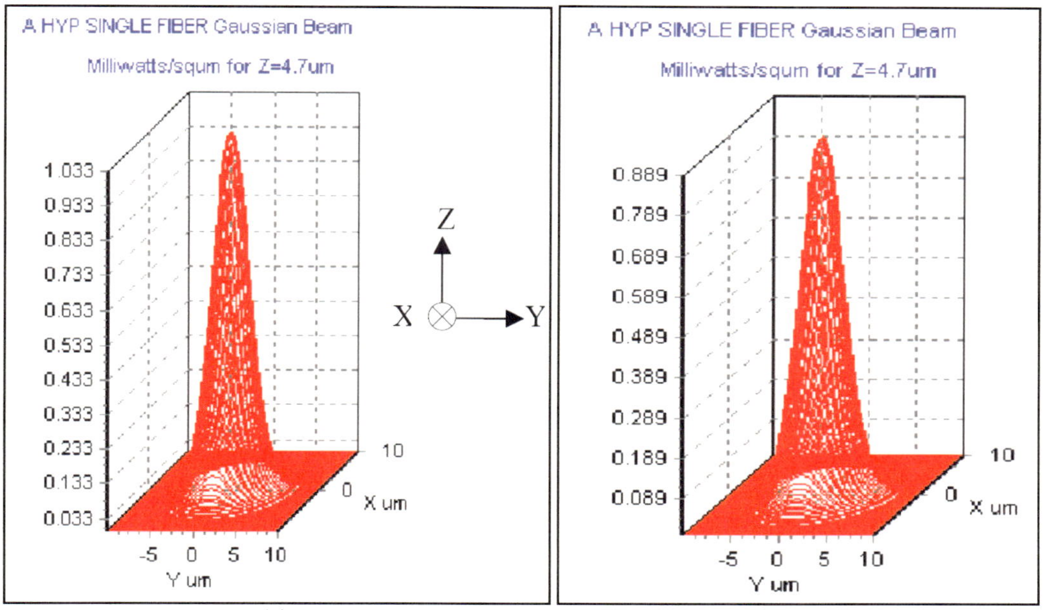


(a) (b)

Fig. 9. Optical intensity of the hyperbolic-shaped microlenses 4.7 μm from the tip of the microlens in the optical waveguide. The strengths were (a) 0.933 a.u. with curvature radius of 9.4 μm and (b) 0.789 a.u. with a curvature radius of 15μm.

Fig. 10(a)–(c) show the variation of wavefronts at the different distances from the tip of microlens in the optical waveguide, of 1.1471, 4.7, 50 μm, respectively. The wavefronts of the hemispherical-end microlenses aren't in accordance with the Gaussian shape in the optical waveguide due to the existence of radiation modes and a greater Fresnel's reflections on the endface of the lens tip. From numerical results with a radius curvature of 9.4 μm, the hyperbolic-shape microlenses have less normalized phase aberration than the hemispherical-end microlenses with wavelength of 1.55 μm, as shown in Fig. 11. It was evidenced in the phase matching after microlens's transformation of the hyperbolic-shape microlens which was not that of the hemispherical-end microlens. From both of the simulations and experiments, the hyperbolic-shape microlenses can be proved with a much better coupling efficiency than the hemispherical-end microlenses, due to phase matching between the Gaussian type laser source and the fiber mode. In addition to the wavefront matching between the propagating laser beam and the fiber mode, the spot-size matching of the Gaussian field distribution is a major reason of improving the coupling efficiency for the hyperbolic-shape microlenses. Compared to the near field intensity patterns, 1 μm of near the tip of lens in free space, both the hyperbolic-shape microlens and the hemispherical-end microlens with a radius curvature of 9.4 μm are shown in Figs. 12(a) and (b). Obviously, the hyperbolic-shape microlenses, as shown in Fig. 12 (b), have much stronger optical intensity and a larger extent of an elliptical field than the hemispherical-end microlenses.

We have compared the optical transformation of the hyperbolic-shaped microlens versus the hemispherical-end microlens for efficient coupling to an elliptical Gaussian laser diode of aspect ratio 1:1.5. The coupling efficiencies for both the tapered hyperbolic-shaped microlens and a tapered hemispherical-end fiber microlens were calculated based on the Fresnel diffraction theory and the results are in good agreement with the measurements. From the theoretical calculations, we show that uncoated tapered hyperbolic-shaped microlenses, affected solely by Fresnel's reflections, suffer less than 0.2 dB loss when coupled to an elliptical Gaussian mode laser of aspect ratio 1:1.5.

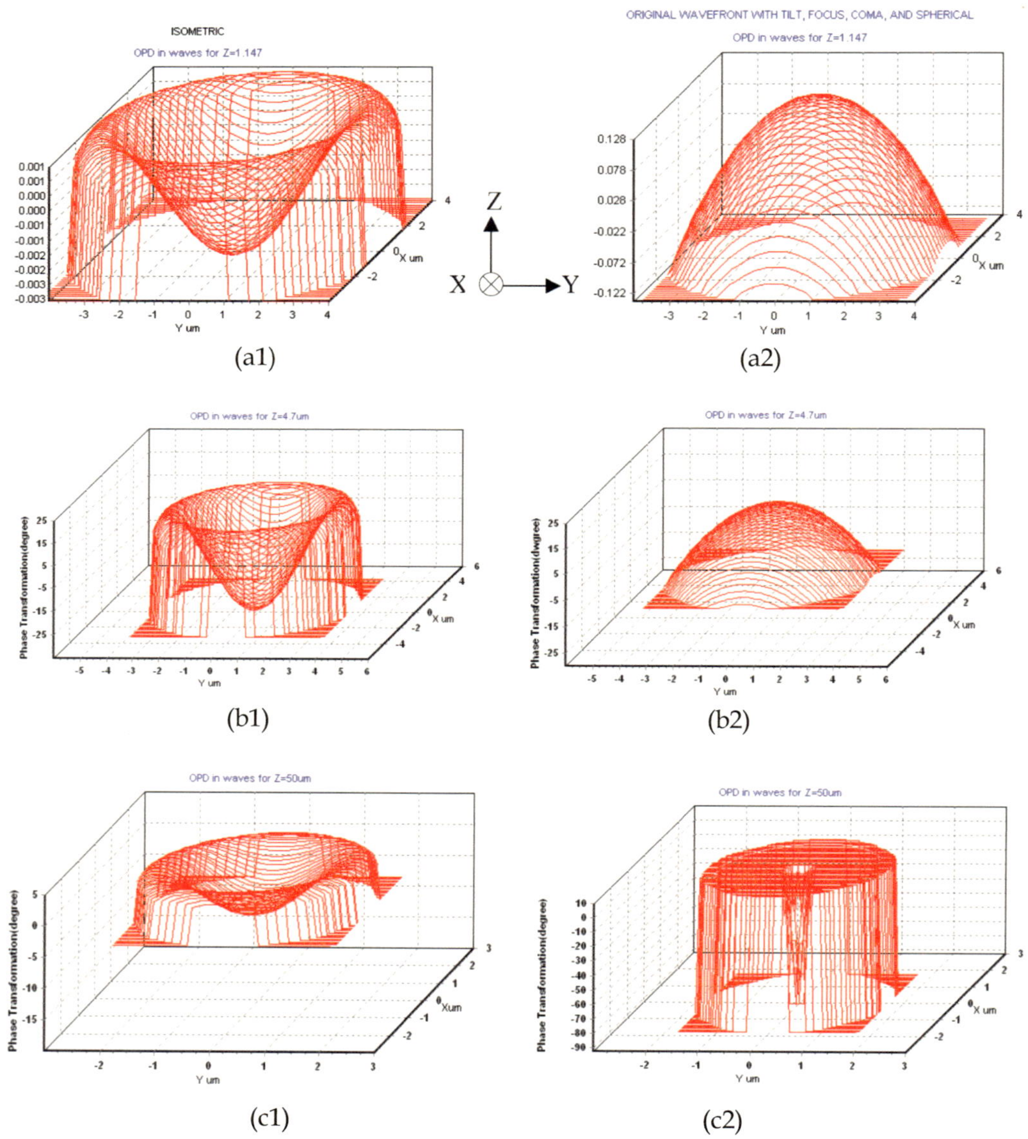


Fig. 10. The variation of wavefronts at the different distances from the tip of microlens in the optical waveguide at (a)1.1471 , (b) 4.7 and (c) 50 μm, respectively. (a-1), (b-1) and (c-1) are for hemispherical-end microlenses. (a-2), (b-2) and (c-2) are for hyperbolic-shape microlenses.

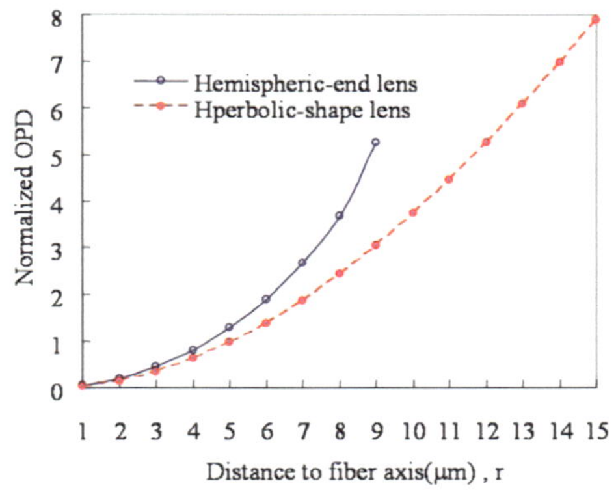


Fig. 11. For the hemispherical-end and the hyperbolic-shaped microlenses of curvature radius 9.4 μm, the hyperbolic-shaped microlenses have a less normalized phase aberration at wavelength 1.55 μm.

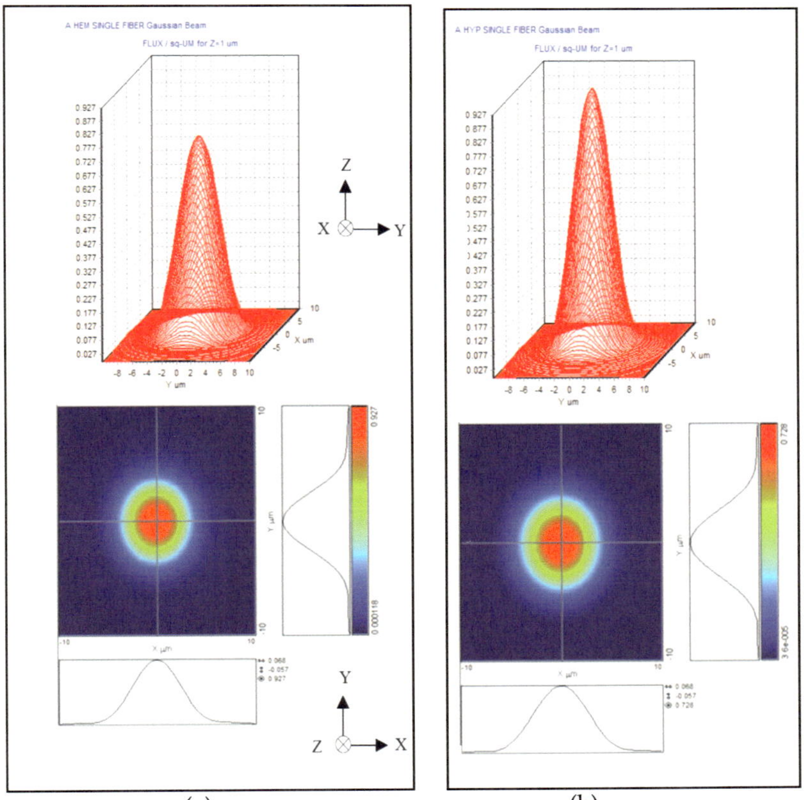


(a) (b)

Fig. 12. Near field intensity patterns for: (a) hemispherical-end microlens and (b) hyperbolic-shaped microlens 1 μm near the tip of the lens. Both the hyperbolic-shaped microlens and the hemispherical-end microlens are of curvature radius 9.4 μm. The hyperbolic-shaped microlens shows a larger extent of the elliptical field.

## 3. Manufacturing process of fiber microlenses

The fibers used in this discuss were Corning step-index single-mode fibers. The glass composition was $SiO_2$-$GeO_2$-$P_2O_5$ for the core and pure $SiO_2$ for the cladding with low impurity content. The fiber core diameter, the beam spot size, and the refractive index difference were 8 μm, 5 μm, and 0.3%, respectively, at a wavelength of 1.55 μm. To obtain efficient coupling between the diode laser and fiber in a previous study, hemispherical fiber lenses with a short taper length and a large taper angle with a small radius of curvature were required. One of the convenient methods of producing short taper lengths and large taper angles in hemispherical fiber lenses has been reported.

In this chapter, the THEF was redesigned with a longer taper length and a smaller taper angle with a small radius of curvature than the hemispherical fiber lens. The THEFs were fabricated by etching the fiber end in a 55% HF solution placed in a teflon beaker, with a thin layer of oil floating on top, as shown in Fig. 13. Fig. 13 is a sketched diagram of the etching process; up to 200 fibers may be etched simultaneously in the experiment. The etching process at room temperature in the HF/oil solution was terminated after 25 min. The oil on top of the HF solution has the effect of reducing the fiber etching above the bath surface from the HF vaporization. The higher-density oil floating on the surface of the HF solution reduces the fiber etching caused by the re-deposition of the evaporated HF molecules. In this process, the longer taper length and the smaller taper angle of the fiber end were achieved by controlling the density of the oil floating on top of the HF solution.

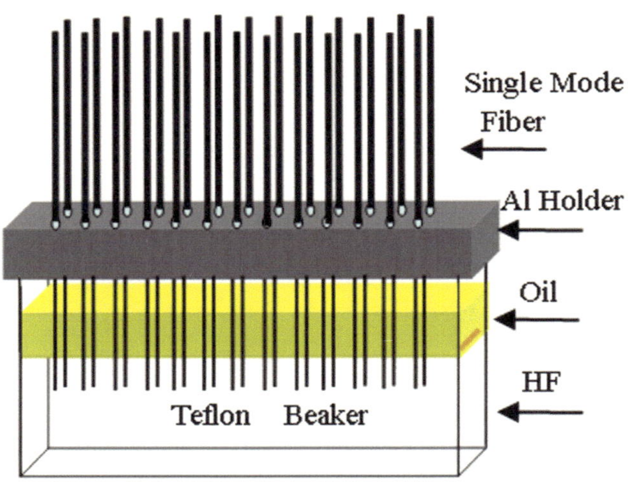


Fig. 13. Setup of the THEF fabrication diagram with the HF solution, oil layer, holder, and the fibers.

The experimental measurements and numerical calculations for the dependence of taper angle on oil density for the THEF are presented. The etching process at room temperature in HF with oil solution was terminated at 25 min. Different densities of oil floating on the HF solution were investigated. The oil density was defined by weight per volume (g/cm3). Fig. 14 shows the taper angle as a function of the oil density. This result shows that the taper angle is dependent on the oil density. The oil floating on the HF with an oil solution with lower density and much etching effect caused by HF deposition due to vaporization exhibits a smaller taper angle. In order to create hyperbolic microlenses for high-coupling efficiency, straight tapers without lateral offsets were required. This was achieved by exactly aligning the fiber vertically with the surface of the HF solution in the fiber holder, as shown in Fig. 13. After the fiber was etched to the desired tapered shape, a hyperbolic microlens for the THEF was formed by heating the fiber tip in a fusion splicer. Only one tapered fiber can be processed with a splicing machine at a time to form a lensed fiber. The duration of the arc discharge and the offset distance between the tapered fiber tip and the arc discharge line were optimized to obtain the proper melting temperatures. In addition, the arcings were performed twice and oriented orthogonally from each other to achieve a symmetrically hyperbolic microlens. A 300-μm-long InGaAsP/InP multi-quantum-well, ridge-waveguide Fabry–Perot laser diode wire bonded on a sub-carrier was used as the gain medium for the fiber gating laser. The FP laser diode, which has a 90% HR coating at the back facet and a low-cost AR-coated front facet of 0.5%, was mounted on a TE cooler to control the substrate temperature. The FP laser beam has a far-field divergence of 20°×30°(lateral × vertical).

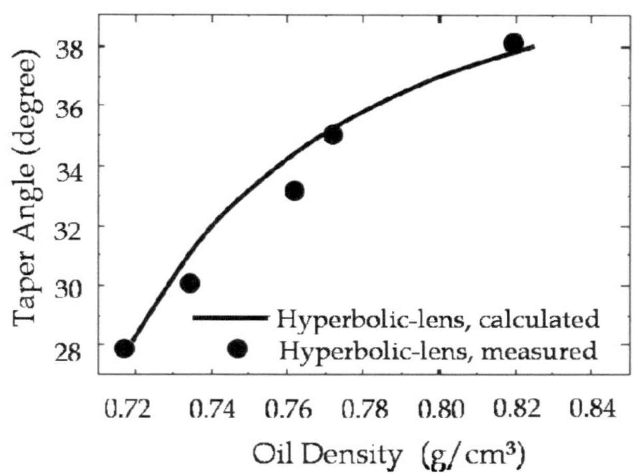


Fig. 14. The taper angle as a function of the local oil density. The taper angle is dependent on the oil density.

### 3.1 Measurements of hyperbolic-end and hemispheric-end microlenses

Fig. 15 shows the hemispherical-end microlens and hyperpolic-end microlens with singular tapered angle by different etching process and fusion parameters. The taper angle of the hemispherical-end microlens and hyperpolic-end microlens are $2\theta_{hs}$ and $2\theta_{hp1}$, respectively. The feature of the hyperpolic-end microlens has a smaller taper angle, i.e. $2\theta_{hp1} < 2\theta_{hs}$, and $2\theta_{hp2}$ larger than $2\theta_{hp1}$ is necessary. In Figs. 15(a) and (b), both microlenses have a radius of curvature of 9.4 μm. The coupling efficiency of the hemispherical-end microlens and hyperpolic-end microlens as a function of the lenses radius of curvature is shown in Fig. 4. The distance *L* between the laser and the fiber lens was optimized at every point for a maximum coupling efficiency. The high coupling performance of microlenses with a hyperbolic profile was due to the improved mode matching and the better wavefront matching between the laser and the fiber compared with the currently available hemispherical microlenses, which have a typical coupling efficiency of 50%. The radius of the hyperbolicend microlens will determine the collimated beam size for mode-size matching, and the hyperbolic shape will remove the hemispherical aberration for wave front matching between the propagating laser beam and the fiber mode. This hyperbolic-end microlens exhibited a better coupling performance compared to other microlenses fabricated by chemical etching and drawn-tapered fiber techniques, which showed a maximum coupling efficiency of 65% for a laser diode with an aspect ratio of 1:1.5. Some previous studies used complicated processes. The lens shape of a perfect THEF was modeled by a hyperbolic curve, is expressed as Eq.(4).

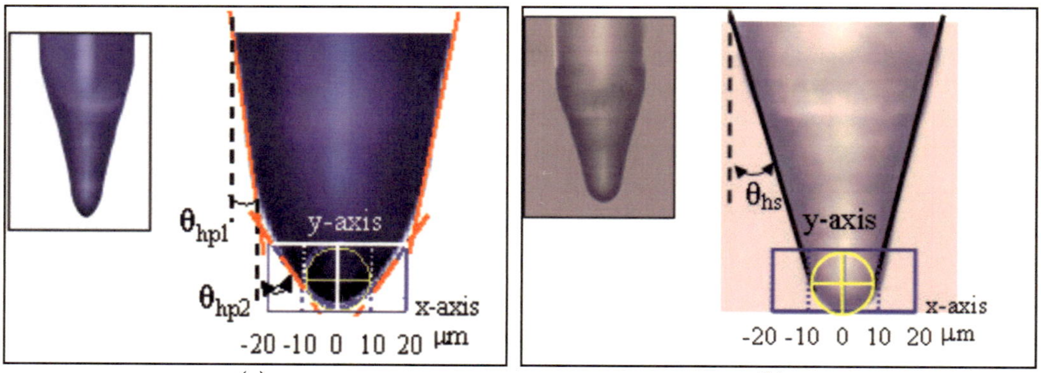


(a) (b)

Fig. 15. (a) Hyperbolic-end microlens, and (b) hemispherical-end microlens with different tapered angles. Both microlenses have a radius of curvature of 9.4 μm.

### 3.2 An emperical model for taper length depending on oil density

The HF with oil solution can be divided into three layers: the bottom layer of uncontaminated HF solution, the interface layer of HF with oil solution, and the top-most layer of oil, as shown in Fig. 16(a). Fig. 16(b) shows the etching processes of tapered fiber. The oil floating on top of the HF solution decreases the vaporization effect from the HF solution onto the fiber surface above the HF solution during etching. The local oil density, $\gamma(\varepsilon)$, is assumed to vary in the e-direction.

According to Fig. 16(a), the layers I, II, and III are the fiber dipped in the uncontaminated HF solution, the interface layer between the oil layer and the HF solution, and oil layer, respectively. Therefore, the local oil density in layers I, II, and III are zero, a function of e, and a constant (depending on the oil), respectively. The fiber in layer I can be etched to a long round stick, while the fiber in layer II can be etched to form a longer taper length and a smaller taper angle. The etching process in both layers I and II is terminated at room temperature after 25 min. The local oil density, $\gamma(\varepsilon)$, in three different layers can be expressed as

$$\gamma(\varepsilon)=\begin{cases} \text{constant} & \varepsilon_1 < \varepsilon \\ [p + q\,\gamma(\varepsilon)](\varepsilon - \varepsilon_0) & \varepsilon_0 < \varepsilon < \varepsilon_1 \\ 0 & \varepsilon < \varepsilon_0 \end{cases} \tag{16}$$

where the $p$ and $q$ are constants. Based on Fig. 16, the taper angle, $\theta_t$, is given by

$$\theta_t = 0.5 \tan^{-1}\left(d_f/2\varepsilon\right) \tag{17}$$

where the $d_f$ is the fiber diameter of 125 μm. From Eqs. (16) and (17) yields

$$\theta_t = 0.5 \tan^{-1}\left\{d_f\left[p+q\gamma(\varepsilon)\right]/2\gamma(\varepsilon)\right\} \tag{18}$$

where the constants $p$ and $q$ are $-0.113$ and 0.169, respectively. The constants $p$ and $q$ are from the experimental values of the taper angle related to the oil density. To complete the etching process, the long round stick in layer I disappeared and left a longer taper fiber, as shown in Fig. 17. Figs. 17(a) and 17(b) show the pictures of different etching time, 18 min. and 25 min., respectively. Figure 17(c) shows a tapered fiber picture after completing an etching process to be enlarged by 500 times by SEM (Scanning Electron Micrograph Image). The taper angles are dependent on etching time and oil thickness, as shown in Fig. 18. In the interface region, the fiber was subjected to the influence of adhesive force (horizontal direction) except for surface tension (vertical direction). The activity of the horizontal direction and the vertical direction were $r_{OH}\sin\theta$ and $r_{OF} = r_{HF} - r_{OH}\cos\theta$, respectively. The dependence of the taper length and the taper angle between the hemispherical microlenses and THEFs is shown in Fig. 19.

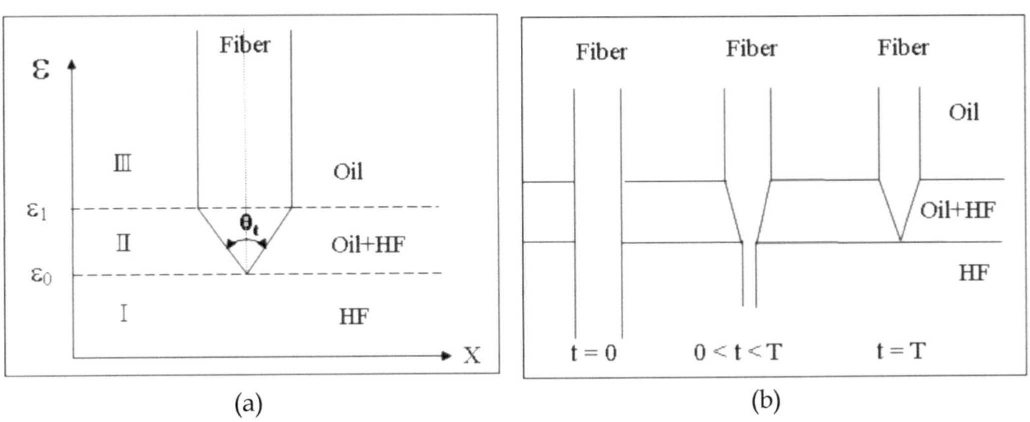


Fig. 16. (a) HF with oil solution divided into three layers; layerI: the fiber dipped in the uncontaminated HF solution, layer II:the interface layer between the oil layer and the HF solution, and layer III: oil layer. (b) Etching processes of fiber microlensat different etching time.

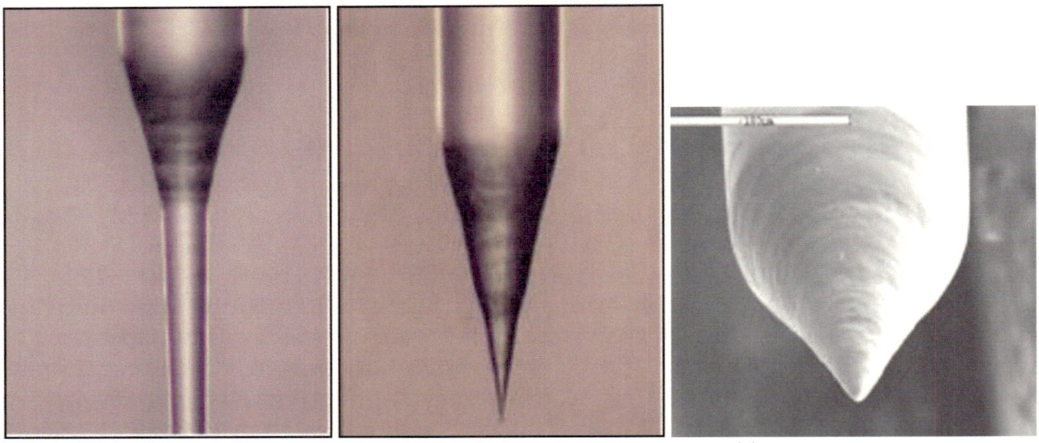

(a) (b) (c)

Fig. 17. Pictures of different etching time, (a) 18 min. (b) 25 min., and (c) enlarged by 500 times taper fiber after etching process, photo by SEM (Scanning Electron Micrograph Image).

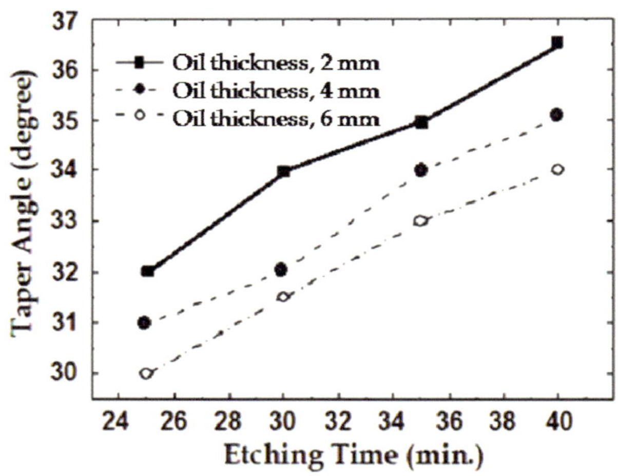


Fig. 18. The taper angles are dependent on etching time and oil thickness.

## 4. Fabrication of fiber gratings laser with uncoated microlenses

In densely spaced wavelength-routed and large-capacity optical network systems, single-mode lasers with narrow linewidth and low temperature sensitivity are of particular importance as the light sources in transmitter modules. Using a temperature-insensitive and low-chirp fiber Bragg grating reflector as the external cavity for the semiconductor gain element, fiber grating external cavity lasers (FGECLs) have been developed for 2.5 Gbit/s or 10 Gbit/s WDM with a low bit error rate penalty. It has been shown that the side-mode suppression ratio (SMSR) of the non-AR-coated FGECL exhibits a current-dependent oscillation. To avoid the formation of an intra cavity and increase the SMSR, Morton et al.

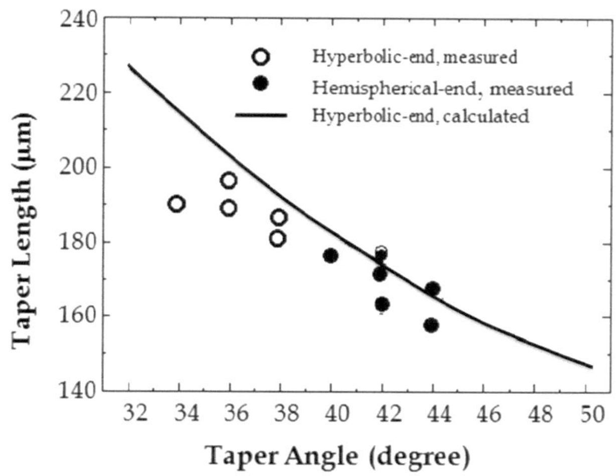


Fig. 19. The dependence of taper length and taper angle between the hemispherical microlenses and THEFs.

have demonstrated an FGECL configuration by applying antireflection (AR) coatings to both the fiber tip and the diode laser front facet (facet closest to fiber tip). However, both the fiber tip and the diode laser front facet required low AR coatings($1\times10^{-5}$). This makes the fabrication process complicated and causes an increase in the cost of the production. Therefore, compact integration of the diode laser and the fiber grating external cavity, which would allow the ease of optical coating and the reduction of back reflection from the fiber tip, is of great research interest for optimizing the FGECL performance. A fiber microlens which provides an efficient coupling mechanism to match the spot size of the diode laser to the fiber is commonly used for optical alignment in FGECL modules.

This paragraph will elucidate the fabrication of a fiber grating external cavity laser (FGECL) module with a low cost while still maintaining a good performance using a low-cost AR(Anti-reflection)-coated($5\times10^{-3}$) laser and a tapered hyperbolic-end fiber (THEF) microlens. Previously, high-performance FGECL modules have only been available by using a complicated AR-coated($1\times10^{-5}$) laser process that leads to a high packaging cost. Our FGECL consisted of a HR/AR-coated diode laser, a THEF microlens, and a fiber grating. The results showed that the FGECL module exhibits a side-mode suppression ratio (SMSR) higher than 44 dB, a higher output power of more than 2 mW, and a larger operation current range of over 50 mA. In addition, excellent wavelength stability and a low-penalty directly modulated 2.488 Gbit/s were also obtained for FGECL modules. The THEF microlens demonstrated up to 86% coupling efficiency for a laser with an aspect ratio of 1:1.5. Low-cost FGECL modules with good performance were achieved primarily owing to the THEF microlens having a high coupling efficiency (typically 75%) which enhanced the feedback power from the fiber grating external cavity to the HR/AR-coated laser compared with the currently available hemispherical microlens which has low coupling efficiency (typically 50%). Therefore, the packaged FGECL module is suitable for use in low cost 2.5 Gbit/s lightwave transmission systems.

### 4.1 Fiber grating external cavity laser structure

The fiber grating used in the FGECL comprises a single-mode fiber grating, and an uncoated THEF microlens. They are fabricated on the same photosensitive single-mode fiber with a mode-field diameter of 9.6 μm at λ = 1.55 μm. The fiber Bragg grating is firstly written on the photosensitive single-mode fiber using 248 nm UV light. The 6-mm-long grating is formed in the core of germanium-doped silica glass. To stabilize the grating formation, thermal annealing at 140°C was performed for 40 h. After the preparation and characterization of the fiber grating, the THEF microlens was processed at the tip of the single-mode fiber with the grating formation. To conveniently fabricate the high coupling efficiency THEF microlens, a unique chemical etching and fusing technique was developed. According to the abovementioned, the THEF were fabricated by symmetrically tapering the fiber during etching and hyperbolically lensing the tip during fusing process. For a high coupling efficiency between the diode laser and the single-mode fiber grating, the THEF microlenses are required to have a small radius of curvature and a low lateral offset between the microlens center and the fiber axis.

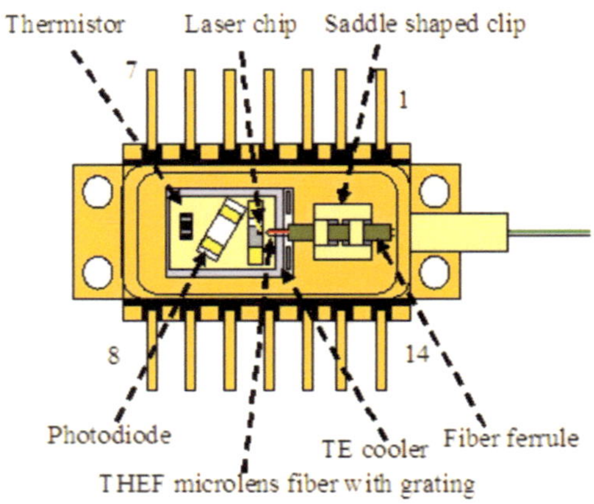


(a)

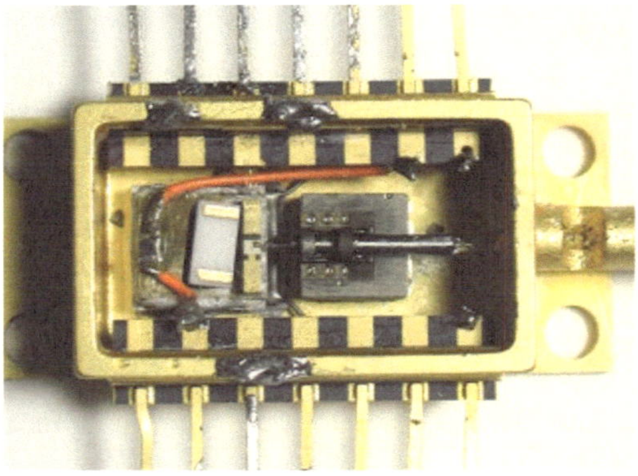

(b)

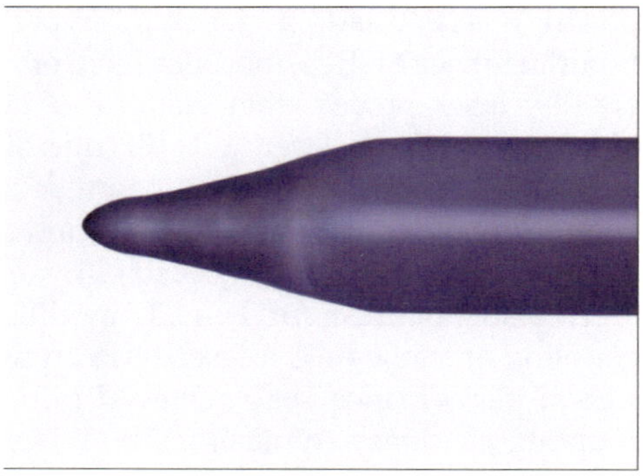

(c)

Fig. 20. (a) AR-coated FGECL module, (b) laser-welded butterfly-type FGECL module, and (c) THEF microlens.

Fig. 20 shows the FGECL module configuration. The FGECL module consisted of an HR/AR-coated diode laser, a THEF microlens, a fiber grating, a photodiode, and a TE cooler, as shown in Fig. 20(a). A laser-welded butterfly-type FGECL module package is shown in Fig. 20(b). A 300-μm-long InGaAsP/InP multi-quantum-well, ridge-waveguide Fabry-Perot laser diode wire-bonded on a sub-carrier was used as the gain medium for the FGECL. The FP laser diode which has a 90% HR coating at the back facet, and a low-cost AR-coated front facet of 0.5%, was mounted on a TE cooler to control the substrate temperature. The FP laser beam has a far-field divergence of 20°×30° (lateral × vertical). The fiber grating with the THEF microlens was assembled in a fiber ferrule, and then the FGECL module was packaged by laser welding technique, as shown in Fig. 20(b). Fig. 20(c) shows a THEF microlens.

### 4.2 Performance of the fiber grating external cavity laser

For each THEF, the maximum coupling efficiency between the laser diode and the fiber was measured. The coupling efficiency is defined as the ratio of the optical power from a 2-m-long fiber pigtail with a THEF to that of the FP laser output power. When the maximum coupling was achieved, the distance between the front facet of the laser and the tapered hyperbolic-end fiber tip was 9 μm. The 3 dB lateral alignment tolerance between the laser and the fiber was ±1.0 μm. The apodized fiber grating has a reflectivity of 50% at a center wavelength of 1539.15 nm, and a 3 dB bandwidth of 0.2 nm. The spectra of non-AR-coated and AR-coated FP laser diodes are shown in Figs. 21(a) and 21(b), respectively, at a substrate temperature of 25°C. The non-AR-coated and AR-coated Fabry-Perot laser diodes, and FGECL module typically exhibit linear L-I curves up to a pumping current of more than 70 mA, as shown in Fig. 22(a). In the Fig. 22(a), the threshold current ($I_{th}$) of the non-AR-coated laser, AR-coated laser, and FGECL were 12 mA, 32 mA, and 18 mA, respectively, at a substrate temperature of 25°C. At an operation current of 70 mA, the output powers of the non-AR-coated laser, AR-coated laser, and FGECL were more than 14 mW, 7 mW, and 5 mW, respectively. In the Fig. 22(a), the FGECL module has a lower threshold current and slope efficiency than the AR-coated laser. The spectral characteristics of the FGECL module can be analyzed as a single-cavity laser formed by a HR back facet and an effective front facet from the reflectivity and coupling efficiency of the fiber grating external cavity.

The $m$th mode effective reflectivity $R_{2s}$ at the diode laser front facet, including all the reflection components from the output end of the FGECL is given by

$$R_{2s} = |r_{2s}|^2 = \left| r_2 + \left(1 - |r_2|^2\right) r_{2ext} \exp(-j2\pi \nu_m \tau_{ext}) \right|^2 \tag{19}$$

where $r_2$, $r_{2ext}$, $\nu_m$, and $\tau_{ext}$ are the reflection coefficient of the diode laser front facet, the external cavity reflection coefficient, the optical frequency, and the round-trip delay through the length of the external cavity, respectively. Here, the external cavity length is the distance from the laser diode front-facet to the center of the fiber grating. For the longitudinal mode of the wavelength located within the 3 dB bandwidth of the fiber grating reflection, the value of $r_{2ext}$ can be estimated by taking into account the fiber grating reflectivity $R_g$ and the coupling efficiency $\eta$, and is given by

$$R_{2ext} = |r_{2ext}|^2 = \eta^2 R_g \tag{20}$$

The effective reflectivity $R_{2s}$ has a higher value for the case of higher $R_g$ and $\eta$. However, for the FGECL, $R_{2s}$ is lower than the reflectivity of the natural cleavage facet. Therefore, the mirror loss of the equivalent cavity is higher. The lower slope efficiency of the FGECL may be induced by the internal refractive loss of the fiber grating, due to the imperfect refractive index modulation and uncoated section of the grating. Fig. 22(b) shows the lasing spectrum with an SMSR of more than 44 dB at a substrate temperature of 25°C and an injection current of 40 mA.

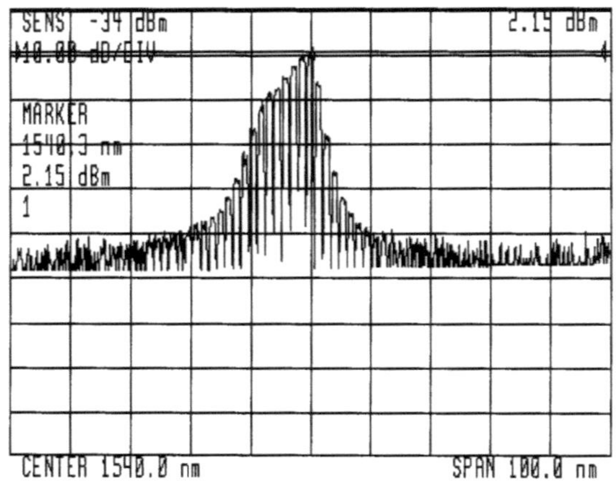


(a)

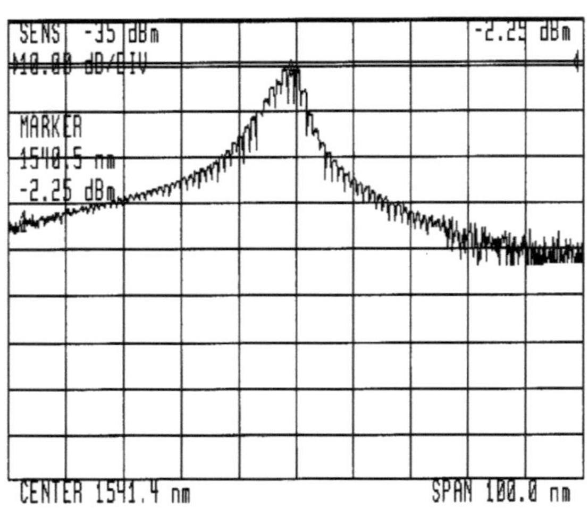


(b)

Fig. 21. Spectra of (a) non-AR-coated and (b) AR-coated FP laser diodes at substrate temperature of 25°C.

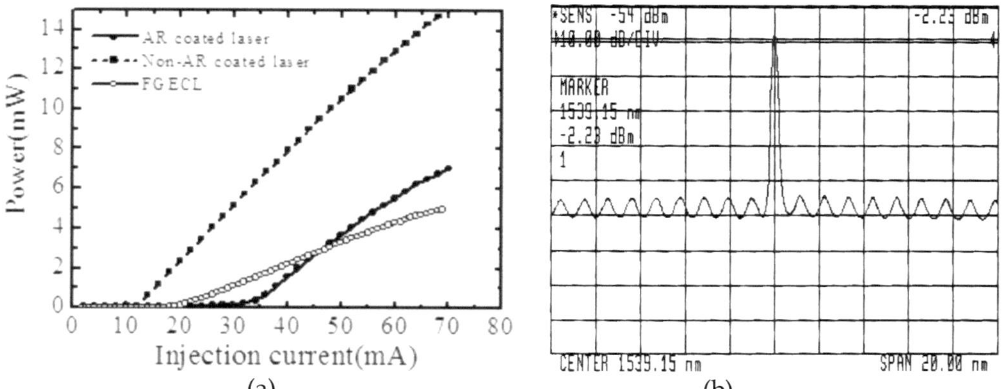


(a) (b)

Fig. 22. (a) Typical L-I characteristics of AR-coated laser, non-AR-coated laser, and FGECL module at substrate temperature of 25°C and (b) lasing spectrum of FGECL module with SMSR of more than 44 dB at substrate temperature of 25°C and an injection current of 40 mA.

The lasing peak wavelength ($\lambda_o$) and the SMSR of the AR-coated FGECL as a function of current (I) is shown in Fig. 23. In the range of current from 18 to 70 mA at an ambient temperature of 28°C, $\lambda_o$ of the AR-coated FGECL is almost fixed at 1539.15 nm, which is consistent with the center wavelength of the fiber grating. This indicates that for wavelength stability, the operation current range of the AR-coated FGECL is tenfold that of the non-AR-coated FGECL. At a substrate temperature of 25°C, the AR-coated FGECL maintains a single longitudinal mode operation with the SMSR mostly greater than 44 dB at operation currents from 30 to 60 mA, as shown in Fig. 23. The high values of the SMSR indicate that the main mode is strongly locked by the fiber grating's external cavity.

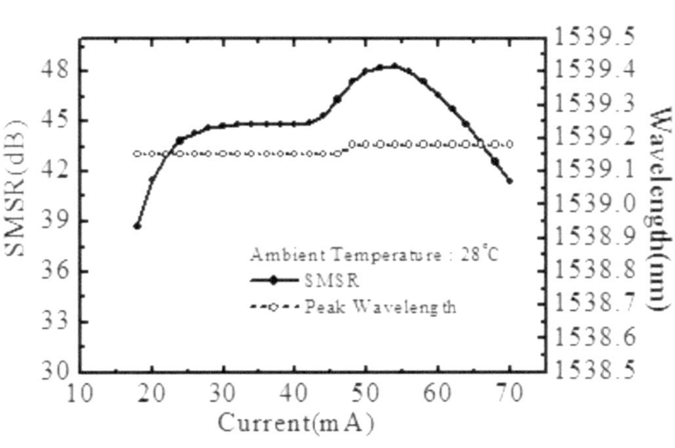


Fig. 23. Lasing peak wavelengths ($\lambda_o$'s) and SMSR of AR-coated FGECL as a function of pumping current at substrate temperature of 25°C.

The lasing peak wavelength ($\lambda_o$) and SMSR of the AR-coated FGECL as a function of temperature (T) are shown in Fig. 24. The lasing peak wavelength is strongly locked by the fiber grating and is dominated by the fiber grating. The $\lambda_o$ is dependent on lasing wavelength of fiber grating temperature at ambient temperature of 28°C. The temperature of the laser chip is controlled using a TE cooler. To determine the effect of temperature on the SMSR and $\lambda_o$, the data was measured at I = 40 mA. The AR-coated FGECL had a high SMSR at temperatures from 20 to 40°C. At 20 to 40°C, the SMSR varies from 38.5 to 47.4 dB, as shown in Fig. 24. This change of SMSR was caused by the temperature-induced red shift of the diode laser gain spectrum. The FGECL module showed excellent single-longitudinal-mode characteristics. In the range of temperatures from 20 to 40°C, little variation of $\lambda_o$ was observed. The average variation of $\Delta\lambda/\Delta T$ was ~ 0.0042 nm/°C, and the shift of $\lambda_o$ was within the 3 dB window of the fiber grating.

On the other hand, for the temperature dependence of the non-AR-coated FGECL, the peak wavelength ($\lambda_o$) has more significant variations between 1536 to 1545.8 nm at 18 to 36°C, as shown in Fig. 25. The FGECL with a non-AR-coated laser operates less effectively in a single mode from the measured SMSRs. This is due to the mode hopping caused by the strong intra-cavity resonance of the FP diode laser. While the lasing mode of the FP diode laser is outside the 3dB window of the fiber grating, there is no more mode selection by the external cavity. Fig. 26 shows a plot of the SMSR and $\lambda_o$ as a function of the pumping current of the non-AR-coated FGECL at a substrate temperature of 25°C. The $\lambda_o$ showed current-dependent variation, and weak wavelength selection by the fiber grating external cavity. The SMSR exhibited current-dependent oscillations with a short operation current range from 17 to 22 mA. Without AR coating on the diode laser's front facet, the FGECL shows a nonlinear L-I curve and a low SMSR, indicating strong mode hopping competed by the diode-laser FP cavity and the external cavity.

Fig. 27 shows the BER performance of the AR-coated and non-AR-coated FGECL modules with a transmission rate of 2.488 Gbit/s. The measured conditions were a NRZ format, $2^{23}$-1 pseudo-random binary sequence (PRBS) pattern, with a modulation current of 20 mA, and an operation current of 40 mA at substrate temperature of 25°C. The power penalties of the AR-coated and non-AR-coated FGECL modules for 25 km SMF (ITU G.652) were measured at 0.8 and 1.42 dB, respectively. The AR-coated FGECL module shows much better BER performance than the non-AR-coated FGECL module. Therefore, the packaged FGECL module is suitable for use in low cost 2.5 Gbit/s lightwave transmission systems.

## 5. Discussion and conclusions

In general, a high-performance FGECL requires a low AR-coated($1\times10^{-5}$) laser. However, such packages require complicated processes, which make their cost too high. In this chapter, the experimental results demonstrate the possibility fabricating FGECL modules at low cost while still maintaining a good performance by employing a low-cost AR-coated ($5\times10^{-3}$) laser and a THEF microlens. Compared with previous studies, new low-cost FGECL modules exhibited a better SMSR (>44 dB), a good wavelength stability (almost fixed from 18~70 mA), and an excellent temperature stability ($\Delta\lambda/\Delta T$~0.0042 nm/°C). In a previous study, an FGECL showed a higher SMSR (>55 dB) and a better wavelength stability ($\Delta\lambda/\Delta I$~0.0062 nm/mA from 8~250 mA), however, both the fiber tip and the diode laser's front facet required a low-AR coating ($1\times10^{-5}$). This makes the fabrication process complicated and causes a rise in the cost of the production. In this chapter, low-cost FGECL modules with good performance were achieved primarily owing to the THEF microlenses having a high coupling efficiency (typically 75%) which enhances the feedback power from

the fiber grating external cavity to the HR/AR-coated laser compared with the currently available hemispherical microlens which has a low coupling efficiency (typically 50%).
In addition, the experimental results demonstrated a good performance of the BER and a low penalty of directly modulated 2.488 Gbit/s for FGECL modules with HR/AR-coated diode lasers and uncoated THEF microlenses. This clearly indicates that the packaged FGECL modules are suitable for use in low-cost 2.5 Gbit/s lightwave transmission systems, such as gigabit passive optical networks (GPONs), metropolitan area networks (MANs), and fiber-to-the-home (FTTH) applications.

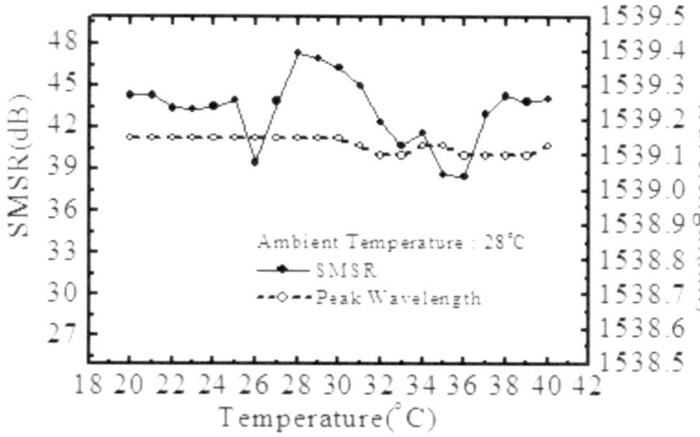


Fig. 24. Lasing peak wavelengths ($\lambda_o$) and SMSR of AR-coated FGECL module as a function of temperature at the injection current of 40 mA. The peak wavelength is dependent on lasing wavelength on fiber grating temperature at ambient temperature of 28°C.

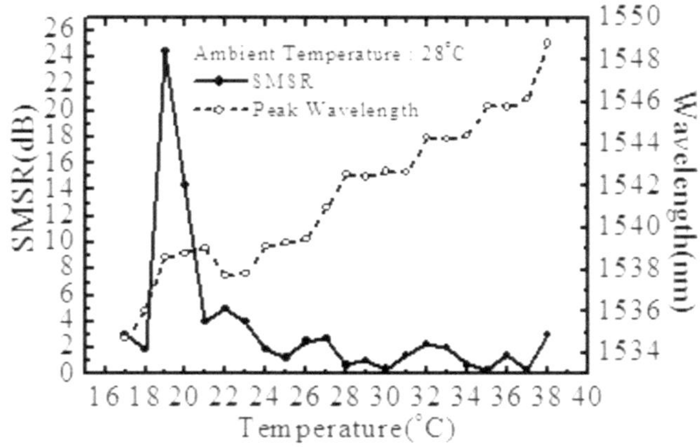


Fig. 25. Lasing peak wavelength ($\lambda_o$) and SMSR of non-AR-coated FGECL module as a function of temperature at the injection current of 40 mA.

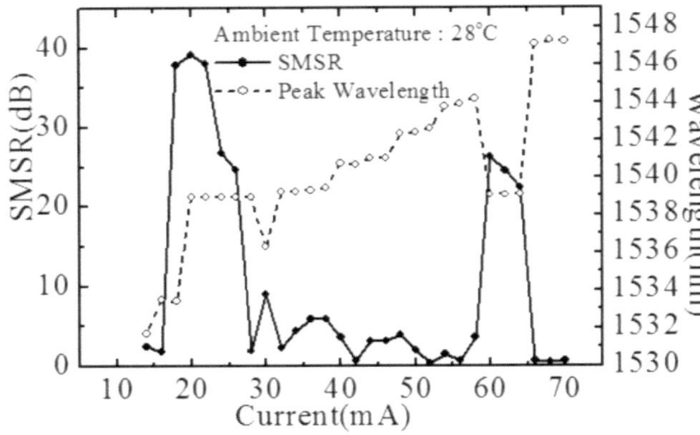


Fig. 26. Lasing peak wavelength ($\lambda_o$) and SMSR of non-AR-coated FGECL module as a function of injection current at substrate temperature of 25°C.

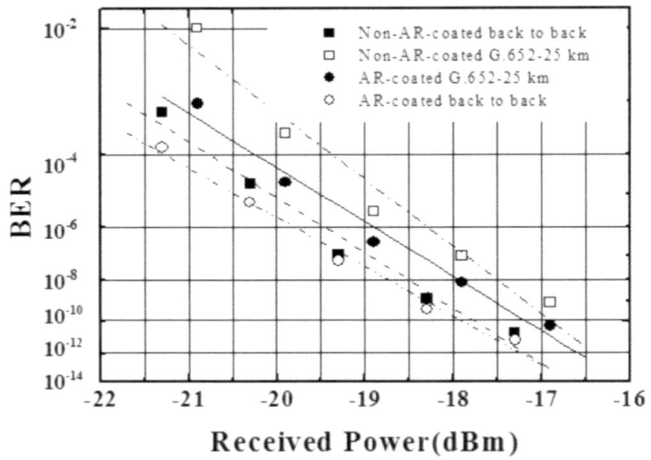


Fig. 27. BER performance of AR-coated and non-AR-coated FGECL modules with a transmission rate of 2.488 Gbit/s.

## 6. References

ASAP™ (Advanced Systems Analysis Program), Breault Research Organization, Inc.,Tucson, Arizona USA

Yang, H. and et al. (2010). Investigation of the efficient coupling between a highly elliptical Gaussian profile output from a laser diode and a single mode fiber using a hyperbolic-shaped microlens, *Optics & Laser Technology,* Vol.42, No. 6, pp.918–926

Hillerich, B. and Guttmann J. (1989). Deterioration of taper lens performance due to taper asymmetry, *J. Lightwave Technol.* No. 7, pp. 99-104

Mathyssek, K. and et al. (1985). Fabrication and investigation of drawn fiber tapers with spherical microlenses, *J. Opt. Commun.* No.6, pp. 142-146

Edwards, C. and et al. (1993). Ideal microlenses for laser to fiber coupling, *J. Lightwave Technol.* No.11, pp. 252-257

Gangopadhyay, S. and Sarkar, S. (1997). *ABCD* matrix for reflection and refraction of Gaussian light beams at surfaces of hyperboloid of revolution and efficiency computation for laser diode to single-mode fiber coupling by way of a hyperbolic lens on the fiber tip, Appl. Opt. Vol. 36, pp. 8582-8586

Presby, H. and et al. (1990). Laser micromachining of efficient fiber microlenses, *Appl. Opt.* Vol. 29, pp. 2692-2695.

Goodman, J. (2005). Introduction to Fourier optics, Roberts & Company Publishers, pp. 63-126. ISBN 0-9747077-2-4, Colorado, USA

Yariv, A., (1996). Optical Electronics in Modern Communications, fifth edition, Oxford Univ. Press, pp. 52-53, ISBN 0-19-510626-1, New York, USA

Kogelink, H., (1964). Coupling and conversion coefficients for optical modes in quasioptics, Microwave Research Institute Symposia Series, Polytechnic Press, New York, Vol.14, pp. 333-347

Mathyssek K, and et al. (1985). Fabrication and investigation of drawn fiber tapers with spherical microlenses, *J. Opt. Commun.* No. 6, pp. 142-146

Hillerich, B. and Guttmann, J. (1989). Deterioration of taper lens performance due to taper asymmetry, J. Lightwave Technol., Vol. 7, pp. 99-104

Lay, T., and et al. (2002). 1.55-µm Non-anti-reflection-coated Fiber Grating Laser for Single-longitudinal Mode Operation, Opt. Quantum Electron.Vol. 34, pp. pp. 687-696

Morton, P. and et al. (1994). Stable single mode hybrid laser with high power and narrow linewidth, *Appl. Phys. Lett.*, Vol. 64, pp. 2634-2636

Yang, H. and et al. (2004). High coupling tapered hyperbolic fiber microlens and taper asymmetry effect, *IEEE J. Lightwave Technol.* Vol. 22, No. 5, pp. 1395-1401

Petermann, K. (1988). *Laser Diode Modulation and Noise,* Kluwer Academic Publishers, Chap. 2&9, ISBN 90-277-2672-8, Tokyo, Japan

Hill, K., and et al. (1993). Bragg gratings fabricated in monomode photosensitive optical fiber by UV exposure through a phase mask, *Appl. Phys Lett.* vol. 62, pp. 1035-1037

Lemaire, P. and et al. (1993). High pressure H2 loading as a technique for achieving ultrahigh UV photosensitivity and thermal sensitivity in GeO2 doped optical fibres, *Electron Lett.,* Vol. 29, no. 13, pp. 1191-1193

Lay, T., and et al. (2003). 1.55-µm Fiber grating laser utilizing an uncoated tapered hemispherical-end fiber microlens, *J. Appl. Phys.* Vol. 42, Part 1, No. 2A, pp. 1-3

Sigg, J. (1993). Effects of optical feedback on the light-current characteristics of semiconductor lasers, *IEEE J. Quantum Electron.,* Vol. 29, No. 5, pp.1262-1270

Kallimani, K. and O'mahony, K. (1998). Calculation of optical power emitted from a fiber grating laser, *IEE Pro-optoelectron,* Vol. 145, No. 6, pp.319-324

Annovazzi-Lodi, V. and et al. (2000). Power efficiency of a semiconductor laser with an external cavity, *J. Optical Quantum Electron.,* Vol. 32, pp. 1343-1350

Timofeev, F. and et al. (1995). Spectral characteristics of a reduced cavity singlemode semiconductor fiber grating laser for applications in dense WDM systems, *Proc. 21st European Conf. Optic. Commun. (ECOC'95),* Brussels, Belgium, paper Tu.P.26, pp. 477–480

Huang, S. and et al. (2000). High coupling optical design for laser diodes with large aspect ratio, *IEEE Trans. Adv. Packag.,* Vol. 23, pp. 165-169

Lin, C. (1989). *Optoelectronic technology and lightwave communications systems*, New York: Van Nostrant Reinhold, pp.175-194, ISBN 04-422-6050-4, New York, USA

Mobarhan, K. and et al. (2002). Laser diode packaging technology: 980 nm EDFA pump lasers for telecommunication applications, Application Note, No. 3, pp. 1-7, Newport Corp., CA, USA

Yang, H. and et al. (2001). An optimum approach for fabrication of tapered hemispherical-end fiber for laser module packaging," *J. Electron. Mater.*, Vol. 30, pp. 271-274

Lee, T. and et al. (1982). Short-cavity InGaAsP injection lasers: dependence of mode spectra and single longitudinal mode power on cavity length," *IEEE J.Quantum Electron.*, Vol. QE-18, pp.1101~1112

Shiraishi, K. and et al. (1995). A fiber lens with a long working distance for integrated coupling between laser diodes and single-mode fibers, *IEEE J. Lightwave Technol.*, Vol. 13, pp. 1736-1744

Lee, T. and et al. (1982). Short-cavity InGaAsP injection lasers: dependence of mode spectra and single longitudinal mode power on cavity length, *IEEE J. Quantum Electron.*, Vol. QE-18, pp.1101~1112

Yang, H. and et al. (2007). Low-cost fiber grating laser module package employing a hyperbolic fiber microlens, *J. Appl. Phys.*, Vol. 46, No. 3A, pp. 1016-1020

Yariv, A. (1973). Coupled-mode theory for guided-wave optics, *IEEE J. Quantum Electron.*, Vol. QE-9, pp. 919~933

Timofeev, F. and et al. (2000). Experimental and theoretical study of high temperature-stability and low-chirp 1.55μm semiconductor laser with an external fifer grating, *Fiber and Integrated Optics*, Vol. 19, pp. 327-353

# Generation of Few Cycle Femtosecond Pulses via Supercontinuum in a Gas-Filled Hollow-Core Fiber

Walid Tawfik Mohamed[1,2], Jungkwuen An[1] and Dong Eon Kim[1]
[1]*Department of Physics & Center for Attosecond Science and Technology (CASTECH), Pohang University of Science and Technology (POSTECH), Pohang,*
[2]*National Institute of laser Enhanced Sciences, Cairo University, Cairo,*
[1]*Korea*
[2]*Egypt*

## 1. Introduction

The interaction of intense short nano- and picosecond laser pulses with plasma leads to reach variety of important applications, including time-resolved laser induced breakdown spectroscopy LIBS (Walid Tawfik et al., 2007), soft x-ray lasers (X. F. Li et al., 1989), and laser-driven accelerators (W. P. Leemans et al., 1989). In most cases the useful output - whether it be photons or accelerated particles - increases with the distance over which the laser-plasma interaction occurs, at least up to some upper limit which might, for example, be set by the onset of phase mismatching. The strength of any such interaction will depend on the intensity of the driving radiation, which for the short pulse lengths reached higher power densities and started the race for ever shorter laser pulses as a unique tool for time-resolved measurements of ultrashort processes (R. A. Cairns et al., 2009).
The progress in generating of femtosecond down to sub-10 fs optical pulses has opened a door for scientists with an essential tool in many ultrafast phenomena, such as femto-chemistry (Zhong et al., 1998), high field physics (Li Y T et al., 2006) and high harmonic generation HHG (John et al., 2000).The advent of high-energy laser pulses with durations of few optical cycles provided scientists with very high electric fields, sufficient to suppress the Coulomb potential in atoms, accelerate electrons up to kilo-electron-volt energies, and produce coherent intense UV and extreme ultraviolet (XUV) radiation with durations of attoseconds ($10^{-18}$ second) (Hentschel et al., 2001, Christov et al., 1997) which allow for the investigation of molecular dynamics with 100-as resolution (Baker et al., 2006). Ultrashort pulsed laser amplifiers can easily provide pulse energies in millijoule and even joule levels but with pulse durations longer than 20 fs. Amplification of pulses with shorter durations even to the millijoule level is a difficult task requiring elaborate setups (Ishii et al., 2005). Durations down to the few-cycle regime can be achieved through pulse compression schemes. Typically, such methods start in the picosecond or femtosecond region, i.e. already in the regime of ultrashort pulses. These methods can be grouped into two categories: linear pulse compression and nonlinear pulse compression (Shank et al., 1982).

In linear pulse compression; when pulses are chirped, their duration can be reduced by removing (or at least reducing) this chirp, i.e. by flattening the spectral phase (Treacy et al., 1969). Dechirping can be accomplished by sending the pulses through an optical element with a suitable amount of chromatic dispersion, such as a pair of diffraction gratings, a prism pair, an optical fiber, a chirped mirror, a chirped fiber Bragg grating or a volume Bragg grating (Martinez et al., 1984, Fork et al., 1984). The smallest possible pulse duration is then set by the optical bandwidth of the pulses, which is not modified by dispersive compression. In the ideal case, bandwidth-limited pulses are obtained. On the other hand, in the nonlinear pulse compression; in a first step, the optical bandwidth is increased, typically with a nonlinear interaction such as a supercontinuum generated by self-phase modulation (SPM) in nonlinear medium (Tomlinson et al., 1984). In most cases, this leads to chirped pulses, often with a longer duration which can be strongly reduced by linear compression (Fork et al., 1984).

The main challenge in pulse compression using supercontinuum generation lies in effective dispersion compensation and hence compression of the generated bandwidths to yield an isolated ultrashort optical pulse (Schenkel et al., 2003). In recent years significant progress has been made in the generation of supercontinuum in the visible and near-infrared spectral region from hollow and microstructured fibers. Recently, a high-energy supercontinuum extending a bandwidth exceeding 500 THz was generated with two gas-filled hollow fibers in a cascading configuration (Nisoli et al., 2002). The supercontinuum at the output of a single gas-filled hollow fiber has been compressed to 4.5 fs by use of a combination of chirped mirrors and thin prisms (Nisoli et al., 1997).

Hollow fiber pulse compression is currently the most widespread method for generating high-power, few-cycle pulses, with pulse durations as short as 2.8 fs and pulse energies up to 5.5mJ being achieved in this manner (Nisoli et al., 1996), with the energy throughput limited by damage and plasma formation at the fiber entrance and self-focusing in the waveguide. In this technique, laser pulses are spectrally broadened in a gas contained within a long (~ 1m, 50–500 μm inner diameter) hollow, glass fiber, that acts as a dielectric waveguide for the light, allowing significantly longer interaction lengths than possible in an unguided interaction. When an intense laser pulse propagates through a gas in the fiber it induces continuum generation due primarily to self-phase modulation (SPM). To maximize energy throughput, it is important to efficiently couple the pulse into the fiber. This is achieved by carefully matching the focal spot of the laser to the guided mode of the fiber. In a typical setup, the fiber is entirely contained within a single chamber which is statically filled (SF) with gas. In a differentially pumped (DP) fiber, gas is injected into the exit end of the fiber, and pumped away at the fiber entrance. The broadened pulses can be temporally compressed to durations shorter than the input pulses by using appropriate dispersive optical elements, e.g. prism compressors or chirped mirrors (Nisoli et al., 1996, Sartania et al., 1997).

In this chapter we represent a review on the generation of few-cycle fs light pulses using a gas-filled hollow-core fiber. Also, we report our experimental results of reaching 3.8-fs light pulses with energies up to 0.5 mJ using a laser of 0.9 mJ energy at reputation rate of 3 kHz and pulse duration of 28 fs. The pulse compressing achieved by the supercontinuum produced in neon gas-filled hollow fibers while the dispersion compensation is achieved by five pairs of chirped mirrors. In these experiments, we demonstrated the broadening variation with static pressures of neon gas. The applied technique allows for a straightforward tuning of the pulse duration via the gas. Since the focusing into the fiber is no longer dependent on pressure and intensity, consistent coupling conditions are achieved for a broader range of pressures and input pulse parameters.

## 2. Supercontinuum generation and ultrashort pulses

Supercontinuum (SC) generation or white-light generation, is a complex physical phenomenon where the interplay between different nonlinear processes and dispersion leads to a significant spectral broadening of laser pulses propagating in a nonlinear medium. White light generation by laser radiation was first reported by Alfano and Shapiro who observed a spectral broadening of a 5 mJ picosecond second-harmonic output pulses of a Nd:YAG laser using a bulk of borosilicate glass (Gersten et al., 1980). The major nonlinear phenomena participating in SC generation are self-phase modulation (SPM), cross phase modulation (XPM), four wave mixing (FWM), intrapulse Raman scattering (IPRS) and soliton self-frequency shift (SSFS) (Alfano et al., 1987). If an intense laser pulse propagates through a medium, it changes the refractive index, which in turn changes the phase, amplitude, and frequency of the incident laser pulse. Changing the phase can cause a frequency sweep within the pulse envelope which called self-phase modulation (SPM). Nondegenerate four-photon parametric generation (FPPG) usually occurs simultaneously with the SPM process. Photons at the laser frequency parametrically generate photons to be emitted at Stokes and anti-Stokes frequencies in an angular pattern due to the required phase-matching condition. When a coherent vibrational mode is excited by a laser, stimulated Raman scattering (SRS) occurs which is an important process that competes and couples with SPM. The interference between SRS and SPM causes a change in the emission spectrum resulting in stimulated Raman scattering cross-phase modulation (SRSXPM) (Gersten et al., 1980). A process similar to SRS-XPM occurs when an intense laser pulse propagates through a medium possessing a large second-order $\chi^{(2)}$ and third-order $\chi^{(3)}$ susceptibility as a nonlinear effect. Sometimes both of second harmonic generation SHG and SPM occur simultaneously and can be coupled together thus the interference between them alters the emission spectrum that is called second harmonic generation cross-phase modulation (SHG-XPM) (Alfano et al., 1987). When a weak pulse at a different frequency propagates through a disrupted medium whose index of refraction is changed by an intense laser pulse according to another nonlinear process called induced phase modulation (IPM) at which the phase of the weak optical field can be modulated by the time variation of the index of refraction originating from the primary intense pulse (Alfano et al., 1986).

Recent development of ultra-short femtosecond lasers has opened the way to the investigation of ultrafast processes in many fields of science. An important milestone in the generation of femtosecond pulses was posed in 1981, with the development of the colliding pulse mode-locked (CPM) dye laser (Alfano et al., 2007). The first 27 fs pulses were generated in 1984 using a prism-controlled CPM laser (Agrawal, 1980). By using chirped mirrors to control the intracavity dispersion of Ti:sapphire oscillator ; pulses as short as 7.5 fs have been generated (Fork et al., 2007) and by with the additional use of broadband semiconductor saturable absorber mirror; sub-6-fs pulses were generated (Valdmanis et al., 1985). In 1985, the first chirped-pulse amplification technique (CPA) was found which introduce amplification of ultrashort pulses as short as 20 fs to become available with extremely high power levels up to terawatt peak power at kilohertz rates (Donna Strickland et al., 1985, Xu et al., 1996, Sutter et al., 1999 ). Compression techniques represent an alternative way to generate ultrashort pulses. Nakatsuka and co-workers introduced an optical compression technique based on the interplay between self-phase modulation (SPM) and group velocity dispersion (GVD) occurring in the propagation of short light pulses in single-mode optical fibers (Nakatsuka et al., 2007). It is found that nonlinear propagation induces both of spectral broadening and chirping of the laser pulses then a subsequent propagation in an appropriate optical dispersive delay line can provides compression of the chirped pulse. Increasing the spectral bandwidth of the output pulse leads to the generation of a compressed pulse shorter in duration than the input one. Using this technique with a prism chirped-mirror Gires–Tournois interferometer compressor, 4.6 fs compressed pulses were obtained from a 13-fs cavity dumped Ti:sapphire laser but with a few with only nanojoules due to the low-intensity threshold for optical damaging of the single-mode optical fibers (Backus et al., 2007). By introducing a novel technique, based on spectral broadening in a hollow fiber filled with noble gases, the pulse compression extended to millijoule energy range which presents the advantage of a guiding element with a large diameter mode and of a fast nonlinear medium with high damage threshold(Nakatsuka et al., 1997). By controlling of group-delay dispersion (GDD) in the compressor stage over an ultrabroadband (650–950 nm) spectral range, the hollow fiber technique could delivered a high-throughput up to 4.5-fs with multigigawatt (Nakatsuka et al., 2007). Advances in the design of chirped multilayer coatings led to the adequate dispersion control without the need for prisms which opened the way to scaling sub-10-fs hollow fiber-based compressors to short pulses of 5 fs with substantially higher peak power up to 0.11 TW were generated at 1 kHz repetition rate (Nisoli et al., 1996).

## 3. Nonlinear pulse propagation in optical fibers

The Kerr-nonlinearity-induced intensity-dependent additive to the refractive index is one of the key physical factors in supercontinuum generation. This nonlinearity is due to the anharmonic motion of bound electrons under the influence of an applied field that causes also the induced polarization P from the electric dipoles to be nonlinear with the electric field E, and follow the nonlinear relation (Shen et al., 1984)

$$P = \varepsilon_0 \left[\chi^{(1)} E + \chi^{(2)} E^2 + \chi^{(3)} E^3 + \ldots\right] \tag{1}$$

where $\chi^{(i)}$ (i = 1, 2, ...) is susceptibility for the order the i-th and $\varepsilon_0$ is the vacuum permittivity. Here, the dominant contribution to P is the linear susceptibility $\chi^{(1)}$ while the second-order susceptibility $\chi^{(2)}$ is responsible for nonlinear processes such as sum-frequency generation (SFG) and second-harmonic generation (SHG) (Shen et al., 1984). However, for media exhibiting an inversion symmetry at the molecular level, $\chi^{(2)}$ is zero. Therefore, optical fibers do not normally show second-order nonlinear effects since $\chi^{(2)}$ vanishes for silica glasses ($SiO_2$ is isotrop). So that the lowest order nonlinear effects in optical fibers originate from the third-order susceptibility $\chi^{(3)}$ which is responsible for phenomena such as third-harmonic generation (THG), four-wave mixing (FWM), and nonlinear refraction. However, the processes are not efficient in optical fibers unless special efforts are made to achieve phase matching. Therefore, most of the nonlinear effects in optical fibers originate from nonlinear refraction due to the intensity dependence of the refractive index resulting from the contribution of $\chi^{(3)}$. That means, the refractive index of a medium with a Kerr nonlinearity is written as (Shen et al., 1984)

$$n = n_0 + n_2 I(t) \tag{2}$$

where $n_0$ is the field-free nonperturbed refractive index of the medium, and $I(t)$ is the laser radiation intensity, and $n_2 = (2\pi/n_0)^2 \chi^{(3)}(\omega; \omega, \omega, -\omega)$ is the nonlinear refractive index at the frequency $\omega$, $\chi^{(3)}(\omega; \omega, \omega, -\omega)$ is the third-order nonlinear-optical susceptibility.
For the case of short laser pulses, the intensity-dependent additive to the refractive index gives rise to a physically significant phase modulation of the laser field, the so-called self-phase modulation (SPM). We used formula (2) to represent the nonlinear phase incursion acquired by a laser pulse over a distance L in a medium with a Kerr nonlinearity in the form

$$\Phi(t) = \frac{\omega}{c} n_2 I(t) L \tag{3}$$

It can be seen from expression (3) that the intensity dependence of the refractive index of the medium maps the temporal profile of the field intensity in a laser pulse on the time dependence of the nonlinear phase shift, which, in turn, gives rise to a time-dependent frequency deviation across the laser pulse:

$$\Delta\omega(t) = \frac{\omega}{c} n_2 L \frac{\partial I}{\partial t} \tag{4}$$

The maximum SPM-induced spectral broadening of the laser pulse can then be estimated as

$$\Delta\omega(t) = \frac{\omega}{c} n_2 L \frac{I_0}{\tau} \tag{5}$$

where $I_0$ is the peak intensity of the laser pulse, and $\tau$ is the pulse width. SPM is responsible for spectral broadening of ultrashort pulses and the existence of optical solitons in the anomalous-dispersion regime of fibers (Hasegawa et al., 1973). The third-order susceptibility controls the nonlinear effects which seem to be elastic in the sense that no energy is exchanged between the electromagnetic field and the dielectric medium. Another class of nonlinear effects results from stimulated inelastic scattering due to the optical field transfers part of its energy to the nonlinear medium.
Stimulated Brillouin scattering (SBS) and stimulated Raman scattering (SRS) represent two important nonlinear effects in optical fibers which are related to vibrational excitation modes of silica (Stolen et al., 1972, Ippen et al., 1972 ). The main difference between the two is that acoustic phonons participate in SBS while optical phonons participate in SRS. In a simple quantum-mechanical picture applicable to both SBS and SRS, a photon of the incident field (the pump) is annihilated to create a phonon with the right energy and momentum to conserve the energy and the momentum and a photon at the downshifted Stokes frequency. Of course, if a phonon of right energy and momentum is available it can creates a higher-energy photon at the so-called anti-Stokes frequency via the inverse process. The FWM generates an anti-Stokes photon, where two pump photons annihilate themselves to produce Stokes and anti-Stokes photons provided the conservation of the total momentum. The conservation of momentum requirement leads to a phase-matching condition, that must be satisfied for FWM to occur. In single-mode fibers, that phase-matching condition is not easily satisfied, thus the anti-Stokes wave is rarely observed during SRS. Even though SBS and SRS are very similar in their origin, different dispersion relations for optical and acoustic phonons lead to some basic differences between the two. A

fundamental difference is that in optical fibers SRS dominates in the forward direction whereas SBS occurs only in the backward direction. Although a complete description of SBS and SRS in optical fibers is quite involved, the initial growth of the Stokes wave can be described by a simple relation for SRS, as follows;

$$\frac{dI_S}{dz} = g_r I_p I_s \tag{6a}$$

where $g_r$ is the Raman-gain coefficient and, $I_p$ is the pump intensity and, $I_s$ is the Stokes intensity. Since the growth of the anti-Stokes wave is negligible, it is not discussed here. A similar relation holds for SBS as follows;

$$\frac{dI_S}{dz} = g_B I_p I_s \tag{6b}$$

where $g_{rB}$ is the Brillouin-gain coefficient and, $I_p$ is the pump intensity and, $I_s$ is the Stokes intensity. The Raman-gain spectrum is measured to be very broad extending up to ~30 THz while the Brillouin-gain spectrum is extremely narrow with a bandwidth of only ~10 MHz (Stolen et al., 1972, Ippen et al., 1972 ). The maximum value of Brillouin-gain decreases by a factor of $\Delta\upsilon_p/\Delta\upsilon_B$ for a broad-bandwidth pump, where $\Delta\upsilon_p$ is the pump bandwidth and $\Delta\upsilon_B$ is the Brillouin-gain bandwidth. So that for short pump pulses (< 50 ns), SBS is negligible. An important feature for both SBS and SRS is that they exhibit a threshold-like behavior. The later means that a significant conversion of pump energy to Stokes energy occurs only when the pump intensity exceeds a certain threshold level. For example, for SRS the threshold pump intensity typically is ~ 10 MW/cm$^2$ (Hasegawa et al., 1973).

## 4. Supercontinuum generation in hollow fibers

Ultrashort pulses can be obtained not only by linear pulse compression, but also by nonlinear pulse compression techniques as descript earlier. In nonlinear pulse compression, the pulses are spectrally broadened by propagation through a suitable nonlinear medium and subsequently compressed in a dispersive delay line. In 1981, a new method based on spectral broadening by SPM was introduced for optical pulse compression. With this technique 6 fs pulses at 620 nm have been generated due to the propagation of short pulses in single-mode optical fibers and by using a prism-pair for external dispersion compensation (Nakatsuka et al., 1981). Furthermore, a 4.5 fs pulses at 800 nm have been achieved using an improved ultrabroad-band dispersion compensation scheme broadened by propagation through a suitable nonlinear medium (Baltuska et al., 1997). However, the use of single-mode optical fibers limits the input pulse energy to a few nanojoules range. Therefore, the need for new spectral broadening techniques was born with the availability of high-energy (mJ) femtosecond pulses from solid-state laser amplifiers. One possibility is to achieve spectral broadening in bulk materials but due to the very short interaction length, limited by Rayleigh lengths, high intensities are needed to achieve the necessary nonlinearity for spectral broadening (Rolland et al., 1988). These high intensities can lead to damage and spatial beam quality problems due to multiphoton ionization. In 1996, another particularly suitable technique for high-energy ultrashort pulses was introduced. This technique depends on spectral broadening by SPM in a hollow cylindrical fused silica fiber filled with a noble gas under constant pressure (Nisoli et al., 1996). That technique enables pulses shorter than 5 fs to be generated at multigigawatt peak powers (Nisoli et al., 1996).

Hollow fiber is suitable for large pulse energies since it provides a guiding element with a large-diameter single mode. Furthermore, the use of noble gases as nonlinear medium offers several important advantages compared to optical fibers.
Light propagation in a hollow waveguide is a well-studied topic, which was developed when long-distance communication in standard optical fibers was still inaccessible (Marcatili et al., 1964). Electromagnetic radiation propagates in hollow fibers by grazing incidence reflections; only leaky modes are supported because of power losses through the fiber walls. Three propagation modes can be excited: transverse circular electric ($TM_{0m}$) modes, with the electric field directed radially; $TE_{0m}$ modes, in which the electric field lines are transverse concentric circles and centered on the propagation axis; hybrid modes ($EH_{pm}$, with $p \geq 1$) at which all field components are present, but axial components are so small that such modes can be thought as transverse. For fiber diameters sufficiently larger than the optical wavelength, $EH_{1m}$ modes appear linearly polarized and can be efficiently coupled to a laser beam. The radial intensity profile of $EH_{1m}$ modes is given by $I_c(r) = I_{c0}\, J_0^2\,(\upsilon_m\; r/a\,)$ where $J_0$ is the zero-order Bessel function, $a$ is the capillary radius, $\upsilon_m$ is the m th zero of $J_0(r)$, $I_{c0}$ is the peak intensity. The complex propagation constant $\beta(\omega)$ of the $EH_{1m}$ mode is given by (Shen et al., 1984)

$$\beta(\omega) = \frac{\omega\eta(\omega)}{c}\left[1 - \frac{1}{2}\left(\frac{\upsilon_m c}{\omega\eta(\omega)a}\right)^2\right] + \frac{i}{a^3}\left(\frac{\upsilon_m c}{\omega\eta(\omega)}\right)^2 \frac{\upsilon^2(\omega)+1}{\sqrt{\upsilon^2(\omega)-1}} \tag{7}$$

where $\eta(\omega)$ is the refractive index of the gas, $\omega$ is the laser frequency, and $\upsilon(\omega)$ is the ratio between the refractive indexes of the external (fused silica) and internal (gas) media. The refractive index $\eta(\omega)$ can be calculated at standard conditions by tabulated dispersion relations then the actual refractive index can be easily determined in the operating conditions used for pulse compression (Lehmeier et al., 1996). Many modes can be excited when the laser beam is injected into the capillary. However single-mode operation is generally required for pulse compression and mode selection must be exists. This aim can be achieved by optimal coupling between the input laser beam and the fundamental fiber mode $EH_{11}$. By assuming a Gaussian linearly polarized input beam, it is possible to determine the equation for the coupling efficiency between the input beam and the capillary modes. Propagation along hollow fibers can be occurred at grazing incidence reflections of the dielectric inner surface. This grazing angle reduces the losses caused by these multiple reflections by suppresses the higher order modes. Thus, only the fundamental mode can propagate, in a sufficiently long fiber (Lehmeier et al., 1996). On the other hand, with the proper mode matching for an optimum value of $w_0/a \approx 0.65$, where $w_0$ is the spot size at the fiber entrance and $a$ is the capillary radius, the coupling efficiency of the $EH_{11}$ mode with the laser beam is ~ 98 %, while higher-order modes show a value lower than 0.5%. Thus, the incident radiation intensity profile as a function of the radial coordinate r is given by

$$I_0(r) = I_0\, J_0^2\,(2.405\;\, r/a\,) \tag{8}$$

where $I_0$ is the peak intensity and Jo is the zero-order Bessel function (Marcatili et al., 1984). It is worth pointing out that even if higher-order modes were excited, mode discrimination would be achieved anyway, owing to the higher loss rate of $EH_{1m}$ with respect to fundamental mode. Mode discrimination in the capillary allows one to perform a spatial filtering of the input beam. By applying the same mode on the complex propagation constant $\beta(\omega), \beta$ the real phase constant of Eq. (8) , and imaginary, $\alpha/2$ (field attenuation constant), parts of the propagation constant are given by:

$$\beta = \frac{2\pi}{\lambda}\left[1 - \frac{1}{2}\left(\frac{2.405\,\lambda}{2\,\pi\,a}\right)^2\right] \tag{9a}$$

$$\frac{\alpha}{2} = \left(\frac{2.405}{2\,\pi}\right)^2 \frac{\lambda^2}{2\,a^3}\,\frac{v^2+1}{\sqrt{v^2-1}} \tag{9b}$$

where $\lambda$ is the laser wavelength in the gas medium. Assuming Gaussian pulse profile and neglecting dispersion and self-focusing, the maximum broadening spectrum after propagating a length of $l$ can be written as

$$\delta\omega_{max} = 0.86 \int_0^l \gamma\,(z)P_0\;\xi\;e^{-\alpha z}dz/T_0 \tag{10}$$

where $z$ is propagating distance, $\alpha$ is given by Eq. (9b), $P_0$ is the peak power; $T_0$ is the half-width (at the 1/e intensity point) of the pulse; $\gamma$ is the nonlinear coefficient and is given by $\gamma = n_2\;p(z)\,\omega_0/c\,A_{eff}$ [$n_2$ is given by Eq. (2)], where $n_2$ is the nonlinear index coefficient, $\omega_0$ is the laser central frequency; $c$ is the light speed in vacuum, $\xi$ is the coupling efficiency, $A_{eff}$ is the effective mode area (Nisoli et al., 1996). In the statically gas-filled case, the pressure is constant along the fiber. While in the differentially pumped case, the pressure is chosen to be a minimum (0 bar) at the entrance and gradually increases along the fiber. This leads to the pressure distribution

$$p(z) = \left[p_0^2 + \left(\frac{z}{L}\right)(p_L^2 - p_0^2)\right]^{1/2} \tag{11}$$

where $p_0$ and $p_L$ are the pressure at the entrance and the exit, respectively. Then the bandwidth broadened in both the cases can be expressed as

$$\Delta\omega_{SF} = 0.86\omega_0 n_2\,p_L\,P_0\;\xi\;(1 - e^{-\alpha z})\,/\alpha\;c\;T_0\;A_{eff}\,, \tag{12}$$

$$\Delta\omega_{DP} = 0.86\;\omega_0 n_2\;p_L\xi\;P_0 \int_0^l \frac{\sqrt{z}\;e^{-\alpha z}}{\sqrt{L}\;c\;T_0\;A_{eff}}dz. \tag{13}$$

So that the bandwidth broadening depends on both the input laser characteristics and the gas used in the hollow fiber. The nonlinear coefficient $\gamma$ takes the values 7.4 × $10^{-25}$ $m^2/W$ bar, 9.8 × $10^{-24}$ $m^2/W$ bar, and 2.78 × $10^{-23}$ $m^2/W$ bar for neon, argon, and krypton respectively (Nisoli et al., 1997, Robinson et al., 2006). For short pulses (20-fs), it is found that a much larger spectral broadening is obtained than long pulses (140-fs). This is because at shorter pulse duration, gas dispersion, in addition to SPM, also plays an important role during pulse propagation while for longer pulses (140 fs) a purely SPM-broadened highly modulated spectrum is observed (Nisoli et al., 1996). The higher order nonlinear effect of Self-steepening has to be taken into account as well for the short input pulses. This higher order nonlinear effect is due to the intensity dependence of the group velocity and leads to an asymmetry in the SPM-broadened spectra with a larger broadening on the blue side (blue-shift) (Baltuska et al., 1997).

In the following, we reviewed recent results of some working groups in that field;

In 2006 Joseph Robinson et al., group used 700-μJ, 30-fs laser pulses with a 50-nm bandwidth centered at 800 nm and 1-kHz repetition rate 15 mm diameter to generate SC for few cycle generation (Robinson et al., 2006). The beam was focused into a 1-m-long hollow fused-silica fiber and reach intensity of about $2\times10^{14}$ $Wcm^{-2}$. The fiber is mounted inside a larger fused-silica support capillary, with an outer diameter of 6 mm and filled with neon pressures of 0–3 bar, in both the SF and DP cases. They found that the transmission remains constant with increasing pressure for the DP case, in contrast to the SF case, which exhibits a sharp drop in transmission with increasing pressure. They conclude that performance is due to defocusing at the fiber entrance, which causes an increase in the focal spot size and beam aberrations and consequently a reduced coupling efficiency. In these experiments, they observed a spectra broaden in both cases up to a pressure of 1.5 bar, with the DP bandwidth lagging that of the SF case, due to the reduced average gas density for the DP case. At higher pressures, while the DP spectra continue to broaden, the SF spectra begin to reduce in bandwidth due to the reduced energy throughput at these pressures. finally, a Spectral broadening of $\Delta\omega = 7\times10^{14}$ rad s−1 with an energy throughput of 40% and corresponding output pulse energies of 290 μJ and subsequent compression to 6.5 fs has been achieved in the DP fiber filled with neon at 3 bar.

In 2008 Katsumi Midorikawa et al., group used 800 nm, 5 mJ, 25-fs laser pulses 1-kHz repetition rate to generate SC for few cycle generation . The beam was focused into a 220 cm-long hollow fused-silica fiber and reach intensity of about 2.4 $\times10^{14}$ $Wcm^{-2}$. The fiber is mounted inside a larger fused-silica support capillary, and filled with neon in gradient pressures of 0 - 1.6 bar (Samuel et al., 1996). They observed the spectral broadening for various neon pressures on the exit side as shown in Fig. 1(a) (Fig. 1(a), (b) is reproduced with permission from Akira Suda (Samuel et al., 1996)). Then they used the same parameters as those used in the experiments to make simulations for these broadenings based on a method described elsewhere (Nurhuda et al., 2003) in order to understand the mechanism driving the spectral broadening as shown in fig.1 (b). A good agreement was found between the experimental and the theoretical simulations. By investigating the simulation data in details they recognized that the spectral broadening is predominantly driven by the self-phase modulation. Moreover, they concluded that the pressure gradient method prevents the beam from collapsing due to self-focusing as well as from plasma defocusing, while allowing the pulse to undergo self-phase modulation during its passage through the hollow fiber. However, the spectral broadening of more than 300 nm was obtained at a pressure of 1.6 atm. Under these conditions, they could obtain 5 fs laser pulse after compression.

In 2009 Chang H. Nam et al., group used a Ti:sapphire, 5 mJ, 29-fs laser pulses 1-kHz repetition rate to generate transform-limited 3.5 fs pulse (Park et al., 2009). The beam was focused into a 100 cm-long hollow fused-silica fiber. The fiber is mounted inside a larger fused-silica support capillary, and filled with neon in gradient pressures of 0 - 2.0 bar. They observed the spectral broadening for various neon pressures on the exit side as shown in Fig. 2 (Fig. (2),(3) are reproduced with permission from Chang H. Nam (Park et al., 2009)). The output beam from the pulse compressor was recollimated by a concave silver mirror, and two sets of chirped mirrors. The first pair of the chirped mirrors has GDD of - 80 $fs^2$ and TOD of +150 $fs^3$ (510 nm to 920 nm), and the second pair has -90 $fs^2$ and +1270 $fs^3$, respectively, (600 nm to 1000 nm). They used SHG FROG (second-harmonic generation frequency-resolved optical-gating) for temporal characterization of the compressed output pulses. They found that the shortest pulse measured was 4.1 fs at neon pressure of 1.6 bar and also at 1.8 bar. For pressures in excess of 1.8 bar, the pulse duration increased due to the high-order chirp induced by self-focusing and ionization. The FROG measurement showed

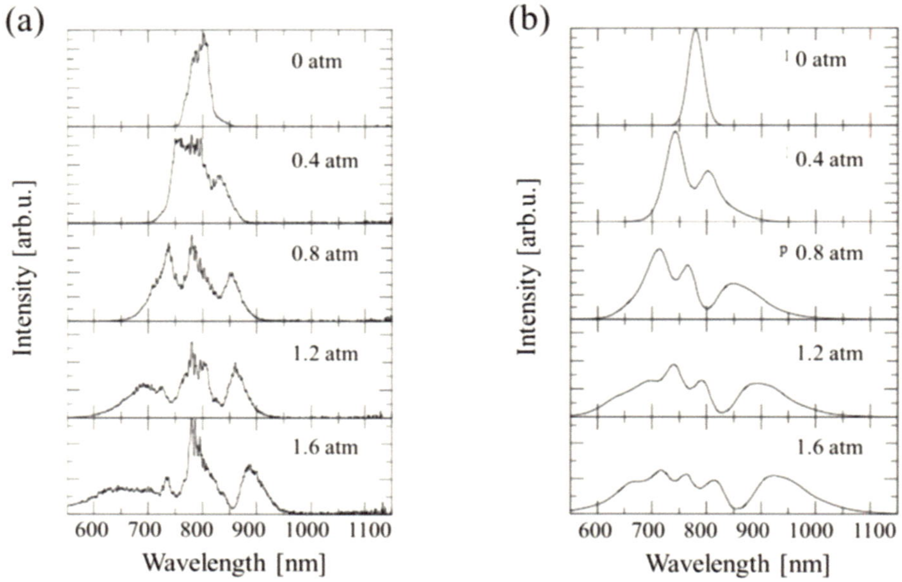


Fig. 1. Comparison of spectral broadening at various neon gas pressures. (a) Experimental data and (b) Simulated results.

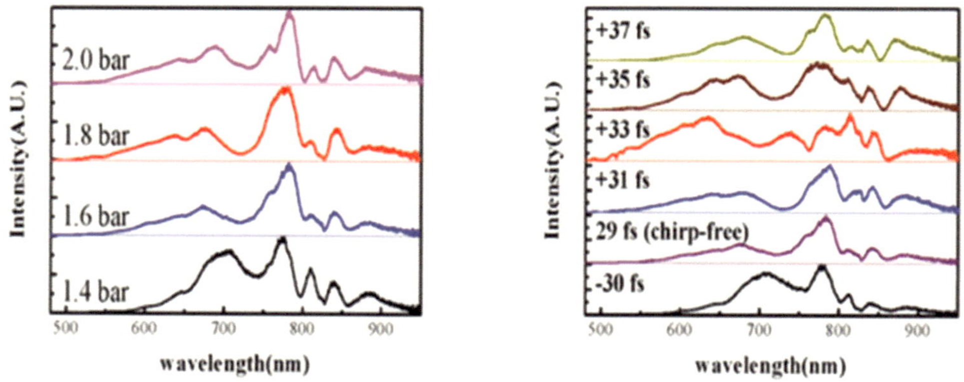


(a) (b)

Fig. 2 Spectral broadening in a differentially pumped hollow-fiber pulse compressor measured while changing (a) Ne pressure with chirp-free 29-fs pulses and (b) laser chirp with 1.6-bar neon.

that the GDD increase with neon pressure was much larger than that theoretically estimated for the increase of neon pressure. In these studies they also studied the effect of laser chirp on spectral broadening. That was done by changing the grating distance in the pulse compressor of the kHz CPA Ti:Sapphire laser as shown in Fig. 3 where the output spectra changed very delicately to laser chirp. Their obtained results showed that the spectral broadening was sensitive to the direction of the laser chirp. With negatively chirped pulses, spectral broadening was reduced, while the broadening was effective with positively chirped pulses. Furthermore, they observed 3.4-fs pulses in the transform-limited case using positively chirped 31-fs pulses, while the negatively chirped 30-fs pulses resulted in a longer transform-limited duration of 4.4 fs. They found that while the compressor cell was evacuated, the energy transmission was 62 % (3.1 mJ), but it decreased to 28% (1.4 mJ) with 1.6-bar neon.

In 2010, Wei Zhi-Yi group compared the spectral broadening from statically neon-filled and differentially pumped hollow fibers (Wei Zhang et al., 2010). They used 800 nm Ti: sapphire 25fs 800 μj laser which is focused into a fused silica 1 m hollow fiber filled with neon. The spectral broadening was obtained at different pressures, as shown by the spectra in Fig. 4 (Fig. (4),(5) are reproduced with permission from Wei Zhi-Yi (Wei Zhang et al., 2010)). They found that for the statically filled case, the spectrum broadens obviously as the gas pressure increases, even at very low pressure. For the optimum pressure of 1.5 bars, the broadening bandwidth covers the range from 460nm to 930nm with transmission efficiency of 50%. However, at pressure 2.0 bars they observed a narrow bandwidth with lower transmission and significant spot splitting. They concluded that this was due the energy loss due to ionization and defocusing at the fiber entrance at very high pressure. For the differentially pumped case, it was noticed that the bandwidth is almost unchanged at the pressure lower than 1 bar. However, the spectrum continuously broadens with nearly constant transmission efficiency with the increase of the pressure. The pressure was limited by the windows pressure maximum capacity at 2.5 bars. At the later pressure, the broadest spectrum was obtained in region from 420 nm to 960 nm with transmission efficiency of 68.8% for 0.55 mJ pulse energy. In this case, both of the bandwidth and the transmission efficiency are much higher than the statically filled case. These results show the obvious advantages of the differentially pumped hollow fiber, which supports 3.3 fs transform limited pulses with higher transmission efficiency. Based on a set of negative dispersion chirped mirrors and _ne adjustment of the small angle wedges, laser pulses as short as 4.4 fs are obtained, as shown in Fig. 3. It corresponds to only 1.6 cycles at the central wavelength. The compressed pulse duration obtained is 1.3 times the Fourier transform limitation; the reason is that the higher order dispersions cannot be compensated perfectly.

In general, the same equations used to describe pulse propagation in optical fibers can be used for gas-filled hollow fibers. Both of the dispersion and the weights relative of SPM can be evaluated using characteristic parameters such as the dispersion length$L_D$, and the nonlinear length $L_{NL}$ where

$$L_{NL} = 1/(\gamma P_0) \tag{14}$$

$$L_D = T_0^2/|\beta_2| \tag{15}$$

where $P_0$ is the peak power of the pulse and $T_0$ is the half-width (at the 1/e-intensity point) of the pulse *and* $\beta_2 = \frac{d^2\beta}{d\omega^2}$ is the GVD (Group-velocity dispersion) of the fiber filled with gas. The fiber length L is a very important factor to affecting the nonlinear processes that if L exceeds the dispersion length $L_D$ *and* the nonlinear length $L_{NL}$, both of SPM and dispersion will play an important role in pulse propagation through the fiber. The optimum fiber length $L_{opt}$ for best pulse compression, i.e. for interplay between GVD and SPM for the generation of linearly chirped pulses, is approximated by $L_{opt} \approx (6\, L_{NL} L_D)^2$ (Tomlinson et al., 1984). For the case of higher pulse energies there are two considerations set the limit

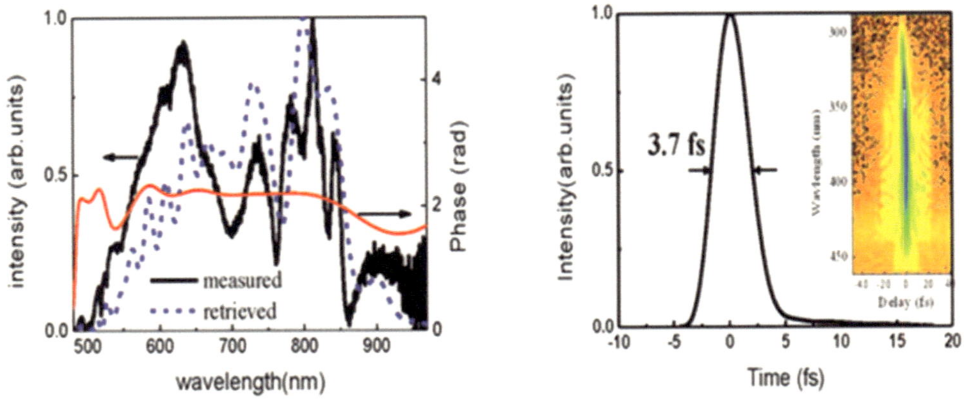


Fig. 3 (a) Measured (black solid) and retrieved (blue dotted) spectra and spectral phase of the laser output obtained with 1.6-bar neon and positively chirped 33-fs laser pulse and (b) temporal profile of the laser pulse measured using the SHG FROG method. The inset shows the FROG trace.

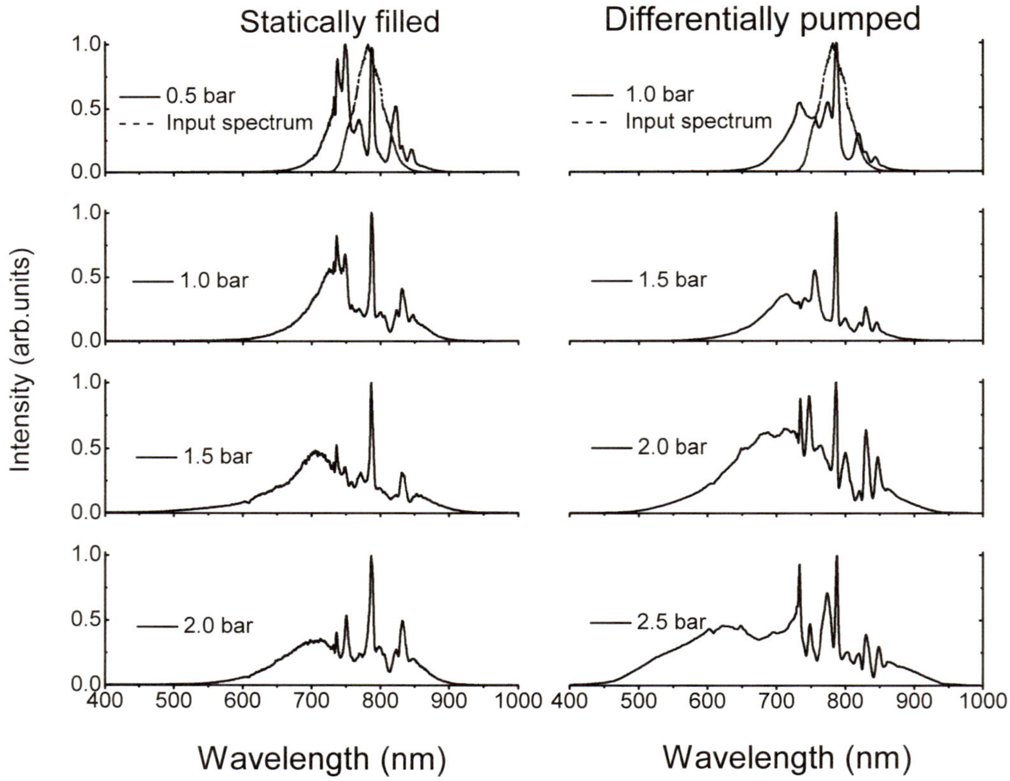


Fig. 4. Spectral broadening in the hollow fiber for the statically gas-filled (a)-(d) and differentially pumped (e)-(h) cases as a function of gas pressure. The dashed lines show the spectrum of the input pulse.

when scaling this approach to supercontinuum generation. First, the input laser peak power must be smaller than the $P_c$ critical power for self-focusing $P_c = \lambda^2/2\pi n_2$ (for a Gaussian beam) which constraints the type of noble gas used and its pressure. Second, to avoid ionization, the input laser peak intensity should be smaller than the multiphoton ionization threshold which constraints the hollow fiber diameter and the type of gas used. The second constraint can be reduced by using shorter pulses, since the threshold for multiphoton ionization increases with decreasing pulse duration.

In the following Sections, we will demonstrate our experimental findings for the generation of few-cycle pulses using supercontinuum generation in hollow fiber filled with neon gas.

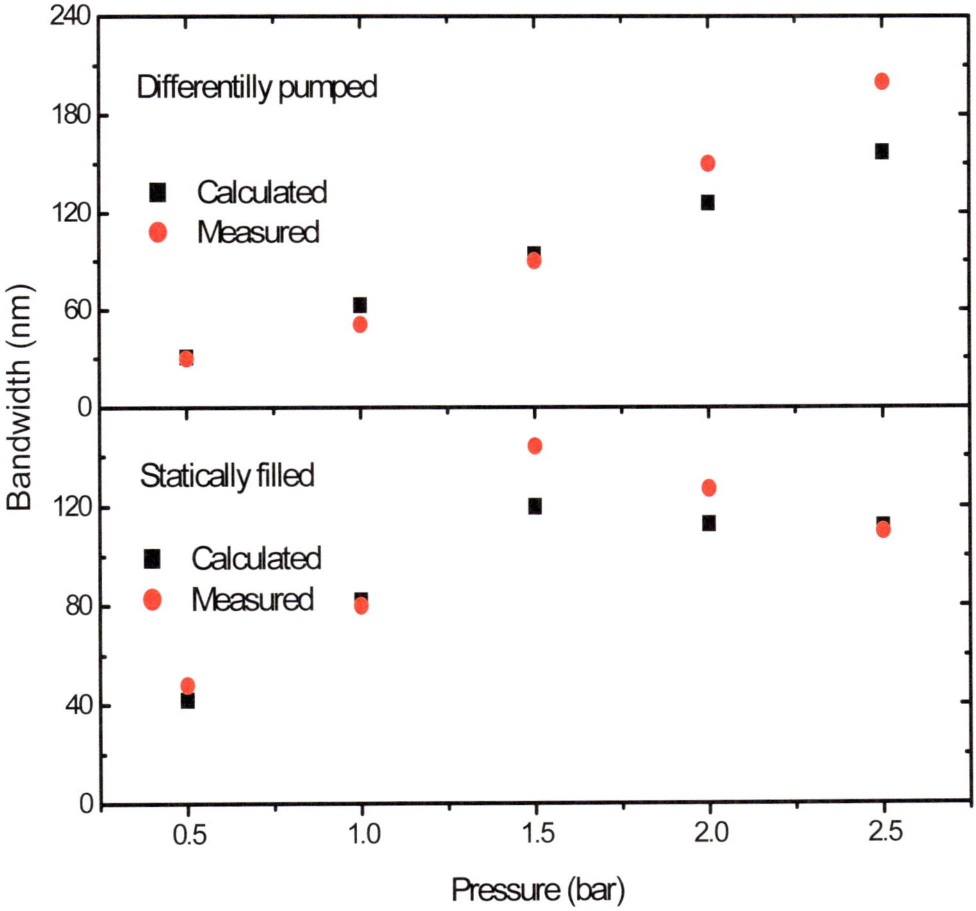


Fig. 5. Calculated and measured bandwidths broadening at the full-width-half-maximum (FWHM) for both the differentially pumped (upper panel) and statically gas filled (lower panel) cases.

## 5. Experimental setup

Figure 6 the experimental setup used for SC generation and pulse compression. A typical chirped-pulse amplifier (CPA) based on a commercial multipass Ti:sapphire system (Femtopower Pro, Femtolasers GmbH) delivers pulses 900-μJ, 28-fs laser pulses with a 55-nm bandwidth centered at 800 nm and 3-kHz repetition rate (Rolland et al., 1988, Nisoli et al. 1997, Cerullo et al. 2000, Marcatili et al. 1964). A Ti:sapphire oscillator (Rainbow; Femtolasers GmbH) produces sub-10 fs seed pulses with 2.4 nJ energy at 76 MHz. The oscillator was pumped by a continuous wave cw diode-pumped frequency doubled Nd:YVO4 laser (Coherent Verdi) at 532 nm and 3.10 watt. The oscillator output pulses are then stretched by 30 mm of SF57-glass before coupling them into the amplifier.

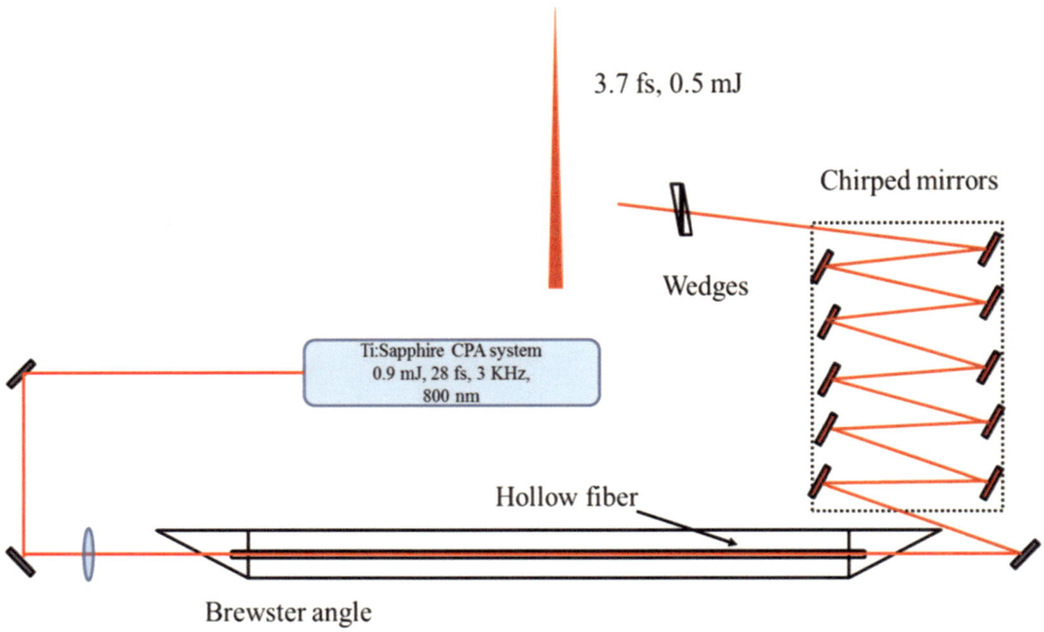


Fig. 6. Schematic of the experimental setup where the laser beam is focused into the statically pumped hollow fiber by a lens with focal length of 1.5m and then compensated for by chirped mirrors and a pair of thin wedges.

A nine-pass amplifier with three-mirror configuration was used to boost the pulse energy to the millijoule level. This amplifier consisted of a pair of dielectric-coated spherical mirrors (R/C=800, 500 cm) and a flat silver mirror overcoated with $MgF_2$. A 5-mm-long Brewster-cut 0.25% titanium-doped-sapphire crystal system was placed at the confocal position of the two spherical mirrors and was pumped by 527 nm, 25.6 watt Nd:YLF laser (Quantronix 527DQ) (Quantronix). About 90% of the focused pump total energy was absorbed by the Ti:sapphire crystal. To reduce thermal lensing effects, the crystal was cooled to -100 °C by using immersion cooling. After the fourth pass through the crystal the beam is coupled out by mirror and fed into the single pulse selecting system. This system consists of a Pockels cell between two crossed polarizers, a Berek compensator and an acousto-optic programmable dispersive filter (DAZZLER) (FASTLITE), which improves the pre pulse suppression ratio of the single pulse selection, provides fine dispersion control, and preventing gain narrowing before coupling back to the amplifier. After amplified in another four passes, the pulse passes through the 8th-pass periscope followed by a telescope which adapts the beam for efficient gain saturation across the entire pumped volume for the last two passes through the crystal. The beam is finally picked out of the multipass system by the 9th-pass PICK-OFF prism. The output of the amplifier was more than 1.1 mJ with hundreds of picosecond duration and 3 kHz repetition rate. After amplification, the beam sent into a grating compressor. The compressor consists of two high-efficiency transmission gratings and a vertical retro-reflector to get 23 fs with about 80% throughput efficiency of 0.9 mJ.

The beam from this laser system was focused into a fused silica hollow fiber with a length of 1 m (250 μm inner diameter, and 750 μm outer diameter) by a lens with 1.5 m focal length. The focusing was chosen to be 65% with respect to the hollow fiber diameter to satisfy the maximum coupling criterion, resulting in an intensity of $2\times10^{14}$ $Wcm^{-2}$.The fiber was kept on a V-groove and then placed inside a tube chamber made of stainless steel with Brewster-cut input and output windows (0.5-mm-thick fused silica), making small enclosures at both ends of the hollow fiber. Stainless steel tubes are attached to both ends of the support capillary using o-ring seals. A CCD camera was used to image the entrance of the fiber, enabling qualitative measurement of the focal spot size. Pulse energy and spectral broadening were measured at the output of the fiber for neon pressures of 0–3 bar, using static pressure. The vacuum chamber was filled with neon with pressure of about 2 bars. Initially, the chamber is vacuumed to 2 x $10^{-2}$ torr using a mechanical vacuum pump. Furthermore, the beam from the hollow fiber was collimated using a silver concave mirror with curvature radius of 4 m. The dispersion induced by gas in the hollow fiber, air and optical components such as lens, two windows, was compensated for by a set of ten broadband chirped mirrors and a pair of fused silica wedges. Each pair of chirped mirrors had group delay dispersion (GDD) of −40 $fs^2$ for the wavelength range from 510 nm to 920 nm. The compressed pulse duration is measured with an autocorrelator.

## 6. Results and discussions

The result of supercontinuum relies on nonlinear phase shift due to self-phase modulation and self-steepening, it is crucial to optimize the dispersion of the input laser pulse for obtaining broadband supercontinuum so that the laser pulse maintains high peak power and steep temporal shape. Dispersion controlled supercontinuum spectra are presented at Fig. 7(a) and their 2D spectrogram at Fig. 7 (b).

To control the dispersion of the input pulse, group delay dispersion (GDD) was adjusted by the DAZZLER for the range from -200 to +160 $fs^2$. It's clearly shown that the dependency of supercontinuum bandwidth on the dispersion of the laser pulse exists. The optimal GDD is about -50 $fs^2$, which indicates this initial negative chirp compensates the positive dispersion from both the focusing lens and hollow fiber. The maximum bandwidth covers single octave from 450 nm to 950 nm wavelength. The transform limited pulse duration of 2.9 fs FWHM can be achieved using this spectrum as shown in Fig. 7(c). Output pulse energy is about 0.5 mJ and no significant dependence on the dispersion is observed.

Not only the dispersion of the pulse affects the spectral broadening but also the pressure of noble gas plays an important role on supercontinuum generation. In general, by increasing the pressure of Ne gas in hollow fiber, more spectral broadening occurs due to the

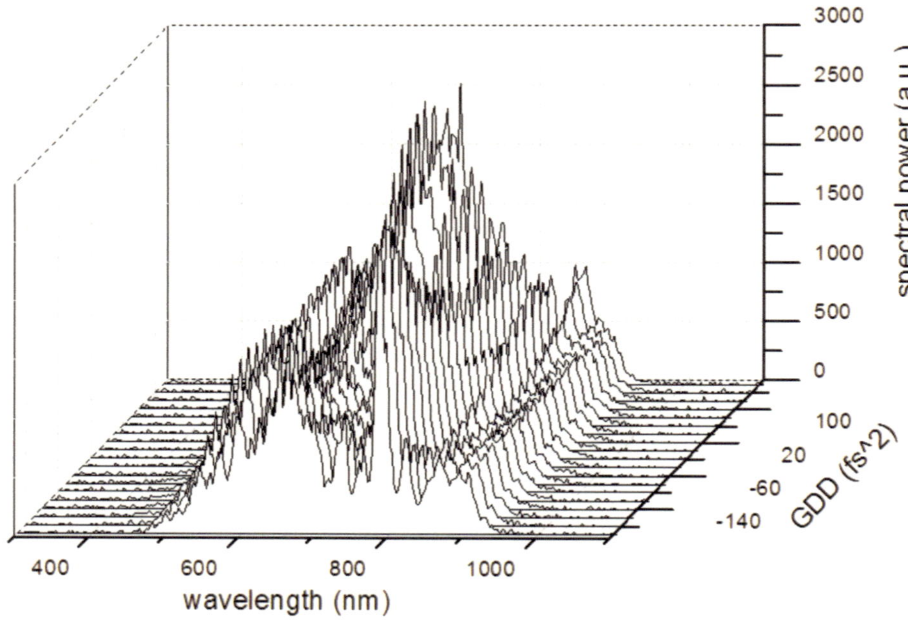


(a)

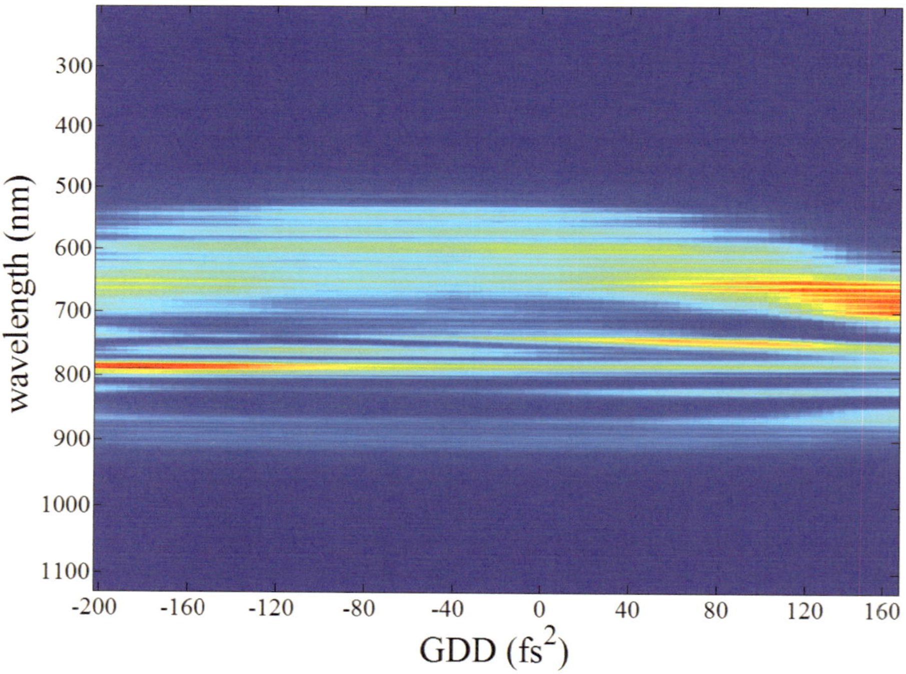


(b)

Fig. 7. (a) super-continuum spectra vs group delay dispersion (GDD) (b) super-continuum spectrogram; red:intense, blue:no-light.

increasing nonlinear susceptibility $\chi$. Fig. 8 shows the measured spectra of supercontinuum for various Ne pressures from 0.0 bar to 2.6 bar. Below the pressure of 2 bar, the spectral bandwidth increases with the increasing pressure of Ne gas, while the intensity of supercontinuum decreases over the pressure of 2 bar due to the absorption in the gas medium. One of the bandwidth limitations of supercontinuum from hollow fiber is energy lose due to leaky mode from mode matching condition for marginal wavelength. For fixed core diameter, only single wavelength satisfies the $EH_{11}$ mode matching condition which allows the best coupling efficiency while other wavelength components suffer energy loss during propagation. So as shown in Fig. 8, the spectral bandwidth does not really cover shorter than 400 nm and longer than 1000 nm wavelength. To overcome this limitation one should consider about multiple stages of supercontinuum generation in parallel or in serial with different parameters such as length and core diameter of hollow fiber, type of gas and gas pressure. Indeed, the serial supercontinuum generation has been successfully demonstrated to enhance the supercontinuum bandwidth (Schenkel et al., 2003), and parallel supercontinuum generation also can be implemented for multi-channel waveform synthesis (Goulielmakis et al., 2007).

The supercontinuum output is sent to compressor which consists of 5 pair of double-angle chirped mirrors for residual dispersion compensation (Pervak et al., 2007). The chirped mirror is originally designed for the wavelength range from 450 to 900 nm and reflectivity drops below 400 nm while reflectivity for other wavelength is above 98%. So after compressor the pulse energy drops to 0.4 mJ depending on how much spectrum is lost below 450 nm.

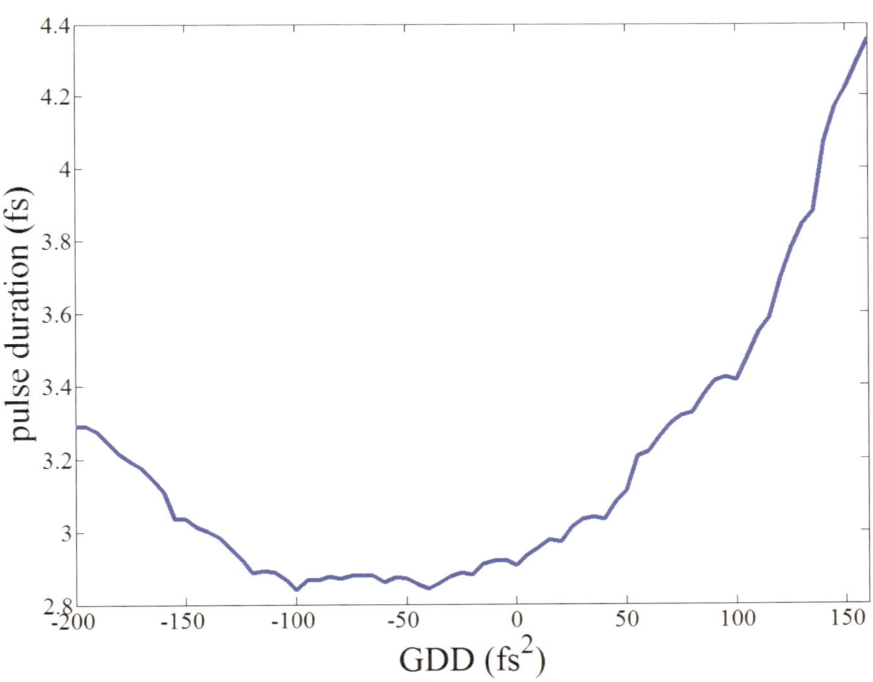


Fig. 7. (c) The transform limited pulse duration corresponding to the spectrum for each GDD.

Compressed pulse duration of 3.8 fs is measured by using second harmonic autocorrelator and the autocorrelation trace is shown in Fig. 9 (a). Precise chirp compensation can be made by either adjusting wedge thickness or slightly changing the pressure of Ne gas which introduces positive material dispersion. The pulse duration for various gas pressures is shown in Figure 9 (b). The shortest pulse duration is achieved as 3.7 fs at 2.4 bar pressure.

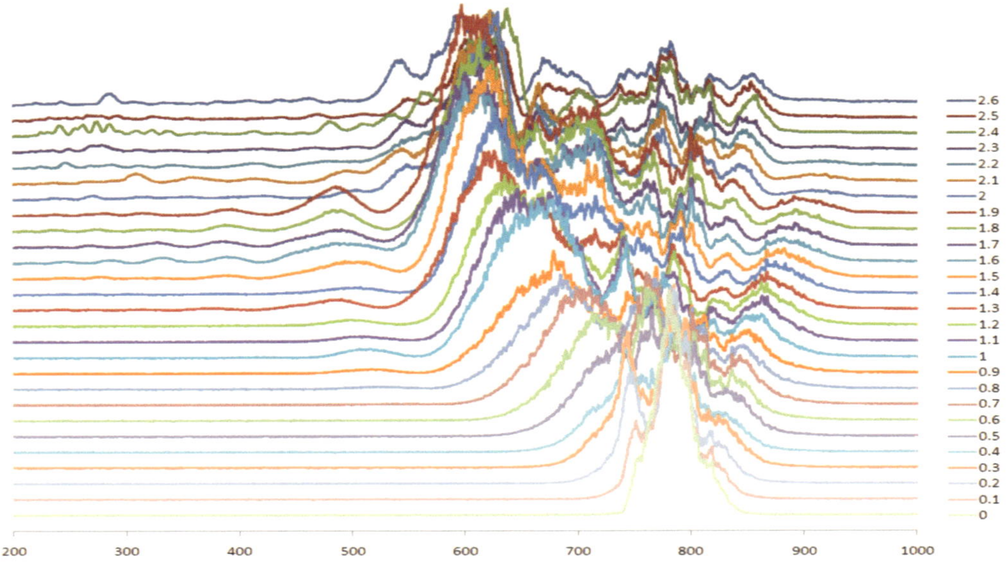


Fig. 8. Supercontinuum spectra vs Ne pressure (0 bar ~ 2.6 bar).

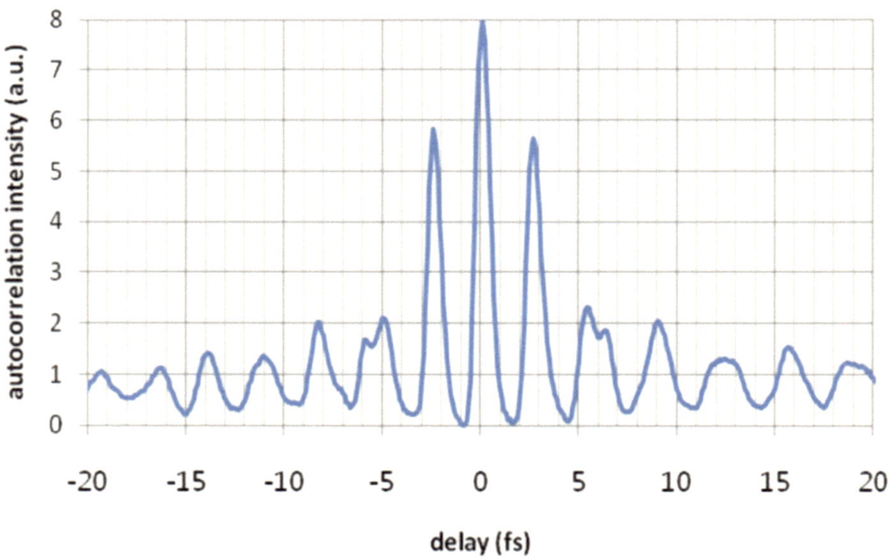


Fig. 9. (a) the measured pulse duration of 3.8 fs.

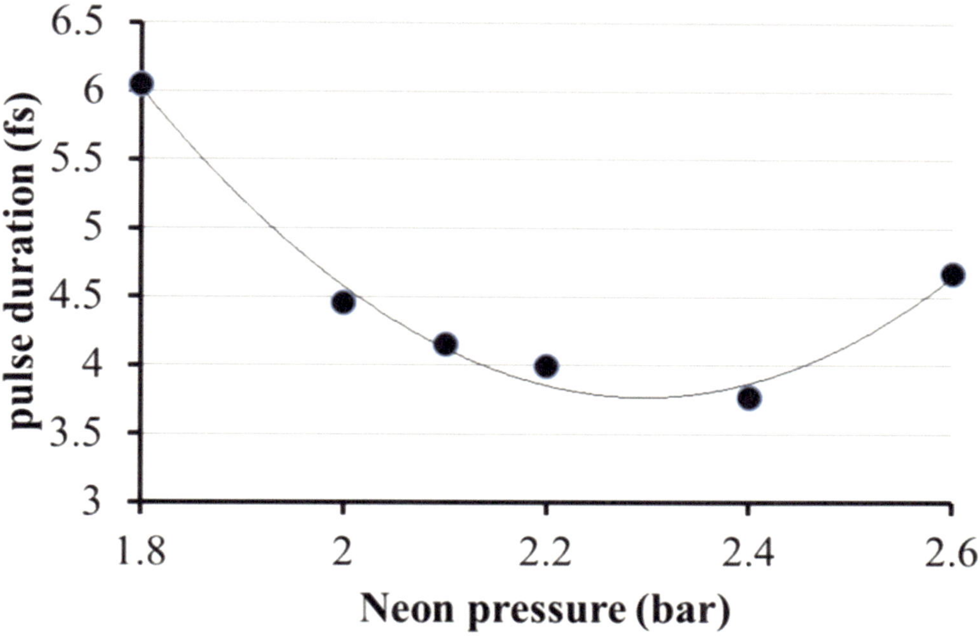


Fig. 9. (b) The variations of the pulse duration with the neon gas pressure.

## 7. Conclusion

Compression of high-energy pulses down to the few cycles regime is a well-established technology nowadays, essentially based on the hollow fiber technique. Using this tool, significant successes have been performed in ultrafast physics and nonlinear optics, in particular in the field of attosecond pulses.

Nonlinear-optical interactions of ultrashort laser pulses can efficiently generate an artificial white light with a controlled and short pulse duration, unique spectral properties, and a high spectral brightness. We have reviewed the basic theory of supercontinuum generation especially in nonlinear hollow fiber optics. In that review, we demonstrated a comparison between the spectral broadening in both SF and DF using hollow fiber filled with gas (usually neon). Also, we have showed our experimental results for the generation of few cycle pulses using different static gas pressures. We successfully observed 3.8 fs, 0.5 mJ laser pulses at 3 kHz with high through output.

In future prospective, ultrashort pulses are expected to lead to breakthroughs in communication and optical computing. Optical computing based on ultrafast logic units has the potential for revolutionizing the field of computers. On the other hand, using SPM more spectrally wider supercontinuum pulses could be achieved and hence compression down to 1 fs and possibly even into attosecond regions may be possible over the next decade using uv pulses.

## 8. Acknowledgment

This research has been supported in part by Global Research Laboratory Program [Grant No 2009-00439] and by Leading Foreign Research Institute Recruitment Program [Grant No 2010-00471] through the National Research Foundation of Korea(NRF) funded by the Ministry of Education, Science and Technology (MEST).

## 9. References

Agrawal G. P., (1995), Nonlinear Fiber Optics , Academic Press, Inc.

Agrawal G.P. (2007), "Nonlinear Fiber Optics", 4th ed., Academic Press.

Alfano R R, and Shapiro S L (1970) "Emission in the Region 4000 to 7000 Å Via Four-Photon Coupling in Glass" Phys. Rev. Lett. 24 584.

Alfano R R, and Shapiro S L (1970) "Observation of Self-Phase Modulation and Small-Scale Filaments in Crystals and Glasses" Phys. Rev. Lett. 24 592.

Alfano, R.R., Li Q., Jimbo T., Manassah J., and Ho P. (1986) Induced spectral broadening of a weak picosecond pulse in glass produced by an intense ps pulse. Opt. Lett. 11, 626–628.

Alfano, R.R., Wang Q.Z., Jimbo T., and Ho P.P. (1987) "Induced spectral broadening about a second harmonic generated by an intense primary ultrashort laser pulse in ZnSe crystals" Phys. Rev. A 35, 459–462.

Backus S., Peatross J., Huang C. P., Murnane M. M., Kapteyn H. C. (1995) "Ti:sapphire amplifier producing millijoule-level, 21-fs pulses at 1 kHz" Opt. Lett. 20, 2000.

Baker S., Robinson J.S., Haworth C.A., Teng H., Smith R.A., Chirila C.C., Lein M., Tisch J.W.G., Marangos J.P. (2006),"Probing Proton Dynamics in Molecules on an Attosecond Time Scale," Science 312, 424.

Baldeck P. L. and Alfano R. R., "Intensity effects on the stimulated four photon spectra generated by picosecond pulses in optical fibers," IEEE J. Lightwave Technol. LT-5, 1712–1715.

Baltusˇka A., Wei Z., Pshenichnikov M. S., Wiersma D. A. (1997) "Optical pulse compression to 5 fs at a 1-MHz repetition rate" Opt. Lett. 22, 102.

Baltuska A., Fuji T., Kobayashi T. (2002), "Visible pulse compression to 4 fs by optical parametric amplification and programmable dispersion control," Opt. Lett. 27, 306.

Baltuska A., Wei Z., Pshenichnikov M. S., and Wiersma D. A. (1997), "Optical pulse compression to 5 fs at 1 MHz repetition rate," Opt. Lett. 22, 102.

Cairns R. A. , Bingham R., Jaroszynski D. A., and Michael Baron (2009) "Laser-Plasma Interactions" Taylor & Francis.

Cerullo G., Silvestri S. D., Nisoli M., Sartania S., Stagira S., and Svelto O. (2000), "Few-Optical-Cycle Laser Puises: From High Peak Power to Frequency Tunability," IEEE J. Sel. Topics Quant. Electronics 6, 948.

Christov I. P., Murnane M. M., and Kapteyn H. C. (1997), "High-Harmonic Generation of Attosecond Pulses in the "Single-Cycle" Regime" Phys. Rev. Lett. 78, 1251.

Corkum P. B. and Rolland Claude (1986) "Supercontinuum Generation in Gases" Phys. Rev. Lett. 57, 2268–2271.

D. Strickland and G. Mourou (1985), "Compression of amplified chirped optical pulses", Opt. Commun. 56, 219.

Dalgarno A., and Kingston A. E. (1966) "The refractive indices and Verdet constants of the inert gases" Proc. R. Soc. London Ser. A 259, 424.

DeMartini F., Townes C. H., Gustafson T. K., and Kelley P. L. (1967), "Self-Steepening of Light Pulses," Phys. Rev. 164, 312.

Donna Strickland and Gerard Mourou, (1985) "compression of amplified chirped optical pulses ", Opt. Commun. 56,3, 219.

FASTLITE 45-47 rue croulebarbe 75013 Paris France, http://www.fastlite.com/

Fork R. L., Brito Cruz C. H., Becker P. C., and Shank C. V. (1987), "Amplification of femtosecond optical pulses using a double confocal resonator" Opt. Lett. 12, 483.

Fork R. L., Cruz C. H. B., Becker P. C., and Shank C. V. (1987), "Compression of optical pulses to six femtoseconds by using cubic phase compensation," Opt. Lett. 12, 483.

Fork R. L., Greene B. I., and Shank C. V. (1981), "Generation of optical pulses shorter than 0.1 psec by colliding pulse mode locking," Appl. Phys. Lett. 38, 671.

Fork R. L., Martinez O. E., and Gordon J. P. (1984), "Negative dispersion using pairs of prisms", Opt. Lett. 9 (5), 150.

Fork R. L., Shank C. V., Hirlimann C., Yen R., and Tomlinson W. J. (1983), "Femtosecond white-light continuum pulses" Opt. Lett. 8, 1.

Goulielmakis, E., Yakovlev, V. S., Cavalieri, A. L., Uiberacker, M., Pervak, V., Apolonski, A., Kienberger, R., Kleineberg, Krausz, U., F., (2007)," Attosecond Control and Measurement: Lightwave Electronics", Science 317, 769.

Gersten J., Alfano R., and Belic M. (1980) "Combined stimulated Raman scattering in fibers" Phys. Rev. A 21, 1222–1224.

Grassl H.P. and Maiera Max (1979) "Efficient stimulated Raman scattering in silicon" Opt. Commun. 30, 253.

Grischkowsky D., Courtens E., and Armstrong J. A. (1973), "Observation of Self-Steepening of Optical Pulses with Possible Shock Formation," Phys. Rev. Lett. 31, 422.

Hasegawa A. and Tappert F. (1973), "Transmission of stationary nonlinear optical pulses in dispersive dielectric fibers. I. Anomalous dispersion," Appl. Phys. Lett. 23, 142.

Hentschel M., Kienberger R., Spielmann C., Reider G. A., Milosevic N., Brabec T., Corkum P. B., Heinzmann U., Drescher D., and Krausz F. (2001), "Attosecond metrology," Nature 414, 509.

Ippen E. P., and Stolen R. H. (1972), "Stimulated Brillouin scattering in optical fibers," Appl. Phys. Lett. 21, 539.

Ishii N., Turi L., Yakovlev V. S., Fuji T., Krausz F., Baltuska A., Butkus R., Veitas G., Smilgevicius V., Danielius R., and Piskarskas A. (2005), "Multimillijoule chirped parametric amplification of few-cycle pulses" Opt. Lett. 30, 567.

John W. G. Tisch (2000), "Phase-matched high-order harmonic generation in an ionized medium using a buffer gas of exploding atomic clusters," Phys. Rev. A 62, 041802(R).

Leemans W. P., Volfbeyn P., Guo K. Z., Chattopadhyay S., Schroeder C. B., Shadwick B. A., Lee P. B., Wurtele J. S., and Esarey E. (1998), "Laser-driven plasma-based accelerators: Wakefield excitation, channel guiding, and laser triggered particle injection" physics of plasma, 5, 5, 1615-1623.

Lehmeier H. J., Leupacher W., and Penzkofer A. (1985) "Nonresonant third order hyperpolarizability of rare gases and N2 determined by third harmonic generation" Opt. Commun. 56, 67.

Li Y T, Yuan X H, Xu M H, et al. (2006), Observation of a fast electron beam emitted along the surface of a target irradiated by intense femtosecond laser pulses. Phys Rev Lett, 96, 165003.

Li,X. F. ;Huillier A. L', Ferray M., Lompré L. A., and Mainfray G. (1989)" Multiple-harmonic generation in rare gases at high laser intensity", Phys. Rev. A 39, 5751.

Lin C, and Stolen R H (1976) "New nanosecond continuum for excited-state spectroscopy" Appl. Phys. Lett. 28 216.

Marcatili E. A. J., and Schmeltzer R. A. (1964), "Hollow metallic and dielectric waveguides for long distance optical transmission and lasers," Bell Syst. Tech. J. 43, 1783.

Martinez O. E., Gordon J. P., and Fork R. L. (1984), "Negative group-velocity dispersion using refraction", J. Opt. Soc. Am. A 1 (10), 1003.

Nakatsuka H., Grischkowsky D., and Balant A. C. (1981), "Nonlinear picosecond-pulse propagation through optical fibers with positive group velocity dispersion," Phys. Rev. Lett. 47, 910.

Nisoli M, De Silvestri S, Svelto O. (1996) Generation of high energy 10 fs pulses by a new pulse compression technique. Appl Phys Lett, 68, 2793.

Nisoli M., De Silvestri S., Svelto O., Szipöcs R., Ferenz K., Spielmann C., Sartania S., and Krausz F. (1997), "Compression of High Energy Laser Pulses Below 5 fs," Opt. Lett. 22, 522.

Nisoli M., Sansone G., Stagira S., De Silvestri S., Svelto O., and Vozzi C. (2002), "Ultra-broadband continuum generation by hollow-fiber cascading" Appl. Phys. B 75, 601.

Nisoli M., Silvestri S.D., and Svelto O. (1996), "Generation of high energy 10 fs pulses by a new pulse compression technique", Appl. Phys. Lett. 60, 2793.

Nisoli M., Stagira S., Silvestri S. D., Svelto O., Sartania S., Cheng Z., Lenzner M., Spielmann C., and Krausz F. (1997), "A novel high-energy pulse compression system: generation of multi-giga watt sub-5-fs pulses," Appl. Phys. B 65, 189.

Nisoli M., Stagira S., Silvestri S. D., Svelto O., Sartania S., Cheng Z., Tempea G., Spielmann C., and Krausz F. (1998), "Toward a terawatt-scale sub-10-fs laser technology," IEEEJ. Sel. Top. Quantum Electron. 4, 414.

Nurhuda M., Suda A., Midorikawa K., Hatayama M., and Nagasaka K. (2003), "Propagation dynamics of femtosecond laser pulses in a hollow fiber filled with argon: constant gas pressure vs differential gas pressure," J. Opt. Soc. Am. B 20, 2002-2011.

Park Juyun, Lee Jae-hwan, and Nam Chang Hee (2009) "Generation of 1.5 cycle 0.3 TW laser pulses using a hollow-fiber pulse compressor" Opt. Lett. 34 (15), 2342.

Pervak, V., Ahmad, I., Trubetskov, M. K., Tikhonravov, A. V., and Krausz, F. ,(2009), "Double-angle multilayer mirrors with smooth dispersion characteristics" Opt. Exp., 17, 10, pp. 7943-7951.

Quantronix, 41 Research Way, East Setauket, NY 11733, USA, http://www.quantronix.com.

Robinson J S, Haworth C A, Teng H, Smith R A, Marangos J P and Tisch J W G (2006), "The generation of intense, transform-limited laser pulses with tunable duration from 6 to 30 fs in a differentially pumped hollow fiber" Appl. Phys.B 85 525.

Rolland C. and Corkum P. B. (1988), "Compression of high-power optical pulses" J. Opt. Soc. Am. B 5, 641.

Samuel Bohman, Akira Suda, Masanori Kaku, Muhammad Nurhuda, Takuya Kanai, Shigeru Yamaguchi, and Katsumi Midorikawa1 (2008), "Generation of 5 fs, 0.5 TW pulses focusable to relativistic intensities at 1 kHz", Opt. Express 16 (14), 16684.

Schenkel, B., Biegert, J., Keller, U., Vozzi, C., Nisoli, M., Sansone, G., Stagira, S., Silvestri, S. De, and Svelto, O., (2003), "Generation of 3.8-fs pulses from adaptive compression of a cascaded hollow fiber supercontinuum" Opt. Lett., 28, 20, pp. 1987-1989.

Sartania S., Cheng Z., Lenzner M., Tempea G., Spielmann C., Krausz F., and Ferencz K. (1997), "Generation of 0.1-TW5-fs optical pulses at a 1-kHz repetition rate" Opt. Lett. 22, 1562.

Schenkel B., Biegert J., and Keller U., Vozzi C., Nisoli M., Sansone G., Stagira S., De Silvestri S., and Svelto O. (2003) "Generation of 3.8-fs pulses from adaptive compression of a cascaded hollow fiber supercontinuum", Opt. Lett. 28, 20.

Shank C. V., Fork R. L., Yen R., Stolen R. H., and Tomlinson W. J. (1982) "Compression of femtosecond optical pulses", Appl. Phys. Lett. 40, 761.

Shen Y. R. (1984), "The Principles of Nonlinear Optics" John Wiley & Sons, New York.

Stingl A., Lenzner M., Spielmann C., Krausz F. and Szipöcs R. (1995),"Sub-20-fs, kilohertz-repetition-rate ti:sapphire amplifier" Opt. Lett., 20, 602.

Stolen R. H., and Lin C. (1978), "Self-phase-modulation in silica optical fibers," Phys. Rev. A 17, 1448.

Stolen R. H., Ippen E. P., and Tynes A. R. (1972), "Raman Oscillation in Glass Optical Waveguide," Appl. Phys. Lett. 20, 62.

Strickland D. and Mourou G. (1985), "Compression of amplified chirped optical pulses", Opt. Commun. 56, 219.

Suda A, Hatayama M, Nagasaka K and Midorikawa (2005) "Generation of sub-10-fs, 5-mJ-optical pulses using a hollow fiber with a pressure gradient" Appl. Phys. Lett. 86, 111.

Sutter D. H., Steinmeyer G., Gallmann L., Matuschek N., Morier-Genoud F., Keller U., Scheuer V., Angelow G., Tschudi T. (1999) "Semiconductor saturable-absorber mirror assisted Kerr-lens mode-locked Ti:sapphire laser producing pulses in the two-cycle regime" Opt. Lett. 24, 631.

Takahashi Eiji J., Nabekawa Yasuo, Mashiko Hiroki, Hasegawa Hirokazu, Suda Akira, and Midorikawa Katsumi (2004),"Generation of Strong Optical Field in Soft X-Ray Region by Using High-Order Harmonics", IEEE J. Select. Topics Quantum Electron., 10, 1315.

Tomlinson W. J., Stolen R. H., and Shank C. V. (1984), "Compression of optical pulses chirped by self-phase modulation in fibers", J. Opt. Soc. Am. B 1 (2), 139.

Treacy E. B. (1969), "Optical pulse compression with diffraction gratings", IEEE J. QE-5, 454.

Valdmanis J. A., Fork R. L., Gordon J. P. (1985)"Generation of optical pulses as short as 27 femtoseconds directly from a laser balancing self-phase modulation, group-velocity dispersion, saturable absorption, and saturable gain" Opt. Lett. 10, 131.

Walid Tawfik Y. Mohamed and Mahmoud Abdel-Aty (Editor) (2007), " Recent advances in laser induced breakdown spectroscopy as elemental analytical technique for environmental applications and space exploration" in " Aspects of Optical Sciences and Quantum Information", Research Signpost 37/661 (2), Fort P.O., Trivandrum-695 023, Kerala India, ISBN: 81-308-0147-7, p. 195-229.

Wei Zhang, Hao Teng, Chen-Xia Yun, Xin Zhong, Xun Hou, Wei Zhi-Yi (2010), "Generation of Sub-2 Cycle Optical Pulses with a Differentially Pumped Hollow Fiber"Chin. Phys. Lett. Vol. 27, No. 5.

Werncke W, Lau A., Pfeiffer M., Lenz K., Weigmann H. J. and Thuy C. D. (1972) "An anomalous frequency broadening in water" Opt. Commun. 4 413.

Xu L., Spielmann Ch., Krausz F., Szipoecs R. (1996) "Ultrabroadband ring oscillator for sub-10-fs pulse generation" Opt. Lett. 21, 1259.

Yu Wet, Alfano R.R., Samb C.L. and Seymourb R.J. (1975) "Spectral broadening of picosecond 1.06 μ pulse in KBr" Opt. Commun. 14 344.

Zhong D P, Zewail A H. (1998) "Femtosecond real-time probing of reactions 23 studies of temporal, velocity, angular, and state dynamics from transition states to final products by femtosecond-resolved mass spectrometry J Phys Chem A, 102, 23,4031.

# 8

# Evanescent-Wave Pumped and Gain Coupled Whispering-Gallery-Mode Fibre Laser

Xiao-Yun Pu, Yuan-Xian Zhang and Li Feng
*Department of Physics, Yunnan University, Kunming, Yunnan, China*

## 1. Introduction

A novel Whispering–Gallery–Mode (WGM) fibre laser is demonstrated by pumping and gain coupling with evanescent-waves in this chapter. The properties of the fibre laser, including energy threshold, produced length and polarization of lasing emission have been investigated. Two important applications of the fibre lasers on optoelectronics, linearly polarized three-color lasing emission and single WGM lasing emission, are also demonstrated in this chapter. The chapter is composed by four parts. In part one, an evanescent wave pumped and gain coupled micro-cavity fiber laser is demonstrated by inserting a bare fused quartz fiber into a glass capillary filled with Rhodamine 6G dye solution, and its energy threshold properties, including the energy threshold varied with the refractive index of the dye solution for different fiber diameters and the produced length of lasing emission, are then investigated. In part two, the polarization properties of an evanescent-wave pumped WGM fibre laser are investigated. It is found that there are two kinds of polarization lasing beam emitting from the evanescent-wave pumped WGM fibre laser under two different pump conditions. In part three, WGM fibre laser, emitting linearly polarized three-color light, is demonstrated by pumping and gain coupling with evanescent waves. In part four, a coupled cylinder-cavity structure, which is used to increase the free spectral range of a WGM fiber laser, is demonstrated by binding two bare optical fibres together, and single WGM lasing emission is realized.

## 2. Part-I - Threshold property of Whispering-Gallery-Mode fiber lasers pumped by evanescent-waves

In Part I, the threshold properties of Whispering–Gallery–Mode (WGM) fiber lasers pumped by evanescent-waves are investigated. Evanescent-wave gain coupled circular microcavity lasers, such as cylindrical [1-6], spherical[7,8], capillary [9,10] and fiber knot [11] lasers, have generated much interest in recent years, which are due mainly to the fact that the gain media are distributed around their resonators and their potential applications in integrated optics, optoelectronics and optofluidics [12,13] etc. As the evanescent field of a WGM in a circular cavity extends into the gain medium, the gain can be coupled into the WGM that provides optical feedback required by lasing oscillations. To optically excite the gain molecules around a circular cavity, laser beams are usually pumped from the outside of the gain medium [1-5], this pumping configuration, called side-pumping scheme, suffers a low pump efficiency, because the pump energy is absorbed by all of excited molecules, and only a small number of molecules residing in the WGM evanescent field contributes to the optical gain. To solve this problem, evanescent-wave pumping scheme has been used in circular cavities in recent years, such as bare optical fiber [6], capillary[9], and microfiber knot[11], which reduces lasing threshold from 200 μJ /pulse in the side-pumping scheme[1] to 9.2 μJ /pulse [9] and further to 100 nJ /pulse[11]. Optical gain in the evanescent-wave pumping scheme is produced by evanescent-waves on a fiber surface, the waves are created by a pump light that propagates within the fiber by total internal reflection. Due to a strong spatial overlap between the gain profile and the WGM evanescent field, a high pump efficiency and a long gain distance along the axis of the circular cavity have been achieved [6], which has been successfully used to invent a WGM fiber laser emitting three-wavelength-range lasing beams in a single optical fiber [14].

In the evanescent-wave pumping scheme, laser resonator is surrounded by a dye solution of low refractive index (RI), which acts not only as the gain medium, but also the cladding material for the pumping and circulating light in a microcavity. The penetration depth of pump light into the dye solution determines the gain value, the smaller RI difference between the cavity material and dye solution is, the larger gain is produced by evanescent-wave pumping scheme. Further more, the quality factor (Q value) of a resonator determines the cavity loss, a small RI difference between the cavity material and dye solution means a large cavity loss, which is not benefit to lasing oscillation. Therefore, the threshold property of an evanescent-wave pumped microcavity laser is affected strongly by the RI of cladding solution. However, few works, either the simulations or the experiments, have been reported to explore this important effect of the microcavity laser.

In this part, firstly, the optical gain of a microcavity fiber laser in the evanescent-wave pumping scheme is analyzed by introducing a Gaussian distribution function of pump light, a gain formula is derived. The energy losses related to a circular cavity, including cavity absorption, light scattering and leakage are considered, which leads to a quality-factor equation. Secondly, assuming the derived gain is equal to the energy loss, the energy threshold formula is achieved that is convenient to be compared with our experimental results. Based on the characteristics of frustrated totally internal reflection (FTIR) of pump light traveling along an optical fiber, an attenuated factor is introduced in the threshold energy formula, to achieve the equation for determining the produced length of lasing emission along the axial of the optical fiber. Thirdly, an evanescent-wave pumped and gain coupled WGM fiber laser is fabricated by inserting a bare fused quartz fiber (no cladding) into a glass capillary filled with Rhodamine 6G dye solution, the energy threshold properties of the laser, including the energy threshold varied with the RI of the dye solution for different fiber diameters and the produced length of lasing emission along the fiber axis, have been experimentally investigated. Fourthly, the experimental results are compared with the theoretical calculations, a summary of this work is given. Finally, the analytical formula for resonant line-width of WGM in a circular cavity is derived, which is arranged as the appendix at the end of this chapter.

### 2.1 Derivation of the energy threshold and the produced length of lasing emission

#### 2.1.1 Optical gain in evanescent-wave pumping scheme

An optical fiber without cladding is immersed in a dye solution, longitudinally pumped by laser light along the fiber axis, pump beams inside the fiber are the meridian beams [15] which can be drawn in XZ-plane as shown in Fig.I-1. When a pump beam experiences a total internal reflection at the interface of the fiber and the dye solution, its evanescent-wave ($E_p$) excites dye molecules and produces an optical gain around the fiber surface. Because fiber cross-section is a circular microcavity that supports WGM oscillation, the optical gain produced by the $E_p$ is within the WGM evanescent-wave ($E_w$), a strong spatial overlap between two evanescent-waves of $E_p$ and $E_w$ leads to a high pump efficiency in the evanescent-wave pumped WGM fiber laser.

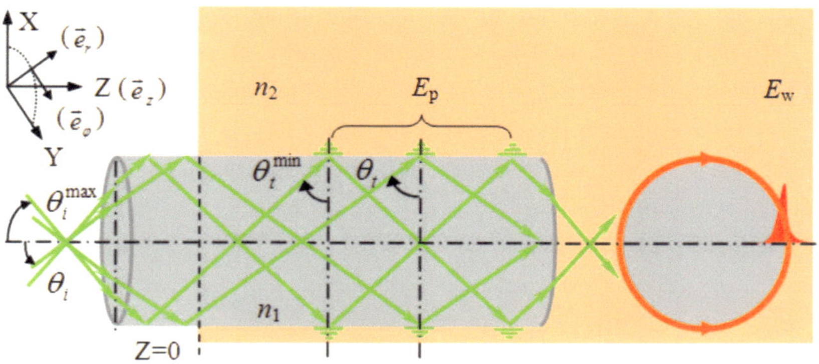


Fig. I-1. The schematic diagram of a WGM fiber laser pumped by an evanescent-wave. Pump light propagating within a bare fiber in the form of meridian beams, $\theta_i$: the entrance angle of a beam on the fiber end face, $\theta_t$: the incidence angle of the beam on the fiber side face, $E_P$ and $E_w$ : the evanescent-fields of a pump light and a WGM.

The evanescence-wave intensity of the pump light can be calculated by [16]

$$I_p(r) = E_p^2(r) = I_0 \exp[-2k(\beta_1 + \beta_2)(r-a)]\,,\ (r \ge a), \tag{I-1}$$

Where $a$ is the fiber radius, $k = 2\pi / \lambda_p$ is the wave vector of the $E_p$, $\lambda_p$ is the wavelength of the pump light in vacuum space, $\beta_1 = [n_1^2 \sin^2 \theta_t - n_2^2]^{1/2}$ is the attenuated factor induced by total internal reflection, $n_1$ and $n_2$ are the RIs of fiber core and cladding solution hereafter ($n_1 > n_2$), $\theta_t$ is the incident angle of the pump light on the fiber interface that is larger than the critical angle $\theta_c = \sin^{-1}(n_2 / n_1)$, $\beta_2 = \lambda_p \alpha_{abs}^{p,out}$ is the attenuated factor due to the absorption of $E_p$ by dye molecules, $\alpha_{abs}^{p,out}$ is the absorption coefficient of dye solution at the wavelength $\lambda_p$. Assuming the concentration of dye molecules to be $N_0$, the optical gain coefficient in the small signal condition can be written as

$$g(r) = C(\zeta, \lambda_c, n_2) N_0 \varepsilon_{p0} \exp[-2k(\beta_1 + \beta_2)(r-a)]\,,\ (r \ge a). \tag{I-2}$$

Where pump energy $\varepsilon_{p0}$ is used to substitute of the intensity of $I_0$ for the convenience of comparing the calculation with the experiment, $C(\zeta, \lambda_c, n_2)$ is a coefficient related to coupling efficiency of pump energy ($\zeta$), the fluorescence-quantum efficiency of dye molecules, which is a function of center wavelength of lasing emission ($\lambda_c$) and the RI for the clad solution ($n_2$). After integrating Eq. (2) along radial direction, the total optical gain can be expressed as

$$G=\int_a^{\infty} g(r)dr=\int_0^{\infty} C(\xi,\lambda_c,n_2)N_0\varepsilon_{p0}\exp[-2k(\beta_1+\beta_2)R]dR$$

$$=\frac{C(\zeta,\lambda_c,n_2)N_0\lambda_p\varepsilon_{p0}}{4\pi[(n_1^2\sin^2\theta_t-n_2^2)^{1/2}+\lambda_p\alpha_{abs}^{p,out}]}. \tag{I-3}$$

Since the pump-light source is usually a tightly focused laser beams [6, 9, 11], each beam coupled into the fiber incidents upon the fiber side face with different angle $\theta_t$. $\overline{\sin^2\theta_t}=\int_{\theta_t^{min}}^{\pi/2} f(\theta_t)\sin^2\theta_t d\theta_t$ is used to substitute the term of $\sin^2\theta_t$ in Eq. (3), where the $\theta_t^{\min}$ is the minimum incident angle at fiber side face that relates to the maximum incident angle $\theta_i^{\max}$ at fiber end face as shown in Fig.I-1, $f(\theta_t)$ is the distribution function of incident angle that is defined as the fraction of pump intensity per unit angle. In order to get the $f(\theta_t)$, the intensity distribution of pump light on the fiber end face is assumed to be a Gaussian function, $I(\theta_i)=I_0\exp[-\tan^2\theta_i/\tan^2\theta_i^{\max}]$. The $f(\theta_t)$ thus can be expressed as

$$f(\theta_t)=f_0\exp[-\frac{\cos^2\theta_t(1-n_1^2\cos^2\theta_t^{\min})}{\cos^2\theta_t^{\min}(1-n_1^2\cos^2\theta_t)}], \tag{I-4}$$

Where $f_0$ is the normalized factor that can be calculated by the numeral integration. For the purpose of comparison, a homogeneous distribution function $f(\theta_t)=1/(\pi/2-\theta_t^{\min})$ has been used to calculate the term of $\sin^2\theta_t$, the calculated optical gains for both distribution functions are shown in Fig.I-2. Let $\sin^2\theta_t$ be equal to $\sin^2\theta_t^{\max}$, the calculated gain curves are also shown in Fig.I-2.

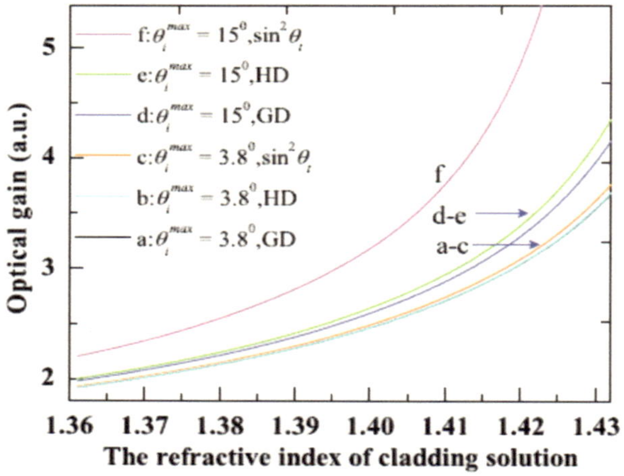


Fig. I-2. The calculated optical gains varied with the RI of cladding solution. The curves a-c: $\theta_i^{\max}=3.8^\circ$ ($\theta_t^{\min}=87.4^\circ$), the curves d-f: $\theta_i^{\max}=15^\circ$ ($\theta_t^{\min}=79.8^\circ$). HD: homogeneous distribution, GD: Gaussian distribution. In the calculations, $n_1$=1.458 for the quartz fiber used in our work, $\lambda_p$=532 nm, and $\beta_2=\lambda_p\alpha_{abs}^{p,out}=0.062$ for the dye concentration of $4\times10^{-3}$ M/L.

The calculations indicate that (1) the optical gain increases with the increase of $n_2$ value and the incident angle $\theta_i$, (2) when the incident angle $\theta_i$ is small, the effect of distribution function on the optical gain is not obvious, however, the effect can not be neglected with the increase of the incident angle.

### 2.1.2 The cavity energy losses

Assuming the $Q_{abs}$, $Q_{leak}$ and $Q_{sca}$ to be the quality factors of a circular cavity related to the energy loss coefficients $\alpha_{abs}$, $\alpha_{leak}$ and $\alpha_{sca}$, which are caused by cavity absorption, light leakage and scattering, respectively, the total quality factor ($Q_{tol}$) of a WGM in a cylindrical cavity can be expressed as

$$\frac{1}{Q_{tol}} = \frac{1}{Q_{abs}} + \frac{1}{Q_{leak}} + \frac{1}{Q_{scat}} \tag{I-5}$$

The absorption loss ($\alpha_{abs}$) includes the losses inside the cavity ($\alpha_{abs}^{w,in}$, related to the inner field of a WGM) and outside the cavity ($\alpha_{abs}^{w,out}$, related to the evanescent field of a WGM). To calculate it, we assume an occupation factor ($\eta_w$) of a WGM in a circular cavity, which is defined as the ratio of the evanescent-field volume to that of the whole WGM [1]. The occupation factor can be written as $\eta_w = \int_a^{\infty} |E(r)|^2 rdr \Big/ \int_0^{\infty} |E(r)|^2 rdr$. For a TE wave in a cylindrical coordinate system shown in the top left of Fig.I- 1, the EM field of a WGM in a cylindrical cavity can be written as [17],

$$\vec{H}(r) = A_1 J_n(n_1 k_n^l r)\vec{e}_z, \quad (0 \le r \le a), \tag{I-6a}$$

$$\vec{H}(r) = A_2 H_n^{(1)}(n_2 k_n^l r)\vec{e}_z, \quad (r \ge a), \tag{I-6b}$$

$$\vec{E}(r) = D_1 \left[ \frac{n}{n_1^2 r} J_n(n_1 k_n^l r)\vec{e}_r - \frac{k_n^l}{n_1} J_n'(n_1 k_n^l r)\vec{e}_\phi \right], (0 \le r \le a), \tag{I-6c}$$

$$\vec{E}(r) = D_2 \left[ \frac{n}{n_2^2 r} H_n^{(1)}(n_2 k_n^l r)\vec{e}_r - \frac{k_n^l}{n_2} H_n^{'(1)}(n_2 k_n^l r)\vec{e}_\phi \right], (r \ge a). \tag{I-6d}$$

Where $A_1$, $A_2$, $D_1$ and $D_2$ are constants; $a$ is the radius of the cavity; $J_n$ and $H_n^{(1)}$ are the $n^{th}$ Bessel and Hankel functions of the first kind, the derivative of the function denoted by a prime is respect to its argument hereafter; $k_n^l = 2\pi a / \lambda_l^n$ is the wave vector in vacuum for the $n^{th}$ angular mode number and $l^{th}$ radial mode order of a WGM. The occupation factors for $(l, n)$ = (1,745), (1, 1530) and (1, 2256), which correspond to fiber diameters ($2a$ = 93 ± 1μm, 196 ± 1μm and 296 ± 1μm) used in our work and the resonant wavelengths at 576.94, 583.14 and 583.72 nm, have been calculated as the function of $n_2$ value, the calculated results are shown in Fig.I-3 as the violet dotted lines. The absorption coefficient is $\alpha_{abs} = (1-\eta_w)\alpha_{abs}^{w,in} + \eta_w \alpha_{abs}^{w,out}$, and the quality factor related to the $\alpha_{abs}$ thus can be written as

$$Q_{abs} = (\frac{1}{Q_{abs}^{in}} + \frac{1}{Q_{abs}^{out}})^{-1} = [\frac{\alpha_{abs}^{w,in} \lambda_l^n (1-\eta_w)}{2\pi n_1} + \frac{\alpha_{abs}^{w,out} \lambda_l^n \eta_w}{2\pi n_1}]^{-1} . \quad \text{(I-7)}$$

In the wavelength range of 560 to 580 nm, $\alpha_{abs}^{w,in} = 2 \times 10^{-4}$ cm$^{-1}$ in our calculation, $\alpha_{abs}^{w,out}$ =20.5 to 2.7 cm$^{-1}$ for the dye concentration of 4×10$^{-3}$ M/L, and 41.0 to 5.4 cm$^{-1}$ for the concentration 8×10$^{-3}$ M/L. The calculated $Q_{abs}$ s varied with $n_2$ value are shown in Fig.I-3 as the red circular points.

For the TE wave, there are two boundary conditions at surface ($r = a$) of a cylindrical cavity, $A_1 J_n(n_1 k_n^l a) = A_2 H_n^{(1)}(n_2 k_n^l a)$ and $\frac{D_1}{n_1} J_n^{'}(n_1 k_n^l a) = \frac{D_2}{n_2} H_n^{'(1)}(n_2 k_n^l a)$, which can be deduced from Eq. (6). As $A_1 D_2 = A_2 D_1$, the characteristic Eq.(6) for TE wave is written as

$$n_2 \frac{J_n^{'}(n_1 k_n^l a)}{J_n(n_1 k_n^l a)} = n_1 \frac{H_n^{'(1)}(n_2 k_n^l a)}{H_n^{(1)}(n_2 k_n^l a)} . \quad \text{(I-8)}$$

Let $x_{l,n}$ =$2\pi a/ \lambda_l^n$ be the size parameter of a circular cavity of radius $a$, based on Eq. (8) , the $Q_{leak}$ for a TE wave of a WGM can be calculated with the analytic asymptotic formula [18] shown in Eq. (9).

$$Q_{leak} \cong \frac{\pi}{4}(n_1^2 - n_2^2) x_{l,n}^2 \left| H_n^{(1)}(n_2 x_{l,n}) \right|^2 \cdot \left[ (\frac{n}{n_1 x_{l,n}})^2 + (\frac{Y_n^{'}(n_2 x_{l,n})}{Y_n(n_2 x_{l,n})})^2 \right] . \quad \text{(I-9)}$$

The $Q_{leak}$ s of ($l$, $n$) = (1,745), (1, 1530) and (1, 2256), have been calculated as the function of $n_2$ value, the calculated results are shown in Fig.I- 3 as the green solid lines.

The value of $Q_{scat}$ is of the order of 10$^{12}$ for an ultrahigh-Q microsphere with smooth surface [19]. The $Q_{scat}$ s may be degraded to the order of ~10$^8$ for the fibers used in our work, which corresponds to a scattering loss $\alpha_{scat}$ ~2×10$^{-3}$ cm$^{-1}$, and can be neglected by comparing with $\alpha_{abs}$ that is in the range of 2 to 0.2 cm$^{-1}$ within the wavelength range of 560 to 580 nm. Therefore, $1/Q_{tol} \cong 1/Q_{abs} + 1/Q_{leak}$, and the total cavity energy loss is expressed as

$$\alpha_{tol} \cong \frac{2\pi n_1}{\lambda_l^n} \left[ \frac{1}{Q_{abs}} + \frac{1}{Q_{leak}} \right] . \quad \text{(I-10)}$$

The $Q_{tol}$ s have been calculated as the function of $n_2$ value, the calculated results are shown in Fig.I-3 as the blue solid lines.

The calculated results indicate that (1) for the fiber with a larger diameter (Fig.I- 3c, $2a$ = 296 μm), the $Q_{leak}$ is always much larger than the $Q_{abs}$ , the $Q_{tol}$ is determined mainly by the $Q_{abs}$, as the result, the cavity loss increases smoothly with the $n_2$ value; (2) for the fiber with a small diameter (Fig.I- 3a , $2a$ = 93 μm), as the $n_2$ value increase, the $Q_{tol}$ is also determined by the $Q_{abs}$ if $n_2$ is smaller than 1.385, however, the $Q_{tol}$ is determined mainly by the $Q_{leak}$ if $n_2$ is larger than 1.385, which leads to a sharp increase of the cavity loss; (3) for the fiber with a medium diameter (Fig.I- 3b, $2a$ = 196 μm), the situation is between the cases (1) and (2).

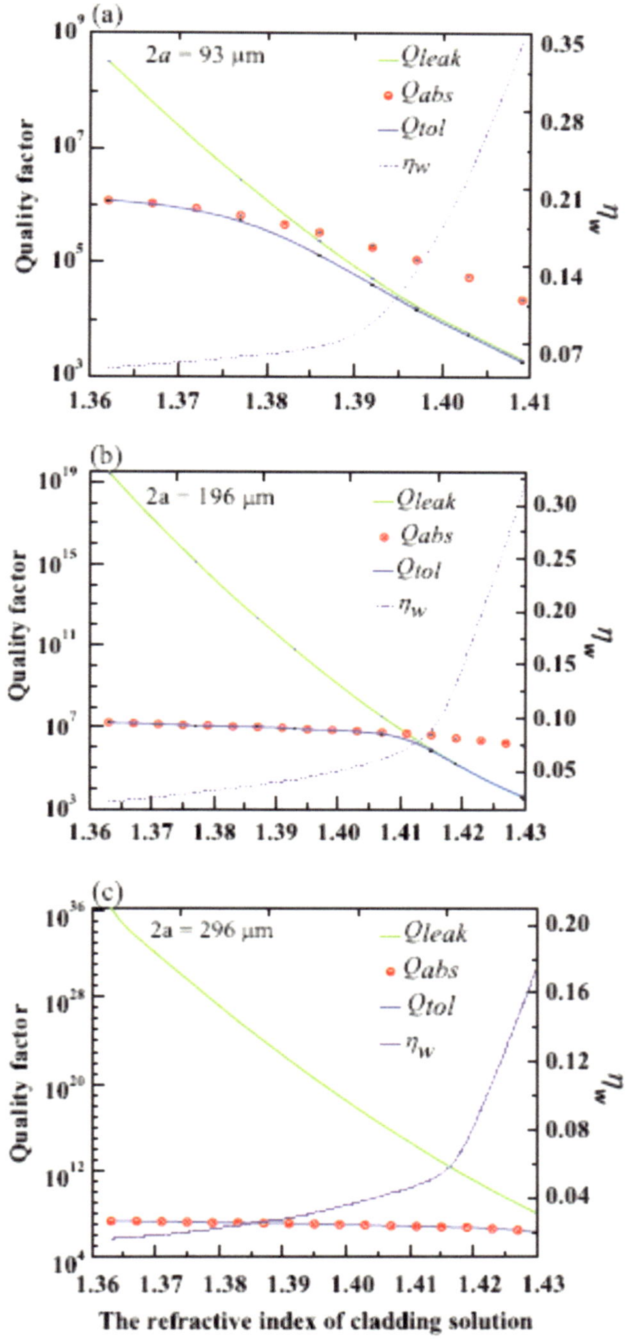


Fig. I-3. Quality factors ($Q_{\mathrm{tol}}$ in the blue solid line, $Q_{\mathrm{leak}}$ in the green solid line and $Q_{\mathrm{abs}}$ in red the circles points) and occupation factor ( $\eta_w$ in the violet dotted line) varied with $n_2$ value. The calculated fiber diameters are 93 µm in Fig.I-3a, 196 µm in Fig.I-3b and 296 µm in Fig.I-3c.

### 2.1.3 The formulas of energy threshold and produced length of lasing emission

The threshold condition of lasing action is the gain equaling to the loss. Let Eq. (3) be equal to Eq. (10), the threshold energy can be expressed as

$$\varepsilon_{p0}^{th} = \frac{4\pi[(n_1^2\sin^2\theta_t - n_2^2)^{1/2} + \lambda_p \alpha_{abs}^{p,out}]}{C(\zeta,\lambda_c,n_2)N_0\lambda_p}\alpha_{tol}$$

$$= \frac{C'(\zeta,\lambda_c,n_2)[(n_1^2\sin^2\theta_t - n_2^2)^{1/2} + \lambda_p \alpha_{abs}^{p,out}][Q_{abs}+Q_{leak}]}{N_0\lambda_p\lambda_l^n Q_{abs}Q_{leak}} \quad \text{(I-11)}$$

Due to the cladding solution is an absorption medium, the pump energy is gradually attenuated along the fiber axis (Z-axis). The energy attenuation is consistent with Beer's law, $\varepsilon_p(Z) = \varepsilon_{p0}\exp(-\alpha_{abs}^p Z)$, where $\alpha_{abs}^p$ is the absorption coefficient of pump light that consists of two parts, the fiber absorption ($\alpha_{abs}^{p,in}$) and the cladding's absorption ($\alpha_{abs}^{p,out}$). Therefore, the optical gain in Eq. (3) must be written as $G(Z) = G(0)\exp(-\alpha_{abs}^p Z)$ at the Z point. To calculate the $\alpha_{abs}^p$ value, similar to the definition of the $\alpha_{abs}$ for a WGM, we define an occupation factor $\eta_p$ for the pump light, which is the fraction of the evanescent-field intensity to that of the whole pump light. The $\alpha_{abs}^p$ is then written as $\alpha_{abs}^p = (1-\eta_p)\alpha_{abs}^{p,in} + \eta_p\alpha_{abs}^{p,out}$.

The produced length of lasing emission along the Z-axis increases with the pump energy when the energy is larger than the threshold. For a given pump energy, there is a maximum produced length $Z_{max}$, which corresponds to the optical gain at $Z = Z_{max}$ equaling to the cavity loss, that is $G(Z_{max}) = \alpha_{tol}$. The threshold energy at the position $Z = Z_{max}$ is thus written as $\varepsilon_{p0}^{th}(Z_{max}) = \varepsilon_{p0}^{th}(0)\exp(\alpha_{abs}^p Z_{max})$, and the produced length of lasing emission along the fiber axis can be expressed by the logarithmic function as

$$Z_{max} = \frac{\ln[\varepsilon_{p0}^{th}(Z_{max})] - \ln[\varepsilon_{p0}^{th}(0)]}{(1-\eta_p)\alpha_{abs}^{p,in} + \eta_p\alpha_{abs}^{p,out}}. \quad \text{(I-12)}$$

Eqs. (11, 12) will be used to compare with the experimental results in Section 2.2.3 of this part.

## 2.2 Experimental results and discussion

### 2.2.1 Experimental setup

The experimental setup is shown in Fig.I-4 schematically. The laser beams (532.0 nm), generated by a frequency doubled and Q-switched Nd:YAG laser, was used as a pump light. The required pump energy was obtained by changing the polarization direction of a polarizer ($P_1$), and its power was monitored by a power meter (PM) through a beam splitter (BS). The polarizer ($P_2$) was used to determine the polarization state of the pump light. The lenses $L_1$ and $L_2$ were used to compress the size of the pump beams. A bare fiber ($F_1$, RI=1.458) was inserted into a long glass capillary (C, 1 mm in inner diameter, 120 mm in length), the open space between $F_1$ and C was filled with ethanol and ethylene-glycol mixed solution of the Rhodamine 6G dye with a concentration of $4\times10^{-3}$ M/L. The RI of the mixed solution, acting as the cladding solution, was varied from 1.361 to 1.430 (measured by an

Abbe refractometer) by adjusting the volume ratio of the two solutions. The pump beam was longitudinally coupled into $F_1$ along the fiber axis by a lens ($L_3$) (focal length =75 mm) with a conical angle of $2\theta_i$ =7.6°. The beam would propagate within $F_1$ by FTIR if the entrance angle $\theta_i$ was smaller than the critical angle $\theta_{ic}$, which was 15.9° in our experiments. To make sure the pump beams within $F_1$ were all meridian beams, the incident direction of the pump beams was set to be along the axial direction of $F_1$.

The WGM lasing emission ($L_w$) from the rim of $F_1$ was recorded by a spectrometer (Spectrapro 500i) with an ICCD detector (PI-Max 1024RB) via an optical fiber $F_2$, which had a 0.05 nm spectral resolution when a grating of density 2400 g/mm was used. The intensity of $L_w$ was detected by a photo detector (PD, DSi200) after the lasing emission passing through an analyzer $P_3$ positioned on the Y-Z plane, the polarization state of the $L_w$ was checked by rotating $P_3$. The produced length of the $L_w$ along Z-axis was measured by a ruler, and the fiber sizes were measured by a reading microscopy with an accuracy of 1μm.

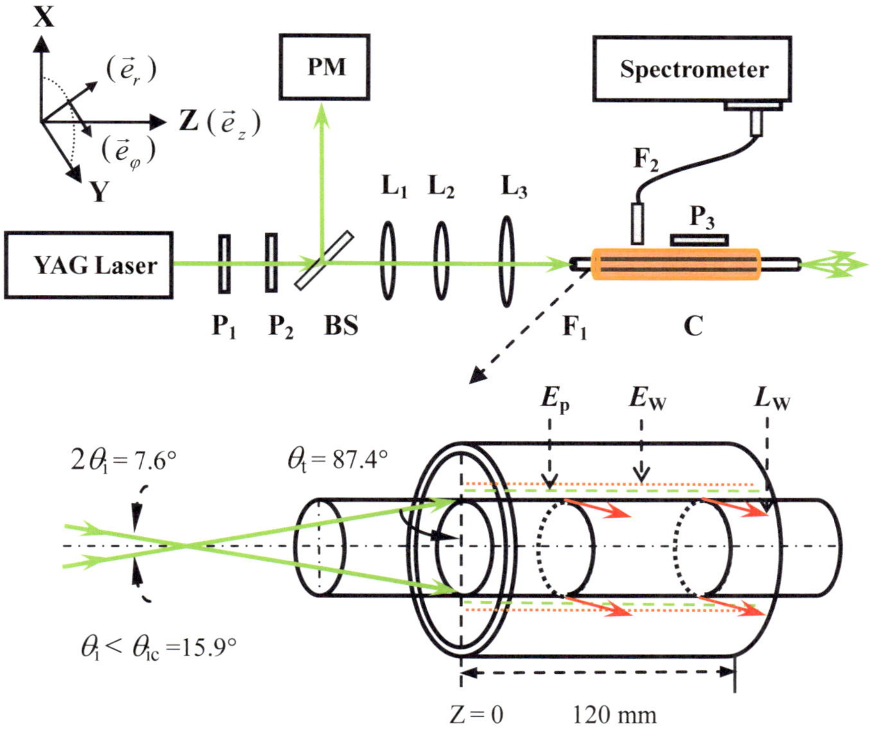


Fig. I-4. Schematic illustration of the experimental set up. $P_1$, $P_2$ and $P_3$: polarizer, BS: beam splitter, PM: Power meter, $L_1$, $L_2$ and $L_3$: lens, $F_1$: bare quartz fiber, $F_2$: optical fiber, C: glass capillary, $E_P$: evanescent field of pump light, $E_w$: evanescent field of WGM, $L_w$: WGM laser radiation.

### 2.2.2 Experimental results

As shown in the insert of Fig.I-5, the output intensities of a WGM fiber laser versus pump energies were measured at the position Z = 0 mm (fiber diameter 196 μm, $n_2$ = 1.402). When the pump energy was lower than 12.8 μJ/pulse, only a weak spontaneous fluorescence was observed around $F_1$, once the pump energy was higher than 12.8 μJ/pulse, a strong deep-yellow light emitting from the rim of $F_1$ was observed in the direction vertical to the $F_1$'s axis (X-Y plane), the energy (12.8 μJ) was the lasing threshold corresponding to the onset of the deep-yellow light. The intensity of lasing emission was found to reach its maximum (or minimum) when the polarization direction of $P_3$ are vertical (or parallel) to the $F_1$'s axis, meaning that electrical vectors of the lasing emission are perpendicular to the fiber axis, and the evanescent-wave pumped WGM lasing emission is a typical TE wave. When the pump energy was fixed at 100 μJ/pulse, the lasing spectrum recorded by the spectrometer was shown in the Fig. I-5. The interval between any pair of adjacent lasing peaks is about 0.39 nm, slightly larger than 0.38 nm that is calculated by $\Delta\lambda = \lambda^2/2\pi a n_1$ (free spectral range). In the calculation of the $Q_{tol}$ in Section 2, the values of $\lambda_l^n$ and ($l$, $n$) should be known in advance. Based on asymptotic formula (Eq. (13)) for the resonant positions of WGMs in cylindrical cavity [20-21], the modes in the lasing spectrum of TE wave in Fig.I-5 were assigned by the pair of numbers ($l$, $n$), where $a_l$ is the root of Airy function.

$$\frac{2\pi a_1}{\lambda_n^l} n_1 \approx n + 2^{-1/3} a_l n^{1/3} + \frac{3}{10} 2^{-2/3} a_l^2 n^{-1/3} + 0^{-2/3}. \quad \text{(I-13)}$$

For the fiber of diameter 196 μm and the resonant wavelength $\lambda_l^n$ =583.14 nm, $l$ and $n$ have been assigned by (1, 1525) as indicated in Fig.I-5. Using the same method, for the fibers of diameter 93 μm ( $\lambda_l^n$ =576.94 nm) and 296 μm ( $\lambda_l^n$ =583.72 nm), $l$ and $n$ have been assigned by (1, 745) and (1, 2256), respectively.

The threshold energies varied with the $n_2$ value were measured for three fibers with different diameters at the position Z = 0 mm, and the results were shown in Fig.I-6. For the fiber of diameter 296 μm, as shown in Fig.I- 6 in green square points, threshold energies decrease monotonously with the increase of $n_2$ value, which is 25.0 μJ (the maximum) if $n_2$ = 1.361, and 13.4 μJ (the minimum) if $n_2$ = 1.430. For the fiber of diameter 93 μm, as shown in Fig.I-6 in the blue triangle points, threshold energies are not sensitive to the change of the cladding solutions when $n_2$ value is between 1.361 and 1.385, but increase sharply when $n_2$ value is larger than 1.385. For the fiber of diameter 196 μm, as shown in Fig.I-6 in the red circle points, the threshold energies decrease slowly with the increase of $n_2$ value when $n_2$ is between 1.361 and 1.410, beyond the value 1.410, the threshold energies increase sharply and there is minimum threshold value (12.0 μJ) when $n_2$ = 1.412.

The output intensities versus pump energies at the positions Z = 50 and 80 mm were measured as shown in Fig.I-7 (cladding solution is the same as that in Fig.I-5). The threshold energy is 100.0 μJ/pulse at Z = 50 mm (Fig.I-7a), while it is 190.0 μJ/pulse (Fig.I- 7b) at Z = 80 mm, both values are much larger than 12.8 μJ/pulse measured at Z = 0 mm (Fig.I-5). The results indicate clearly that the threshold energy increases with the measured position (Z), which yields two important issues: first, what experimental law does the pump energy and produced length of lasing emission along the fiber axis follow? Second, which factors affect the produced length of lasing emission?

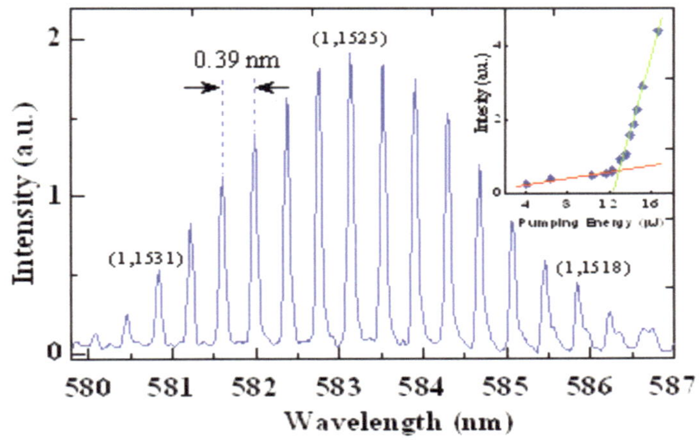


Fig. I-5. Lasing spectrum from the WGM fiber laser of a fiber diameter 196 μm. The output intensities of the laser versus pump energies are shown in the insert of the figure. The measured free spectral range is 0.39 nm indicated by a pair of arrows, and the assigned modes for each lasing peaks are indicated by brackets.

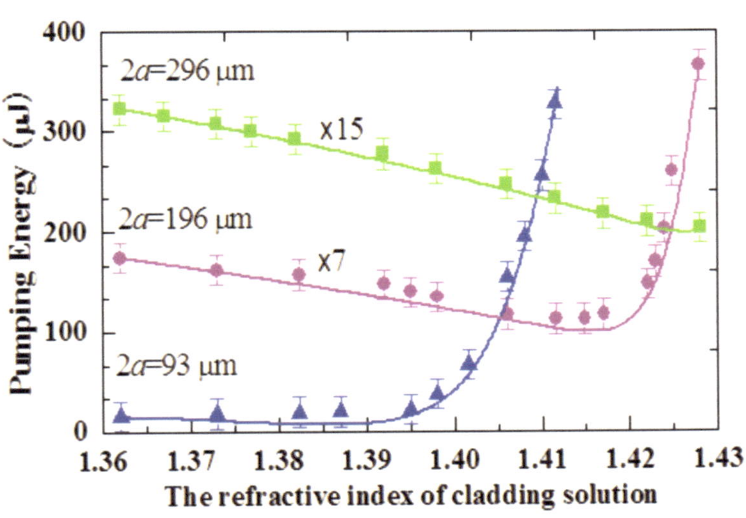


Fig. I-6. Threshold energies varied with the refractive index of cladding solution at the position of Z = 0 mm. Green square points: experimental data for the fiber of diameter 296 μm, red circle points: experimental data for the fiber of diameter 196 μm, blue triangle points: experimental data for the fiber of diameter 93 μm. Solid curves: theoretically calculated results.

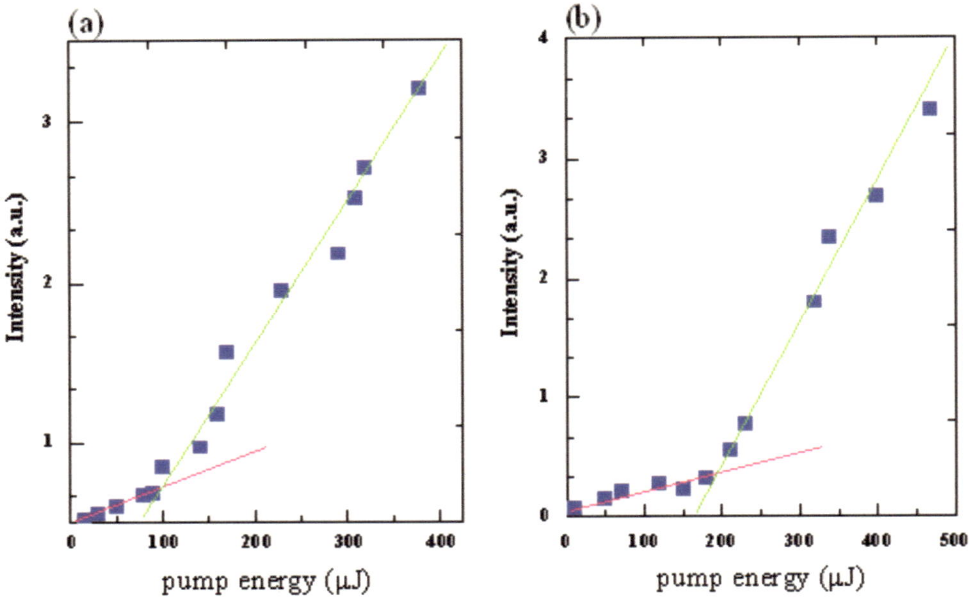


Fig. I-7. The output intensities of a fiber laser versus pump energies at different fiber position (Z). Z = 50 mm in Fig.I-7*a*, and Z = 80 mm in Fig.I-7*b*.

To address the problems mentioned above, a contrastive experiment was carried out by using the fiber of diameter 196 μm. First, different dye concentrations were used with a fixed RI of cladding solution (ethanol, $n_2$=1.361), and the produced lengths versus the pump energies was measured as shown in Fig.I-8a ($c$ = 4×10$^{-3}$ M/L) and Fig.I-8b ($c$ = 8×10$^{-3}$ M/L). Second, the dye concentration is the same as that in Fig.I- 8a, but the $n_2$ value is increased to 1.402, the produced lengths versus the pump energies was measured as shown in Fig.I-8c.
The experimental data (red triangle points) in Fig.I-8 shows that the produced length increases dramatically when the pump energy is just larger than the threshold energy at Z = 0 mm, however, the increasing trend become slow when the produced length reaches a certain value, beyond this value, the produced length is almost saturated with the pump energy. Comparing Fig.I-8a with Fig.I-8b, the energy threshold at Z = 0 mm is shifted from 49.0 μJ/pulse in Fig.I-8a (low dye concentration) to 22.0 μJ/pulse in Fig.I-8b (high dye concentration), however, the produced length is decreased from 120 to 89 mm when the pump energy is fixed at 400 μJ/pulse. Comparing Fig.I- 8a with Fig.I-8c, the energy threshold at Z = 0 mm is lowered from 49.0 μJ/pulse in Fig.I-8a ($n_2$=1.361) to 12.8 μJ/pulse in Fig.I-8c ($n_2$=1.402), however, the produced length is decreased from 120 to 97 mm when the pump energy is fixed at 400 μJ/pulse. The observed phenomena shown in Fig.I-6 and Fig.I-8 can be well explained by the theory described in Section 2.1.3.

### 2.2.3 Comparing experimental results with the theoretical calculation

The Eq. (11) is used to calculate the threshold energy varying with $n_2$ value at the position Z = 0 mm. Each parameter is known except the coefficient $C'(\zeta,\lambda_c,n_2)$ in the Eq. (11). $C'(\zeta,\lambda_c,n_2)$ is treated as $C'(\lambda_c,n_2)$ for a fiber with certain diameter, since the coupling efficiency of pump energy is unchangeable, and $C'(\lambda_c,n_2)$ is approximately treated as

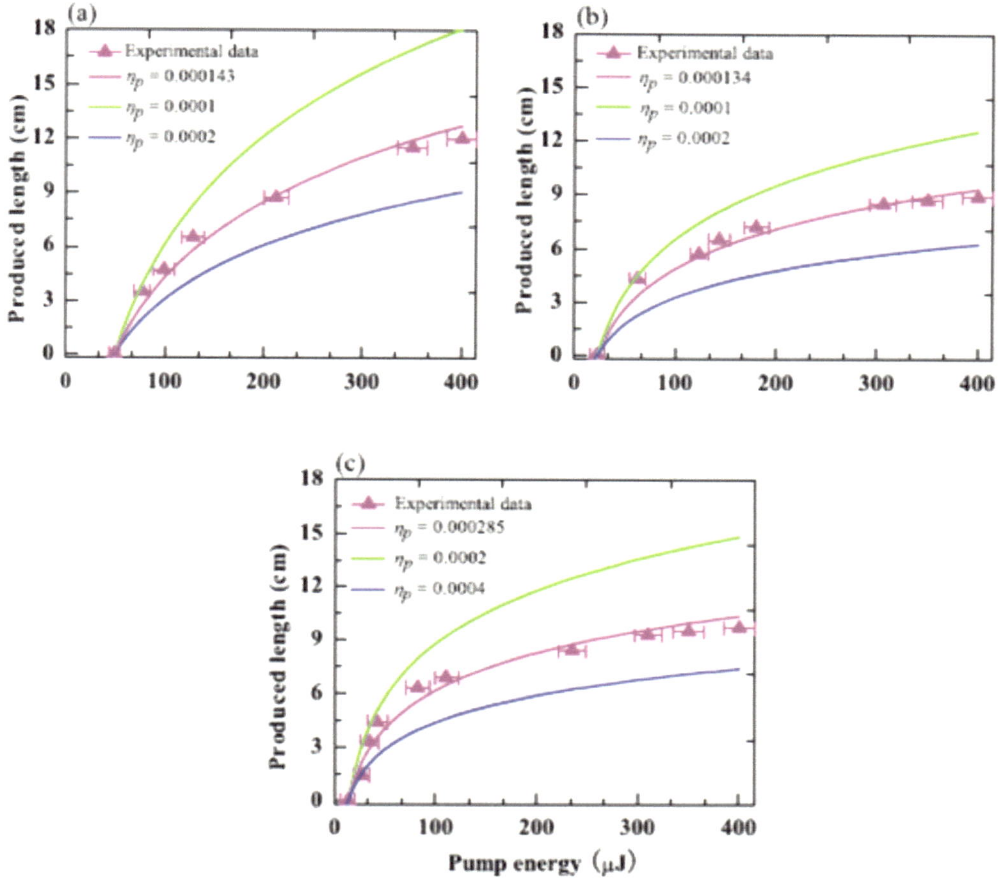


Fig. I-8. The produced length of lasing emission versus pump energy. The $n_2$ value is fixed at 1.361, but the dye concentrations are 4×10$^{-3}$ M/L in Fig.I- 8a and 8×10$^{-3}$ M/L in Fig.I- 8b. The dye concentration is the same as that in Fig.I- 8a, but the $n_2$ values is 1.402 in Fig.I- 8c. The red triangle points with error bars are experimental data, the solid curves are theoretically calculated results for various fitting parameters ( $\eta_p$ ).

$C^{'}(\lambda_c)$ when solvent effect in dye molecules is negnected. Pairs of experimental data $(n_2, \varepsilon_{p0}^{th})$ and the related $Q_{tol}$ value are substituted into the Eq. (11) to determine $C^{'}(\lambda_c)$ . For the fiber of diameter 2$a$ = 296 μm at central wavelength $\lambda_c$ ~583.72 nm, $(n_2, \varepsilon_{p0}^{th}, Q_{tol})$ = (1.361, 21.5μJ, 2.21×10$^{7}$), the determined $C^{'}$ =6.07×10$^{20}$(μJ. m$^{-1}$). For the fiber of diameter 2$a$ =196 μm at central wavelength $\lambda_c$ ~ 583.14 nm, $(n_2, \varepsilon_{p0}^{th}, Q_{tol})$ = (1.361, 25.0 μJ, 1.66×10$^{7}$), the determined $C^{'}$ =5.30×10$^{20}$ (μJ. m$^{-1}$). For the fiber of diameter 2$a$ = 93 μm, three values $(n_2, \varepsilon_{p0}^{th}, Q_{tol})$ = (1.361, 15.5 μJ, 1.19×10$^{6}$), (1.386, 20.0 μJ, 1.36×10$^{5}$) and (1.409, 256.1 μJ, 1.85×10$^{3}$) are chosen to suit the spectral shift with $n_2$ value, the determined $C^{'}$ ($\lambda_c$ = 576.94nm) = 2.33×10$^{19}$, $C^{'}$ ($\lambda_c$ = 569.91 nm) = 3.86×10$^{18}$ and $C^{'}$ ($\lambda_c$ = 564.92 nm) = 7.82×10$^{17}$ (μJ. m$^{-1}$), respectively. The calculated energy thresholds shown in Fig.I- 6 fit the experimental data very well after the $C^{'}$ values are determined.

The dependence of energy threshold on $n_2$ value for the different fiber diameters can be understood when the properties of optical gain and cavity loss are considered simultaneously. Both optical gain and cavity loss increase with $n_2$ value, the increment of energy threshold can be expressed by Eq. (11) as

$$\Delta\varepsilon_{p0}^{th}(n_2) = g(n_2)\Delta\alpha_{tol}(n_2) - |\Delta g(n_2)|\alpha_{tol}(n_2), \tag{I-14}$$

where $g(n_2) = 4\pi[(n_1^2 \sin^2\theta_t - n_2^2)^{1/2} + \lambda_p \alpha_{abs}^{p,out}] / CN_0\lambda_p$ is the factor related to optical gain. If $|\Delta g| / g > \Delta\alpha_{tol} / \alpha_{tol} = |\Delta Q_{tol}| / Q_{tol}$, that is the relative increment of optical gain being larger than that of cavity loss, thus $\Delta\varepsilon_{p0}^{th}(n_2) < 0$, otherwise $\Delta\varepsilon_{p0}^{th}(n_2) > 0$.

For the fiber of diameter $2a$ = 296 μm, $|\Delta g| / g$ is always larger than $\Delta\alpha_{tol} / \alpha_{tol} = |\Delta Q_{tol}| / Q_{tol}$ when $n_2$ varying from 1.361 to 1.430, thus $\Delta\varepsilon_{p0}^{th}(n_2) < 0$, meaning that the energy threshold decreases monotonously with the increase of $n_2$, which is shown in Fig.I- 6 as the green square points. For the fiber of diameter $2a$ = 93 μm, $|\Delta g| / g$ is slightly larger than $\Delta\alpha_{tol} / \alpha_{tol}$ when $n_2$ varying from 1.361 to 1.385, thus $\Delta\varepsilon_{p0}^{th}(n_2) \sim 0$, meaning that the increased optical gain compensates the increased cavity loss and the energy threshold keeps stable; however, $|\Delta g| / g << \Delta\alpha_{tol} / \alpha_{tol}$ after $n_2$ > 1.385, thus $\Delta\varepsilon_{p0}^{th}(n_2) >> 0$, meaning that more pump energy is required to compensate the cavity loss and the threshold energy increases with the increase of $n_2$ value, which is shown in Fig.I- 6 as the blue triangle points. For the fiber of diameter $2a$ = 196 μm, $n_2$ =1.412 is a critical RI that corresponding to $|\Delta g| / g \sim \Delta\alpha_{tol} / \alpha_{tol}$. When $n_2$ < 1.412, $|\Delta g| / g > \Delta\alpha_{tol} / \alpha_{tol}$, but $|\Delta g| / g < \Delta\alpha_{tol} / \alpha_{tol}$ when $n_2$ >1.412. Therefore, there is a minimum threshold energy at $n_2$ =1.412, which is shown in Fig.I- 6 as the red solid points.

Eq. (12) is used to calculate the produced length of lasing emission varied with pump energy. Fiber absorption coefficient ($\alpha_{abs}^{p,in}$) is 0.0002 cm$^{-1}$, and the cladding absorption ($\alpha_{abs}^{p,out}$) is 1156 cm$^{-1}$ (2312 cm$^{-1}$) at $\lambda_p$ =532 nm for dye concentration of 4×10$^{-3}$ M/L(8×10$^{-3}$ M/L). The occupation factor for pump light ($\eta_p$) is still an unknown value that is used as a fitting parameter in Eq. (12), the $\eta_p$ value for fitting the experimental data best is considered as the true $\eta_p$. The fitting results are shown in Fig.I- 8. For the three experimental conditions of (1) $c$ = 4×10$^{-3}$ M/L and $n_2$ = 1.361 in Fig.I- 8a, (2) $c$ = 8×10$^{-3}$ M/L and $n_2$ = 1.361 in Fig.I- 8b, and (3) $c$ = 4×10$^{-3}$ M/L and $n_2$=1.402 in Fig.I- 8c, Eq. (12) fits the experimental data best if $\eta_p$ = 1.43×10$^{-4}$, 1.34×10$^{-4}$, and 2.85×10$^{-4}$, respectively. It is clear that the experimental data of $Z_{max}$ and $\varepsilon_p^{th}(Z_{max})$ follow Eq. (12) well.

Comparing Fig.I- 8a with Fig.I- 8b in the same $n_2$ value, a large dye concentration in Fig.I- 8b will attenuate more evanescent field of the pump light, and reduce the penetration depth of the pump light into the cladding solution, leading the $\eta_p$ value to lower slightly down from 1.43×10$^{-4}$ in Fig.I- 8a to 1.34×10$^{-4}$ in Fig.I- 8b. Based on $\alpha_{abs}^p = (1-\eta_p)\alpha_{abs}^{p,in} + \eta_p \alpha_{abs}^{p,out}$, the calculated $\alpha_{abs}^p$ values are 0.17 cm$^{-1}$ in Fig.I- 8a and 0.31 cm$^{-1}$ in Fig.I- 8b. A large absorption coefficient $\alpha_{abs}^p$, of cause, will lead to a short produced length of lasing emission for a given pump energy. The energy threshold at the position Z = 0 mm is determined mainly by the dye concentration, the dye concentration in Fig.I- 8b is twice as that in Fig.I- 8a, therefore, the energy threshold at Z = 0 mm is reduced from 49 ($c$ = 4×10$^{-3}$ M/L in Fig.I- 8a) to 22 μJ/pulse ($c$ = 8×10$^{-3}$M/L in Fig.I- 8b).

Comparing Fig.I- 8a with Fig.I- 8c in the same dye concentration, a large $n_2$ value in Fig.I- 8c implies a long penetration depth of the pump light into the cladding solution, which leads $\eta_p$ value to rise up from 1.43×10$^{-4}$ in Fig.I- 8a to 2.85×10$^{-4}$ in Fig.I- 8c, corresponding $\alpha_{abs}^{p}$ values to increase from 0.17 cm$^{-1}$ to 0.33 cm$^{-1}$. Based on Eq. (12), the produced length of lasing emission in Fig.I- 8c will be shorter than that of Fig.I- 8a for a given pump energy. The energy threshold at the position Z = 0 mm is determined mainly by $n_2$ value at this time, a large $n_2$ value implies a long penetration depth of pump light into cladding solution, and thus a large occupation factor $\eta_p$. When $\eta_p$ value increasing from 1.43×10$^{-4}$ in Fig.I-8a ($n_2$=1.361) to 2.85×10$^{-4}$ in Fig.I-8c ($n_2$=1.402), more dye molecules are excited by the given pump energy, it is reasonable that the energy threshold is reduced from 49 to 12.8 μJ/pulse.

## 3. Part II - Polarization properties of an evanescent-wave pumped Whispering-Gallery-Mode fibre laser

In Part II, the polarization properties of an evanescent-wave pumped Whispering Gallery Mode (WGM) fibre laser evanescent-wave pumping scheme are investigated. The dye gain is produced by the evanescent field of the pump light, therefore, the polarization state of the evanescent field determines the vibration states of excited molecules, consequently the polarization properties of lasing emission of the evanescent-wave pumped WGM fibre laser. We find that when the pump beam is strictly along the axial direction of an optical fibre, the lasing emission is transverse electric (TE) wave and the electrical vector of emitting point is along the radial direction of the optical fibre, which forms a special radial polarized radiation; while the pump beam deviates from the axial direction of the optical fibre, both TE and transverse magnetic (TM) waves exist in the lasing emission, which forms a mixed polarization radiation with radial and axial polarized states. The radial polarization laser is useful in the application of photo-etching [22] and high resolution microscope [23], etc. We report on our experimental results and give the explanation for observed phenomenon in this part.

### 3.1 Experimental setup

The experiment setup is shown in Fig.II-1. A pulsed Nd:YAG laser (λ=532 nm, pulse width=7 ns) is used as a pump source. The pump laser beams pass two polarizers $P_1$ and $P_2$. $P_2$ is used to determine polarized state of the pump beams, while $P_1$ is used to change pump energy by rotating its polarization direction. Lenses $L_1$ and $L_2$ are used to shrink the pump beams size from original 8 mm to 1.5 mm. $F_1$ of diameter 196 μm is a barely optical fibre with the refractive index of 1.458. After being focused by lens $L_3$ (focal length =75 mm), the pump beams are end-fired into $F_1$ with an incident angle $\theta_i$ =1.2°. To make sure the pump light is meridian beams [15] when propagating in $F_1$, the axis of $F_1$ is adjusted strictly along the z-axis. The pump light will be skew beams [15] in $F_1$ when the angle between the axis of $F_1$ and the z-axis is about 10°.

The fibre $F_1$ is inserted into a glass tube D with the inner diameter of 2 mm. A mixed solution of ethanol and ethylene glycol doped with Rhodamine 6G dye molecules is filled into the glass tube, and the concentration and refractive index of the dye solution are 3×10$^{-3}$ mol/L and 1.422, respectively. The pump beams will propagate within $F_1$ by TIR if the incident angle $\theta_i$ is smaller than 37.6°. Evanescent field $E_p$ of the pump beams excite the dye molecules in the dye solution, photons in evanescent field $E_{WGM}$ that belongs to a WGM of circular cavity built in $F_1$ stimulate the excited molecules, the $E_{WGM}$ also couples dye gain into the circular cavity [1, 2], and then forms lasing oscillation in the circular cavity with the support of the WGM. Lasing energy $L_{WGM}$ radiates from the rim of $F_1$ in the plane of xy, an optical fibre $F_2$ transmits part of the $L_{WGM}$ to the entrance slit of the spectrometer (Spectrapro 500i, ICCD：PI－MAX). The intensity of $L_{WGM}$ also can be detected by a photo detector PD (DSi200) after the lasing emission passing the polarizer $P_3$, which is positioned on the yz plane. The angle between $P_3$ and the y-axis is defined as Φ which is zero as when $P_3$ is along in the y-axis direction. Φ can be adjusted by rotating $P_3$.

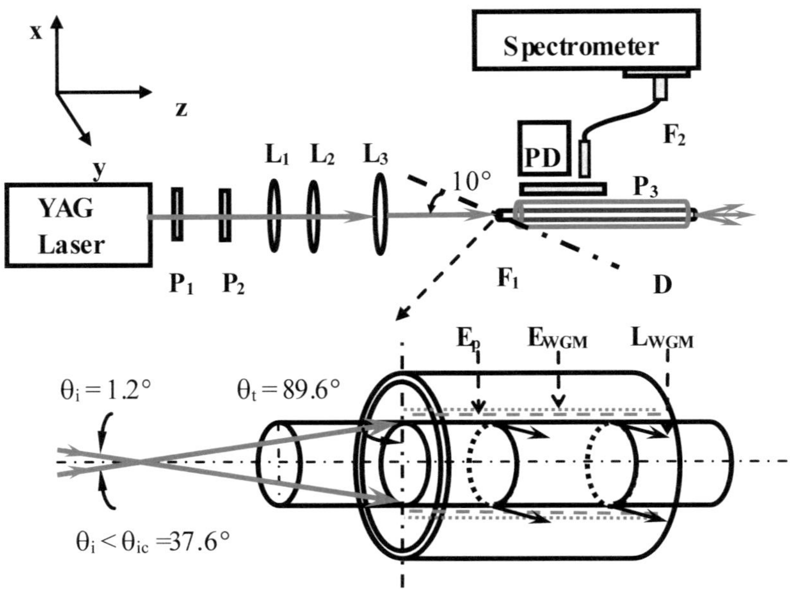


Fig. II-1. Experimental setup

### 3.2 Experimental results and discussion

#### 3.2.1 Pump along the optical fibre's axis

When pump beams are strictly in the axial direction of $F_1$, polarizer $P_2$ is set along the y-axis direction at first. Polarization properties of lasing beams are checked and recorded by PD together with another polarizer $P_3$, the background subtracted and normalized signals are shown in Fig.II-2. Fig.II-2 shows that the intensity of lasing emission (solid points with error bar) reaches its maximum when $P_3$ is set vertically to the $F_1$'s axis (Φ=0°,180° and 360°), while it reaches its minimum when $P_3$ is set along the $F_1$'s axis (Φ=90° and 270°). The curve in Fig.II-2 is drawn by Malu' s Law, $I = \cos^2 \Phi$, the consistency between the curve and experimental data indicates that, when the pump beams are strictly along the $F_1$'s axis, the electrical vector of lasing emission is vertical to the fibre's axis, and the evanescent-wave pumped WGM laser is of a typical TE wave lasing emission. To check the dependence of the lasing emission on the polarization of pump beams, the polarizer $P_2$ is reset along the x-axis and the middle between x and y-axis, respectively. The experimental results obtained are the same as that when $P_2$ is set along the y-axis direction, which indicates that the polarization of evanescent-wave pumped WGM fibre laser is independent on the polarized state of the pump beams.

Removing out the polarizer $P_3$, the lasing spectrum is recorded with the spectrometer of grating density 2400 g/mm and is shown in Fig.II-3, which shows that the spectrum is consisted of a series of peaks with equal wavelength interval. The mode-assignment results (see text 2.2) also indicate that the spectral lines are the TE wave with the same radial mode order $l=1$ but different angular mode number $n$[24].

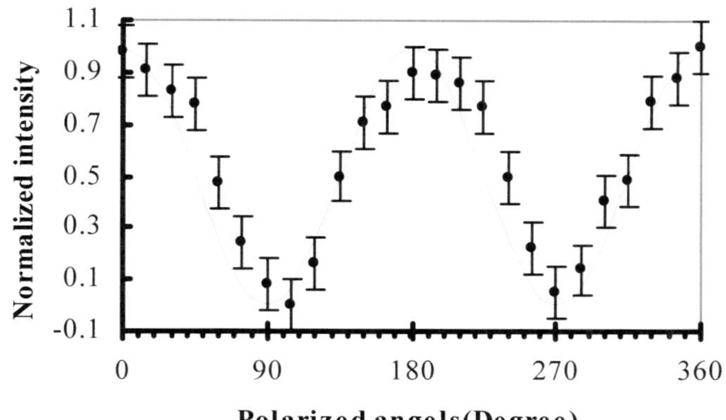


Fig. II-2. The law of WGM laser's intensity varied with polarized angels of polarizer $P_3$.

### 3.2.2 The form of radial polarization laser

When pump beams are in the fibre $F_1$'s axial direction, as shown in Fig.II-4-A, the beams are meridian beams that propagate within the fibre by TIR. Hereinafter, the pump beam in xy plane is taken as an example for discussing the polarization properties of the pump beams. When the pump beam travels to the interface of $F_1$, if the entrance angle $\theta$ is larger than the critical angle $\theta_c$, the beam will experience a Goos-Hänchen shift $d_g$ [25] that is a length about a wavelength of the pump light $\lambda_p$, and then is totally reflected into the fibre by the interface. In the distance of $d_g$, the pump beam tunnels the fibre out a distance of $d_p$ that is also a length of $\lambda_p$, and then an evanescent field of the pump beam $E_p$ is formed outside $F_1$. $E_p$ is a traveling wave along the z-axis, but it can not travel along the x-axis [25], therefore, there is no electric field of $E_p$ existing in the z-axis and this characteristic is independent on the polarization of the pump beam. For the evanescent-wave pumped laser, since no electric field of $E_p$ exists in the z-axis, it is impossible to produce stimulated photons polarizing in the direction of z-axis. After dye gain is coupled into $F_1$, the WGM lasing oscillation and emission, of course, will lack the polarization element of z-axis, so the lasing radiation belongs to TE wave, of which electric field is strictly vertical to the axis of $F_1$.

WGM within a circular cavity is a kind of "Quasi-normal Mode" [26], meaning that part of energy in a WGM leaks tangentially out of the surface of the cavity in the way of evanescent-wave, which forms lasing emission in our situation as shown in the solid arrows of Fig.II-4-C. As both lasing emission and its electric vector are vertical to the axis of $F_1$, based on a simple geometrical consideration, all electric vectors of lasing radiation that emit from the same big circle of $F_1$ , as shown in the dashed arrows of Fig.II-4-C, cross at the centre of the big circle. Therefore, when pump beam travels strictly along the axis of $F_1$, the lasing emission from evanescent-wave pumped WGM fibre laser is a special axial polarized emission of which emitting direction is vertical to the axis of $F_1$.

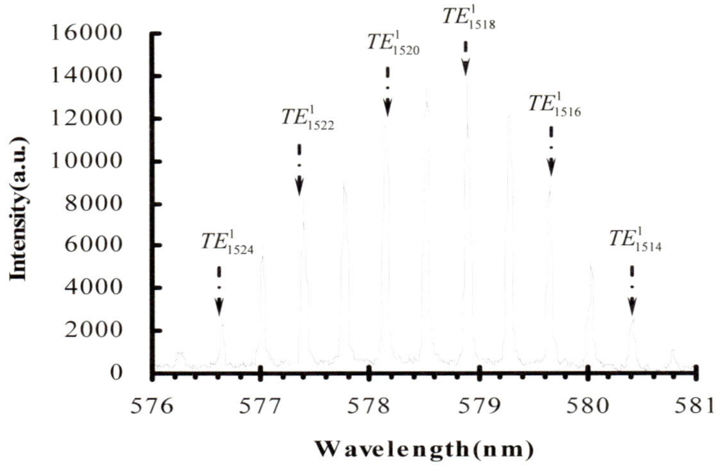


Fig. II-3. WGM lasing spectrum of TE wave acquired by pump beam being along a fibre's axis.

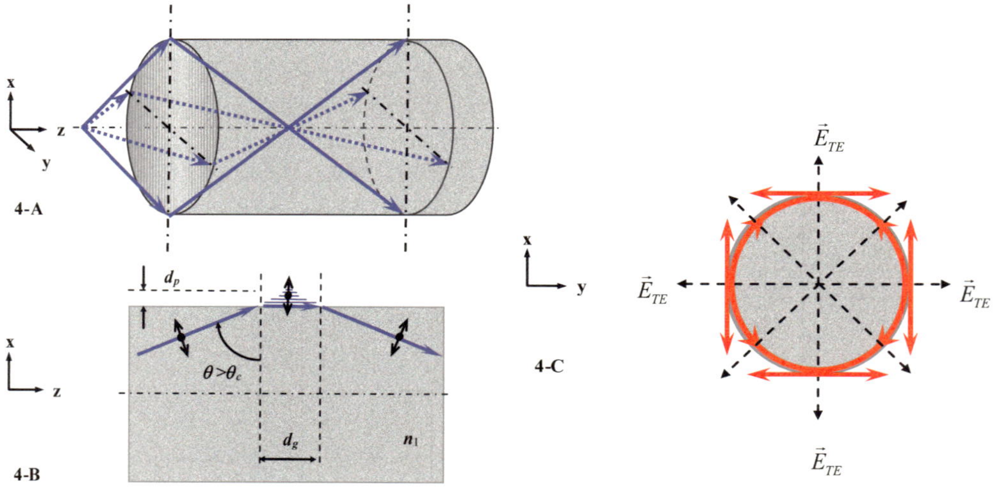


Fig. II-4. Diagram for the formation of radial polarization laser.

### 3.2.3 Pump deviating from the optical fibre's axis

When pump beams deviate from the axial direction of $F_1$, the lasing spectrum acquired by removing out the polarizer $P_3$ is shown in Fig.II-5-A. The spectrum is consisted of a group of weaker spectral lines and a group of stronger spectral lines, and each group is of the same wavelength interval. The lasing spectrum got by adding $P_3$ in the direction of y-axis is shown in Fig.II-5-B. The group of weaker lines vanishes completely in the spectrum, and the TE-wave spectrum remains. The lasing spectrum obtained by adding $P_3$ in the direction of z-axis is shown in Fig.II-5-C. The group of stronger lines almost vanishes in the spectrum, and the TM-wave spectrum remains. Fig.II-5 indicates that, when pump beam deviates from the fibre $F_1$'s axis, both TE and TM waves exist in the lasing emission.

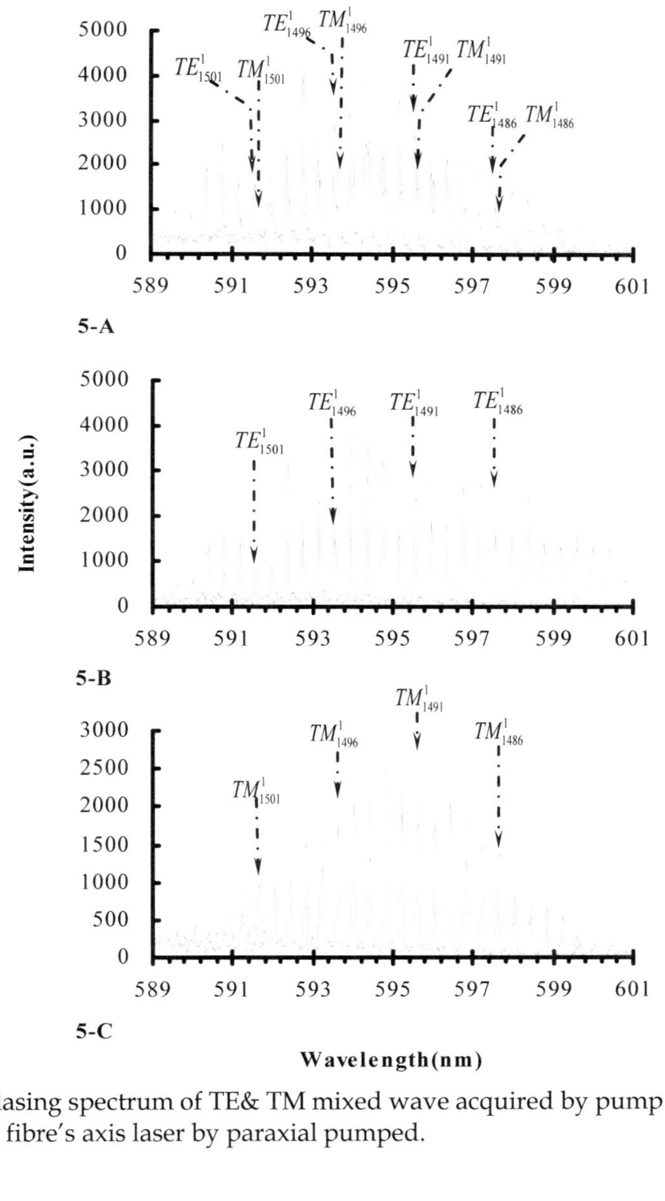


Fig. II-5. WGM lasing spectrum of TE& TM mixed wave acquired by pump beam deviating from the optical fibre's axis laser by paraxial pumped.

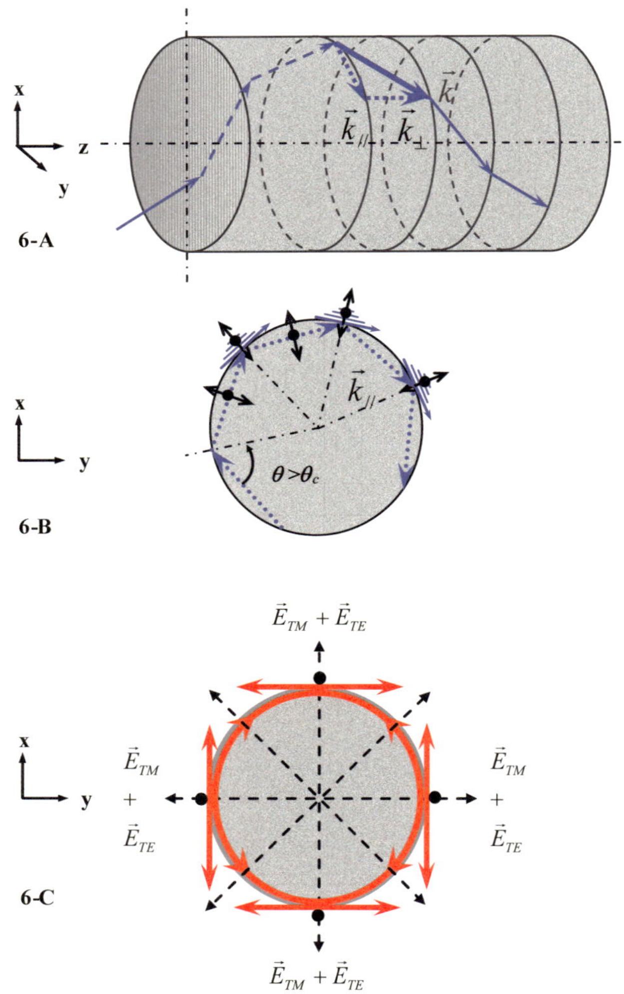


Fig. II-6. Diagram for the formation of mixed polarization laser.

### 3.2.4 The form of radial and axial mixing polarization laser

When pump beams are deviated from the optical fibre's axis, as shown in Fig.II- 6-A, the beams propagate as skew beams in the fibre $F_1$ along the z-axis. To analyse the property of the beams in the fibre, one of the beams with wave vector $\vec{k}$ is taken as an analytic sample. The vector $\vec{k}$ can be decomposed into $\vec{k}_{//}$ and $\vec{k}_{\perp}$, as shown in Fig.II-6-B in dashed arrow,

$\vec{k}_{//}$ is a component that expresses a light beam propagating in the xy plane by TIR; while $\vec{k}_{\perp}$ is a component that expresses a light beam propagating along the z-axis. For the beam expressed by $\vec{k}_{//}$, both electric vectors, the vector being parallel to z-axis and the vector being vertical to z-axis, exist simultaneously in the fibre $F_1$, when the beam experiences TIR on the fibre's interface, as shown in Fig.II-6-B, both electric vectors also exist in the evanescent field of pump beam. Because the dye molecules in the evanescent field of pump beam are excited by two different electric vectors, two kinds of stimulated photons, polarizing in the z-axis and in the xy plane, therefore, exist simultaneously in the evanescent field of WGMs of the circular cavity built in the optical fibre $F_1$. After the dye gain is coupled into the circular cavity, the WGM lasing oscillation and emission, of course, contain TE wave that is polarizing in the xy plane and TM wave that is polarizing along the z-axis. Combining with the analysis at the end part of this text 2.2, as shown in Fig.II-6-C, it can be conluded that the electric field of TE wave crosses at the centre of the circular cavity, while the electric field of TM wave is parallel to the z-axis; when pump beams are deviated from the axis of the optical fibre, the lasing emission from evanescent-wave pumped WGM fibre laser is a mixed polarization emission.

## 4. Part III - Linearly polarized three-color lasing emission from an evanescent wave pumped and gain coupled fibre laser

In Part III, an evanescent-wave pumped and gain coupled WGM fiber laser is introduced by inserting a bare quartz fiber (no cladding layer) into a glass capillary filled with dye molecules in a lower refractive index solution. Once the bare fiber is inserted into three glass capillaries filled with Rhodamine 6G, Rhodamine 610 and Rhodamine 640 dye solutions, respectively, WGM laser oscillations at the wavelengths of 567-575, 605-614 and 656-666 nm occur simultaneously, and a linearly polarized three-color lasing emission is achieved in a single optical fiber.

### 4.1 Experimental setup

The experimental setup is shown in Fig.I-4 schematically. An bare fiber ($F_1$, 196±1 μm in diameter, refractive index 1.458) was inserted into a long glass capillary (C, 1 mm in inner diameter), the residual space between $F_1$ and C was filled with ethanol and ethylene-glycol mixed solution doped by Rhodamine 6G dye with a concentration of $3\times10^{-3}$ M/L. The refractive index of the mixed solution, acting as the cladding solution of $F_1$, was 1.395 measured by an Abbe refractometer. The pump beam was longitudinally coupled into $F_1$ along the fiber's axis by a lens (L) (focal length =75 mm) with a conical angle of $\theta_i$ =9.2°. The beam would propagate within $F_1$ by TIR if the conical angle $\theta_i$ was smaller than the critical entrance angle $\theta_{ic}$, which was 35.6° in our experiments.

The evanescent field of the pump light ($E_p$) excites dye molecules in the mixed solution, thus the photons in evanescent field of WGM ($E_{WGM}$) of a circular cavity stimulate the excited dye molecules, and the $E_{WGM}$ also couples dye gain into the circular cavity at the same time. Supported by the WGMs, lasing oscillations occur in the circular cavity. The WGM lasing emission ($L_{WGM}$) from the rim of $F_1$ can be recorded by a spectrometer (Spectrapro 500i) equipped with an ICCD detector (PI-Max 1024RB) via an optical fiber $F_2$, it gives 0.05 nm spectral resolution when a grating of density 2400g/mm is used. The intensity of $L_{WGM}$ is detected by a photo detector (PD, DSi200) after the lasing emission passing through an analyzer $P_3$ positioned on the *yz* plane.

### 4.2 Experimental results and discussion

#### 4.2.1 Three-color WGM lasing emission

It is found that the generating length of lasing emission along the $F_1$ increases with pumping energy by arranging pump configuration as described in the part I, in which the fiber $F_1$ was replaced by a thick fiber (288 μm in diameter) and the direction of the incident beam was still strictly along the axial direction of $F_1$. When the pumping energy equaled 20, 50 and 100 μJ/pulse, the lengths of the lasing emission were 40, 90 mm and over 120 mm, respectively. Based on this, a longer gain length in evanescent-wave pumped WGM fiber laser, of course, can be utilized to build a multicolor WGM fiber laser [27]. That is to say, the fiber $F_1$ was inserted into three short glass capillaries (15 mm long) in series and their intervals between the capillaries were 5 mm. Rhodamine 6G, Rhodamine 610 and Rhodamine 640 dye molecules were dissolved in ethanol and ethylene-glycol mixed solution to obtain concentrations $5\times10^{-4}$, $2\times10^{-3}$ and $1\times10^{-2}$ M/L, respectively. The prepared dye solutions with an index of 1.422 were filled in the residual space between the $F_1$ and the capillaries. Evanescently pumped by 532 nm line of the pulsed YAG laser along the fiber's axis, only a weak fluorescent light emitted out from the fiber when the pumping energy was 20 μJ/pulse. The fluorescent spectrum was recorded by a spectrometer with a grating of density 150g/mm, which is the grey curve shown in Fig. III-1A. When pumping energy was 80 μJ/pulse, we observed three-color (yellow, orange and red) WGM lasing emissions simultaneously in experiments, the emitting wavelengths from the fiber laser were in the ranges of 567-575 nm (Rhodamine 6G), 605-614 nm (Rhodamine 610) and 656-666 nm (Rhodamine 640) as shown in Fig. III-1A with the black curve. Since three-color WGM lasing emission from a single optical fiber was realized, a novel evanescent-wave pumped and gain coupled WGM fiber multicolor laser was demonstrated eventually.

To show the fine spectral structure of the three-color WGM lasing emission, the spectra were recorded with a grating of density 2400g/mm, and shown in Figs. III-1B, 1C, and 1D in the wavelengths 567-575 nm, 605-614 nm and 656-666 nm, respectively. The measured spectral intervals between two adjacent peaks are 0.246 nm (Fig.III-1B), 0.282 nm (Fig.III-1C) and 0.335 nm (Fig.III-1D), which can be compared with the calculated free spectral widths of WGMs in the used circular cavity. For a larger diameter fiber (288 μm in our case), the wavelength intervals (free spectral width) can be approximately calculated by [27] $\Delta\lambda = \lambda^2/2\pi a n_1$, where $a$ and $n_1$ are the radius and the refractive index of the used fiber, respectively. The calculated wavelength intervals are 0.247, 0.281 and 0.332 nm in the wavelength ranges of 567-575 nm, 605-614 nm and 656-666 nm, which accord well with the measured values.

#### 4.2.2 Spacial overlap between evanescent wave of WGM and lasing gain

To explain the characteristic of the longer gain in an evanescent-wave pumped WGM fiber laser, the radial intensities of the WGMs in the used circular cavity and the evanescent field of pump light are calculated. For a TE wave in a cylindrical coordinate system shown in the top left of Fig.I-4, the electric field of a WGM in a cylindrical cavity can be written as (I-6c, 6d). For TM wave with radial mode number $l$ and angular mode number $n$, the WGM is assigned by $TM_n^l$ $(l, n)$, while it is assigned by $TE_n^l$ $(l, n)$ for TE wave in this part. Two lasing lines at the wavelengths of 591.59 and 589.43 nm have been mode assigned [20,28] as $TE_{1500}^1$ and $TE_{1490}^2$, which are chosen to calculate the distribution of radial intensity. Let $a$ = 98 μm, $n_1$ =1.458, and $n_2$ =1.395, the radial intensities $I_{WGM}(r)$ for the two WGMs are calculated and shown in Fig.III-2.

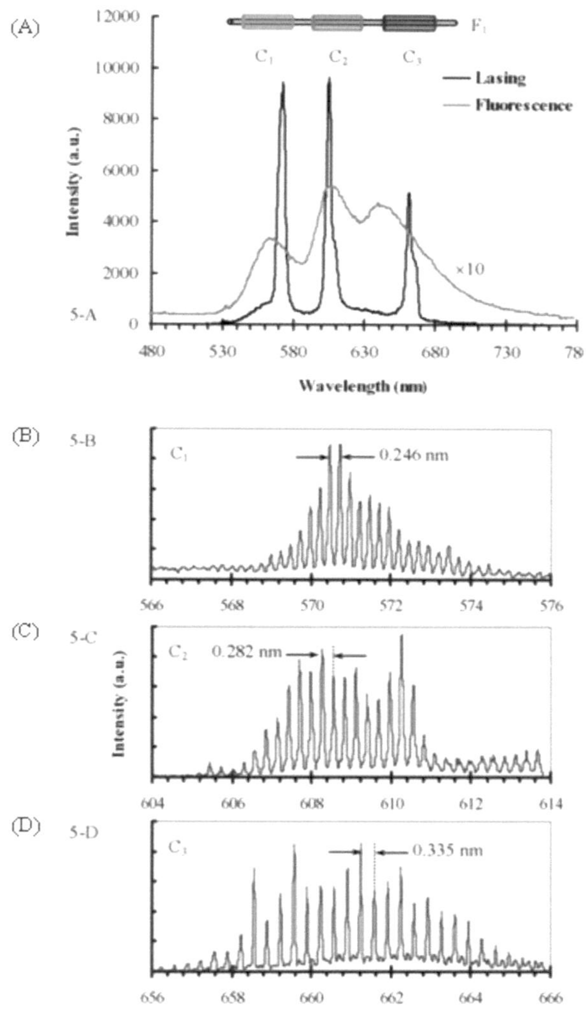


Fig. III-1. Lasing spectra recorded by gratings with densities: 150g/mm (A) and 2400g/mm (B), (C) and (D). Panel (A) shows fluorescent (grey curve) and lasing (black curve) spectra emitting from a three-colour evanescent-wave fiber laser, and the laser structure with the capillaries $C_1$, $C_2$ and $C_3$ being filled with different dye solutions. Panels (B), (C) and (D) indicate lasing spectra emitted from $C_1$, $C_2$ and $C_3$ filled with Rhodamine 6G, 610 and 640 dye solutions, respectively.

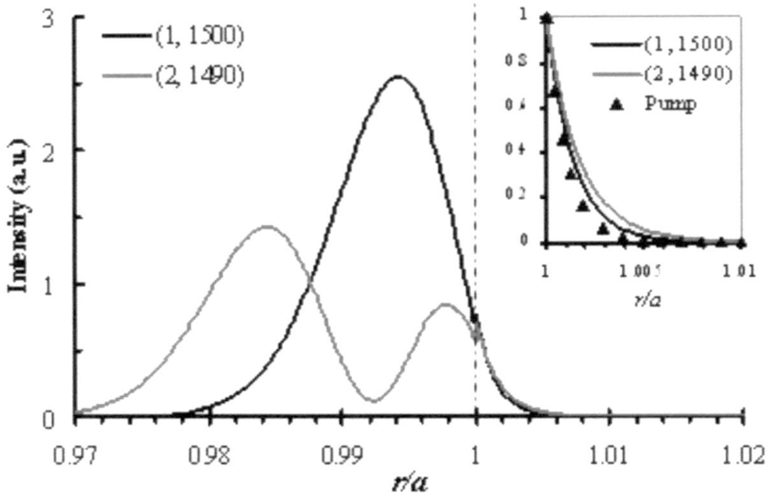


Fig. III-2. Radial intensity distributions calculated for two WGMs assigned by $TE^{1}_{1500}$ (1, 1500) and $TE^{2}_{1490}$ (2, 1490), where the bumps at $r = a$ are caused by the discontinuity of TE wave, and the insert shows the intensities of evanescent waves for the pump light (solid triangles) and WGMs (black and grey curves)

The evanescence-wave intensity of the pump light is also calculated from [16]

$$I_p(r) = I_0 \exp[-2k\beta(r / a - 1)a], \quad (r \geq a), \tag{III-1}$$

where $\beta = [n_1^2 \sin^2\theta / n_2^2 - 1]^{1/2}$, and $k = 2\pi n_2/\lambda_p$. Letting $I_0$=1, $\theta$ =86.8°, $\lambda_p$ = 532 nm, the calculated result is shown in the insert of Fig. 6 by solid triangles, for a convenient comparison, the normalized evanescent-wave intensities of the modes $TE^{1}_{1505}$ and $TE^{2}_{1490}$ are also presented. The curves in the insert show clearly that in the radial direction of the cylindrical cavity ($r \geq a$), the intensity distributions of the two WGMs overlap well with the distribution of evanescence-wave of the pump light. Because the lasing gain in the evanescent-wave pumped WGM fiber laser is excited by the evanescent field of the pump light, the spatial distribution of the lasing gain is the same as that of the pump light. Therefore, it is reasonable to attribute the longer gain distance and the higher pumping efficiency in evanescent-wave pumped WGM fiber laser to the good spatial overlap between the lasing gain and the evanescent field of WGMs in a circular cavity.

## 5. Part IV - Broadening free spectral range of an evanescent-wave pumped Whispering-Gallery-Mode fibre laser by Vernier effect

In Part IV, evanescent-wave pumping scheme is used in a WGM fibre laser, which is formed by inserting a piece of bare quartz fibre (without cladding) into a glass capillary filled with a dye solution of low refractive index, as shown in Fig.IV-1*a*. When the pumping light propagates along the fibre's axis by total internal reflection, its evanescent-field excites a lasing gain around the fibre, and WGM lasing emission occurs as the pumping energy exceeds the threshold. To increase the FSR of WGM fibre laser, two optical fibres of different radii are bound together with their axes parallel to make a coupled cylinder-cavity structure (CCCS). Due to the Vernier effect [29-31], lasing occurs only when both cavities are in resonance at the same frequency, and the FSR of the CCCS is thus greatly enlarged.

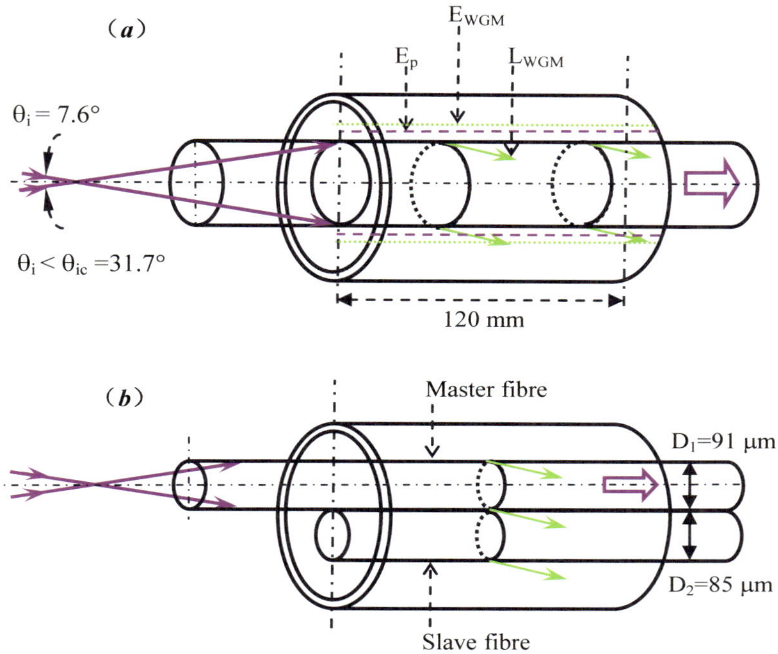


Fig. IV-1. Schematic of evanescent-wave pumped WGM fiber laser is shown in (*a*), where $E_p$ and $E_{WGM}$ indicate the evanescent-fields of the pump light and WGMs, $L_{WGM}$ indicates the WGM lasing emission. The coupled cylinder-circular cavity is shown in (*b*), the master fiber provides the gain and make up a CCCS with the slave fibre. Note that the vertical and horizontal directions are drawn with different scales

## 5.1 Experiment and results

The experimental setup is shown in Fig.IV-1a. A laser beam (355 nm), generated by a frequency tripled and Q-switched Nd:YAG laser, was used as the pump light. A bare fibre (refractive index 1.46) was inserted into a long glass capillary (1 mm inner diameter), and the open space between the fibre and the capillary was filled with ethanol solution doped by Coumarin 500 (LC 5010, Lambda Physik) dye with a concentration of $4\times10^{-3}$ Mol/L. The refractive index of the dye solution, acting as the cladding layer of the bare fibre, is 1.36 measured by an Abbe refractometer. The pump beam was coupled into the bare fibre along its axis with a conical angle $\theta_i$ =7.6°. The beam would propagate within the bare fibre by total internal reflection if the conical angle $\theta_i$ was smaller than the critical entrance angle $\theta_{ic}$, that was 31.7° in our experiments. The evanescent field of the pump light ($E_p$) excites dye molecules in the dye solution, thus the photons in evanescent field of WGM ($E_{WGM}$) of a circular cavity (formed by any cross section of the fibre) stimulate the excited dye molecules, supported by the WGMs, lasing oscillations occur in the circular cavity. It is worth noting that the evanescent-wave pumping scheme used in this work is different from those of references

[30] and [31]. The ring cavity in reference [30] is composed by gain material and the gain is obtained by directly pumping the cavity, while the side-pumping scheme is used in reference [31]. The WGM lasing emission ($L_{WGM}$) from the rim of the fibre can be recorded via a detecting fibre by a spectrometer (Spectrapro 500i) equipped with a CCD detector (PI-Max 1024RB), which gives 0.05 nm spectral resolution when a grating of density 2400g/mm is used.

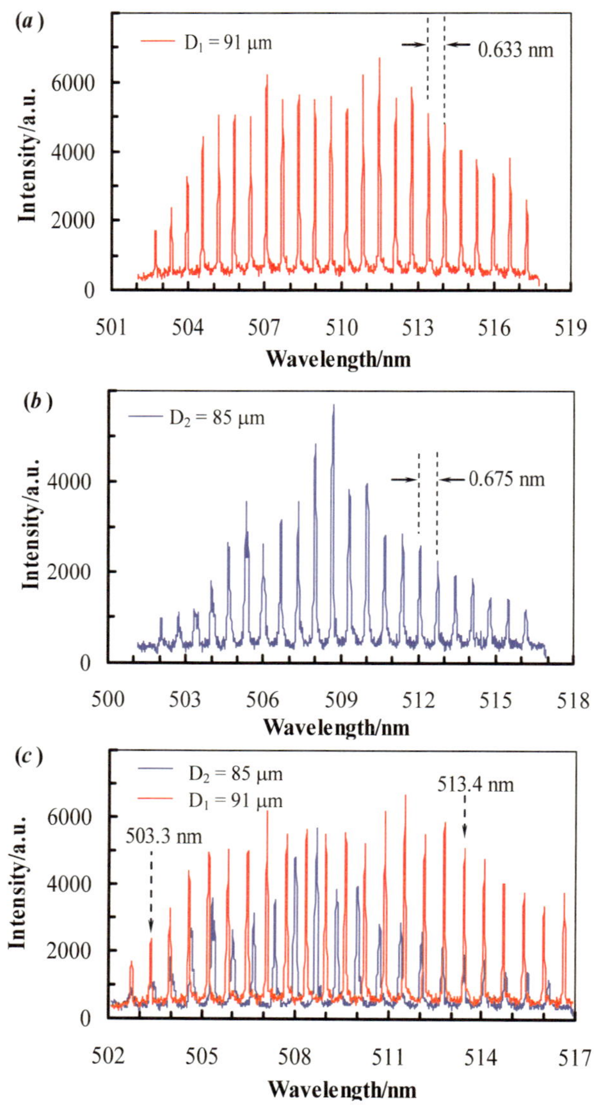


Fig. IV-2. WGM fiber lasing spectrum from a bare optical fibre of diameter $D_1 = 91$ μm is shown in (*a*), the measured FSR is 0.633 nm as indicated by the arrows. The spectrum from the bare optical fibre of diameter $D_2 = 85$ μm is shown in (*b*), the measured FSR is 0.675 nm as indicated by the arrows. The combination of spectra (*a*) and (*b*) in the same wavelength region is shown in (*c*). Two arrows in (*c*) indicate the overlapping positions of the WGM resonant wavelengths of the two fibers.

Two types of fibres with diameter $D_1$ = 91 μm and $D_2$ = 85 μm, measured by a microscope with 1 μm uncertainty, were prepared by etching two bare quartz fibres of original diameter 100 μm in a diluted HF-acid solution. After supersonic cleaning in ethanol, one of the fibres was inserted in the glass capillary filled with ethanol dye solution. The fibre was evanescently pumped by the 355 nm laser beam along its axis, and WGM lasing spectra were investigated. For the fibre with a diameter $D_1$ = 91 μm ($D_2$ = 85 μm), lasing emission was observed when the pumping energy was larger than 25 μJ (30 μJ), and the lasing spectrum was recorded as shown in Fig.IV-2a (Fig.IV- 2b) with the pumping energy fixed at 100 μJ. The measured average FSR in the spectral range from 504 to 516 nm was 0.633 nm for the fibre of $D_1$ = 91 μm, while it was 0.675 nm in the spectral range from 506 to 513 nm for the fibre of $D_2$ = 85 μm. When the two spectra (Fig.IV-2a and 2b) are merged together in the same wavelength region, as shown in Fig.IV-2c, it is found that two pairs of peaks corresponding to WGM resonant wavelengths overlap, and the overlapping wavelengths are 503.3 nm and 513.4 nm as indicated in Fig.IV-2c by the dashed arrows.

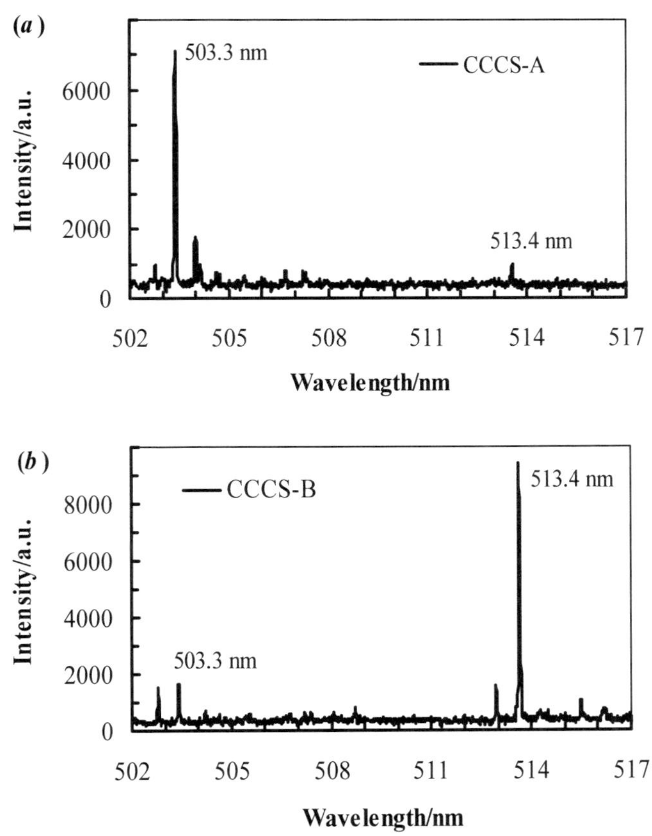


Fig. IV-3. Lasing spectra from the CCCS. The spectra (*a*) and (*b*) were acquired by setting a detecting fibre at different positions along the axis of the CCCS. The wavelength values shown in the figure are the same as the overlapping positions of the WGM resonant wavelengths in Fig.IV- 2(c).

After the single fibre experiments, the two fibres (we called the fibre of $D$ = 91 μm in diameter as master fibre, and the fibre of $D_2$ = 85 μm in diameter as slave fibre) were bound together with their axes parallel by two fine hairs to form a CCCS. The CCCS was then inserted in the same glass capillary as shown in Fig.IV-1b, evanescently pumped by the 355 nm laser beam along the axis of the master fibre (pumping energy was fixed at 100 μJ), the lasing emissions were observed along the rims of both fibres, and the lasing spectra recorded were shown in Fig.IV-3. The CCCS suppresses most of WGM resonant peaks shown in Fig.IV-3. However, a strong lasing peak appears at 503.3 nm and a weak peak appears at 513.4 nm, where two pairs of peaks overlap together as shown in Fig.IV-2c. When the detecting fibre was moved to another position along the master fibre's axis, as shown in Fig.IV-3b, we obtained another lasing spectrum similar to that shown in Fig.IV-3a from the CCCS, in which most of WGM resonant peaks have been suppressed, such that a strong lasing peak appears at 513.4 nm and a weak peak appears at 503.3 nm. The wavelength difference between the strong and the weak peaks is 10.1 nm. This value, corresponding to FSR of the CCCS, is 16 and 15 times larger than that of the master and slave fibres, respectively. That is FSR is broadened by using the CCCS. The lasing peaks around 503.3 and 513.4 nm are possibly caused by the partial overlap of resonant frequencies between the WGMs of the main and the slave fibres.

### 5.2 Discussion and conclusion

Let $D$ and $m$ be the diameter and the refractive index of a circular cavity, respectively, the WGM resonant condition can be approximately written as $\pi Dm = N\lambda$ , where $\lambda$ is resonant wavelength, $N$ is an integer. For the master and the slave fibres, the resonant conditions are given by,

$$\pi D_1 m = N_1 \lambda_1 , \tag{IV-1}$$

$$\pi D_2 m = N_2 \lambda_2 \tag{IV-2}$$

For a CCCS composed by two circular cavities, based on Vernier effect, the resonance occurs only at the same wavelength of both cavities satisfies resonant condition, which leads to $\pi(D_1 - D_2)m = (N_1 - N_2)\lambda_{CCCS}$, where $\lambda_{CCCS}$ is the resonant wavelength of the CCCS. Therefore, the FSR of the CCCS can be written as,

$$FSR_{CCCS} = \delta\lambda_{CCCS} = \frac{\lambda_{CCCS}^2}{\pi(D_1 - D_2)m} \cdot \tag{IV-3}$$

Instead of using the measured diameters of the two fibres, we utilize the spectral diameter defined by $D^s = \lambda^2 / \pi m \delta\lambda$ to calculate $\delta\lambda_{CCCS}$, where $\delta\lambda$ is the measured FSR from the lasing spectrum shown in Fig.IV- 2. For the master and the slave fibres, the spectral diameters are $D_1^s$ = 89.6 μm ($\delta\lambda$ = 0.633 nm) and $D_2^s$ = 84.0 μm ($\delta\lambda$ = 0.675 nm), respectively. Substituting the values of $D_1^s$ and $D_2^s$, together with $m$ = 1.46 into equation (3), the calculated $\delta\lambda_{CCCS}$ is 10.28 nm for $\lambda_{CCCS}$ = 513.4 nm, and 9.87 nm for $\lambda_{CCCS}$ = 503.3 nm. The average value of $\delta\lambda_{CCCS}$ is 10.08 nm, which is very close to the measured FSR of the CCCS.

A broad FSR can also be obtained by using a single fibre of small diameter. For example, the diameter must be reduced to about 6 μm to obtain the same FSR as the CCCS used in this work. However, a fibre of small diameter means a low efficiency to couple the pumping energy into the fibre, as well as a decrease in its quality factor due to a larger leakage loss, especially when the cavity is immersed in a liquid solution.

The difference between the geometric diameters ($D_1$ = 91 μm, $D_2$ = 85 μm) and the spectral diameters ($D_1^s$ = 89.6 μm, $D_2^s$ = 84.0 μm) is mainly caused by the property of intensity distribution of WGMs. Most of photons in a WGM don't travel strictly along the inner surface of a circular cavity, but slightly inside the surface. Therefore, the spectral diameter ($D_i^s$) calculated by the measured FSR of a circular cavity is always smaller than its geometric diameter ($D_i$).

## 6. Chapter summary

Threshold property of the novel WGM fiber lasers pumped by an evanescent-wave has been investigated . The laser has been fabricated by inserting a bare fused quartz fiber into a glass capillary filled with dye solution, and its energy threshold properties, including the energy threshold varied with the RI of the dye solution for different fiber diameters and the produced length of lasing emission, are experimentally studied. We find that the energy threshold is very sensitive to the change of the refractive index of dye solutions and the fiber diameter, and the produced length of lasing emission is dependent on dye concentration and the refractive index of dye solutions. Based on the characteristic of evanescent-wave pumping scheme, we derive the threshold energy formula of the WGM fiber lasers from the general theory of WGM microcavity laser. Based on the characteristic of frustrated totally internal reflection of light traveling along the fiber, we obtain the equation. to determine the produced length of lasing emission according to the characteristic of FITR of light traveling along the fiber. The calculated results with our theory fit the experimental data very well.

The polarization property of lasing emission, for the evanescent-wave pumped WGM fibre laser, is related closely to the light-pump conditions. When pump beams travel strictly along the axis of the optical fibre, only TE wave exists in lasing emission, and the electric vector of the lasing emission goes along the radial direction of the fibre, which forms a special radial polarization emission whose emitting direction is vertical to the fibre's axis. When pump beams are deviated from the axis of the optical fibre, both TE and TM waves exist in lasing emission, which forms a special emission mixed with radial and axial polarization emission, and the emitting direction is also vertical to the fibre's axis.

For an evanescent-wave pumped and gain coupled WGM fiber laser, the dye gain is excited by the evanescent wave of the pumped light, distributed around the fiber and coupled into WGMs of a circular cavity. A good spatial overlap between the dye gain and the evanescent fields of WGMs leads to a high pumping efficiency and a longer gain distance along the fiber, and linearly polarized three-color lasing emissions from a single optical fiber can be realized by immersing the fiber into three different dye solutions, where may be found a potential application in multicolor display.

Evanescent-wave pumping scheme has been successfully used to excite WGM fibre laser. A coupled cylinder-cavity structure has been fabricated by binding two bare optical fibres together, and the measured FSR of the laser emission from the CCCS is effectively broadened. This structure can be used to obtain a single mode lasing emission in an evanescent-wave pumped WGM fibre laser.

## 7. Acknowledgments

We acknowledge support from the National Science Foundation of China (NSFC) through grant 60877037、11164033 and the program of IRTSTYN in University of Yunnan Province.

## 8. References

[1] H. J. Moon, Y .T. Choung, and K. An, "Cylindrical micro-cavity laser based on the evanescent-wave-coupled Gain,"Phys. Rev. Lett. 85, 3161-3164 (2000).
[2] M .Kazes, D. Y. Lewis, Y. Ebenstein, T. Mokari and U. Banin, "Lasing from semicon ductor quantum roads in a cylindrical micro-cavity," Adv. Mater. 14, 317-321 (2002).
[3] H. J. Moon, G. W. Park, S. B .Lee, K. An, and J. H. Lee, "Waveguide mode lasing via evanescent-wave-coupled gain from a thin cylindrical shell resonator," Appl. Phys. Lett. 84 4547-4550(2004).
[4] N. K. Chen, L. Y. Zhang, K. C. Hsu, L. L. Hu, S .Chi, Y. C. Lai, S .M .Tseng, and J .T. Shy，"CW-pumped evanescent amplification based on side-polished fiber with heavily Er3+-doped glass overlay," Jpn. J. Appl. Phys . 45, 6328-6330 (2006).
[5] J. D. Suter, W. Lee, D. J. Howard, E. Hoppmann, I. M. White, and X. Fan, "Demonstration of the coupling of optofluidic ring resonator lasers with liquid waveguides," Opt. Lett. 35, 2997-2999(2010).
[6] X. Y. Pu, N. Jiang, D. Y. Han, Y. L. Feng, and Y. T. Ren, "Linearly polarized three-colour lasing emission from an evanescent wave pumped and gain coupled fibre laser," Chin. Phys. B 90, 054207_1-054207_7(2010).
[7] H. Fujiwarra, and K. Sasaki, "Lasing of a microsphere in dye solution," Jpn. J. Appl. Phys. 28, 5101-5104(1999).
[8] S. B. Lee, M. K. Oh, J. H. Lee, and K. An, "Single radial-mode lasing in a submicron-thickness spherical shell microlaser," Appl. Phys. Lett. 90, 201102_1-201102_4 (2007).
[9] A. Shevchenko, K. Lindfors, S. C. Buchter, and M. Kaivola, "Evanescent wave pumped cylindrical micro-cavity laser with intense output radiation," Opt. Commun. 245, 349-353(2005).
[10] S. Lacey, I. M. White, Y. Sun, S. I. Shopova, J. M. Cupps, P. Zhang, and X. Fan, "Versatile opto-fluidic ring resonator lasers with ultra-low threshold," Opt. Express 15, 15523-15530 (2007).
[11] X. Jiang, Q. Song, L. Xu, J. Fu, and L. Tong, "Microfibre knot dye laser based on the evanescent-wave-coupled Gain," Appl. Phys. Lett. 90, 233501_1-233501_3 (2007).

[12] C. Monat, P. Domachuk, and B. J. Eggleton, "Integrated optofluidics: A new river of light," Nat. Photonics 1, 106-114(2007).

[13] Y. Sun, S. I. Shopovab, C. S. Wu, S. Arnold, and X. Fan, "Bioinspired optofluidic FRET lasers via DNA scaffolds," PNAS 37, 16039-16042(2010).

[14] X .Y. Pu, N. Jiang, R. Bai, W. L . Xiang , Y. X. Zhang, and D. Y. Han, "A novel micro-cavity fiber laser with three-colorWGM lasing emission in a single optical fiber," Chinese Patent No.ZL200810058304 (30 Sep. 2009).

[15] K. M. Djafar, and L. S. Lowell, *Fibre-optic Communications Technology*(Science Press, Beijing, 106-108(2002)).

[16] M. Born, and E. Wolf, *Principles of Optics*(Cambridge University Press, Cambridge, 1997).

[17] C. F. Bohren, and D. R. Huffman, *Absorption and Scattering of Light by Small Particles*(John Wiley & Sons,New York, 1998).

[18]Y. X. Zhang, X.Y. Pu, K. Zhu, and L. Feng, "Threshold property of Whispering- Gallery-Mode fiber lasers pumped by evanescent-wave," J. Opt. Soc. of Am. B. 28, 2048-2056 (2011)

[19] Y. S. Choi, H. J. Moon, K. Y. An, S. B. Lee, J .H. Lee, and J. S. Chang, "Ultrahigh-Q Microsphere Dye Laser Based on Evanescent-Wave Coupling," J. Korean Phys Soc. 39, 928-931(2001).

[20] C. C .Lam, P. Y .Leung, and K. Yang , "Explicit asymptotic formulas for the positions, widths, and strengths of resonances in Mie scattering," J. Opt. Soc. of Am. B. 9, 1585-1592(1992).

[21] D. L. Wang, N. Jiang, L. Q. Jiang, and X. Y. Pu, "The precise assignment of whispering gallery modes for lasing spectra emitting from cylindrical micro-cavities (in Chinese)," Spectroscopy Spectr Anal.28, 2749-2753(2008).

[22] D. G. Hall, "Vector-beam solutions of Maxwell's wave quation," Opt. Lett. 21, 9-11 (1996).

[23] Q. Zhan, J. R. Leger, "Microellipsometer with radial symmetry," Appl. Opt. 41, 4630-4637(2002).

[24] P. W. Barber, S. C. Hill, *Light scattering by particles: computational methods*. (World Scientific Publishing Co Pte Ltd, Singapore, 25-77(1990)).

[25] J. D. Jackson, *Classical Electrodynamics*. (Advanced Education Press, Beijing, 306-308 (2001)).

[26] E. S. C. Ching, P. T. Leung, K. Young, *The role of Quasi-normal Modes-Optical Processes in Microcavities*, edited by R. K. Chang and A. J. Campillo. (World Scientific Publishing Co Pte Ltd, Singapore, 1-75(1996)).

[27] X. Y. Pu, R. Bai, W. L. Xiang, F. Du and N. Jiang , "Two-wavelength-range whispering-gallery-mode fiber laser pumped by evanescent wave , " Acta Phys. Sin. 58, 3923-3928(2009).

[28] R. Yang, W. H. Yu, Y. Bao, Y. X. Zhang and X. Y. Pu, "whispering-gallery modes based on evanescent field in cylindrical micro-cavity," Acta Phys. Sin. 57, 6414-6418 (2008).

[29] K. Oda, N. Takato, and H. Toba, "A wide-FSR waveguide double-ring resonator for optical FDM transmission systems," IEEEJ. Light wave Technol. 9, 728-736 (1991).

[30] L. Shang, L. Liu, and L. Xu, "Single-frequency coupled asymmetric microcavity laser,"Opt. Lett. 33, 1150-1152 (2008).

[31] X. Wu, Y. Z. Sun, J. D. Suter, and X. D. Fan, "Single mode coupled optofluidic ring resonator dye lasers," Appl. Phys. Lett. 94 241109_1-241109_3 (2009).

# Part 2

# Medical, Imaging, Spectroscopy and Measurement Applications

# 9

# Optical Fiber Near Infrared Spectroscopy for Skin Moisture Measurement

Ahmad Fairuz Omar and Mohd Zubir MatJafri
*School of Physics, Universiti Sains Malaysia*
*Malaysia*

## 1. Introduction

Skin covers the entire human body and function as shield from various types of external stimuli, damage and also from loss of moisture. The main characteristic factors for protecting the body and assisting in motion are the softness and pliability of skin (Obata and Tagami, 1990). These factors are dependent on the amount of moisture available in the stratum corneum, which is the outermost layer of skin. Modification in the amount of water content of the stratum corneum may lead to significant consequences to the functional properties of human skin. It is essential to retain sufficient moisture in the stratum corneum for healthy skin since the water level in this superficial layer of the human skin are of the utmost importance in determining many of its properties (Suh *et al*, 2005). Many instruments have been developed for studying skin physiology and among these applications, measurements of moisture in stratum corneum or sometimes also referred as hydration is one of the fundamental in the study of biophysical properties and function of the skin barrier (Fluhr et al, 1999). The most well-established technique to measure water content in skin is based on measuring electrical properties such as capacitance and alternating current conductivity on the skin surface (Arimoto *et al*, 2005). However, optical fiber near infrared spectroscopy technique has emerge as a popular substitute to conventional measuring methodology in various field of study including environmental monitoring, agricultural and food product quality analysis and also in medical, particularly skin health analysis. This mainly due to the technique which is seen to be able to produce measuring instruments that are non-destructive, portable, low cost, fast and easy operation besides having high precision and reproducibility. The integration of optical fiber probe into the spectroscopy system has added the flexibility of measurement and the fiber design itself can serve into raising the efficiency of measurement. For skin moisture measurement, one of the advantages of applying optical measuring technique is that the interface of optical fiber probe does not necessarily have to be made to contact the skin surface. Therefore, non-occlusive measurements can be made. To cater for the increasing interest in the development of optical fiber spectroscopy system, this chapter will present the existing application of upper range of NIR (1100-2500nm) and the possible application of lower range of NIR (700-1100nm) particularly 970nm in the measurement of skin moisture content.

## 2. Overview on skin pathology and moisture content

Human skin consists of epidermis and dermis and covers the entire body and function as protection from various types of external stimuli, damage and from moisture loss as well as assisting in motion through the skin softness and pliability (Obata and Tagami, 1990, Woo *et al*, 2001). These characteristics are reliant on the amount of moisture available in the stratum corneum, the outermost layer of skin (epidermis) and are controlled by the barrier function that maintains adequate water content in the skin layer. Therefore, changes in the water content of the stratum corneum have important effects on the functional properties of human skin. To sustain a healthy skin, it is very important to maintain sufficient moisture in the stratum corneum (Suh *et al*, 2005). Stratum corneum is about 10-40μm thick and is composed of partially flattened and keratinized layers, except on palms and soles. It is also moderately dehydrated cells in a lipid matrix. Below the stratum corneum is the epidermis with about 100-200microns thickness. Below epidermis is the dermis with about 2-4mm thickness. Skin is more hydrated at the deeper layers. In general, the increase in tissue hydration rate can be influenced by the increase in relative humidity (Martin, 1993). If the health of the stratum corneum is not maintained during environmental changes, the efficiency of the barrier and moisture-maintaining functions of the skin may drop. Consequently, the skin will become easily dried, roughened and even more at risk to infection. Therefore, it is crucial to sustain adequate moisture in the stratum corneum for healthy skin (Woo *et al*, 2001). It is not necessarily to characterize dry skin as lacking in moisture. Dry skin is more often considered to have a rough and uneven surface that efficiently scatters light, leading to a dry and matte appearance of the skin. Climate, cleansing age or heredity may lead to normal dry skin. Water loss in skin can be reduced by applying moisturizer that function by creating a barrier to surface evaporation. This will create a smoother, softer feel to the skin and to improve the appearance of the skin (Martin, 1993). Skin moisture and thickness may vary according to anatomical body site, age, gender and sun exposure. The thickness of the skin for adults may vary from a few millimetres at the eyelid until up to a centimetre at the foot sole. Majority of the current methods used for evaluating skin and diagnosing skin diseases are subjective, time consuming, expensive and invasive. Improvement in non-invasive and objective diagnostic methodology would be valuable from an economic point of view as well as to quality in health care (Boden *et al.*, 2008).

Abnormal changes in the skin are common indicator of many diseases. There is a growing interest in the development of non-invasive methods for diagnosing various illnesses through skin measurements (Boden *et al*, 2008). Since the last three decades, skin bioengineering is a growing field in cutaneous research. Various instruments have been introduced for skin physiological analysis and in experimental trials. The measurement of stratum corneum hydration is one of the primary importances in the study of the biophysical properties and function of the skin barrier (Fluhr *et al*, 1999). Two non-invasive skin characterization techniques suggested for the diagnosis and monitoring of various markers of diseases are near infrared (NIR) spectroscopy and skin impedance (IMP). These two techniques have been applied for diagnosis of neuropathy, blood glucose levels, microcirculation in patients with diabetes and radiotherapy induced erythema (Boden *et al*, 2008).

NIR spectroscopy application for the measurement of water content in skin has long been developed. For clinical diagnostics and for the evaluation of the efficacy of cosmetics products, an exact water content measurement technique is required. For instance, an easy measurement method of water content is essential for atopic cases (Arimoto *et al*, 2005). Hansen and Yellin (1972) have applied NMR and IR to determine that at water content lower than 10%, the water present was tightly bound, most likely due to the polar sites of the proteins. At water contents in between 10-40%, they found less tightly bound water and suggested it was hydrogen-bonded to the protein-bound water. For water content greater than 50%, the water resembled the bulk liquid. The effectiveness of skin moisturizers is generally verified through indirect measurements of hydration such as high frequency electrical conductivity, TEWL (transepidermal water loss), biomechanical measurements and subjective clinical evaluations. However, all these techniques suffer from low precision and no well-understood relationship to water content (Martin, 1993).

## 3. Skin moisture measurement

There are several techniques that have been conventional used as skin moisture measuring devices. According to Suh *et al*, 2005, a various techniques have been developed for measuring stratum corneum water content, including electric conductance such as by transepidermal water loss (TEWL) which measures the rate of evaporation of water from the skin surface, attenuated total reflectance (ATR) fourier transform infrared or ATR-FTIR and confocal raman spectroscopy. However, TEWL is very sensitive towards environmental changes and requires several minutes to retrieve stable readings. This can be seen through sample measurement conducted using DermaLab TEWL module (Cortex Technologies, Hadsund, Denmark) as shown in Figure 1. The probe comes together with humidity and temperature sensor. The measurement will stop and the final result will be displayed when the value of standard deviation (SD) reaches the user-defined value which for this example is 0.3.

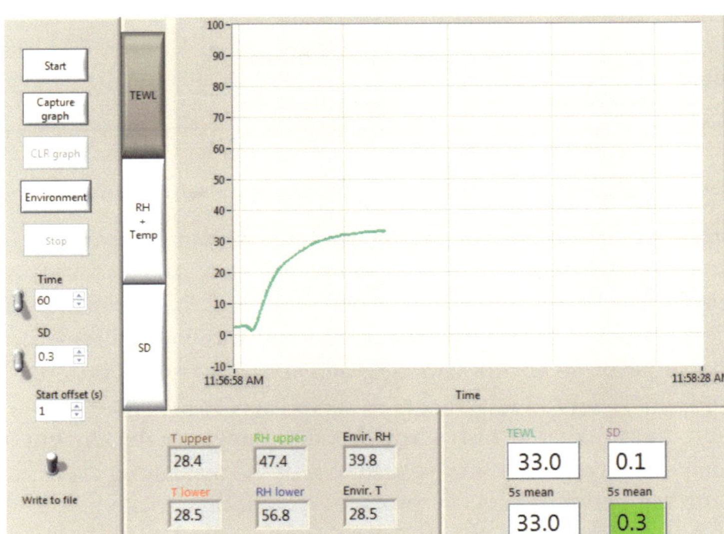


Fig. 1. TEWL measurement.

ATR-FTIR measurements on the other hand rely on the ambient conditions and are limited to the uppermost stratum corneum. ATR-FTIR method is neither rapid nor portable (Suh *et al*, 2005). ATR measurements on the other hands require occlusion of the skin, which can

affect the measurement of water content. The magnitude of contact between the internal reflectance element and the skin is not constant, especially after the skin is treated with a lotion which may alter the refractive index of the skin upon hydration, hence affecting the depth of penetration of radiation (Martin, 1993).
Arimoto *et al* (2005) stated that the most established technique to measure water content in skin is through the measurement of electrical properties such as capacitance and alternating current conductivity on the skin surface. Table 1 shows the current available commercial instrument for skin moisture measurement and its parameters.

| Instrument | Technique | Frequency | Unit |
|---|---|---|---|
| Corneometer CM 820 | capacitance | 40-75 kHz | Arbitrary Units (AU) |
| Corneorneter CM 825 | capacitance | mean frequency of 1 MHz (1.15 MHz - very dry medium; 0.95 MHz - very hydrated medium) | Arbitrary Units (AU) |
| Skicon 200 | conductance | 3.5 MHz | psiemens (pS) |
| Nova DPM 9003 | impedance based capacitance | up to 1 MHz | Arbitrary Units (AU) |
| DermaLab - Moisture Module | impedance-based capacitance | 100 kHz | µsiemens (µS) |

Table 1. Parameters of different instruments for moisture measurement (Fluhr *et al*, 1999).

The skin moisture sensor design consists of two metal plates which are isolated by an isolating medium, called a dielectric, as a capacitor. When a voltage source is connected to the capacitor, electrons will start to flow from one plate over the terminal to the other plate. The capacitor will store the electric charge. The quantity of charge stored is called capacitance. The capacitance of the capacitor will increase when material is introduced between the capacitor plates. Vacuum has a dielectric constant less than 7 while water dielectric constant is approximately 81. Therefore, the changes in the amount of water in the measured skin lead to a modification of capacitance of the measuring capacitor. For CM 825 system from Courage and Khazaka, its probe is electrically isolated from the measuring electronics and hence eliminating the influence from ground capacitance and salty skin surface. Total electrical opposition to the flow of an alternating current is defined as impedance. Resistance ($R_x$), capacitance ($C_x$) and frequency ($f$) contribute to the impedance (Z) through the following equation (Fluhr *et al*, 1999):

$$Z=(R_x^2 + [1/2\pi f C_x]^2)^{1/2} \quad (1)$$

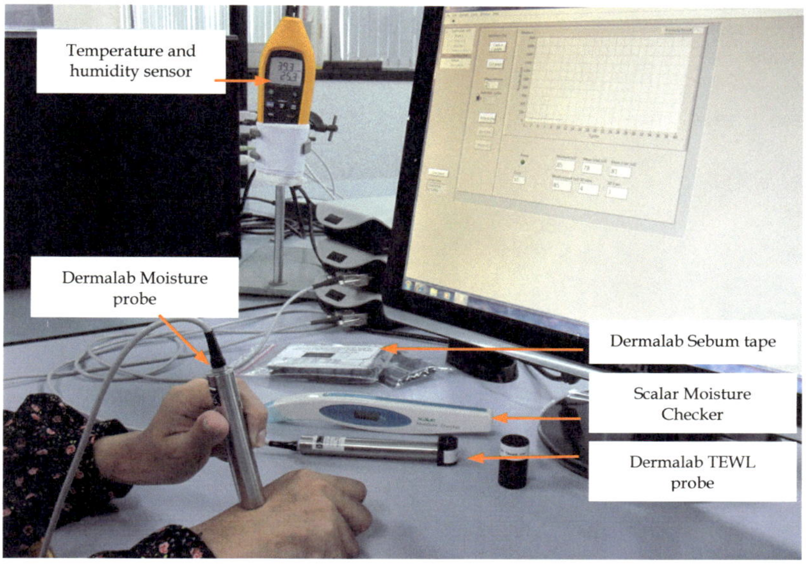


Fig. 2. Skin moisture measurement technique and other skin analysis modules in Spectroscopy Laboratory. Universiti Sains Malaysia.

Electrical conductance is measured when a constant frequency alternating current is applied to skin. The skin moisture is then calculated from the electric conductivity that is dependent on the water content of the skin (Woo et al, 2001). Conductance is identified to correlates well with the superficial portion of the stratum corneum even the electrical field on the stratum corneum is non homogeneous (Tagami, 1994). Nonetheless, this technique has limitation. Moisture measurement using electric conductance devices are easily affected by the amount of electrolytes in the skin and by the contact area of the probe's surface on the skin. These devices are also influenced by external temperature and humidity. This requires the devices to be kept at a constant temperature and humidity. Besides, the amount of electrolyte that the skin contains can alter the conductance value with no regard to water content (Woo *et al*, 2001). Figure 3 shows the skin moisture distribution across a person (a) right upper limb and (b) face measured at room temperature of 22°C and relative humidity of 44.1% using Dermalab moisture module from Cortex Technologies (Hadsund, Denmark). Cortex Technologies suggested moisture content below 150μS as very dry, between 150-300 μS as dry and above 300μS as skin with sufficient amount of moisture.

There is a non linear relationship between electrical conductance and water content. However, this is depending on the binding state of water molecules to the keratin chainsthat is described by the water sorption isotherm. Berardescal stated that there has been regular in the literature to define three types of water according to their strength of binding to the keratin namely *"tightly bound water"* for water contents from in between 0% and 7%, *"bound water"* between about 7% and 35% and *"free water "* which is above 35%. This division is generally helpful and can be considered simplistic on the basis of more detailed theory. Due to the variation in water binding strength, there is no direct proportionality between total water content and electrical conductance. Substances or treatments that interact with the keratin-water network may modify conductance without changing the water content of the sample (Berardescal, 1997). Berardescal (1997) disagree with the approach to refer to the electrical *"capacitance"* of the skin without specifying the stimulating frequency and other

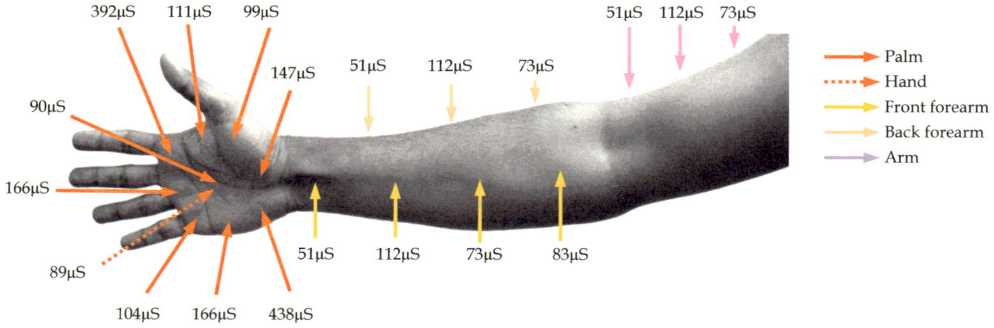


(a)

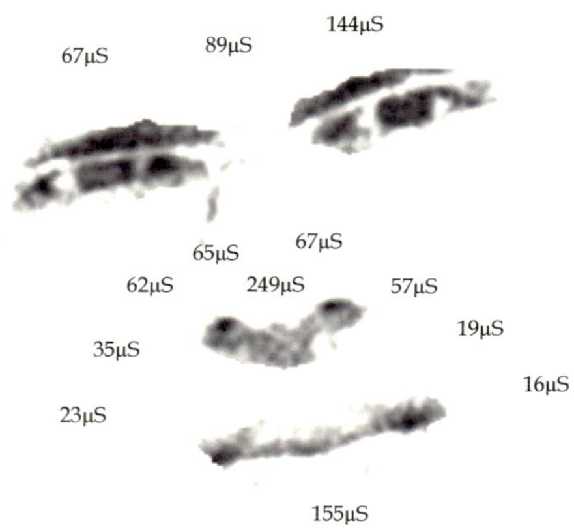


(b)

Fig. 3. Skin moisture distribution across a person (a) right upper limb (b) face.

experimental settings used to make the estimate. *"Capacitance"* term generally referred by many literatures in similar study is not actually electrical capacitance in the usual scientific and engineering sense since it is frequency dependent, in contrary to the true electrical capacitance.

Dermalab moisture module has the capability of showing the result of moisture measurement in two ways, either a single instantaneous data or continuous measurement. For the continuous measurement where the probe is pressed onto the skin for a few cycle of measurement, the measured data will show increment through time as shown in Figure 4. This is due to occlusion which enhances stratum corneum hydration (Zhai and Maibach, 2001). The application of NIR spectroscopy through optical fiber probe has shown a promising solution to this problem since the measurement of moisture can be performed without the need for the probe to be in contact with the skin surface or with very minimum contact pressure.

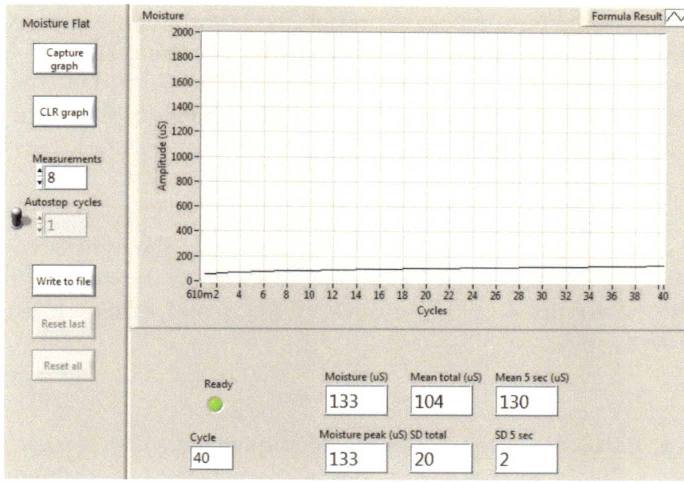


Fig. 4. Occlusion during continuous skin moisture measurement.

## 4. Optical fiber NIR spectroscopy technique (1100-2000nm)

Near Infrared (NIR) spectroscopy technique are capable of providing information on constituent's concentration. NIR spectroscopy was first developed in the 1980s by Norris *et al.* in analysis of agricultural products (Williams and Norris, 2001). Measuring oxygen saturation of hemoglobin is one of the most successful NIR spectroscopy applications in the field of biomedical engineering. There are also large interests of NIR application in non-invasive blood glucose measurement even though most of them are not clinically reliable until today (Arimoto *et al*, 2005). NIR region are the most prominent for water absorption bands. It allows direct moisture measurement at a certain range of wavelength. This is due to overtones and combinations of the fundamental vibrations that are active in the NIR range and taking place from hydrogen covalent bonds (Suh *et al*, 2005). Water spectrum dominating NIR spectra with overtone bands of the O-H bonds with peak absorption at 760 nm, 970 nm (due to the second overtone of the O-H stretching band), 1190nm (the combination of the first overtone of the O-H stretching and the O-H bending band), 1450 nm first overtone of the OH-stretching band and a combination band), and 1940 nm (combination of the O-H stretching band and the O-H bending band) (Luck, 1974, Nicolai *et al.* 2007).

Spectroscopic measurements are directly related to water content and can be represented by a classical Beer's Law relationship through the absorption of the hydroxyl moieties. The basic working relationship of the light attenuation can be stated by the exponential law of absorption. The differential absorption can be expressed as:

$$dI = - \alpha I_o \, dx \tag{2}$$

Where α is the absorption coefficient and is a measure of the rate of loss of light from the direct beam due to the dissolved and suspended substance within the water and the water itself. It is an inherent optical property. Upon integration from 0 to x, gives the exponential law of absorption:

$$I = I_o e^{-\alpha x} \tag{3}$$

0 is the starting point of the light passage through the absorbing medium. X is the length of the medium or the distance of light travel through the medium. $I_o$ is the light intensity at point 0. Since the medium is a solution, the concentration, c is included. The absorbance of a sample is proportional to both concentration and the path length that light travels (Martin, 1993). Therefore, the equation becomes:

$$I = I_o e^{-\alpha x c} \tag{4}$$

However, based on the above mentioned, the attenuation of the light is not entirely due to the absorption of light energy, but to some extent, it is also as the result of light scattered to the other side by the particles of the solution (Jenkins and White, 1976). Therefore, the equation can be rewritten as:

$$I = I_o e^{-[\alpha a + \alpha s]xc} \tag{5}$$

One of the advantages of the usage of optical techniques for water content measurement is its flexibility. Non-occlusive measurements can be made using NIR technique since it is not necessarily for the interface of probe light to be made in contact with the skin surface. Moreover, a single measurement point technique can be expanded to a two-dimensional area by using an image sensor with a series of spectral filters (Arimoto *et al*, 2005). Eun-Jung Suh (2005) concluded the following advantages that NIR spectroscopy has for biomedical applications:

i. it is a non-invasive and non-destructive analytical technique.
ii. One can use fiber optics for *in vivo* measurements.
iii. It is possible to monitor not only the surface of biological tissues but also their insides because NIR light penetrate into the tissues.

An *in vitro* spectroscopic experiment using porcine skin has been reported by Walling and Dabney (1989). Study on the differences of the absorption spectra between free and bound water has been conducted by Martin (1995). Martin has successfully showed that the absorption spectra can distinguish four types of water in skin. Those are water associated with the lipid phase within the stratum corneum, bulk water below the stratum corneum, secondary water of hydration on stratum corneum keratin and primary water of hydration on stratum corneum keratin. Also in the work, Martin (1995) has experimentally obtained profiles of the measurement depth in diffuse reflectance spectroscopy. Attas *et al* (2002) have conducted an experiment using near infrared spectroscopic imaging with wavelength between 960nm and 1700nm for in-vivo skin hydration measurements. The system is the combination between a near infrared camera with a liquid-crystal tunable filter (LCTF) to acquire spectral images at multiple narrow wavelength bands. The proposed system produces two-dimensional skin hydration mapping system. This technique enables the reading of absorption distribution over the skin surface. The use of mid-infrared range of wavelength in attenuated total reflectance (ATR) spectroscopy is also an efficient technique to assess skin water content. The penetration depth of mid-infrared signal is much shallower than the NIR diffuse reflectance. This is because the ATR measurement utilizes evanescent wave that localizes on the surface of the ATR crystal (Arimoto et al, 2005). Good correlation between NIR absorbance and water content has been showed empirically through *in vitro* and *in vivo* technique. This is an additional advantage of NIR analysis over other measurements.

Suh *et al* (2005) have conducted an experiment to evaluate water content in stratum corneum using a FT-NIR spectrometer with a fiber-optics probe. This technique is considered rapid and non-destructive. This serve as an advantage over other conventional NIR instruments with integrating sphere or moving grating that are hard to move and handle for practical use. The use of optical fiber probe is valuably being used for clinical diagnostics. The correlation between skin water contents and the NIR values have been investigated by using the conventional capacitance method as the reference. In order to find a robust model for skin moisture measurement, they have used three regions of NIR wavelengths which are 1130-1830nm, 1200-1670nm and 1380-1600nm. The calibration results of their experiment are shown in Table 2. The results presented also include NIR spectral treatment through first and second derivatives techniques. The first derivative will removes the baseline or offset produced by the scattering effects. On the other hand, the second derivative eliminates the gradient of the spectrum (Owen, 1995; Dabakk *et al*, 1999).

| Spectral Range (nm) | Spectral Treatment | No. Factors | SEC (%) | SEP (%) |
|---|---|---|---|---|
| | None | 7 | 3.82 | 3.98 |
| 1130-1830 | 1D | 2 | 5.52 | 6.17 |
| | 2D | 2 | 4.65 | 7.21 |
| | None | 6 | 4.35 | 4.90 |
| 1200-1670 | 1D | 3 | 5.06 | 7.35 |
| | 2D | 3 | 4.56 | 9.20 |
| | None | 2 | 7.43 | 6.53 |
| 1380-1600 | 1D | 1 | 6.95 | 7.50 |
| | 2D | 3 | 4.90 | 9.63 |

Table 2. Calibration results for water content measurement using FT-NIR conducted by (Suh *et al*, 2005).

In the near infrared application on skin conducted by Suh *et al* (2005), water molecules show two clear absorption bands at 1450 nm and 1940 nm. The peaks absorbances are sufficiently high and can easily be identified in the spectrum of the human skin. The absorption band around 970nm is considered weak, produces higher light scattering effect which is mainly due to short wavelength and there is too much penetration into the inner skin that make it not valuable for the measurement of water content in the stratum corneum. Starting from 1100nm, the NIR detector is changed from silicon to a lead sulphide (PbS) detector. The development of a calibration model is suggested at the spectral range above 1150nm.

Woo *et al* (2001) have presented a comparative evaluation between the performances of conventional scanning-type spectrometer using NIRSystems model 6500 (Foss NIRSystems Inc., Silver Spring, MD) and portable NIR system, a newly integrated system equipped with an optical fiber using microchip technology. The fiber-optic probe used with the portable NIR system has a total of nine reflectance fiber-optic bundles. Eight surrounding bundles are used for illuminating the sample while one at the center of the bundle for receiving the light reflected from the sample. A 0.3-mm gap was maintained between the fiber terminal and the skin surface in order to receive the light reflected light effectively and to avoid contacting the skin surface directly. This is done by using a holder that could strongly support the fiber-optic bundle to obtain a stable spectrum. They identified in the experiment that the calibration results efficiency were fairly similar between both spectrometers. The calibration results for water content measurement using portable NIR system is shown in

Table 3. It is also identified that better calibration result was acquired when second derivative spectra is applied on the 1150-1650nm wavelength range, with SEC value of 5.1% and SEP value of 5.6% (Woo *et al*, 2001).

| Spectral Range (nm) | Spectral Treatment | No. Factors | SEC (%) | SEP (%) |
|---|---|---|---|---|
| 1150-1650 | None | 13 | 4.5 | 4.9 |
| 1150-1650 | 1D | 9 | 4.7 | 5.0 |
| 1150-1650 | 2D | 8 | 4.7 | 5.3 |
| 1350-1500 | None | 8 | 6.6 | 6.7 |
| 1350-1500 | 1D | 11 | 5.9 | 6.9 |
| 1350-1500 | 2D | 10 | 6.3 | 7.4 |

Table 3. Calibration results for water content measurement using Portable NIR System conducted by Woo *et al* (2001).

## 5. Optical fiber NIR analysis at 970nm

### 5.1 NIR reflectance spectroscopy

The range of NIR signal between 1100nm to 2500nm is commonly being suggested for biochemistry application, typically for quantitative analysis. Nonetheless, lower NIR region in between 700nm to 1100nm have also been used widely in physical analysis as well as in analytical chemistry. For instance, International Organization for Standardization, ISO 7027 (1990) has suggested LED with peak wavelength at 860nm to be used as illuminating radiation in water turbidity measurement. In fruits quality analysis, range of NIR wavelength in between 700-1100nm has widely been applied for the measurement of soluble solids content and acidity (Nicolai *et al*, 2007). This section presents an overview on spectroscopy skin analysis using lower range of NIR wavelength and the focus will be given particularly for wavelength at 970nm.

The NIR spectral representation for different parts of body may be different due to the variation of skin thickness and surface roughness that will define the magnitude of diffuse and specular reflectance. To understand this further, spectroscopic implementation can be conducted on a person by collecting spectral signature from different part the person body. The spectroscopic instrumentations used for the measurement in this example are from Ocean Optics (Dunedin, Florida, USA) as demonstrated in Figure 5. The value of reflectance was measured using Spectrometer (650-1100nm). Other custom setup prior to the measurement includes integration time = 10ms, spectra averaged = 30 and boxcar smoothing = 1. Light source used was HL-2000 tungsten halogen lamp with spectral emission between 360nm to 2000nm and colour temperature of 2960K. The reflectance spectrum of the halogen lamp was calibrated using WS-1-SL, a white diffuse reflectance standard with above 99% reflectivity from 400-1500nm and above 96% reflectivity from 250-2000 nm. Reflectance probe used in this measurement is R600-7-SR-125F, a standard reflectance/backscattering probe with 6 illumination fibers around 1 read fiber. Each fiber has a core diameter of 600μm.

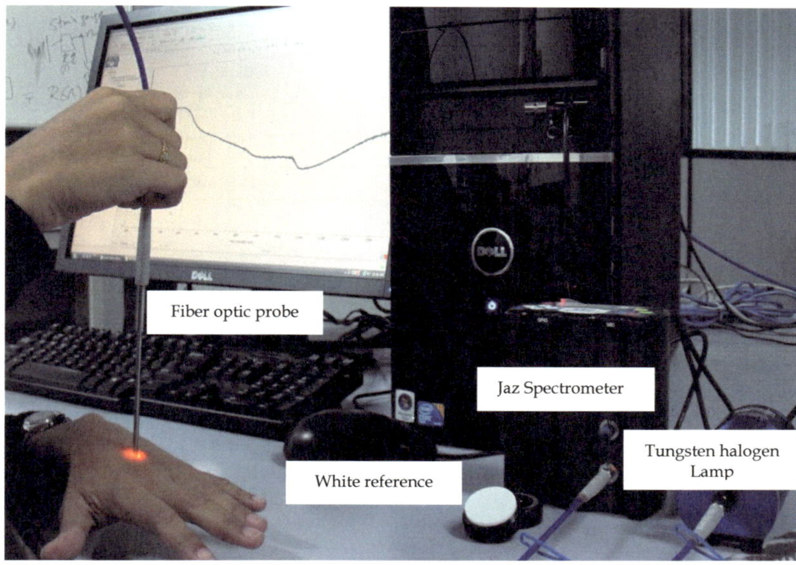


Fig. 5. Spectroscopy experimental setup.

Figure 6 (a) shows the NIR spectra of a person right upper limb and Figure 6 (b) shows the spectra of a person face. The patterns of spectra for every parts of the upper limb appear to be the same except at wavelength above 1050nm where the reflectance increases in its intensity for palms, front forearms and arms. While the reflectance of hands and back forearms drop further beyond 1050nm.

From the measurement of reflectance spectra shown in Figure 6 (a) and (b), it is identified that the reflectance intensity varies naturally according to different body parts. The high reflectance variation may greatly hide the absorbance magnitude that comes from moisture content of the skin. Therefore, it is important to localise the calibration equation of the skin moisture measurement based on body part, unless if the optical fiber probe is designed in such away that the penetration depth is only within the stratum corneum.

### 5.2 Skin optical simulation through ASAP - RSM

ASAP (Advanced Systems Analysis Program) is an optical engineering software developed by Breault Research Organization, Inc. that can be used to simulate optical system with 4 major steps in its application; build the system, create sources, trace the rays and perform analysis. RSM (Realistic Skin Modelling) provides interactive scripts for modelling light propagation, absorption and scattering in skin. RSM accurately replicates the absorption and scattering properties of human skin. The absorption and scatter characteristics of each layer of tissue are calculated separately. This is done by taking into consideration the different chromophore concentrations in each layer such as melanin, hemoglobin, water, bilirubin and beta carotene. RSM can function with the range of source wavelength between 250nm and 1000nm (Breault Research Organization, Inc, 2008).

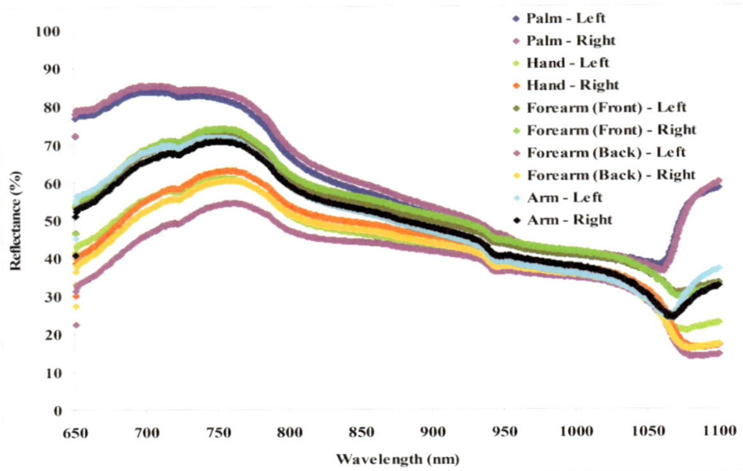


(a)

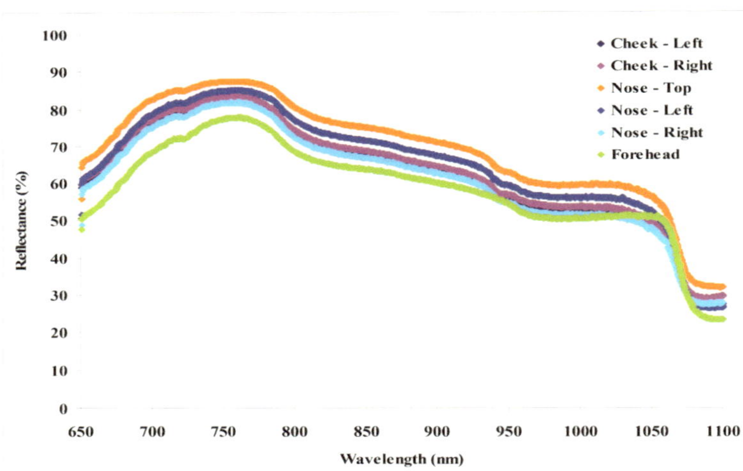


(b)

Fig. 6. NIR spectra from a person (a) upper limb (b) face.

ASAP-RSM software can be used to address the issue related to low absorptivity, high scattering and deep penetration of NIR wavelength at 970nm. Figure 7 shows the results of simulation when a circular beam with 1000000 rays are illuminated perpendicularly onto a skin surface with variation in the amount of water content in the stratum corneum with 15μm of thickness. The relative volume absorption of NIR radiation increases with higher amount of water content in the stratum corneum. The peak responsivity is identified at 970nm, indicated by the graph steepest gradient produced if compared to other neighbouring wavelengths of 950nm, 960nm, 980nm and 990nm. Figure 8 shows the image of the magnitude of volume absorbance at 970nm, taken at 10μm depth from the skin surface.

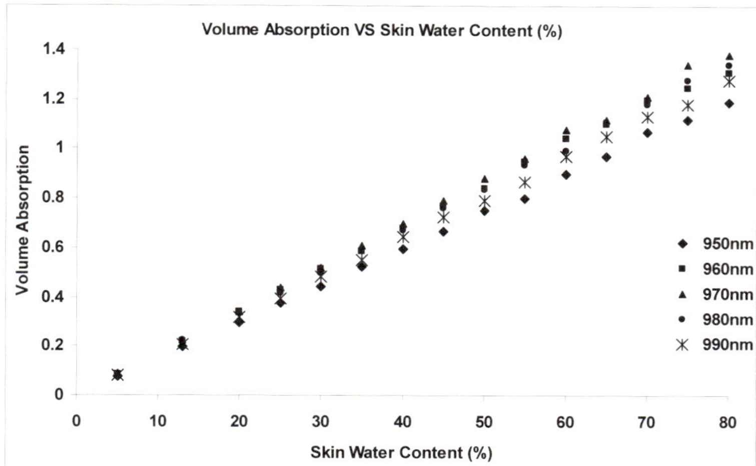


Fig. 7. Relationship between relative volume absorption and water content in stratum corneum for 5 different near infrared wavelengths.

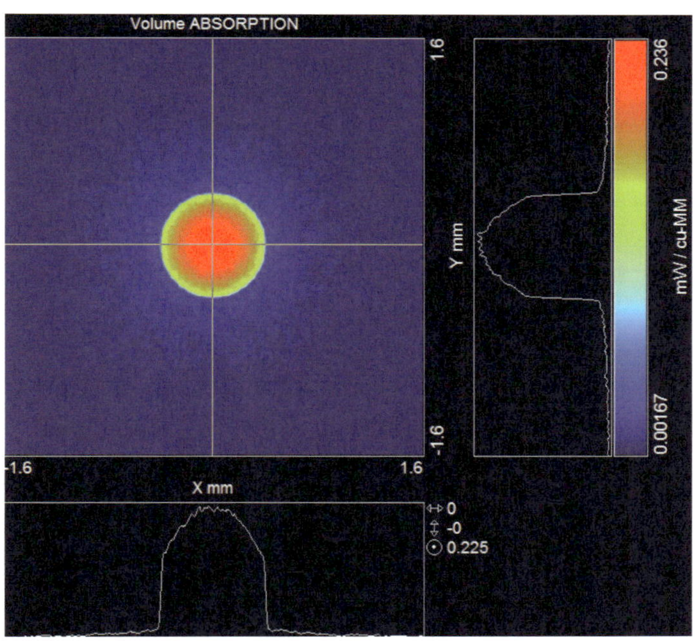


Fig. 8. Volume absorbance at 970nm, taken at 10μm depth from the skin surface.

Figure 9 shows in details the peak absorbance of water content for wavelength from 700nm to 1000nm. As has been proven by Figure 7, 970nm produces the peak absorbance for water content in stratum corneum. Wavelength at 760nm has a very weak peak absorbance for water content and can be seen slightly when compared with wavelength 700nm and 800nm.

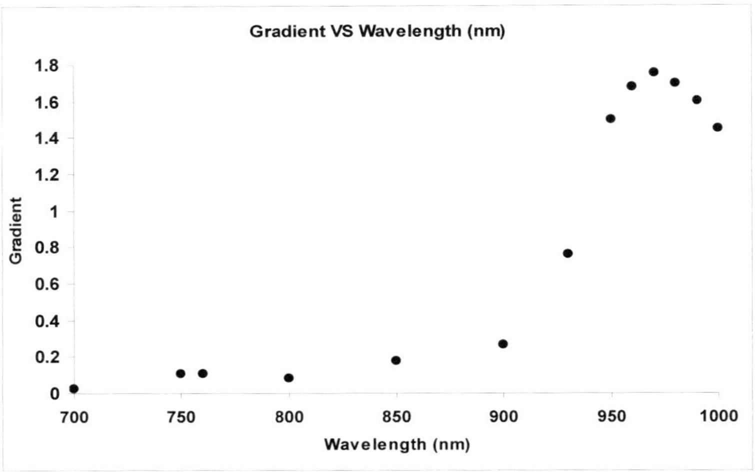


Fig. 9. Relationship between gradient (responsivity) and wavelength.

As has been stated by Suh *et al* (2005), the absorptivity of moisture at 970nm is very low while the scattering volume is high. Due to small thickness of stratum corneum, it is difficult to directly quantify its moisture content using a typical fiber probe configuration. Table 4 shows the absorption coefficient, $\mu_a$ ($mm^{-1}$) for 6 different percentages of moisture content in stratum corneum. In the simulation through RSM as shown in the example within this chapter, the scattering coefficient, $\mu_s$ for stratum corneum is 8.8794$mm^{-1}$ and anisotropy factor (g), the average directional cosine of the scattered light is kept constant at 0.9, indicating that the tissue is highly forward scattering.

| **Moisture content (%)** | 5 | 10 | 15 | 20 | 25 | 30 |
|---|---|---|---|---|---|---|
| **Absorption Coefficient, $\mu_a$ ($mm^{-1}$)** | 0.00226 | 0.00453 | 0.00679 | 0.00906 | 0.01132 | 0.01359 |

Table 4. Moisture absorption coefficient.

The typical implementation of spectroscopy analysis is fundamentally based on detection of broad spectrum from ultraviolet to near infrared. This can be done by illuminating the sample with light source with broad spectral wavelength. The resultant spectrum will then be detected by spectrometer that consist of grating, to disperse the incoming signal into its respective wavelength and optically allowing it to fall on a single or two photodiode array, depending on the effective range of wavelength that the instrument is required to interpret. Fiber optic probe is commonly being used as light transferring medium between the light source and spectrometer. As has been discussed in the earlier section of this chapter, for a single spectroscopy application such as for the measurement of skin moisture content, only a single wavelength that usually response the best. To apply a large range of non-contributed wavelength for a simple spectroscopy measurement has unnecessarily increased the cost of operation. A single wavelength sensitivity spectroscopy system that making use of LED with a narrow wavelength with peak emission that corresponds the best with the composition of interest and photodiode with spectral responsivity matches the LED spectral emission will significantly lowering the cost of operation and directly suit the application.

Figure 10 illustrates the conceptual design of optical fiber sensor for skin moisture measurement using ASAP-RSM. The reflectance fiber probe is arranged in bifurcated configuration with each cable having a diameter of 1mm. The LED is 5mm type and illuminates its radiation from a cubical die.

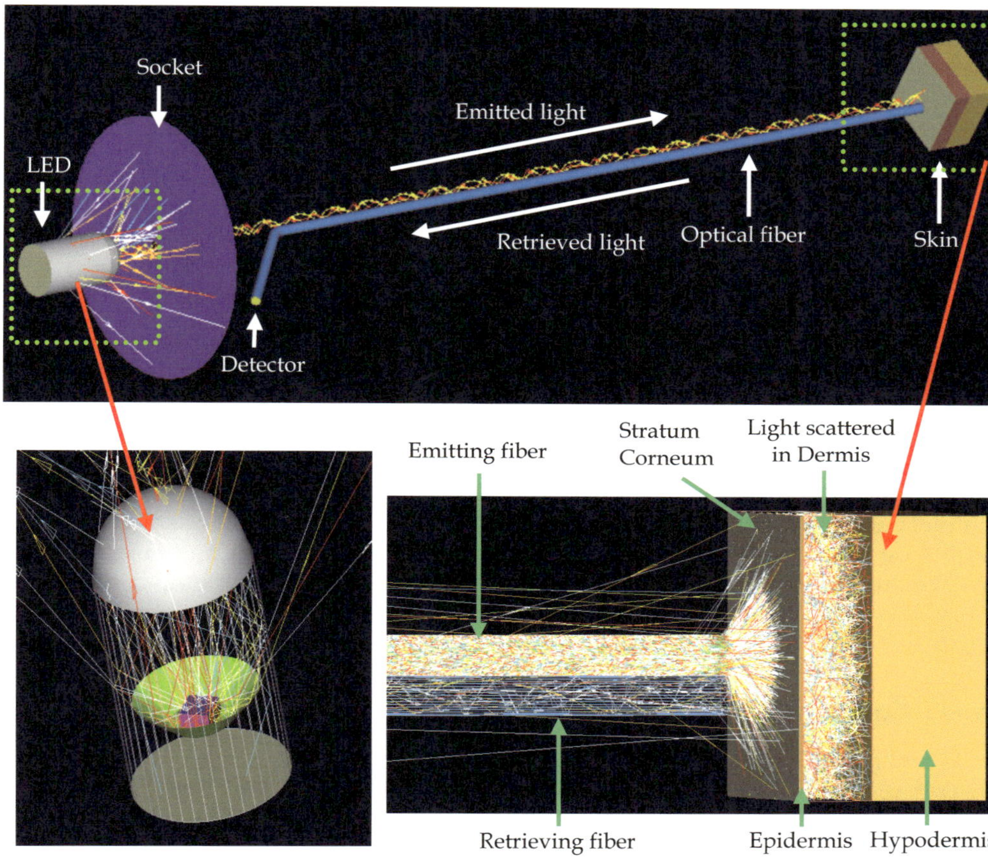


Fig. 10. Conceptual design of optical fiber sensor for skin moisture measurement.

In the application of ASAP-RSM, to obtain good results for analysis, highest possible number of rays traced during simulation is desirable. This will require a powerful computer with high processing speed and memory capacity. With smaller number of rays to be traced in the optical system, the small value of absorbance will drown within high intensity of light scattering, thus, no meaningful results can be generated. An optical model has been developed with one retrieving fibers allocated in the middle of two emitting fibers. All fibers used are having core diameter of 1mm. Light source (wavelength=970nm) with emitting diameter of 1mm and $45^{o}$ angle of emission has been directed through both emitting fibers with 3 millions total number of rays. The probe is placed right on top of the skin surface. The thickness of stratum corneum is set to 40μm. Using the existing optical parameters for all skin layers, no valuable results can be generated. When the absorption coefficients of water content are increased 100 times, then some useful results can be observed. This is presented in Figure 11 where fluctuation of intensity still occur despite a good correlation is achieved with low root mean square error (RMSE) of 2.63%.

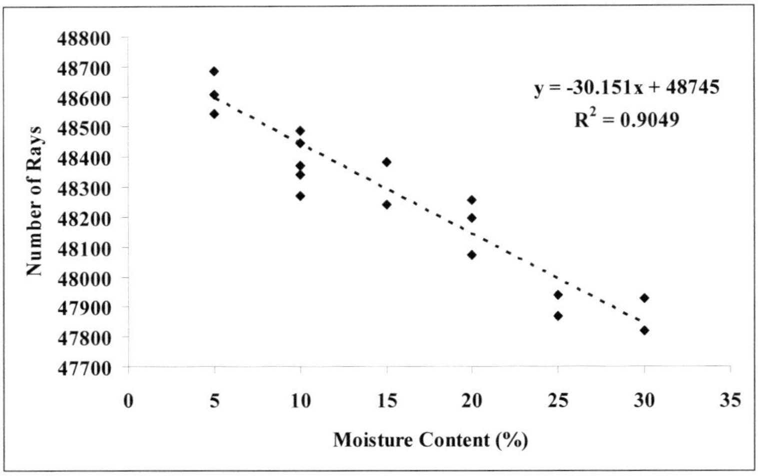


Fig. 11. Number of rays VS moisture content (absorption coefficient X100).

This simulation analysis has presented an early overview on possible application of 970nm in skin moisture measurement. The real challenge is to be able to produce an optical fiber probe that can minimize the high scattering effect from the radiation and ensuring very minimal amount of light penetrating deeper into the skin, beyond the stratum corneum. To put this implementation into being will enable the development of low cost optical fiber NIR sensor for skin moisture measurement.

## 6. Conclusion

This chapter has presented multiple experimental designs that are commercially available in the market such as impedance based design, within the clinical trial such as long NIR wavelength (1100-2500nm) and those that has possible venture into implementation such as lower range of NIR wavelength (970nm). The main aim of this continuous research is to be able to produce an optical fiber instrument that can minimise or to some extend eliminate any setback in the conventional implementation. Optical fiber NIR sensor has always be a promising technology as quantitative instrumentation for analytical science.

## 7. Acknowledgment

This project is sponsored by Universiti Sains Malaysia Research University Grant (Grant no. 1001/PFIZIK/811153) and Research University Postgraduate Research Grant Scheme (Grant No. 1001/PJJAUH/843028).

## 8. References

Arimoto, H., Egawa M. and Yamada, Y. (2005). Depth profile of diffuse reflectance near-infrared spectroscopy for measurement of water content in skin, *Skin Research and Technology*, 11, pp. 27–35

Attas, M., Schattka, T.P.B., Sowa, M., Mantsch, H. and Zhang, S. (2002). Long-wavelength nearinfrared spectroscopic imaging for in-vivo skin hydration measurements. *Vibrat Spectrosc*, 28, pp. 37-43.

Berardescal, E. (1997). EEMCO guidance for the assessment of stratum comeurn hydration: electrical methods, *Skin Research and Technology*, 3: pp. 126-132.

Breault Research Organization, Inc. *Realistic Skin Model (RSM): BIO Toolkit interactive script for ASAP*, ASAP Technical Publication, 2008).

Boden, I., Nilsson, D., Naredi, P. and Lindholm-Sethson, B. (2008). Characterization of healthy skin using near infrared spectroscopy and skin impedance, *Med Biol Eng Comput*, 46, pp. 985–995.

Dabakk, E., Nilsson, M., Geladi, P., Wold, S. and Renberg, I. (1999), Sampling Reproducibility and Error Estimation in Near Infrared Calibration of Lake Sediments for Water Quality Monitoring, *J. Near Infrared Spectrosc*, 7, pp. 241–250.

Fluhr, J.W., Gloor, M., Lazzerini, S., Kleesz, P., Grieshaber R. and Berardesca, E. (1999). Comparative study of five instruments measuring stratum corneurn hydration (Comeometer CM 820 and CM 825, Skicon 200, Nova DPM 9003, DermaLab). Part I. *In vitro. Skin Research and Technology*, 5, pp. 161-170

Hansen, J.R. and Yellin, W. (1972). NMR and Infrared Spectroscopic Studies of Stratum Corneum Hydration, in Water Structure at the Water-Polymer interface H. H. G. Jellinek, Ed. (Plenum Press New York, 1972), pp. 19-28.

International Standard ISO 7027. *Water Quality–Determination of Turbidity*. ISO: Geneva, Switzerland. 1990.

Jenkins, F.A. and White, H.E. (1976), *Fundamentals of Optics*, McGraw-Hill, Inc. pp. 455-473.

Luck, W.A.P. (1974). Infrared overtone region. In W. A. P. Luck (Ed.), *Structure of water and aqueous solutions*, Weinheim: Verlag Chemie. pp. 248–284.

Martin, K.A. (1993). Direct measurement of moisture in skin by NIR spectroscopy, *J. Soc. Cosmet. Chem.* 44, 249-261.

Martin K. (1995). Depth-profiling of skin by near infrared spectroscopy. In: Davies AMC, Williams P, eds. Near infrared spectroscopy: the future waves. *The Proceedings of the 7th International Conference on Near Infrared Spectroscopy*. Montreal: NIR Publications, pp. 328–333.

Nicolai, B.M., Beullens, K., Bobelyn, E., Peirs, A., Saeys, W., Theron, K.I. and Lammertyn, J. (2007). Nondestructive measurement of fruit and vegetable quality by means of NIR spectroscopy: A review, *Postharvest Biology and Technology*, 46, pp. 99–118.

Owen, A.J. (1995). *Uses of Derivative Spectroscopy, UV-Visible Spectroscopy*, Application Note, Agilent Technologies, pp. 1-8.

Suh, E.J., Woo, Y.A. and Kim, H.J. (2005). Determination of Water Content in Skin by using a FT Near Infrared Spectrometer, *Arch Pharm Res*. 28(4), pp. 458-462.

Tagami H. (1994). Hardware and measuring principle: skin conductance. In: Elsner P, Berardesca E, Maibach HI, eds. Bioengineering of the skin: water and the stratum comeum. Boca Raton: CRC Press, pp. 197-203.

Walling, P.L. and Dabney, J.M. (1989). Moisture in skin by nearinfrared reflectance spectroscopy. *J Soc Cosmet Chem*, 40, pp. 151–171.

Williams, P. and Norris, K. (2001). Near-Infrared Technology, in the Agricultural and Food Industries. Amer Assn of Cereal Chemists, 2nd edition, 2001.

Woo, Y.A., Ahn, J.W., Chun, I.K. and Kim, H.J. (2001). Development of a Method for the Determination of Human Skin Moisture Using a Portable Near-Infrared System, *Anal. Chem*, 73, pp. 4964-4971

Zhai, H. and Maibach, H.I. (2001), Skin occlusion and irritant and allergic contact dermatitis: an overview, *Contact Dermatitis, 2001*, 44, pp. 201–206

# 10

# Live Cells as Optical Fibers in the Vertebrate Retina

Andreas Reichenbach[1], Kristian Franze[1,2], Silke Agte[1,3], Stephan Junek[4], Antje Wurm[1], Jens Grosche[1], Alexej Savvinov[5], Jochen Guck[2] and Serguei N. Skatchkov[5]
[1,3]*Leipzig University;*
[2]*Cambridge University;*
[4]*Max Planck Institute for Brain Research, Frankfurt/M.;*
[5]*Universidad Central del Caribe, Bayamon;*
[1,3,4]*Germany*
[2]*UK*
[5]*Puerto Rico*

## 1. Introduction

The vertebrate eye is equipped with an inverted-type retina; that means, light must pass through all proximal retinal layers before it arrives at the photoreceptor cells which are aligned at the back of the tissue. Though it is often stated that the transparency of the intact vertebrate retina is 'almost total' (Enoch and Glisman, 1966), it contains numerous structures which differ in size and refractive index. These differences should lead to significant scattering. Accordingly, it has been pointed out that the situation in the inverted retina 'is equivalent to placing a thin diffusing screen directly over the film in your camera' (Goldsmith, 1990). In fact, many current digital cameras posses such a "diffusing screen", in order to prevent artifacts due to the discrete and periodic sampling of the image: the anti-aliasing filter. As the layout of the retina exhibits a similar sampling strategy, one might hypothesize that the cell layers in front of the photoreceptors act as an anti-aliasing filter. However, the fact that aliasing artifacts can be observed in the vertebrate eye (Coletta et al. 1990) argues against such a hypothesis. On the contrary, these reports confirm the high resolution provided by the vertebrate retina, close to its physical limits. We will discuss here whether this apparent discrepancy is resolved by the presence of cellular light guides within the retinal tissue.

## 2. The inverted vertebrate retina: optical properties of the tissue

Before going into more detail, the basic structure, phylogenetic origin, and embryonic development of the vertebrate retina will briefly be introduced.

### 2.1 Basic morphology of the vertebrate retina

The retina has a well-organized structure with seven main layers (Fig. 1). Three layers contain the cell bodies with the cell nuclei (outer nuclear layer, ONL; inner nuclear layer,

INL; ganglion cell layer, GCL), two layers, connecting the nuclear layers, contain cellular processes and the neuronal synapses (the outer plexiform layer, OPL, and the inner plexiform layer, IPL), the innermost layer contains the axons of the ganglion cells on their course towards the optic nerve head (nerve fiber layer, NFL), and the outermost layer is formed by the photoreceptor segments (PRL). The neural retina can be divided into an inner and outer part. The inner retina includes the NFL, GCL, IPL, and INL, and the outer retina consists of the OPL, ONL, and PRL (Fig. 1).

There are three types of neurons that lie in series and mediate the forward transmission of the visual information from the photoreceptor outer segments (where light is absorbed by the photosensitive pigments) at the outer surface of the neural retina to the axons at the inner surface of the retina running towards the optic nerve. These are, (i) the photoreceptor cells (the first-order neurons of the retina), (ii) the bipolar cells (the major second-order neurons, which are also called interneurons), and (iii) the ganglion cells (the third-order neurons, also called projection neurons). Photoreceptor cells consist of three parts: the soma of the cell, containing the cell nucleus, is located in the outer nuclear layer; the sensory process that consists of an inner segment (containing the energy-producing mitochondria) and an outer segment (containing the molecular machinery that captures the photons and converts the light energy into neuronal activity); and an axon which forms synapses with the bipolar cells (and other interneurons) in the outer plexiform layer. There are two basic types of photoreceptor cells, rods (responsible for the low contrast, 'black-and-white', but high sensitivity vision at low light intensities; scotopic vision) and cones (responsible for high contrast and color vision at bright daylight; photopic vision).

In addition to these neurons, elongated 'radial' glial cells span the entire thickness of the tissue; these are called Müller cells. As shown in Figure 1, a distinct set of retinal neurons is aligned along each Müller cell, thus forming a so-called columnar unit (Reichenbach et al., 1994; Reichenbach and Robinson, 1995). Basically, a vertebrate retina can thus be considered as being composed of a large number (almost 10.000.000 in the human eye) of such repetitive columnar units, each contributing its part to the visual information collected by the retina. The constituents of a columnar unit are remarkably constant across a given retina and largely independent of local topographic specializations (there are some exceptions such as the primate fovea centralis). However, the cellular composition of the columnar units differs considerably among the diverse vertebrates, even among mammalian species; this depends mainly on the diurnal *vs.* nocturnal lifestyle, i.e., the photopic *vs.* scotopic specialization of the retina (cf. section 4.3). In all mammals studied so far, every unit contains, in addition to the 'core Müller cell', (i) about 1 cone plus a variable number of rods, (ii) at least three interneurons of the INL, and (iii) one or two ganglion cell(s) in photopically specialized retinas (or less than one in the other retinas).

### 2.2 Phylogenetic origin and embryonic development of the vertebrate retina

During both phylogenesis and embryonic development, the retina of all vertebrates emerges from the same primitive neural tissue (the 'neural plate') as the brain does; it is thus considered as an 'externalized' part of the brain. To understand the inversion of the vertebrate retina, it is essential to keep in mind that vertebrates belong to the deuterostomian animals which means that our ancestors belong to the relatives of recent starfish and sea urchins. The starfish possesses an epithelial nervous system. It locally constitutes the outer surface of the body, where it directly contacts the seawater as a fluid environment, with a basal lamina deliminating the epidermal tissue at the inner side. It

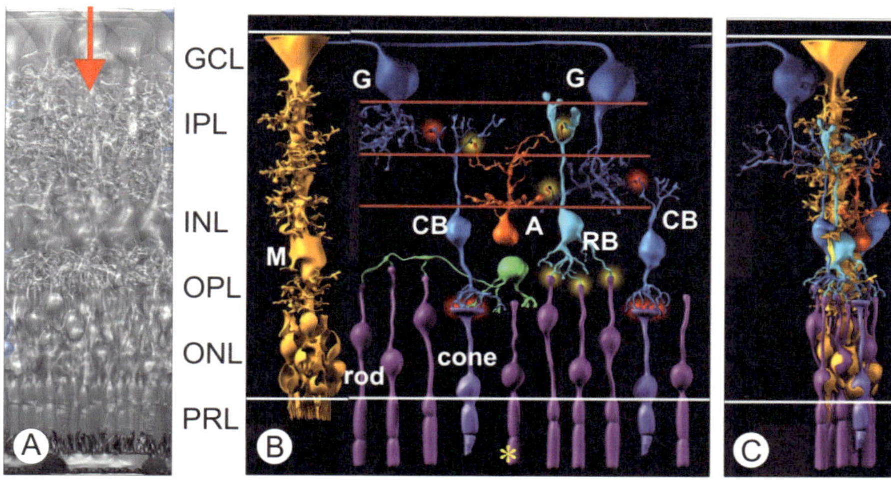


Fig. 1. Structural composition of a typical vertebrate retina; artists view. (A) Radial section through a mammalian retina with the typical layers. (B) The concept of columnar units in the vertebrate retina. The tissue is built up by a large number of repetitive groups of cells which are arranged in columns (C). The center of each column is constituted by a Müller cell (M) which extends lateral side branches ensheathing the adjacent neurons of the unit. G, retinal ganglion cell; B, bipolar cells; A, amacrine cell. Each column spans all retinal layers (GCL, ganglion cell layer; IPL, inner plexiform layer; INL, inner nuclear layer; OPL, outer plexiform layer; ONL, outer nuclear layer). The inner and outer segments of the photoreceptor cells reside in the photoreceptor layer (PRL). Original.

contains so-called supporting cells - which can be considered as the ancestors of radial glial cells (Reichenbach and Robinson, 1995) and two principal types of neurons. The sensory processes of the sensory cells extend into the maritime environment as the source of the stimuli to be monitored, whereas their axons reach towards the ganglion cells where the information is processed. Notably, this polarity is obviously 'correct' and easily comprehensible.

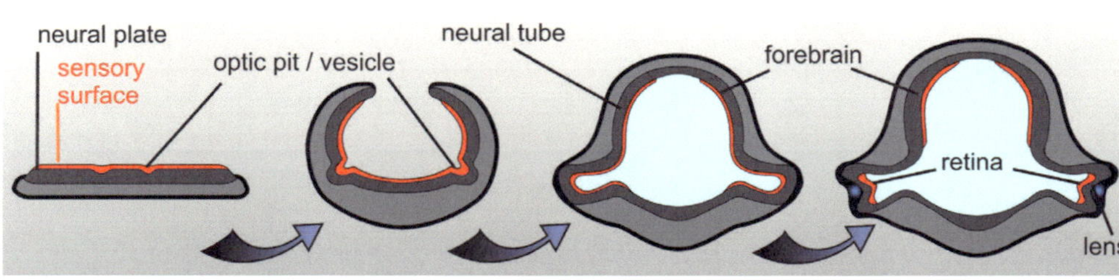


Fig. 2. Developmental mechanism causing the inversion of the vertebrate retina. Embryonic development of the retina occurs by evagination of the eye anlage (optic pit / vesicle) from the neural epithelium (from *left* to *right*). Later on, the outer wall of the optic vesicle becomes invaginated by the developing lens. This outer wall differentiates into the neural retina whereas the inner wall becomes the retinal pigment epithelium; the stalk is transformed into the optic nerve. Initially the 'sensory surface' (*red*) constitutes the outer surface of the neural plate (and the embryo); however, it becomes translocated into the inner surface by the invagination of the neural tube.

In the further course of evolution, the epithelial nervous system was maintained as such, but was enrolled into a tube and moved below the surface of the body by the overfolding or overgrowing skin and subepidermal layers. Similar events occur during our embryogenesis when the – originally superficial - neural plate is enrolled and overlayed in a process called neurulation (Fig. 2). Inevitably, this mechanism is accompanied by an inside-out turn of the polarized epithelium: the sensory cells which had faced the environment at the surface of the body now extend their sensory processes into the lumen – i.e., the inner surface - of the neural tube. This also explains for the 'odd' orientation of our retina (Fig. 1), with the photoreceptor cells directed away from the light. During the evagination of the optic vesicle from the neural tube the 'sensory surface' remains at the inner, 'wrong' surface (Fig. 2). Thus, the normal developmental mechanisms of our eyes inevitably lead to an inverted retina.

### 2.3 General optics of the retinal tissue

It has long been known that there is considerable light scattering in the retina which constitutes a 'turbid medium' (Boehm, 1940b) contributing up to 40% to entopic light scatter (Vos & Bouman, 1964), and allowing for a light transmission of only 85% at 500 nm (Hammer et al., 1995). Whereas a freshly isolated retina appears virtually transparent when viewed from top (i.e., orthogonally to its surface; see *arrow* in Fig. 3B) it looks 'milky' if one looks obliquely on the tissue (which tends to enrol) (Fig. 3).

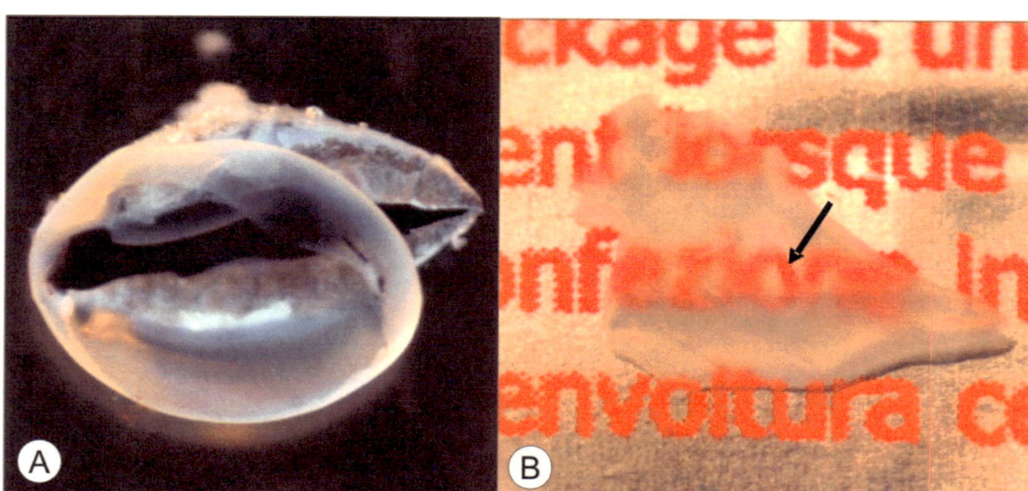


Fig. 3. Limited transparancy of the vertebrate retina. (A) Freshly dissociated guinea-pig retina in a buffer-filled Petri dish over a black background. (B) A piece of the same retina over some text. Only where the tissue is flat over a letter, the letter can be read easily (*arrow*). Wherever the retinal tissue is somehow enrolled and light incidence is thus not normal to the surface, the image is blurred.

It has also become evident that light scattering is inhomogenous among the retinal layers; particularly strong light scatter has been located in the inner plexiform layer (Martins-Ferreira & de Oliveira Castro, 1966) and in axon bundles of the nerve fiber layer (Knighton et al., 1989). Much of this scatter is due to small subcellular elements with dimensions in the order of the wavelengths of visible light (Boehm, 1940; Knighton et al., 1989; de Oliveira Castro et al., 1985) such as neurofilaments, neurotubuli, and other organelles which are enriched in the nerve fiber and plexiform (i.e., synaptic) layers. This scattering is, for example, used to visualize (some of the) individual retinal layers in intact eyes of human patients by optical coherence tomography (OCT) (e.g., Puliafito et al., 1995) (Fig. 4).

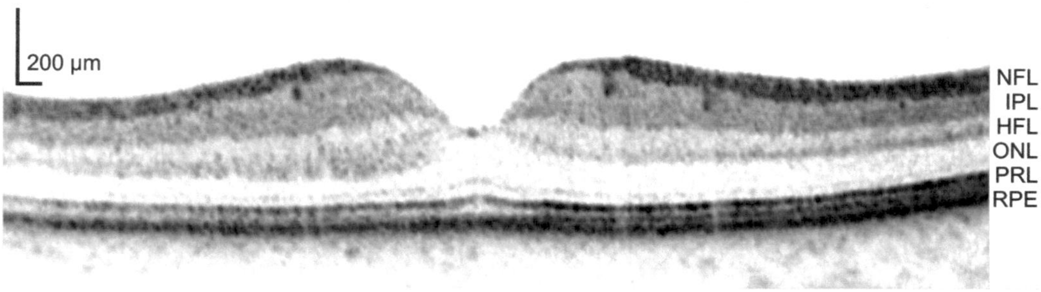


Fig. 4. Optical coherence tomography (OCT) demonstrates differential backscattering of light from human retina.

When a human retina in the intact patient's eye is studied by OCT, most of the retinal layers are easily identified because of the different amount of their backscattering of the laser. The cell process-containing layers (NFL, IPL, and HFL) display more backscattering (dark) than the cellular layers. NFL, nerve fiber layer; IPL, inner plexiform layer; HFL, Henle's fiber layer; ONL, outer nuclear layer; RPE, retinal pigment epithelium.

In fact, the refractive index of the retinal tissue is higher (about 1.36) (Chen, 1993; Valentin, 1879; Nordenson, 1934; Ajo, 1949) than that of the vitreous (1.336) (Nordenson, 1934). This causes a specular reflection from the retinal surface (Millodot, 1972; Charman, 1980; Gorrand, 1986) which may contribute to about half of the light leaving the eye, and which causes the 'fundus reflexes' familiar to ophthalmoscopists. It can be summarized that the retina is all but perfectly transparent; later we will show data revealing even different refractive indices among retinal cell types (sections 3.1, 4.1, 4.2).

### 2.4 Performance of the vertebrate retina

Despite these considerations, most vertebrates including humans can reliably detect a light signal when only a few photons enter the eye (Pirenne, 1967). Moreover, our visual acuity - even outside the *fovea centralis* where the inner retinal layers are missing – is as good as the spacing of two neighboring cone photoreceptor cells allows, resulting in the highest possible resolution (Livingstone and Hubel, 1988; Williams et al., 1993). This has stimulated the search for a mechanism that may minimize intraretinal light scatter.

## 3. Methods to study cell and tissue optics of the vertebrate retina

In order to understand the pathway of light through the inverted vertebrate retina it is essential to collect precise data about the optical properties of the retinal tissue and of its constituents, the retinal cells. Whereas first measurements of retinal optics date back to the late 19th century, many methods only became available in recent years. The following sections are aimed at providing an overview of 'traditional' and current techniques that have been employed, or may be employed, to study retinal optics down to the cellular level.

### 3.1 Assessing the refractive index of single cells and retinal tissue

There are different indirect and direct approaches to determine the *refractive index n* (RI) of living tissues and individual cells.

In historical experiments, Valentin used a refractometer to measure the RI of different types of biological liquids and tissues of different species including retinae (1.3411 (rabbit) $\leq n \leq$ 1.3461 (young chicken) (Valentin, 1879)). However, this technique assesses the refraction angles at parallel interfaces, which are rarely found in biological tissues. Nerve tissue, for example, had to be crushed to a mushy mass in order to provide reproducible results. Furthermore, this method only yields an average value for the RI of the whole sample.

The RI of biological samples can be taken as a measure of the total organic solids present (which is mostly determined by the protein concentration) (Barer, 1954). Based on their total dry mass concentration, Chen estimated the RI of individual retinal layers to be between 1.369 and 1.358 (Chen, 1993). The ratios of the refractive indices of neighboring retinal layers were similar.

Furthermore, the RI of individual, acutely isolated cells can, for example, be determined with an index matching method. Here, cells are sequentially exposed to a series of isotonic superfusates of graded RI, which are adjusted to a value at which most of the cells become virtually invisible in phase contrast microscopy. One way of obtaining solutions with higher RI is by adding different concentrations of bovine serum albumin (BSA). This technique is comparable to the immersion method established in mineralogy and crystallography. However, it is difficult to raise the RI of the medium to higher values without damaging vital cells, and all of the above-mentioned approaches suffer from relatively poor resolution.

More sophisticated setups to measure the RI of biological samples with high spatial resolution include Jamin-Lebedeff microscopes (Franze et al., 2007) and digital holographic microscopy (DHM) (Charriere et al., 2006). These methods exploit interference microscopy, which can be used to quantitatively determine optical pathway differences. The optical path difference depends on both the RI and the thickness of the sample (the phase profile represents the multiplication of RI differences and sample thickness). Thus, the refractive index itself cannot be measured without determining the thickness of the cells separately. This determination is often subject to considerable errors. Methods to decouple the RI from the cell thickness include a combination of interference microscopy with confocal laser scanning microscopy or atomic force microscopy (the latter methods are used to measure the sample height) as well as measuring the phase profiles of the same sample in media with different RIs. In case of cells in solution, exact three-dimensional tomographical image can be obtained with a dual-beam laser trap-based optical cell rotator (Kreysing et al., 2008).

### 3.2 Measuring light propagation through individual cells

Light transmission through retinal cells can be observed *'in situ'*, i.e., while the cells are within the intact retinal tissue. This way, light exiting the distal light path, i.e., the photoreceptor outer segments, can be visualized. Pioneering experiments by Enoch and

Tobey in the 1960s and 1970s suggested that the outer segments of photoreceptor cells act as "light collectors" (Tobey et al., 1975) and wave guides. Light transmission through these outer segments *in situ* occurs in modal patterns as known from optical fibers (Enoch, 1961). The most commonly seen modes included $HE_{11}$, $TE_{01}$, and $TM_{01}$ (Enoch, 1961) (Fig. 5). In some of the observed cells, mode changes occurred with changing wavelength.

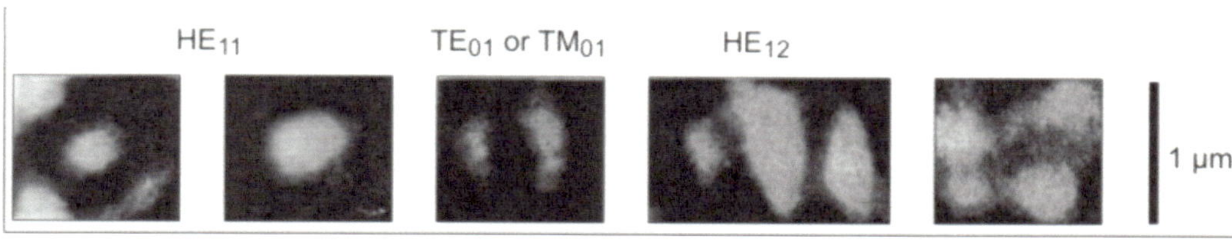


Fig. 5. Modal patterns of light exiting photoreceptor cell outer segments of rhesus macaque monkeys. Scale bar: 1 µm. (adapted from Enoch, 1961).

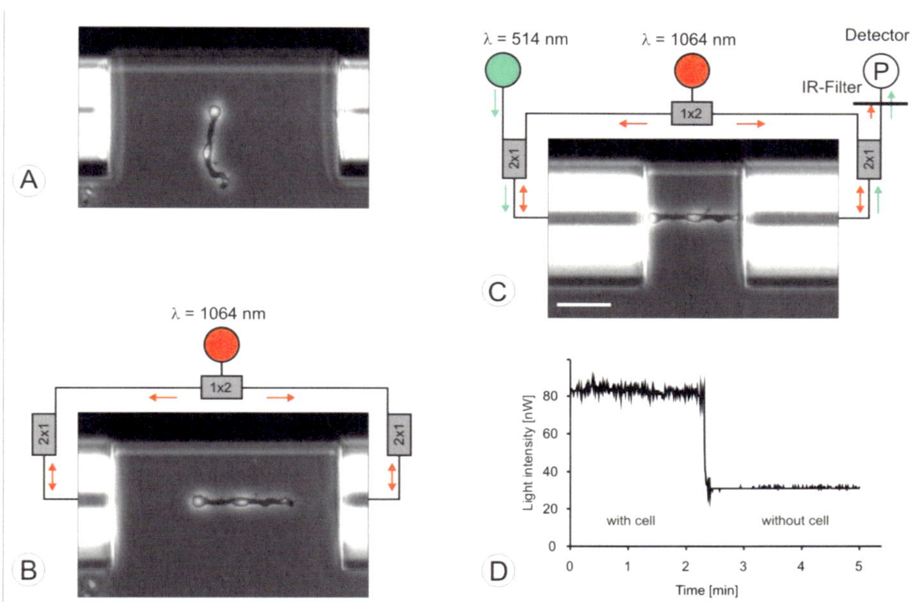


Fig. 6. Individual Müller cells as optical fibers. Demonstration of light guiding properties of individual Müller cells measured in a modified dual-beam laser trap. **A,** A cell is floating freely between the ends of two optical fibers, which are aligned against a backstop visible at top. **B,** The Müller cell is trapped, aligned, and stretched out by two counter-propagating infrared (IR)-laser beams ($\lambda$ = 1064 nm; not visible in the image) diverging from optical fibers. **C,** The fibers are brought in contact with the cell. Blue light ($\lambda$ = 514 nm) emerges from the left optical (input) fiber, and is collected and guided by the cell to the right (output) fiber. The fraction of blue light re-entering the core of the output fiber is measured by a powermeter, while the IR light is blocked by an appropriate cut-off filter. Scale bar, 50 µm. **D,** Typical time course of the power of the blue light measured. When the cell is removed from the trap, only a fraction of the blue light, which is no longer confined to the cell but diverges freely, is measured. The ratio $P_{\mathrm{cell}} / P_{\mathrm{no\ cell}}$ defines the relative guiding efficiency.

Franze et al. developed a setup to investigate light transport through individual cells that are located proximally to the photoreceptors (i.e., closer to the inner retinal surface) (Franze et al., 2007) (Fig. 6). Enzymatically dissociated retinal cells were optically trapped between the ends of two opposing optical fibers (single-mode at 1064 nm) by counter-propagating infrared laser beams (Ytterbium fiber laser, λ = 1064 nm). The refractive index of the cell medium has been elevated to that of retinal neurons (n ~ 1.36) by adding BSA (cf. chapter 3.1). Additionally, visible laser light (Argon Ion laser, λ = 514 nm) was coupled into one of the fibers, and the intensity of light coupling into the opposite fiber was measured. When the optical fibers are moved apart, only a fraction of the visible light emanating from one fiber couples into the opposing one, because of its divergence (Bass, 1995). However, when a structure capable of guiding light and thus preventing it from diverging is trapped, the intensity loss is minimized (Fig. 6).

### 3.3 Measuring the transmission and reflection of light on retinal tissue

Obtaining reliable quantitative data on light transmission and reflection of intact retinal tissue is a difficult task. Experiments have been carried out on bovine retina samples in a special setup (Fig. 7) with subsequent mathematical evaluation of the data (Hammer et al., 1995). Basically, the setup consisted of two integrating spheres (coated with barium oxide) between which the retinal sample was placed in the path of monochromatic light. Four detectors measured the light intensities at the start and end of the light path, and the scattered light collected within the integrating spheres (Fig. 7).

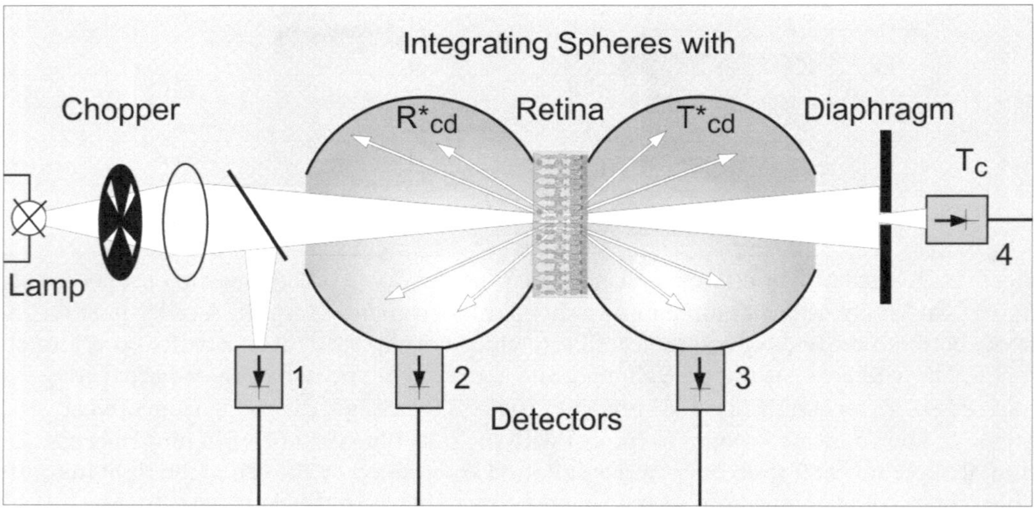


Fig. 7. Setup used to measure light transmission through and reflection at retinal tissue (modified from Hammer et al., 1995).

The calculation of optical tissue parameters involved three steps including an inverse Monte Carlo simulation. This approach revealed that, at $\lambda$=500nm (i.e., in the visible range of light), the retina transmits less than 90% of incident light, due to about 7% absorption and about 5% reflection (Hammer et al., 1995).
The retinal sample is placed between two integrating spheres which allow to collect light lost from straight transmission by diffuse reflectance ($R^*_{cd}$) and diffuse transmittance ($T^*_{cd}$). $R^*_{cd}$ and $T^*_{cd}$ were measured by two photodiodes (detectors 2 and 3, respectively). Furthermore, the incident light and the collimated transmittance (Tc) were measured by detectors 1 and 4, respectively. Light was provided by a xenon-high pressure lamp with monochromator.

### 3.4 Visualization of backscattering of light within retinal tissue by confocal microscopy

The retina contains numerous cellular structures (phase objects) with varying RI and sizes on the order of the wavelength of visible light. These structures should lead to significant scattering. Back-scattering can be visualized by detecting the reflection of (monochromatic) light that is sent through the tissue. Confocal microscopy offers a convenient way for such a measurement (Franze et al., 2007). An additional advantage of this approach is the possible three-dimensional reconstruction of optical slices, revealing the distribution of intensities of back-scattered light along the light path (cf. Fig. 11). Another technique that exploits the back-scattered light within a retina is optical coherence tomography (cf. Fig. 4).

### 3.5 Visualization of light propagation through the retinal tissue

To simulate the physiological light path in the inner retina, Franze et al. illuminated the retinal surface by light exiting a multimode optical fiber, which was inserted into a freshly dissected eye cup (Fig. 8A) (Franze et al., 2007). At the opposing side, the sclera, the choroid, the pigment epithelium and the photoreceptor cells were surgically removed, which allowed access to the end of the retinal light path with a confocal microscope. This approach revealed that the light transport through the inner retina is inhomogeneous (Fig. 8A).
To study the light transmission through the whole retina, a single mode optical fiber was mounted vertically above the inner surface of a retina (Fig. 8B). A micromanipulator was used to move the fiber in defined steps along the retinal surface, and the position of illuminated photoreceptor groups at the opposing side of the retina was recorded and compared to the fiber position (Fig. 8B) (Agte et al., 2011).
For a direct visualization of the light path within the retina, confocal microscopy of retinal cross-sections was combined with an optical fiber setup (Fig. 8C). Retinal slices were adhered to a nitrocellulose membrane, and Müller glial cells stained with a fluorescent dye. Green laser light exiting a single mode optical fiber was used to illuminate individual Müller cell endfeet. The fluorescence signal of Müller cells was recorded by one channel, the forward-scattered light was detected by another channel of the confocal microscope. Transmitted light was scattered in the filter membrane and could thus also be detected. An overlay of the channels allowed the detection of changes in light scattering as a function of changes in the fiber position relative to Müller cell positions (Fig. 8C). These experiments suggested that light hitting a Müller cell is guided more efficiently through the retinal tissue.

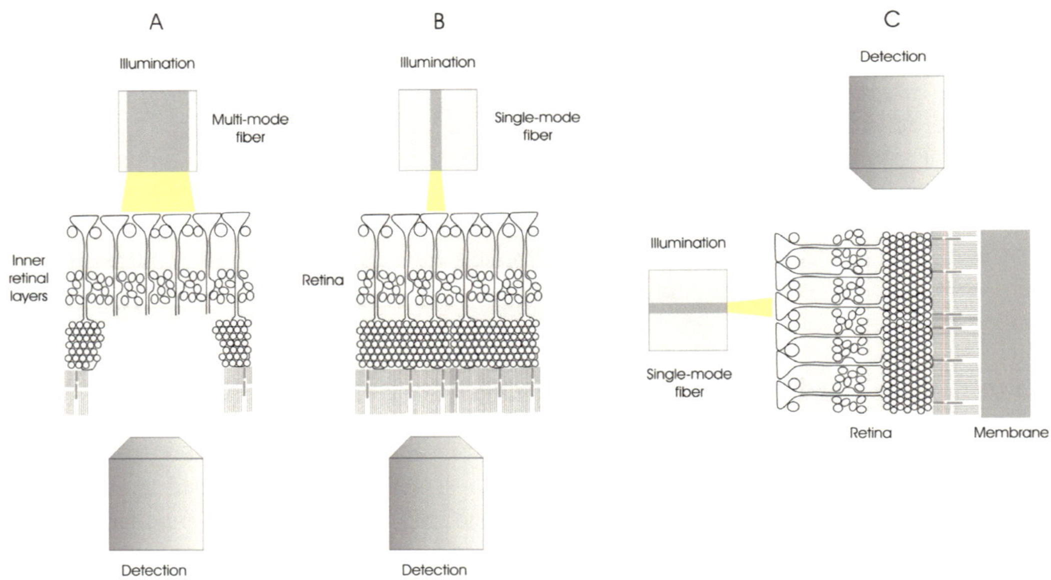


Fig. 8. Schemata of the illumination methods used to study the light transport through the retina. (A) The retina is placed on the stage of an inverted confocal microscope after surgically removing the photoreceptor cells. A multimode optical fiber provides a wide-field illumination of the retinal surface. The confocal microscope scans a plane close to the outer plexiform layer and monitors the light transmission through the inner retina. (B) To investigate the light distribution through the whole tissue, a single mode optical fiber is used to simulate the physiological illumination of individual Müller cells. An objective of an inverted microscope detects the light after passing all retinal layers. (C) A retinal slice is fixed on a nitrocellulose membrane and placed on the stage of an upright confocal microscope. A fluorescent staining visualizes Müller cells. The single mode fiber is positioned in front of individual Müller cells. Light scattering within the retinal cross-section is detected by the microscope. Light transmitted through the retina is visualized by monitoring the scattering in the membrane to which the retina is attached.

### 3.6 Physical and mathematical simulation

Assuming that total light reflection occurs in (outer and inner segments of) photoreceptor cells, light transport through the cells can be simulated by employing model 'cells' magnified into the microwave spectrum (Jean and O'Brian, 1949). Although only indirect evidence can be obtained, this magnification and the use of custom-made models allow a wide variation of parameters, and provide basic insights into the mechanisms involved.

The light path through cells and tissue compartments can also be simulated based on experimental data and appropriate physical models. Assuming that (certain) retinal cells are wave guides, the $V$ parameter can be used to estimate the efficiency of light guidance. This method has been applied to (outer and inner segments of) photoreceptor cells (Enoch and Tobey, 1978) as well as to Müller glial cells (Franze et al., 2007). This parameter of a circular dielectric waveguide is given by the expression

$$V = (\pi d/\lambda)\ (n_1^2 - n_0^2)^{1/2},$$

where $d$ is the diameter of the guide, $\lambda$ is the wavelength in vacuum, and $n_1$ and $n_0$ are the indices of refraction of the inside (cell) and outside (surrounding) of the fiber, respectively. Generally, a cell can be considered as an efficient waveguide if $V > 2$ (cf. Fig. 10).

Using the same basic optics (and assuming that light propagates in the cells by total reflection), the light collecting (or light radiating) properties of non-cylindrical (paraboloid of revolution, cone-like) cell structures can be estimated (Winston and Enoch, 1971; Miller and Snyder, 1973) (cf. Fig. 9). It should be noted however, that this older theoretical view of light propagation with ray optics in mind is not really appropriate to describe the light transport through structures that are comparable in size to the wavelength of light propagating through them.

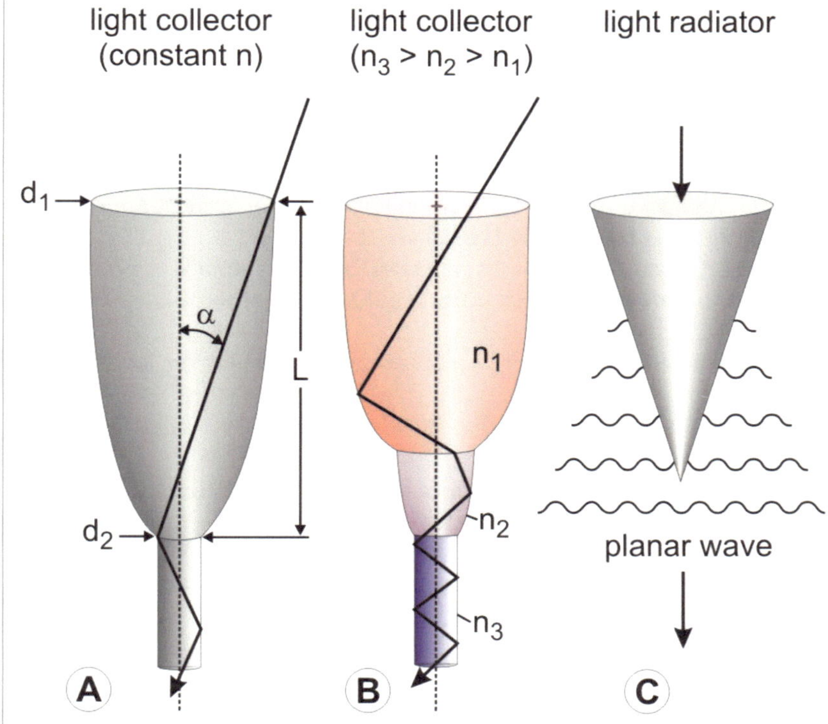


Fig. 9. Models for light collection (**A**, **B**) and light radiation (**C**) by (cone) photoreceptor cells. Incident light may be collected by a parabolic structure if its refractive index (n) exceeds that of its microenvironment (light collector: (**A**)). The accepted input semi-angle **α** can be estimated from morphological data (length **L** and larger - $\mathbf{d_1}$ - and smaller - $\mathbf{d_2}$ - diameter) and from refractive index differences ($\mathbf{n}_{\text{object}} - \mathbf{n}_{\text{environment}}$). The light collector can be 'improved' (i.e., **α** can be increased) if **n** increases along the light path (**B**). Both types of light collection have been ascribed to cone inner segments (ellypsoids) (Miller, 1981). By contrast, tapering cylinders may function as light radiators from which light spreads as a planar wave (**C**). It has been suggested that this applies to peripheral human cone outer segments which thereby may help to distribute light to the rod outer segments (Miller and Snyder, 1973). Modified from Winston and Enoch, 1971 (light collectors) and Miller and Snyder, 1973 (light radiator).

Numerical finite-difference time-domain and an analytic Mie theory approaches have been applied to model light transport through photoreceptor cell nuclei of diurnal and nocturnal animals in two and three dimensions (Solovei et al., 2009, Kreysing et al. 2010). These simulations revealed that, while the specific chromatin arrangements have little impact on the far-field scattering cross-section, scattering in the near field (which is the relevant regime inside the retina) shows a significant difference between retinae of diurnal and nocturnal animals. The "inverted" photoreceptor cell nuclei of nocturnal mammals have been suggested to act as collection lenses, which could improve night vision (cf. chapter 4.3).
Finally, Labin and Ribak recently used a direct three-dimensional numerical solution of the Helmholtz equation (the fast Fourier transform split-step beam propagation method of the global third order) to obtain a description of a propagating electromagnetic (EM) field along retinal cells and their vicinity (ref.). They confirmed that Müller glial cells preserve the acuity of images in the retina (Franze et al., 2007).

## 4. Experimental evidence for the presence of cellular optical fibers in the retina

The application of the above-mentioned experimental and theoretical approaches to retinal tissue samples and isolated retinal cells has shown that

- light propagation through – and light reflection by – the retina is non-homogenous;
- pronounced back-scattering of light occurs in the two plexiform layers, and in the nerve fiber layer;
- all layers are spanned by regularly arranged 'tubes' displaying low backscattering of light;
- these tubes were shown to be Müller cells, by vital dye filling and immunohistochemistry;
- isolated vital Müller cells display higher refractive indices than retinal neurons;
- these Müller cells provide efficient laser light guidance if interposed between glass fibers;
- in intact retinal tissue, Müller cells effectively guide light from vitread retinal surface to the photoreceptor cells;
- photoreceptor cell nuclei of nocturnal animals likely serve as densely packed arrays of microlenses;
- if hit by light, the photoreceptor cells themselves provide efficient light guidance towards their outer segments where the photosensitive molecules are located.

The following sections will highlight particular aspects of this knowledge, as it was accumulated over the years.

### 4.1 The 'classics': photoreceptor outer (and inner) segments are wave guides

It has been shown already in the mid 20th century that vertebrate photoreceptor cells display the properties of wave guides. In a comparative study, Sidman (1957) showed that various vertebrate rod outer segments display solid concentrations of 40-43%, and refractive indices of 1.4056 to 1.4106 whereas these values in cone outer segments were slightly lower (29.7 to 34.9% and 1.365 to 1.3958, respectively). These data suggested that photoreceptor cells can function as 'classical' wave guides, and light transmission through these cells occurs in modal patterns (Enoch, 1961; Tobey et al., 1975) (cf. Fig. 5).

Similar direct evidence for the wave guide properties of photoreceptor inner segments is missing, but the results of physical and mathematical simulations (cf. section 3.6) strongly suggest that light guidance also occurs in these inner segments. Moreover, based on mathematical simulations it was suggested that cone inner segments are light collectors, with an entrance aperture $\theta_{max}$ of 13° (Winston and Enoch, 1971) (cf. Fig. 9A). Taking into account that cone inner segments consist of several consecutive compartments (myoid, paraboloid) with different distinct refractive indices, a more complex simulation indicated that these inner segments of cones are ideal light-collectors with an even increased acceptance angle (Winston and Enoch, 1971) (cf. Fig. 9B).

In summary, it is now generally accepted that photoreceptor outer (and inner) segments are wave guides, and that (parts of cone) inner segments are light collectors funneling the light into the outer segments. Until very recently, however, it was unclear how sharp images can arrive at the 'apertures' of the inner segments.

### 4.2 The 'news': Müller glial cells are light guides that bridge the light-scattering elements in the inner retina

Whereas the healthy vertebrate retina appears very transparent if viewed from top (Fig. 4), it becomes turbid and acquires a 'milky' appearance when the Müller cells swell during 'spreading depression' (a condition involving extreme changes in intra- and extracellular ion concentrations) (Gouras, 1958; Mori et al., 1976; van Harreveld 1984; De Oliveira Castro et al., 1984) and when Müller cells lose their typical alignment, shape, und (ultra-) structure under pathological conditions such as posterior uveitis (Eberhardt et al., 2011). These observations suggest that Müller cells may be involved in intraretinal light guidance. More direct arguments stem from different experimental approaches as mentioned in the following.

If light hits the inner (vitread) surface of the retina, it enters the Müller cell endfeet which constitute the innermost retinal layer (Fig. 1). The inner and outer stem processes of these columnar cells then extend throughout the thickness of the retinal tissue (Fig. 1B) and could thus provide a substrate for wave guidance towards the photoreceptor inner segments. Assuming total internal reflection and using measured data of cell diameters and refractive indices, V parameter values of >2 (up to 4) have been calculated (Franze et al., 2007) (Fig. 10).

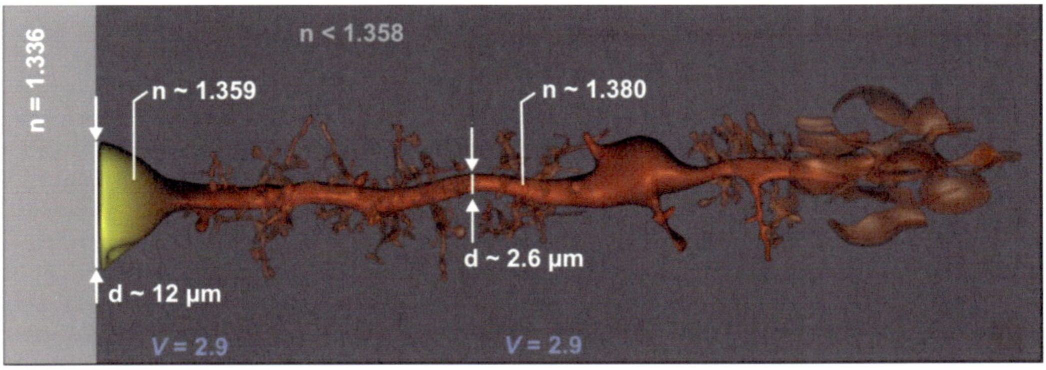


Fig. 10. Estimation of the *V* parameter along the different sectors of a typical guinea-pig Müller cell, modified from Franze et al., 2007.

While the diameter **d** of the Müller cell decreases from endfoot (left sinde; d ~ 12μm) to the stem processes (d ~ 2.6 μm), the local refractive index **n** increases from endfoot (mean n ~ 1.359; yellow) to the process (n ~ 1.380; red); thus, the ***V*** parameter remains fairly constant (2.9 for long [red] wavelengths, up to 4 for shorter [blue] wavelengths). The Müller cell endfeet constitute the boundary between the vitreous body (bright grey, n = 1.336) and the retinal neuropile (dark grey; mean n < 1.358).

Using the same assumptions and data sets, a simulation according to that shown in Figure 9B shows that Müller cell endfeet can be considered as light collectors, with a half-angle of acceptance close to 13° (similar to the cone inner segments). Moreover, the comparatively low refractive index of the innermost part of the endfeet (Fig. 10) provides index matching between the vitreous fluid and the retina, and reduces loss of light be reflection at the inner retinal surface. More detailed simulations of light propagation through human Müller cells suggest that (i) a significant increase of the intensity at the photoreceptors is obtained, (ii) for pupils up to 6 mm width, the coupling between neighboring cells is only a few percent, and (iii) low cross talk over the whole visible spectrum also explains the insensitivity to chromatic aberrations of the eye (Labin & Ribak, 2010).

In addition to these simulations, several different lines of experimental data support the view that Müller cells are light-guiding elements in the vertebrate retina. First, isolated individual cells were shown to effectively conduct light between two opposing glass fibers (cf. section 3.2 and Fig. 6). Second, the 'tubes' of low reflection spanning the vital retina, which were identified by confocal reflection microscopy, (cf. section 3.4), were shown to be the Müller cells (Fig. 11).

To understand the natural path of light through the vital retina, multiple experiments were performed in which light was applied onto the vitread surface. At this surface the Müller cell endfeet form a smooth plane covering the entire retina before they are funnelled into thick cell processes (e.g. guinea pig 2-3 μm). The tubular shape of the processes is retained until they are connected with the photoreceptor layers. Here, the Müller cell processes branch into thin cytoplasmatic arms, whose diameter is smaller than the wavelength of the visible light (Fig. 10).

In a first experiment, the inner part of the retina - or the pre-photoreceptor light path - was studied by using a multi mode optical fiber to illuminate a wide area of the retinal surface. The image, recorded at a plane close to the outer plexiform layer displays a pattern of bright spots on a dark background (Fig. 12B). This pattern represents the negative image of the image detected in the backscattering experiments. Thus, the 'tubes', which were identified as Müller cells (3.4.), transmit more of the incident light and scatter less then their surrounding structures.

To investigate the interaction between Müller cells and photoreceptor cells (i.e. the distribution of light from one Müller cell to the attached light-sensitive photoreceptor cells) the whole retina was analyzed regarding its light transmission. These experiments require a small light source to allow an illumination of individual Müller cell endfeet. For this reason, the multi mode fiber previously used was replaced by a single mode fiber. The beam waist was small enough to shine a light spot on single Müller cells. The illuminated photoreceptor cells were observed at the opposite surface (Fig. 12 C, D). These experiments revealed that the light, which enters one Müller cell, is clearly assigned to a small group of adjacent photoreceptor cells (Fig. 12C, red). When the fiber was moved along the retinal surface, the receptor groups moved in the same direction (Fig. 13). However, several fiber steps were required to achieve one 'jump' of the illuminated spot at the receptor layer. Sometimes sideward-deviations' occurred as well as even 'backward jumps' which can be explained by the orientation of Müller cells in the tissue. In some cases, more than one group of photoreceptors were illuminated simultaneously, probably reflecting the positioning of the beam over two or three adjacent Müller cell endfeet (Fig. 13, green and 12D).

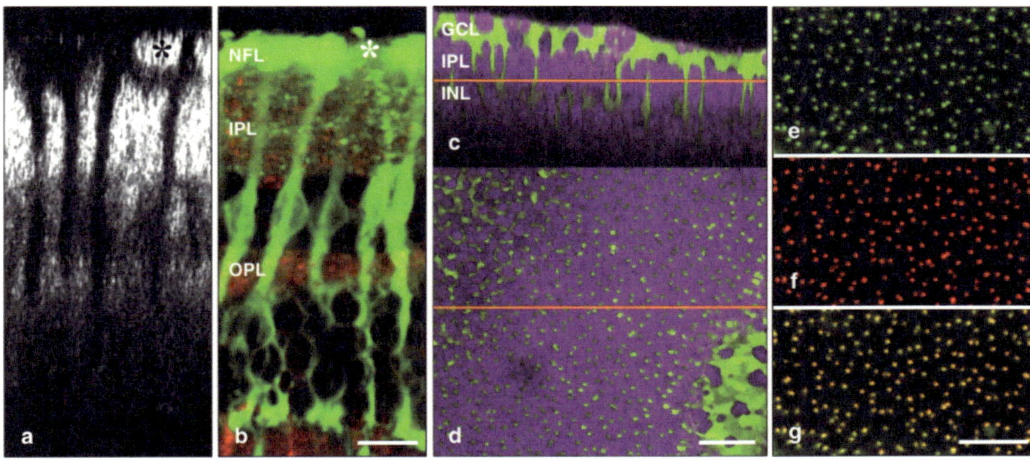


Fig. 11. Structures of low light reflection in the inner retina are Müller cells. (*a*) Z-reconstruction of reflection images of a living retina. Main scattering elements (bright) are the axon bundles and both plexiform layers. Low-reflecting tubular structures are spanning the entire retina. (*b*) Living retinal slice preparation, visualizing Müller cells (green) and synaptic elements in both plexiform layers (IPL, OPL) (red) (20). The levels of the IPL, OPL, and NFL (nerve fiber layer) are the same as in (*a*). The asterisks indicate axon bundles in the NFL. Scale bar, 10 μm, also valid for (*a*). (*c, d*) Overlay of light detected in reflection mode (purple) and the fluorescence signal of Müller cells. (*c*) Z-reconstruction of a confocal image stack. (*d*) Oblique optical section at the level of the red horizontal line in (*c*). The dye-filled irregularly shaped Müller cell somata of the inner nuclear layer (INL) are visible in the left upper corner. The central area shows Müller cell cross-sections in the inner plexiform layer (IPL). In the lower right part, Müller cell endfeet are visible, which enclose the ganglion cell somata in the ganglion cell layer (GCL). The lack of merging of the two colors, which would result in white areas, demonstrates that the dye filled exclusively those structures that showed low light reflection. (*e-g*) Confocal image at the IPL of a retinal wholemount fixed in 4% paraformaldehyde, after exposure to the green vital dye and immunocytochemical labeling of vimentin (red), which in the retina is specific to Müller cells (17, 22). (*e*) Fluorescence of the vital dye. (*f*) Vimentin immunofluorescence. (*g*) Overlay of (*e*) and (*f*). Colocalization of the red and green dyes results in yellow labeling. The observed complete colocalization means that the vital dye-filled and the immunoreactive cells are identical, and thus identifies the low-reflecting tubular structures as Müller cells. Scale bars in (*c-g*), 25 μm. Adapted from Franze et al. 2007, Copyright (2007) National Academy of Sciences, USA.

To further investigate this jumping behaviour of light, retinal cross-sections were observed while the retinal surface was again exposed to laser light emanating from a single mode optical fiber. Müller cells were fluorescently labelled, and retinae were attached to a membrane. This membrane served as a 'screen' to visualize the transmitted light. In these

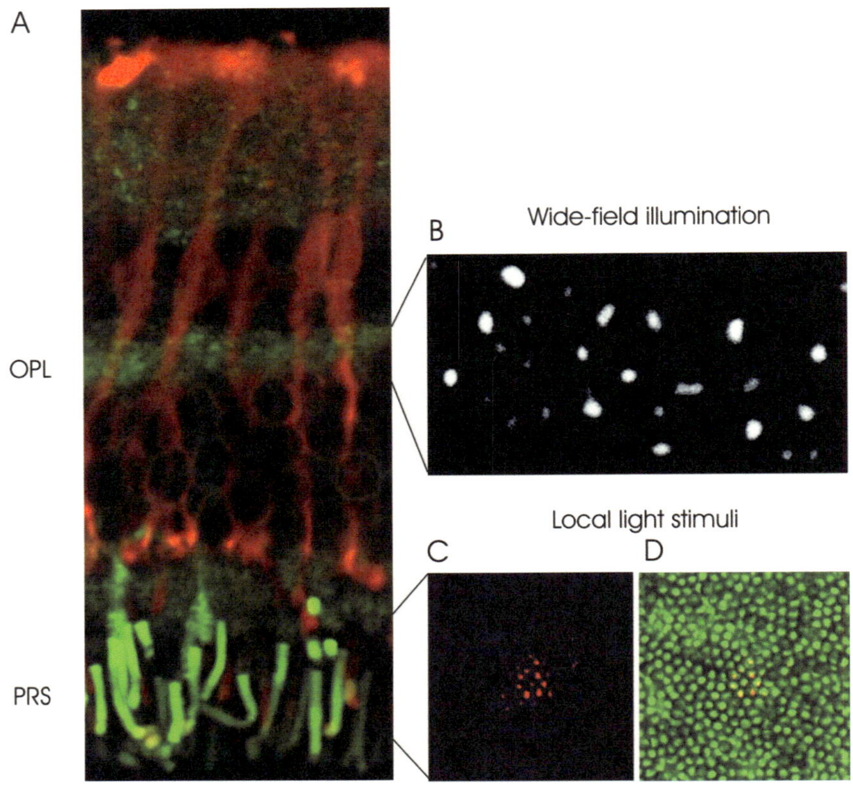


Fig. 12. Detection of light guiding structures in the retina. (A) Fluorescent image of the vital retina. Fluorescently labeled Müller cells (red) span the entire tissue. The endfeet are funnelled into the Müller cell processes, which are split in thin arms surrounding the photoreceptor nuclei. The light-sensitive segments of the photoreceptor cells are visualized by another fluorescent dye (green). (B) Wide-field illumination, provided by a multimode fiber, is used to illuminate a relatively large area of the retinal surface. At the outer plexiform layer, a pattern of brighter spots is visible. (C, D) Arrangement of the photoreceptor outer segments (green) illuminated by a multimode fiber (D). For selective illumination of individual Müller cells, the multi mode fiber is replaced by a single mode fiber. Light entering a small group of photoreceptor segments (red) is detected (C). (OPL: outer plexiform layer, PRS: photoreceptor segment layer).

experiments, (i) the position of the glass fiber, (ii) the Müller cells in the living retina, (iii) the light scattering inside the tissue, and (iv) the transmitted light were simultaneously observed. When the laser beam was not directly illuminating a Müller cell, intraretinal scattering occured mainly in the inner plexiform layer (IPL). The transmitted light at the membrane appeared as a wide spot (Fig. 14, bottom). If , however, the beam entered a Müller cell endfoot, most of the previously detected scattering inside the retina disappeared and a much narrower, bright light spot was transmitted to the photoreceptors (Fig. 14, top). Taken together, all experiments indicated that each Müller cell is optically coupled to a distinct small area in the photoreceptor layer. Further analysis has shown that the distance between the intensity maxima of these 'illuminated receptor fields' corresponds to the average distance between the axes of two neighbouring Müller cells. This suggests that every Müller cell serves as 'optical fiber' for 'its own' group of photoreceptor cells (Agte et al., 2011). A quantitative evaluation of the photoreceptors in 'typical' mammalian retinae such as human (outside the fovea centralis), guinea pig, and others, showed that a group of 1 cone and about 10 rods belong to 1 Müller cell (Reichenbach and Robinson, 1995), suggesting that Müller cells play a central role in vision at both daylight and dawn. During night, the small amount of light is efficiently guided to the low-light sensitive rod photoreceptors. At daylight, the resolution is limited by the cone spacing. Every cone photoreceptor is equipped with its own wave guide, which thus provides high-contrast vision at daylight (Agte et al., 2011).

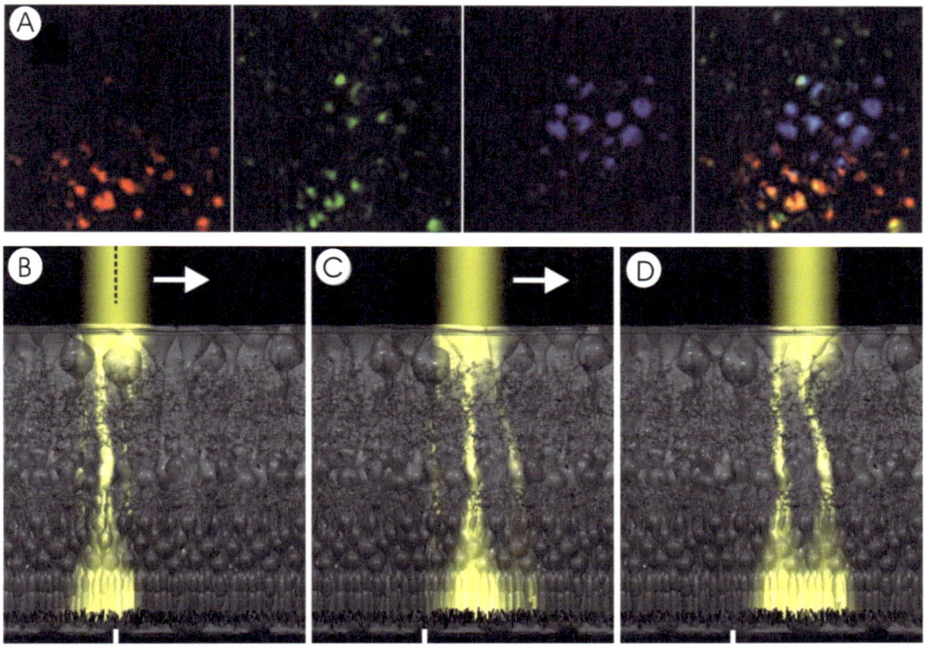


Fig. 13. Each Müller cell provides light for its own 'illuminated receptor field'. (A) When light is sent through an individual Müller cell, a distinct group of bright dots appears at the level of the outer segments (red). Movement of the light source relative to the inner retinal surface results in a 'jumping' of discrete illuminated fields at the backside of the tissue. (B, C, D) Schemata of the light behaviour inside the retina explaining the effect observed in (A). (B) The light beam illuminates a single Müller cell endfoot, which results in a single 'illuminated receptor field'. (C) A fiber step moves the beam in front of another Müller cell and another group of receptors is addressed. In this position the coupling of light into adjacent Müller cells is less efficient. (D) The beam above the edge of two neighbouring Müller cells results in a simultaneous detection of two receptor fields.

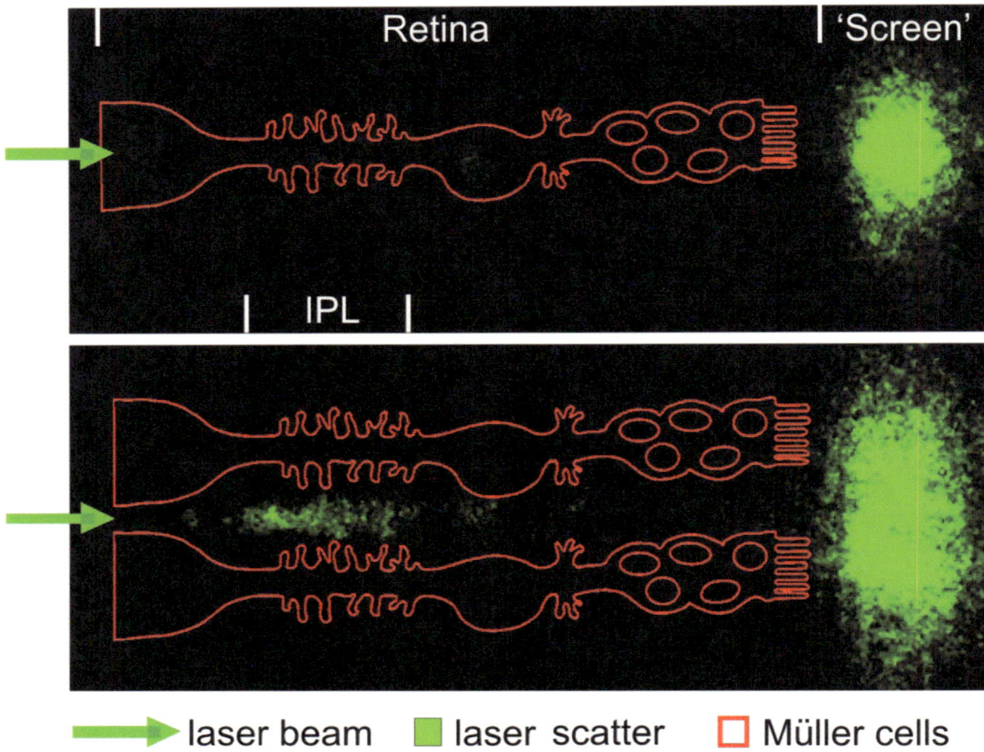


Fig. 14. Visualization of the light path within vital retinal slices. The retina is attached to a nitrocellulose membrane. A micromanipulator horizontally positions a single mode fiber in front of the vitread surface. This assembly is placed under the objective of an confocal microscope to detect the light scattering inside the retinal tissue as well as the transmitted light at the membrane. The membrane serves as a 'screen' for the light after passing the retina. In a position where the fiber is placed between two Müller cell processes (bottom) light scattering occurs particularly in the inner plexiform layer (IPL). The transmitted light spot at the membrane covers a wide area. For a better impression of the situation, Müller cell positions are schematically indicated (red line). If the laser beam hits a Müller cell endfoot (top) the scattering of light in the retina decreases and the light spot at the membrane shows rather sharp boundaries as well as a smaller spot size, suggesting that Müller cells relay light through the inner retina.

### 4.3 The 'special case': chains of cell nuclei act as 'microlense arrays' in the outer retina of nocturnal mammals

Nocturnal mammals such as mice and rats possess retinae with very high rod photoreceptor cell densities which increase light sensitivity by an optimal absorption probability of photons at low intensity levels. However, it requires the piling of rod somata, which mainly contain the nuclei, in many rows (10 or more). An extraordinary high degree of heterochromatin condensation reduces the size of nuclei and somata. In contrast, retinae of mammals with diurnal or crepuscular life style (such as those of horses, guinea pigs, and rabbits) display lower rod densities; their ONL contains few layers of somata / nuclei (2-5) and their rod nuclei appear less dense. The thick ONL in retinae of nocturnal mammals appears to be a particular challenge for effective light propagation, because the Müller cell processes in the ONL are thin and irregularly shaped such that they probably cannot act as wave guides. Recently, it has been shown that the disadvantage of an elongated light path may be compensated by improved optical properties of the condensed nuclei, acting as perfect chains of lenses (Solovei et al., 2009) (Fig. 15). Also in this instance have the experimental findings be supported by thorough theoretical modelling, taking into account the wave nature of light (Kreysing et al., 2010). Noteworthy, Müller cells may play an indirect role in this mechanism because they help to align the stacks of photoreceptor somata (Fig. 1).

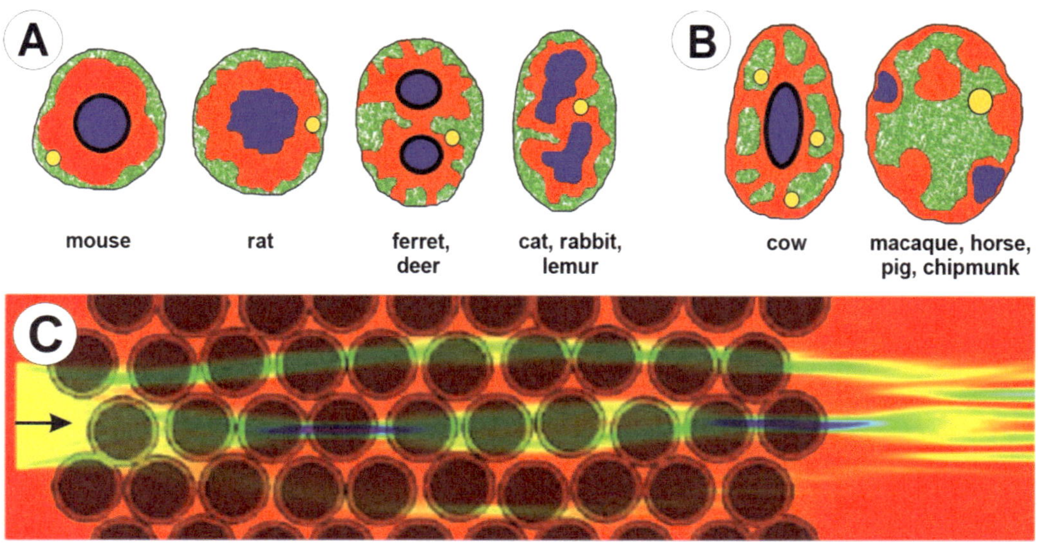


Fig. 15. Stacks of rod photoreceptor nuclei may act as light-guiding chains of lenses in nocturnal mammals. A, B: Distribution of euchromatin (green) and heterochromatin (red & blue) in rod photoreceptor cell nuclei of several mammals: A: nocturnal species, B: diurnal species; The nucleoli are labeled in yellow. C: Simulation of light propagation through the columnar stacks of rod cell nuclei in the outer nuclear layer, considered as chains of optical lenses. In this simulation, a light beam is entering from left side (arrow), i.e. from the outer plexiform layer; the inner and outer segments of the photoreceptors would be illuminated at the right side. Modified from Solovei et al. (2009).

### 4.4 The vertebrate retina consists of several 'stacks' of arrays of light-guiding elements

Taken together, the vertebrate retina apparently consists of several successive arrangements of vital light-guiding live 'optical fibers' or light-guiding elements that compensate the disadvantages of the inverted structure. To give an impression of the full-length light path through the retina, Figure 16 summarizes the available data and hypotheses. It has long been accepted that both the outer and the inner segments of the photoreceptors display the properties of wave guides (section 4.1). More recently, the Müller cells (section 4.2) and the nuclei of the photoreceptor cells (section 4.3) were suggested to contribute to intraretinal light propagation. This leaves us with not less than 4 subsequent components of a light-

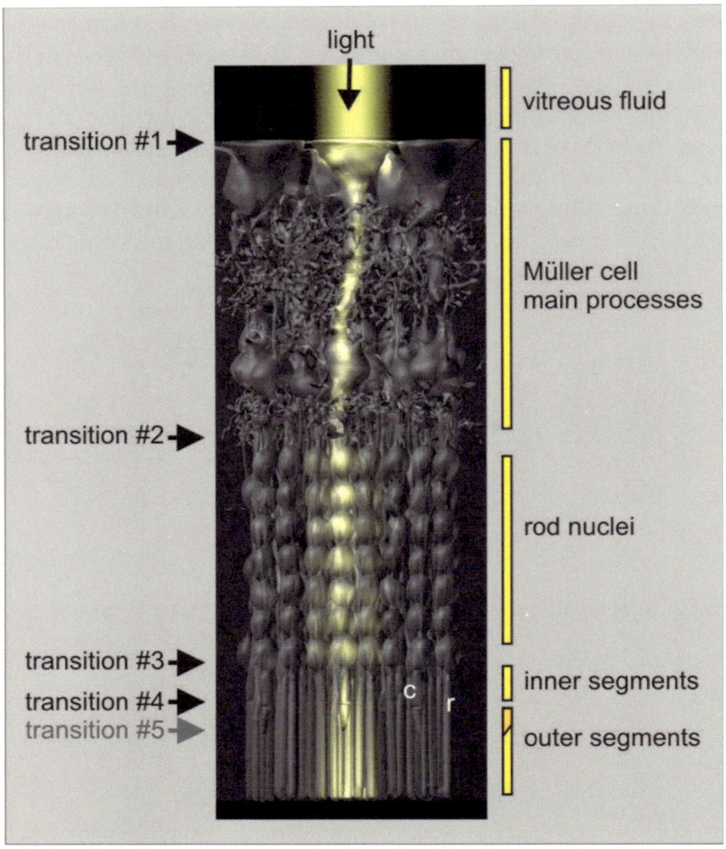


Fig. 16. Artist's view of the light path through the vertebrate retina. After being projected through the vitreous body, light is further transmitted by (i) the Müller cell stem processes, (ii) the rows of photoreceptor cell nuclei, (iii) the inner segments, and (iv) the outer segments of photoreceptor cells. This involves four transition sites, (1) from vitrous to Müller cell endfeet, (2) from Müller cell outer stem process to rod nuclei, (3) from photoreceptor cell nuclei to inner segments, and (4) from inner to outer segments. An additional transition from (peripheral) cone to rod outer segments (#5) has been proposed. c, cone; r, rod.

guiding pathway from the inner retinal surface to the end of the photoreceptor outer segments (Fig. 16). In addition to these 4 components (which differ in their light-guiding parameters and even mechanisms), the scenario involves the occurrence of (at least) 4 'transition sites'.

At first place, light arriving via the vitreous fluid must enter the retina. As the inner retinal surface (below a basal lamina) is formed by the endfeet of the Müller cells, these structures must constitute the site of this transition (see also section 4.2). The next transition is supposed to occur at the border between the OPL and the ONL, i.e., between the tapering Müller cell stem processes and the rod nuclei. Not much is known about the detailed (ultra-) structure of this transition zone, and about possible mechanisms of light transport at this site. Subsequently, at the third transition zone light must pass from the nuclear stacks to the inner segments of photoreceptor cells, close to the OLM. It has been argued that the inner

segments of photoreceptors (particularly, of the cones) are well-suited to function as light-collectors (Winston and Enoch, 1971) which may facilitate this transition. The fourth 'obligatory' transition occurs between the inner and outer segments of the photoreceptors. These structures are connected by a very thin cilium (well below the wavelength of visible light) which certainly cannot function as a light-guiding fiber with total internal reflection. This transition has never been discussed in detail. It should be noted, however ,that considerable backscattering of light seems to occur at the level of the cilia, in optical coherence tomography of the human retina (Anger et al., 2004).
Finally, an additional ('facultative') transition has been proposed by Miller and Snyder (1973). These authors suggested that light that was not absorbed in the short, tapering outer segments of the cones may spread from there towards the (longer) outer segments of the rods. This pathway was explained by the 'light radiator' properties of the peripheral cone outer segments (see Fig. 9C, *light radiator*).

## 5. Open questions

Although the above-mentioned experiments and models provided some indication for the presence of 'live optical elements' in the vertebrate retina, some open questions remain. For instance,

- there might be other mechanisms involved than classical wave guidance in optical fibers, such as for instance, microcrystal guidance;
- it remains to be shown to what degree does light transport by retinal cells improve vision / how efficient is light transport;
- light transport across the 'transition zones' between the successive arrangements of light-guiding live 'optical fibers' remains to be explained (cf. Fig. 16);
- the Müller cell population has been proposed to constitute a natural 'optical fiber plate' (Franze et al., 2007) but it remains to be clarified whether the ocular optics (cornea, lens, etc) focus the images onto the inner retinal surface or onto the photoreceptor layer;
- Müller cell light guidance as discussed above is supposedly in conflict with the small-eye effect (Glickstein and Millodot, 1970) and the Stiles-Crawford effect (Enoch & Tobey, 1981); this remains to be resolved (also requiring a careful reinterpretation of the original experimental evidence from the older studies in the light of the new findings;
- it remains to be shown whether all Müller cells in all vertebrates are light guides; in many (non-mammalian, or nocturnal mammalian) species the Müller cells, or parts of them, are very thin (< 0.3 µm in diameter) and thus appear not to be suitable as wave guides, at least for 'classical' total reflection.
- cells are active systems: can cells in the retina regulate light transport?;

These (and other) open questions must remain subject of future studies.

## 6. Conclusion

Often the retina is considered as a transparent tissue through which light passes without loss or scattering. However, this is not true; all cells and their processes and organelles are phase objects which differ in their refractive indices, which means that they must scatter the light. In particular, the synapses in the two plexiform layers have diameters close to 500nm, i.e., within the range of visible light (ca. 400-700nm). Indeed, light scattering by the retinal tissue layers is evidenced by the mere fact that optical coherence tomography delivers images of these layers in the living eye (cf. Fig. 4). Thus, the inverted structure of the vertebrate retina has prompted the question how it is possible that we see sharp, unblurred images at daylight, and can detect a few photons entering our eye (Pirenne, 1967). The presence of light-scattering elements in front of our photoreceptor cells should decrease both visual acuity and light sensitivity. An explanation can be provided by the existence of light-guiding elements that traverse all retinal layers, and directly transfer the light (and the image of the environment) towards the light-absorbing outer segments of the photoreceptor cells. Here we summarize evidence supporting the view that the vertebrate retina apparently consists of several successive arrangements of such 'optical elements' that compensate for the disadvantages of the inverted structure (Fig. 16). There are 4 subsequent components of a light-guiding pathway from inner retinal surface to the end of the photoreceptor outer segments, i.e., Müller cells, stacks of photoreceptor cell nuclei (only in nocturnal mammals?), and the inner and outer segments of the photoreceptor cells. These 4 components, which differ in their light-guiding parameters and the underlying mechanisms, must be 'bridged' by at least 4 'transition sites'. Although many open questions remain, it appears to be very likely that cellular light guidance is a major factor in vertebrate vision.

## 7. References

Agte, S., Junek, S., Matthias, S., Ulbricht, E., Erdmann, I., Wurm, A., Schild, D., Käs, J.A., & Reichenbach, A. (2011). Müller glial cell-provided cellular light guidance through the vital mammalian retina. *Biophys J* accepted

Ajo, A. (1949). On the refractive index of the retina. *Acta Physiol. Scand.,* Vol. 13, No. 1-2, (Feb. 1947), pp. 130-149, ISSN 1748-1708

Anger, E.M., Unterhuber, A., Hermann, B., Sattmann, H., Schubert, C., Morgan, J.E., Cowey, A., Ahnelt, P.K, & Drexler, W. (2004). Ultrahigh resolution optical coherence tomography of the monkey fovea. Identification of retinal sublayers by correlation with semithin histology sections. *Exp. Eye Res.,* Vol. 78, No. 6, (Jun. 2004), pp. 1117-1125, ISSN 0014-4835

Barer, R. (1954) Refractometry of living cells. *Q. J. Microsc. Sci.,* Vol. 95, No. 4, (Dec. 1954), pp. 399-423, ISSN 0370-2952

Bass, M. (1995) Handbook of Optics, Volume I: Fundamentals, Techniques, and Design (2nd ed.), McGraw-Hill, ISBN 9780070477407, USA

Boehm, G. (1940). Über ein neues entoptisches Phänomen im polarisierten Licht. „Periphere" Polarisationsbüschel. *Acta Ophthalmol.,* Vol. 18, No. 2, (Sep. 1940), pp. 143-169, ISSN 1755-375X

Charman, W.N. (1980). Reflection of plane-polarized light by the retina. *Bri.t J. physiol. Opt.,*Vol. 34, (1980) pp. 34-41., ISSN 0007-1218

Charriere, F., Marian, A., Montfort, F., Kuehn, J., Colomb, T., Cuche, E., Marquet, P., & Depeursinge, C. (2006). Cell refractive index tomography by digital holographic microscopy. *Opt. Lett.,* Vol. 31, No. 2, (Jan. 2006). pp. 178-180, ISSN 0146-9592

Chen, E. (1993). Refractive indices of the rat retinal layers. *Ophthalmic Res.,* Vol. 25, No. 1, (Jan./Feb. 1993), pp. 65-68, ISSN 0030-3747

Coletta, N.J., Williams, D.R., & Tiana, C.L.M. (1990). Consequences of spatial sampling for human motion perception. *Vision Res.,* Vol. 30, No. 11, (1990), pp. 1631-1648, ISSN 0042-6989

De Oliveira Castro, G., Martins-Ferreira, H., & Gardino, P.F. (1985). Dual nature of the peaks of light scattered during spreading depression in chick retina. *An Acad. bras Cienc.*, Vol. 57, No. 1, (Mar. 1985), pp. 95-103, ISSN 0001-3765

Eberhardt, C., Amann, B., Feuchtinger, A., Hauck, S.M., & Deeg, C.A. (2011). Differential expression of inwardly rectifying K+ channels and aquaporins 4 and 5 in autoimmune uveitis indicates misbalance in Müller glial cell-dependent ion and water homeostasis. *Glia*, Vol. 59, No. 5, (May 2011), pp. 697-707, ISSN 0894-1491

Enoch, J.M. (1961). Visualization of wave-guide modes in retinal receptors. *Am. J. Ophthalmol., Vol.* 51, (May 1961), pp. 1107-1118, ISSN 0002-9394

Enoch, J.M., & Tobey, F.L. (1978). Use of the waveguide parameter V to determine the difference in the index of refraction between the rat rod outer segment and the interstitial matrix. *J. Opt. Soc. Am.*, Vol. 68, No. 8, (Aug. 1978), pp. 1130-1134, ISSN 1084-7529

Enoch, J.M., & Tobey, F.L. (1981). Vertebrate Photoreceptor Optics (1st ed.), Springer Verlag, ISBN 9783540105152, Berlin-Heidelberg-New York

Enoch, J.M., & Glismann, L.E. (1966). Physical and optical changes in excised retinal tissue. Resolution of retinal receptors as a fiber optic bundle. *Invest. Ophthamol. Vis. Sci., Vol.* 5, No. 2 (Apr. 1966), pp. 208-221, ISSN 0146-0404

Franze, K., Grosche, J., Skatchkov, S.N., Schinkinger, S., Foja, C., Schild, D., Uckermann, O., Travis, K., Reichenbach, A., & Guck, J. (2007). Müller cells are living optical fibers in the vertebrate retina. *Proc. Natl. Acad. Sci. U S A*, Vol. 104, No. 20, (May 2007), pp. 8287-8292, ISSN 0027-8424

Glickstein, M., & Millodot, M. (1970). Retinoscopy and eye size. *Science*, Vol.168, No. 931, (May 1970), pp. 605-606. ISSN 0036-8075

Goldsmith, T.H. (1990). Optimization, constraint, and history in the evolution of eyes. *Quarterly Rev. Biol.*, Vol. 65, No. 3, (Sep. 1990), pp. 285-287, ISSN 0033-5770

Gorrand, J-M. (1986). Separation of the reflection by the inner limiting membrane. *Ophthalmic Physiol. Opt.*, Vol. 6, No. 2, (Apr. 1986) pp. 187-196, ISSN 0275-5408

Gouras, P. (1958). Spreading Depression of Activity in Amphibian Retina. *Am. .J Physiol.*, Vol. 195, No. 1, (Oct. 1958), pp. 28-32, ISSN 0002-9440

Hammer, M., Roggan, A., Schweitzer, D., Müller, G. (1995). Optical properties of ocular fundus tissues-an in vitro study using the double-integrating-sphere technique and inverse Monte Carlo simulation. *Phys. Med. Biol.*, Vol. 40, No. 6, (Jun. 1995), pp. 963-978, ISSN 0031-9155

Jean, J., & O'Brian, B. (1949). Microwave test of a theory of the Stiles-Crawford effect. *J. Opt. Soc. Amer., Vol.* 39, No. 9, (1949), p. 1957, ISSN 1084-7529

Kreysing, M., Kießling, T., Fritsch, A., Dietrich, C., Guck, J., & Käs, J. (2008). The optical cell rotator. *Opt. Express* , Vol. 16, No. 21, (Oct. 2008), pp. 16984-16992, ISSN 1094-4087

Kreysing, M., Boyde, L., Guck, J., & Chalut, K.J. (2010). Physical insight into light scattering by photoreceptor cell nuclei. *Opt. Lett.*, Vol. 35, No. 15, (Aug. 2010), pp. 2639-2641, ISSN 0146-9592

Knighton, R.W., Jacobson, S.G., & Kemp, C.M. (1989). The spectral reflectance of the nerve fiber layer of the macaque retina. *Invest. Ophthalmol. Vis. Sci.*, Vol. 30, No. 11, (Nov. 1989), pp. 2393-2402, ISSN 0146-0404

Labin, A.M., & Ribak, E.N. (2010). Retinal glial cells enhance human vision acuity. *Phys Rev Lett.*, Vol. 104, No. 15, (Apr. 2010), pp. 158102, ISSN 0031-9007

Livingstone, M., & Hubel, D. (1988). Segregation of form, color, movement, and depth: anatomy, physiology, and perception. *Science*, Vol. 240, No. 4853, (May 1988), pp. 740-749, ISSN 0036-8075

Martins-Ferreira, H., & De Castro, O.G. (1966) Light-scattering changes accompanying spreading depression in siolated retina. *J. Neurophysiol.*, Vol. 29, No. 4, (Jul. 1966), pp. 715-726, ISSN 0022-3077

Miller, W.H., & Snyder, A.W. (1973). Optical function of human peripheral cones. *Vision Res.*, Vol. 13, No. 12, (Dec. 1973), pp. 2185-2194, ISSN 0042-6989

Millodot, M. (1972). Reflection from the fundus of the eye and its relevance to retinoscopy. *Atti Fond. G. Ronchi., Vol.* 27, No. 1, (1972), pp. 31-50, ISSN 0391-2051

Mori, S., Miller, W.H., & Tomita, T. (1976). Müller cell function during spreading depression in frog retina. *Proc. Natl. Acad. Sci. U S A*, Vol. 73, No. 4, (Apr. 1976), pp. 1351-1354, ISSN 0027-8424

Nordenson, J.W. (1934). Über den Brechungsindex der Netzhaut. *Acta Ophthalmol. Copenhagen,* Vol. 12, (1934), pp. 171-175, ISSN 0001-639X

Pirenne, M.H. (1967). Vision and the Eye (2nd ed.), Chapman and Hall Ltd., ISBN 9780412042201, London, (and references therein)

Puliafito, C.A., Hee, M.R., Lin, C.P., Reichel, E., Schuman, J.S., Duker, J.S., Izatt, J.A., Swanson, E.A., & Fujimoto, J.G. (1995). Imaging of macular diseases with optical coherence tomography. *Ophthalmol.*, Vol. 102, No. 2, (Feb. 1995), pp. 217-229, ISSN 0161-6420

Reichenbach, A., & Robinson, S. (1995). Phylogenetic constraints on retinal organization and development. *Progr. Retin. Eye Res., Vol.* 15, No. 1, (1995), pp. 139-171, ISSN 1350-9462

Sidman, R.L. (1957). The structure and concentration of solids in photoreceptor cells studied by refractometry and interference microscopy. *J. Biophys. and Biochem. Cytology*, Vol. 3, No. 1, (Jan. 1957), pp. 15-30, ISSN 0095-9901

Solovei, I., Kreysing, M., Lanctôt, C., Kösem, S., Peichl, L., Cremer, T., Guck, J., & Joffe, B. (2009) Nuclear architecture of rod photoreceptor cells adapts to vision in mammalian evolution. *Cell, Vol.* 137, No. 2, (Apr. 2009), pp. 356-68. ISSN 0092-8674

Tobey, F.L., Enoch, J.M., Scandrett, J.H. (1975). Experimentally determined optical properties of goldfish cones and rods. *Invest. Ophthalmol, Vol.* 14, No. 1, (Jan. 1957), pp. 7-23, ISSN 0020-9988

Valentin, G. (1879). Ein Beitrag zur Kenntniss der Brechungsverhältnisse der Thiergewebe. *Arch. ges. Physiol.*, Vol. 19, No.1 (Dez. 1879), pp. 78-105, ISSN 0365-267X

van Harreveld, A. (1984) Visual concomitants of retinal spreading depression. *Ann. Acad. Brasil. Ciênc.*, Vol. 56, No. 4, (Dec. 1984), pp. 519-524, ISSN 0001-3765

Vos, J.J., & Bouman, M.A. (1964). Contribution of the retina to entoptic scatter. *J. Opt. Soc. Am.*, Vol. 54, No. 1, (Jan. 1964), pp. 95-100, ISSN 1084-7529

Williams, D., Sekiguchi, N., & Brainard, D. (1993). Color, contrast sensitivity, and the cone mosaic. *Proc. Natl. Acad. Sci. U S A.*, Vol. 90, No. 21, (Nov. 1993), pp. 9770-9777, ISSN 0027-8424

Winston, A., & Enoch, J.M. (1971). Retinal cone receptor as an ideal light collector. *J. Opt. Soc. Am.*, Vol. 61, No. 8, (Aug. 1971), pp. 1120-1122, ISSN 1084-7529

# 11

# New Window on Optical Brain Imaging; Medical Development, Simulations and Applications

Chemseddine Mansouri[1] and Nasser H. Kashou[2,3,4]
[1]*CRAN UMR 7039 CNRS-UHP-INPL, Nancy University*
*Vandoeuvre-les-Nancy,*
[2]*Department of Radiology, Children's Radiological Institute,*
*Nationwide Children's Hospital*
[3]*Department of Radiology, Department of Ophthalmology,*
*The Ohio State University Medical Center*
[4]*Department of Biomedical, Industrial and Human Factors Engineering,*
*Wright State University,*
[1]*France*
[2,3,4]*USA*

## 1. Introduction

As telecommunications advances optical fiber technology, the costs decreases and signal quality increases. This trend is advantageous for the medical industry specifically for neurological applications and research. In general, the field of optics is based on light behavior. The addition of biological tissue interaction with this light introduces and extra factor. This opens up a field of modeling and optimization of the optical fibers based on scattering, absorption and paths traversed by these photons. This modeling includes simulation and signal reconstruction in order to quantify chromophore concentrations in biological tissue, in this case the human head.
Functional near-infrared spectroscopy (fNIRS) allows the ability to monitor brain activation by measuring changes in the concentration of oxy- and deoxy-hemoglobin (Hb) by their different spectra in the near-infrared range (Arridge 1995; Arridge and Lionheart 1998; Arridge and Schweiger 1995; Benaron et al. 2000; Benaron and Stevenson 1993; Brazy et al. 1985; Chance et al. 1998; Cope 1991; Cope and Delpy 1988; Everdell et al. 2005; Ferrari et al. 2004; Franceschini 2000, 2003; Gandjbakhche et al. 1994; Gratton et al. 1995, 1997; Gratton and Fabiani 2003; Greisen 2006; Hintz et al. 1998, 1999, 2001; Hoshi 2003; Jasdzewski et al. 2003; Kato et al. 1993; Kusaka et al. 2004; Meek et al. 1995, 1998; Miki et al. 2005; Obrig and Villringer 2003; Obrig et al. 2000; Plichta et al. 2006; Schroeter et al. 2004, 2006a, 2006b; Strangman et al. 2002, 2003; Taga et al. 2003; Toronov et al. 2000; Villringer and Chance 1997; Wilcox et al. 2005; Wolf et al. 2003; Wyatt et al. 1988; Xu et al. 2005). Functional imaging with NIR light is made possible in a spectrum window that exists within tissues in the 600–950 nm NIR region, in which photon transport is dominated by scattering rather than absorption. For more than two decades, the

single channel measurement technique of NIR spectroscopy has been successfully used to measure the haemodynamic response to brain activity in both adults and neonates (Ferrari et al. 2004; Hoshi 2003; Obrig et al. 2000). In a typical design the photon can have three possible paths: scattering, absorption or transmission. Its resolution is intrinsically limited by the diffusive nature of NIR light in tissue. More recent studies using the multi channel NIRS technique, NIR optical topography (OT), have improved spatial and temporal resolution in both adults (Kennan et al. 2002) and infants (Taga et al. 2003).

NIRS measurement accuracy can vary depending on the design of the detector head. Constant optode distance is crucial; if head circumference changes even by a fraction of a millimeter, the trends are significantly biased (Greisen 2006). This dependency has been shown in both experiments and simulation. A comprehensive study was performed on factors affecting the accuracy of NIR concentration calculations and found that the wavelength selection and optode placement to be important factors in reducing error (Strangman et al. 2003). A limitation of NIR is its low spatial resolution (Gandjbakhche et al. 1994). On the other hand compared to other imaging methods, optical approaches have an excellent temporal resolution (Pouratian et al. 2003; Villringer and Chance 1997) that enables analysis in the frequency domain.

In this chapter we hope to give a technological review of near-infrared light and systems, discuss optode design considerations including background on the fiber design as it relates to this field and finally touch on current trends and applications. For the latter, we will focus on diffusion theory and simulation of photon propagation using a head model. We will follow this with concluding remarks.

## 2. Technology review

### 2.1 Light-tissue interaction

Interaction between light and the biological system consists of the absorption of light by a molecular species and the redirection of incidental light over a range of angles due to refractive index mismatch in tissue microstructure, i.e. scattering. The energy change measured by absorption is defined by

$$\Delta E = hv = h(c/\lambda) \tag{1}$$

Where c is the speed of light, $\lambda$ is the wavelength and $h$ is Planck's constant 6.626 x 10-34 J S. From this we can derive a relationship between the absorption of chromophores and the wavelength of light. **Fig. 1** shows the absorption spectrum of two chromophores that are particularly critical for NIRS, oxy- and dexoy- hemoglobin. As will be seen later, the choice of wavelengths for different systems is dependent upon this. We highlight 690 and 830 nm because of the difference in absorption for oxy- and deoxy- hemoglobin. From this figure we also see a point where the absorptivity is the same for both, about 805 nm. It is advantageous to include this additional wavelength as a quality and validation check for instruments.

This property of tissue will have a direct effect on the light pathway and behavior which leads us into another fundamental equation/law known as the Beer Lambert Law, **Equation 2**.

This states that the attenuation of light (I) as it travels through an absorptive tissue is directly proportional to the product of the concentration of the chromophore ([C]) and the optical path length (L), **Fig. 2**.

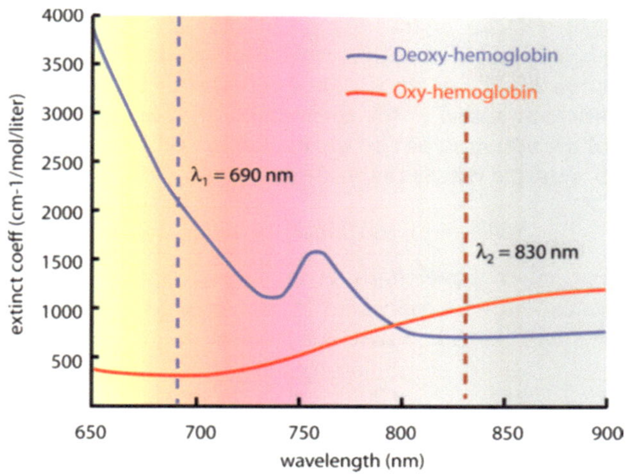


Fig. 1. Absorption spectrum of oxyhemoglobin (HbO) and deoxyhemoglobin (Hb) chromophores. Two or more wavelengths are needed to differentiate between these two chromophores, 690 and 830 nm are commonly used.

$$I = I_o\, e^{-\varepsilon[C]L} \tag{2}$$

Taking the log and inverting the output over input yields the attenuation

$$\mathrm{A} = \log\left(\frac{I_o}{I}\right) = \varepsilon[C]L \tag{3}$$

Again, note that absorption spectrum is dependent on the wavelength of light. Thus, a more appropriate formulation is

$$A_{\lambda_i} = \varepsilon_{\lambda_i}[C]L \tag{4}$$

Where $\lambda_i$ is the i$^{th}$ wavelength for the extinction coefficient ($\varepsilon$) based on **Fig. 1**. In the case of NIRS, two chromophores are unknown so we would need a minimum of two equations; meaning we need two wavelengths to find the solution, e.g. 690 and 830 nm. This formulation is common in pulse oximetry and follows the simple model shown in **Fig. 2(a),** where the light is transmitted at one end of the tissue and collected at the back end. For example, by placing a light source on our finger nail and a light detector directly opposite of that, on our finger. If we wanted to measure the light on the same side as our finger nail or concentration changes of a chromophore from more than one detector across a period of time then we will need to work with a slightly different equation as is discussed in the next subsection.

### 2.2 Near Infrared Light (NIR)

Near infrared light ranges from 600-950*nm* and is strongly absorbed by two chromophores, namely, oxyhemoglobin (HbO) and deoxyhemoglobin (Hb). As a result of this interaction, the modified Beer-Lambert Law (MBLL), seen in **Equation 5**, can be used to quantify changes in chromophore concentrations (Cope 1991; Cope and Delpy 1988).

$$\Delta\mathrm{OD} = -ln(I_{final}/I_{initial}) = \varepsilon\Delta\mathrm{CLB} \tag{5}$$

where $\Delta OD = OD_{final} - OD_{initial}$ is the change in optical density, $I_{final}$ and $I_{initial}$ are the measured intensities before and after the concentration of the chromophores change, $\Delta C$ and replace $I$ from **Equations 2 & 3**. $L$ is the distance between incident light and detected light, $\varepsilon$ is the extinction coefficient, and $B$ is the differential path-length factor (DPF), **Fig. 2**. The general one chromophore equation can be further expanded for brain functional monitoring applications in which multiple wavelengths are used.

$$\Delta OD^{\lambda} = (\varepsilon^{\lambda}_{HbO}\Delta[HbO] + \varepsilon^{\lambda}_{Hb}\Delta[Hb])B^{\lambda}L \qquad (6)$$

Where $\lambda$ indicates particular wavelength. After rearranging the mathematical terms the concentration changes can now be computed by measuring the change in optical density at two different wavelengths as seen in **Equations 7 and 8**.

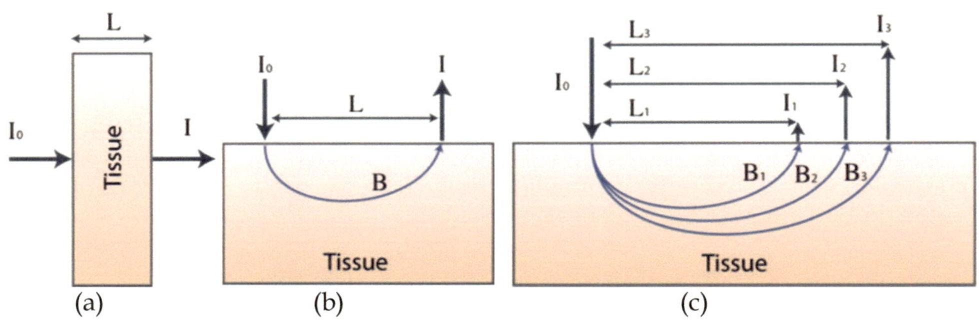


Fig. 2. (a) Light propagated through tissue ($I_o$) and transmitted at the output (I) traversing a distance L allows for chromophore concentration [C] calculations using Beer Lamberts Law. (b) Modified Beer Lamberts Law is used in this case as the photon traverses a path of distance (B). (c) Increasing the number of detectors and varying the distance allows for spatially resolved spectroscopy and several depths of penetration.This leads us to the modified Beer Lambert Law.

$$\Delta[Hb] = \frac{\left[\left(\varepsilon^{\lambda 2}{}_{HbO}\left(\Delta OD^{\lambda 1} / B^{\lambda 1}\right)\right) - \left(\varepsilon^{\lambda 1}{}_{HbO}\left(\Delta OD^{\lambda 2} / B^{\lambda 2}\right)\right)\right]}{\left[\left(\varepsilon^{\lambda 1}{}_{Hb}\varepsilon^{\lambda 2}{}_{HbO} - \varepsilon^{\lambda 2}{}_{Hb}\varepsilon^{\lambda 1}{}_{HbO}\right)L\right]} \qquad (7)$$

$$\Delta[HbO] = \frac{\left[\left(\varepsilon^{\lambda 1}{}_{HbO}\left(\Delta OD^{\lambda 2} / B^{\lambda 2}\right)\right) - \left(\varepsilon^{\lambda 2}{}_{HbO}\left(\Delta OD^{\lambda 1} / B^{\lambda 1}\right)\right)\right]}{\left[\left(\varepsilon^{\lambda 1}{}_{Hb}\varepsilon^{\lambda 2}{}_{HbO} - \varepsilon^{\lambda 2}{}_{Hb}\varepsilon^{\lambda 1}{}_{HbO}\right)L\right]} \qquad (8)$$

The basic process is now to shine two near infrared lights through the tissue, one which is more sensitive to oxyhemoglobin and the other to deoxyhemoglobin at two different time periods and take the difference of these concentrations. In this case all the other variables in **Equations 7 and 8** are defined by the design. This is possibly the simplest way to do these measurements. Next we will introduce the types of systems in use.

### 2.3 Experimental techniques

Time domain (TD), frequency domain (FD), and continuous-wave (CW) systems are currently being used to solve the above implementation of NIRS (Boas et al. 2002). For this section we will discuss these three techniques separately and highlight the advantages and disadvantages of each.

### 2.3.1 Time domain

TD systems (Benaron and Stevenson 1993; Chance et al. 1988; Cubeddu et al. 1996; Grosenick et al. 1997; Hebden et al. 1997) introduce light impulses into tissue on the order of picoseconds which after passing through different layers such as skin, skull, cerebrospinal fluid (CSF), and brain become broadened and attenuated (Boas et al. 2002). The output of this pulse after being transmitted through the highly scattered medium is known as the temporal point spread function (TPSF)(Hebden et al. 1997), **Fig. 3**.

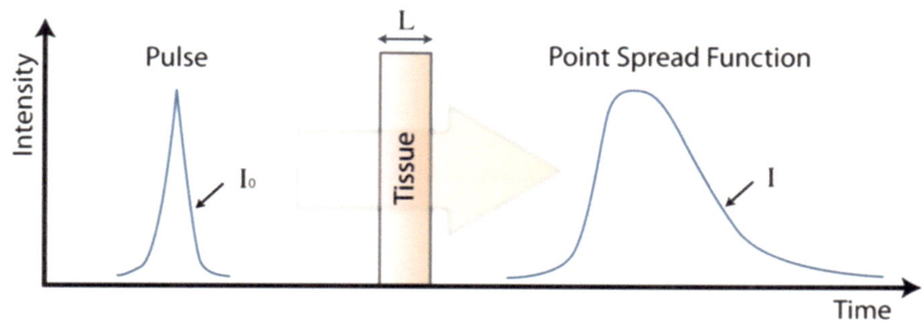


Fig. 3. In time domain systems an impulse (picoseconds) of light is sent through the tissue and the area of the point spread function is measured.

An advantage of TD systems over the other two is that theoretically it can obtain the highest spatial resolution and can accurately determine absorption and scattering (Boas et al. 2002). However, it has the disadvantage of lower temporal resolution as a direct result of trying to achieve an adequate signal to noise ratio, as well as the need to mechanically stabilize the instrument, the large dimensions and high cost of the necessary ultra fast lasers (Hebden et al. 1997).

### 2.3.2 Frequency domain

In FD systems, (Gratton et al. 1997; Jiang et al. 1995; Pogue et al. 1997; Pogue and Patterson 1994) the light source is continuously on but is amplitude modulated at frequencies on the order of tens to hundreds of megahertz, **Fig. 4**.

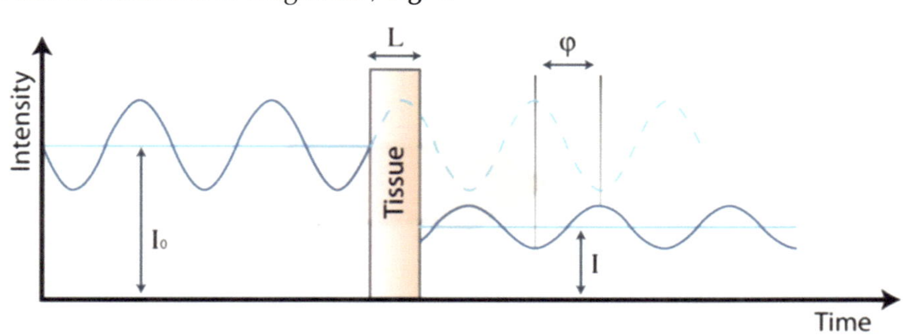


Fig. 4. In frequency domain systems a continuous intensity modulated source is sent through the tissue as a result and amplitude reduction and phase shifted signal is measured.

FD systems have the advantage of achieving higher temporal resolution than TD systems. One disadvantage is the noise in the measurements of scattering effects (Boas et al. 2002). Currently, FD is not as developed as the other two.

### 2.3.3 Continuous wave

In CW systems, (Maki et al. 1996; Nioka et al. 1997; Siegel et al. 1999) light sources emit light continuously at a constant amplitude or modulated frequencies not higher than a few tens of kilohertz (Boas et al. 2002). The most highly developed application of CW imaging technology is the study of hemodynamic and oxygenation changes in superficial tissues, and of the outer (cortical) regions of the brain; particularly using optical topography (Gibson et al. 2005). An advantage of CW systems over the other two is their cost. However, the main disadvantage is the inability to uniquely quantify the effects of light scattering and absorption (Arridge and Lionheart 1998). Other drawbacks are that intensity measurements are far more sensitive to the optical properties of tissues at or immediately below the surface than to the properties of localized regions deeper within the tissue (Gibson et al. 2005). This is due to the characteristic 'banana' shape **(seen in Fig. 2 (a) and (b))** of the volume of tissue over which the measurement is sensitive, which is narrow near the source and detector and very broad in the middle (Arridge 1995; Arridge and Schweiger 1995). Additionally, the detected intensity is highly dependent on surface coupling, meaning if an optical fiber is moved slightly or pressed more firmly on the skin, it can result in a significantly large change in measurement (Gibson et al. 2005). **Fig. 5** illustrates the CW system.

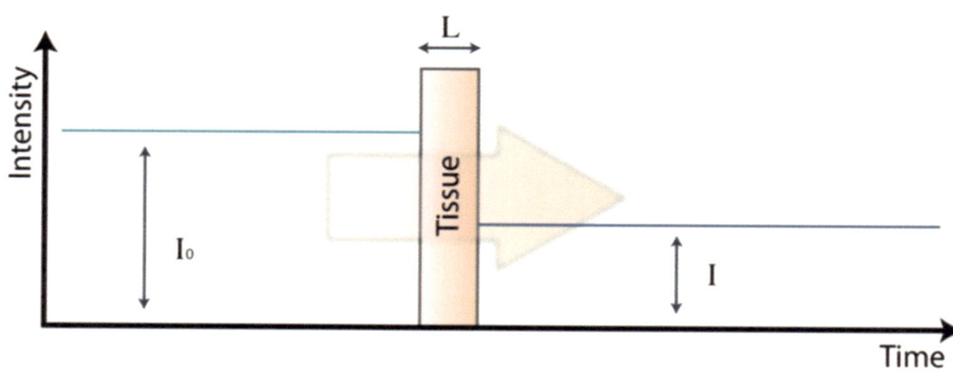


Fig. 5. In continuous domain systems, the light with constant is continuously emitted through the tissue and an attenuated signal is measured.

A more comprehensive discussion of these three systems will be beyond the scope of this chapter. A brief introduction was given so that a better understanding and ability to distinguish the protocols of current applications can be possible. Next we will discuss the optical fiber and configurations.

### 2.4 Optical fibers

Ideally, in transmitting light via optical fibers, we would achieve total internal reflection in the fiber as long as the incident angle is not too large, otherwise light will escape. Originally, bare fibers were used to transmit light. The limitation of using bare fibers is that the light guiding surface is exposed. For example, light could leak from finger prints, hence, fibers currently consist of two layers; the core, for light transmittance and the cladding or barrier to keep the light from escaping, **Fig.6**. These layers have different indices of refraction.

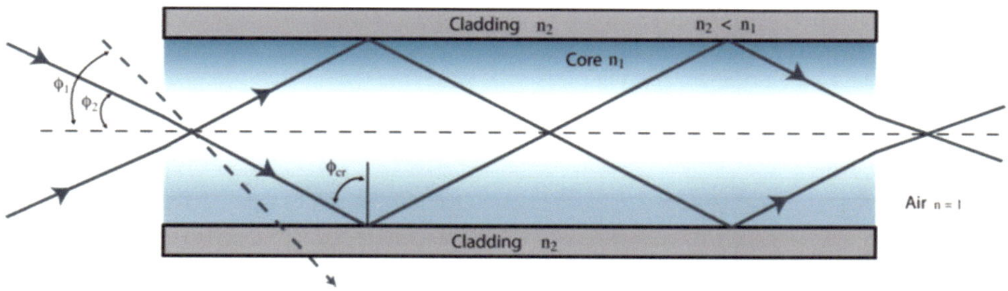


Fig. 6. Standard optical fiber light path. There is a slight refraction as the light enters the fiber from air. Cladding is used with a different index of refraction ($n_2$) to minimize light from escaping and entering. Based on the critical angle $\Phi cr$, the light either refracts into the cladding barrier or reflects back into the core as related to the angles of incidence $\Phi_1$ and$\Phi_2$.

The light is transmitted at incident angles which are given by Snell's law

$$n_1 \sin \Phi_1 = n_2 \sin \Phi_2 \tag{9}$$

where $n$ and $\Phi$ are the refractive index and angle of incidence respectively. The light is refracted slightly inward as it passes from the air ($n$=1) to the optical fiber ($n_1$>$n$). Depending on the angle of incidence, the light may reflect back into the core or refract out into the cladding. The maximum angle of incidence allowable for some light to be refracted is called the critical angle ($\Phi_{cr}$). Angles above this will result in total internal reflection when $n_2 < n_1$.

$$\Phi_{cr} = \arcsin(\sin \Phi_2) \tag{10}$$

## 2.5 Optode considerations

In optimizing near infrared spectroscopy signals several considerations should be taken into account from the optode perspective. There are two types of materials that are used in designing the optical fibers; glass and plastic. The latter is recommended for prototyping because of the reduced cost; however, glass yields better signal transmission and detection. Additionally, glass fibers are strong but the fiber breaks with little stretching. Plastic fibers are cheap and more flexible than glass. However, they have much higher attenuation and a limited temperature range. Thus they are commonly used in designing sensor heads but once they are finalized, it is optimal to switch to glass. The size of the area of measurement will determine the fiber diameters both for sources and detectors. The placement of the sources and detectors for in-vivo measurements is highly dependent on the region of interest and based on the particular application. Proper understanding of photon propagation allows for an educated arrangement to come up with spatially resolved spectroscopy **(refer to Fig.2 (c))**. Illustrated are several modified configurations designed for the visual cortex from previous work (Kashou et al. 2007), **Fig. 7**. By varying the distance between the source and detectors we can manipulate the depth of the signal propagation. Optimizing the optode layout is highly dependent on the application and region of interest. The sensor heads with two detectors were designed to allow for the option of using only the inner sources, the outer sources or both simultaneously. The one detector sensor head was designed to be more flexible by allowing the same option as the previous ones but with the addition of spatially varying the sensors (horizontally) and the detector (vertically) in order to manipulate the penetration depth of the signal.

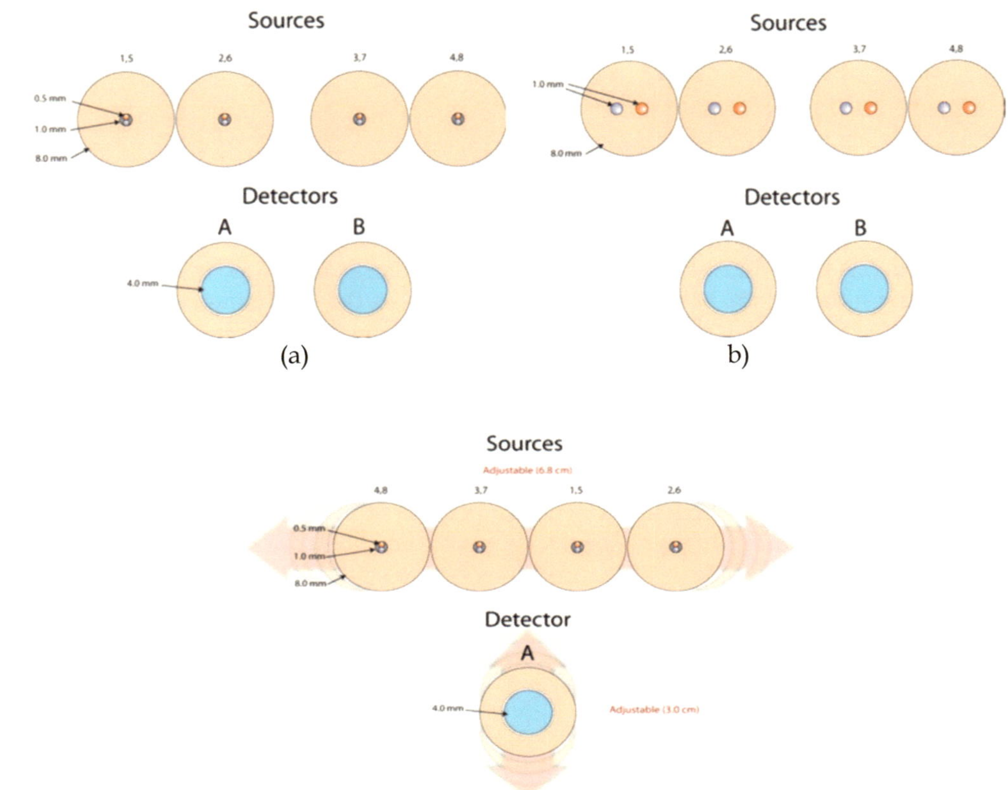


(c)

Fig. 7. Three optode layouts for visual cortex measurements using two wavelengths 690 (orange, sources 1-4) and 830 (gray, sources 5-8), transmitted through optical fibers and collected by the detector (s) (blue). (a) Co-located wavelengths with diameter of 0.5mm each and two detectors, (b) separated wavelengths with diameter of 1.0mm each and two detectors, (c) co-located wavelengths with diameter of 0.5mm each and one detector which are spatially adjustable.

**Table 1**, lists the approximate distances for the configurations illustrated in **Fig. 7 (a) & (c)**. Note, in configuration 1, the sensor head was not exactly symmetric and thus needs to be considered in the analysis of the signals. A more accurate methodology of fabrication and design of the sensor heads needs to be incorporated for the future as research labs move forward in this field. These numbers are important in the process of validating and optimizing the sensor head. They can also be used in the photon propagation simulation for correlation with experimental data. We will discuss simulation in detail in Section 3.

| Wavelength | Source | Configuration 1 | | Configuration 2 |
|---|---|---|---|---|
| | | Distance from A | Distance from B | Minimum Distance from A |
| 830 nm | | | | |
| | 1 | 3.2 cm | 4.5 cm | 2.0 cm |
| | 2 | 3.0 cm | 3.7 cm | 2.5 cm |
| | 3 | 3.7 cm | 3.0 cm | 2.0 cm |
| | 4 | 4.8 cm | 3.3 cm | 2.5 cm |
| 690 nm | | | | |
| | 5 | 3.2 cm | 4.5 cm | 2.0 cm |
| | 6 | 3.0 cm | 3.7 cm | 2.5 cm |
| | 7 | 3.7 cm | 3.0 cm | 2.0 cm |
| | 8 | 4.8 cm | 3.3 cm | 2.5 cm |

Table 1. The source detector distances from a one and two detector sensor head design for the visual cortex.

### 2.6 Optode configurations

In optical tomography, several approaches for optode configurations have been evaluated. Most of the clinical studies have been performed using two parallel glass plates which compress the tissue (such as the breast). This source-detector imaging configuration has the advantage to reduce the attenuation of the transmitted light through a well-known thickness of the interrogated structure. However, the compression technique induces changes in optical properties of the tissue (Gibson et al. 2005), and it is generally difficult to achieve depth resolution from off-axis transmission data acquired in parallel plane geometry. 3D image reconstruction can be ideally generated by surrounding the tissue with multiple sources and fibre-detectors. This circular configuration does not require compression, but the detected light intensities at the tissue boundaries notably decrease as the size of the tissue increases (Ntziachristos et al. 2002).

### 2.7 Applications

In this section, a more detailed look at the methods and techniques of selected studies is described, in order to gauge reader interest in the possibilities of future work. Since traditionally, imaging modalities in children usually involves higher risk and acceptance, NIR has and is proving to have potential in this population. As a result, new methods have been introduced for functional studies in infants with the hopes of carrying on to adults. An advantage of starting with infants is mainly because of anatomy and the distance the light has to traverse.

Single channel NIRS (Ferrari et al. 2004; Hoshi 2003; Obrig et al. 2000) has been used in the past to record brain activity. The typical design would have three possible photon paths. The first possible path is that the photon undergoes a number of scattering events to reach the detector. The second possibility is that the photon is absorbed after a number of scattering events. The third is that the photon leaves the head without being detected. Unfortunately, using a single channel NIRS technique has the major constraint of spatial resolution; hence multiple NIRS systems have been used for optical topography (Gratton et al. 1995). The first report of optical topography on premature babies included the use of nine sources and four detectors using a CW system (Chance et al. 1998). In a more recent study, optical topography in awake infants (Taga et al. 2003) was demonstrated by using the

Hitachi system. This multi-channel system used a CW source and generated two wavelengths (780 and 830*nm*). Twenty laser diodes and eight avalanche photodiode detectors were used and separated into individual light sources by 48 lock-in amplifiers. **Table 2** gives a non-exhaustive list of studies involving NIRS mainly for visual cortex monitoring.

| Reference | Source | Detector | Wavelength | Type | Stimuli |
|---|---|---|---|---|---|
| Taga et al., 2003 | 20 | 8 | 780, 830 | CW | Checkerboard |
| Wilcox et al., 2005 | | 2 | 680, 830 | | Ball & Box |
| Chance et al., 1998 | 9 | 4 | | CW | |
| Toronov et al., 2000 | 16 | 2 | 758, 830 | TD | |
| Franceschini et al., 2003 | 16 | 16 | 690, 830 | CW | Finger tap |
| Franceschini et al., 2000 | 8 | 2 | 758, 830 | FD | |
| Everdell et al., 2005 | 16 | 8 | 785, 850 | FD | |
| Xu et al., 2005 | 9 | 4 | 808 | CW | |
| Meek et al., 1995 | 4 | | 779, 821, 855, 908 | TD | Red, blue, green discs |
| Jasdzewski et al., 2003 | 2 | 4 | 682, 930 | CW | Checkerboard |
| Plichta et al., 2006 | 17 | 16 | 695±20, 830±20 | CW | Checkerboard |
| Wolf et al., 2003 | 4 | 2 | 670, 830 | FD | Checkerboard |
| Gratton & Fabiani 2003 | 8 | 2 | 750 | FD | Gratings (CP) |
| Schroeter et al., 2006a | | | 775, 810, 850, 910 | | Checkerboard |
| Schroeter et al., 2004 | 8 | 7 | 780, 830 | FD | Checkerboard |
| Miki et al., 2005 | | | 780, 805, 830 | | Monocular Flashes |
| Kusaka et al., 2004 | 8 | 8 | 776, 804, 828 | CW | Flashing light |
| Obrig et al., 2000 | 4 | 1 | 775, 825, 850, 905 | | Checkerboard |

Table 2. The details of the system and configurations used by some research groups.

## 3. Photon diffusion equation

We discussed MBLL earlier however a more efficient way of modeling and simulating the light path through tissue is by the photon diffusion equation. The well-known photon diffusion equation derived from the transport equation is widely used to describe photon migration in biological tissues:

$$\frac{1}{v}\frac{\partial}{\partial t}L(r,\vec{s},t)+\nabla.L(r,\vec{s},t).\vec{s}=-(\mu_a+\mu_s)L(r,\vec{s},t)+\mu_s\int_{4\pi}L(r,\vec{s}',t)P(\vec{s}',\vec{s})d\vec{s}'+S(r,\vec{s},t) \quad (11)$$

## 4. Photon simulation and reconstruction

### 4.1 Monte Carlo (MC) method

The 2D MC code proceeds in the following way: First, a photon packet is launched into the medium at the defined source position with an initial survival weight $w$ that indicates the probability of a photon's successively propagation without absorption. Snell's law and Fresnel reflection formulas are then applied at each boundary. After each scattering event,

a new direction of the photon is determined using the Henyey-Greenstein scattering function. The effects of absorption in each region can be obtained by $exp(\mu_a L_m)$ where $L_m$ is the partial path length of photon through region $m$. The survival weight and path length in each layer is recorded when the photon was scattered out of the head model, to obtain the photon trajectory.
The same rules used previously for the 2D MC code, to describe photon migration are also adopted for the 3D MC code. The tissue is divided into voxels, each of which may have different optical properties. In our simulation, before the photons started to emit, we created a matrix to record the spatial arrangement of tissue constituents. Each voxel in the matrix was assigned an integer to represent the type of tissue in the voxel. As the photon is propagated from one scattering event to the next, a check was made every 1 grid element of 1 mm spacing to see if the scattering or absorption coefficient has changed.
We performed 2D and 3D MC simulations based on various head models. We modified Monte Carlo code (Wang et al. 1995) to calculate the spatial sensitivity profiles *SSP* ("banana shaped"). The mean optical path length travelled by the detected light within each region $m$ of the tissue types is defined as:

$$L_m = \frac{\partial A}{\partial \mu_{a,m}} \tag{12}$$

Where $A = -\ln(I/I_0)$, which is the attenuation change in the detected intensity $I$ and $I_0$ is the incident radiation. **Equation 13** can be used to calculate the radiation distribution profile (spatial sensitivity profile, SSP) which depends on the probability of photons traveling within a particular region $m$ at position $r$ from the source optode to the detector optode:

$$SSP(r) \equiv L_m(r) = \frac{\partial A}{\partial \mu_{a,m}(r)} \tag{13}$$

### 4.2 Head models

#### 4.2.1 Two-dimensional head model

A three-layer semi-infinite slab having various scattering and absorption optical properties with the single optode spacings $\rho$ = 28 mm, as shown in figure 1, is adopted. The layers start with the scalp-skull region of 7.9 mm, then the cerebrospinal fluid (CSF) of 1.5 mm and the brain being infinitely extended. The optical properties and thicknesses of each layer used for this model were chosen based on reports from literature.

#### 4.2.2 Three-dimensional homogeneous head model

The sphere had dimensions of 40 x 40 x 40 mm with the single source optode positioned at (20,20,0) and 19 detectors depicted in **Fig. 8**, having each 1 mm of diameter, placed in various positions on the surface.

#### 4.2.3 Three dimensional realistic head model

We used segmented MRI data (3D-SPGR, T1-weighted images) of an adult human head that we employed for 3D MC simulations. **Figure 8** shows the axial slice of the anatomical MRI, segmented into five tissue types (air, scalp, CSF, and gray/white matter). The whole volume was contained within 512x512x120 voxels. The optical properties used for each tissue type were reported from literature. The source and detector were placed on the transaxial plane.

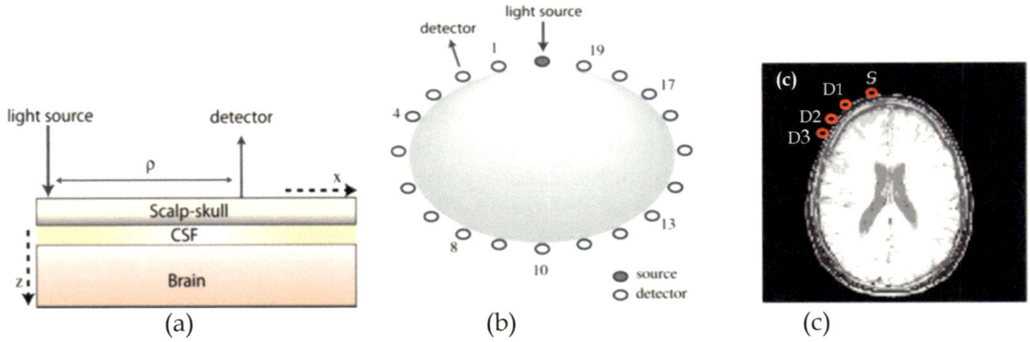


Fig. 8. (a) Two-dimensional heterogeneous, (b) Three-dimensional homogenous (symbols full and open circles indicate the position of the source and the detectors respectively) and (c) anatomical MRI, adult head models used in MC simulations (symbols S and D1-D3 indicate the position of the source and the detectors respectively).

## 5. Results and discussion

We simulated the radiation distribution so-called "banana-shape" in order to determine the contribution of the photons scattered from the different regions within the head model as function of separation distance between the source light and the detector location. We ran $10^8$ photons for all MC simulations to achieve an appropriate SNR. The spatial sensitivity distributions were calculated for each head model. **Fig. 9** shows a cross section of the "banana shaped" across the 2D three-layer model **(Fig. 8(a))** for a source-detector separation of 28 mm, which is commonly used for NIRS. The effects of the CSF layer can clearly be seen as a significant reduction in the depth penetration of light. The ratio of photons scattering through the brain decreases, such that most of the light is detected from scalp-skull layer. A significant part of photons was guided along the CSF layer, which has the effect of concentrating measurement sensitivity to the more superficial layer (scalp-skull) and may change the depth sensitivity profile.

The sensitivity measurements at various source-detector spacings **(Fig. 9)** was investigated by considering a 3D homogeneous head model **(Fig. 8 (b))**. The migrating of all $10^8$ photons was performed in a single 3D MC simulation for all 19 detectors at same time. The sensitivity to deeper region seems to vary between different source-detector spacings, as shown in figures **(Fig. 10 (a)-(d))**. For the smaller optode spacings, the minimum depth is obtained **Fig. 10(a)-(b)**. As the separation between the source and the detector increases, the depth of penetration by the detected light also increases but intensity of light decreases more strongly, as shown in **Fig. 10 (c)-(d)**. This model can be useful in investigating the propagation of light in the infant brain. A reference homogeneous phantom (Hebden et al. 2002) was previously used in representing the premature infant brain which consisted of a balloon filled in a solution of intralipid.

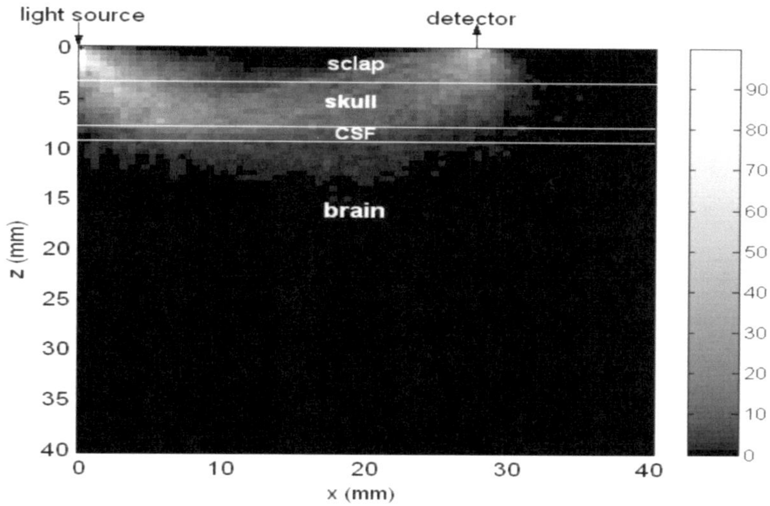


Fig. 9. Spatial sensitivity profile of photons in 2D three-layer head model as given by MC simulation for a single source- detector separation. Light in the near-infrared range (700-900 nm), compared to the visible light range penetrates tissues more deeply. The photons change direction multiples times as they scatter, and they follow a generally curved path called the "banana shaped" curve between the light source and the detector location at the surface (Mansouri and L'Huillier, 2007).

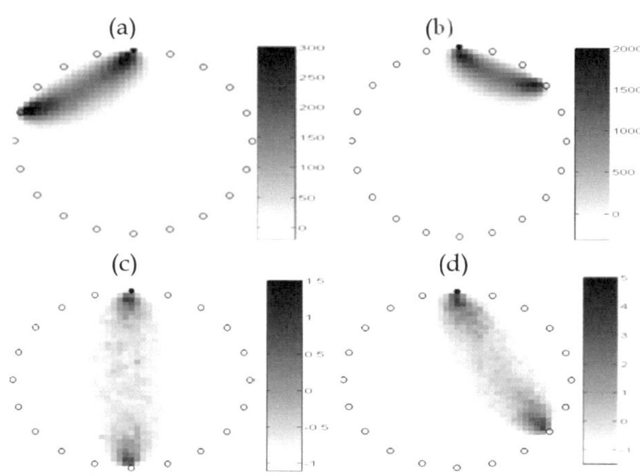


Fig. 10. ("banana shaped") photon distributions through the homogeneous 3D head model between the single source and various detectors 4, 17,10, and 13 from (a) to (d) respectively.

**Fig. 11** shows a cross section of the "banana shaped" across the 3D segmented MR image **(Fig. 8 (c))**. The light penetration between the source and the detector is not along any major gray matter, however, a significant part of the light is detected from scalp and skull. This suggests that the contribution of light that travels within the superficial layers of the scalp and skull to the whole detected light is rather large. This indicates that the number of photons that travel through the brain surface are the largest among the detected photons penetrating into the brain tissue which spreads near the brain surface. Thus, the fundamental question that arises here: is the NIR signal sensitive to changes associated with functional brain activation? While NIRS is increasingly gaining attention in studies on brain function, the sensitivity of NIRS signals to functional brain activation remains to be discussed.

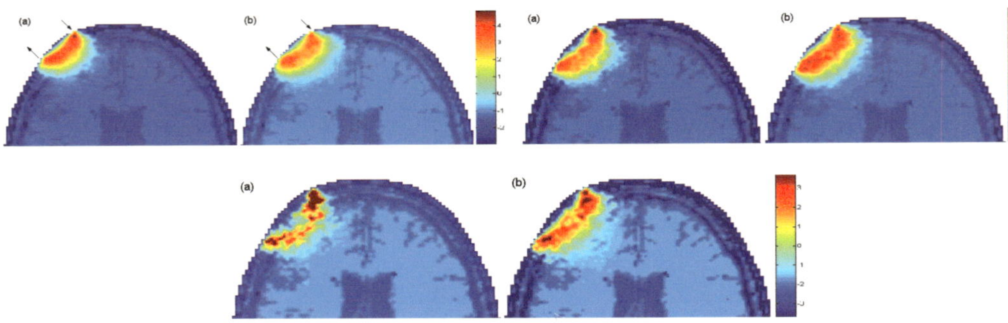


Fig. 11. Spatial sensitivity profiles of photons in the realistic 3D head model. Depth sensitivity is limited by the presence of endogenous chromophores such as haemoglobin, water and cytochrome. The optical path travelled by photons between the source and the detector is strongly dependent to the thickness and the optical properties of the superficial tissues and the 'void' region, which is the CSF of the head. Data measurements from multiple source-detector separations provide overlapping measurement from a limited area. These are used for tomographic image reconstruction utilising 3D models (Mansouri et al., 2010).

The present study has demonstrated that the path length is strongly dependent on the thickness and the optical properties of the superficial tissues and the CSF in the head model. This indicates that the contribution of the extracerebral layers to the radiation distribution is large, while the contribution of the brain is small. Therefore, the estimation of the optical path travelled by photons between the source and the detector is crucial for evaluating the sensitivity of NIRS signals to the brain.

## 6. Conclusion

In this chapter, we presented a technical overview of NIRS as it relates to the medical field. NIRS has the potential to be the wide spread standard of care in hospitals. Mainly because it is low cost, non-invasive, and portable. The number of research groups both preclinical and clinical using this technology is on the rise. This is because the nature of the light propagation and the depth of penetration of NIRS is ideal for the infant population. As seen in this chapter, there is still room for improvement and even though not discussed here, there are studies trying to measure neuronal responses using NIRS. Other groups have tried to combine NIRS with various modalities. There are still challenges which are being tackled, such as sensor head design and optimization. However, within the next several years it is expected that there will be much more advancement to this promising modality.

## 7. References


Arridge, S. R. (1995). "Photon-measurement density functions. Part I: Analytical forms." *Appl Opt* 34(31): 7395-7409.

Arridge, S. R. and W. R. Lionheart (1998). "Nonuniqueness in diffusion-based optical tomography." *Opt Lett* 23(11): 882-884.

Arridge, S. R. and M. Schweiger (1995). "Photon-measurement density functions. Part 2: Finite-element-method calculations." *Appl Opt* 34(34): 8026-8037.

Benaron, D. A., S. R. Hintz, et al. (2000). "Noninvasive functional imaging of human brain using light." *Journal of Cerebral Blood Flow and Metabolism* 20(3): 469-477.

Benaron, D. A. and D. K. Stevenson (1993). "Optical time-of-flight and absorbance imaging of biologic media." *Science* 259(5100): 1463-1466.

Boas, D. A., M. A. Franceschini, et al. (2002). Noninvasive Imaging of Cerebral Activation with Diffuse Optical Tomography, *In vivo optical imaging optical imaging of brain function*. R. D. Frostig, CRC Press.

Brazy, J. E., D. V. Lewis, et al. (1985). "Noninvasive Monitoring of Cerebral Oxygenation in Preterm Infants - Preliminary-Observations." *Pediatrics* 75(2): 217-225.

Chance, B., E. Anday, et al. (1998). "A novel method for fast imaging of brain function, non-invasively, with light." *Optics Express* 2(10): 411-423.

Chance, B., J. S. Leigh, et al. (1988). "Comparison of Time-Resolved and Time-Unresolved Measurements of Deoxyhemoglobin in Brain." *Proceedings of the National Academy of Sciences of the United States of America* 85(14): 4971-4975.

Cope, M. (1991). The Application of Near-Infrared Spectroscopy to Non Invasive Monitoring of Cerebral Oxygenation in the Newborn Infant. *Department of Medical Physics and Bioengineering*. London, University College London. PhD.

Cope, M. and D. T. Delpy (1988). "System for Long-Term Measurement of Cerebral Blood and Tissue Oxygenation on Newborn-Infants by near-Infrared Trans-Illumination." *Medical & Biological Engineering & Computing* 26(3): 289-294.

Cubeddu, R., A. Pifferi, et al. (1996). "Time-resolved imaging on a realistic tissue phantom: mu(s)' and mu(a) images versus time-integrated images." *Appl Opt* 35(22): 4533-4540.

Everdell, N. L., A. P. Gibson, et al. (2005). "A frequency multiplexed near-infrared topography system for imaging functional activation in the brain." *Review of Scientific Instruments* 76(9): -.

Ferrari, M., L. Mottola, et al. (2004). "Principles, techniques, and limitations of near infrared spectroscopy." *Canadian Journal of Applied Physiology-Revue Canadienne De Physiologie Appliquee* 29(4): 463-487.

Franceschini, M. A., S. Fantini, et al. (2003). "Hemodynamic evoked response of the sensorimotor cortex measured noninvasively with near-infrared optical imaging." *Psychophysiology* 40(4): 548-560.

Franceschini, M. A., V. Toronov, et al. (2000). "On-line optical imaging of the human brain with 160-ms temporal resolution." *Optics Express* 6(3): 49-57.

Gandjbakhche, A. H., R. Nossal, et al. (1994). "Resolution Limits for Optical Transillumination of Abnormalities Deeply Embedded in Tissues." *Medical Physics* 21(2): 185-191.

Gibson, A. P., J. C. Hebden, et al. (2005). "Recent advances in diffuse optical imaging." *Physics in Medicine and Biology* 50(4): R1-R43.

Gratton, E., S. Fantini, et al. (1997). "Measurements of scattering and absorption changes in muscle and brain." *Philosophical Transactions of the Royal Society of London Series B-Biological Sciences* 352(1354): 727-735.

Gratton, G., P. M. Corballis, et al. (1995). "Shades of Gray-Matter - Noninvasive Optical-Images of Human Brain Responses during Visual-Stimulation." *Psychophysiology* 32(5): 505-509.

Gratton, G. and M. Fabiani (2003). "The event-related optical signal (EROS) in visual cortex: Replicability, consistency, localization, and resolution." *Psychophysiology* 40(4): 561-571.

Greisen, G. (2006). "Is near-infrared spectroscopy living up to its promises?" *Seminars in Fetal & Neonatal Medicine* 11(6): 498-502.

Grosenick, D., H. Wabnitz, et al. (1997). "Time-resolved imaging of solid phantoms for optical mammography." *Applied Optics* 36(1): 221-231.

Hebden, J. C., S. R. Arridge, et al. (1997). "Optical imaging in medicine: I. Experimental techniques." *Physics in Medicine and Biology* 42(5): 825-840.

Hebden, J. C., A. Gibson, et al. (2002). "Three-dimensional optical tomography of the premature infant brain." *Physics in Medicine and Biology* 47(23): 4155-4166.

Hintz, S. R., D. A. Benaron, et al. (2001). "Bedside functional imaging of the premature infant brain during passive motor activation." *Journal of Perinatal Medicine* 29(4): 335-343.

Hintz, S. R., D. A. Benaron, et al. (1998). "Stationary headband for clinical time-of-flight optical imaging at the bedside." *Photochemistry and Photobiology* 68(3): 361-369.

Hintz, S. R., W. F. Cheong, et al. (1999). "Bedside imaging of intracranial hemorrhage in the neonate using light: Comparison with ultrasound, computed tomography, and magnetic resonance imaging." *Pediatric Research* 45(1): 54-59.

Hoshi, Y. (2003). "Functional near-infrared optical imaging: Utility and limitations in human brain mapping." *Psychophysiology* 40(4): 511-520.

Jasdzewski, G., G. Strangman, et al. (2003). "Differences in the hemodynamic response to event-related motor and visual paradigms as measured by near-infrared spectroscopy." *Neuroimage* 20(1): 479-488.

Jiang, H., K. D. Paulsen, et al. (1995). "Simultaneous reconstruction of optical absorption and scattering maps in turbid media from near-infrared frequency-domain data." *Opt Lett* 20(20): 2128-2130.

Kashou, N. H., R. Xu, et al. (2007). "Using FMRI and FNIRS for localization and monitoring of visual cortex activities." *Conf Proc IEEE Eng Med Biol Soc* 2007: 2634-2638.

Kato, T., A. Kamei, et al. (1993). "Human visual cortical function during photic stimulation monitoring by means of near-infrared spectroscopy." *J Cereb Blood Flow Metab* 13(3): 516-520.

Kennan, R. P., S. G. Horovitz, et al. (2002). "Simultaneous recording of event-related auditory oddball response using transcranial near infrared optical topography and surface EEG." *Neuroimage* 16(3): 587-592.

Kusaka, T., K. Kawada, et al. (2004). "Noninvasive optical Imaging in the visual cortex in young infants." *Human Brain Mapping* 22(2): 122-132.

Maki, A., Y. Yamashita, et al. (1996). "Visualizing human motor activity by using non-invasive optical topography." *Front Med Biol Eng* 7(4): 285-297.

Mansouri, C. and J. P. L'Huillier (2007). "Time-resolved photon migration through an adult head model: Comparison between Finite Element and Monte Carlo calculations -

art. no. 66320L." *Therapeutic Laser Applications and Laser-Tissue Interaction III* 6632: L6320-L6323.

Mansouri, C., J. P. L'Huillier, et al. (2010). "Depth sensitivity analysis of functional near-infrared spectroscopy measurement using three-dimensional Monte Carlo modelling-based magnetic resonance imaging." *Lasers in Medical Science* 25(3): 431-438.

Meek, J. H., C. E. Elwell, et al. (1995). "Regional Changes in Cerebral Hemodynamics as a Result of a Visual Stimulus Measured by near-Infrared Spectroscopy." *Proceedings of the Royal Society of London Series B-Biological Sciences* 261(1362): 351-356.

Meek, J. H., M. Firbank, et al. (1998). "Regional hemodynamic responses to visual stimulation in awake infants." *Pediatric Research* 43(6): 840-843.

Miki, A., T. Nakajima, et al. (2005). "Near-infrared spectroscopy of the visual cortex in unilateral optic neuritis." *American Journal of Ophthalmology* 139(2): 353-356.

Nioka, S., Q. Luo, et al. (1997). "Human brain functional imaging with reflectance CWS." *Oxygen Transport to Tissue Xix* 428: 237-242.

Ntziachristos, V., J. Ripoll, et al. (2002). "Would near-infrared fluorescence signals propagate through large human organs for clinical studies? (vol 27, pg 333, 2002)." *Optics Letters* 27(18): 1652-1652.

Obrig, H. and A. Villringer (2003). "Beyond the visible - Imaging the human brain with light." *Journal of Cerebral Blood Flow and Metabolism* 23(1): 1-18.

Obrig, H., R. Wenzel, et al. (2000). "Near-infrared spectroscopy: does it function in functional activation studies of the adult brain?" *International Journal of Psychophysiology* 35(2-3): 125-142.

Plichta, M. M., M. J. Herrmann, et al. (2006). "Event-related functional near-infrared spectroscopy (fNIRS): Are the measurements reliable?" *Neuroimage* 31(1): 116-124.

Pogue, B., M. Testorf, et al. (1997). "Instrumentation and design of a frequency-domain diffuse optical tomography imager for breast cancer detection." *Optics Express* 1(13): 391-403.

Pogue, B. W. and M. S. Patterson (1994). "Frequency-Domain Optical-Absorption Spectroscopy of Finite Tissue Volumes Using Diffusion-Theory." *Physics in Medicine and Biology* 39(7): 1157-1180.

Pouratian, N., S. A. Sheth, et al. (2003). "Shedding light on brain mapping: advances in human optical imaging." *Trends Neurosci* 26(5): 277-282.

Schroeter, M. L., M. M. Bucheler, et al. (2004). "Towards a standard analysis for functional near-infrared imaging." *Neuroimage* 21(1): 283-290.

Schroeter, M. L., M. M. Bucheler, et al. (2006a). "Circadian variability is negligible in primary visual cortices as measured by fNIRS." *International Journal of Psychophysiology* 62(1): 9-13.

Schroeter, M. L., T. Kupka, et al. (2006b). "Investigating the post-stimulus undershoot of the BOLD signal - A simultaneous fMRI and fNIRS study." *Neuroimage* 30(2): 349-358.

Siegel, A., J. J. Marota, et al. (1999). "Design and evaluation of a continuous-wave diffuse optical tomography system." *Optics Express* 4(8): 287-298.

Strangman, G., D. A. Boas, et al. (2002). "Non-invasive neuroimaging using near-infrared light." *Biological Psychiatry* 52(7): 679-693.

Strangman, G., M. A. Franceschini, et al. (2003). "Factors affecting the accuracy of near-infrared spectroscopy concentration calculations for focal changes in oxygenation parameters." *Neuroimage* 18(4): 865-879.

Taga, G., K. Asakawa, et al. (2003). "Brain imaging in awake infants by near-infrared optical topography." *Proceedings of the National Academy of Sciences of the United States of America* 100(19): 10722-10727.

Toronov, V., M. A. Franceschini, et al. (2000). "Near-infrared study of fluctuations in cerebral hemodynamics during rest and motor stimulation: Temporal analysis and spatial mapping." *Medical Physics* 27(4): 801-815.

Villringer, A. and B. Chance (1997). "Non-invasive optical spectroscopy and imaging of human brain function." *Trends in Neurosciences* 20(10): 435-442.

Wang, L., S. L. Jacques, et al. (1995). "MCML--Monte Carlo modeling of light transport in multi-layered tissues." *Computer Methods and Programs in Biomedicine* 47(2): 131-146.

Wilcox, T., H. Bortfeld, et al. (2005). "Using near-infrared spectroscopy to assess neural activation during object processing in infants." *Journal of Biomedical Optics* 10(1): -.

Wolf, M., U. Wolf, et al. (2003). "Fast cerebral functional signal in the 100-ms range detected in the visual cortex by frequency-domain near-infrared spectrophotometry." *Psychophysiology* 40(4): 521-528.

Wyatt, J. S., M. Cope, et al. (1988). "The Response of Cerebral Blood-Volume to Changes in Arterial Carbon-Dioxide Tension Measured by near-Infrared Spectroscopy in Newborn-Infants." *Early Human Development* 17(2-3): 291-291.

Xu, C., D. Wang, et al. (2005). "Continuous-wave Near-infrared Topography System and Its Clinical Application." *Conf Proc IEEE Eng Med Biol Soc* 1: 196-198.

12

# Novel Conductive and Transparent Optical Fiber Probe for Multifunctional Scanning Tunneling Microscopy

Guo Xinli[1,2] and Fujita Daisuki[3]
[1]*School of Materials Science & Engineering, Southeast University, Nanjing, Jiangsu,*
[2]*Jiangsu Key Laboratory for design and Manufacture of Micro-Nano Biomedical Instruments*
[3]*Advanced Nanocharacterization Center, National Institute for Materials Science, Tsukuba, Ibaraki,*
[1,2]*P. R. China*
[3]*Japan*

## 1. Introduction

It is well known that progress in modern science is impossible without reliable tools for characterization of structural, physical, and chemical properties of materials and devices at the micro-, nano, and atomic scale levels. The structural information of materials can be obtained by the 1st generation of optical microscopy, and second generation of scanning and transmission electron microscopy. While characterization of local electronic properties became possible after the 3rd generation of Scanning Tunneling Microscopy (STM) based on the quantum mechanical tunneling between an atomically sharp metallic tip and conductive surface by G. Binning and H. Rohrer in 1981 at IBM Zurich, which earned the Nobel Prize in Physics five year later [Binning et al, 1982, 1983]. STM has become the first instrument to directly probe local geometries of various nanostructures with an atomic resolution and has triggered development of new classes of STM-related techniques and a rapid growth of a variety of scanning probe microscopy (SPM) techniques over the last two decades, which include atomic force microscopy (AFM) based on the mechanical detection of the Van der Waals forces between the tip and the sample surface using a pliable cantilever[Binning et al, 1986, magnetic force microscopy (MFM), electrostatic force microscopy (EFM), scanning capacitance microscopy (SCM), near-field scanning optical microscopy (NSOM) etc to provide capability to access local electrical, magnetic, chemical, mechanical, optical, and thermal properties of materials on the nanometer scale. The SPM technique has demonstrated not only imaging, but also manipulation, control and modification of the local structure and material functionality at the nano-and atomic level. The rapid development of nanoscience and nanotechnology was really started and strongly stimulated by the generation and availability of SPM techniques, and in turn constantly stimulates development of novel SPM techniques [Sergei & Alexei, 2007]. Typically, one of the advancement of STM under the stimulation of nanometer-scale science and technology is to detect and analyze not only tunneling electrons but also photons induced by tunneling electrons [Berndt, 1998]. This kind of STM-induced photon emission, or STM-induced luminescence (STML) is a local energy-dissipation process where a part of the energy carried by a tunneling electron is transferred to a radiative process, it has the potential ability to unveil local electronic structures and chemical states with high spatial resolution up to atomic scale, and the fabrication of nanoscale light source. Up to date, STML has been successfully observed on various surface nanostructures including metal surfaces [Berndt et al, 1993a, 1995; Uehara et al, 1999], indirect-gap semiconductor surfaces [Downes & Welland, 1998; Thirstrup et al, 1999], direct-gap semiconductors [Sasaki & Murashita, 1999; Fujita et al, 2004; Guo et al, 2007], fluorescence molecules [Berndt et al, 1993b; Fujita et al, 2001; Dong et al, 2004; Guo et al, 2003, 2004], conjugated polymer films [Alvarado et al, 1997 & Lidzey et al, 1997], quantum nanostructures [Evoy et al, 1999 & Tsuruoka et al, 2003] etc. The STML is believed to be realized by the excitation of highly localized tunneling current of a STM through following two ways as schematically shown in figures 1 (a) and (b).

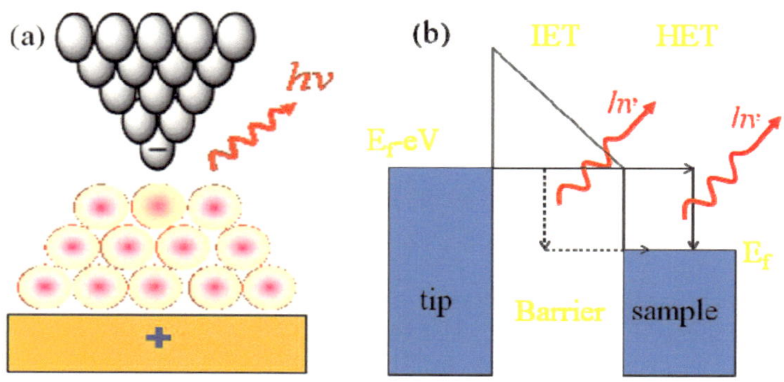


Fig. 1. Schematic drawing of the STML from molecular surface in STM tunneling junction (a), and (b) band gap in STM tunneling junction to show the two types of excitation process of STML. i.e: Inelastic Electron Tunneling (IET), and Hot Electron Thermalization (HET).

The inelastic tunneling (IET) process in figure 1 (b) shows that photon emission is triggered by an inelastic process during electron tunneling through the barrier. The STML process from metal and indirect band-gap semiconductor surfaces can be well explained by the IET process i.e.: the tunneling electrons excite a localized surface plasmon or a tip-induced plasmon (TIP) whose radiative decay processes cause light emission (Downes et al., 1998; Johansson, 1998, 1990). While the "hotelectron thermalization (HET)"process in figure 1(b) shows that energetic electrons injected into the sub-surface layers lose their energy in the bulk by a so-called "thermalization" process, and finally recombine with positive holes that emit photons. The STML process from the direct bandgap semiconductors and their quantum nanostructures can be well explained by the HET process i.e.: a part of tunneling electrons injected from an STM tip combine with major carriers (positive holes) to form excitons at low temperatures, which subsequently decay radiatively or non-radiatively. In addition, the electron-hole recombination process is also a reasonable explanation for the STML of organic semiconductor thin films (Alvarado et al., 1997; Lidzey et al., 1997), which has been studied with relation to the industrial applications for organic thin-film lightemitting devices (Friend et al., 1999; Lidzey et al., 1998).
There are two types of photon detection system for STML measurements as show in figures 2(a) and (b).

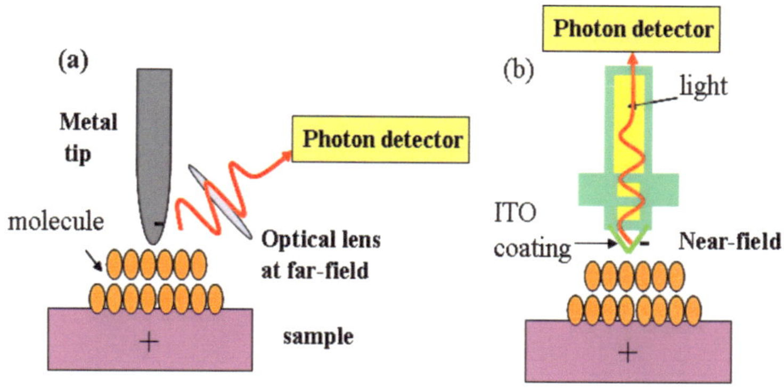


Fig. 2. Schematic drawing of the two types of photon detection systems for STML measurements: (a) optical far-field photon detection by setting optical lens of fibers near to the STM tunneling junction; (b) optical near-field photon detection by using a conductive and transparent optical fiber probe.

In Figure 2(a), the emitted photons in STML measurements are normally collected at optical far-field using lenses (Dong et al., 2002; Hoffmann et al., 2002), fibers (Arafune et al., 2001; Thirstrup et al., 1999), or ellipsoidal reflectors (Nishitani et al., 1998; Suzuki et al., 1999) to focus light from a tip to a detector outside. This system have been widely used due to its relatively easy instrumentation (Arafune et al., 2001; Dong et al., 2002; Hoffmann et al., 2001, 2002; Nazin et al., 2003; Nishitani et al., 1998; Suzuki et al., 1999; Thirstrup et al., 1999). The other detection system is shown in figure 2 that the emitted photons are collected within the optical near-field into a conductive and transparent optical fiber probe. This system is first developed by Murashita (1999), which has a great similarity with near-field scanning optical microscopy (NSOM) operating in collection mode, where an optical fiber probe is used for collection of light (Garcia-Parajo et al., 1994). This near-field photon detection offers a unique capability of near-field detection and potential high detection efficiency.

In this chapter, we introduce the application of a Novel Conductive and Transparent optical fiber probe for multifunctional scanning tunneling microscropy. This kind of novel probe have the functions for high-quality scanning tunneling microscope (STM) imaging, near-field excitation and detection of high-intensity STM-induced electroluminescence (STML). The probe is fabricated by coated a sharpened an optical fiber with an uniform ITO layer with the thickness in the range of ~50 nm to ensure the conductivity and improve the tip-apex geometry of the fiber probe High-quality STM image, and high-intensity STML corresponding to a maximum quantum efficiency of $7\times10^{-4}$ and $1\times10^{-3}$ photons/electron was obtained respectively on Au(111)/mica and p-GaAs(110) surface at room temperature (RT) by using of this fiber probe. This kind of novel conductive and transparent optical fiber probe provided a new way for equipping the STM not only with the functions of conventional STM but also with the functions of near-field excitation and detection of high-intensity STML.

## 2. Experimental procedure

### 2.1 Novel optical fiber probes for STML

Figures 3(a) and (b) show the SEM images of a novel optical fiber probe used for photon detection and STM imaging.

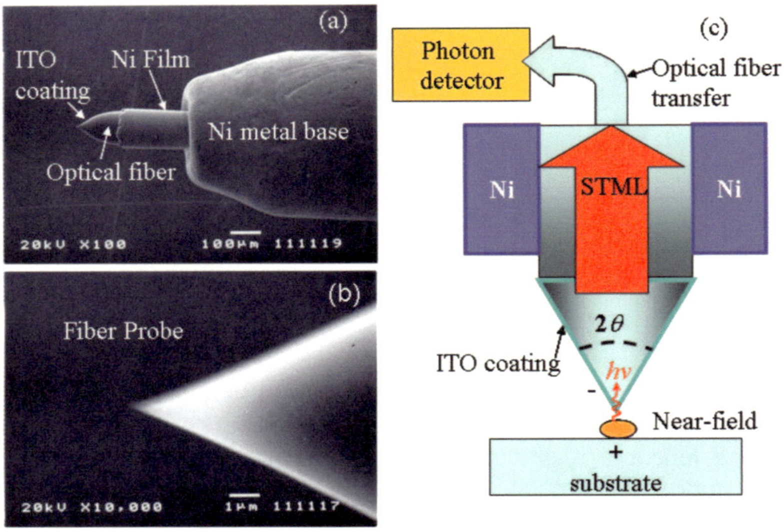


Fig. 3. (a)-(b): SEM images of a conductive and transparent STM fiber probe taken at different magnifications. The fiber probe is coated with 100-nm-thick indium-tin-oxide (ITO) thin film. (c): Schematic drawing of the optical near-field detection geometry for STML using the conductive and transparent fiber probe and its light transfer.

The optically transparent STM probes consisting of a $GeO_2$-doped multimode silica optical fiber with a core diameter of 100 μm and a sharpened tip apex are fabricated by NTT-AT Ltd., Japan. The a conductive indium-tin oxide (ITO) film is famous for its good electrical conductivity and optical transparency in the wide wavelength regions ranging from ultraviolet (UV), visible, to near infrared (NIR). The apex of the probe is coated with a conductive indium-tin oxide (ITO) thin film of 100-nm thickness with an optical transmission ~75%. Typical curvature of the tip apex of the optical fiber probe is 50–100 nm. As shown in Figure 3(c), photons emitted just under the probe apex are collected at near-field into the core of the optical fiber probe and transferred to the detectors outside. The cone solid angle with apex angle of 2θ is the area of a spherical cap on an unit sphere: $\Omega = 2\pi(1-\cos\theta) \sim 0.842$ [sr]. In addition,

## 2.2 STM equipped with photon detection system for STML

The STML is realized by the combination of a STM system with a photon detection system as schematically shown in Figure 4.

For the STM system in Figure 4, we use an ultrahigh vacuum system (UHV) with a base pressure ~3× $10^{-9}$ Pa, which consists of a low temperature (LT)-STM chamber, a sample preparation chamber, and a deposition chamber. The whole LT- STM system manufactured by Unisoku Ltd., Japan, is installed on an air-dumping system. The STM head is suspended under a liquid helium/nitrogen cryostat with an internal eddy-current damping system. The sample can be cooled down to about 10 K by liquid helium in a tank through a bundle of copper wires for thermal connection. The lowest temperature of 8.7 K at the STM head has been achieved by pumping liquid helium in the tank. This practical UHV-LT-STM is shown in figure 5.

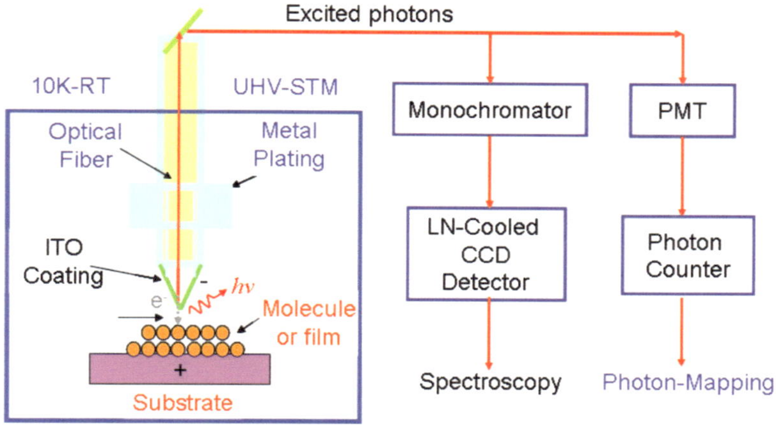


Fig. 4. Schematic drawing of the experimental set up for STML detection using an ITO-coated optical fiber probe based on the combination of a STM system with a photon detection system.

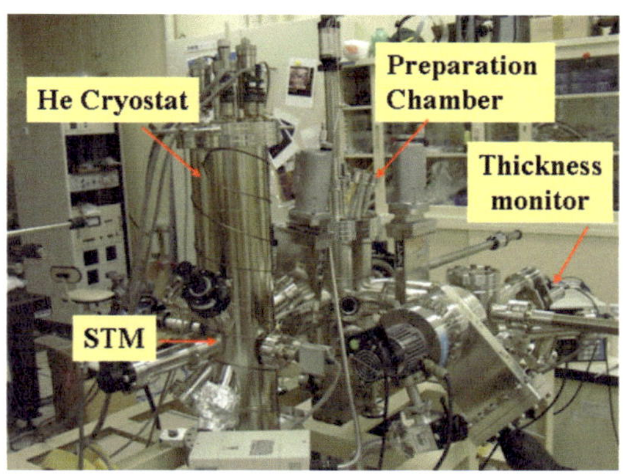


Fig. 5. UHV-LT-STM equipped with the facilities for sample preparation, sputtering and heating. Base pressure is ~~3× $10^{-9}$ Pa, the lowest temperature is ~8.7 K.

For the photon detection system in figure 4, STML experiments were performed with the conductive optical-fiber probes at RT and at about 80 K using liquid nitrogen. Photons induced by injecting electrons to the surface are collected into the STM probe and transferred to the outside photon detectors through several optical fiber connections. Several measurement modes of STML have been developed, i.e.: photon mapping, fixed-point optical spectroscopy, spectrum mapping spectroscopy (Nishitani et al., 1998), and spectroscopic imaging (Hoffmann et al., 2002). The latter two methods are only possible in very sensitive setups. In this study, two kinds of photon detection systems have been developed. One is the low-noise single photon counting system for quantitative photon mapping and the other is a fixed point optical spectroscopy. The single photon detection system is composed of a Peltier-cooled photomultiplier tube (PMT; Hamamatsu Photonics, R943-02), a discriminator, and a pulse counter. The PMT is set in the cooling system, which can not only cool it down to around -30°C but also acts as an effective electromagnetic shield. The photons transferred through the optical fibers are focused on the center of the GaAs(Cs) photocathode surface using a condenser lens system. Although the PMT has a sufficiently high sensitivity over a wide spectral range from UV to NIR (160 to 930 nm), the effective cooling system can realize significantly low dark counts of ~3 counts/s (cps). A single TTL pulse is generated from a single voltage pulse caused by a single photon injection into the PMT. The pulse counter measures the number of the generated TTL pulses and outputs the analog voltage signal, which is proportional to the number of the pulses detected per a gate time. Connecting the analog output of the photon counter to the external input of the STM controller, simultaneous photon mapping with STM scanning is realized. Typical gate time for a photon mapping is 50 ms.

To perform STML spectroscopy, photons emitted from a fixed position are introduced to a multichannel spectrometer composed of an optical monochromator and a liquid-nitrogen cooled CCD detector. The monochromator is a Czerny-Turner type triple grating monochromator with a focal length of 300 mm (Acton Research Corp., SpectraPro-300i). The liquid nitrogen cooled CCD detector is operated at the fixed temperature of -110°C and has a back-illuminated CCD array with large pixels of 1,100×330 (Princeton Instruments, LN/CCD-1100PB). STM images presented here are all taken in topographic mode at a constant tunneling current. The tunnel bias is defined as the voltage applied to the sample. The practical set up of this photon detection system is shown in figure 6.

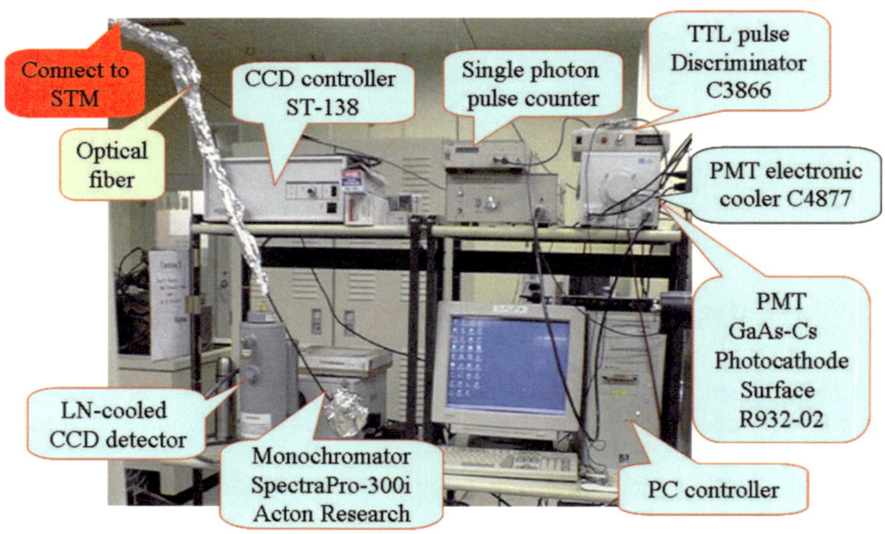


Fig. 6. Photon detection system for the measurements of STML.

### 2.3 Photon detection efficiency of STML

In order to perform a quantitative photon mapping of STML, light-detection efficiency of the photon detection system should be properly estimated. A schematic illustration of transmission and detection of STM-induced photons is shown in figure 7.

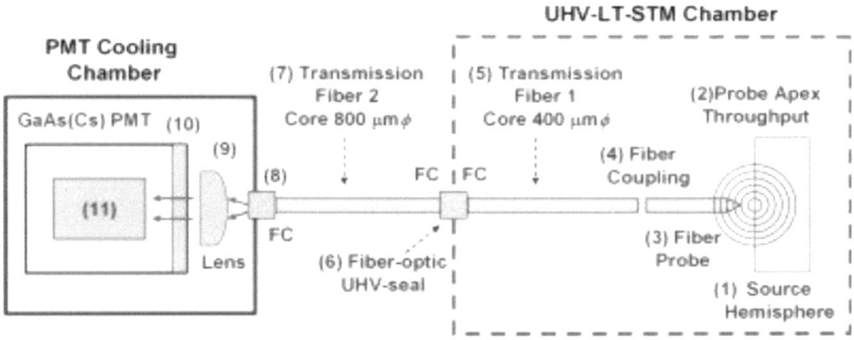


Fig. 7. Schematic illustration of transmission and detection of STML. From the photon source to the photocathode surface of the PMT, photons are transmitted through an optical fiber probe (core: 100 μmϕ), a first transmission fiber (core: 400 μmϕ), a fiber-optic UHV-seal with FC-type receptacles, a second transmission fiber (core: 800 μmϕ), a condenser lens with a FC-receptacle, and the PMT input window.

Firstly, the light source is located just under the probe apex. If the photons are uniformly emitted in space, approximately half of the photons can propagate to the vacuum-side hemisphere (1). Part of light escaped from the surface is collected into the probe with a specific throughput (2). From the probe apex to the photocathode surface (11) of the PMT, photons are transmitted through an optical fiber probe (3: numerical aperture, NA = 0.2, core 100 μmϕ), a fiber coupling (4), a first transmission quartz fiber (5: NA = 0.2, core: 400 μmϕ) in UHV, a fiber-optic UHV-seal with FC-type receptacles (6), a second quartz transmission fiber (7: Na = 0.2, core: 800 μmϕ), a FC-receptacle (8), a condenser lens to make a parallel light (9), and the PMT input window (10). Each optical path or connection has certain reflection and transmission losses. The total throughput of the light transferred from the tip apex to the photocathode of the PMT is approximately ~30% using the previously reported values by Murashita (1999). The quantum efficiency of the PMT (11) is 10–20% over a wide spectral range from UV to NIR (200 to 850 nm). The last task is to estimate collection yield of the fiber probe. In the case of an optical fiber with a flat surface, collection yield can be evaluated as follows. When light emitted from a point source is collected with an optical fiber located at optical far-field, the solid angle of collection is expressed by $2\pi(1-\cos(\sin^{-1}(NA)))$ if the fiber end is sufficiently close to the source (Arafune et al., 2001). Assuming NA = 0.2 for a quartz optical fiber, the solid angle of collection is 0.13 steradian, which corresponds to ~2 % of the hemisphere solid angle ($2\pi$). With fiber-probe collection, on the other hand, luminescence is collected at optical near-field into the apex of the probe located very close to the sample surface. Macroscopic shape of the probe apex is a cone, whose apex angle ($2\theta$) is ~60° for our case, as shown in Figure 8(a). In this case, the collection solid-angle of the emitted photons can be evaluated by the cone solid angle $\Omega = 2\pi(1-\cos\theta)$. Since $\theta$ is ~30°, the collection solid angle is ~0.84 steradian, which corresponds to ~13% of $2\pi$. In order to propagate within an optical fiber by total internal reflection, incident light should be injected within the limited acceptance cone. Although the gradually tapered shape can collimate the injected light into the acceptance cone with half angle $\theta_a$ = 11.5°, a fraction of the photons may propagate out of the fiber due to imperfect total internal reflection. From a microscopic point of view, the probe apex is not a real cone but has a rounded apex with a curvature of ~100 nm as shown in figure 8(b).

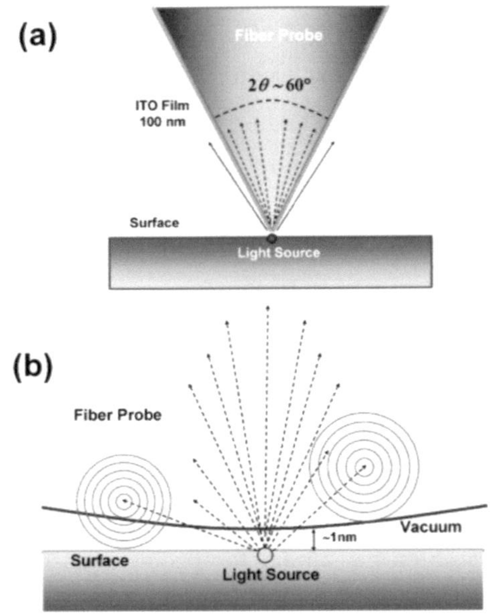


Fig. 8. (a): Schematic macroscopic representation of photon collection of STML with an ITO-coated optical fiber probe. The shape of the fiber probe can be approximated to a circular cone with half angle (θ) of around 30°. (b): Schematic microscopic illustration of near-field interaction between an optical fiber STM probe and a photon source at surface. Most of the emitted photons are injected into the fiber probe and some are transferred to propagating waves by scattering with nanometer-scale objects located in optical near-field.

Since the photons are generated in the nanometer-scale region at the surface and the distance between the probe and the surface is typically within one nanometer, most of the photons emitted to the vacuum are first captured into the probe apex within the optical near-field. As a result, enhanced collection yield of light into the fiber probe is expected due to the near-field detection (Murashita, 1999). Here we have assumed that the near-field enhancement can roughly compensate for the propagation loss caused by the imperfect total internal reflection. Taking into account the transmission efficiency of the ITO thin film (~75%), the collection yield of the optical fiber STM probe can be roughly estimated as ~10 %. Compared with reported throughput values (from ~$10^{-5}$ to ~$10^{-1}$) of NSOM fiber probes with flat apertures (Ohtsu, 1998), the estimated collection yield of the fiber probe corresponds roughly to the best values of NSOM probes. Compared with the conventional far-field detection configurations, the collection yield ~10% of the fiber probe is as good as one of the best values (~9%) of the state-of-the-art lens detection system (Hoffmann et al., 2002). As will be shown later, the estimated collection yield of the fiber probe should be validated experimentally. By multiplying the above estimated values for the light collection efficiencies of the STM-luminescence detection system, total detection efficiency of the emitted photons (200–850 nm) is estimated to be ~0.15–0.3%.

### 2.4 Samples used for the measurement of STML

Samples prepared for the STML experiments were (1) a Zn-doped p-type GaAs (001) wafer with carrier concentration of ~$1 \times 10^{19}$ $cm^{-3}$, (2) a Si-doped n-type GaAs (001) wafer with carrier concentration of ~1 × $10^{18}$ $cm^{-3}$ , (3) an Au(111) film on freshly cleaved mica substrate. In order to produce a clean GaAs (110) surface, the samples were cleaved in ultrahigh vacuum in the sample preparation chamber. Immediately after the cleavage, the samples were transferred to the LT-STM chamber for the STML measurements. The Au(111) film with a thickness ~200 nm on freshly cleaved mica substrate is fabricated by the use of DC magnetron sputtering followed by an in-situ annealing at ~300 °C for 20 min. Cleaning of the Au(111) films were performed by a couple of cycles of Ar ion sputtering (~1 keV) and subsequent annealing at about ~600 K.

## 3. Results and discussions conclusion

### 3.1 High-resolution STM image using an ITO-coated fiber probe

Topographic STM images of a cleaved p-type GaAs(110) surface taken with a conventional tungsten tip at 100 K and with the optical fibre probe are compared in figure 9.

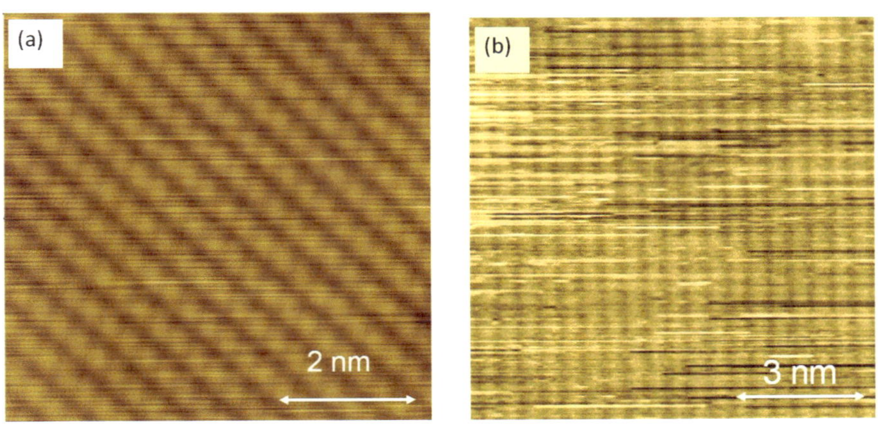


Fig. 9. STM topography images of cleaved GaAs(110) surfaces taken with (a) a W tip at RT (7.6×7.6 $nm^2$, -1.8 V, 0.2 nA); (b) an ITO-coated optical fiber probe at RT (10× 10 $nm^2$, -1.5 V, 0.2 nA).

Both images show the typical features of the relaxed GaAs(110) surface, where buckled dimer rows consisting of Ga atoms and As atoms are running along the ($\bar{1}$10) direction. The overall features of both images are consistent with the previously reported atomic-resolution STM images on GaAs(110) surfaces [Yokoyama & Takiguchi 2001; Ohtsu 1998]. The observed STM imaging quality using the optical fibre probe is a little bit worse than that with a conventional electrochemically sharpened tungsten tip [Ohtsu 1998].

### 3.2 STML from p-GaAs(110) surface using an ITO-coated fiber probe

#### 3.2.1 The bias dependence of STML

Figures 10(a) and (b) show the bias dependencies of the STML intensity on the p-type GaAs(110) surface observed by the ITO-coated optical fiber probe at RT and 80 K, respectively.

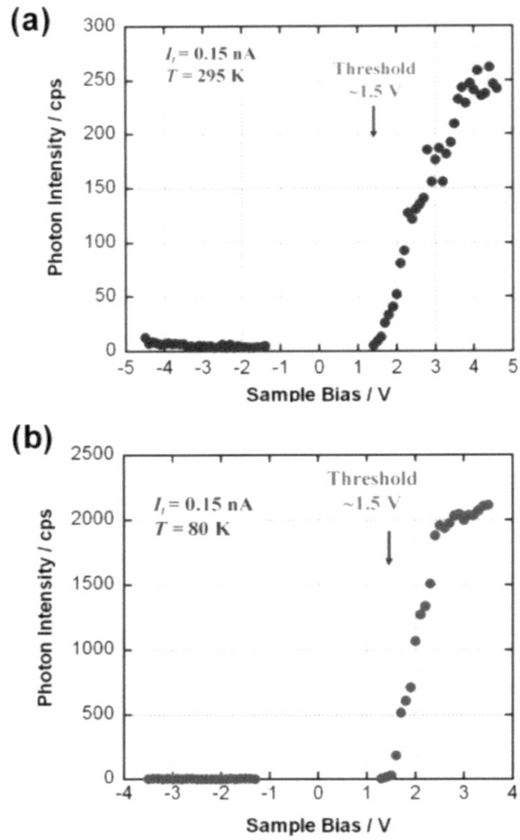


Fig. 10. Bias dependencies of STML intensity on a p-type GaAs(110) surface (Zn-doped) observed by an ITO-coated optical fiber probe at 295 K (a) and at 80 K (b). In both cases, threshold bias voltage of STML is at 1.5 V, which corresponds to a band-gap of GaAs. Photon intensity saturates around the sample bias voltage of ~3.5–5.0 V, where the quantum efficiencies at 295 and 80 K are 1.8 $\times 10^4$ photons/electron and 1.5 $\times 10^3$ photons/electron, respectively. Tunneling current $I_t$ =0.15 nA.

The tunneling current was kept constant at 0.15 nA during the measurements by a feedback control. Much more intense STML has been observed for positive sample bias than for negative polarity. At the positive sample biases, threshold of STML is observable at ~1.5 V and saturates around the sample bias voltages of ~3 V. Typical maximum intensities of STML observed at 295 and 80 K are ~1.7$\times$ $10^3$ and 1.4 $\times 10^4$ cps/nA, respectively. The corresponding maximum quantum efficiencies (the number of emitted photons per an electron) at 295 and 80 K are ~1.8 $\times 10^4$ and ~1.5 $\times 10^3$ photons/electron, respectively. The quantum efficiency value (~1.8 $\times 10^4$) at 295 K shows a good agreement with the previously reported value (~1 $\times$ $10^4$ photons/electron) of STML from p-type GaAs regions of cleaved (110) GaAs/Al-GaAs heterostructures at RT (Abraham et al., 1990). Therefore, the total detection efficiency (about 0.15%) of the single photon counting system is estimated quite reasonably. In the positive polarity, the observed threshold voltage (~1.5 V) of STML corresponds closely to the direct band gap energies (1.424 and 1.493 eV) of GaAs at RT and

80 K (Madelung, 1996). In the case of p-doped semiconductors, the Fermi level is located around the top of the valence band and the major mechanism of the STML is radiative recombination of majority holes in the valence band and the injected electrons in the conduction band. To inject electrons into the conduction band, application of certain bias voltage is required, which almost corresponds to the band gap energy of GaAs in this case. On the contrary, in the negative polarity case, the STML intensity is much lower than the positive polarity case. The possible mechanism is direct recombination of electron-hole pairs created by the impact ionization of injected electrons.

### 3.2.2 The spectroscopy of STML

We performed spectroscopic studies not only at RT but also at low temperatures. Figure 11 is a comparison of the optical spectra of STML from a p-GaAs(110) surface obtained at three different temperatures (295, 85 and 18 K).

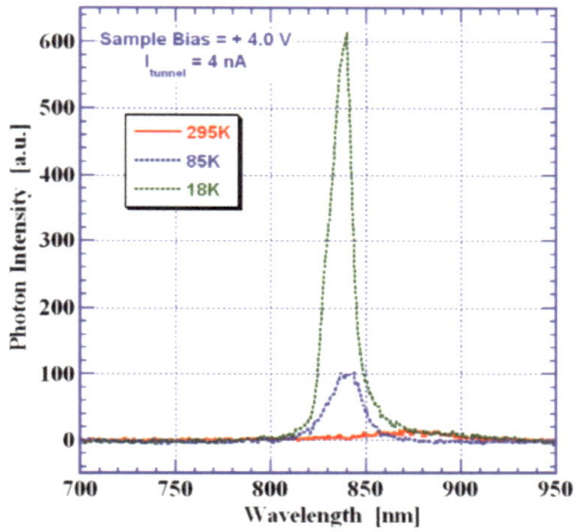


Fig. 11. A plot of near-field optical spectra of STML from a cleaved p-type GaAs(110) surface obtained with an ITO-coated optical fibre probe at 295, 85 and 18 K. The spectra were taken with a fixed sample bias (+4.0 V) andwith a constant tunnelling current of 4 nA.

The spectra were taken with positive sample bias (+4.0 V) and constant tunnelling current of 4 nA. Both a significant increase in the number of emitted photons and a sharpening of the peak shapes are observed with decreasing temperature. Also the peak position shift towards lower wavelength (blue shift) with decreasing temperature. The direct band gap of GaAs at 85 and 18 K can be empirically calculated as 1.507 eV (823 nm) and 1.518 eV (817 nm), respectively [Madelung,1996]. At sufficiently low temperatures, majority carriers (holes) are frozen on impurities [Yu & Cardona, 1996]. Taking into account the binding energy of a Zn acceptor (30.6 meV), the radiation energies of the exciton decay in GaAs at low temperatures can be estimated as 1.487 eV (834 nm) at 18 K and 1.476 eV (840 nm) at 85 K. The observed peak positions are 837 nm at 18 K and 840 nm at 85 K, which show good agreement with the estimated photon energies as above. Thus the dominant mechanism of STM-PEat lowtemperatures ($T$ < 85 K) can be attributed to the radiative recombination of the injected free electrons with the majority holes trapped on the acceptors, which is known as a free-to-bound transition [Yu & Cardona, 1996].

### 3.2.3 The spatial distribution of STML

Spatial distribution of the STML can be visualized by multi-channel recording of STM-related data (topography, current, conductance, etc.) and photon intensity as a function of lateral position of the STM probe. Simultaneously obtained STM topography images and near-filed photon maps are obtained from the cleaved p-type GaAs(110) surface using the ITO coated probe. The experiment has been performed at 80 K keeping the tunneling current constant at 0.15 nA. At low temperatures, major photon emission process in p-GaAs can be attributed to the radiative recombination of injected electrons and major holes. By using this photon mapping technique, we have observed a significant decrease of STML intensity around the heavily-doped Zn accepter atoms located at the sub-surface regions of the GaAs(100) plane. Figures 12 (a) and (b) show an STM topography image and an STML map obtained simultaneously on the p-GaAs(110) surface. The protrusions appearing on the STM topography image correspond to the Zn dopants located at the sub-surface layers (de Raad et al., 2001; Zheng et al., 1994). The apparent height of the protrusions is related to the depth of the Zn dopants in the sub-surface layers.

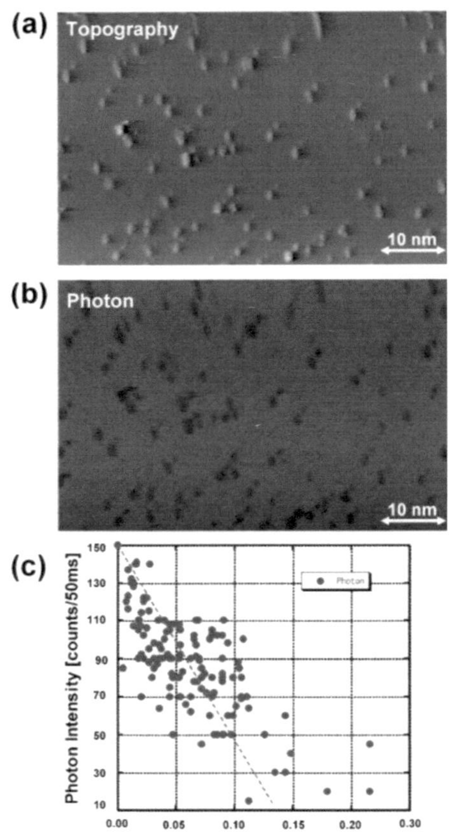


Fig. 12. Simultaneously obtained STM image (a) and a photon map (b) on a p-type GaAs(110) surface at 80 K using an ITO-coated optical fiber probe. ($V_{sample}$ =+2.5 V, $I_t$ =0.15 nA, scan size =59 ×39 nm). (c) : Relation between the apparent height and the STML intensity at the Zn dopant positions.

The protrusion with the highest apparent heights can be assigned to the Zn dopants in the first sub-surface layer. The protruding nature of the Zn dopants in the constant-current imaging is the result of enhanced tunneling near the core of Zn acceptor in GaAs (Zheng et al., 1994).

What is more interesting here is the spotty dark re- gions appearing on the near-field photon map in figure 12 (b). Those spots correspond to the locations of the Zn dopants. It should be noted that the individual dopants has been resolved by near-field STML photon mapping for the first time. Figure 12(c) shows the relation between the STM apparent heights and the STML intensities observed at the core positions of the Zn dopants. It is found that the STML intensity decreases significantly with the increase of the apparent height. Since the optical spectroscopy of the STML on the Zn dopants and on the normal terraces shows the same spectral features in shape and peak position, the basic luminescence mechanism itself should be the same, that is, the radiative recombination of the electron-hole pairs. Thus the decrease of the emitted photons can be attributed to the decrease of the number of injected electrons into the conduction band or to the decrease of the number of majority holes in the valence band.

### 3.2.4 The mechanism of STML

Figure 13 shows a schematic band diagram that illustrates the electron tunnelling process between the ITOcoated fibre probe and a p-GaAs(110) surface with positive sample bias. Since the sample is a heavily doped p-type GaAs(110), the Fermi level lies just above the valence band maximum (VBM). By applying a positive bias voltage *(Vs)* to the sample, the tunnelling electrons can get the injection energy of $eVs$ measured from the VBM [Murashita & Tateno, 2001]. When the injection energy becomes larger than the direct band gap *(Eg)* of the sample *(eVs > Eg)*, it is possible to inject the tunneling electrons into the conduction band. At RT the injected electrons in the conduction band lose their energy through thermalization until they reach the conduction minimum (CBM). Then direct recombination process of the injected electrons and majority carriers (holes) causes the emission of photons.

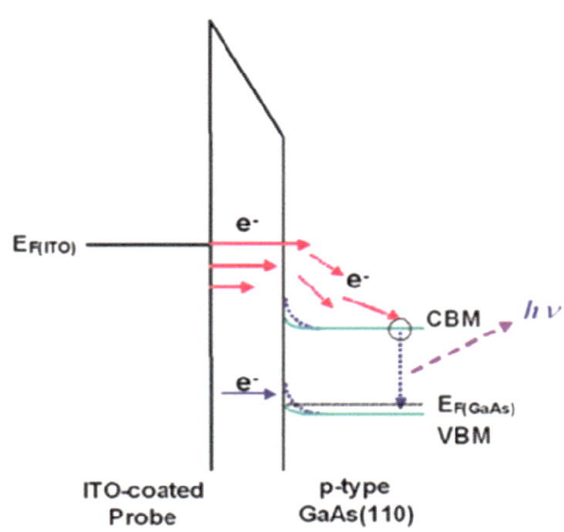


Fig. 13. A schematic energy band diagram illustrating the electron tunnelling process between the ITO-coated fibre probe and a p-type GaAs(110) surface with positive sample bias.

### 3.3 STML from n-GaAs(110) surface using an ITO-coated fiber probe

#### 3.3.1 The bias dependence of STML

Figures 14 (a) and (b) show the tunnel-bias dependence of the STMEL intensities from the surface of n-type GaAs(110) induced by the ITO-coated optical fibre probe at RT and 80 K, respectively.

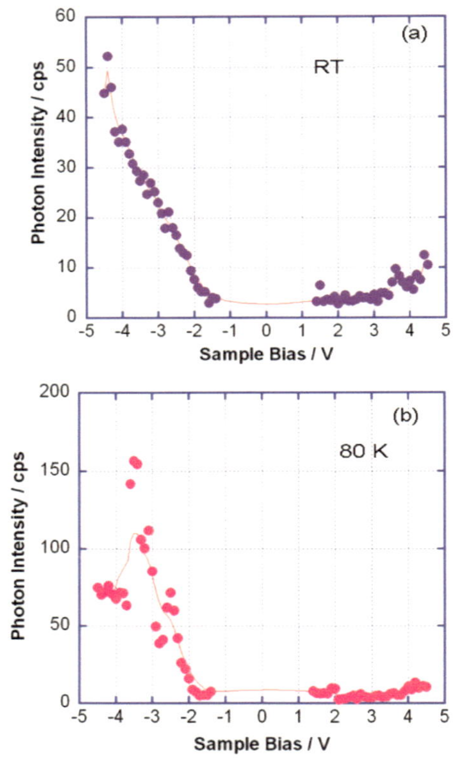


Fig. 14. Bias dependence of STML intensities from surface of n-type GaAs(110) at RT (a) and 80 K (b). At both temperatures, the STML becomes observable when the applied sample bias exceeds −1.5 V and shows higher intensity at negative bias than that at positive bias (tunnelling current = 0.15 nA). The dotted lines are drawn by a polynomial fit to the experimental data, which shows the continuous changes in the STML intensities versus bias.

The tunnelling current is kept at 0.15 nA by a feedback control. At both temperatures, the STML becomes observable when the applied sample bias exceeds −1.5 V in the negative bias region, and saturates around a sample bias of −5.0 to −6.0 V. The direct band gaps of GaAs at RT and 80 K can be calculated empirically as 1.424 and 1.507 eV, respectively [Evoy et al,1999; Alvarado et al, 1991;Rossi et al, 1970], which are in good agreement with the above observed threshold 'turn on' voltage (−1.5 V) shown in figures 14(a) and (b). In the positive bias region, a weak photon emission can be observed when the bias exceeds~+3.5 V. At RT the maximum value of the STML intensity at negative sample bias is about five times higher than that at positive sample bias, as shown in figure 14(a), while at 80 K the maximum value of STML intensity at negative sample bias is about 15 times higher than that at positive sample bias, as shown in figure 14(b). This can be explained by the different recombination process of hole- electron pairs, which will be discussed below, along with the optical spectroscopy data. The large difference in photon emission intensity at RT and 80 K can probably be attributed to the fact that majority carriers (electrons) are frozen on impurities at sufficiently low temperatures [Yu & Cardona,1996], which leads to the increase in radiation efficiency of the exciton decay in GaAs. The maximum quantum efficiency (the number of emitted photons per electron or hole) at RT and 80 K can be calculated as ~$3\times10^{-5}$ and ~$1 \times 10^{-4}$ (photons/electron), respectively.

### 3.3.2 The mechanism of STML

Figure 15 is a schematic energy band diagram illustrating the STML process on the surface of n-type GaAs(110) at both polarities.

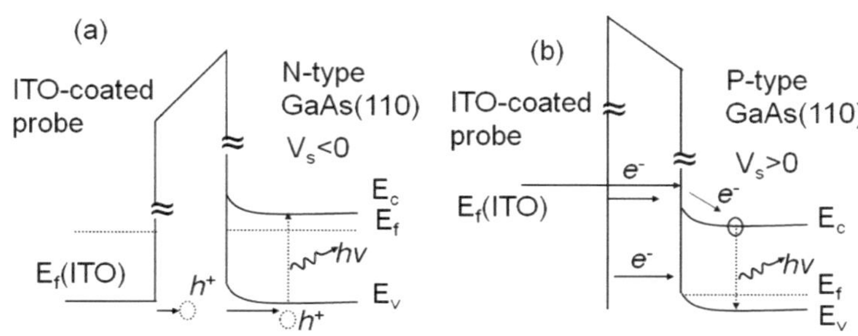


Fig. 15. Schematic energy band diagrams illustrating STML process on the surface of n type GaAs(110), showing two kinds of radiative recombination process of bulk hole–electron pairs: (a) direct recombination of injected holes from the tip with majority carriers (electrons) into n-type GaAs at negative sample bias, and (b) recombination of extra bulk hole–electron pairs created by an impact ionization effect of energetic electrons injected from the tip into the surface of n-type GaAs at positive sample bias.

In figure 15(a), at a sufficient negative sample bias, the tip Fermi level is aligned with the valence band, and holes are injected from the tip side into n-type GaAs (electrons are tunnelled out). Injected at low biases (<−1.5 V in this experiment), carriers are trapped at the surface. Injected at higher biases (>−1.5 V), carriers may reach the bulk and recombine radiatively with readily available majority carriers, emitting light of an energy equal to the bulk band gap energy, $Eg = h\nu$. The STML generated at positive sample bias rarely occurs for n-type semiconductors at positive sample bias. The most likely reason is probably due to an impact-ionization effect, which is shown in figure 15(b). In figure 15(b), under a sufficient positive sample bias (~+3.5 V in this experiment), the electrons injected from the tip side into the conduction band can cause an impact ionization effect that creates extra holes. The radiative recombination of the extra holes with the majority carriers (electrons) generates a relatively weak STML due to its large threshold voltage (~+3.5 V) in generating extra bulk hole–electron pairs. Both above cases will generate carriers which ultimately recombine over the band gap and thus yield a peak at ~1.5 eV.

## 3.4 Development of Au/ITO dual-layer-coated optical fiber probe

In the case of HET-type STM-luminescence such as the case of p-GaAs, the electronic property of the tip materials does not significantly affect the luminescence efficiency. However, in the case of TIP-type STM-luminescence, proper selection of the tip material is very important to enhance the quantum efficiency since the electronic properties of both tip and surface contribute the TIP-type STM-luminescence (Johansson, 1998; Johansson et al., 1990). Furthermore, the near-field photon detection using an ITO-coated fiber probe has a higher and more stable photon detection efficiency of STML than that of far-field photon detection. However, the electron injection efficiency, STM image resolution and tunneling stability of the semiconductor ITO-coated fiber probe are generally lower than those of metal tip. Because the ITO layers for the commercially available ITO-coated fiber probes are normally very crystalline or smoothly coated on the fiber probe with a symmetrical geometry and sharp tip apex, they can easily crash with low durability and change the tunneling condition during scanning process. Therefore, it is necessary to coat a thin metal layer such as Au, Ag to improve the performance of ITO-coated fiber probe. In this research, the thin Au/ITO dual layer is deposited by DC magnetron sputtering on the surface of the tip apex of the ITO-coated fiber probe to improve its tip geometry, imaging quality and electron injection efficiency. Sputtering of the ITO layer was performed using a ceramic high-density ITO target (5 wt% $SnO_2$) at ~$1 \times 10^{-1}$ Pa in a pure $Ar/O_2$ gas mixture with the distance between the target and tip holder fixed at 10 cm. The DC power was maintained at 100 W for all the deposition. The base pressure of the chamber was ~$1 \times 10^{-4}$ Pa. Sputtering of the Au layer was performed using an Au metal target (purity: 99.95%) in a pure Ar gas (~$1 \times 10^{-1}$ Pa) environment after depositing the ITO layer.

Figure 16(a) shows a SEM image of the fabricated Au/ITO dual-layer-coated fiber probe. The deposited Au/ITO dual layer around the tip apex looks smooth and the coated tip apex looks transparent by naked eye. The outer Au layer is confirmed by the corresponding AES Au map and spectrum analysis, as shown in figures 16(b) and (c), respectively. The Au map in figure 16(b) matches very well with the SEM image in figure 16(a) and shows that the Au layer uniformly covers the tip apex. The Au NOO, $M_5NN$, $M_4NN$ peaks in figure 16(c) appeared constantly when randomly examined at points around the tip apex region; this also indicates the uniform coverage of the Au layer.

The thickness of the Au layer is a critical parameter to obtaining a better yield of high electron injection efficiency and near-field photon detection efficiency. We measured the optical transmission with UV-2450 spectrophotometers, Shimadzu, Japan and the electric resistivity by the four-probe method for Au films of various thickness, uniformly deposited on a glass substrate by sputtering, to determine the appropriate thickness of the Au layer to realize a better combination of conductivity and transparency. The measured electric resistivity and optical transmission values versus film thickness are shown in figures 17(a) and (b), respectively.

Figure 17(a) illustrates a spread of conductivity at low thickness values (<25 nm). This is because the resistivity of metal deposits at low thickness is complicated by the dynamics of film nanostructure [Lidzey et al, 1997]. The electric resistivity of the Au films is almost constant with variation of film thickness up to 1000 nm, whereas the optical transmission decreases monotonously with the film thickness. The appropriate thickness value of the Au film is determined as 13 nm, where both a resistivity ~$1.5 \times 10^{-5}$ $\Omega$cm and a high optical transmission ~48% are achieved. This conductive value is enough for tunneling and getting a better yield of high electron injection efficiency in the STM tunneling junction. Based on the above results, the optical collection efficiency of the fabricated Au/ITO dual-layer-coated fiber probe is estimated as ~10%, which is similar to that of the standard optical tip [Tsuruoka et al, 1998].

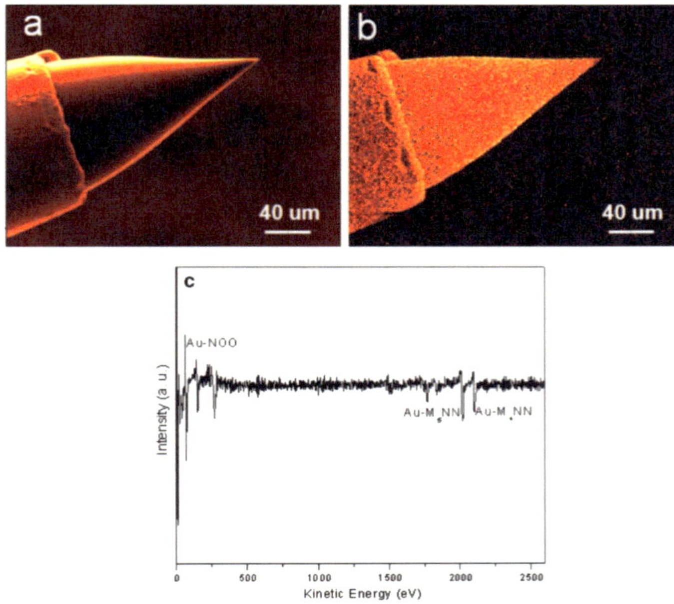


Fig. 16. (a) SEM image of the fabricated Au/ITO-coated fiber probe; (b) corresponding Au Auger map which shows a uniform coverage of the Au layer around the tip apex; (c) typical AES spectrum randomly examined at the points around the tip apex region.

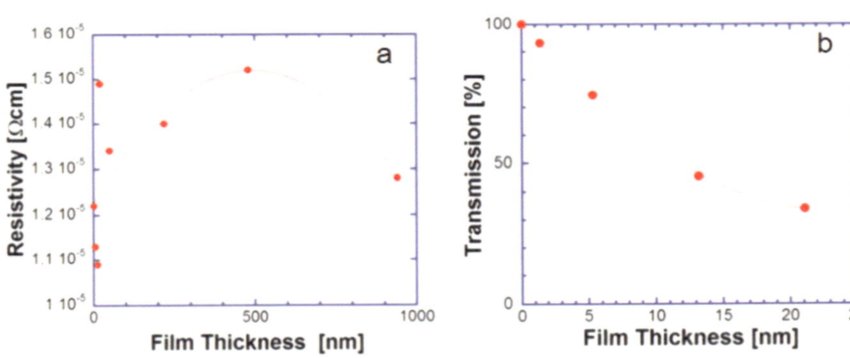


Fig. 17. (a) Resistivity versus Au film thickness measured by the four-probe technique; (b) optical transmission versus Au film thickness. A best combination point of resistivity with transmission is determined at ~13 nm in thickness.

High-quality STM imaging with a herringbone reconstruction on a Au(111) surface on mica is clearly observed by using this Au(13 nm)/ITO(50 nm) dual-layer-coated fiber probe, as shown in figure 18.

On the contrary, the herringbone reconstruction is difficult to observe using the pristine ITO-coated fiber prober without the Au/ITO dual coating layer.

For comparison, the STML is investigated on the surface of Au(111)/mica using both the Au/ITO dual-layer-coated fiber probe and the pristine ITO-coated fiber probe at RT. Figures 19(a) and (b) are the measured photon intensity versus the sample bias on Au(111)/mica by use of these two kinds of fiber probes, respectively.

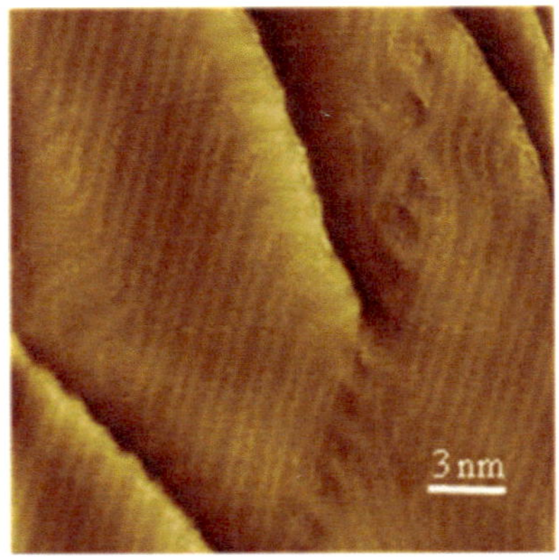


Fig. 18. STM image of a Au(111) surface using the Au (13 nm in thickness)/ITO-coated fiber probe, which shows a clear herringbone reconstruction structure. (Sample bias = −1.7 V, tunneling current = 0.15 nA, 20 nm × 20 nm).

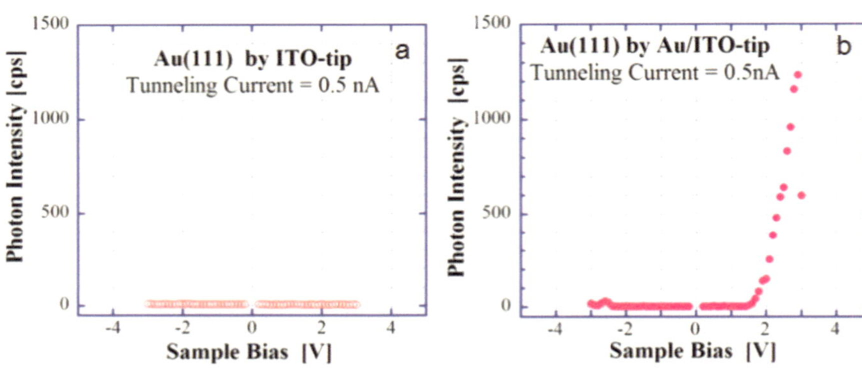


Fig. 19. Bias dependence of STML intensity from the Au(111) surface at RT (tunneling current = 0.5 nA): (a) using the ITO-coated fiber probe; (b) using the Au/ITO-coated fiber probe. The STML is generated only in the case of (b), with a maximum quantum efficiency ~7 × 10−4 photon/electron at a sample bias of +3 V, whereas the STML is too weak to be observed in case (a).

The surface plasmon-mediated emission is hard to generate without Au coating (see figure 19(a)) whereas the maximum quantum efficiency of ~$7 \times 10^{-4}$ photons per electron was obtained at a sample bias of ~2.5 V using the Au/ITO dual-layer-coated fiber probe. The Au layer shows an effective enhancement of the intensity of Plasmon mediated STML from the surface of Au(111)/mica at positive sample bias. Because the plasmon-mediated STML is generated by inelastic tunneling [Hoshino & Yamamoto 2003], it is reasonable to believe that such a kind of enhancement is attributed to the improvement of the outside Au layer on the sharpness and electron injection efficiency of the ITO-coated fiber probe. It is worth noting that the plasmon-mediated STML at negative sample bias is usually very weak (as can be seen from the weak emission peak at ~−2.8 V in figure 19(b)) compared to that at positive sample bias. The main reason is probably attributed to the symmetry of the tip apex and the difference of crystallinity between the Au layer on the surface of the fiber probe and the Au(111) on mica substrate.

## 4. Conclusion

An ultrahigh-vacuum low-temperature scanning tunneling microscope with a conductive and optically transparent probe was employed to perform detailed analyses of STML from a cleaved Zn-doped p-type GaAs(110) surface and Si-doped n-type GaAs(110), respectively. Photons generated in a nanometre-scale area just under the STM probe can be collected effectively within the optical near-field region through the apex of the optical fibre tip. We observed a strong photon emission at positive sample biases where electrons are injected into the p-GaAs(110) surface. While under both bias polarities, we observed nanoscale STML from the surface of n-type GaAs(110) at RT and 80 K. The developed Au/ITO dual-layer-coated optical fiber probe for STML exhibits high performance in further improving the quality, stability and reproducibility of STM imaging and STML, and enhancing the electron injection efficiency from the tip apex to generate high-intensity plasmon-mediated STML. This novel conductive and transparent optical fiber probe provides an effective way to construct a multifunctional STM and for the fabrication of highly efficient nanoscale light sources.

## 5. Acknowledgment

We would like to thank for the help of Ms. N. Niori in the experiment work, and the supports from National Natural Science Foundation of China (21173041 from National Natural Science Foundation of China (21173041), National Basic Research Program of China (973 plan, 2012CB619401), Open project of Jiangsu Key Laboratory for design and Manufacture of Micro-Nano Biomedical Instruments, China(JSNBI201101) and Pairs of Talent Foundation of Jiangsu, China in writing this chapter.

## 6. References

Abraham DL, Veider A, Scho¨nenberger Ch, Meier HP, Arent DJ, Alvarado SF. (1990). Luminescence in scanning tunneling microscopy on III-V nanostructures. *Appl Phys Lett* No. 56, pp.(1564–1566)

Arafune R, Sakamoto K, Meguro K, Satoh M, Arai A, Ushioda S. (2001). Multiple-fiber collection system for scanning tunneling microscope light emission spectroscopy. *Jpn J Appl Phys* No. 40, pp.(5450–5453)

Alvarado S F, Renaud P, Abraham D L, Schonenberger C, Arent D J, Meier H P. (1991). *J. Vac. Sci. Technol.* B No.9, pp.( 409-413)

Binning G, H. Rohrer, Ch. Gerber, E. Weibel. (1982). *Physics. Rev. Lett.* No. 49, pp.(57-60)

Binning G, H. Rohrer, Ch. Gerber, E. Weibel. (1983). *Physics. Rev. Lett.* No. 50, pp.(120-123)

Binning G, C. F. Quate, Ch. Gerber. (1986). *Physics. Rev. Lett.* No. 56, pp.(930-933)

Berndt R, (1998). Scanning probe microscopy and spectroscopy. New York: Springer. pp. (97–134)

Berndt R, Gimzewski JK, Johansson P. (1993a). Electromagnetic-interactions of metallic objects in nanometer proximity. *Phys Rev Lett* No. 71, pp.(3493–3496).

Berndt R, Gaisch R, Schneider WD, Gimzewski JK, Reihl B, Schlittler RR, Tschudy M. (1995). Atomic-resolution in photon-emission induced by a scanning tunneling microprobe. *Phys Rev Lett* No. 74, pp.(102–105)

Berndt R, Gimzewski JK. (1993). Photon emission in scanning tunneling microscopy: Interpretation of photon maps of metallic systems. *Phys Rev B* No. 48, pp.(4746–4754)

Berndt R, Gaisch R, Schneider WD, Gimzewski JK, Reihl B, Schlittler RR, Tschudy M. (1993b). Photon-emission from adsorbed $C_{60}$ molecules with subnanometer lateral resolution. *Appl Phys A* No.57, pp.(513–516)

Daisuke Fujita, Keiko Onishi and Noriko Niori. (2004). Light emission induced bytunnelling electrons from a p-type GaAs(110) surface observed at near-field by a conductive optical fibre probe, *Nanotechnology* No. 15 , pp.(S355-S361)

de Raad G J, Bruls D M, Koenraad P M, Wolter J H. (2001). *Phys. Rev.* B, No. 64, pp.( 075314-075320)

Dos Sandos DA, Bredas JL, Logdlund M, Salaneck WR. (1999). Electroluminescence in conjugated polymers. *Nature* No. 397, pp(121–128)

Downes A, Welland ME. (1998). Photon emission from Si(111)-(7x7) induced by scanning tunneling microscopy: atomic scale and material contrast. *Phys Rev Lett* No.81, pp.(1857–1860)

Downes A, Taylor ME, Welland ME. (1998). Two-sphere model of photon emission from the scanning tunneling microscope. *Phys Rev B* No. 57, pp.(6706–6714)

Dong Z. C, Guo X. L, A. S. Trifonov, P. Dorozhkin, K. Miki, K. Kimura, S. Yokoyama, S. Mashiko. (2004). Vibrationally Resolved Fluorescence from Organic Molecules near Metal Surfaces Induced by a Scanning Tunneling Microscope. *Phys. Rev. Lett.*, No.92, pp. (0868011-0868014)

Dong Z-C, Kar A, Zou Z-Q, Ohgi T, Dorozhkin P, Fujita D, Yokoyama S, Terui T, Yamada T, Kamikado T, Zhou M, Mashiko S, Okamoto T. (2002). Light emission from porphyrin molecules induced by a scanning tunneling microscope. *Jpn J Appl Phys* No. 41, pp.(4898–4902)

Evoy S, Harnett CK, Craighead HG, Keller S, Mishra UK, DenBaars SP. (1999). Low-temperature scanning tunneling microscope-induced luminescence of an InGaN/GaN multiquantum well. *Appl Phys Lett* No. 74, pp.(1457–1459)

Evoy S, Craighead H G, Keller S, Mishra U K , DenBaars S P. (1999). *J. Vac. Sci. Technol.* B No. 17, pp.( 29-33)

Fujita D, Ohgi T, Deng W, Ishige K, Okamoto T, Yokoyama S, Kamikado T, Mashiko S. (2001). Photon emission induced by tunneling electrons from a Cu(100) surface covered with porphyrin molecules. *Surf Sci* No.493, pp.(702–707)

Friend RH, Gymer RW, Holmes AB, Burroughes JH, Marks RN, Taliani C, Bradley DDC, Garcia-Parajo M, Cambril E, Chen Y. (1994). Simultaneous scanning tunneling microscope and collection mode scanning near-field optical microscope using gold coated optical fiber. *Appl Phys Lett* No.65, pp.(1498–1500)

Guo X. L, D. Fujita, N. Niori, K. Sagisaka, and K. Onishi. (2007). Nanoscal electroluminescence from n-type GaAs(110) in tunnel junction, *Nanotechnology* No.18, pp.(195201-195204)

Guo X. L, Dong Z. C, A. S. Trifonov, S. Mashiko, T. Okamoto. (2003). Tunneling-induced molecular luminescence from nanoscale organic molecular layers on metal substrate *Appl. Phys. Lett.*, No.84, pp. (969-972)

Guo X. L, Dong Z. C, A. S. Trifonov, S. Mashiko, T. Okamoto. (2003). Role of molecules in tunneling current induced photon emission from the surface of a perinone derivative molecular monolayer on Au(100) , *Phys. Rev. B.* No. 68, pp.(113403-113405)

Guo X. L, Dong Z. C., A. S. Trifonov, K. Miki, Wakayama, D. Fujita, K. Kimura, S. Yokoyama, S. Mashiko. (2004). Nanoscale organic electroluminescence from tunnel junctions, *Phys. Rev. B*, No. 70, pp. (233204-233207)

Guo X. L, Fujita D, Niori N, Sagisaka K and Onishi N. (2007). Scanning tunneling Microscopy luminescence from nanoscale surface of GaAs(110), *Surf. Sci.* No.601, pp. (5280-5284)

Hoffmann G, Kliewer J, Berndt R. (2001). Luminescence from metallic wells in a scanning tunneling microscope. *Phys Rev Lett* No.87, pp.(176803–176806)

Hoffmann G, Kro¨ger J, Berndt R. (2002). Color imaging with a low temperature scanning tunneling microscope. *Rev Sci Instrum* No. 73, pp.(305–309)

Hoshino M and Yamamoto N. (2003). *Mater. Res. Soc. Symp. Proc.* No.738, pp.( 149-153)

Johansson P. (1998). Light emission from a scanning tunneling microscope: Fully retarded calculation. *Phys Rev B* No.58, pp.(10823–10834)

Johansson P, Monreal R, Apell P. (1990). Theory for light-emission from a scanning tunneling microscope. *Phys Rev B* No. 42, pp.(9210–9213)

Lidzey DG, Alvarado SF, Seidler PF, Bleyer A, Bradley DDC. 1997. Electroluminescence from a soluble poly(p-phenylenevinylene) derivative generated using a scanning tunneling microscope. *Appl Phys Lett* No.71, pp.(2008–2010)

Lidzey DG, Bradley DDC, Skolnick MS, Virgili T, Walker S, Whittaker DM. 1998. Strong exciton-photon coupling in an organic semiconductor microcavity. *Nature* No. 395, pp.(53–55)

Madelung O. (1996). Semiconductors: basic data. Berlin: Springer. p.101.

Murashita T. (1999). Optical system for tunneling-electron luminescence spectro/ microscopes with conductive-transparent tips in ultrahigh vacuums. *J Vac Sci Technol B* No.17, pp.(22–28)

Murashita T, Tateno K. (2001). Direct measurement of sub-10 nm-level lateral distribution in tunneling-electron luminescence intensity on a cross-sectional 50-nm-thick AlAs layer by using a conductive transparent tip. *Appl Phys Lett* No.78, pp.(3995–3997)

Murashita T, Tateno K. (2001). *Appl. Phys. Lett.* No. 78, pp.(3995–4002)

Nishitani R, Umeno T, Kasuya A. (1998). A study of metal films using a spectrally resolved photon map obtained by spectrum mapping measurements of STM-induced light. *Appl Phys A* No.66, pp.(S139–S143)

Ohtsu M. 1998. Near-field nano/atom optics and technology. Tokyo: Springer. p 71.

Rossi J A, Wolfe C M , Dimmock J O. (1970). *Phys. Rev. Lett.* No.25, pp.( 1614-1618)

Sasaki S, Murashita T. (1999). Crossover energy where cathodoluminescence becomes comparable to electroluminescence in p-type GaAs observed by injecting tunneling electrons. *Jpn J Appl Phys* No. 38, pp.(L4–L6)

Sergei Kalinin, Alexei Gruverman, (2007). Scanning Probe Microscopy, Volume I, Springer,ISBN-10 : 0-387-28667-5, Springer Science+ Business Media, LLC, 233 Spring Street, New York, NY 10013, USA

Suzuki Y, Minoda H, Yamamoto N. (1999). STM light emission from Ag/Si(111). *Surf Sci* No. 438, pp.(297–304)

Thirstrup C, Sakurai M, Stokbro K, Aono M. (1999). Visible light emission from atomic scale patterns fabricated by the scanning tunneling microscope. *Phys Rev Lett* No. 82, pp.(1241–1244)

Tsuruoka T, Ohizumi Y, Ushioda S. (2003). Light intensity imaging of single InAs quantum dots using scanning tunneling microscope. *Appl Phys Lett* No.82, pp.(3257–3259

Tsuruoka T, Ohizumi Y, Ushioda S, Ohno U and Ohno H. (1998). *Appl. Phys. Lett.* No.73, pp.( 1544-1547)

Uehara Y, Fujita T, Ushioda S. (1999). Scanning tunneling microscope light emission spectra of Au(110)-(2 ×1) with atomic spatial resolution. *Phys Rev Lett* No. 83, pp.(2445–2448).

Yokoyama T, Takiguchi Y. (2001). *Surf. Sci.* No. 482–485, pp.( 1163–1168)

Yu P Y and CardonaM. (1996). *Fundamentals of Semiconductors* (Berlin: Springer) p 339

Zheng ZF, Salmeron MB, Weber ER. (1994). Empty state and filled state image of ZnGa acceptor in GaAs studied by scanning-tunneling- microscopy. *Appl Phys Lett* No.64, pp.(1836–1838)

13

# Applications of Optical Fibers to Spectroscopy: Detection of High Explosives and Other Threat Chemicals

Natalie Gaensbauer[1], Madeline Wrable-Rose[1], Gabriel Nieves-Colón[2], Migdalia Hidalgo-Santiago[2], Michael Ramírez[2], William Ortiz[2], Oliva M. Primera-Pedrozo[2], Yahn C. Pacheco-Londoño[2], Leonardo C. Pacheco-Londoño[2] and Samuel P. Hernandez-Rivera[2]

*ALERT DHS Center of Excellence for Explosives*
*[1]Northeastern University, Boston, MA,*
*[2]Department of Chemistry, University of Puerto Rico, Mayagüez,*
*[1]USA*
*[2]Puerto Rico*

## 1. Introduction

In the last decades, research worldwide in areas of national defense and homeland security has focused on the search for new technologies for the detection and characterization of explosives, chemical and biological threats (CBT) and narcotics in different environments and scenarios. In the case of *in situ* field detection of explosives, the technique generally used is ion mobility spectrometry (IMS). The major advantages of IMS are its sensitivity in the picogram range, its continuous real time monitoring capability, reasonable price due to instrumental simplicity and ease of automation (Steinfeld and Wormhoudt, 1998). However, in general terms, IMS has a limited linear range and cannot be used for quantitative analysis. Also, analyte and background responses exhibit variations that occur with different reactive gas compositions and sample compositions. Moreover, it is relatively easy to overload an IMS. Therefore, sample mass and size must be limited.

Optical spectroscopy is routinely used for the measurement of many different species at trace levels. However, optical spectroscopy has not been extensively applied in the highly energetic materials (HEM) detection arena. This is due, in part, to physical constraints such as low vapor pressure, limited sample size, concealment, interferences and in part, to the spectroscopic characteristics of the compounds themselves (Steinfeld and Wormhoudt, 1998). Spectroscopic techniques have the potential to provide the best selectivity for explosives and offer an information-rich fingerprint that allows for near unambiguous identification. In the 1990's, direct detection by infrared absorption spectroscopy was not possible because of the limited sensitivity of this method. In addition, the test materials had to be placed physically within the spectrometer's sample compartment for measurement. The production of optical fibers cables (OFCs) that transmit in the mid-IR range made possible the development of a range of spectroscopic probes for *in situ* analysis (Melling and

Shelley, 2002); (Mehta et al., 2002). FTIR spectroscopy can now be effectively used outside the confinement of the sample compartment, making it available for field work. The establishment of RAIRS as an active area in the trace detection of chemicals is the result of a combination of high extinction coefficients in the MIR and the optical advantages of working at the grazing angle (Umemura, 2002). By combining a grazing angle head or probe (GAP) with a OFC that transmits in the MIR, the instrumentation becomes a platform for developing methodologies for real time, remotely sensed, *in situ* analysis (Mehta et al, 2002). GAP-FTIR operating in RAIRS mode with a sensing probe coupled to fiber optics has been used for the detection of active pharmaceutical ingredients (APIs) on metals (Mehta et al, 2003) and on glass surfaces (Hamilton et al., 2005). In these reports, combined teams from academia and the private sector demonstrated that the technique is an excellent alternative for the validation of cleanliness of metal walls of pharmaceutical reactors. Low limits of detection (LOD) ranging from 10 to 50 ng/cm$^2$ of single API were achieved (Mehta et al, 2003). In addition, the methodology was applied to quantify APIs in mixtures and was demonstrated to not depend solely on the reflective properties of metallic surfaces, making detection on other surfaces such as glass and plastic surfaces viable (Person et al., 2007). These characteristics of the methodology offer clear advantages over the traditional, time consuming, laborious and operator driven, swab-based HPLC method of analysis routinely used by the pharmaceutical industry in the validation of the cleanliness of batch reactors and other vessels and pipes.

Over the last ten years our group has dedicated many of its research efforts to the development and use of coupling optical fibers (OF) to spectroscopic instrumentation for applications in CBT and explosives detection: from near field close to the sample to far field at sample to detector distances over 100 m. From transmitting the excitation source energy to collecting the sample emitted/scattered energy from the ultraviolet (UV) to the mid-infrared (MIR) regions of the electromagnetic spectrum, optical fibers have led the way to doing experiments outside the traditional sample compartments, thus taking the experiments to the sample rather than the sample to the instrument. In this chapter, three applications in which optical fibers play the central role of interfacing the samples with the sensors are reviewed. First, MIR transmitting optical fibers used to couple an FTIR interferometer to a grazing angle probe and measure the reflectance/absorption IR (RAIS) spectra of highly energetic materials (HEM) in neat form, in mixtures of HEM and in quantification and discrimination experiments. In the second application, bundles of optical fibers and single strands of fibers were use to guide laser beams to act as Raman excitation sources. Another set of fibers were used to collect the scattered Raman signal to the guide it to the entrance slit/plane of a high throughput optical spectrometer. In the third application discussed, optical fibers were used to couple visible and UV reflective telescopes to imaging spectrometers in order to implement a remote Raman system for standoff detection to 140 m of hazardous chemicals.

## 2. Optical fiber coupled grazing angle probe reflectance-absorption infrared spectroscopy

The analysis of traces of solid phase highly energetic materials (HEM) deposited onto substrates is essential in homeland security and national defense applications (Gillen et al, 2004). Moreover, the preparation of HEM standards is critical during system detection design for field applications and post-delivery instrument operation validation (MacCrehan, 2004). The practical development of high-quality solid-on-solid standards depends on the material type, on the physical properties of the deposited analyte, such as its vapor pressure and solubility in an appropriate solvent, on HEM adhesion forces to the test surface, on temperature, on the solvents used for mass transfer and on the sample preparation method. Several analytical techniques have been used to ensure reliable and reproducible *in situ* trace HEM detection. Among these techniques, ion mobility spectrometry (IMS) (Hernandez et at., 2004; Hallowell, 2001; Ewing et al., 2001; Phares et al., 2000), secondary ion mass spectrometry: SIMS (Gillen et al., 2004) and optical fiber coupled grazing angle probe/Fourier transform reflection absorption infrared spectroscopy: OFC-GAP/FT-RAIRS (Pacheco-Londoño et al., 2007; Primera-Pedrozo el at., 2008; Primera-Pedrozo et al., 2009) have been used the most frequently. In the first two techniques, chemical analysis takes place after sample collection. This is typically used for handbags and carry-on luggage detection in airports (Ewing et al., 2001). In OFC-RAIRS, the mid IR (MIR) beam is directly focused on the sample area, and the measurements are done *in situ* without sample consumption or destruction (Primera-Pedrozo el at., 2008). Remote detection using remote infrared spectroscopy (RIRS) and remote Raman spectroscopy (RRS) systems have become important for detecting highly energetic materials as well as their formulations and mixtures on surfaces (Pacheco-Londoño et al., 2009). Thus, in the development of new techniques and instrumentation for trace detection, the incorporation of methodologies for production of high-quality samples and standards is required. For proper functioning of sensors, reliable standards are essential. Calibration curves can then be generated to quantify known analytes, sensors can be trained with standards to detect samples of unknown identity with chemical/physical properties similar to compounds in the data base and thus detection methods can be significantly enhanced. The applications of these specially prepared samples include use in experiments that require fine control of the distribution of loadings of analytes on surfaces. Another important application is in the establishment of reliable standards that may support an instrument response validation program.

The sample analysis setup is schematically presented in Fig. 1. A Remspec MIR grazing angle probe was used to collect the spectra. The grazing-angle head uses carefully aligned mirrors to deliver the beam to the sample surface at the grazing angle (approximately 80° from normal), to collect the reflected beam, and to return it to a detector (liquid nitrogen cooled MCT detector).

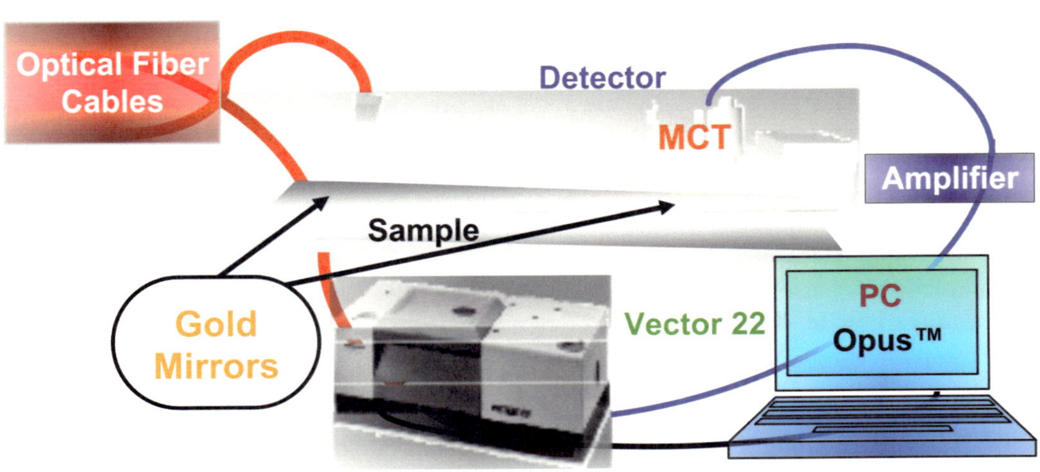


Fig. 1. Experimental setup for Fiber Optic Coupled-Grazing Angle Probe FTIR.

The signal is delivered from the spectrometer to the head by IR transmitting optical fiber cables. The grazing angle accessory is connected to the external beam port of the Bruker Vector 22 spectrometer by a 1.5 m, 19–fiber chalcogenide glass optical bundle in the As-Se-Te system, which transmits throughout the MIR with the exception of a strong H-Se absorbance band at 2200 $cm^{-1}$. The IR footprint produced by the grazing angle probe is elliptical with the intensity decaying from the middle towards the edges. The specially configured head illuminates a large spot on the sample surface. The spot is an ellipse 1 inch by six inches that is defined by a Gaussian distribution with a center spot about 1/8 inch by an inch. Electric signal from the MCT is delivered to the FTIR using an amplifier.

Sample smearing (Soto-Feliciano et al., 2006; Primera-Pedrozo, 2007; Primera-Pedrozo el at., 2008; Primera-Pedrozo et al., 2009) aerosol spray deposition, thermal inkjet deposition (Primera-Pedrozo et al, 2005; Wrable et al. 2010), deposition by rotary evaporation (MacCrehan, 2009), and direct transfer deposition using micropipettes (Primera-Pedrozo, 2005) have been reported previously as methods for sample and standards preparation on metals, silica, glass and plastic surfaces. Also, trinitrotoluene (TNT) was deposited on cloth and planar surfaces with post removal by air jets. The results proved that the particles that were less efficiently removed from the polycarbonate surface (Fletcher et al, 2008). Stickiness constitutes another method for producing plastic explosives deposits (Heimerl, 1999). A standard deposit on fibrous substrates method was developed by Phares and collaborators (Phares et al, 2000), in which the explosive suspension is first transferred to the Teflon with post drying. Then, the dried explosive deposit is pressed onto the tested substrate and mechanically transferring the sample without the associated liquid.

Piezoelectric and bubble-jet printing technologies have been used for microfabrication of biological samples with a rapid preparation of a large number of printed arrays at extremely low cost (Gillen et al. 2004). Both technologies have enabled deposition of femtomoles of analytes at an inkjet printer dot. These results make sample printing technologies excellent means for trace hazardous chemicals samples and standard manufacture. Thermal inkjet (TIJ) printers offer the advantages of being easy to wash and load and transfer of analytes using small amounts of sample (~ 20 μL). However, TIJ would be inappropriate for delivery of proteins on account of possible denaturation caused by exposure to a high temperature gradient. Inkjet printers can be used to fabricate microarrays containing biological materials (proteins, enzymes, etc). Thermal inkjet printers offer advantages compared to piezoelectric ones because they are easier to wash and reload and have the capability of depositing small volume samples. However, piezoelectric inkjet are preferred for protein and other biological molecules deposition because of heat stimulated denaturalization problems (Allain et al., 2004). In TIJ printing, a thin film resistor superheats less than 0.5% of the fluid in the chamber to form a gas bubble. This bubble rapidly expands (in less than 10 μs) and forces a drop to be ejected through a very small orifice (Beeson and Skip, 2000). The loading concentration of the sample on the surface can be controlled by varying parameters, such as the number of passes, the dispensing frequency, the applied energy and the pen architecture. Precise delivery of a known number of droplets with known mass and concentration is achieved. Also, by proper selection of the dispensing solution concentration, only one solution is required to dispense a broad range of surface concentrations, avoiding the preparation of several dilutions that could lead to analytical errors. In previous work, it was demonstrated that TIJ technology can be used for depositing 2,4,6-trinitrotoluene (TNT) and cyclotrimethylenetrinitramine (RDX) onto stainless steel surfaces, and TIJ-based deposition has been shown to be a superior method for further detection using OFC-GAP/FT-RAIRS (Primera-Pedrozo et al., 2005). Visualization of deposits, using optical microscopy images, showed that there was a close dependence between the number of passes and the crystallization of the explosives on the surface. Even at a low (1.25 μg/cm$^2$) loading concentration, the surface was practically covered by crystals. Below this surface concentration, the formation of a metastable form of TNT was observed (Primera-Pedrozo et al., 2005).

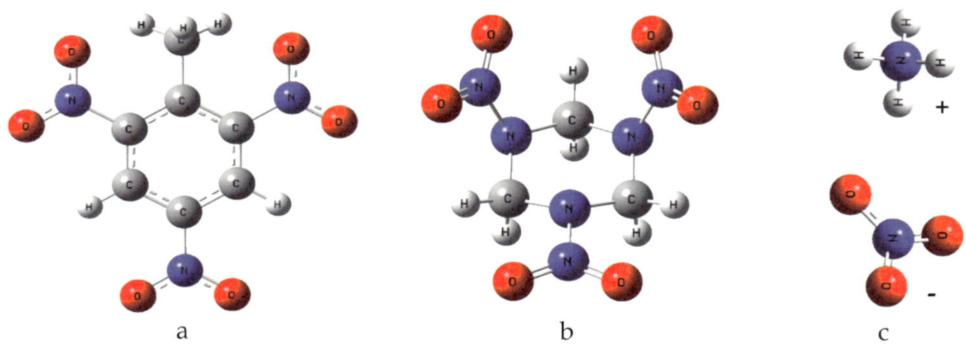


Fig. 2. Chemical structures of some of the highly energetic materials studied in OGC-GAP-RAIS experiments. (a) TNT; (b) RDX; (c) AN.

As demonstrated by Thundat and co-workers (Van Neste et al., 2009), the residence time of an analyte on a substrate depends on the surface energies, on the analyte-surface interaction forces and on the physical properties of the analyte deposited on the surface. Highly energetic materials, such RDX and PETN, take days to months to desorb from the surface. Thus, the physical properties of the deposited analytes and the substrate characteristics, such as vapor pressure and analyte-surface interactions, influence the preparation schemes of samples and standards. This report presents a broad view, from start to completion, of two schemes that can be used to establish efficient, reproducible and representative methodologies for preparing samples and standards of common contamination compounds. In this research, the following three HEMs were selected for the experiments: TNT, RDX and ammonium nitrate. Their chemical structures are shown in Fig. 2. Through the testing of various variables from materials to methods, a tailored approach will be apparent. This paper follows the timeline in which the experiments were performed, allowing the reader to easily follow why and how certain materials (solvents) and methods were modified based on the deposition characteristics desired. Each HEM studied required a unique mixture of solvents to prepare the optimum solutions for TIJ dispensing. The solvents used were all of HPLC grade and included methanol (MeOH, $CH_3OH$), acetonitrile (ACN, $CH_3CN$) and isopropyl alcohol, (IPA, $(CH_3)_2CHOH$)). These solvents were purchased from Fisher Scientific International. Distilled, deionized water was purified by reverse osmosis, followed by deionization and filtration through four stages of cartridges to remove organic and inorganic impurities and by exposure to far UV light to remove bacterial impurities (Nanopure Diamond, Barnstead, Thermo-Fisher Scientific, Inc., Waltham, MA).
A thermal inkjet (TIJ) X-Y-Z dispenser, model IIS-300S manufactured by ImTech, Inc. (Corvallis, OR), was used for printing the HEM onto the target surfaces. To begin the experiment, a bitmap (.bmp) image file of the desired design (1, 2 or 3 blocks) was chosen and loaded. The user-coordinates (print zone coordinate system) were as follows: x = 3.5 in,

y = 10 in. and z = 0.7 in. For the print head commands, the following settings were used: fire pulse = 2 μsec, warm pulse = 0.5 μsec, dead pulse = 0 μsec, voltage = 0 volts and the maximum frequency was 12 kHz. The IIS acceleration was as follows: x = 100 in/sec$^2$, y =100 in/sec$^2$ and z = 100 in/sec$^2$. The velocity was x = 10 in/sec, y=10 in/sec, z =10 in/sec and the IIS cap location was x = 1.90 in, y = 1.83 in, and z = 1.30 in. The solutions were dispensed using zero-dot spacing script (space between drops using HP inkjet) at a printing resolution of 600 dots per inch (dpi). Once the above specifications were set, the bitmap image was loaded and selected to print on the TIJ dispenser. The total printer area per substrate ("chip") was 1.0 cm × 0.8 cm (area = 0.8 cm$^2$). The bitmap images had three 0.8 cm$^2$ chips to be deposited per image, allowing 3 substrates to be printed at the same time. The total dispensing time for the surface area of 0.8 cm$^2$ was 14 s for a total of 56 s for three substrates. To allow enough drying time, a waiting period of 60 s was employed between printing passes. This waiting period was particularly important for high loading substrates that involved many passes. To carry out shelf life experiments, FOC-GAP/FT-RAIRS was used as a non destructive technique during a period of 15 hr. Details of the setup have been described previously (Primera et al. 2009).The most intense peak of the IR spectrum for each HEM was selected for the kinetics studies. In the case of RDX, the experiments were performed over one year. The experimental conditions were co-addition of 64 scans, 4 cm$^{-1}$ resolution and a spectral range of 900–4,000 wavenumbers (cm$^{-1}$). In terms of consecutive measurements, the repeat setting was 100 times, and the delay between measurements was 100 s. Background spectra from clean test substrates were acquired using the same instrumental conditions as the sample spectra prior to each sample measurement session. All spectra were recorded in reflectance mode to facilitate data processing. To compare the influence of the surface type on the shelf life, experiments were performed on gold-on-silicon, glass, and stainless steel (SS) plates.

White light images of samples were acquired in order to maintain a record of the information concerning the loading surface distributions. This allowed comparisons between the two sample preparation methods and between the solvent mixtures used. An Olympus America, Inc. (Center Valley, PA) model BH2-UMA high resolution optical microscope designed for mineralogy studies and equipped with 10-250× magnification and a 6.0 MB PAX-Cam image capturing CCD camera controlled by the PAX-it!™ Software (Midwest Information Systems, Inc., Villa Park, IL) was used to capture white light images (micrographs) of the substrates with HEM loadings on them. Most of the images captured were taken with a 10× objective, but some micrographs required the use of infinity-corrected ultra long working distance Olympus objectives of 100× and 250× magnification. Nine images per sample were obtained by dividing the CCD chip into three rows and three columns.

For chemical analysis of the surface loadings of RDX and TNT, a model 1100 high performance liquid chromatograph (HPLC) from Agilent Technologies (Santa Clara, CA) equipped with a C-18 column (platinum, 100 A, 5 μm, 150 mm x 4.6 mm, Eclipse XDB Alltech), and a variable wavelength detector was used. The HPLC conditions employed were a flow of 1.0 mL/min, a stop time of 5 min and 3 min for TNT and RDX, respectively, a solvent mixture of 50% water and 50% methanol and a pressure limit of 400 bar. The chromatographic column was operated at 40 °C. Other conditions included a detector wavelength of 224 nm and 254 nm for TNT and RDX, respectively, with a pump time of 5 min, a peak width (response time) of less than 0.1 min, post time = off and an analysis time of 5 min. A model 732 IC ion chromatograph (Metrohm, Riverview, FL) was used for surface loading analysis of AN. IC conditions included a retention time of 3.94 min, a mobile phase of 0.1 mM L-tartaric acid and 0.1 mM picolinic acid, an injection volume of 10 μL and a flow rate of 1 mL/min. Experiments were performed in cationic mode, and ammonium cations were detected.

The two methods tested for preparation of homogeneous samples and standards of solids/traces deposited onto surfaces were sample smearing (Primera-Pedrozo et al., 2008) and TIJ sample delivery (Primera-Pedrozo et al., 2005). Both methods were compared in the study. Two important parameters for dispensing analytes using TIJ are the solvent viscosity and surface tension. The value of the solvent viscosity must be between 2 and 6 x $10^{-2}$ g cm$^{-1}$ s$^{-1}$ (centipoises, cp), whereas the surface tension of the solution must be between 30 and 40 mN/m. IPA has a high viscosity compared to methanol and acetonitrile (2.8 cp), resulting in the formation of uniform spherical deposits (Van Neste et al., 2009). However, due the slight solubility of TNT and RDX in methanol and IPA, it was necessary first to dissolve them in acetonitrile and then to prepare a mixture of methanol and 2-propanol in order to have the adequate viscosity to dispense. The composition of the solvent was 10% acetonitrile, 20% methanol and 70% isopropanol. The use of 100% IPA as the solvent results in the formation of non-uniform droplets on silicon wafer surface and spread-out is observed. AN is highly soluble in water, and this constitutes a problem while dispensing. Thus, for the preparation of AN solutions, a small amount of the reagent was added to 3 mL of methanol and heated to 35°C until dissolved. This mixture was then transferred to a flask, and the vial was rinsed three times and transfer to the flask with IPA. The solvent ratio was changed until uniform, spherical deposits were found on the substrate images. Once the solutions were prepared, the analytes were deposited onto gold-on-silicon, glass or stainless steel substrates. The latter two substrates were used for comparison and as a means of obtaining information on analyte-substrate interactions. HEM solutions were prepared as explained in the previous section and, using a syringe, were transferred into one of the printer cartridges that was then used to print directly onto the substrates studied using the TIJ dispenser (Fig. 3). Low concentrations were used to avoid HEM nucleation and subsequent crystallization in the jet nozzles.

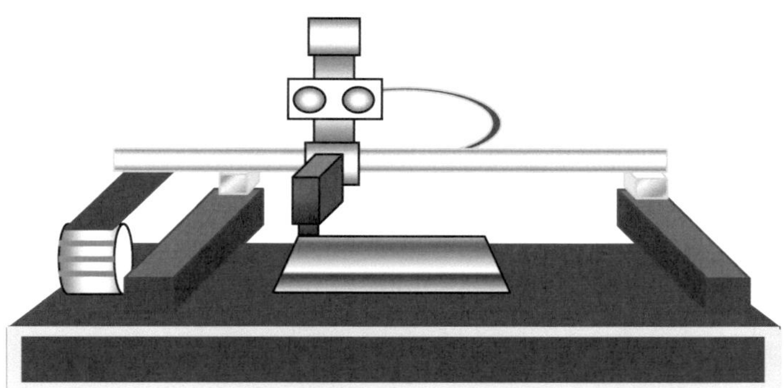

Fig. 3. Deposition method: thermal inkjet printer used for sample dispensing.

To obtain higher surface concentrations as 20-50 μg/cm$^2$, several print layers of solutions (passes) were applied. The cartridges were filled completely with 40 mL of HEM solution and then sealed shut with a small glass bead secured in place using adhesive tape. A vacuum was then applied to the nozzle of the cartridge to ensure that the solution was ready to dispense. The cartridge was then placed in the printer head, and a personal computer (PC) was used to control the deposition program for dispensing by TIJ printing. Before dispensing by TIJ, a bitmap (*.bmp) image was required to determine precisely where to place the substrates. An MS Word™ (Microsoft Corporation, Redmond, WA) file was used to scale the target boxes until the dimensions matched those of the substrate. A width of 2.25 in and height of 2.00 in was found to produce an image that is 1.0 cm × 0.8 cm. Three samples were printed at one time (Fig. 4). These small squares were printed out in black to clearly show where the surfaces needed to be placed to ensure proper deposition.

Fig. 4. Substrate deposition area for TIJ dispensing. Surface coverage was not complete, leaving 1 mm of surface uncoated on two sides for handling purposes.

The HEM loadings or surface concentrations (μg/cm$^2$) deposited by TIJ or smearing were determined by HPLC or IC. To determine the total mass deposited onto the substrates, the latter were rinsed with an appropriate solvent and volume in which the HEM were highly soluble. Acetonitrile was used for rinsing RDX and TNT, but water was used for rinsing AN samples due to the high solubility of this HEM in water. Rinsing is the first and most critical step for determining how much explosive was originally deposited using both deposition methods. The substrates were printed onto an area of 0.8 cm × 1.0 cm, with 0.1 cm left free (Fig. 4) on each side to enable handling the substrate. To ensure that the analyte was rinsed entirely from the surface with as little loss as possible, tweezers were used to clamp the substrates by their corners, and a Pasteur pipette was used to rinse the substrate with the appropriate solvents. Care was taken to ensure that as much analyte as possible was removed. Once completed, the rinses were collected, transferred to small 1.0-mL vials, stirred to ensure uniform concentrations and analyzed for total mass collected using HPLC or IC. In the case of smeared samples, the total mass recovered was converted to percent recovery; however, in the case of TIJ deposition, this was not possible because the initial mass deposited was not readily known.

TIJ technology offers several advantages to smearing deposition. The method is less subject to human errors and provides more uniform target surface coverage. Moreover, the surface loading concentration can be varied by changing the numbers of passes delivered to the sample, the dispensing frequency, the applied energy and the dispensing pen architecture. Also, the method includes precise delivery of a number of droplets with well-characterized mass and concentration. Furthermore, only one solution is required, avoiding the need for serial dilutions that can increase the analytical errors caused by human intervention.

To ensure the feasibility of the method, the prepared dispensing solutions were also run in HPLC or IC to determine their exact concentration. Table 1 shows the results of the calibration curves used for determining the concentration of the HEM studied using HPLC or IC. RDX and TNT calibration curves were prepared using acetonitrile as the solvent, while ultra high purity water was used for AN. Limits of detection (LOD) and limits of quantification (LOQ) for TNT and RDX using HPLC analysis and for AN using IC analysis are also reported. TNT had the lowest LOD and LOQ compared to RDX and AN. The exactly concentration of HEM solutions was also determined by HPLC or IC. The concentrations of these HEM solutions transferred to the ink cartridge were higher than the concentrations in the calibration curves (See Table 1), and thus it was necessary to prepare 1:10 (v/v) dilutions in acetonitrile to ensure the data fell into the proper concentration range.

| HEM | Concentration range (ppm) | Method of analysis | $R^2$ | LOD (ppm) | LOQ (ppm) | Significance F-value |
|---|---|---|---|---|---|---|
| TNT | 0.1-1.5 | HPLC | 0.9990 | 0.05 | 0.15 | $1.30x10^{-05}$ |
| RDX | 0.1-1.5 | HPLC | 0.9978 | 0.07 | 0.23 | $4.43x10^{-05}$ |
| AN | 1.0-5.0 | IC | 0.9601 | 0.2 | 0.6 | $3.25x10^{-06}$ |

Table 1. HPLC and IC calibration curve results for the highly energetic materials (HEM) studied.

A different solvent was used for the rinsing step of each HEM on account of their distinct physical properties. Many factors had to be taken into account such as HEM-solvent affinity (in terms of solubility), deposition methodology and HEM vapor pressure. The first step consisted of finding the right solvent for each HEM. TNT and RDX were soluble in methanol, but AN was soluble in water. The inorganic AN is a highly hygroscopic chemical, and it was difficult to avoid the absorption of water during the preparation of standards. To remove as much water as possible, AN was dried at room temperature in a vacuum oven and maintained in a desiccator for 24 hr prior to use. The solubility of TNT and RDX is high in acetonitrile and low in isopropanol. According to Primera-Pedrozo and co-workers, using methanol as the solvent for sample smearing gave the most uniform surface concentrations as determined by micro-FTIR surface analysis (Primera-Pedrozo et al., 2009). For TNT and RDX, a solvent mixture of 10% ACN, 20% MeOH and 70% IPA was used, whereas a mixture of 12% MeOH and 88% IPA was used in the case of AN.

The second step of the methodology was to find a solvent mixture compatible with each deposition method. Smearing requires a solution that can be easily spread over the substrate and dries quickly in order to achieve a uniform coverage and surface concentration. In addition, the size of the substrate where the deposition takes places influences the distribution and the readiness of the smearing process. When the size of the substrate is relatively large (> 10 $cm^2$), it is very easy to dispense the mixture containing the analyte with the Teflon applicator, resulting in a uniform surface coverage (Fierro-Mercado et al., 2010). Another parameter that plays an important role in the nature of the coverage of the substrate is analyte-surface interactions, and this effect will be discussed later. TIJ deposition requires the use of a moderately viscous solution for high quality printing. Thus, the solvent solutions used were tailored to each specific deposition method. Depending on the speed of drying of the solution containing the dispensed analyte, the mix was modified with a higher or lower vapor pressure adsolvent. If the solution took too long to dry, then the percentage of MeOH in the solvent was increased. It was found that HEMs studied were soluble in methanol. The high vapor pressure of MeOH resulted in quick evaporation and uniform surface loading (Primera-Pedrozo et al., 2009). However, if the solvent mix required higher viscosity, then the percentage of IPA was increased to compensate for the lower viscosity of the adsolvent added. These modifications to the solvent mix tailored the vapor pressure and physical characteristics of the mix as well as the deposition characteristics of the HEMs studied.

Fig. 5 contains white light micrographs of RDX deposits on gold coated silicon substrates obtained after a single thermal inkjet pass. The dispensing experiments were performed using an 812 ppm RDX stock solution in 10% ACN, 20% MeOH and 70% IPA. Figure 5a shows that the central portion of the substrate is covered by a nearly uniform distribution of RDX deposits. The micrographs of the borders (Figure 5b, inner part of border) showed the presence of small RDX crystals, while the micrograph included in Figure 5c (outer part of the border) shows the formation of elongated, larger RDX crystals. The onset of crystallization of the deposited RDX sample was observed immediately, even after a single TIJ pass. RDX crystal formation was observed only at the edges (inner and outer rim) of the deposited sample on the substrate. Although crystallization was observed to a small degree and only at the substrate edges, the solvent mixture used was considered adequate for the depositions and was left unchanged. The idea of decreasing the dispensing solution concentration and increasing the numbers of passes to increase the surface loading concentration also led to crystallization, as was demonstrated by Manrique-Bastidas and collaborators for TNT (Manrique-Bastidas et al. 2004a; Manrique-Bastidas et al., 2004b). However, this behavior is not general, and each case depends on the solvent mixture/HEM combination. For example, the high affinity of AN for water made its deposition onto surfaces a challenge.

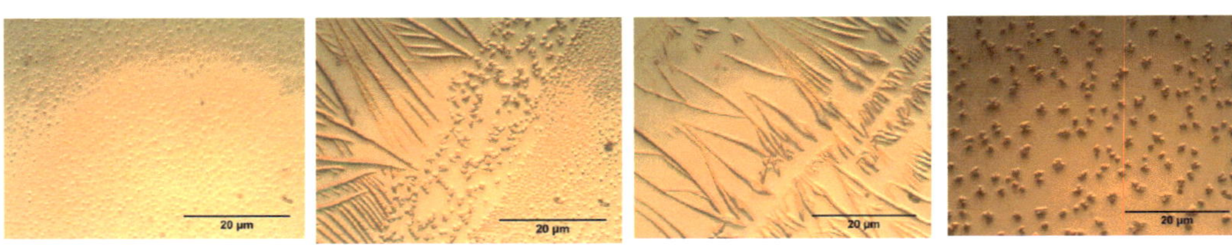

a. Center: 1 pass　b. Inner Edge　c. Outer Edge　d. Center: 5 passes

Fig. 5. White light microscope images of TIJ-deposited RDX on a gold-silicon substrate: (a) center section of the substrate; (b) inner edge of the substrate; (c) outer edge of the substrate. (d) center section of the substrate after five TIJ pass. The conditions included a stock solution of 892 ppm RDX and a solvent consisting of 10% ACN, 20% MeOH and 70% IPA. The image was acquired at an optical magnification of 100×.

Although water is the optimum solvent for TIJ because of its high surface tension and low viscosity, attempts to dispense aqueous solutions of AN resulted in poor spreading of the deposited analyte/solvent mix over the surface; instead of spreading evenly, the mixture remained in isolated droplets on the surface. Fig. 6 depicts the micrographs of the results of dispensing AN on gold-silicon substrate using a solvent mixture of 10% $H_2O$ and 90% IPA. The results were similar after one TIJ pass (Fig. 6a) and five TIJ passes (Fig. 6b). The lack of homogeneity of the distributions obtained can be clearly observed. Instead forming of a uniform layer of analyte on the surface, the formation of dispersed, isolated droplets is favored. When initially deposited, the droplets were much more finely dispersed, but moisture in the air was enough to induce the formation of larger droplets on the substrates. Because of the problems related with using water as the solvent for depositions of AN, a solvent mixture was prepared containing 12% MeOH and 88% IPA. The appearance of AN in this solvent mixture after dispensing onto the gold-coated silicon slides is depicted in Fig. 6. The difference in comparison to the water-based solvent mix can be clearly seen as a collection of fine and uniformly dispersed deposits, especially along the edges. While absorption of water from the atmosphere was still taking place in the center of the substrate, this process occurred to a much smaller degree than in the samples deposited with AN/water solutions only. The conditions included a stock solution of ~ 812 ppm AN and a solvent consisting of ~ 10% $H_2O$ and 90% IPA. Optical magnificationused was 10×. White light 100× magnification micrographs of TIJ deposits of 1059 ppm TNT in MeOH after 25 passes: (c) edges; (d) center.

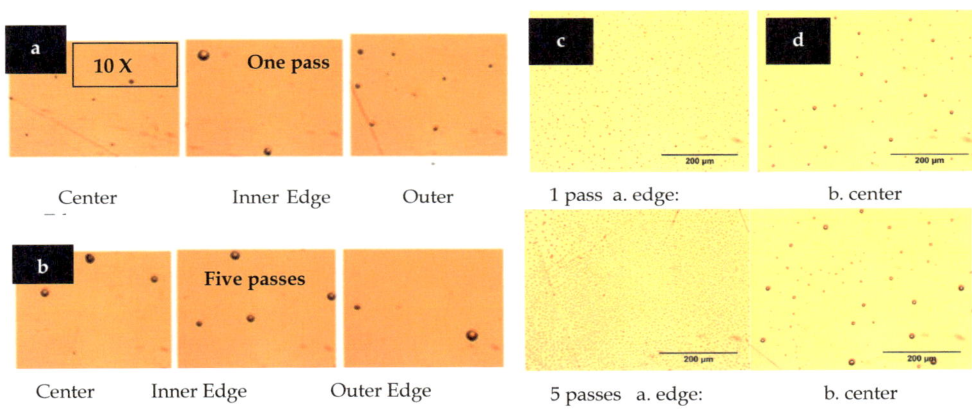


Fig. 6. White light micrographs of TIJ deposits of AN onto a gold-silicon substrate. (a) one TIJ pass; (b) five TIJ passes. The conditions included a stock solution of ~ 812 ppm AN and a solvent consisting of ~ 10% H2O and 90% IPA. The optical magnification is ~ 10×. (c) one TIJ pass; (d) five TIJ passes. The conditions included a stock solution of ~ 812 ppm AN and a solvent containing ~ 12% MeOH and 88% IPA. The optical magnification is ~ 10×.

Smearing deposition has been used as a method for sample preparation of solid compounds on substrates. Samples and standards of nitrocompounds [7, 9] and cyclic organic peroxides as well as pharmaceutically active ingredients and excipients have been prepared (Primera-Pedrozo et al., 2004). These studies centered on covering relatively large surface areas (~43 $cm^2$) on stainless steel, aluminum or glass (Hamilton et al, 2005; Perston et al., 2007). In all mentioned applications, smearing proved to be a high-quality deposition technique as judged by the excellent agreement between the amount aimed to be deposited, the actual amount deposited, coverage uniformity and the percent of analyte recovered after substrate rinsing. The introduction of TIJ to design new methodologies for preparation of samples and standards required a comparison with the well established sample smearing technique. As already mentioned, samples were prepared by both deposition methods using the same solvent mixtures to make a real comparison. After sample preparation, the substrates were rinsed and analyzed using protocols based on HPLC or IC in order to determine the total mass collected in the solution. Results obtained from the chemical analyses were then expressed in terms of surface loadings (μg/cm$^2$).

The images acquired after each carrying out each method were excellent tools in assessing the performance of the methods. In general, smearing was a somewhat inconsistent process because it relies on human intervention to be carried out. At times, even the environment in which the method took place altered the reproducibility. Examples including wind drifts from a fan or high lab humidity altered the way in which the solution adhered, spread out or dried on the substrate surface. The smearing process was more exposed to the local environment than the TIJ method because surface depositions using the latter technique took place within a custom acrylic built covering with a controlled atmosphere. Images were acquired at several locations. Another important observation during smearing depositions on these substrates was the poor adhesion of the HEM studied to the gold coated silicon surface, in contrast to the strong adhesion experienced by analytes (including nitrocompounds and cyclic peroxide HEMs) to other test surfaces such glasses, stainless steel, aluminum and plastics. Solid-phase residues of HEM were often left on the Teflon applicator after smearing was performed in one or two passes. This residue was considered an indication that the surface used in the experiments was very inert and thus rejected the adhesion of the analytes.

Images acquired of thermal inkjet depositions showed entirely different results compared to smearing. TIJ reduced significantly the direct operator interaction during the deposition process. Thus, reproducible results with uniform surface concentrations were obtained for TIJ depositions. Figure 5 shows an even distribution of TNT/10% ACN, 20% MeOH and 70% IPA droplets on the right edge and center of the substrate after dispensing. It is possible to observe slight differences in various areas depending on the location of the deposited sample on the substrate, but overall, even surface concentration dominated the depositions. An interesting finding was that the droplets were in a metastable state (similar to super cooled water in a refrigerator) instead of in the assumed crystalline state (Manrique-Bastidas et al. 2004a; Manrique-Bastidas et al., 2004b). As a result, the images captured show the TIJ method of deposition to yield more favorable results than smearing with respect to achieving consistent homogeneous surface concentrations.

Both deposition methods have positive and negative aspects. Smearing is fast and easy to use. Typically, it has a relatively high susceptibility to human error, at times leading to irreproducible results, and requires large amounts of time for dispensing a wide range of concentrations. Moreover, it produces poor surface loading concentrations over surface areas of ~ 1.0 cm$^2$ and smaller on highly inert surfaces such as gold coated silicon. As can be observed in Table 2, only between 29% and 70% of the deposited material could be recovered after deposition by rinsing and subsequent HPLC analysis. The highest deposited surface concentration tested in this work (50 μg/cm$^2$) was very difficult to achieve because of the low adhesion of TNT to the small sample area of the surface and the high affinity of TNT for the Teflon applicator. These characteristics explain why the lowest percent recovery (37% - 55%) was obtained for this surface loading. Deposition of extremely low concentrations of approximately 1.0 μg/cm$^2$ failed because the small volume of HEM in solution was not enough to cover the entire surface, without modifying the protocol established. This could be attributed to the fact that the pressure applied to the substrates or the speed of spreading are both operator-dependent. Also, because the deposition takes place in the open air, explosives with high vapor pressures would sublimate quickly during this procedure. Thus, smearing was prone to a large degree of human variability during its preparation stage.

| Deposited (μg/cm²) | Recovered (μg/cm²) | Recovered (%) |
|---|---|---|
| 50 | 29.8 - 18.8 | 37.6 - 55.7 |
| 10 | 6.1 - 7.1 | 60.6-70.9 |
| 5.0 | 2.9 - 3.6 | 57.8 - 71.4 |
| 1.0 | 0.3 – 0.7 | 29.8 - 66.3 |

Table 2. Comparison of the theoretical amount of TNT deposited using smearing and the real surface concentration detected after rinsing.

TIJ, however, has a small margin for operator error, resulting in reproducible and even surface loadings. TIJ also has the possibility of becoming even more useful in conjunction with simple programming. Different shapes, different surface loadings on the same substrate and the spacing between deposited areas can all be optimized using this technique, whereas sample smearing has fewer degrees of freedom available for improvement. Thus, with respect to overall ease of use and functionality, TIJ was demonstrated to be the superior method of deposition when compared to sample smearing. Sample deposition using TIJ was time consuming for high surface loadings, even for a substrate of only 0.8 cm². Each successive pass required the operator to manually push a button that alerted the software to perform another pass. Thus, although depositions with TIJ can take a significant amount of time, particularly for high surface concentrations, the operation results in high mass transfer yields and there is a finer control over the place where the dispensing takes place. The TIJ dispenser took roughly 14 s to complete a single pass on the 0.8 cm² surface area, and, depending on the behavior of the solutions used, some depositions required a delay of up to 60 s between passes to allow the sample to dry.

The main advantage offered by TIJ is that, in principle and in practice, only one concentration is required for dispensing a broad range of surface loadings, reducing significantly the errors associated with operator intervention and with the preparation and use of multiple dilutions. In this work, it was estimated that HEMs stock solutions of concentrations in the range of 800 to 1100 ppm would result in loadings of the desired surface concentrations so that each TIJ pass added an easily traceable amount. The devised protocol proved to be very simple and limited the number of steps in the methodology. In addition, the number of TIJ passes controlled the surface loading concentration very efficiently, as shown in Table 3 for TNT. For example, after performing one TIJ pass using a 1059 ppm TNT stock solution, 1.04 μg/cm² were deposited onto the gold-silicon surface. In Figure 6, the relationship between the number of TIJ passes and the surface loadings for the HEMs studied is represented. The deposition of RDX and TNT exhibited linear behavior with respect of the number of TIJ passes applied to the substrates, but the deposition of AN stock solutions with TIJ was not.

This non-linear behavior of AN was attributed to the affinity of the analyte for water in the form of ambient humidity; as time passed, AN absorbed additional water from the air. This phenomenon would explain why, when more AN was available, the surface concentration began to increase exponentially, but the precise reason behind this behavior is currently unknown. These results demonstrate that, using the TIJ method, it is possible to generate samples and standards of HEMs with more uniform coverage and surface loading concentrations that can be varied by changing the numbers of TIJ passes without the need for serial dilutions. Attempts to prepare standards lower than 5 μg/cm² for AN were unsuccessful because that solution, after rinsing, was outside of the lower value of the IC calibration range (1.0 ppm). As explained above, for TNT there was a direct relationship in the number of passes and loading concentration. For TNT, one pass resulted in a surface loading of 1.0 μg/cm², and 10 passes led to a surface concentration of 9.2 μg/cm². For RDX, one pass led to 0.55 μg/cm², five passes generated a loading of 3.2 μg/cm² and 10 passes resulted in a deposition of 5.4 μg/cm². This result confirms that the amount dispensed by TIJ is markedly dependent on the substance to be deposited and also on the solvents and surfaces used for the depositions.

| Number of passes | Rinsing Vol. (mL) | Loading concentration (μg/cm²) |
|---|---|---|
| 1 | 5 | 1.04 |
| 2 | 10 | 2.08 |
| 5 | 10 | 5.58 |
| 7 | 10 | 7.88 |
| 10 | 10 | 9.19 |

Table 3. HPLC analysis of TNT loading concentrations obtained by varying the number of TIJ passes.

Even though TIJ promises to be a good method for the preparation of samples and standards on surfaces, maintenance of the TIJ print is a potential problem, and care must also be exercised when re-using solutions stored in the cartridges. If the cartridges are not properly sealed after use, then the solutions could leak out of the nozzles and solidify, clogging the pores of the inkjet cartridge. This clog eventually could lead to disruptions in the depositions of the desired surface concentrations upon further use.

The shelf life is a critical parameter to evaluate during the production of samples and standards of solid analytes deposited onto substrates. Materials with high vapor pressure will not stay on the surface for a long time. However, the sublimation of the material depends on the material-surface interactions. TNT, which has a higher vapor pressure than RDX, will sublimate faster than RDX. As will be shown, RDX stays on the stainless steel surface for almost an entire year without any significant changes of the FTIR band intensities (expressed as peak areas). Another significant parameter to be considered is the photodecomposition of analytes by exposure to light (Irrazabal et al., 2007) leading to the formation of new products. To perform comparisons of surface effects, samples were dispensed using TIJ onto gold coated silicon, glass and stainless steel (SS) substrates. Three samples of TNT were prepared on each surface using 10 passes, leading to a final surface concentration of approximately 10 μg/cm². The samples were then analyzed by OFC-GAP/FT-RAIRS for 900 min (15 hr) to examine sublimation from the test surfaces. In the case of RDX, experiments were performed over 11 months after smearing on a SS substrate. Figure 7 shows the RAIRS spectra of 10 μg/cm² TNT on glass, SS and gold-silicon surfaces. The vibrational signatures of the HEM can be clearly observed (1096 $cm^{-1}$, 1181 $cm^{-1}$, 1354 $cm^{-1}$, 1559 $cm^{-1}$) (Lin-Vien et al., 1991). The nitro symmetric stretching vibration of TNT band appeared at 1354 $cm^{-1}$, and the nitro asymmetric stretching vibration was found at 1559 $cm^{-1}$. The locations of persistent IR bands were almost the same on the different test surfaces, with only a few changes in intensity. The presence of a strong band at about 1267 $cm^{-1}$ can be attributed to Si-O vibration in the glass. The spectra of TNT on gold-silicon are shown at 0 hr and at 15 hr, after completing the sublimation studies. At 15 hr, the TNT vibrational signatures disappeared from the spectrum, indicating that the compound had sublimated from the surface.

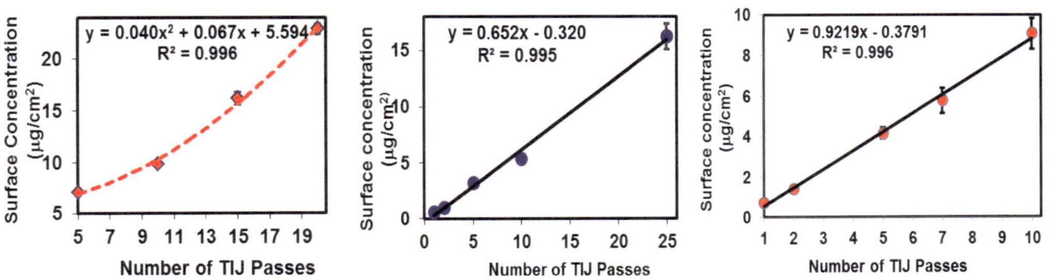


(a) (b) (c)

Fig. 6. Surface loadings obtained by variation of the number of TIJ passes dispensed to gold-coated silicon substrates. (a) AN, 892 ppm; (b) RDX, 812 ppm; (c) TNT, 1059 ppm.

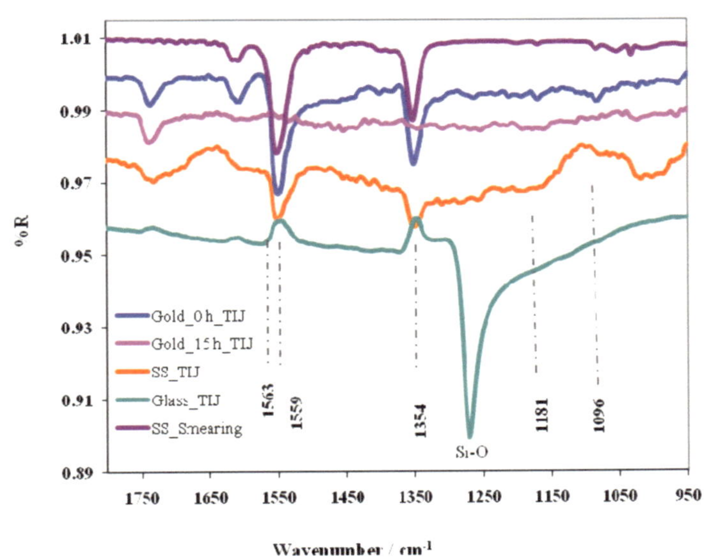


Fig. 7. FTIR spectra using a grazing angle objective for 10 μg/cm² RDX (smearing) and TNT (TIJ): (a). RDX/SS; (b) TNT/Au@Si, 0 h; (c). TNT/Au@Si, 15 h; (d) TNT/SS; TNT/glass.

Sublimation rate studies were performed for TNT deposited onto the mentioned surfaces. Peak areas were calculated for the IR band centered at 1354 $cm^{-1}$ using the OPUS™ Bruker Optics software (Billerica, MA) for all the spectra recorded over 15 hr. Peak areas were calculated by integration using straight lines that connected the wavenumber limits of the peak envelopes as baselines. These results are shown in Figure 8 for TNT on SS, glass and gold-silicon. In the same graph, the sublimation behavior of RDX (in days) is observed (the *x* axis is on the top of Figure 8).

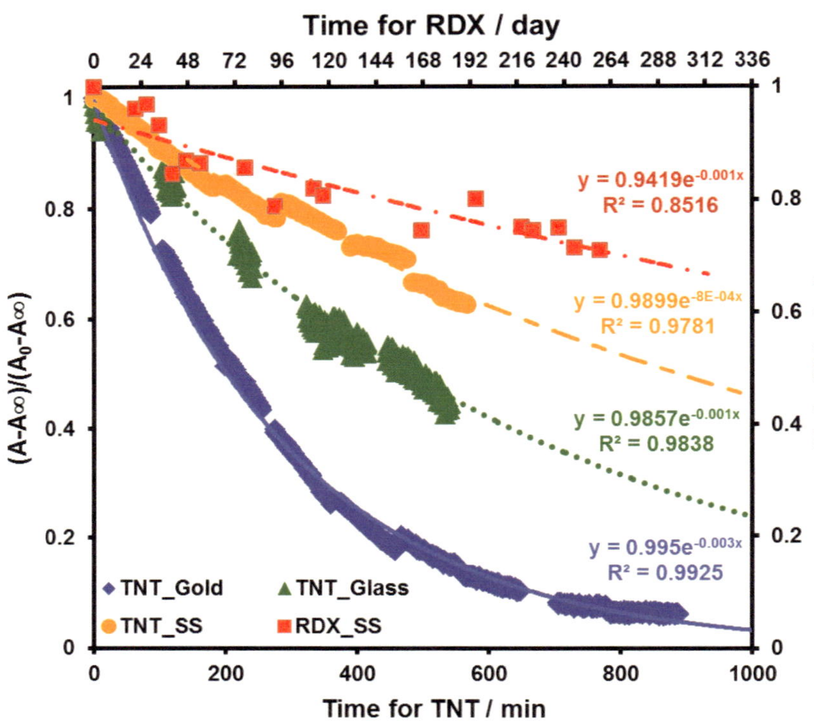


Fig. 8. Plots of $(A-A_\infty)/(A_0-A_\infty)$ vs. time for surface sublimation for: TNT on gold-silicon (♦); TNT on glass (▲), TNT on stainless steel (●) and RDX on stainless steel (■).

According to standard procedures in kinetics and dynamics measurements, a physical property can be taken to be directly proportional to the concentration of a given species in the concentration range used. In the present studies, integrated IR band areas were taken to be directly proportional to the surface concentrations. In Equation 1, $A_\infty$ is the area at infinite time, $A_t$ is the peak area at time $t$ (given in seconds) and $k$ is the surface sublimation constant. The decay of peak areas was monitored as a function of time for the different surfaces, and the data were plotted as ln $(A_t-A_\infty)$ vs. $t$. The surface sublimation constant was obtained from the value of the slope for the linear regression trend.

$$ln\,(A_t - A_x) = kt \quad (1)$$

The data were adjusted to Equation 1, and the value of the constant parameter $k$ was calculated. The results are shown in Table 4. The residence time of the explosive on the surfaces (gold on-silicon, glass and SS) is affected by the sublimation process, vapor pressure (Phares et al., 2000) and HEM interaction with the surface. The residence time (RT) can be described as the time the material will persist on a surface after exposure to ambient conditions. The results indicate that TNT on SS has a residence time longer than the other surfaces studied. The half life values ($t_{50}$) for surface residence are also shown in Table 4 for TNT on gold-silicon, glass and SS surfaces and for RDX on SS. Residence time half life values for TNT on glass and SS are higher than on gold-silicon. These results are closely dependent of the material type. Gold on silicon is a very inert surface compared to glass and SS surfaces. This caused the poor adhesion of the material on the gold-silicon indicating that the interaction between the HEM and substrate is stronger in glass and SS than gold on silicon. However, for the gold coated silicon surface, this interaction should be minimal, due to small value of $t_{50}$ = 195 min. The high value of 630.1 days for $t_{50}$ of RDX is an indication of strong interactions between the HEM and the stainless steel surface compared to gold on silicon, and it is closely related to the low vapor pressure of RDX.

| Kinetics Parameters | TNT @ Gold (min) | TNT @ Glass (min) | TNT @ SS (min) | RDX @ SS (day) |
|---|---|---|---|---|
| k | 0.00356 | 0.00142 | 0.00081 | 0.0011 |
| $t_{1/2}$ | 195 | 488 | 857 | 630 |

Table 4. Kinetics parameters of the sublimation of TNT on different surfaces and RDX on SS.

The results presented above confirm that physical properties such vapor pressures and surface type influence the lifetime of the highly energetic materials on the substrates and the practical limits of substrates for preparation of samples and standards. Although AN was not tested in sublimation studies, it is estimated that it will remain on the surfaces for a long time due to its ionic character and low vapor pressure. However, the high affinity for water constitutes a practical limitation, particularly if the samples/standards are meant to be used for IR detection applications.

## 3. Detection of hazardous liquids concealed in commercial products by OFC-Raman spectroscopy

In August 2006 a terrorist plot to destroy aircrafts on transatlantic flights was discovered and timely stopped in London. The plan involved the use of liquid explosives stored in beverage bottles that would pass check points without being detected. The liquids were going to be mixed in flight, generating an improvised explosive device (IED) and were going to be left in the aircraft and detonated remotely after the aircraft landed (CNN news report, 2006). Immediately, airport security agencies at UK and USA established a ban to all liquids except from medicine and infants food beyond checkpoint. The position has been changed several times to allow certain amount of liquid or gel based products. However, when or how terrorists would try to pass hazardous liquids into an aircraft to create a threat, explosive or chemical is uncertain. For this reason it is important to develop a methodology through which it would be possible to differentiate between common products and compounds that can be combined for terrorisms intends.

The use of hazardous liquids as tools for terrorist intentions is not a new modality. Terrorists have used hazardous chemicals that are liquids at room temperature in many occasions. Some examples are the bombing to the World Trade Center in 1993, the attempt to a Philippine Air flight in 1994 and the deployment of the CWA Sarin in a Japan subway in 1995. Some of these attempts involved the use of liquid explosives like nitroglycerin and other nitro compounds. Peroxide based explosives are also easily prepared from common liquids such as acetone. Other extremely hazardous liquids are chemical warfare agents (CWA) and Toxic Industrial Compounds (TIC). TICs are chemicals toxic to humans that are widely used in manufacturing or primary material processing. They have received more attention in recent years because ease of accessible in large quantities by potential terrorists. Chemical warfare agents (CWA) differ from TICs in that they are intended to immediately incapacitate as many soldiers as possible when released against an enemy in war. To do this, the CWA must be toxic enough to cause an instant response when it is inhaled or comes into contact with the skin. Table 1 contains examples of some hazardous liquids and their volatility. Highly volatile materials can be easily deployed just by opening the container. Along with the volatility of these chemical is the Immediately Dangerous to Life or Health (IDLH) level and is defined by the US National Institute for Occupational Safety and Health. This value is the concentration in air that would cause immediate or delayed adverse health effects after 30 min of unprotected exposure. If these chemical are used in large quantities toward civilians, the amount of casualties will be huge. However even small quantities of a toxic chemical or a small IED can cause chaos specially in closed environments like an aircraft or a train or high transit areas like a building or a transportation terminal.

| Chemical | Description | Volatility ($mg/m^3$) | IDLH (ppm) |
|---|---|---|---|
| GB ($C_4H_{10}FO_2P$) | CWA: Sarin, organophosphate nerve agent. | 16,091 | .03 |
| AC (HCN) | CWA: hydrogen cyanide, blood agent | 1,080,000 | 50 |
| $HNO_3$ | TIC: nitric acid, a highly corrosive acid | 63,000 | 25 |
| $H_2SO_4$ | TIC: sulfuric acid, a strong mineral acid | 1.3 | 3.7 |
| GD ($C_7H_{16}FO_2P$) | CWA: Soman, a nerve agent | 3,900 | .008 |
| $PCl_3$ | phosphorus trichloride | 130,000 | 25 |

Table 5. Description of liquid CWA and TIC at room temperature.

Various approaches have been taken to detect and identify chemical agents including HPLC/MS GC/MS, Ion Mobility Spectroscopy, Infrared Spectroscopy and Raman Spectroscopy (D'Agostino et al., 2006; Sun and Ong, 2005, Marrs et al., 1996). Vibrational spectroscopy has the advantage that provides chemical information and provide with the sensitivity and selectivity required for Chemical Point Detection systems and has the potential for remote sensing. Specifically, Raman is able to analyze samples though various transparent glass and plastic containers, such as beverage bottles or food containers. This allows the contents of a container to be analyzed without opening the container, minimizing exposure to potentially harmful substances and helping to speed the screening process. In this work concealed liquids scenarios are studied by Raman spectroscopy. Hazardous liquids including CWAs TICs and other prohibited are studied through the walls of commercial drink containers. Fiber optic coupled Raman was used to evaluate the content of plastic and glass containers. Standoff Raman detection was used to obtain information of hazardous liquids and mixtures from 20 feet distance. The objective of this work is to study, in real field conditions, the detection of CWAs and TICs by point detection or with a as a remote surveillance tool.

Raman spectroscopy is one of the promising tools under consideration (Fraquharson et al., 2005; Pearman and Fountain, 1999). Since Raman scattering brings information on vibrational modes of molecules it can be used as a specific mean of detection. These days, Raman systems are portable, sensitive, flexible tools that are used in the field and the laboratory (Marrs et al., 1996). Raman based systems are able to analyze samples through various transparent glass and plastic containers (US NRC, 2004). This allows the non-invasive, non-destructive inspection of the content of a container without opening, minimizing exposure to potentially harmful substances and helping to speed the screening process. Also water does not present a strong signature in Raman measurements therefore Raman based systems result attractive to characterize aqueous based commercial products.

In this study Optical Fiber Coupled Raman spectroscopy (OFCRS) was used to characterize samples where a hazardous liquid is concealed either in a commercial product container or mixed with consumer products. Several commercial consumer products such as: mouthwash, sodas, juices and liquors were mixed with liquid explosives, flammables and CWA's. FOCRS was used to evaluate the content in plastic and glass containers. The dependence on liquid color, container material was also studied. The results suggested that the technique can be used to discriminate if the commercial liquids are the intended or a hidden hazardous liquid.

Raman spectroscopy presents various strengths that make it a potential technique for detection of chemical agents in the field. Since Raman scattering brings information on vibrational modes of molecules it can be used as a specific mean of detection. Water does not present a strong spectra making possible to analyze a variety of samples where water can be interference. In 1999 Christensen reported the use of a portable Raman system to characterize chemical agents sealed on glass containers. Samples were part of a library for military training in chemical agent identification. (Christesen et al. 1999). Optic fiber coupled Raman spectroscopy (OFCRS) was used to characterize liquid explosives and commercial liquids (Alvarez-Rivera, 2002). The results suggested that the technique can be used to discriminate if the commercial liquids are the intended or a hazardous liquid. Eliasson and colleagues (Elliasson et al., 2007) have reported the detection of drugs and liquid explosives concealed in dense colored plastic containers. The technique consists of a variation of the angle of detection to collect scattered Raman radiation. This approach allowed reducing fluorescence and interferences.

In 2002 Harvey evaluated a portable Raman system for forensic applications (Harvey et al, 2002). The evaluation of chemical agents in glass and plastic vials included library matches and variability due to containers and the use by operators. The third component of the present work is to evaluate Raman as a detection tool for concealed hazardous liquids. Liquid explosives and chemical agents were detected in a variety of consumer product containers either pure or missed with the commercial product.

The hazardous materials used for the detection experiments were: hydrogen peroxide 50% wt. in water, toluene, benzene and 99.5% ethanol from Aldrich chemicals. Also, acetone, dimethylmethyl phosphonate (DMMP) and triethyl phosphate (TEP) from Fisher Scientific were used. Fisher chemicals were obtained from Fisher Scientific International, Chicago, IL. Aldrich chemicals were obtained from Sigma-Aldrich Chemical Company (St. Louis, MO).

DMMP and TEP are structural analogs of chemical warfare agents (CWA) such as TABUN, SARIN, SOMAN, and GF and therefore are commonly used as CWA simulants (CWAs). The commercial liquid products employed in this study were: Gatorade variety drinks (distributed by The Gatorade company, Chicago, USA), Scope mouthwash (distributed by Procter and Gamble, Cincinnati, USA), Dewar's White Label whiskey (imported by Mendez & Co., PR), Ron Bacardi light rum (Produced by Bacardi Corp, PR), V8 vegetable juice (distributed by Campbell Company, NJ, USA), 7Up (distributed by Coca Cola Company, Puerto Rico). The effect of thickness and color of commercial glass and plastic containers were evaluated (See Table 2 for details). Raman spectrum of the original liquid in its container was obtained. Then the liquid was replaced by a hazardous liquid or mixture. Light absorption of containers at different wavelengths was studied.

| Container | Material | Description |
|---|---|---|
| green glass bottle | glass | soda water (Canada Dry®) |
| amber glass bottle | glass | Malt beverage (Malta India®) |
| clear glass bottle | glass | Perfume (Adidas®) |
| clear glass bottle | glass | Juice (Snapple®) |
| blue plastic bottle | Plastic (PET) | Water (Aquacal®) |
| green plastic bottle | Plastic (PET) | Refreshment (7up®) |
| clear plastic bottle | Plastic (PET) | Baby bottle |
| clear plastic bottle | Plastic (PET) | Perfume (Ralph® body spray) |

Table 6. Description of containers used in the OFC-Raman spectroscopy experiments.

A Renishaw RM1000 system microscope with a NIR 785 nm laser was used to obtain the spectra of explosives mixtures. The Raman shift spectra were obtained from 200 to 3200 $cm^{-1}$. The system was calibrated using silicon single crystal sample as an external standard by measuring the vibration at 520.56 $cm^{-1}$. Two portable fiber optic Raman spectrometers (Raman Systems R-3000 HR) were used to evaluate the spectral signature of commercial products. The excitation wavelengths used were 532 nm (green) with 25 mw maximum output power and 785 nm (red) with 250 mw maximum output power. The spectra of the original liquid in its container was collected though the walls. A volume of 30 mL of a hazardous liquid compound was then transferred and analyzed in different commercials containers varying the time and the power. Raman spectra were acquired from 200 $cm^{-1}$ to 1800 $cm^{-1}$. The systems were calibrated using cyclohexane.

The spectra of three liquors were obtained in the same clear glass bottle. The results are presented in Fig. 9a. As expected the rum, gold tequila and whiskey present the same spectra. The yellowish color of the tequila and whiskey presented some fluorescence identified by the shift in the baseline of the spectrum. However it is possible to identify the main peaks associated to the alcohol content. These spectra as well as the spectrum for ethanol, the main component of hard liquors, present peaks at 881(C-C-O stretch), 1048 $cm^{-1}$ (ring vibration), 1186 and 1280 $cm^{-1}$ (ring stretch) and 1456 $cm^{-1}$ ($CH_3$ deformation). Then the content of the clear glass bottle was replaced with acetone and hydrogen peroxide so that the content appears to be clear rum. These products can be easily found at beauty supplies and drugstores. These are commonly found in consumer products such as nail polish remover and hair care products. However together they can be mixed to for acetone peroxide, a known homemade explosive. The Raman spectra of acetone, peroxide and the acetone peroxide explosive are presented in Fig. 9b. Acetone's main peak is located at 789 $cm^{-1}$. For peroxide the strong Raman line at 878 $cm^{-1}$, characteristic O–O stretching mode, is stronger than that the other peaks of the spectra. For the AP the main peaks are located at 588 $cm^{-1}$, and three characteristic peaks at 780 $cm^{-1}$, 890 $cm^{-1}$ and 936 $cm^{-1}$. The position and intensity of the major peaks in acetone and peroxide make them easily detected by Raman and can be used as markers for explosive mixtures.

Fig. 9c presents the Raman spectra of 20% and 60% peroxide in whiskey. As previously mentioned, the spectroscopic signature of peroxide is hidden or masked by the whiskey. Samples from 0% to 90% peroxide in whiskey were prepared without changing the location of peaks. Since the presence of peroxide increased the intensity at 874 $cm^{-1}$ the peaks ratio can be used to discriminate between the original liquid and adulterated liquor. OPUS PLS Package was used to build a model to determine the concentration of peroxide in whiskey. The result of the cross validation is included in Figure 10a. The data from these same samples was correlated just for presence of peroxide. The graph at Figure 10b confirms that peak ratios can be used to discriminate between an original liquid and a concealed hazardous material.

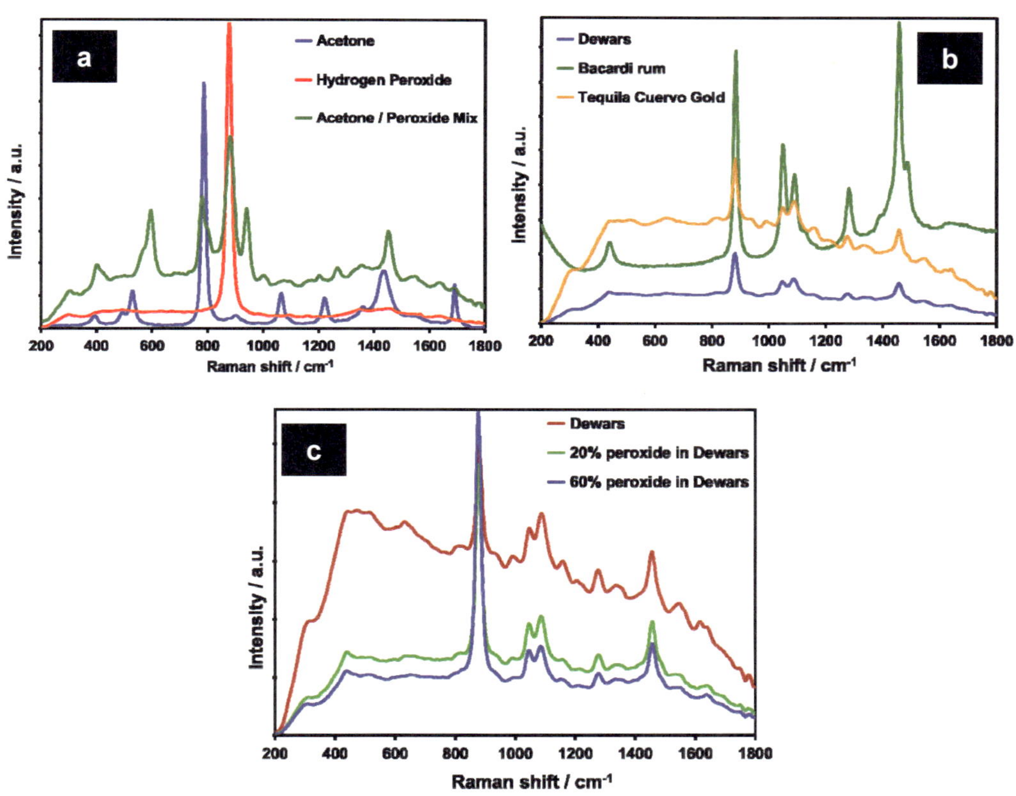


Fig. 9. (a) Raman spectra of liquors in their original containers acquired with optical fiber probe. (b) Raman spectra of acetone, peroxide and acetone peroxide explosive in a clear glass bottle; (c) Raman spectra of a mixture of hydrogen peroxide and Dewar's whiskey. Conditions: 785 nm laser, 200 mW power and 2 s integration time.

The spectra of a perfume and Plax™ mouthwash are shown in Fig. 11. Intensity of the peaks is reduced by the fluorescence of the liquid. Most noticeable peaks for the mouthwash appear at 875 $cm^{-1}$, 1000 $cm^{-1}$, 1080 $cm^{-1}$, and 1450 $cm^{-1}$. For the perfume the sharper peaks are: 798 $cm^{-1}$, 878 $cm^{-1}$, 1260 $cm^{-1}$, 1090 $cm^{-1}$, 1450 $cm^{-1}$ and 1610 $cm^{-1}$. Consumer products such as personal care and cosmetics present a challenge. Commonly these products contain ingredients such as alcohols, peroxides and others with chemical composition related to hazardous chemicals. Also these liquids are usually colored and present fluorescence.

The perfume bottle was emptied and replaced by several toxic industrial compounds. Fig. 11b shows the detection of toxic industrial compounds using a 532 nm laser. These liquids will exhibit strong peaks that will differentiate the material from the original liquid (perfume). The spectra of dimethyl methyl phosphonate (DMMP), a chemical warfare agent simulant is shown in Fig. 12a. This spectrum was collected through the walls of a clear glass juice bottle (Snapple) and was detected using 532 nm laser beam with 10 mW and the acquisition time was 1 s for recording the spectrum. The peaks at about 715 $cm^{-1}$ for DMMP correspond to a stretching mode involving the phosphorus atom. Fig. 12b shows the OFC Raman spectrum of the chemical agent simulant triethylphosphate (TEP).

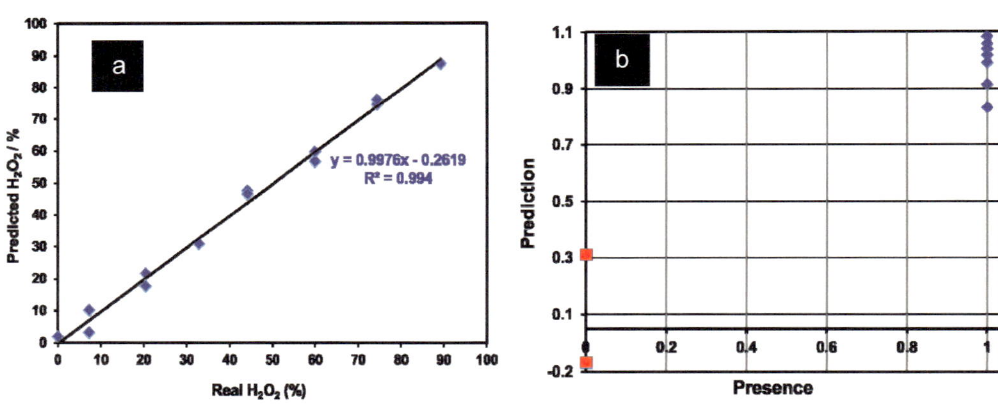


Fig. 10. (a) Prediction of the concentration of peroxide in whiskey; (b) prediction of presence of peroxide in whiskey.

The Raman spectrum of acetone was collected in different bottles using the same acquisition parameters at 532 nm (Fig. 11a) and 785 nm (Figure 11b) laser source. The characteristic peaks of acetone at 780 $cm^{-1}$, 1400 $cm^{-1}$ and 1700 $cm^{-1}$ were observable for all samples. However for 532 nm data the intensity of the peaks decreased in the following order: clear glass > clear plastic > green glass > amber glass. When the 785 nm laser was used the order was: clear glass > clear plastic > amber glass > green glass. It was not possible to collect a spectrum of the chemical in the green glass bottle using 30s of 200 mw of 785 nm laser power.

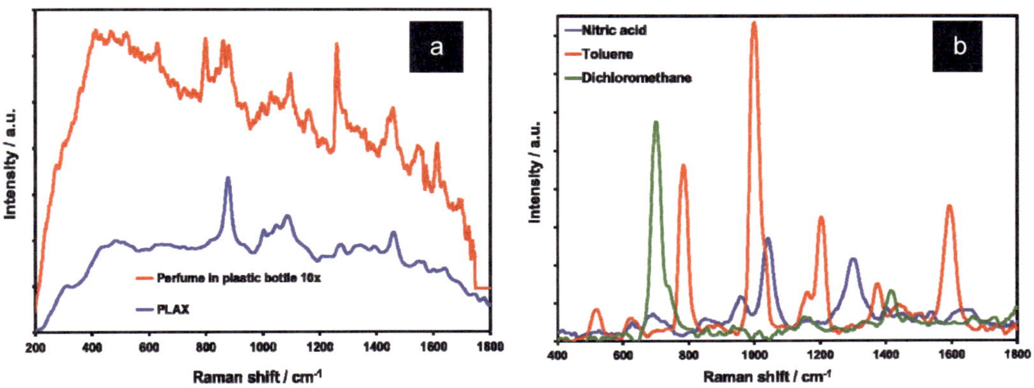


Fig. 11. (a) Raman spectrum of a perfume and mouthwash acquired with optical fiber probe; conditions: 785 nm laser, 100 mW and 1 s of integration time. (b) Raman spectra of toxic compounds within a perfume clear glass bottle at 532 nm, 1 s and a laser power of 12 mW.

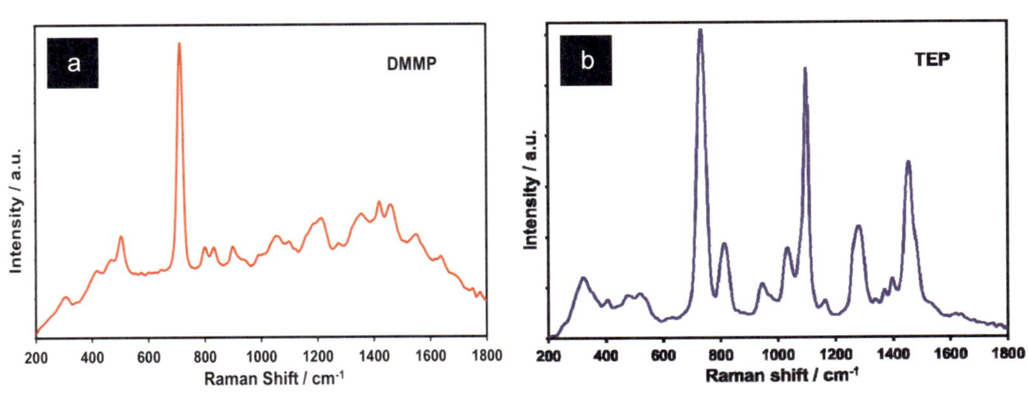


Fig. 12. Raman based detection of chemical agents simulants in a clear glass bottle at 532 nm laser, 1 acquisition, 1 s, 10 mW: (a) DMMP; (b) TEP.

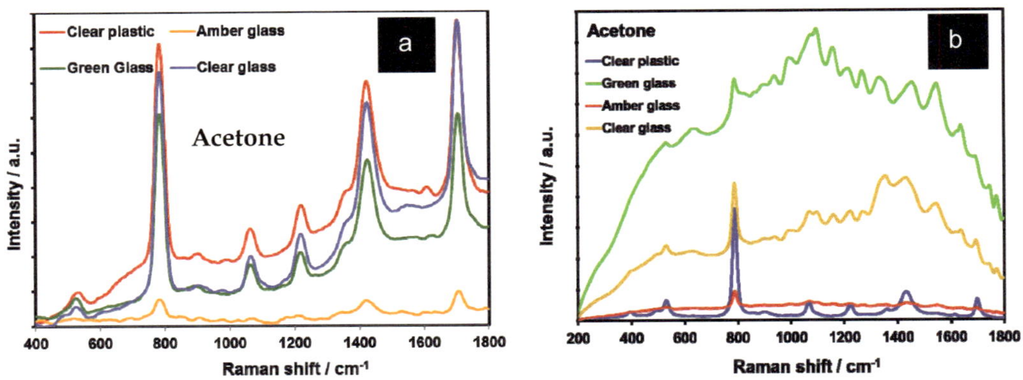


Fig. 13. Raman spectra of acetone in different bottles. (a) 532 nm 20 s, 33 mW; (b) 785 nm 30 s at 200 mW.

Since it was possible to collect a spectrum from the green plastic bottle, the behavior could be associated to thickness of the walls. The thickness of glass and plastic bottles was measured. Table 7 present the results that are similar among glass bottles and differing considerably from a water plastic bottle.

| Product | Material | Color | Thickness (mm) |
|---|---|---|---|
| water | plastic (PET) | clear | 0.22 ± 0.02 |
| fruit juice | glass | clear | 1.92 ± 0.42 |
| malt beverage | glass | amber | 2.28 ± 0.44 |
| beer | glass | green | 2.02 ± 0.47 |

Table 7. Thickness of common containers evaluated with OFC Raman.

Then the region between 700 $cm^{-1}$ and 900 $cm^{-1}$ of acetone was used to study the effect of different collection parameters on the spectra inside different bottles. The acquisition time was changed from 1 to 30 s. The laser power was varied from 11 to 32 mW. The power intensity of the instrument modulating and the output power was measured at the probe with a digital power meter. At fixed laser power of 18 mW the response was linear with increasing acquisition time (Figure 14). For the amber glass bottle the increase in peak area was significantly lower. This suggests that the bottle color is responsible for the fluorescence.

The transmission spectra of different bottles are shown in Fig. 15. These spectra were acquired in the 220 nm to 1200 nm range. The objective of this experiment was to measure how much light is allowed transmitted in a specific bottle at different wavelengths. Solid vertical lines represent the laser wavelengths evaluated in the present work and dashed vertical lines mark the wavelength range for Raman spectra with shifts from 200 to

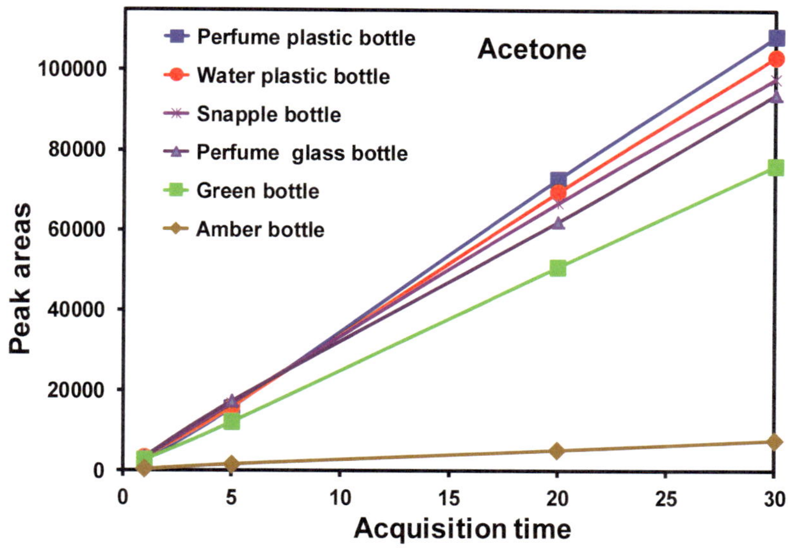


Fig. 14. Detection of acetone at different collection times inside different bottles at 532 nm and 18 mW (1 scan).

1800 $cm^{-1}$. According to this illustration all clear containers (glass and plastic) will allow light of all wavelengths to reach the container as well as allow detecting scattered radiation. However for amber containers the optimum range for interrogation will be 600 to 750 nm. The comparison with green containers suggests that the container material absorbs most of the light but a spectrum of the content would be possible with the appropriate laser power.

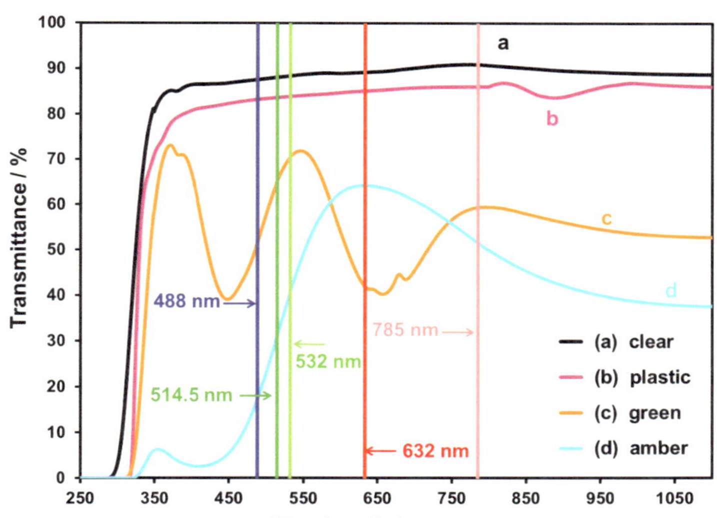


Fig. 15. Transmission spectra of different bottles materials. Solid vertical lines indicate excitation lines in Raman experiments. Dashed lines indicate range for detected scattered radiation in the 200-1800 $cm^{-1}$.

The limits of detection of a CWA simulant in heavily colored liquids were studied by OFC Raman. Triethyl phosphate (TEP) is commonly used as a simulant of Soman (GD), a nerve agent. Fig. 16a presents the spectra of different flavors of Gatorade. The spectra were obtained at the same conditions. Fruit punch presents a higher baseline associated to endogenous fluorescence. TEP was prepared in random concentration from 0 to 100% (V/V). Then the region from 675-855 $cm^{-1}$ was integrated (Fig. 16b). This region was selected because of the presence of the characteristic peaks for phosphates. This region will be the same for the real CWA and related simulants.

Measurements were taken in triplicates from 0 to 100%. Fig. 16c presents average peak areas as a function of TEP concentration in the highly colored solutions concentration. The objective of these graphs is to compare the limit of detection (LOD) and the limit of quantification (LOQ) between colored liquids. Error bars for each data point represent the calculated standard deviation derived from all repetitive measurements each.

| Gatorade flavor | LOD (%) | LOQ (%) |
|---|---|---|
| fruit punch | 6 | 21 |
| lemon | 9 | 30 |
| orange | 9 | 32 |

Table 8. LOD and LOQ for TEP in highly colored Gatorade liquids.

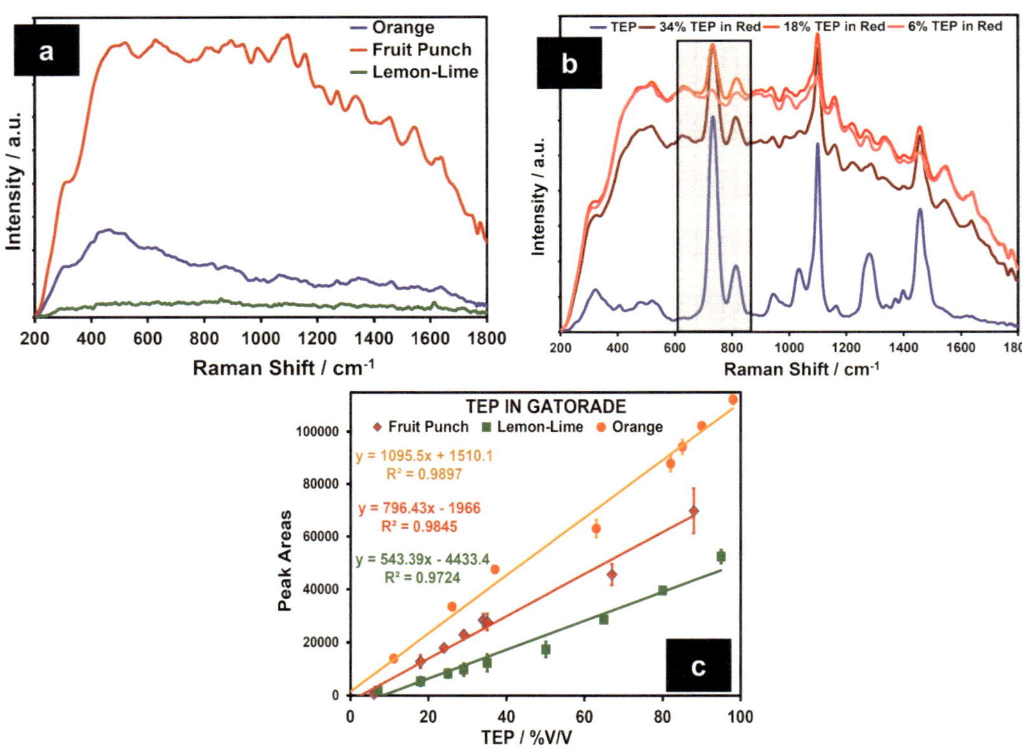


Fig. 16. (a) Raman spectra of Gatorade lemon, orange and fruit punch. (b) Spectrum of neat triethylphosphonate. (c) Peak areas vs. TEP content in the highly colored solutions.

Limits of detection for this analyte have been calculated according to IUPAC by the 3σ criteria (3 times standard deviation of the peak-to-peak noise related to the slope of the linear regression function). When the signal is 3 times as great as the noise, it is ready detectable but still too small for accurate measurement. A signal that is 10 times as great as the noise is defined as the lower limit of quantification (LOQ), or the smallest amount that can be measured with reasonable accuracy. Table 8 presents the results of LOD and LOQ for TEP in Gatorade lemon, orange and fruit punch.
A 488 nm OFC Raman probe was designed, built and tested in quantification of CWA simulants camouflaged as consumer beverages such as purified water and energy drinks.

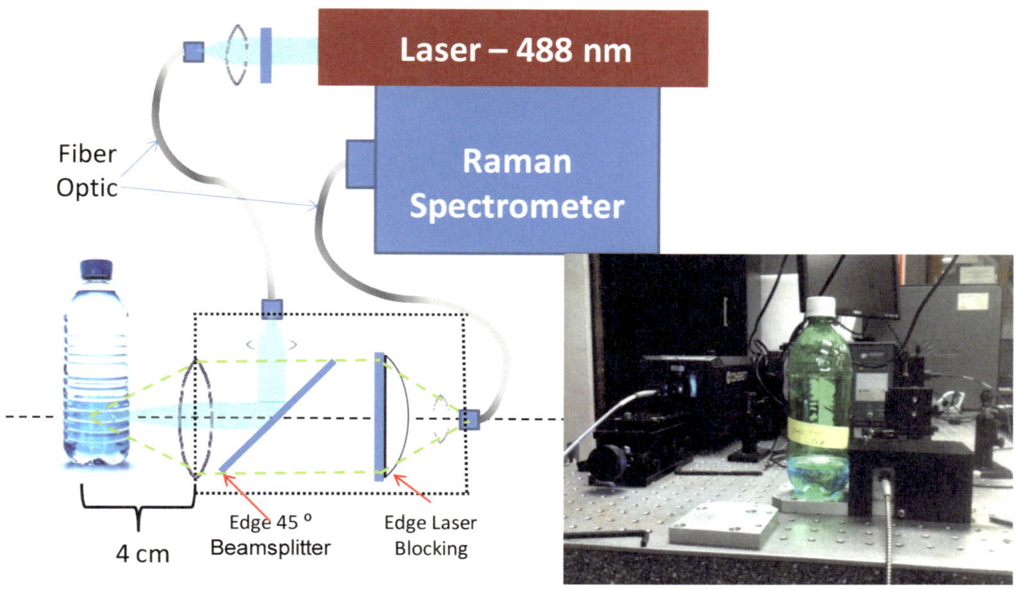


Fig. 17. Optical fiber coupled Raman probe for excitation at 488 nm. Interface to spectrograph was also done with OFC.

Raman experiments were performed using the strong blue line of an argon ion laser INNOVA 310/8 from Coherent, Inc. at 488.0 nm.A single strand optical fiber (non-imaging, 600 μm diameter, model AL 1217, Ocean Optics, Inc.) was used to couple the Raman probe to which a set of laser line filter (to clean satellite lines) and Semrock RazorEdge™ edge filter was used to filter the Rayleigh scattered light. An Andor Technologies spectrograph: Shamrock SR-303i (aperture: f/4; focal length: 303 mm; wavelength resolution: 0.1 nm or 4.2 $cm^{-1}$ at the excitation wavelength) equipped with a 1200 grooves/mm grating was used to analize the Stokes scattered light. A high performance, back thin illuminated CCD camera (Andor Technologies model # DU970N-UVB) with quantum efficiencies of 90% (200 $cm^{-1}$) to 95% (3200 $cm^{-1}$) served as light detector. The probe was designed as a combination of two edge filters: the first works at 45° and acts as a mirror for 488.0 nm light, blocking the backscattered Rayleigh light. The second is an edge filter laser blocking at 0° and blocks of remains of the Rayleigh light and transmits the Stokes component of the inelatically scattered signal. The laser light was guided to the probe by optical fiber and the Stoke Raman signal was also transmitted to the spectrometer by optical fibers. A 4 cm focal length lens focuses the laser light onto of the interior of the plastic/glass bottles. This makes the signal produced at the wall of the bottle minimal.

Four different bottles: clear glass, green glass, brown glass and plastic were used for measure signal Raman of simulant within the container. The simulant used was TEP. Mixtures of simulant and water were used for check the limit of detection. Fig. 18a shows typical OFC Raman spectra of TEP solutions in plastic bottles excited by 488.0 Ar+ line. The calibration curves were obtained by using partial least squares (PLS) regression algorithm of chemometrics and two models were generated. The first model used no preprocessing and the entire spectral region available. For the second model, selected spectral regions were picked by an optimization method, until the best preprocessing was achieved. The optimal regions are shown in Fig. 18b. The optimal preprocessing was vector normalization.

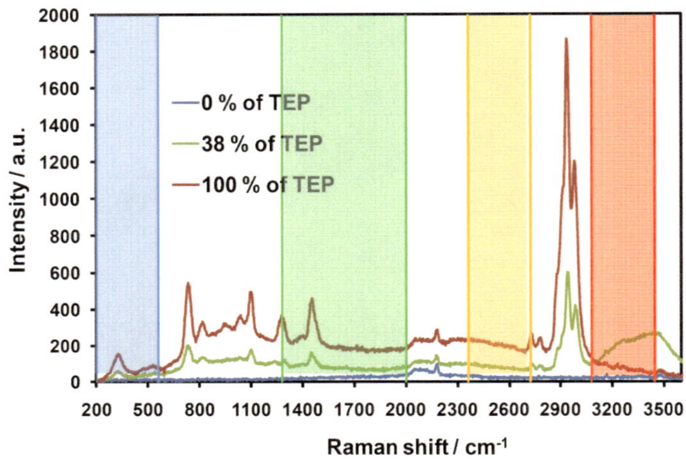


(a)

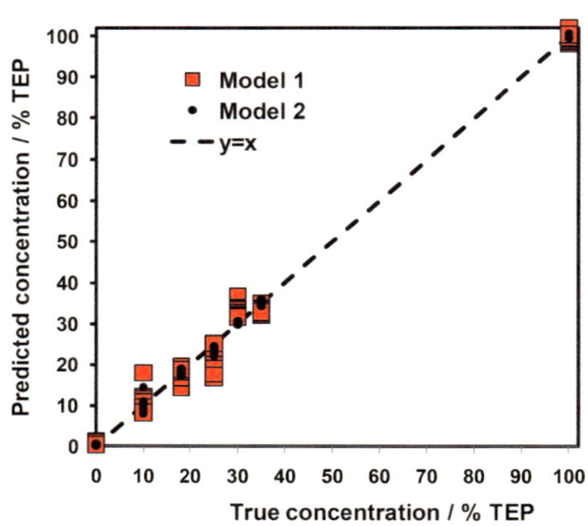


(b)

Fig. 18. (a) OFC Raman spectra of TEP solutions contained in plastic bottles and excited by 488.0 laser line. (b) Predicted vs. true concentrations of TEP in aqueous solutions.

Results of the analysis of the experiments are summarized in Table 9. Both models generated worked well for predicting the TEM concentration, with slightly improved results when vector normalization was applied to the data and individual spectral windows were used instead of the complete spectral range. As can be inferred from Table 9 and using LOD values calculated as 3.3xRMSECV, the detection limit for the model 1 was 7% and for the model 2 was 3%.

| | **Model 1** | **Model 2** |
|---|---|---|
| Region ($cm^{-1}$) | 3600-199 | 3441.5-3080.2; 2721.3-2360; 2001-1278.5; 560.5-199.2 |
| Preprocessing | none | Vector normalization |
| $R^2$ | 99.87 | 99.91 |
| $R^2CV$ | 99.32 | 99.88 |
| RMSEE | 1.14 | 0.914 |
| RMSECV | 2.48 | 1.03 |
| Rank | 6 | 5 |

Table 9. Summary of statistical inferences for the two methods of analysis of OFC Raman spectral data of TEP solutions.

## 4. Optical fibers coupled remote raman detection of chemical warfare agents simulants

Chemical warfare agents (CWA) can be classified as weapons of mass destruction (WMD). They include nerve agents, blister agents, choking agents and blood agents. Nerve agents are a group of particularly toxic chemical warfare compounds (Marrs et al., 1996). They were developed just before and during World War II, and they are chemically related to organophosphorus insecticides. The principal compounds in this group are Tabun (GA), Sarin (GB), Soman (GD) and methylphosphonothioic acid (VX). During last two decades, the world has suffered many terrorist attacks that employed chemical warfare agents (CWA) and other hazardous compounds. Examples of such events were seen during the Iran-Iraq war (Henderson, 1999), and in the terrorist attacks in Matsumoto and the Tokyo subway in Japan (Miyaki et al., 2005). These types of terrible events have motivated many countries to focus their defense and security-related research toward the detection of explosives, hazardous liquids and chemical agents that can be used by terrorist organizations as WMD threats against troops or civilians. The anticipation of future attacks requires a wide array of detection systems for a variety of potential deployment scenarios (Sun et al., 2005; Farquharson et al., 2005). There is a need for more sensitive and selective remote detection techniques for chemical threat compounds operating at ambient conditions *in situ* and on a realistic time scale. Remote Raman Spectroscopy (RRS) provides a method for identifying chemicals in samples located meters from the excitation source. In fact, telescope-based Raman spectroscopy detection methods have been reported for standoff detection of chemicals using both visible and UV laser excitation (Hirschfeld, 1974; Angel et al., 1992; Wu et al., 2000; Sedlacek et al., 2001; Sharma et al., 2002; Thomson and Batchelder, 2002; Sharma et al., 2003; Sharma et al, 2005; Pacheco-Londoño et al., 2009; Ramírez-Cedeño et al, 2010).

Recently, our group has applied Remote Raman and Infrared detection systems to the detection and quantification of military high explosives (HEs) and homemade explosives: HME (Pacheco-Londoño et al., 2009) and also the remote detection of hazardous liquids concealed in commercial products bottles (Ramírez-Cedeño et al, 2010). In this manuscript, we report on the design, assembly and testing of two RRS Systems: one operating with continuous wave (CW) laser lines and the other using a pulsed laser system. The collector telescope was modified to operate with both visible (400 - 700 nm) and near-ultraviolet (NUV: 200 - 400 nm) excitation sources for use in the detection of CWAS: dimethyl methylphosphonate (DMMP), 2-chloroethylethylsulfide (2-CEES) and 2-(butylamino)-ethanethiol (2-BAET). Raman scattering cross sections of the studied CWAS were measured using VIS and NUV excitation lines.

The remote-sensing spectroscopic system was also used to detect toxic industrial compounds (TICs): benzene, chlorobenzene, toluene, carbon tetrachloride, cyclohexane and carbon disulfide. The experiments utilizing this remote system used excitation laser lines at 514.5, 488.0, 363.8 and 351.1 nm and a target-collector telescope at a fixed distance of 6.6 m. Further modification of the reflector telescope for use as a receiver allowed for near-field sensing applications at target-collector distances as close as 2.2 m to the target threat liquid chemicals (CWAS and TIC). The visible CW laser excitation system has been described in previous publications (Pacheco-Londoño et al., 2009; Ramírez-Cedeño et al, 2010). Remote Raman measurements of TIC, which are typically strong inelastic scatterers, were initially carried out using this system without any modifications at a fixed target-collector distance of 6.6 m with visible light excitation. In order to measure remote Raman spectra in the NUV and to measure at closer target-collector distances, several modifications had to be made to the receiver reflector telescope. The modified experimental setup for the prototype RRS-based system is shown schematically in Fig. 17. The spectroscopic analysis system consisted of an Andor Technologies Shamrock SR-303i spectrograph, which was equipped with a high quantum response charge-coupled device detector (CCD, Andor Technologies model Newton™ DU-970N-UVB), the appropriate filters for the rejection of satellite plasma lines (laser line filters), and a laser radiation filter designed to block Rayleigh scattered light (edge filters, obtained from Semrock, Inc. Rochester, NY). The nominal detector efficiencies were 95% (532-570 nm), 93% (514.5 nm), 90% (488 nm) and 35% (350-390 nm).

The other necessary components were a reflective telescope used as a collector or signal receiver, a optical fiber assembly, and a single-line laser system operating at 351.1, 363.8, 488.0, 514.5 and 532 nm (Coherent INNOVA 308, Coherent SABRE 25/7 argon ion laser systems; Coherent VERDI-5 solid state diode laser system). The telescope used was a MEADE ETX-125 Maksutov-Cassegrain design (125 mm clear aperture, 1900 mm focal length $f/15$). The reflecting collector was coupled to the Raman spectrometer with a non-imaging, 600 μm diameter optical fiber (model AL 1217, Ocean Optics, Inc., Dunedin, FL). Two lenses were used to collimate the light from the telescope output, from which the focusing objective was removed, and direct it into the fiber optic assembly. The output of the fiber optic assembly was directly coupled to the Raman spectrometer entrance slit.

The telescope used in the remote detection system was obtained from the manufacturer as a reflective receiver operating in the VIS region only. It was modified to allow for the collection of scattered Raman signals in the near-ultraviolet region, 350-390 nm, by coating the secondary mirror with a thin layer (~ 200 nm) of UV-reflective aluminum. In addition, the minimum focal point where a clear image could be formed was 5 m. An anodized aluminum tube with capabilities of rigidly holding two quartz lenses and moving the lenses to change the focal distance was integrated to the beam path to allow for a reduction of the minimum focal point to 2.0 m adding near-field proximity detection capability. The system was successfully tested at a 2.2 m target-collector distance. Although RRS spectra are not shown, data is but included as part of system performance tests. The changes made to the collector telescope are also illustrated in Fig. 19.

The pulsed laser standoff Raman system used the components of the CW system with the exception of the excitation source and the spectrometer detector. A frequency-doubled 532 nm Nd:YAG pulsed laser system (Quanta Ray INDI Series, Newport-Spectra Physics, Mountain View, CA) was used as the excitation source. The maximum energy/pulse of the laser at 532 nm was 200 mJ, and it operated at a repetition rate of 10 Hz. The pulse width was approximately 5-8 ns, and the beam divergence was less than 0.5 mrad. A gateable, intensified CCD detector (iStar™ ICCD camera, Model DH-720i-25F-03, Andor Technology, Belfast, Northern Ireland) was used as the photon detector. Andor Technology Solis™ software for spectroscopic, imaging and time-resolved studies was used for spectral data acquisition and processing from the intensified and gated CCD detector. Using this software, the data could be acquired in both imaging and spectroscopic modes.

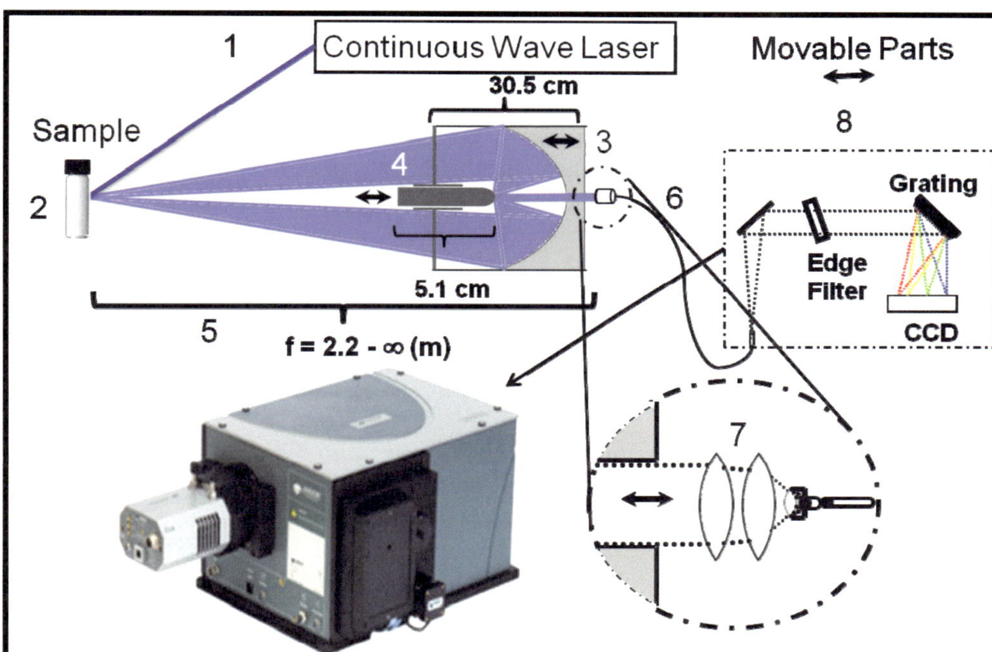


Fig. 19. Design details of the Remote Raman Detection System: (1) laser source; (2) sample; (3) reflective telescope signal collector; (4) variable focus secondary mirror; (5) standoff distance; (6) fiber optic coupling; (7) details of optical coupling; (8) spectrograph. Movable optical elements are represented by double head arrows.

The remote system was tested using TIC and CWAS. The toxic industrial compounds investigated were chlorobenzene, toluene benzene, carbon disulfide, carbon tetrachloride and cyclohexane (all from Fisher Scientific International, Chicago, IL). The chemical warfare agents simulants (CWAS) studied were dimethylmethyl phosphonate (DMMP, 99%, Fisher Scientific International, Chicago, IL), 2-chloroethyl ethyl sulfide (2-CEES, Sigma Aldrich Chemical Company, St. Louis, MO), and 2-(butylamino)-ethanethiol (2-BAET, Sigma-Aldrich). For CW remote detection experiments, pure liquid samples were transferred to 5 mL glass or quartz vials and placed 6.6 m away from the receiver telescope/excitation laser. The spectra of all compounds were collected with the laboratory lights off to avoid background light and mercury lines from fluorescent lamps. The spectra were acquired in the Raman Shift range of 100-1800 $cm^{-1}$ at laser powers (measured at head) ranging from 0.05 to 1 W and a single acquisition with integration time of 1-30 s. Quantification studies of DMMP/water solutions were carried out at a fixed 6.6 m target-collector distance using the CW RRS system only.

Standoff detection experiments on DMMP at 35 m and cyclohexane at 60, 90 and 141 m were carried out using pulsed mode laser RRS system. DMMP was contained in clear glass bottles 2.5 cm in diameter and 5.0 cm high. For acquisition of remote Raman spectra of cyclohexane at the longest distances, the sample was contained in clear glass bottles that were 7.5 cm in diameter and 15.2 cm high. The criterion used for changing vials at longer distances was that the beam diameter at the sample would be smaller than the cross section of the vials containing the samples. Liquids were remotely detected in the spectroscopic range of 500 to 3200 $cm^{-1}$ using 1 to 1000 pulses of 532 nm excitation wavelength, with pulse energy of ~ 200 mJ.

Important industrial solvents and starting materials used for manufacturing in the petrochemical, pharmaceutical, electronic, and other chemical industries are highly flammable, lachrymatory, toxic, mutagenic or carcinogenic. Some of these compounds are used in amounts that constitute potential threats. Thus, it is vital to find ways of monitoring these compounds in air and in their container bottles. Remote Raman systems offer new and simple alternatives to carry out these monitoring processes by remote sensing of the liquid and even the vapor phase (UV Raman). Figures 2 and 3 illustrate the remote CW Raman spectra of organic solvents, some of which are constituents of hydrocarbons found in petroleum-processing plants. All of the spectra were collected at a distance of 6.6 m from the telescope under laboratory conditions using low illumination. Samples were detected using 514.5 nm laser line excitation at 1W (measured at head) using a single acquisition and integration time of 10 s.

The most prominent feature common among all the spectra in Fig. 20a is the aromatic ring-breathing mode at ca. 1000 $cm^{-1}$. This strong Raman band is characteristic of the symmetric stretch mode of the benzene ring at 992 $cm^{-1}$ (Lin Vien et al, 1991; Shrader, 1995; McCreery, 2000). In the case of toluene and chlorobenzene, the same band appeared at 1003 $cm^{-1}$. Another characteristic band of toluene was found at 786 $cm^{-1}$. The C-Cl stretching mode of chlorobenzene was clearly observed at about 700 $cm^{-1}$ in the Raman Shift spectrum of this TIC.

Several organic solvents widely used in industry, teaching, and research labs were also studied. Among these solvents were carbon disulfide, carbon tetrachloride and cyclohexane. Typical Remote Raman spectra are shown in Fig 20b. All of these compounds have characteristic physical and chemical properties. Cyclohexane is typically used as a calibration standard for the Raman Shift axis in dispersive instruments, particularly in Raman studies of liquids (Evans and Bernstein, 1956). In this work, the Remote Raman spectra of these compounds were measured for liquids contained in 5 mL clear glass vials in the 150–1800 $cm^{-1}$ range at a standoff distance of 6.6 m (Fig. 20b).

In Fig. 20b, it is possible to identify carbon disulfide by its very strong (highly symmetric nature) peak at ca. 655 $cm^{-1}$, corresponding to the C-S symmetric stretching mode (Wakabayashi et al., 2007). Carbon tetrachloride shows its three main Raman active peaks. The most important band for $CCl_4$ appears at 461 $cm^{-1}$. This band is attributed to C-Cl symmetric stretching mode (Crain et al., 1992). According to the literature, cyclohexane is characterized by four prominent bands in the 800-1650 $cm^{-1}$ region: a very strong peak at 801 $cm^{-1}$ due to the C–C skeletal breathing mode, the weak peak localized at 1029 corresponds to C–C stretching modes, and other, weak peaks that appear at 1260 and 1445 $cm^{-1}$ due to twisting and scissoring modes (Evans and Bernstein, 1956; Sharma, 2007).
The continuous wave (CW) standoff Raman system was originally designed to measure Raman Shift spectra excited at visible light frequencies (514.5 and 488.0 nm) at a fixed remote distance of 6.6 m. This was based on two main factors: maximum standoff distance due to restrictions based on laboratory space and minimum focal distance of the visible wavelength reflective telescope used as receiver. After successful measurements under the original operating conditions of HE, HME [14] and TICs (the present study), the system was switched operate in the near-ultraviolet region (NUV), and the signal receiver (telescope) was modified to operate as a close field (proximity) detector (1-2 m). After making the necessary changes, the remote detection system was subjected to the challenging task of detecting extremely low cross-section liquids: CWASs.

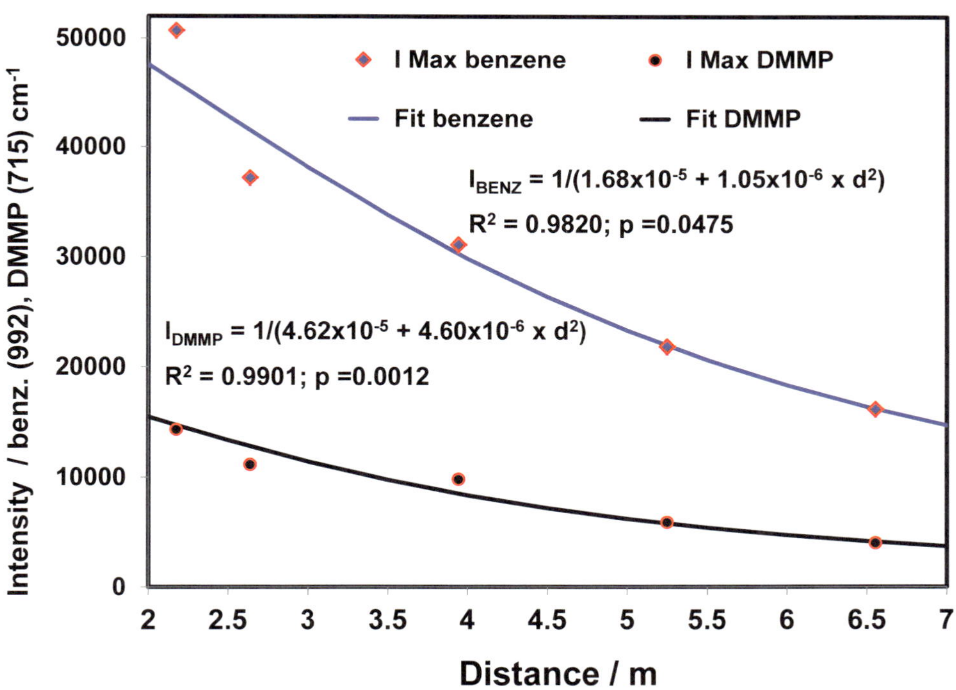


Fig. 20. Dependence of Raman Shift signal (vibrational band intensities) with standoff distance for benzene and DMMP using the 363.8 nm excitation line from a UV-argon ion laser.

Fig. 21 shows results of the remote detection system for the intensities of the Raman bands of benzene (ring breathing mode at 992 $cm^{-1}$) and DMMP (C-P stretch at 715 $cm^{-1}$) at various distances from 2 to 7 m using a CW argon ion laser excitation line at 363.8 nm. The dependence of the band intensities for the most prominent signals of benzene (used as primary external standard) and DMMP were plotted and then a line was fitted to the data as a function of the target-receiver distance using Statgraphics-Centurion™ data analysis and statistical software package (StatPoint, Inc., Herndon, VA). The resulting fits, shown in Figure 4, were non-linear, as expected, and compared favorably with the data.

The decrease of the Raman intensity of the benzene and DMMP peaks with increasing distance is partially attributed to the $1/R^2$ dependence of the remote Raman signal with standoff distance. However, the signal losses are also related to near-field effects and to partial defocusing of the image on the slit of the spectrograph (Chistesen, 1988). The current version of the CW laser remote spectroscopic system is probably less able to detect very long source-target distances (standoff distances) compared to pulsed laser systems because of the higher energy densities in the beam axis of the pulsed systems. However, at close range, working as a near field detector, the CW standoff Raman detection system works equal or better than the pulsed system. The following two cases will provide experimental evidence for this supposition.

Remote Raman spectra were excited using uncollimated CW and pulsed laser beams. Laser power levels of CW lasers were 0.140 to 5.1 W. The calculated values for energy/area taking into account the laser beams spot size at the sample were 1.5 to 8.3 $W/cm^2$ (1.5 - 8.3 $W/cm^3$ energy densities for a sample vial 1-cm in diam). For the pulsed laser experiments, up to 1000 pulses of 200 mJ/pulse (6 ns, 10 Hz) were used, resulting in a maximum average power of 2.0 W. At a standoff distance of 6.6 m, the pulsed laser spot was an ellipse with a major semi-axis of 2.5 cm and minor semi-axis of 1.5 cm. This corresponds to an energy/area of 0.68 $W/cm^2$ (0.68 $W/cm^3$ in a 1-cm diam. vial) or roughly 47% of the minimum value for the steady-state detection system. Table 10 lists the relevant laser beam characteristics of the standoff system. In contrast, typical normal or spontaneous Raman measurements under the microscope use power density values on the order of 12,500 $W/cm^3$ to excite the Raman Shift spectra of samples contained in capillary tubes (100 mW, 10x objective, 10 μm spot size in confocal mode, $8\times10^{-6}$ $cm^3$ interrogation volume). This represents a 1,500 to 8,000-fold higher energy density value for the microscope experiments than the telescope-based CW Raman experiments. Raman scattering cross sections, $(d\sigma/d\Omega)_s$, of CWAS were calculated by performing intensity measurements and relating them according to the treatment of Christesen (1988) via Eq. 1:

$$\frac{\left(\frac{d\sigma}{d\Omega}\right)_s}{\left(\frac{d\sigma}{d\Omega}\right)_r} = \frac{A_s\, N_s\, n_s\, E_s}{A_r\, N_r\, n_r\, E_r} \qquad (1)$$

In the above equation, the subscripts s and r indicate the CWAS and reference values, respectively. $A_s$ and $A_r$ are the sample and reference integrated peak areas, which was measured from the spectra, and $N_i$ is the number of molecules per unit volume, which was obtained from the sample and reference densities. The collection solid angle of the spectrometer is a function of the index of refraction of the liquid ($n_i$) and was accounted for by the ratio ($n_s/n_r$). $E_i$ is an instrument efficiency factor that depends on the wavelength of the scattered light. If the sample and reference bands are close together (as they are typically chosen) the ratio $E_s/E_r$ can be taken as 1. An additional factor would be necessary in Eq. 1 if either the sample or reference absorb at the laser wavelength (Christesen, 1988).

| λ (nm) | Spot Diam. at 6.6 m (cm) | Area ($cm^2$) | Laser Power (Energy) (W / mJ) | Energy/Area ($W/cm^2$) |
|---|---|---|---|---|
| 532.0 (CW) | 1.0 | 0.79 | 5.10 | 6.49 |
| 514.5 | 0.4 | 0.13 | 1.00 | 7.96 |
| 488.0 | 0.4 | 0.13 | 1.00 | 7.96 |
| 363.8 | 0.35 | 0.10 | 0.80 | 8.32 |
| 351.1 | 0.35 | 0.10 | 0.14 | 1.46 |
| 532.0 (pulsed) | 2.5x1.5* | 2.95 | 2.00 / 200 | 0.68 |

* spot is elliptical in shape

Table 10. Remote Raman spectroscopy systems laser beams characteristics.

This method was applied to a comparison between the most prominent Raman band of the CWAS and the 992 $cm^{-1}$ band of benzene and/or the 801 $cm^{-1}$ band of cyclohexane. Benzene and cyclohexane were chosen as references (external primary standards). The cross-section data for the CWAS and reference compounds studied are shown in Table 11. These compounds are relatively weak Raman scatterers in comparison to the reference compounds. For this reason, the standoff detection of CWAS was an important instrument challenge, as suggested by Christesen (Christesen, 1988).
The Remote Raman spectra of three CWASs: 2-BAET, 2-CEES and DMMP, are shown in Fig. 21. All spectra were collected in the visible and NUV at a distance of 6.6 m from the collector telescope. The samples were detected using the strong blue line of the argon ion laser at 488.0 nm and NUV lines at 363.8 and 351.1 nm with 1 W laser power (at head) and a spectral acquisition time of 10 s. Raman peaks at 660 and 1440 $cm^{-1}$ were tentatively assigned to C-S and C-N stretching vibrations, respectively, for the 2-BAET simulant (Fig. 21, top). A strong line appearing at about 700 $cm^{-1}$ was tentatively assigned to C-Cl in 2-CEES (Fig. 5, center). Raman signals located about 660 and 750 $cm^{-1}$ were attributed to the C-S and C-S-C of this simulant, respectively. The peak at 715 $cm^{-1}$ for DMMP (Fig. 21, bottom trace) was assigned to a P-C stretching mode.

| | Cross Sections ($10^{-30}$ $cm^2/sr/molecule$) | | | | |
|---|---|---|---|---|---|
| Exc. line (nm) | benzene (992 $cm^{-1}$) | cyclohexane (801 $cm^{-1}$) | DMMP (715 $cm^{-1}$) | 2-CEES (700 $cm^{-1}$) | 2-BAET (1440 $cm^{-1}$) |
| 532.0 | 13.4 [c] | 3.6 | 3.7 | 2.8 | 2.7 |
| 488.0 | 32.5 [c] | 9.06 [b] | 18.0 [a] | 15.1 | 20.3 |
| 363.8 | 328.4 | 105.6 | 58.0 [a] | 21.4 | 10.3 |
| 351.1 | 160.0 [b] | | 17.2 | 5.4 | 3.3 |

Data adapted from: a ~ (Evans and Bernstein, 1956); b ~ (Wakabayashi et al., 2007); c ~ (Crain et al., 1992).

Table 11. Relative differential Raman scattering cross sections.

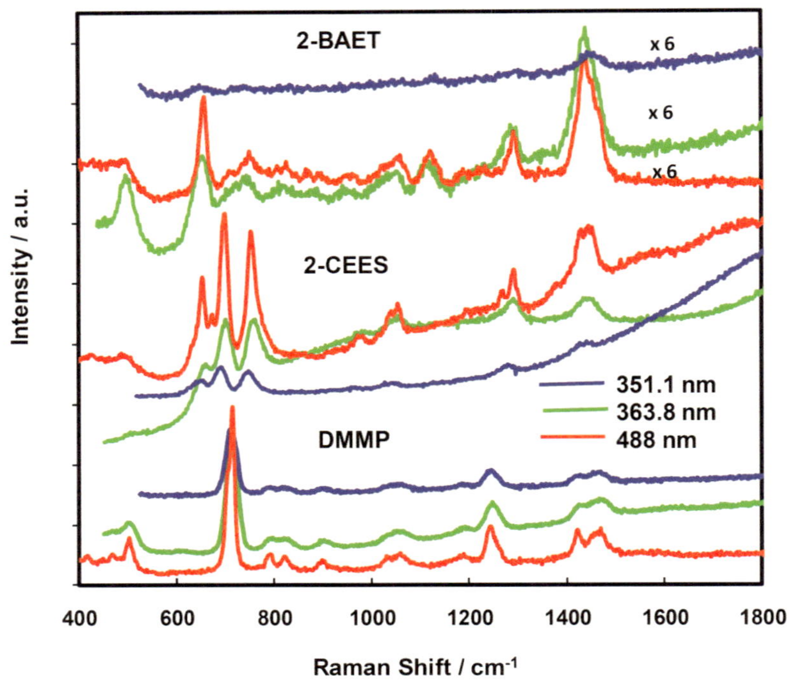


Fig. 21. Remote Raman spectra of CWAS using excitation lines of 488.0, 363.8 and 351.1 nm at 6.6 m target-collector distance, 1 W laser power (at head) and one acquisition time of 10 s integration. Top: 2-(butylamino)-ethanethiol (2-BAET); center: 2-chloroethyl-ethyl sulfide (2-CEES); bottom: dimethylmethyl phosphonate (DMMP).

Pulsed laser Standoff Raman spectra of cyclohexane and DMMP in the spectral range 500 - 3200 $cm^{-1}$ are shown in Figs. 22a and 22b. The measurements were collected at distances of 35 m (DMMP) and 60, 90 and 141 m (cyclohexane) with gated detection using several 532 nm laser pulses (from 1 to 1000 shots) with 200 mJ per pulse. These experiments were performed under "lights on" conditions. The most relevant spectroscopic information for both compounds was presented in the CW laser excitation section. Pulsed mode experiments (Figs. 6 and 7) were obtained using the ICCD camera, which was gated and synchronized with laser pulses to minimize interference from ambient light. The cyclohexane RRS spectra at 60 m and 90 m standoff distances were nearly identical. This similarity was the result of a size beam smaller that the target at both distances. However, the beam diameter at 141 m was significantly larger than the sample, which resulted in a reduced energy density and a lower scattering signal. However, even operating under these conditions, the cyclohexane spectrum at 141 m obtained with 1000 laser pulses had a good signal-to-noise ratio (S/N = 450). The S/N value was determined by dividing the highest intensity Raman band of cyclohexane by the Root-Mean-Square (RMS) noise, calculated by taking a portion of the flattest region of the cyclohexane spectrum (1900-2100 $cm^{-1}$). Both compounds can be detected with a single laser shot, but S/N became statistically significant by averaging the intensity collected after 10 laser shots, as can be seen in Fig. 22a. It is important to note the advantages of a pulsed laser system over CW systems for Raman detection at a distance. When operating in a gated detection mode, the background light signal and fluorescence signal are significantly reduced. Because the detector is acquiring for 400 ns, the fluorescence contributions that dominate at a longer time scale (~ μs) are minimized. One of the main disadvantages of the CW Remote Raman detection system is the necessity of operating under low-illumination conditions. This problem can be readily circumvented by using a gated Intensified-Charge Coupled Device (I-CCD), UV-VIS-capable detector.

An even more challenging application for the standoff Raman spectroscopy detection system was to perform quantification studies of a weak Raman scatterer mixed to less than 10% dilution with water. This task was undertaken using DMMP. In order to carry out the quantification experiments on DMMP, water, solutions of the analyte ranging from 1 to 50 % w/v were prepared and analyzed using CW RRS. For the quantification studies, peak areas of the strongest DMMP Raman signal located about 715 $cm^{-1}$ were used (Fig. 8).

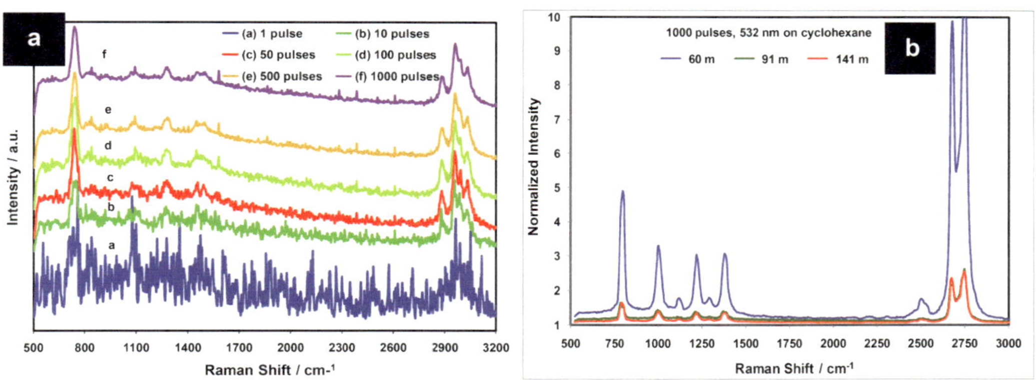


Fig. 22. Left: Remote Raman spectra of DMMP using a 532 nm pulsed laser excitation source at a distance of 35 m, measured with various laser shots in gated mode. Laser: 532 nm, 200 mJ/pulse, 10 Hz. Right: Remote Raman spectra of cyclohexane using a 532 nm pulse laser at standoff distances of 60, 90 and 141 m, measured with 1000 laser shots in gated mode. Laser: 532 nm, 200 mJ/pulse, 10 Hz; gate width 400 ns.

## 5. Conclusion

In this contribution, TIJ and sample smearing were used as deposition methods for the preparation of samples and standards of highly energetic materials deposited as solid traces onto gold coated silicon substrates. Inkjet printing of HEM was demonstrated to have the following important characteristics: precision in sample deposition, drop delivery with non-contact fluid transfer and high reproducibility. These characteristics led to the production of evenly distributed analytes on the test surfaces. The methodology promises to be a good method for development of samples and standards for trace HEM reference materials on substrates. Sample smearing, a technique that had been proven successful for similar operations on stainless steel, glass and plastics, although easy to implement and of inherently low cost, yielded very poor yields on the gold on silicon substrate because of weak interactions between the analytes studied and the test surfaces. Gold coated silicon is probably an excellent test surface for IR reflection studies, but it lacks the sample adhesion forces required to deposit and retain analytes for long enough periods of time to be of value in samples and standards preparation. The residence time of TNT on gold coated silicon was twice as short as TNT on glass and four times shorter than TNT on a stainless steel substrate.

Both methods demanded relatively long preparation and analysis times, depending on the specific sample and substrate requirements, and thus neither methodology stands out as superior from these points of view. Sample deposition by smearing requires higher human intervention, both in the pre-deposition steps and in the actual smearing operation. TIJ requires minimal operator intervention. Deposition by TIJ is a more efficient process than smearing with regard to ease of use, fine control of the amount deposited and the surface loadings achieved. The actual amounts deposited (loadings) had to be determined post-deposition by destructive chemical analysis. This is in contrast to sample smearing deposition, for which the amount deposited is accurately known except in cases of insufficient adhesion such as with gold coated substrates.

The loading concentrations of the highly energetic materials studied (TNT, RDX and ammonium nitrate) were varied by changing the number of passes of the TIJ dispenser. TIJ samples and standards were deposited onto three surfaces, gold-silicon, glass and SS, for surface comparison experiments. Studies are required when searching for substances that stabilize the analytes on the substrate by delaying sample loss through sublimation, which would ultimately interfere with detection applications. The characteristics of the stabilizing agent must not alter or mask the activity and reactivity of the energetic materials with respect to the detection applications.

A standoff technique using an open-path Fourier transform infrared (OP/FTIR) spectrometer has been demonstrated in In another contribution, concealed liquids scenarios are studied by Raman spectroscopy. The Raman spectra of hazardous liquids were differentiated from common drinks and consumer products. A fiber optic coupled Raman probe was applied to the detection of hazardous liquids mixed with consumer products and drinks. The results demonstrated that Raman can be used as tool to quickly characterize if the content of a bottle is the intended commercial product or a hazardous liquid that could be used as a threat to property or human beings. The fluorescence of a red liquid did not affect the detection of a CWA simulant in a clear glass container. The limits of detection for a hazardous liquid in a series of colored liquids were estimated in 6-9 % using standard conditions and no data manipulation. This suggests that trace level detection can be achieved with enhanced experimental setups and statistical analysis of the data (chemometrics).

In the third application of optical fibers in spectroscopy, remote Raman systems have been designed, assembled and tested by coupling a Raman spectrometer with a reflective telescope using fiber optics. The CW RRS system employed 351.1, 363.8, 488.0, 514.5 and 532 nm continuous wave lasers as excitation sources. This prototype system was used in the detection of the chemical warfare agent simulants DMMP, 2-CEES and 2-BAET, and to detect the hazardous industrial solvents and reagents (TICs) benzene, toluene, chlorobenzene, carbon disulfide, carbon tetrachloride and cyclohexane. The operational range of the CW standoff system was tested up to 6.6 m in the laboratory with no background illumination. In addition, quantification studies of DMMP in water were carried out at a standoff distance of 6.6 m using the CW remote Raman detection system. Low limits of detection (LOD) values of 3% w/v were consistently obtained. The pulsed mode RRS system was based on a 532 nm, frequency doubled Nd:YAG laser in lieu the CW excitation lasers of allowed standoff detection experiments of DMMP at 35 m target-collector distances, from single shot to 1000 shots and also allowed detection of cyclohexane at 141 m standoff distance even in single shot mode. The Remote Raman Spectroscopy systems designed in this work should be useful for defense and security applications, for screening hazardous liquids in government installations, airports and seaports and in public installations to improve defense against terrorist attacks.

## 6. Acknowledgments

Parts of the work presented in this contribution were supported by the U.S. Department of Defense, University Research Initiative Multidisciplinary University Research Initiative (URI)-MURI Program, under grant number **DAAD19-02-1-0257**. The authors also acknowledge contributions from Mr. Aaron LaPointe from Night Vision and Electronic Sensors Directorate, Fort Belvoir, VA, Department of Defense, Dr. Jennifer Becker MURI Program Manager, Army Research Office, DoD and Dr. Stephen J. Lee Chief Scientist, Science and Technology, Office of the Director, Army Research Office/Army Research Laboratory, DoD.

Support from the U.S. Department of Homeland Security under Award Number **2008-ST-061-ED0001** is also acknowledged. However, the views and conclusions contained in this document are those of the authors and should not be interpreted as necessarily representing the official policies, either expressed or implied, of the U.S. Department of Homeland Security.

The authors want to acknowledge graduate and undergraduate students from the Departments of Chemistry, Chemical Engineering and Biology for participating in several of the stages of this project. Support personnel from the Department of Chemistry are also gratefully acknowledged, including Aracelis Cardona, Cynthia Ramos, Cacimar Ramos and Rosalie Ramos.

## 7. References

Allain, L.R., Stratis-Cullum, D.N., Vo-Dinh, T. (2004). Investigation of microfabrication of biological sample arrays using piezoelectric and bubble-jet printing technologies. *Anal. Chim. Acta*, 518 (1-2), 77-85.

Alvarez-Rivera, M. (2002). Fiber Optic Coupled Raman Spectroscopy of liquid explosives mixtures. University of Puerto Rico, Mayaguez.

Angel, S.M., Kulp, T.J. & Vess, T.M. (1992). Remote-Raman Spectroscopy at Intermediate Ranges Using Low-Power cw Lasers. *Appl. Spectrosc.*, *46*, 1085-1091.

Beeson, R., Skip, R.D., Thermal inkjet technology -review and outlook. Advanced Research Laboratory, Inkjet Business Unit, Hewlett-Packard Company, Corvallis Oregon, US.

Christesen, S. D., (1988). Raman cross sections of chemical agents and simulants. *Appl. Spectrosc.*, *42*, 318-321.

Christesen, S.; Maciver, B.; Procell, L.; Sorrick, D.; Carrabba, M.; Bello, J., (1999) Nonintrusive Analysis of Chemical Agent Identification Sets Using a Portable Fiber-Optic Raman Spectrometer. *Applied Spectroscopy 53*, 850-855.

Crain, J., Poon, W. C., Cairns-Smith, A., Hatton, P. D., (1992). High-pressure Raman spectroscopic study of cyclohexane $C_6H_{12}$ and $C_6D_{12}$. *J. Phys. Chem.* 96, 8168 - 8173

D'Agostino, P.A.; Hancock, J.R.; Chenier, C.L.; Lepage, C.R., (2006). Liquid chromatography electrospray tandem mass spectrometric and desorption electrospray ionization tandem mass spectrometric analysis of chemical warfare agents in office media typically collected during a forensic investigation. *J Chromatogr A*, 1110 (1-2), 86-94.

Eliasson, C.; Macleod, N.; Matousek, P., (2007). Noninvasive Detection of Concealed Liquid Explosives Using Raman Spectroscopy. *Analytical Chemistry 79* (21), 8185-8189.

Evans, J. C. and Bernstein, H. J., (1956). The effect of Intermolecular interaction on Raman spectrum of carbon disulphide. *Can. J. Chem.* 54, 1127 – 1133.

Ewing, R.G., Atkinson, D.A., Eiceman, G.A., Ewing, G.J. (2001). A critical review of ion mobility spectrometry for the detection of explosives and explosive related compounds. *Talanta.* 54 (3), 515-529.

Farquharson, S., Gift, A., Maksymiuk, P., (2005). Inscore, F. Surface-enhanced Raman spectra of VX and its hydrolysis products. *Appl. Spectrosc.*, *59*, 654-659.

Fierro-Mercado, P.M., Primera-Pedrozo, O.M., Hornedo, A., Hernández-Rivera, S.P., (2010). An In situ FTIR Fiber Optic Method for the Detection of Active Pharmaceutical Ingredients and Excipients on Metallic Substrates, in "Fourier Transform Infrared Spectroscopy: Developments, Techniques and Applications. Nova Science Publishers, Inc. Hauppauge, NY.

Fletcher, R., Briggs, N., Ferguson, E., Gillen, G., (2008). Measurements of air jet removal efficiencies of spherical particles from cloth and planar surfaces. Aerosol science and technology, 42 (12), 1052-1061.

Gillen, G., Zeissler, C., Mahoney, C., Lindstrom, A., Fletcher, R., Chi, P., Verkouteren, J., Bright, D., Lareau, R.T., Boldman, M. (2004). Automated analysis of organic particles using cluster SIMS. Applied Surface Science, 231-232, 186-190.

Hallowell, S.F. (2001). Screening people for illicit substances: a survey of current portal technology. *Talanta.* 4 (3), 447-458.

Hamilton, M.L.; Perston, B.B.; Harland, P.W.; Williamson, B.E.; Thomson, M.A.; Melling, P.J., (2005). *Organic Process Research & Development.* 9, 337–343.

Harvey, S. D.; Vucelick, M. E.; Lee, R. N.; Wright, B. W., (2002).Blind field test evaluation of Raman spectroscopy as a forensic tool. *Forensic Science International,* 125 12–21.

Heimerl, J.M., (1999). A Forensic Workshop Report Concerning (1) The Measurement and Analysis of Energetic Materials and (2) Databasing. A.R. Laboratory (Ed.), Aberdeen Proving Ground, MD.

Henderson, D.A. (1999). The looming threat of bioterrorism. *Science,* 283, 1279-61282.

Hernández, N.M., Rosario, S.V., Hernandez, S.P., Mina, N., (2005). Detection and characterization of smokeless powders with ion mobility spectrometry. *Proc. SPIE Int. Soc. Opt. Eng.* 5778, 607-616.

Hirschfeld, T., (1974). Range Independence of Signal in Variable Focus Remote Raman Spectrometry. *Appl. Opt., 13,* 1435-1437.

Irrazabal, M., Florian, V., Castro, M., Hernandez-Rivera, S.P., Briano, J.G., (2007). Effect of environmental parameters on the chemical signature of TNT in soil. *Proc. SPIE Int. Soc. Opt. Eng.* 6553, 65531N-65541N.

Lin-Vien, Colthup, N.B., Fateley, W.G., Grasselli, J.G. (1991). The Handbook of Infrared and Raman Characteristic Frequencies of Infrared and Raman Characteristic Frequencies of Organic Molecules. Academic Press. San Diego, CA.

McCreery, R. L. *Raman Spectroscopy for Chemical Analysis.* (2000). New York, NY: Wiley-Interscience.

MacCrehan, W.A., (2004). Development of a NIST trace particulate explosives Reference material to evaluate IMS detectors National Institute of Standards and Technology, Gaithersburg, MD 20899-8392 USA.

MacCrehan, W.A. (2009). A NIST Standard Reference Material (SRM) to Support the Detection of Trace Explosives. Analytical Chemistry, 81 (17), 7189-7196.

Manrique-Bastidas, C.A., Castillo-Chara, J., Mina, N., Castro, M.E., Hernandez-Rivera, S.P., (2004). Nucleation and crystallization studies: A vibrational spectroscopy investigation of 2,4,6-TNT. *Proc. SPIE Int. Soc. Opt. Eng.* 5415, 1345-1356.

Manrique-Bastidas, C.A., Primera-Pedrozo, O.M., Pacheco-Londono, L., Hernandez-Rivera, S.P., (2004). Raman microspectroscopy crystallization studies of 2,4,6-TNT in different solvents. *Proc. SPIE Int. Soc. Opt. Eng.* 5617, 429-441.

Marrs, T.C. Maynard, R.L. and Sidell, F.R., (1996). Chemical Warfare Agents: Toxicology and Treatment. John Wiley & Sons Ltd., London, UK.

Melling, P.J.; Shelley, P. (2002). Spectroscopic Accessory for Examining Films and Coatings on Solid Surfaces. U.S Patent 6,3,10,348, United States Patent and Trademark Office, Washington, DC.

Mehta, N.K.; Goenaga-Polo, J.E.; Hernández-Rivera, S.P.; Hernández, D.; Thomson, M.A; Melling, P.J., (2002). Development of an *In-Situ* Spectroscopic Method for Cleaning Validation Using Mid-IR Fiber Optics. *Bio Pharm.,* 15, 36-42.

Mehta, N.K.; Goenaga-Polo, J.E.; Hernández-Rivera, S.P.; Hernández, D.; Thomson, M.A; Melling, P.J., (2003). Development of an *In-Situ* Spectroscopic Method for Cleaning Validation Using Mid-IR Fiber Optics *Spectroscopy,* April.

Meier, B., Weidner, P. and Penzkofer, A., (1990). Double line stimulated Raman scattering in benzene. *Appl. Phys. B, 51,* 404-413.

Mocak, J., Bond, A.M., Mitchell, S. and Scollary, G.A. (1997). Statistical overview of standard (IUPAC and ACS) and new procedures for determining the Limits of detection and quantification: application to voltammetric and stripping techniques. *Pure Appl. Chem., 69*, 297-328.

Miyake, A.; Yamada, N.; Ogawa, T., (2005). Mixing hazard evaluation of organic peroxides with other chemicals. *Journal of Loss Prevention in the Process Industries 18* (4-6), 380-383.

Miyaki, K., Nishiwaki, Y., Maekawa, K., Ogawa, Y., Asukai, N., Yoshimura, K., Etoh, N., Matsumoto, Y., Kikuchi, Y., Kumagai, N. and Omae, K. (2005). Effects of Sarin on the Nervous system of subway workers seven years after the Tokyo subway Sarin attacks. *J. Occup. Health, 47*, 299-304.

National Research Council. Final Report on Existing and Potential Standoff Explosives Detection Techniques. (2004). Published by the National Academy of Sciences.

Pacheco-Londono, L.C., Santiago, A., Pujols, J., Primera-Pedrozo, O.M., Mattei, A., Ortiz, W., Ruiz, O., Ramirez, M., Hernandez-Rivera, S.P., (2007). Characterization of layers of Tetryl, TNB and HMX on metal surfaces using fiber optics coupled grazing angle-FTIR. *Proc. SPIE Int. Soc. Opt. Eng.* 65423K-65410.

Pacheco-Londoño, L.C., Ortiz-Rivera, W., Primera-Pedrozo, O., Hernández-Rivera, S.P. (2009). Vibrational spectroscopy standoff detection of explosives. *Anal. Bioanal. Chem.*, 395 (2), 323-335.

Perston, B.B., Hamilton, M.L., Williamson, B.E., Harland, P.W., Thomson, M.A., Melling, P.J. (2007). Grazing-Angle Fiber-Optic Fourier Transform Infrared Reflection-Absorption Spectroscopy for the in Situ Detection and Quantification of Two Active Pharmaceutical Ingredients on Glass. *Anal. Chem.*, 79, 1231-1236.

Phares, D., Holt, J., Smedley, G., Flagan, R. (2000). Method for characterization of adhesion properties of trace explosives in fingerprints and fingerprint simulations. *J. Foren. Sci.* 45 (4), 774-784.

Police: Plot to blow up aircraft failed. www.cnn.com. Posted August 10, 2006.

Primera-Pedrozo, O.M., Pacheco-Londono, L.C., De La Torre-Quintana, L.F., Hernandez-Rivera, S.P., Chamberlain, R.T., Lareau, R.T., (2004). Use of fiber optic coupled FT-IR in detection of explosives on surfaces. *Proc. SPIE Int. Soc. Opt. Eng.* 5403, 237-245.

Primera-Pedrozo, O.M., Pacheco-Londoño, L., Ruiz, O., Ramirez, M., Soto-Feliciano, Y.M., De La Torre-Quintana, L.F., Hernandez-Rivera, S.P., (2005). Characterization of thermal Inkjet technology TNT deposits by fiber optic-grazing angle probe FTIR spectroscopy. *Proc. SPIE Int. Soc. Opt. Eng.* 5778, 543-552.

Primera-Pedrozo, O.M., Rodriguez, N., Pacheco-Londono, L.C., Hernandez-Rivera, S.P., (2007). Detection of 2,4,6-trinitrotoluene on non-traditional surfaces using fiber optic coupled grazing angle probe: FTIR. *Proc. SPIE Int. Soc. Opt. Eng.*, 6542, 65423J-65410.

Primera-Pedrozo, O., Soto-Feliciano, Y., Pacheco-Londoño, L.C., Hernández-Rivera, S.P., (2008). High Explosives Mixtures Detection Using Fiber Optics Coupled: Grazing Angle Probe/Fourier Transform Reflection Absorption Infrared Spectroscopy. *Sens Imaging: Int. J.* 9 (3), 27-40

Primera-Pedrozo, O., Soto-Feliciano, Y., Pacheco-Londoño, L., Hernández-Rivera, S.P. (2009). Detection of High Explosives Using Reflection Absorption Infrared Spectroscopy with Fiber Coupled Grazing Angle Probe/FTIR. *Sens Imaging: Int. J.* 10 (1), 1-13.

Ramirez-Cedeño, M.L., Ortiz-Rivera, W., Pacheco-Londono, L C. and Hernandez- Rivera, S.P., (2010). Remote Detection of Hazardous Liquids Concealed in Glass and plastic Containers. *IEEE Sensors J., 10*, 693-698.

Soto-Feliciano, Y., Primera-Pedrozo, O.M., Pacheco-Londono, L., Hernandez-Rivera, S.P., (2006). Temperature dependence of detection limits of TNT on metallic surfaces using fiber optic coupled FTIR. *Proc. SPIE Int. Soc. Opt. Eng.* 6201, 62012H-62019.

Sedlacek III, A. J., Ray, M. D., Higdon, N. S. and Richter, D. A. (2001). Short-range noncontact detection of surface contamination using Raman lidar. *Proc. SPIE Int. Soc. Opt. Eng., 4577*, 95-104.

Sharma, S. K., Angel, S. M., Ghosh, M., Hubble, H. W. & Lucey, P. G., (2002). A remote pulsed-laser Raman spectroscopy system for mineral analysis on planetary surfaces at 66 meters. *Appl. Spectrosc, 56*, 699-705.

Sharma, S. K., Lucey, P. G., Ghosh, M., Hubble, H. W., Horton, K. A., (2003). Stand-off Raman Spectroscopic Detection of Minerals on Planetary Surfaces. *Spectrochim. Acta A, 59*, 2391-2407.

Sharma, S.K., Anupam, K. M. & Bhavna, S., (2005). Portable remote Raman system for monitoring hydrocarbon, gas hydrates and explosives in the environment. *Spectrochimica Acta Part A: Molecular and Biomolecular Spectroscopy,* 61, pp. 2404-2412,

Sharma, S. K., (2007). New trends in telescopic remote Raman spectroscopic instrumentation. *Spectrochim. Acta A, 68*, 1008–1022.

Schrader, B. *Infrared and Raman spectroscopy: Methods and applications.* (1995). Schrader, B. Ed. New York, NY: VCH, p. 215.

Shimanouchi, T. (1972). Tables of Molecular Vibrational Frequencies Consolidated. vol. 1, Bethesda, MD: National Bureau of Standards.

Steinfeld, J.I.; Wormhoudt, J., (1998). *Annual Review of Physical Chemistry,* 49, 203–232

Sun, Y. and Ong, K.Y., (2005). Detection Technologies for Chemical Warfare Agents and Toxic Vapors. CRC Press, Boca Raton, FL.

Thomson, G. and Batchelder, D., (2002). Development of a hand-held forensic-lidar for standoff detection of chemicals. *Rev. Sci. Instrum., 73*, 4326-4328.

Umemura, J., (2002). Reflection–absorption spectroscopy of thin films on metallic substrates. In J. M. Chalmers & P. R. Griffiths, Eds., Handbook of vibrational spectroscopy. Chichester, UK: Wiley & Sons; Vol. 2, pp. 982–998.

Van Neste, C.W., Senesac, L.R., Thundat, T. (2009). Standoff Spectroscopy of Surface Adsorbed Chemicals. Analytical Chemistry, 81 (5), 1952-1956.

Wakabayashi, K., Matsumura, T., Nakayama, Y., Koshi, M., (2007). Temporal change of Raman spectra of carbon tetrachloride under laser-driven shock compression. *AIP Conference Proceedings.* 955, Issue 1, 1267-1270.

Wu, M., Ray, M., Fung, K. H., Ruckman, M. W., Harder, D. & Sedlacek III, A. J. Stand-off detection of chemicals by UV Raman spectroscopy. (2000). *Appl. Spectrosc.*, *54*, 800-806.

# 14

# Phase-Shifting Point Diffraction Interferometer Having Two Point Light Sources of Single-Mode Optical Fibers

Oshikane Yasushi, Nakano Motohiro and Inoue Haruyuki
*Graduate School of Engineering, Osaka University*
*Japan*

## 1. Introduction

State-of-the-art research fields such as projection optics for Extreme Ultraviolet Lithography (EUVL) (Otaki et al., 2002a; Stulen & Sweeney, 1999; Goldberg et al., 2004), focusing optics for X-ray microscope (Yamamura & Takai, 2008; Handa et al., 2008; Yumoto et al., 2008; Matsuyama et al., 2008), and Michelson interferometers for detection of gravitational waves (Ando et al., 2001; Sato et al., 1999) require ultra precise mirrors with large surface areas and extremely low surface figure error (in the sub-nanometer scale). In EUVL, a light wave of 13.5 nm is used for photoresist exposure, and such an optical system accepts reflection optics instead of refractive optics owing to the short wavelength. In addition, EUV exposure of a dynamic random access memory (DRAM) pattern with a half pitch (hp) of 45 nm or less requires a numerical aperture (NA) of 0.2 or more. An EUV optical system is designed using Mo/Si multilayer mirrors. The surface figure accuracy of the mirror substrate determines the reflectance of the mirror and the long-range error must be less than 0.2 nm rms. Production of high-precision mirrors with large surface areas is usually supported by highly accurate surface figure measurement methods. For example, the slope integration method (Higashi et al., 2007), three-flat method (Schulz & Schwider, 1967), and point diffraction interferometry (PDI) (Smartt & Steel, 1975; Sommargren, 1996; Medecki et al., 1996; Millerd et al., 2004; Otaki et al., 2002b) have been studied and applied to the measurement of mirrors. The slope integration method determines the surface normal vector at each point on the test surface and the resultant angle of inclination is integrated to obtain the absolute surface figure. However, this method requires highly accurate locations of each measurement point. The three-flat method uses three mirrors including the test mirror and measures the relative surface figure between two of the mirrors at a time using conventional optical interferometry. The relative figure data are solved as a set of simultaneous equations to obtain each surface figure. However, this method returns a line profile of the mirrors rather than a whole surface figure. PDI is a novel absolute surface figure measurement system using an optical diffraction wave from a tiny aperture as the reference surface. A conventional Fizeau interferometer uses a substantial reference surface for the measurement and the machining accuracy of the reference results in measurement accuracy of approximately $\lambda/50$. In contrast, the wavefront accuracy of the optical diffraction wave used in the PDI system has been estimated to be $10^{-5}\lambda$ or less at an NA of 0.2 (Sommargren, 1996).

Therefore, PDI has a much better reference surface and will enable the realization of ultra-high precision measurement of the absolute surface figure. Thus, PDI has been adopted as the phase-shifting (PS) method for the analysis of interference patterns (de Groot, 1995a; de Groot, 1995b). In the PDI system, generation of a diffraction wave is dependent on a tiny pinhole or single-mode optical fiber. Nikon has measured spherical or aspherical concave mirrors using a diffraction wave in the visible region emitted from a tiny pinhole (Otaki et al., 2002a; Otaki et al., 2002b; Ota et al., 2001). On the other hand, EUVL researchers at the Lawrence Livermore National Laboratory (LLNL) have reported surface figure measurement data using the PDI system with a single-mode fiber (Sommargren, 1996). In the two types of PS/PDI systems, a diffraction wave radiating from a single tiny hole serves as both a diagnostic wave and a reference wave, the diffraction wave could not be used entirely as the diagnostic wave. This means that the maximum measurable mirror diameter is approximately half the diameter of the diffraction wave. Therefore, we have developed a PS/PDI system with two optical fibers as point sources for large-aperture optics (Oshikane et al., 2003). One fiber acts as a point source for a diagnostic wave, and the other fiber generates a reference wave. The entire diffraction wave will be applied to the measurement of large optics. PDI is suitable for spherical concave mirrors due to its spherical wavefront as the reference surface. When an aspherical mirror is tested, the diagnostic wavefront gets highly distorted and the resultant PDI interferograms do not reflect the surface figure in PS analysis. Therefore, aspherical surface figure measurement by PS/PDI usually requires (1) the stitching method, which connects all valid PS/PDI data in every part of the surface (Ota et al., 2001; Chen et al., 2006), or (2) numerical restoration of the actual wavefront on the mirror surface from the highly distorted PS/PDI data (Yamaguchi et al., 2006; Iemmi et al., 2005; Nomura et al., 2006). Procedure (2) is also predictably effective for spherical mirror measurement to obtain higher-accuracy data. In this research, we have developed PS/PDI with two optical fibers point sources for absolute surface figure measurement of large-aperture optics. The inevitable wavefront distortion in principle is attempted to be eliminated through an inverse-problem-based numerical reconstruction of the wave, and highly precise measurements of spherical and aspherical mirrors has been realized. Section 2 provides a brief review of PDI research and PS methods. In section 3, a PS/PDI system for large-aperture optics is introduced and the interferogram recording procedure is explained. Section 4 describes the origin of measurement error estimated by interference between diffraction waves from two optical fibers. Effectivity of numerical wavefront restoration is described in section 5. In section 6, typical results of a spherical concave mirror are shown and the measurement accuracy is discussed. To demonstrate aspherical surface measurement, a highly distorted spherical concave mirror is measured and the result is discussed in section 7. An off-axis parabolic mirror surface and Si wafer surface are also measured and discussed in section 7. Measurement error of the PS/PDI for aspherical surfaces is studied. This chapter is concluded in section 8.

## 2. Principle of PS/PDI

### 2.1 Sphericity of diffraction wavefront

A point diffraction interferometer (PDI) was developed by Smartt in the 1970s (Smartt & Strong, 1972). The precursor to the PDI was an interferometer with an optical diffraction wave emitted from a tiny aperture fabricated on a so-called PDI plate, and the diffraction wave was used as a reference spherical wave for interferometry (Smartt & Steel, 1975;

Sommargren, 1996; Medecki et al., 1996; Millerd et al., 2004; Otaki et al., 2002b; Ota et al, 2001). Presently, demand is growing for highly precise optical surfaces, both spherical and aspherical, in the case of mirrors. In this section, (1) the principle of PDI measurement and (2) a phase-shift algorithm (de Groot, 1995a; de Groot, 1995b) for numerical processing of interferogram data are explained. In ordinary optical interferometry, surface distortion between the reference and test surfaces is measured and the measurement accuracy is limited by the accuracy of the substantial reference surface. In the PDI measurement, the reference surface is a diffraction wave emitted from a small aperture. Accordingly, wavefront error of the diffraction wave limits the PDI measurement accuracy. Numerical estimation of the wavefront error can be completed using diffraction theory. The broad theoretical base is the Huygens principle and the Helmholtz-Kirchhoff integral theorem is the strict definition. Moreover, the Fresnel-Kirchhoff diffraction integral is derived from application of the Helmholtz-Kirchhoff integral theorem to a tiny aperture and provides information regarding the diffraction wavefront precision (Born & Wolf, 1959). As an example, fig. 1 shows the angular distribution of optical intensity and phase distortion of a diffraction wave emitted from a single-mode (SM) optical fiber core 4 μm in diameter at a wavelength of 632.8 nm. The resulting sphericity of the diffraction wave is less than $10^{-5}\lambda$ within an NA of 0.2. This means that a spherical concave mirror with a 500-mm diameter and a 1500-mm radius of curvature (ROC) can be measured at an accuracy of less than 6 pm.

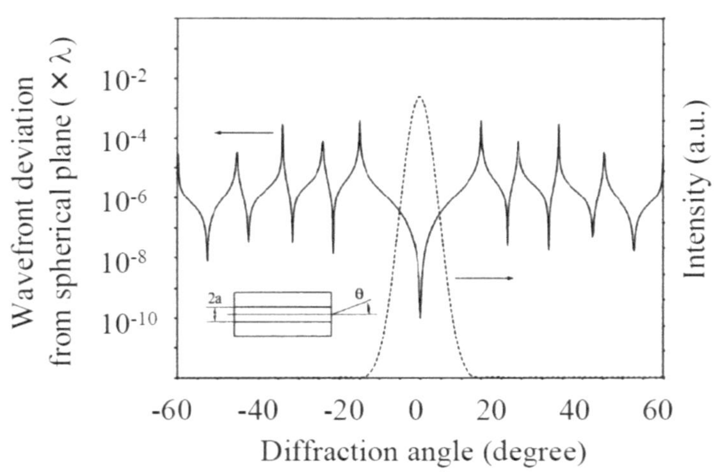


Fig. 1. Intensity and phase distribution of diffraction wavefront from tiny aperture.

## 2.2 Wavefront and surface figure measurement by PDI

With the highly precise diffraction wavefront, vertical spatial resolution of the PDI measurement of the sub-nanometer order, and wavefront estimation using a PDI plate and surface figure measurement with an optical fiber or tiny pinhole are already realized. Fig. 2 explains the principle of PDI measurement using a PDI plate. This configuration is used to evaluate wavefront error in imaging optics. A focused light wave makes an Airy pattern according to the diffraction theory. In the Airy disk, more than 90% of the light wave is converged. The PDI plate is then placed at the focal point. The plate has two apertures; one (hole a) is for the Airy disk and the other (hole b) is for generation of a reference diffraction

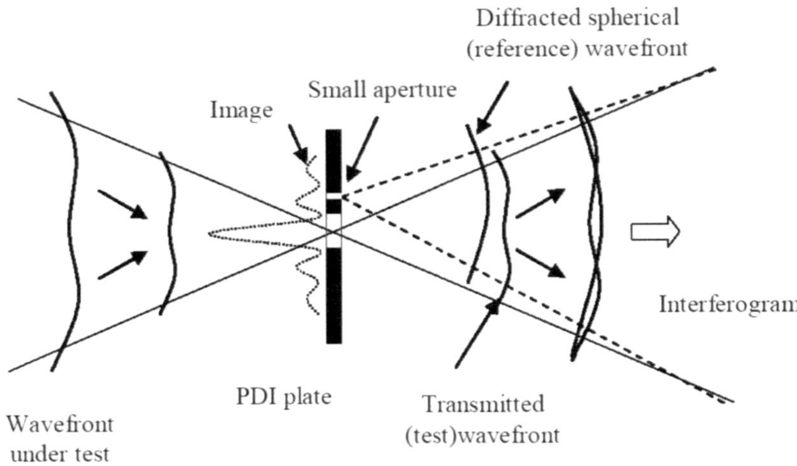


Fig. 2. PDI configuration with PDI plate.

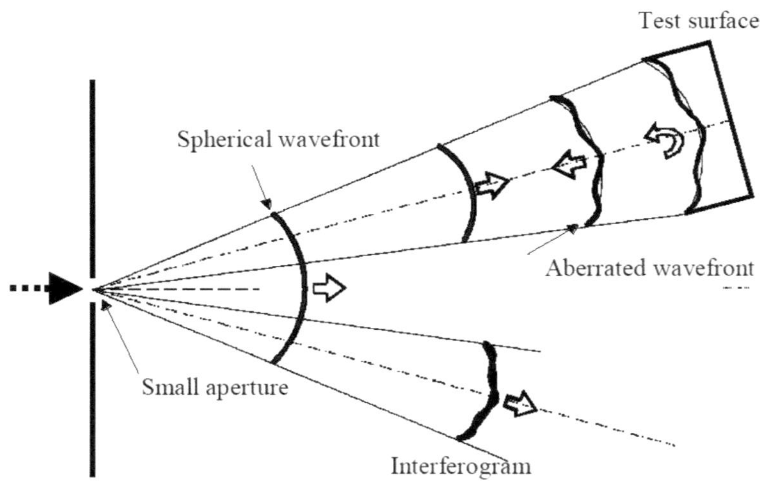


Fig. 3. PDI configuration with a tiny pinhole.

wave. Hole b is located on the first ring pattern of the Airy disk, and the resultant diffraction wave does not include information about the test wave. Hole a gives passage to a component inside the Airy disk. As a result, the wavefront emitted from hole a interferes with the reference wavefront emitted from hole b and optical interference can be accomplished. Fig. 3 shows the PDI configuration for surface figure measurement with a single tiny aperture (Sommargren, 1996). A pinhole or single-mode optical fiber is accepted as the aperture. Part of the diffraction wave generated at the aperture makes a complete trip to the test surface and is reflected around the aperture. The reflected test wave interferes with the rest of the diffraction wave, serving as a reference wave, emitted from the tiny aperture. The resultant interferogram provides surface figure data of the test mirror. However, this configuration has several disadvantages. Reflection around the aperture could distort the test wavefront because of surface roughness or the lack of a reflecting surface. This configuration could not be applied to large test optics because a single diffraction wavefront is divided into a test wave and a reference wave.

### 2.3 Analysis of interferogram by phase-shift method

Optical interferometry treats a series of interferograms for precise measurement. Phase-shift algorithms are of the accurate procedures for interferograms. Fig. 4 shows the principle of the phase-shift procedure. Determining a step-by-step phase difference φ(t) between reference and test surfaces causes sinusoidal changes in optical intensity at point P. A set of discrete data is diagnosed to obtain the initial phase value at point P. Many algorithms exist that determine the initial phase value and the standard procedure is based on Fourier transformation (Freischlad & Koliopoulos, 1990; Larkin & Oreb, 1992). In the following section, the entire experimental dataset is processed using seven bucket algorithms.

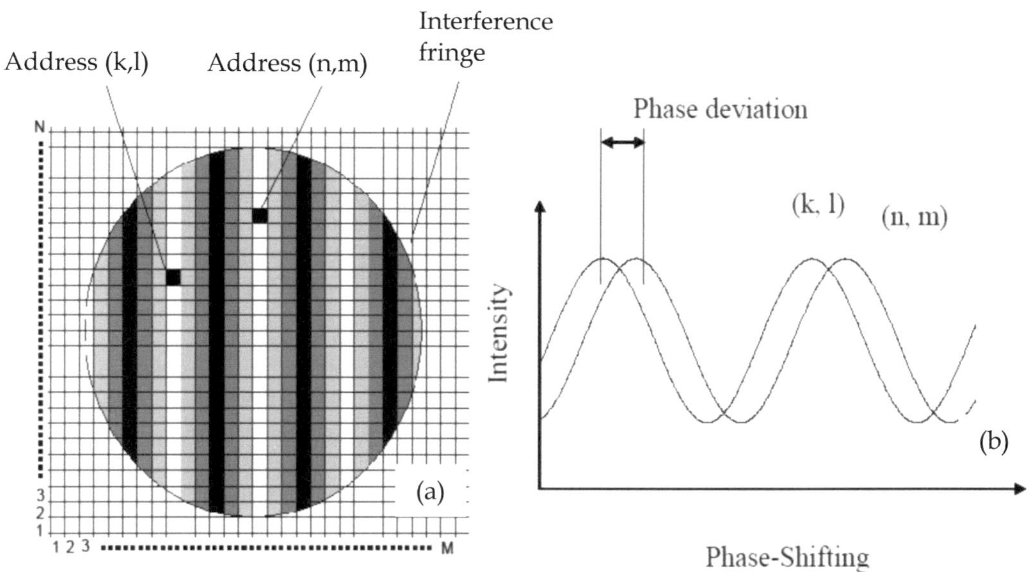


Fig. 4. Principle of phase shift method. (1) CCD pixels and interference pattern, (2) intensity changes in the pixels by phase shift interferometry.

## 3. PDI Configuration with two optical fibers for large-aperture optics

### 3.1 Experimental apparatus

In this section, detailed optical configuration of a PS/PDI system with two optical fibers serving as tiny apertures is described. The recording process for phase-shifting interferograms is also explained. Fig. 5 shows the principle of the PS/PDI configuration with two tiny apertures. A light wave with sufficiently long coherence length is divided into two waves and each wave is introduced into a small aperture. Two independent spherical diffraction waves are then generated. One wave propagates to the test surface and the other serves as the reference wave and propagates in the opposite direction. Therefore, the whole wavefront can be used for surface figure measurement. In the entire optical path, there are

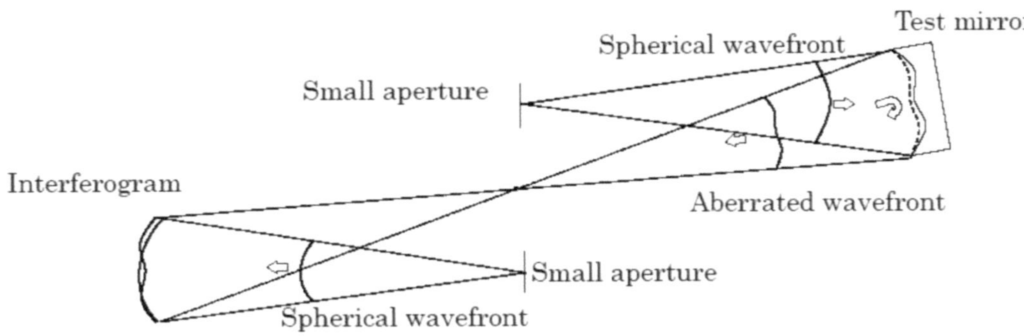


Fig. 5. Optical configuration of PDI with two small apertures.

no additional optics and, in principle, there is no disturbance in the propagating waves. This PDI configuration makes full use of the ultra-high-accurate wavefront as reference. However, this is not a complete common-path system and tilt fringe inevitably occurs due to spacing between the focal point of the test wave (point a) and the origin of the reference wave (point b). Dense tilt fringes are not recorded correctly by a charge-coupled device (CCD) sensor because the width of each fringe is smaller than or comparable to the size of a CCD pixel. To avoid such a consequence, a shorter distance between points a and b and a correction process are needed. A schematic diagram of an actual PS/PDI system is shown in fig. 6 as a means for realizing the above discussed outcome. For temporal stability of the system, the entire optical system is installed on an air-suspension-type vibration-isolation table in a temperature-controlled room. Air circulation in the experiment room causes optical path differences through inhomogeneous refractive index of the atmosphere, resulting in random error in the experimental result. This error is suppressed by a large styrofoam box serving as both an air screen and an adiabatic case. A He-Ne laser light source, which also serves as a heat source, is installed outside the box. The He-Ne laser is linearly polarized and its wavelength is 632.8 nm. In this optical system, light from the components is blocked from the laser using an optical isolator (60 dB) and the laser output power is stabilized. A half-wave plate (HWP) is used to rotate the light polarization to adjust the branching ratio of the laser beam at the polarization beam splitter (PBS). Each branched beam is introduced into the core of an SM optical fiber of 4-μm diameter. Diagnostic and reference waves are thus generated independently. Changing the branching ratio at the PBS optimizes the interferogram contrast generated on the CCD camera. The interferogram is recorded as a 14-bit digital image by the CCD camera with an electronic shutter. In the interferometry system, the phase-shift method is executed by changing the optical path of the diagnostic wave. The aluminum mirror on the single-axis PZT stage in fig. 6 is translated along the optical path and the step-by-step motion is synchronized with the CCD camera shutter. Thus, a series of phase-shifting interferograms are obtained. In this PDI system, separation between the close fiber ends shown in fig. 7 causes a partial diffraction wave due to the opposed fiber end. Therefore, the fiber end is obliquely polished with a pipette grinder to bend the beam output axis away from the fiber axis. Fig. 8 shows the relation between the angle of polish and the output beam direction. For optimal installation configuration, the fiber ends are obliquely polished at 15 degrees from surface normal to fiber axis. As a result, the fiber end serving as the diffraction wave origin has an ellipsoidal section, which may degrade the accuracy of the spherical wavefront. However, as shown in fig. 9, the diffraction wavefront caused by the ellipsoidal fiber core still generates a

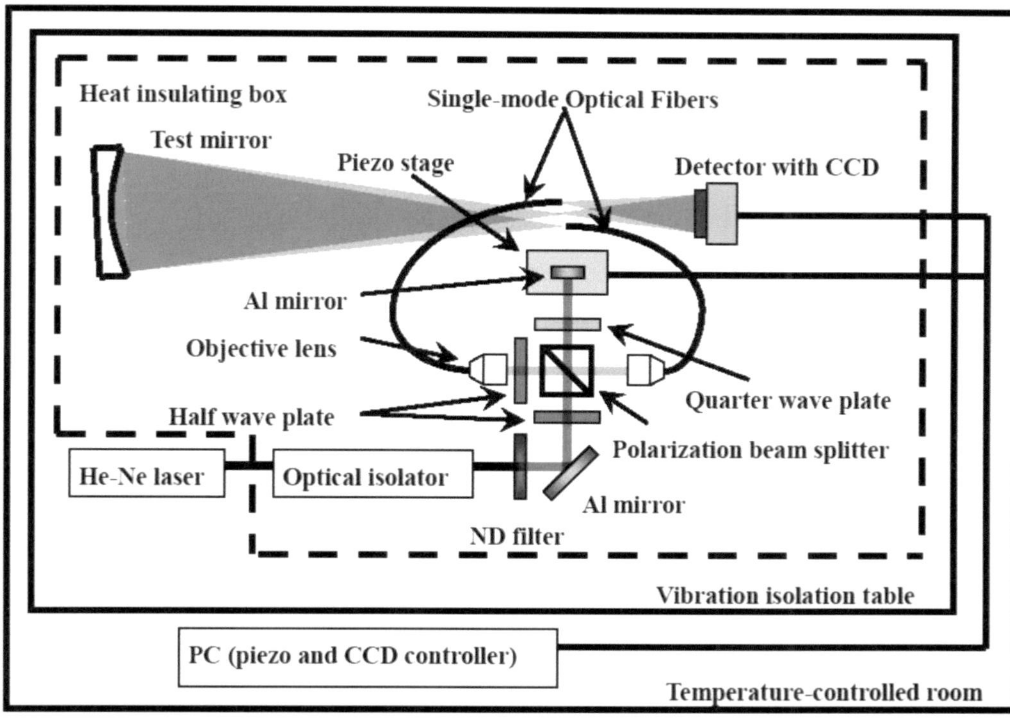


Fig. 6. Schematic diagram of PS/PDI system with two point sources of single mode optical fibers.

microscopic photo of a sharpened and obliquely polished fiber end is shown in fig. 12. The polished section must include both the core and surrounding clad area.

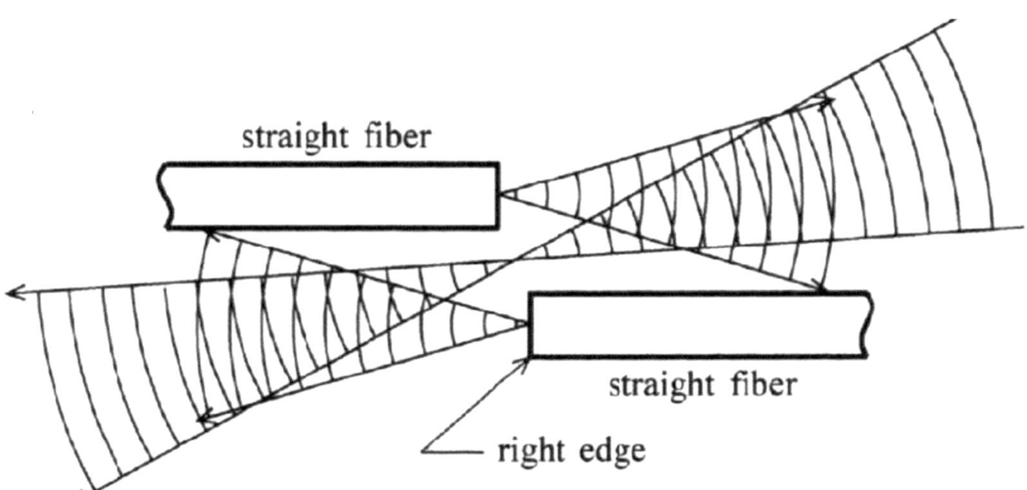


Fig. 7. Two fiber ends arranged in opposite directions and the outgoing diffraction waves.

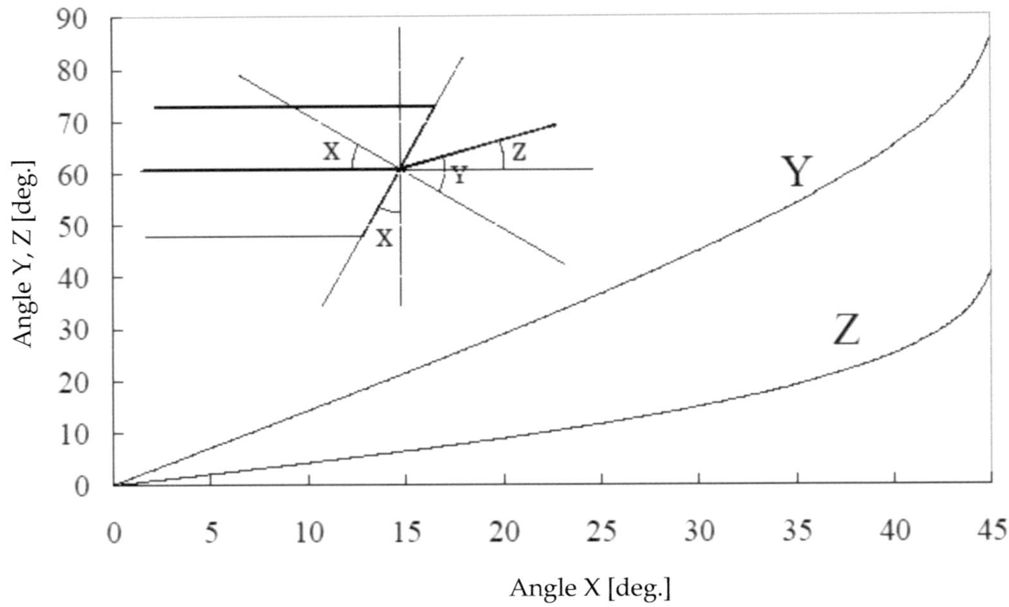


Fig. 8. Change in direction of the outgoing diffraction wave by obliquely polished angle.

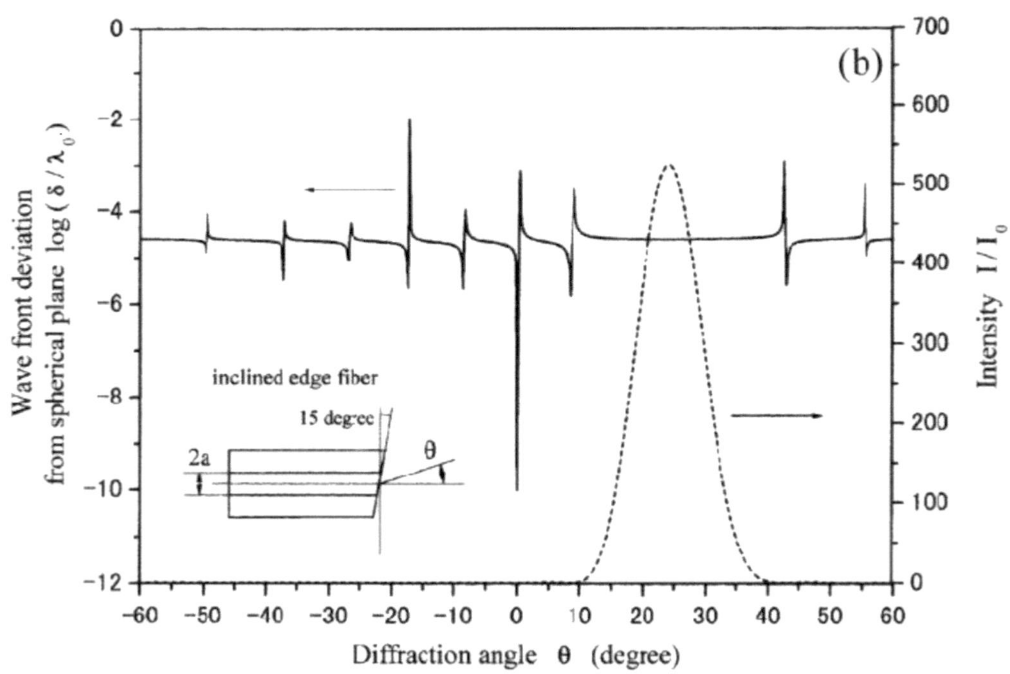


Fig. 9. Intensity and phase distribution of diffraction wavefront emitted from obliquely polished fiber end.

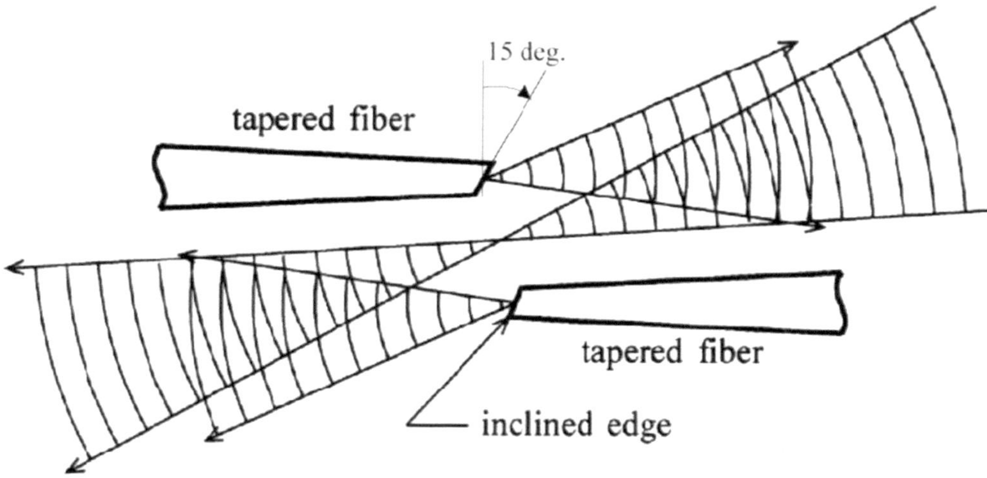


Fig. 10. Resultant configuration of two fiber ends having sharpened apex and obliquely polished ends.

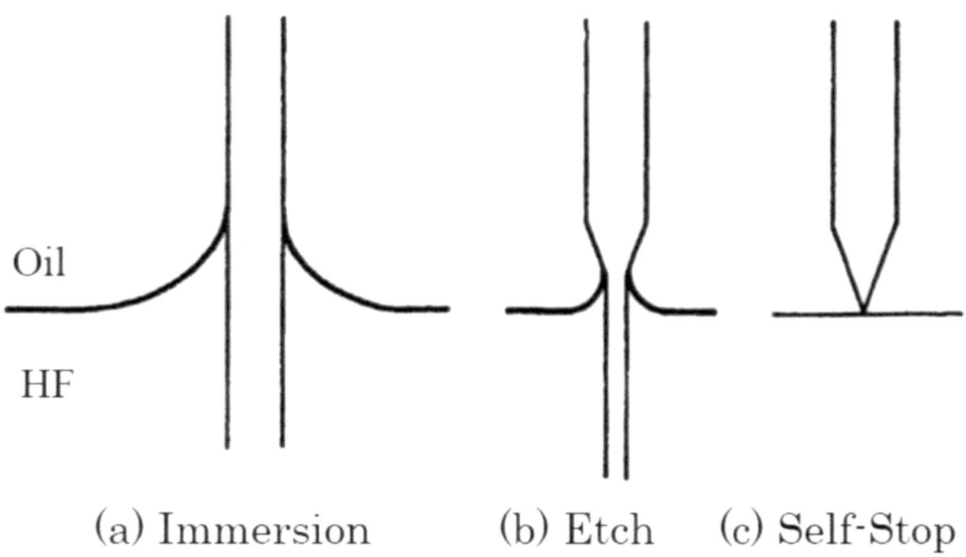


Fig. 11. Chemical etching of fiber end at boundary between oil and conc. HF.

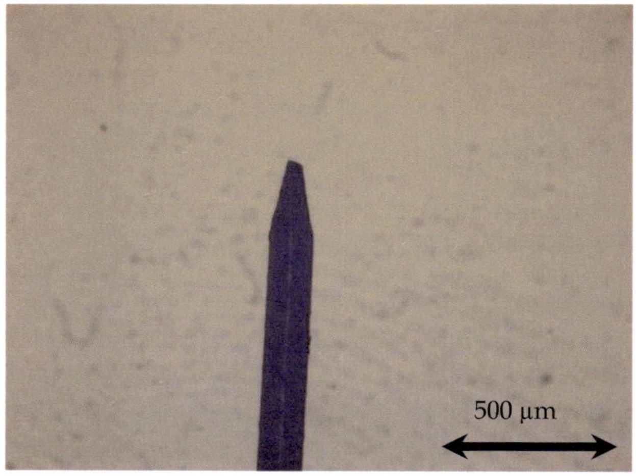


Fig. 12. Microscopic photo of single-mode optical fiber end after chemical etching.

highly spherical wave. Because the fiber end is close to the focal point of the reflected diagnostic wave, the fiber end must have a sharpened tip. With the sharpened fibers, optical configuration as shown in fig. 10 will be accomplished and the whole diffraction wavefront will be used for large-aperture optics. The sharpened end of the fiber is formed with a mixture of hydrofluoric acid and organic fluids. At the interface between the two liquids, the fiber end is automatically sharpened owing to surface tension as shown in fig. 11. A

### 3.2 Recording of interferograms

For precise data analysis using the phase-shift method, serial interferograms having an equivalent phase difference must be recorded. The CCD shutter and movement of the Al mirror or PZT stage are computer controlled using LabVIEW. Motion of the PZT stage is optimized to restrict mechanical vibration of the mirror. Fig. 13 shows the control system and fig. 14 shows a flow chart of the measurement.

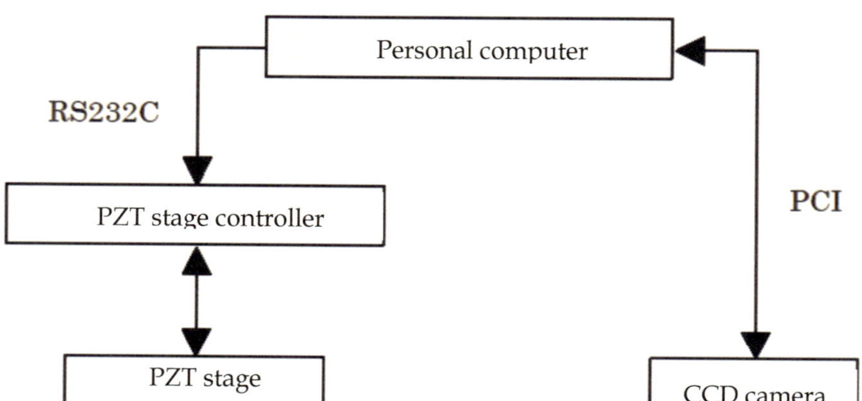


Fig. 13. Diagram for control of the PS/PDI system.

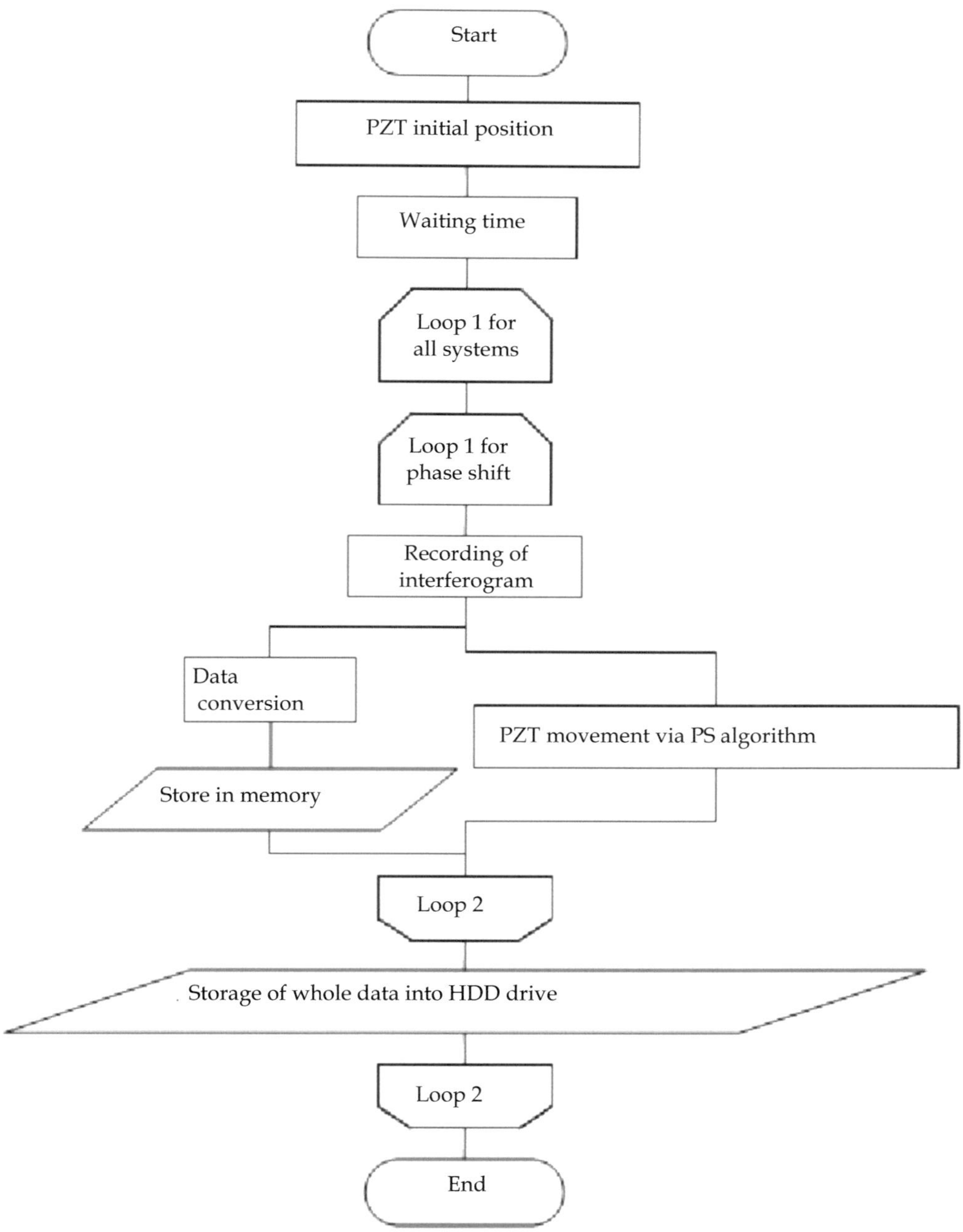


Fig. 14. Flow chart for the PS/PDI measurement.

## 4. Estimation of measurement accuracy by null test

### 4.1 Wavefront analysis procedure

Traditional PS/PDIs analyze interferograms caused by a diagnostic wavefront reflected by the test surface and a reference wavefront to obtain the surface figure of the test mirror (Smartt & Steel, 1975; Sommargren, 1996; Medecki et al., 1996; Millerd et al., 2004; Otaki et al., 2002b). In contrast, the PS/PDI described in this chapter uses two optical fibers as light sources, and the interferogram generated by these two fibers is recorded by a CCD camera. PS analysis of this interferogram provides the accuracy of the spherical wavefronts or measurement error of the PS/PDI system. Two optical-fiber point sources cannot be physically overlapped. As a result, an interferogram having many black-and-white-banded stripes is seen on the screen or CCD surface (Born & Wolf, 1959). We assume an optical system as shown in fig. 15. Points A and B are fiber ends for reference and test waves, respectively, and they are origin symmetrical. Projection plane S, which has a tilt component, is described in the following equation.

$$z=\alpha x+\beta y+d \tag{1}$$

An optical path difference between AP and BP at a point P on S can be written as follows:

$$\begin{aligned}\delta_i &= \overline{AP}-\overline{BP} \\ &= \sqrt{(x_0-x_i)^2+(y_0-y_i)^2+(z_0-z_i)^2}-\sqrt{(-x_0-x_i)^2+(-y_0-y_i)^2+(-z_0-z_i)^2}\end{aligned} \tag{2}$$

Here, P is the i-th pixel on the CCD detector. When the path difference is an even multiple of the half-wavelength of the light, interference fringe becomes bright, and an odd multiple of the half-wavelength causes dark fringe. As an example, numerical simulation of the fringe pattern is shown in fig. 16. In this figure, fiber-to-fiber distance is 60 μm and the projection screen is parallel to the xy plane at z = 150 mm. This interference pattern is a contour having $\lambda/2$ spacing, and an initial phase map will be calculated with this pattern through a phase-shift algorithm. Then, by changing the optical path difference at intervals of $\pi/2$, seven continuously changing interference patterns are created. The seven patterns are processed through seven bucket algorithms with a Von Hann window, as summarized in table 1. Consequently, an initial phase map is formed, as shown in fig. 17. In the phase-shift algorithm, a phase is calculated using an arctangent function and the map has many discontinuous lines, as can be seen in fig. 17. Here, we must unwrap the phase information. After unwrapping, a continuous phase map is generated, as can be seen in fig. 18. The initial phase map shown in fig. 18 indicates a phase difference between the reference wave from a point and the test wave from another point on the projection plane. To obtain the surface figure of test wave, location of both the reference fiber end and projection plane must be determined, and these position coordinates are analytically derived. Locations of every component in fig. 15 can be determined through the least-square method with the phase map on a projection screen because the test and reference waves are assumed to be ideal spherical waves. The optical path difference between AP and BP can be obtained from the schematic diagram of the optics, $\delta_i$, and measurement result, $\Delta_i$. Locations of points A, B, and the projection screen are determined by minimizing s, the square sum of the differences between $\delta_i$ and $\Delta_i$. Here, the measurement result is a relative phase map and the value is described by the following equation with parameter h.

| $P$ | $P'$ | $\alpha$ | $X$ | $T$ |
|---|---|---|---|---|
| 5 | 8 | $\pi/2$ | 2 | $\dfrac{2(g_2 - g_4)}{-(g_1 + g_5) + 2g_3}$ |
| 6 | 8 | $\pi/2$ | 6 | $\dfrac{-(g_0 - g_5) - 3(g_1 + g_4) + 4(g_2 - g_3)}{-(g_0 + g_5) + 3(g_1 - g_4) + 4(g_2 + g_3)}$ |
| 7 | 8 | $\pi/2$ | 8 | $\dfrac{-(g_0 - g_6) + 7(g_2 - g_4)}{-4(g_1 + g_5) + 8g_3}$ |
| 11 | 12 | $\pi/2$ | 16 | $\dfrac{(g_0 - g_{10}) - 8(g_2 - g_8) + 15(g_4 - g_6)}{4(g_1 + g_9) - 12(g_3 + g_7) + 16g_5}$ |
| 8 | 9 | $2\pi/3$ | 20 | $\dfrac{2(g_0 - g_7) - 7(g_1 - g_6) + 17(g_3 - g_4)}{(g_0 + g_7) + 4(g_1 + g_6) - 15(g_2 + g_5) + 10(g_3 + g_4)}$ |
| 8 | 12 | $\pi/3$ | 2.7 | $\dfrac{-g_0 + g_1 + g_3 - g_4 - g_6 + g_7 + 2(g_2 - g_5)}{-(g_0 + g_1 + g_6 + g_7) + 2(g_3 + g_4)}$ |
| 9 | 16 | $\pi/4$ | 2 | $\dfrac{g_1 + g_3 - g_5 - g_7 + 2(g_2 - g_6)}{-g_0 - g_1 + g_3 + g_5 - g_7 - g_8 + 2g_4}$ |

Table 1. Examples of phase shift algorithms based on Von Hann window. Equation T at P=7 is adopted in this study.

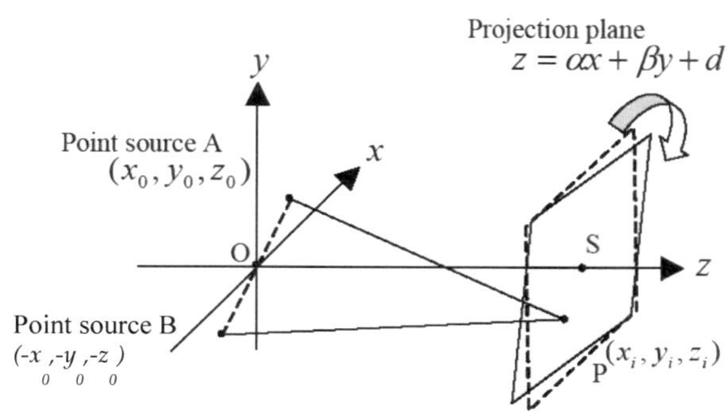


Fig. 15. Coordinates for wave front procedure.

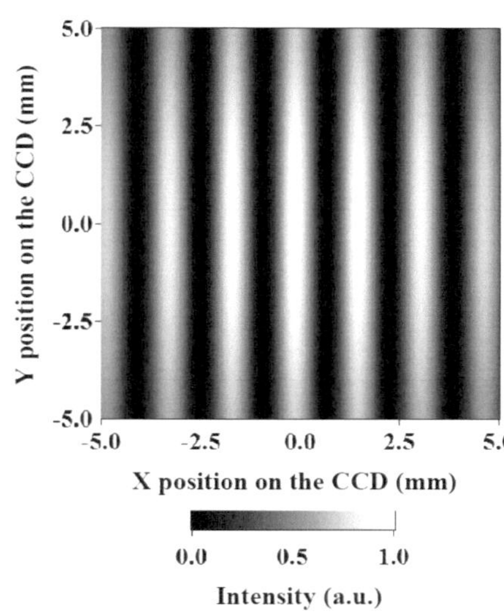


Fig. 16. Computer generated interference pattern.

$$s = \sum_{i} \{(\Delta_i - h) - \delta_i\}^2 \tag{3}$$

As a result, the seven parameters $x_0$, $y_0$, $z_0$, $\alpha$, $\beta$, d, and h are used to determine the locations of the two fiber ends, projection screen, and the offset of phase map. A small change of $\delta_i$ by changing $x_0$, $y_0$, $z_0$, $\alpha$, $\beta$, and d can be expressed as follows.

$$\Delta\delta_i = \frac{\partial\delta_i}{\partial x_0}\Delta x_0 + \frac{\partial\delta_i}{\partial y_0}\Delta y_0 + \frac{\partial\delta_i}{\partial z_0}\Delta z_0 + \frac{\partial\delta_i}{\partial \alpha}\Delta\alpha + \frac{\partial\delta_i}{\partial \beta}\Delta\beta + \frac{\partial\delta_i}{\partial d}\Delta d \tag{4}$$

Parameter s indicated in Eq. (3) is then defined as follows.

$$s = \sum_{i} [(\Delta_i - h) - (\delta_i + \Delta\delta_i)]^2 \tag{5}$$

To minimize parameter s, the following condition is used to form simultaneous equations.

$$\begin{aligned} \frac{\partial s}{\partial \Delta x_0} &= 0 \\ \frac{\partial s}{\partial \Delta y_0} &= 0 \\ \frac{\partial s}{\partial \Delta z_0} &= 0 \\ \frac{\partial s}{\partial \Delta \alpha} &= 0 \\ \frac{\partial s}{\partial \Delta \beta} &= 0 \\ \frac{\partial s}{\partial \Delta d} &= 0 \\ \frac{\partial s}{\partial h} &= 0 \end{aligned} \tag{6}$$

Optical path differences AP and BP are geometrically determined as follows.

$$\begin{aligned} \frac{\partial\delta_i}{\partial x_0} &= \frac{(x_0 - x_i)}{s_1} + \frac{(-x_0 - x_i)}{s_2} \\ \frac{\partial\delta_i}{\partial y_0} &= \frac{(y_0 - y_i)}{s_1} + \frac{(-y_0 - y_i)}{s_2} \end{aligned} \tag{7}$$

$$
\begin{aligned}
\frac{\partial \delta_i}{\partial z_0} &= \frac{(z_0 - \alpha x_i - \beta y_i - d)}{s_1} + \frac{(-z_0 - \alpha x_i - \beta y_i - d)}{s_2} \\
\frac{\partial \delta_i}{\partial \alpha} &= \frac{x_i(z_0 - \alpha x_i - \beta y_i - d)}{s_1} - \frac{x_i(-z_0 - \alpha x_i - \beta y_i - d)}{s_2} \\
\frac{\partial \delta_i}{\partial \beta} &= \frac{y_i(z_0 - \alpha x_i - \beta y_i - d)}{s_1} - \frac{y_i(-z_0 - \alpha x_i - \beta y_i - d)}{s_2} \\
\frac{\partial \delta_i}{\partial d} &= \frac{(z_0 - \alpha x_i - \beta y_i - d)}{s_1} - \frac{(-z_0 - \alpha x_i - \beta y_i - d)}{s_2}
\end{aligned}
\tag{8}
$$

With these equations, each term in Eq. (4) can be described as follows.

$$
\begin{aligned}
\frac{\partial s}{\partial \Delta x_0} &= 2\sum_i \left( -\frac{\partial \delta_i}{\partial x_0} \right) [\Delta_i - h - (\delta_i - \Delta\delta_i)] = 0 \\
\frac{\partial s}{\partial \Delta y_0} &= 2\sum_i \left( -\frac{\partial \delta_i}{\partial y_0} \right) [\Delta_i - h - (\delta_i - \Delta\delta_i)] = 0 \\
\frac{\partial s}{\partial \Delta z_0} &= 2\sum_i \left( -\frac{\partial \delta_i}{\partial z_0} \right) [\Delta_i - h - (\delta_i - \Delta\delta_i)] = 0 \\
\frac{\partial s}{\partial \Delta \alpha} &= 2\sum_i \left( -\frac{\partial \delta_i}{\partial \alpha} \right) [\Delta_i - h - (\delta_i - \Delta\delta_i)] = 0 \\
\frac{\partial s}{\partial \Delta \beta} &= 2\sum_i \left( -\frac{\partial \delta_i}{\partial \beta} \right) [\Delta_i - h - (\delta_i - \Delta\delta_i)] = 0 \\
\frac{\partial s}{\partial \Delta d} &= 2\sum_i \left( -\frac{\partial \delta_i}{\partial d} \right) [\Delta_i - h - (\delta_i - \Delta\delta_i)] = 0 \\
\frac{\partial s}{\partial h} &= 2\sum_i [\Delta_i - h - (\delta_i - \Delta\delta_i)] = 0
\end{aligned}
\tag{9}
$$

As a result, parameter s is determined by six small changes, $\Delta x_0$, $\Delta y_0$, $\Delta z_0$, $\Delta\alpha$, $\Delta\beta$, and $\Delta d$. Additionally, from Eq. (5) and (6), the following expressions are formed.

$$
A = \begin{pmatrix}
\sum_i \left(\frac{\partial \delta_i}{\partial x_0}\right)^2 & \sum_i \frac{\partial \delta_i}{\partial x_0}\frac{\partial \delta_i}{\partial y_0} & \sum_i \frac{\partial \delta_i}{\partial x_0}\frac{\partial \delta_i}{\partial z_0} & \sum_i \frac{\partial \delta_i}{\partial x_0}\frac{\partial \delta_i}{\partial \alpha} & \sum_i \frac{\partial \delta_i}{\partial x_0}\frac{\partial \delta_i}{\partial \beta} & \sum_i \frac{\partial \delta_i}{\partial x_0}\frac{\partial \delta_i}{\partial d} & \sum_i \frac{\partial \delta_i}{\partial x_0} \\
\sum_i \frac{\partial \delta_i}{\partial x_0}\frac{\partial \delta_i}{\partial y_0} & \sum_i \left(\frac{\partial \delta_i}{\partial y_0}\right)^2 & \sum_i \frac{\partial \delta_i}{\partial y_0}\frac{\partial \delta_i}{\partial z_0} & \sum_i \frac{\partial \delta_i}{\partial y_0}\frac{\partial \delta_i}{\partial \alpha} & \sum_i \frac{\partial \delta_i}{\partial y_0}\frac{\partial \delta_i}{\partial \beta} & \sum_i \frac{\partial \delta_i}{\partial y_0}\frac{\partial \delta_i}{\partial d} & \sum_i \frac{\partial \delta_i}{\partial y_0} \\
\sum_i \frac{\partial \delta_i}{\partial x_0}\frac{\partial \delta_i}{\partial y_0} & \sum_i \frac{\partial \delta_i}{\partial y_0}\frac{\partial \delta_i}{\partial z_0} & \sum_i \left(\frac{\partial \delta_i}{\partial z_0}\right)^2 & \sum_i \frac{\partial \delta_i}{\partial z_0}\frac{\partial \delta_i}{\partial \alpha} & \sum_i \frac{\partial \delta_i}{\partial z_0}\frac{\partial \delta_i}{\partial \beta} & \sum_i \frac{\partial \delta_i}{\partial z_0}\frac{\partial \delta_i}{\partial d} & \sum_i \frac{\partial \delta_i}{\partial z_0} \\
\sum_i \frac{\partial \delta_i}{\partial x_0}\frac{\partial \delta_i}{\partial \alpha} & \sum_i \frac{\partial \delta_i}{\partial y_0}\frac{\partial \delta_i}{\partial \alpha} & \sum_i \frac{\partial \delta_i}{\partial z_0}\frac{\partial \delta_i}{\partial \alpha} & \sum_i \left(\frac{\partial \delta_i}{\partial \alpha}\right)^2 & \sum_i \frac{\partial \delta_i}{\partial \alpha}\frac{\partial \delta_i}{\partial \beta} & \sum_i \frac{\partial \delta_i}{\partial \alpha}\frac{\partial \delta_i}{\partial d} & \sum_i \frac{\partial \delta_i}{\partial \alpha} \\
\sum_i \frac{\partial \delta_i}{\partial x_0}\frac{\partial \delta_i}{\partial \beta} & \sum_i \frac{\partial \delta_i}{\partial y_0}\frac{\partial \delta_i}{\partial \beta} & \sum_i \frac{\partial \delta_i}{\partial z_0}\frac{\partial \delta_i}{\partial \beta} & \sum_i \frac{\partial \delta_i}{\partial \alpha}\frac{\partial \delta_i}{\partial \beta} & \sum_i \left(\frac{\partial \delta_i}{\partial \beta}\right)^2 & \sum_i \frac{\partial \delta_i}{\partial \beta}\frac{\partial \delta_i}{\partial d} & \sum_i \frac{\partial \delta_i}{\partial \beta} \\
\sum_i \frac{\partial \delta_i}{\partial x_0}\frac{\partial \delta_i}{\partial d} & \sum_i \frac{\partial \delta_i}{\partial y_0}\frac{\partial \delta_i}{\partial d} & \sum_i \frac{\partial \delta_i}{\partial z_0}\frac{\partial \delta_i}{\partial d} & \sum_i \frac{\partial \delta_i}{\partial \alpha}\frac{\partial \delta_i}{\partial d} & \sum_i \frac{\partial \delta_i}{\partial \beta}\frac{\partial \delta_i}{\partial d} & \sum_i \left(\frac{\partial \delta_i}{\partial d}\right)^2 & \sum_i \frac{\partial \delta_i}{\partial d} \\
\sum_i \frac{\partial \delta_i}{\partial x_0} & \sum_i \frac{\partial \delta_i}{\partial y_0} & \sum_i \frac{\partial \delta_i}{\partial z_0} & \sum_i \frac{\partial \delta_i}{\partial \alpha} & \sum_i \frac{\partial \delta_i}{\partial \beta} & \sum_i \frac{\partial \delta_i}{\partial d} & n
\end{pmatrix}
$$

The resultant simultaneous equations are described as follows.

$$B = \begin{pmatrix} \sum_i \frac{\partial \delta_i}{\partial x_0}(\Delta_i - \delta_i) \\ \sum_i \frac{\partial \delta_i}{\partial y_0}(\Delta_i - \delta_i) \\ \sum_i \frac{\partial \delta_i}{\partial z_0}(\Delta_i - \delta_i) \\ \sum_i \frac{\partial \delta_i}{\partial \alpha}(\Delta_i - \delta_i) \\ \sum_i \frac{\partial \delta_i}{\partial \beta}(\Delta_i - \delta_i) \\ \sum_i \frac{\partial \delta_i}{\partial d}(\Delta_i - \delta_i) \\ \sum_i (\Delta_i - \delta_i) \end{pmatrix} \qquad X = \begin{pmatrix} \Delta x_0 \\ \Delta y_0 \\ \Delta z_0 \\ \Delta \alpha \\ \Delta \beta \\ \Delta d \\ h \end{pmatrix} \qquad AX = B \tag{10}$$

Solution to the system of equations gives $\Delta x_0$, $\Delta y_0$, $\Delta z_0$, $\Delta\alpha$, $\Delta\beta$, and $\Delta d$, and all of the locations for components in fig. 15 will be determined. Using the above procedures and the data shown in fig. 18, the following values are obtained: x = 29.9 μm, y = 0.0 μm, z = 0.2 μm, d = 150.0 mm, α = 0.0, and β = 0.0, and the surface figure deviation of the test surface from an ideal sphere is determined as shown in fig. 19, which is highly spherical. The x and z values are slightly different from the initial values due to computational error.

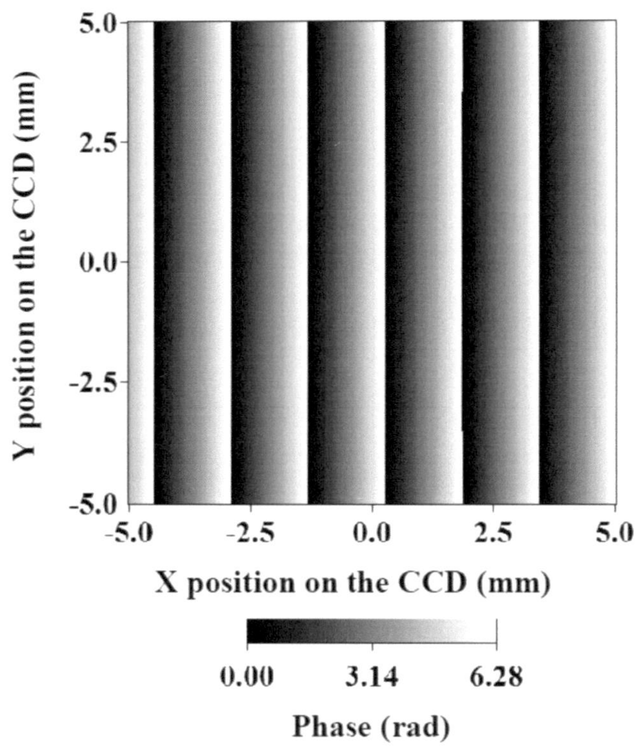


Fig. 17. Wrapped initial phase map with discontinuity.

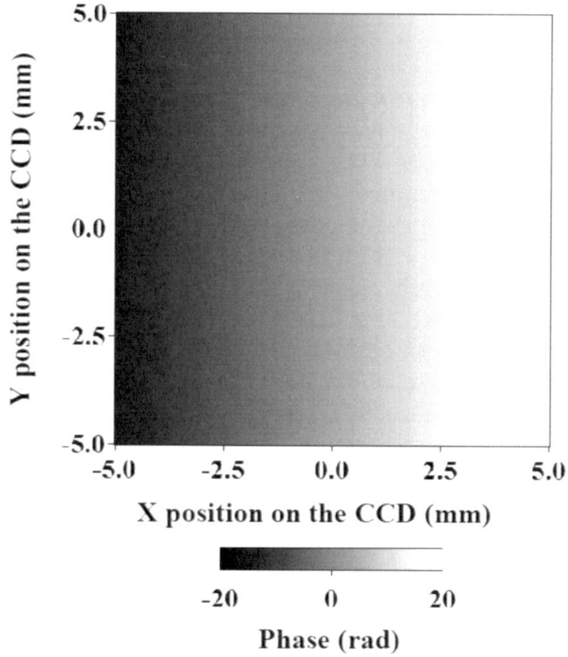


Fig. 18. Unwrapped continuous initial phase map.

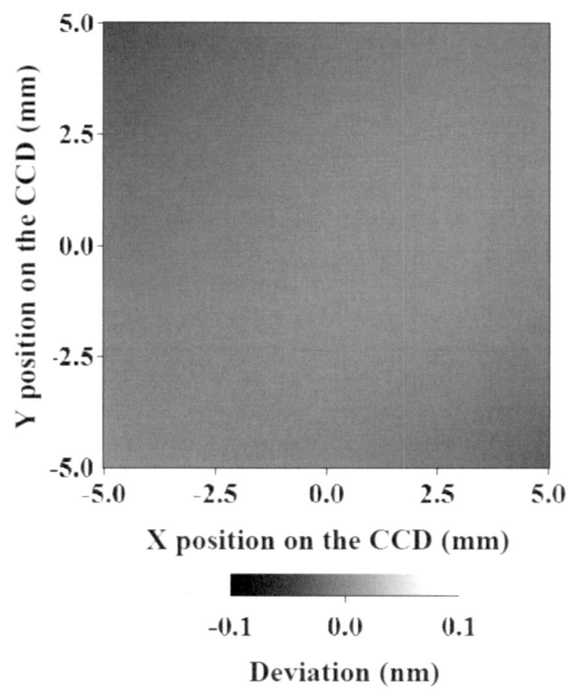


Fig. 19. Deviation of test wavefront from ideal spherical surface.

### 4.2 Measurement error caused by projection plane

At the beginning of error estimation, the interference fringe pattern was recorded using a CCD detector through an imaging lens as shown in fig. 20. However, the two diffraction waves from the fiber ends pass through different parts of the lens and receive different amounts of aberrations. As an example, error caused by a commercial plane-convex lens is estimated by the ray-trace method. The convex lens surface is not hyperboloidal but spherical, and a type of aberration must be included on the diffraction waves. The optical path differences with and without the imaging lens are estimated as shown in fig. 21. This phenomenon causes measurement error as shown in fig. 22. This error may depend on both the optical configuration of the fibers and lens and surface shape of the lens, and cannot be eliminated by post numerical procedure. Consequently, optics should not be inserted in the optical path before the place where interference pattern is formed. Hence, the interference pattern is generated without optics as shown in fig. 23 and transferred to a CCD surface through a relay lens. In this configuration, the relay lens makes a copy of the interferogram on the CCD without adding aberrations to the fringe pattern on the CCD. In general, a rotating ground glass plate is used as the projection plane (Otaki et al., 2002a; Otaki et al., 2002b), but it may generate mechanical vibration. Instead of the ground glass, we tested a Fiber Optic Window (FOW) as the projection plane (Matsuura et al., 2007). The external view and magnified surface image of the FOW are shown in fig. 24. The FOW is a bundle of multi-mode optical fibers, which transfers an optical image from the front side to rear side. This plate has several advantages: it transfers optical intensity distribution correctly to the rear side, the transferred image loses its coherency, and unwanted interference pattern does not occur on the CCD surface. In fig. 25, an experimental interferogram taken by the setup shown in fig. 23 is displayed. The interference pattern is vertically striped because the two fiber ends must be separated physically. Seven consecutive interference patterns are recorded in a $\pi/2$ phase-shift procedure, and the distortion of the test wavefront from the reference wavefront is calculated and shown in fig. 26. This figure should be zero for the entire area without measurement error because the test surface is the same as the reference surface. However, it exhibits long-range or low-frequency figure error across the entire area, meaning the error originated in the FOW. As shown in fig. 24, the fiber cladding of the FOW is thin, and crosstalk between the fibers is not negligible. Hence, the interference pattern is slightly deformed. This error is highly reproducible as exhibited in fig. 27, and subtraction of two results appears to eliminate the error perfectly. The difference between two datasets is shown in fig. 28. The FOW-induced long-range error is reduced adequately, and the vertically striped error pattern having twice the frequency of the interferogram is newly shown clearly. This may be a typical error usually caused by mechanical vibration in the phase-shift measurement (de Groot, 1995c; Creath & Hariharan, 1994). By using seven successive interferograms with phase-shift errors, the same pattern is reproduced as shown in fig. 29. This may be due to the phase-shift error enhanced at the borders between bright and dark fringes where the intensity change is great. Mechanical vibration is a random variation, and the error shown in fig. 29 can be reduced by averaging many datasets. After eliminating the FOW-induced error and averaging 50 consecutive datasets, the difference between the two averaged datasets is shown in fig. 30. The vibrational error shown in fig. 29 disappeared. A small residual error in fig. 30 distributes with the same period as the interferogram, which may be caused by changes in intensity of the interferogram during the phase-shift algorithm (van Wingerden et al., 1991). Fig. 30 is also reproduced by seven interferograms with slightly different intensities. The result is displayed in fig. 31. This error may be caused by relatively large and small intensity changes in bright and dark regions. Intensity fluctuation of the interferogram or laser light is caused by non-rigid components in the optical system, such as the laser-fiber coupler and fiber support. These effects are minimized by a long setup time before measurement. Finally, the difference between the two measurement datasets is shown in fig. 32, which is also the measurement accuracy of PS/PDI with the FOW. The remaining errors may be caused by residual errors of both vibration and intensity fluctuation and random noise in each pixel of the CCD detector. The FOW-induced error depends on the incident angle of light and is non-constant, and eliminating the error is difficult.

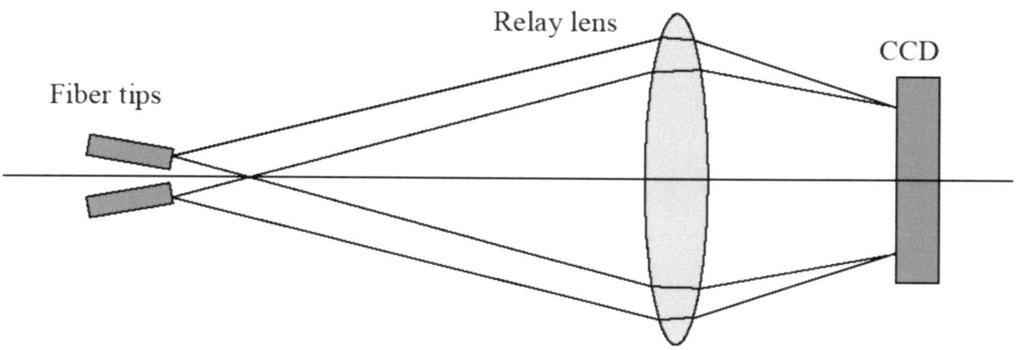


Fig. 20. Recording of interference pattern by two fiber ends with imaging lens.

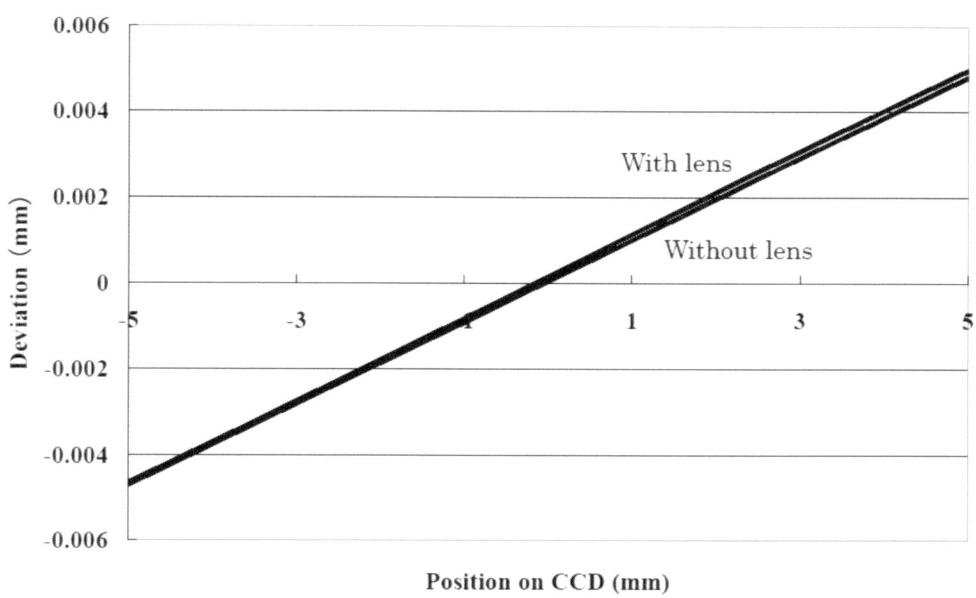


Fig. 21. Change in optical path with and without flat-convex lens.

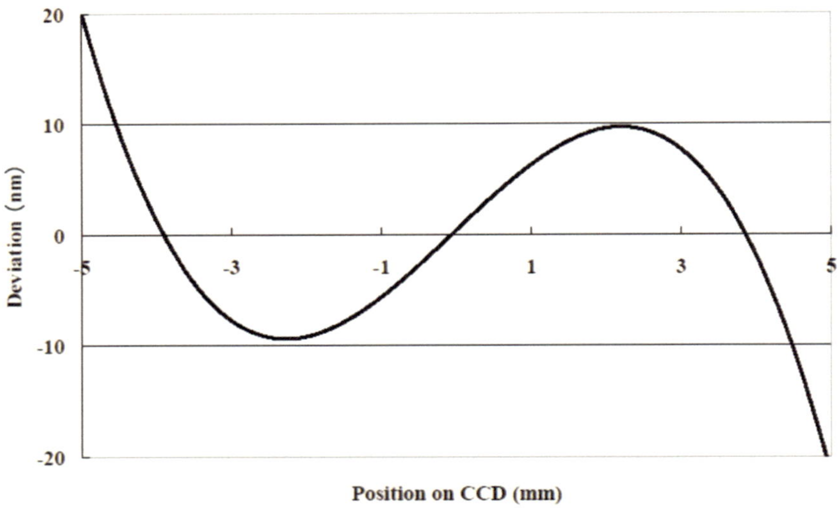


Fig. 22. Error caused by aberration of flat convex lens.

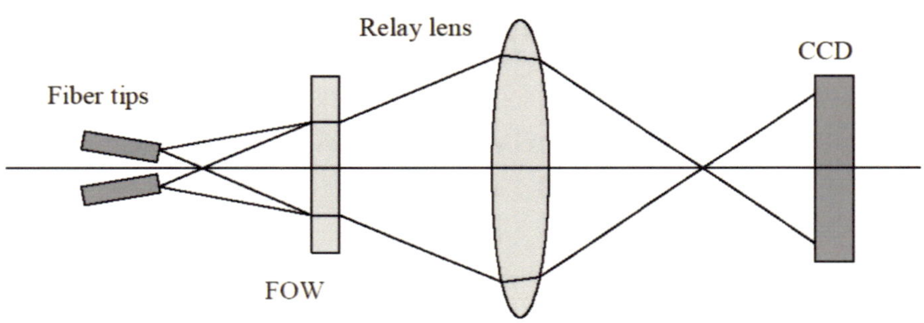


Fig. 23. Recording of interference pattern by FOW.

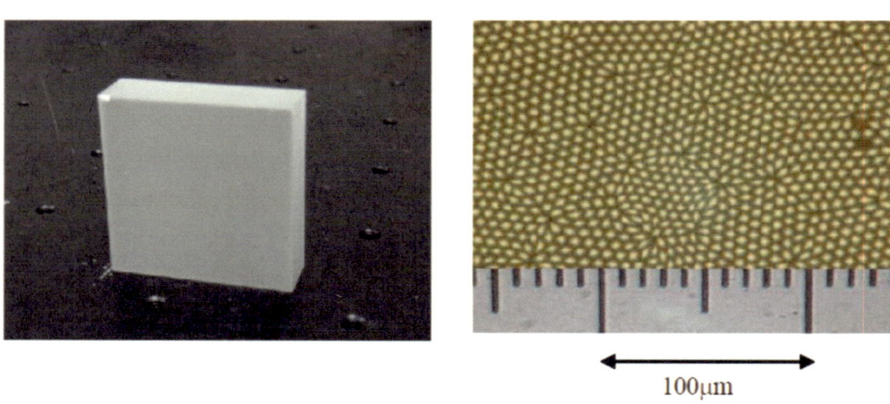


Fig. 24. Photo of the FOW plate and its microscopic surface image.

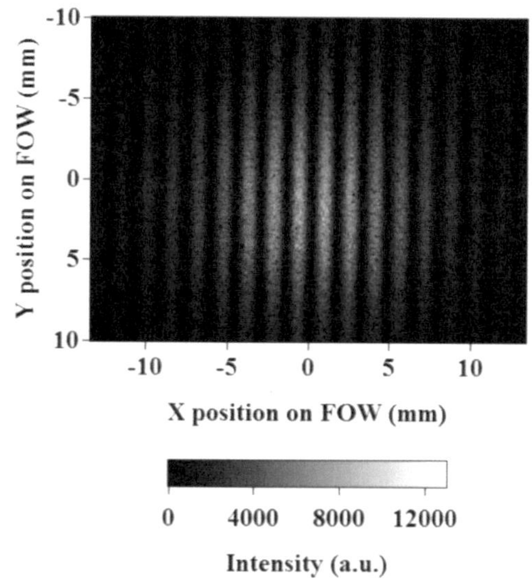


Fig. 25. Typical interference pattern recorded with the FOW plate.

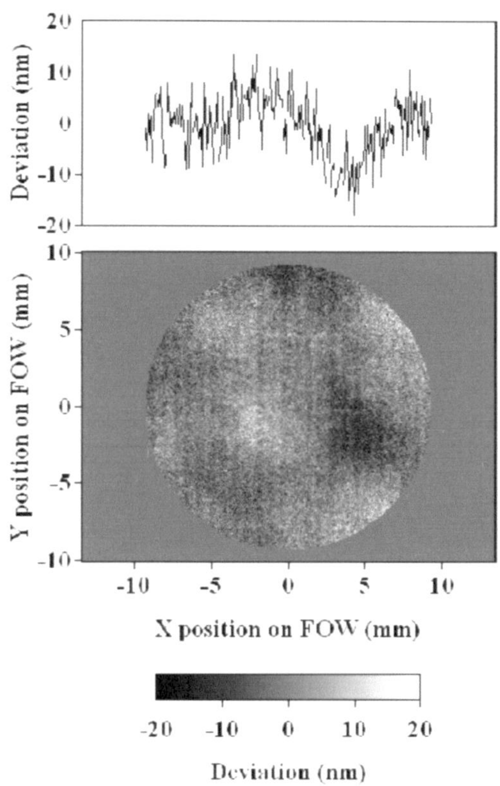


Fig. 26. Deviation of test wavefront from ideal sphere.

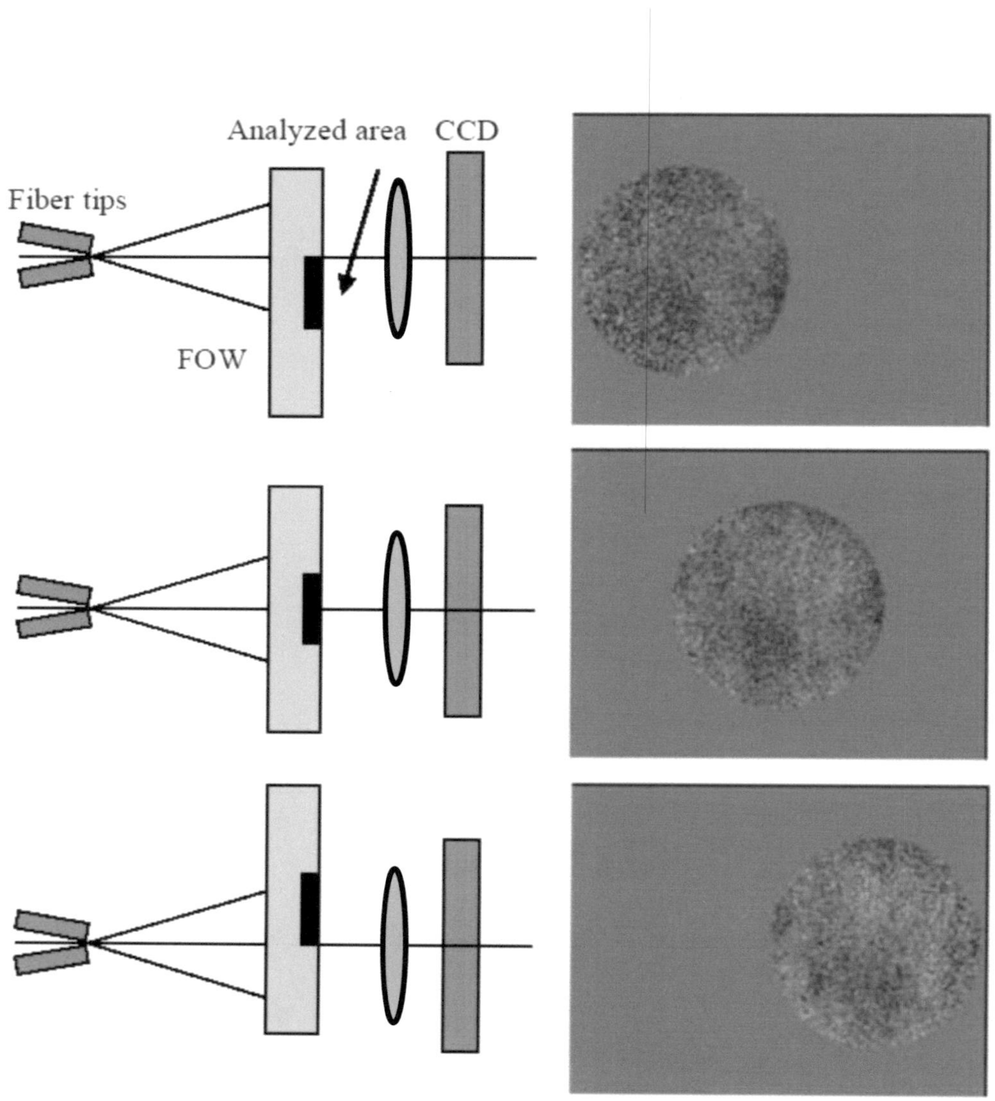


Fig. 27. Low frequency error caused by the FOW depends on its surface position.

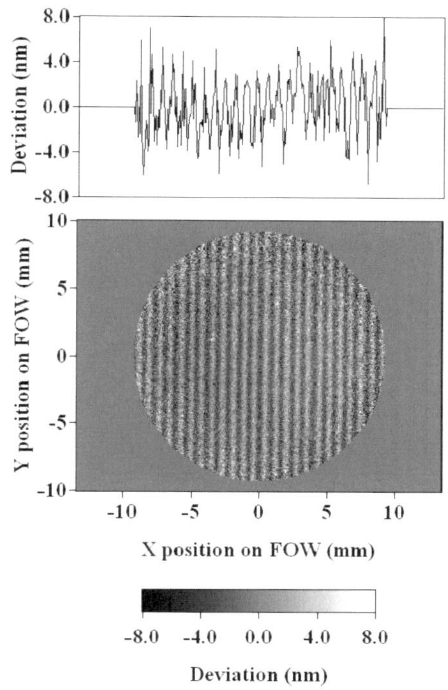


Fig. 28. Difference between two datasets taken with the FOW.

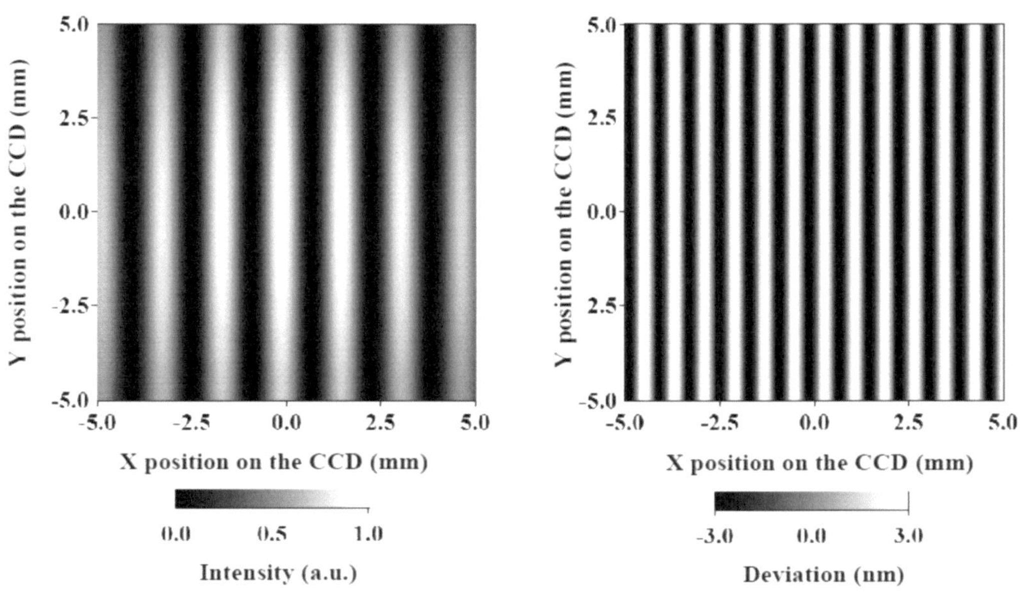


Fig. 29. Estimated phase error (a) in the PS analysis caused by error in movement of PZT stage with ideal interference pattern (b).

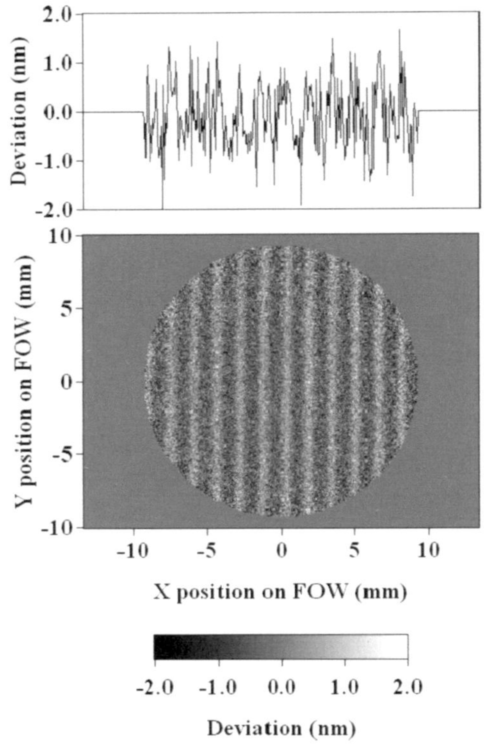


Fig. 30. Difference between two averaged datasets, which are averaged from 50 consecutive datasets.

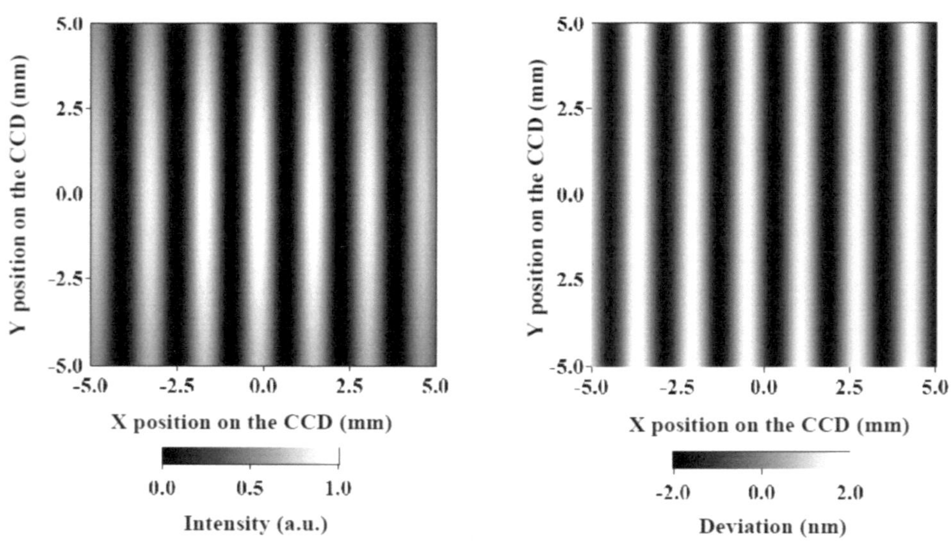


Fig. 31. Estimated phase error (right) in the PS analysis caused by fluctuation in laser light intensity.

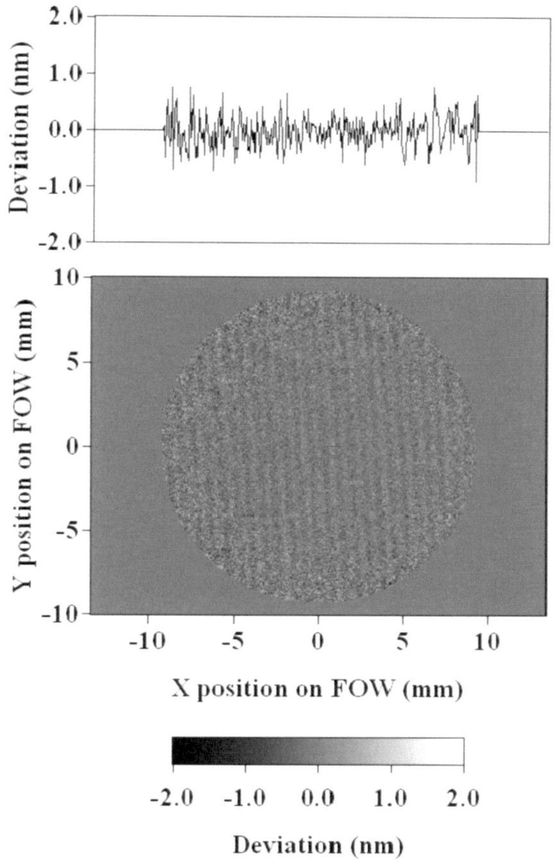


Fig. 32. Difference between two averaged datasets with stabilized laser light, which are averaged from 50 consecutive datasets.

### 4.3 Recording of interference pattern by glassless CCD camera

In the previous section, measurement error caused by FOW is highlighted. Therefore, instead of the FOW, we constructed an alternative optical system without a lens and projection plane as shown in fig. 33. In the new configuration, the interference pattern is directly recorded by a CCD camera. In principle, this configuration requires system stability and random error control by averaging many datasets. In general, a CCD detector is covered by a glass plate to maintain the vacuum and for protection. Such a glass plate may cause internal reflection and unwanted interference inside the plate. Therefore, we adopted a glassless CCD camera without a protective glass plate. This CCD camera has considerable noise because it does not have a cooling system. To constrain the thermal and readout noises, each interference pattern processed by seven bucket algorithms is constructed by averaging 32 interference patterns at the same phase. Movement of the PZT stage, which actualizes phase shift by changing the optical path length accurately, is optimized to minimize the mechanical vibration of the phase-shift mirror. Thus, the movement is more segmentalized. The resultant deviation of the test surface from reference surface is shown in fig. 34(a), and its reproducibility is shown in fig. 34(b). These exhibit the measurement accuracy of this system. In fig. 34(a) and (b), vibrational error still exists. Here, an additional consideration is the condition of the fiber end. When the fiber end is polished poorly, fig. 34(a) becomes fig. 35. To confirm the sphericity of a diffraction wave, after the null test described above, one fiber end is rotated about its optical axis before a second null test. If both null tests return good results, the diffraction waves used in the PS/PDI are highly spherical. The test wave reflected off the test mirror has double the deformation information of the surface figure. Thus, half the distortion of the test wavefront from the ideal sphere corresponds to the mirror surface figure. For this reason, exact measurement accuracy of the PS/PDI is half the value shown in fig. 34. The accuracy is 1.6 nm PV and 0.17 nm rms and the reproducibility is 0.7 nm PV and 0.07 nm rms.

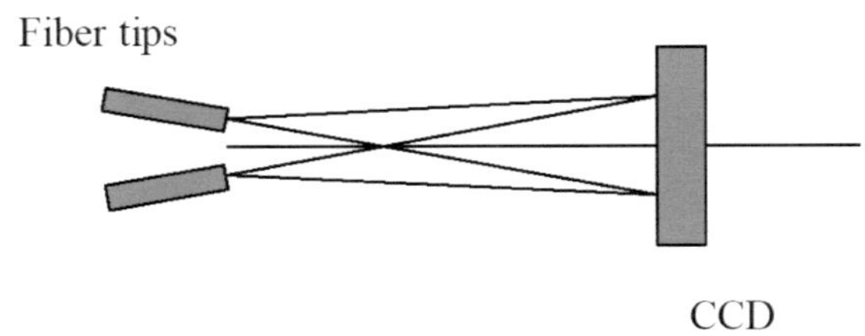


Fig. 33. Recording of interference pattern without any optics.

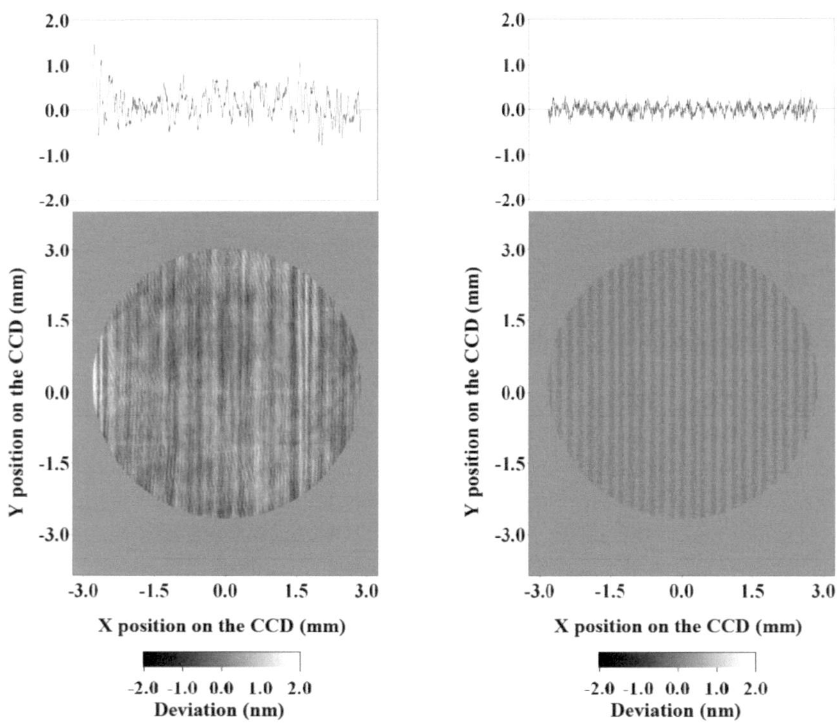


Fig. 34. Measurement accuracy of our PS/PDI system evaluated experimentally. Deviation of wavefront from ideal sphere (left) and the measurement reproducibility (right).

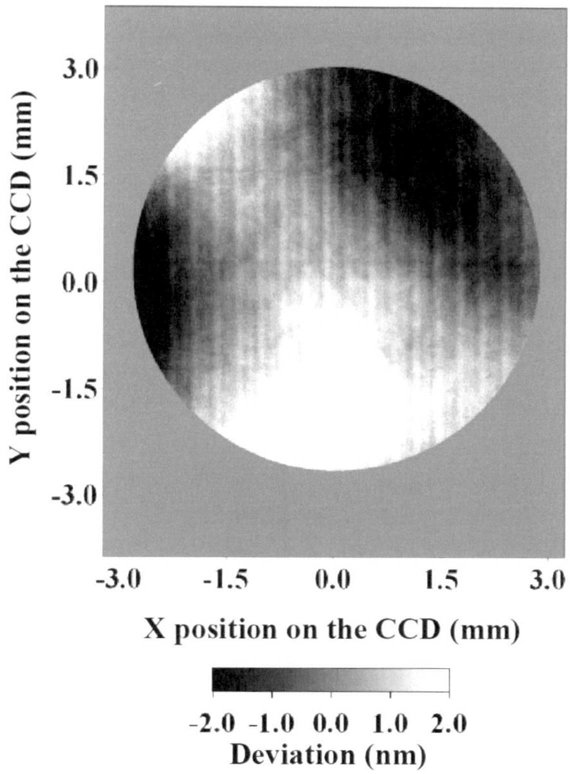


Fig. 35. Deviation of wavefront from ideal sphere caused by inadequate polish at fiber apex.

## 5. Wavefront reconstruction on test mirror surface

### 5.1 Kirchhoff-Helmholtz integral theorem

In the PS/PDI measurement or ordinary optical interferometry, the measurement result is a phase map of a wavefront including surface figure information of a test mirror on a CCD sensor. Here, reflection of light wave on a test mirror surface is considered in geometric optics. As shown in fig. 36, the propagation direction of each part of the reflected wave depends on angle of slope at the test mirror surface. In other words, the surface figure of the reflected wavefront depends on the observation or recording position, and the PS/PDI result with a glassless CCD camera will not agree with the exact surface figure of the test mirror. To obtain the actual mirror surface data, the reflected wavefront should be reconstructed on the test mirror surface using a digital holography technique with both phase and intensity maps computed from the CCD data (Yamaguchi et al., 2006; Matsuura et al., 2008). Fig. 37 shows a schematic diagram of the PS/PDI optics with two optical fibers. Planes P and Q are a spherical concave mirror and a projection plane for the interferogram, respectively. The test wave emitted from one fiber at point O propagates to test mirror P, and the reflected wave converges on point O and emanates to screen Q. The reference wave from another fiber end at point O illuminates screen Q, and it interferes with the light wave from plane P. The resultant interference pattern provides the surface figure of plane P. If surface P is highly spherical, the interference pattern on screen Q will display the deviation of surface P from an ideal sphere. In a practical sense, as explained in section 5.1, precise analysis is necessary for wavefront propagation. Here, a brief description of wave-optics analysis based on the Kirchhoff-Helmholtz (KH) Integral Theorem follows. From the KH theorem, as shown in fig. 38, an optical wave field U(Q), scalar value of electric or magnetic field, at point Q in closed surface S can be expressed with wave field U on closed surface S as Eq.(11) (Born & Wolf, 1959).

$$U(\mathrm{Q})=\frac{1}{4\pi}\iint_{\mathrm{S}}\left[U\frac{\partial}{\partial \mathrm{n}}\left\{\frac{\exp(-\mathrm{i}kr)}{r}\right\}-\frac{\exp(-\mathrm{i}kr)}{r}\frac{\partial U}{\partial \mathrm{n}}\right]\mathrm{d}s \tag{11}$$

Here, $\boldsymbol{r}$ is a vector between point Q and an arbitrary point on S, and $\boldsymbol{n}$ is a unit normal vector at that point on S. In Eq.(11), the wave field on surface P is a reflected spherical wave emitted from point O, and it is expressed as follows with a position vector $\boldsymbol{r}_0$.

$$U = A\frac{\exp\,(\mathrm{i}kr_0)}{r_0} \tag{12}$$

Here, A is amplitude of wave field, and $k$ is wavenumber. Then, the directional derivative of Eq.(12) to the normal direction is as follows.

$$\begin{aligned}\frac{\partial U}{\partial n}&=\frac{\partial U}{\partial r_0}\frac{\partial r_0}{\partial n}\\&=-A\frac{ikr_0\exp(ikr_0)-\exp(ikr_0)}{r_0^2}\quad\left(\because\frac{\partial r_0}{\partial n}=-1\right)\\&=-A\exp(ikr_0)\frac{ikr_0-1}{r_0^2}\\&=-A\frac{ik}{r_0}\exp(ikr_0)\quad(\because kr_0>>1)\end{aligned} \tag{13}$$

The following expression is also true.

$$\begin{aligned}&\frac{\partial}{\partial n}\left\{\frac{\exp(-ikr)}{r}\right\}\\&=\frac{\partial}{\partial r}\left\{\frac{\exp(-ikr)}{r}\right\}\frac{\partial r}{\partial n}\\&=\frac{-ikr\exp(-ikr)-\exp(-ikr)}{r^2}\cos(\mathbf{n},\mathbf{r})\quad\left(\because\frac{\partial r}{\partial n}=\cos(\mathbf{n},\mathbf{r})\right)\\&=-\frac{ikr+1}{r^2}\exp(-ikr)\cos(\mathbf{n},\mathbf{r})\\&=-\frac{ik}{r}\exp(-ikr)\cos(\mathbf{n},\mathbf{r})\quad(\because kr>>1)\end{aligned} \tag{14}$$

Finally, Eq.(11) is modified as follows.

$$U(\mathrm{Q}) = \frac{1}{4\pi}\iint_{\mathrm{S}}\left[-A\frac{\exp(\mathrm{i}kr_0)}{r_0}\frac{\mathrm{i}k}{r}\exp(-\mathrm{i}kr)\cos(\mathbf{n},\boldsymbol{r}) + \frac{\exp(-\mathrm{i}kr)}{r}A\frac{\mathrm{i}k}{r_0}\exp(\mathrm{i}kr_0)\right]\mathrm{d}s$$
$$= \frac{\mathrm{i}k}{4\pi}\iint_{\mathrm{S}} A\frac{\exp(\mathrm{i}kr_0)}{r_0}\frac{\exp(-\mathrm{i}kr)}{r}\{1-\cos(\mathbf{n},\boldsymbol{r})\}\,\mathrm{d}s \tag{15}$$

These are wave analyses in the system of fig. 37. If the optical intensity of a spherical wave is uniform with no phase distortion, the term $A\exp(ikr_0)/r_0$ in Eq.(15) is constant. However, in an actual case, the test surface is aspherical due to a machining error, and intensity distribution of the diffraction wave is non-uniform. These two matters must be considered in the reconstruction procedure of the wavefront on the test mirror surface. Now, as shown in fig. 39, an orthogonal coordinate system is positioned with its origin at an arbitrary point on spherical P, and the x-y plane is a tangential plane of spherical P. Hence, in fig. 39, the normal vector $\boldsymbol{n}$ faces the negative direction of the z-axis. Using this coordinate system, the surface figure of the wavefront near the origin is expressed as follows.

$$z = z_0\,(x, y) \tag{16}$$

Propagation direction of the wavefront may be its normal direction, and a normal unit vector $\mathbf{n}_w$ is introduced at the intersection point between the z-axis and wavefront. Instead of Eq.(12), wave field of the reflected wave is as follows.

$$U = A\exp\left\{ikz_0(x,y)\right\} \tag{17}$$

And the related expressions are derived as follows.

$$\frac{\partial U}{\partial n} = \frac{\partial U}{\partial z_0}\frac{\partial z_0}{\partial n} = Aik\exp(ikz_0)\cos(\mathbf{n},\mathbf{n}_W) \tag{18}$$

$$U(\mathrm{Q}) = \frac{1}{4\pi}\iint_{\mathrm{S}}\left[-A\exp(\mathrm{i}kz_0)\frac{\mathrm{i}k}{r}\exp(-\mathrm{i}kr)\cos(\mathbf{n},\boldsymbol{r}) - \frac{\exp(-\mathrm{i}kr)}{r}A\mathrm{i}k\exp(\mathrm{i}kz_0)\cos(\mathbf{n},\mathbf{n}_{\mathrm{w}})\right]\mathrm{d}s$$
$$= -\frac{\mathrm{i}k}{4\pi}\iint_{\mathrm{S}} A\exp(\mathrm{i}kz_0)\frac{\exp(-\mathrm{i}kr)}{r}\{\cos(\mathbf{n},\mathbf{n}_{\mathrm{w}})+\cos(\mathbf{n},\boldsymbol{r})\}\,\mathrm{d}s \tag{19}$$

Using these equations, if the wave field on surface P is obtained as described in Eq.(17), the wave field on surface Q can be calculated by integration. Finally we develop the expression for the terms cos(**n**,**r**) and cos(**n**, $\mathbf{n}_W$). Referring to fig. 37, cos(**n**,**r**), **n**, and $\mathbf{e}_r$ are expressed as follows.

$$\cos(\mathbf{n}, \boldsymbol{r}) = \mathbf{n} \cdot \mathbf{e}_r \tag{20}$$

$$\mathbf{n} = \frac{(-x_p, -y_p, -z_p)}{\sqrt{x_p^2 + y_p^2 + z_p^2}} \tag{21}$$

$$\mathbf{e}_r = \frac{(x_p - x_q, y_p - y_q, z_p - z_q)}{\sqrt{(x_p - x_q)^2 + (y_p - y_q)^2 + (z_p - z_q)^2}} \tag{22}$$

$$\cos(\mathbf{n}, \boldsymbol{r}) = -\frac{x_p(x_p - x_q) + y_p(y_p - y_q) + z_p(z_p - z_q)}{\sqrt{(x_p^2 + y_p^2 + z_p^2)\{(x_p - x_q)^2 + (y_p - y_q)^2 + (z_p - z_q)^2\}}} \tag{23}$$

And referring to fig. 38, cos(**n**, $\mathbf{n}_W$), $\mathbf{e}_z$, and $\mathbf{n}_W$ are expressed as follows.

$$\cos(\mathbf{n}, \mathbf{n}_w) = -\mathbf{e}_z \cdot \mathbf{n}_w \tag{24}$$

$$\mathbf{e}_z = (0,0,1) \tag{25}$$

$$\mathbf{n}_w = \frac{\left(-\frac{\partial z_0}{\partial x}, -\frac{\partial z_0}{\partial y}, 1\right)}{\sqrt{1 + \left(\frac{\partial z_0}{\partial x}\right)^2 + \left(\frac{\partial z_0}{\partial y}\right)^2}} \tag{26}$$

Finally, Eq.(27) is obtained.

$$\cos(\mathbf{n}, \mathbf{n}_w) = -\frac{1}{\sqrt{1 + \left(\frac{\partial z_0}{\partial x}\right)^2 + \left(\frac{\partial z_0}{\partial y}\right)^2}} \tag{27}$$

Here, after reflection at surface P, the reflected wave propagates in the negative direction of the z-axis because of Eq.(17).

### 5.2 Wave propagation in measurement of perfect spherical surface

Changes in wavefront via propagation are simulated numerically based on the KH theorem as described in the previous section. When the simulation condition is set, light intensity and phase maps of the test wave on the test mirror are determined. As the simulation is conducted, the test wavefront on the CCD sensor is calculated using the KH theorem, and the resultant phase map as an ordinary PS/PDI result is compared with the reconstructed phase map on the test mirror surface. In the calculation, discretization of the integral is

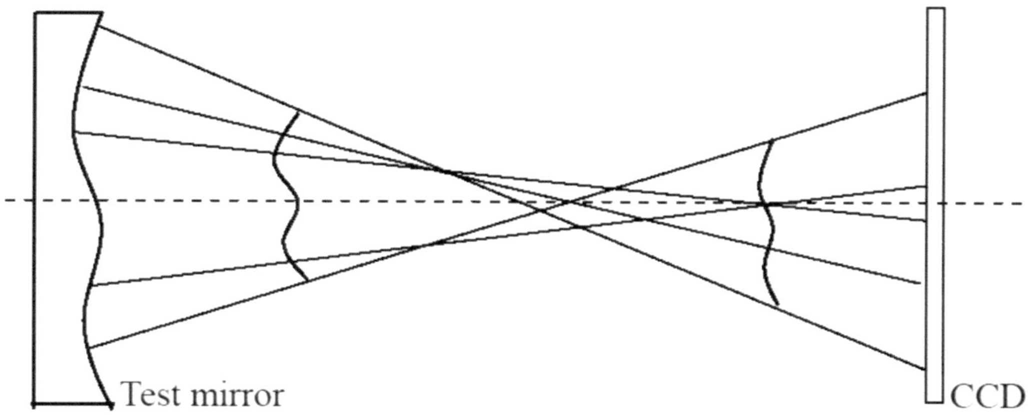


Fig. 36. Correspondence relation between points on test mirror and CCD surface.

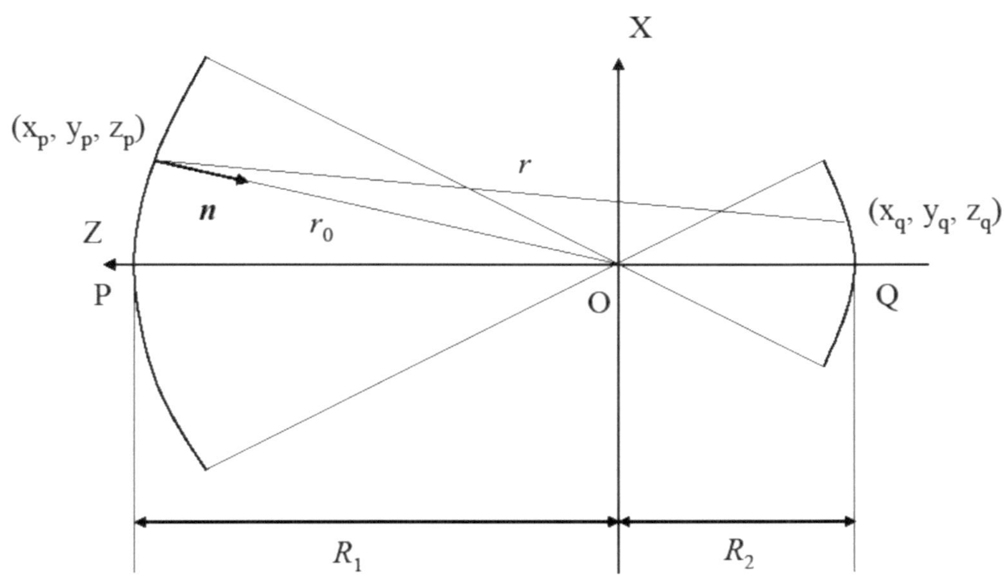


Fig. 37. Mathematical model for the PDI system.

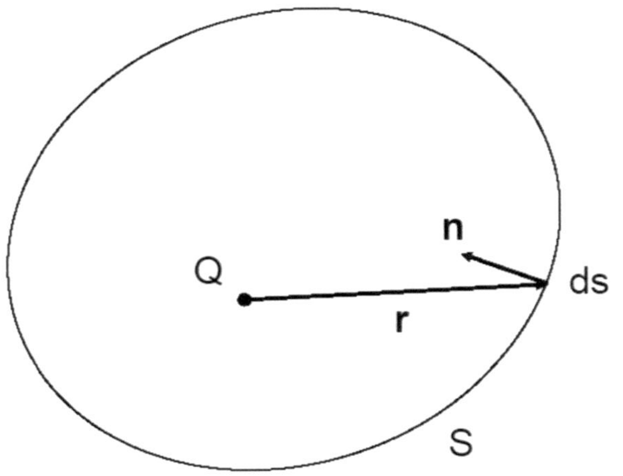


Fig. 38. Domain of integration in KH integral theorem.

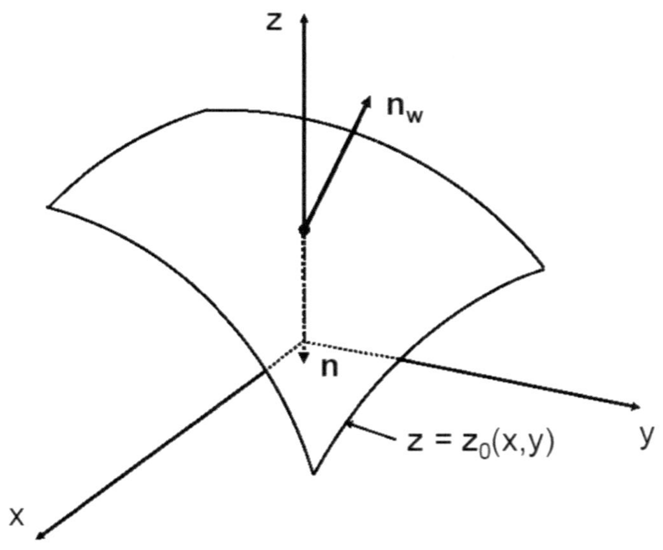


Fig. 39. Wavefront around a point on sphere P.

required, and a sufficient number of discretized integral terms are needed for accurate solution. The actual mirror surface is divided into 5000 x 5000 pixels as the integral terms. In contrast, the CCD surface is divided into 400 x 400 pixels for convenience of calculation time. The reconstruction process for a wavefront at the test mirror requires a sufficient amount of data on the CCD surface, and the 400 x 400 dataset is interpolated. The resultant finer data of 5200 x 5200 pixels are used to reconstruct the wavefront on the test mirror surface in 400 x 400 pixels. The simulation described above is completed with an ideal spherical concave mirror based on the configuration shown in fig. 40. The test mirror is 180 mm in diameter and has a 1500-mm ROC. The origin of the diffraction wave is positioned at the center of curvature of the mirror. The wavelength of the light wave is 632.8 nm. The CCD surface is assumed to be spherical concave with a 150 mm ROC, and it is concentric with the mirror surface. Light intensity distribution on the diffraction wavefront is determined as the Gaussian distribution shown in fig. 41(a). With the uniform phase map on the mirror surface and fig. 41(a), wavefront on the CCD surface is simulated as shown in fig. 41(b). It corresponds to the ordinary PS/PDI result, and is vaulted entirely. The phase map near the mirror edge fluctuates erratically, and an unexpected pattern appeared. Using the resultant intensity distribution and phase maps on the CCD surface, the wavefront was reconstructed on the test mirror surface. As a result, the mirror surface figure or phase map is calculated as shown in fig. 41(c), and the phase map is uniform in most of the area. Huge fluctuations in the phase map appeared near the edge. Next, the same simulation is performed with the more sharpened intensity distribution shown in fig. 42(a). Fig. 42(b) shows the result, which is also vaulted but the phase disturbance near the edge is reduced effectively. The reconstructed wavefront, as shown in fig. 42(c), has a uniform phase map in most area. However, huge fluctuations of the phase appeared near the edge region. In the above simulations, an integral corresponds to superposition of the weighted diffraction waves emitted from everywhere in the integral range. The weighting depends on intensity distribution. Hence, two different results were obtained and are shown in figs. 41 and 42. In fig. 42, contribution of the central part of the wavefront is effective due to its sharpened intensity distribution profile. Large fluctuations in phase maps may be caused by both diffraction at the mirror edge and insufficient discretization of the integral range. In the latter case, light intensity is weak at the mirror edge, and the edge effects are highly suppressed. The phase map in the edge area includes both systematic error from numerical calculations (Greivenkamp, 1987) and Moire fringes caused by grid-like segmentalization in the integral range (Ferraro et al., 2007). The same phenomenon appears in fig. 42. Reconstruction of the low frequency component results in a correct solution, while reconstruction of the high frequency component has considerable errors. In the latter case, the wavefront in only the central square region is correctly restored due to extremely low light intensity near the edge and the domain of integration is confined in the square region. In principle, the KH integral requires a closed surface for its integral domain. However, in these simulations, most of the reflected light propagates inside the square region, and an unhandled component of the wavefront is negligible. Therefore, the wavefront reconstruction processes are successfully achieved as shown in fig. 41 and 42. However, in fig. 42, light intensity in the outer edges of the test mirror is extremely low, and the subsequent lack of wavefront information caused incorrect solution of outer edges in the reconstruction procedure.

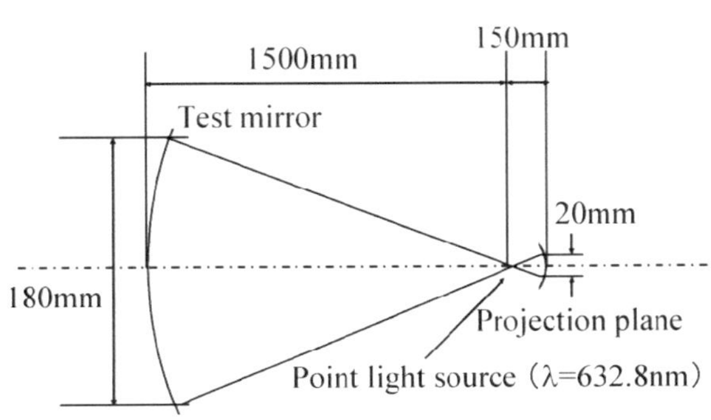


Fig. 40. Mathematical model of PS/PDI for numerical simulation.

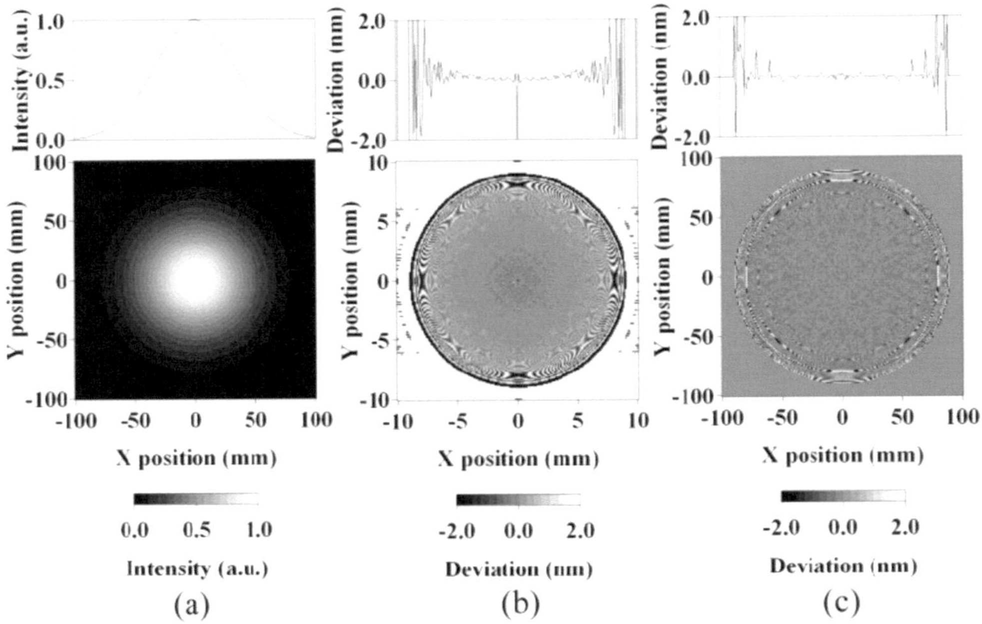


Fig. 41. Wave propagation simulation with ideal spherical concave mirror. (a) light intensity distribution, (b) phase map on CCD, and (c) reconstructed phase map on mirror surface.

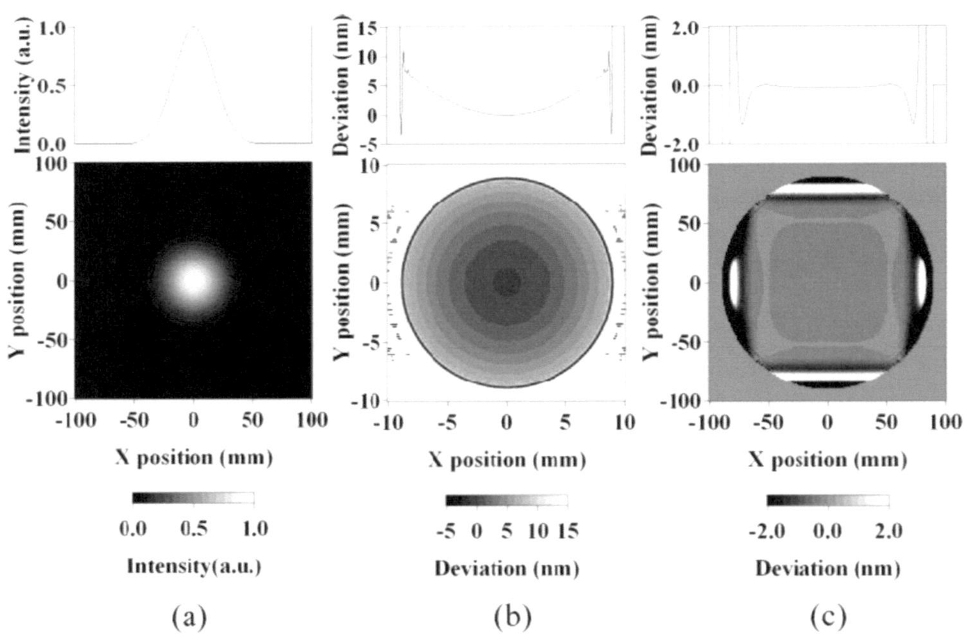


Fig. 42. Wave propagation simulation with ideal spherical concave mirror. (a) sharpen intensity distribution of light, (b) phase map on CCD, and (c) reconstructed phase map on mirror surface.

### 5.3 Wave propagation in measurement of imperfect spherical surface

An actual test mirror may have errors at different spatial frequencies. In this section, wave propagation analysis for test mirrors having surface figure error at both high and low spatial frequencies is conducted, and reconstruction of the wavefront is completed. Here, the reflected wave emanates in diverse directions, and a distant projection plane or CCD surface might not be able to collect most of the reflected light. Therefore the CCD surface is located at a distance of 15 mm from the focal point of the reflected wave. The test mirror surface has a high frequency surface figure deviation from an ideal sphere, similar to a Gaussian profile in the central portion as shown in fig. 43(a) and (d). The maximum deviation is 83.3 nm from ideal surface as shown in fig 43(d), and this deviation profile corresponds approximately to a flat surface 1 mm in diameter. The intensity profile of the diffraction wave is the same as that of fig. 41(a), and the wave propagation analysis provides the test mirror surface figure data as a phase map on the CCD sensor shown in fig. 43(b). A part of the diffraction wave reflected at the central flat area on the test mirror surface continues spreading, and not a trace of the central small area is visible in fig. 43(b). However, the reconstruction process with these data results test the mirror figure as shown in fig. 43(c), and the central cross-section profile shown in fig. 43(e) exhibits approximately the same profile as fig. 43(d). The wavefront reconstruction process succeeded in reproducing the exact surface figure with high frequency error in the numerical simulation. Next, we assume a test surface with low-frequency surface error of sinusoidal fluctuation of 50 nm in amplitude in the entire area as shown in fig. 44(a). With the same intensity distribution as fig. 41(a), the wavefront reconstruction process provides the solution indicated in fig. 44(b) and (c), which are phase maps on the CCD surface and the reconstructed wavefront, respectively. In both cases, low-frequency surface figure error is reproduced successfully. Therefore, an exact surface figure having low frequency error can be obtained only from a phase map on the CCD, according to underlying basic optics. However, an actual mirror having different kinds of frequency components in its error profile must be measured by a PS/PDI system with a wavefront reconstruction procedure.

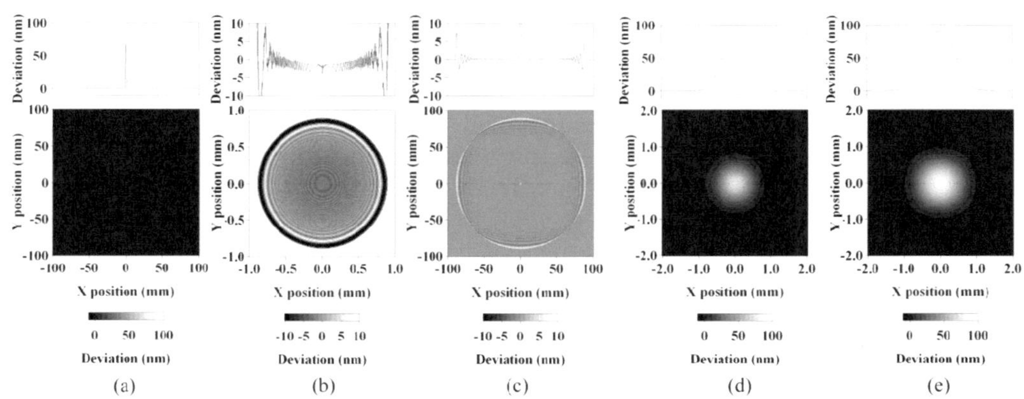


Fig. 43. Wavefront propagation analysis with test mirror surface having figure error in central area at high spatial frequency. (a) test mirror surface figure, (b) wavefront obtained on CCD surface, (c) wavefront reconstructed on test mirror surface, (d) magnified image of surface figure error in (a), and (e) magnified image of reconstructed surface figure error in the analysis.

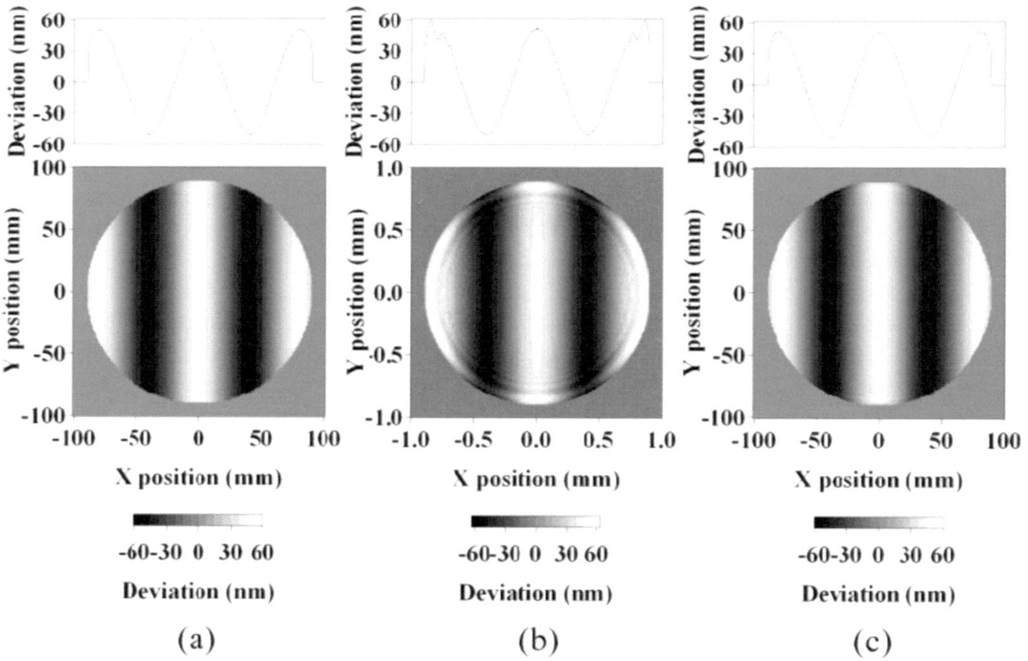


Fig. 44. Wavefront propagation analysis with test mirror surface having figure error in whole surface at low spatial frequency. (a) test mirror surface figure, (b) wavefront obtained on CCD surface, and (c) wavefront reconstructed on test mirror surface.

## 6. Measurement of spherical concave mirror

### 6.1 Procedures for measurement and analysis

The PS/PDI system is advantageous in the measurement of a spherical concave mirror because its reference plane is an ideal sphere for the diffraction wavefront. In this section, trial measurement of a conventional spherical concave mirror 200 mm in diameter, 1500 mm in ROC, and with $\lambda/6$ figure accuracy is reported with reproducibility of the measurement with and without change in the mirror alignment (Matsuura et al., 2010). Firstly, the supporting technique is examined with an ordinary CCD camera and imaging lens. The test mirror described above is epoxied onto an aluminum block (300 mm x 300 mm x 80 mm) and supported by three-point mounting with steel balls for highly rigid supporting. The block is supported by an aluminum thin plate mounted between its top face and supporting rod stands on an optical bench to control mechanical vibration. A photo of the mirror apparatus is shown in fig. 45. On the outer edge of mirror surface, three triangular black papers serving as position markers are attached, and the measurement is completed inside the marks. In figs. 46, 47, and 48, both the interference pattern and phase map on the CCD are shown for three different ambient temperatures of the PS/PDI measurement. Temperature change caused stress at the supporting ball due to the difference in the coefficient of thermal expansion (CTE) between the mirror substrate and aluminum block. Hence, the three phase maps on the CCD are slightly altered. Their differences are indicated in figs. 49 and 50. A temperature change of 2.7 °C caused surface deformation of ±20 nm in the diffraction wavefront. The CTE for aluminum and Pyrex glass are $29\times10^{-6}$[$K^{-1}$] and 3.0-3.6$\times10^{-6}$[$K^{-1}$], respectively, and the difference is enough to cause considerable stress for such a deformation. Next, the test mirror is placed with its side surface resting on three steel balls, and the top edge is touched by the apex of a needle to control mechanical vibration. The mirror supported by this method is shown in fig. 51. To examine the effect of the supporting needle, a series of 60 measurement datasets were taken and each ten datasets were averaged to form an average map, and the resultant six images are displayed in fig. 52. Moreover, each average map is numerically approximated to a plane, and the change in direction of the surface normal vector is plotted in fig. 53. In fig. 52, there is no remarkable change. However, tilt of the mirror surface is clearly changing, especially in the vertical direction, as exhibited in fig. 53. In fact, thermal expansion of the supporting needle causes tilt change of test mirror surface. As a result, the three-point steel ball support system causes not issues with the mirror measurement when it is combined with a thermal control system, such as a temperature-controlled room. Second, in actual PDI measurement, which differs from the null test in section 4, a common-path interferometry will not be realized because only the test wave propagates along the optical path through the measurement atmosphere, and the influence of air fluctuation is experimentally studied by collecting 100 consecutive datasets. Fig. 54 shows change in each averaged dataset, which is calculated from 10 consecutive datasets. A low frequency error pattern is moving from map-to-map, which may be caused by air current. The actual PS/PDI measurement is performed with the random error averaged out. In the following, a glassless CCD camera is used as mentioned in section 4, and the wavefront reconstruction process is employed according to the discussion in section 5. Each interference pattern at a different phase is composed of 32 averaged patterns for reduction of the CCD noise, and the step of phase shift is $\lambda/2$. A frequently used seven bucket algorithm is used to determine an initial phase map, which is averaged over 100 consecutive datasets to eliminate random noise such as air current. With location data of the fibers and CCD determined by pre-experiment analysis, the test wavefront on the CCD surface is calculated. Finally, reconstruction of the wavefront determines the exact surface figure of the test mirror. In figs. 55 and 56, a typical light intensity distribution and an interference pattern are exhibited, respectively. In the images, the lower part is misshapen due to the tip-resistant cover over the mirror surface. Coaxial fringes at the outer edge of the image are caused by edge diffraction. The position markers produced three circular patterns in fig. 55. Tilt fringes or vertical stripes caused by a position gap between the reference fiber end and the focal spot of the test wave are dominant in fig. 56. Location data of the fibers and CCD are derived by the analytical method mention in section 4 since the interference pattern is nearly ideal. The wavefront is shown in fig. 57 as the pattern deviation from an ideal sphere. From 100 consecutive datasets, the difference between the averaged first 50 and second 50 datasets is presented in fig. 58, and they agree within 0.11 nm rms. In section 4, measurement reproducibility was estimated to be 0.07 nm rms, and the lower value may be caused by air current in the optical system. Before the wavefront reconstruction procedure, location data based on the interference between two ideal spherical waves are evaluated by reconstructing the intensity profile at the focal point of the test wave. In fig. 59(a) and (b), the derived profile and an ideal profile are displayed, respectively. The data confirms that the reflected wavefront is focused well and is fairly spherical. In the final step, the exact location of the test mirror surface is required. The position coordinates are derived from the focal position of the test wave and the ROC of the mirror specification data, i.e. 1500 mm. Fig. 60 shows the resultant reconstructed intensity profile and surface figure of the test mirror. The derived profile at Y=0 mm in fig. 60 is compared with the actual intensity profile of a diffraction wave as shown in fig. 61. Within a 170 mm radius of the mirror, the profiles agree well, and the wavefront reconstruction succeeded, while the outer edge data is not good due to a lack of wavefront data outside the CCD sensor.

Fig. 45. Mirror supported by three-point bonding on aluminum block.

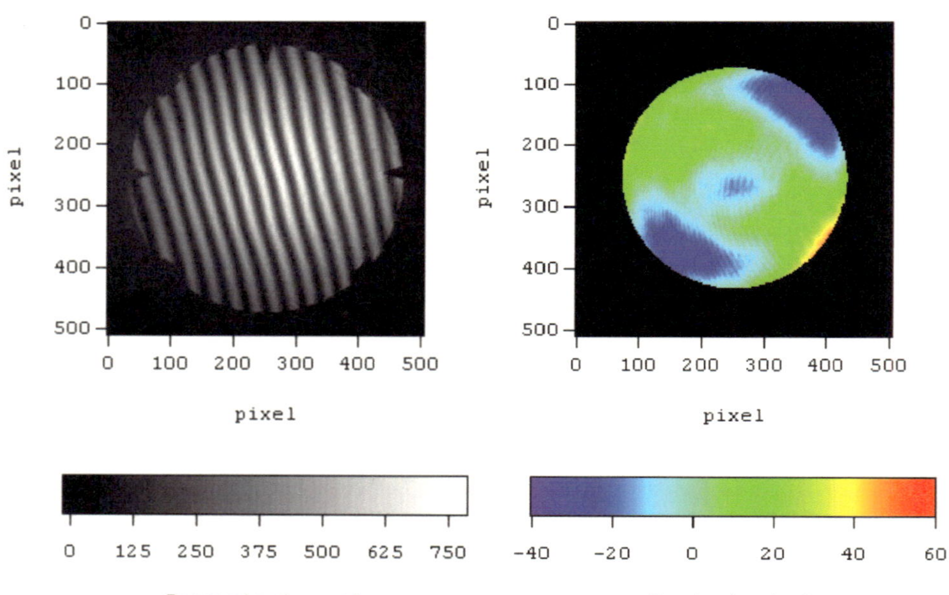


Fig. 46. Interference pattern and measured phase map on CCD surface at ambient temperature of 23.46 degrees Celsius.

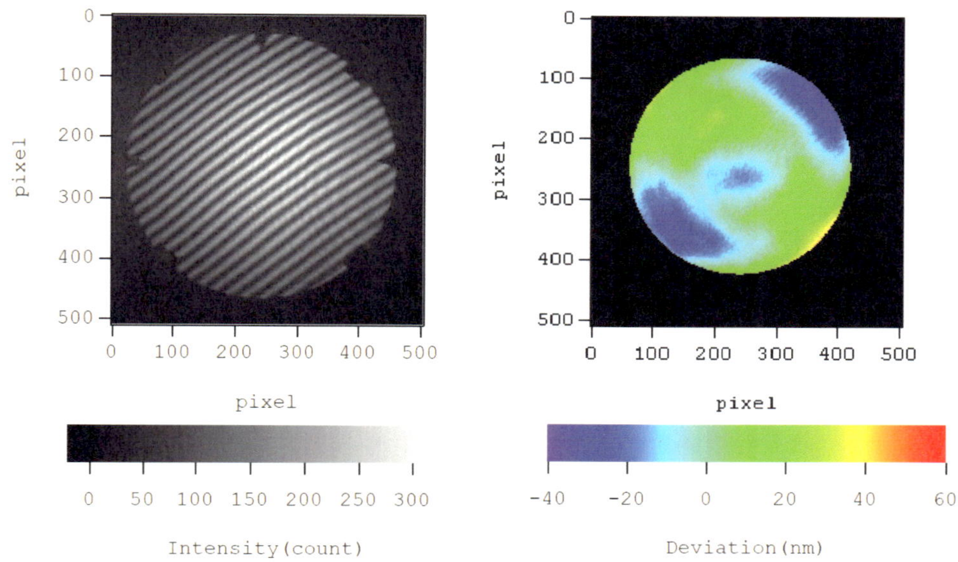


Fig. 47. Interference pattern and measured phase map on CCD surface at ambient temperature of 20.77 degrees Celsius.

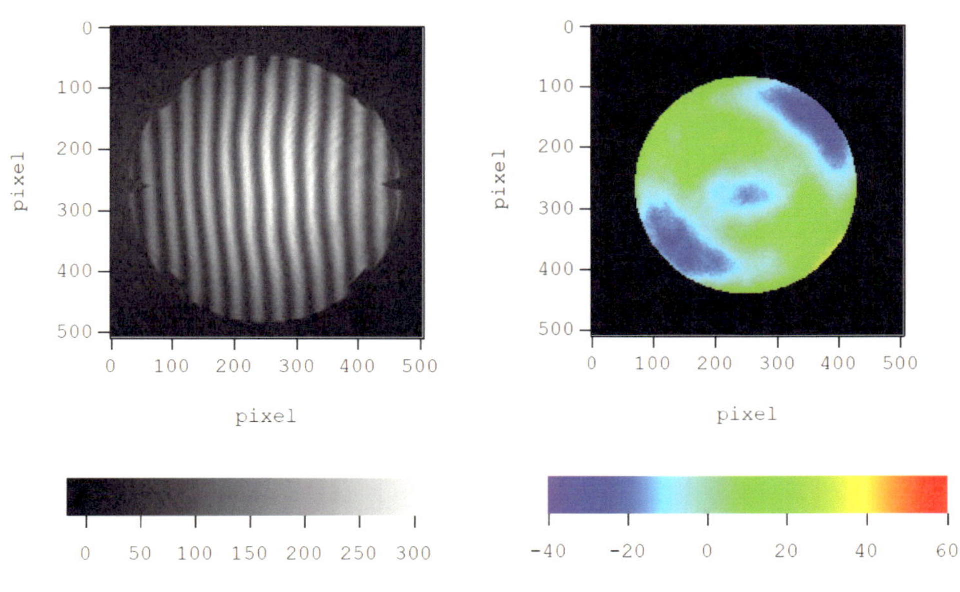


Fig. 48. Interference pattern and measured phase map on CCD surface at ambient temperature of 20.71 degrees Celsius.

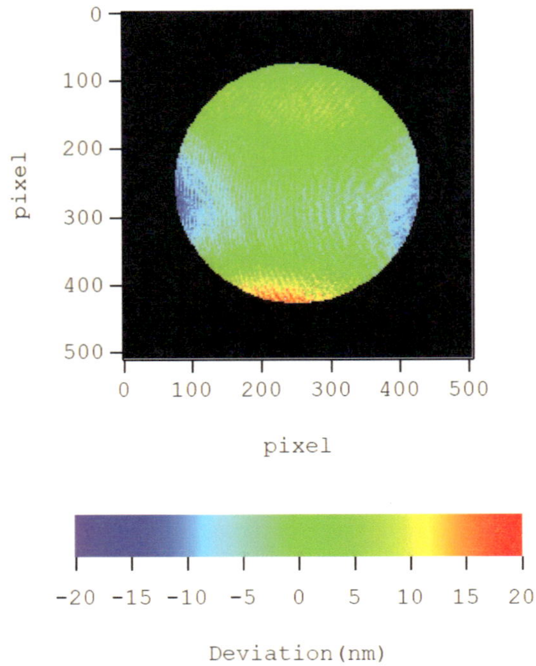


Fig. 49. Relative difference in phase map between fig. 46 and fig. 47.

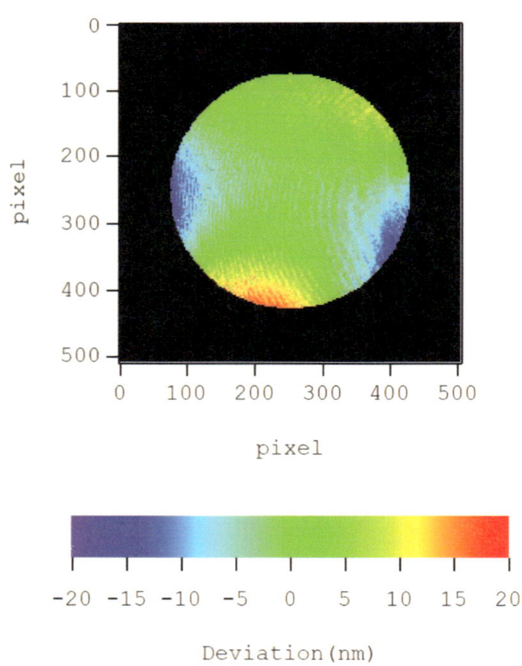


Fig. 50. Relative difference in phase map between fig. 46 and fig. 48.

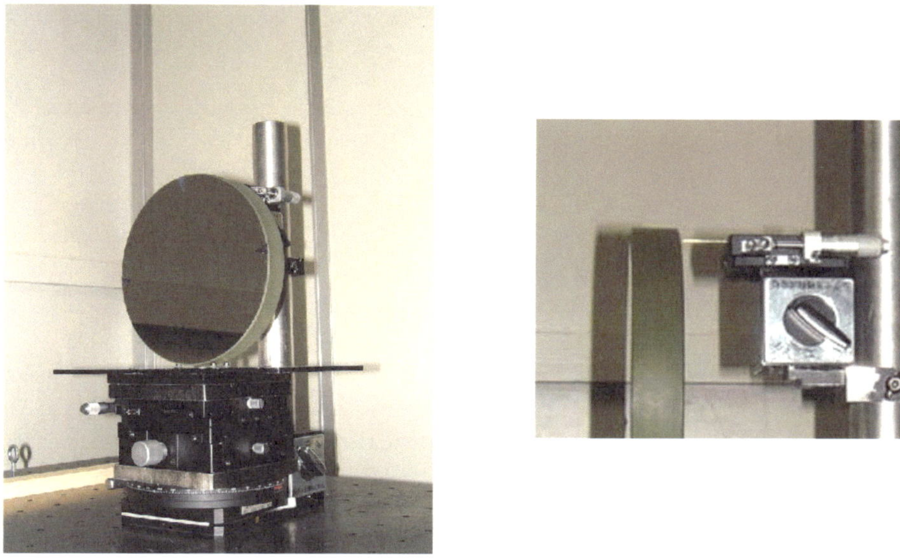

Fig. 51. Mirror supported by three-point suspension of steel balls and single needle.

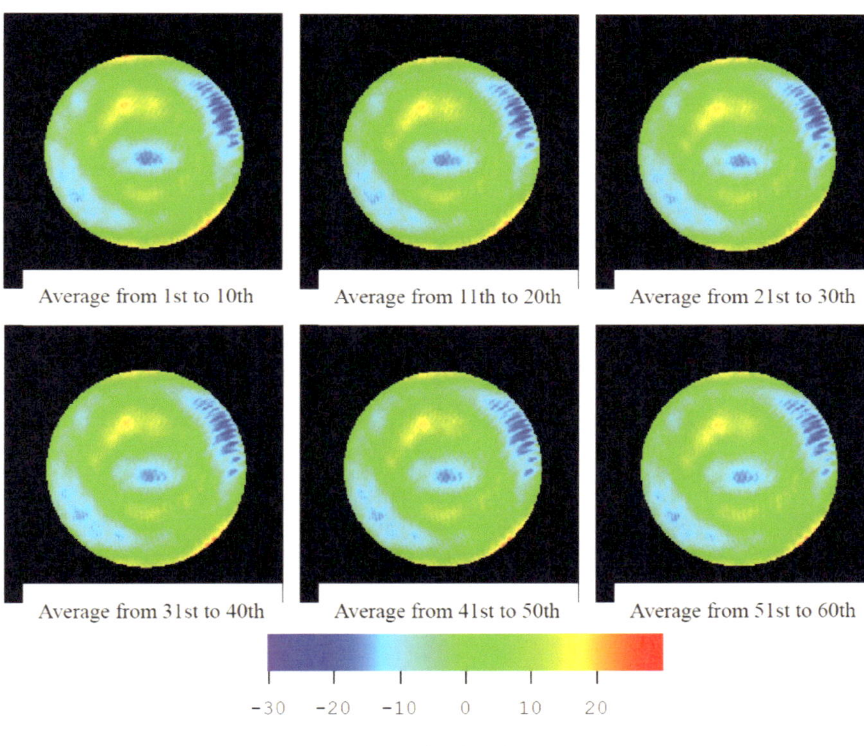


Fig. 52. Averaged images of wavefront for every 10 measurements.

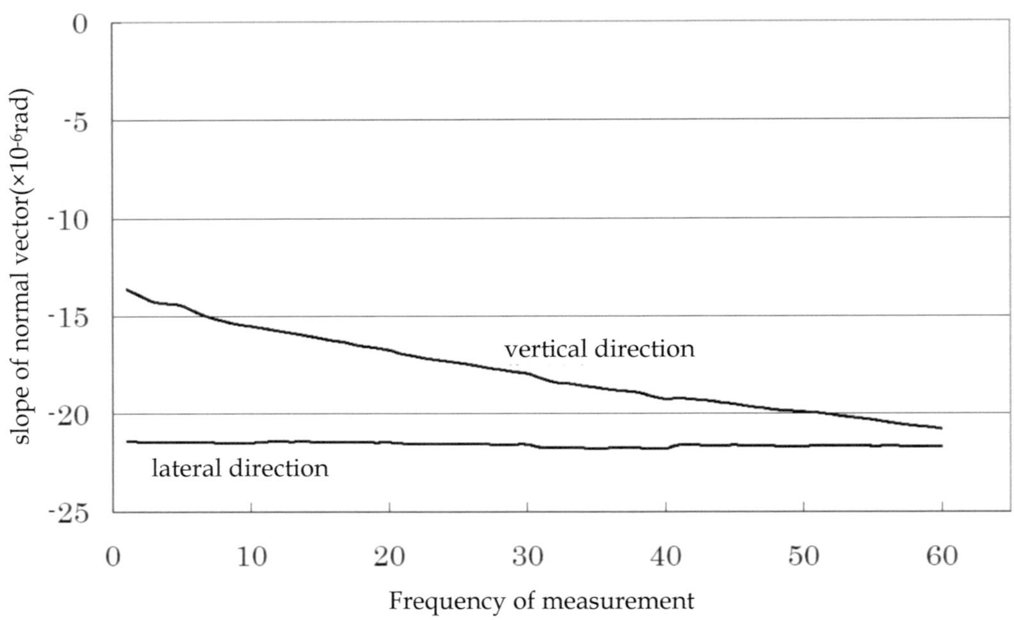


Fig. 53. Temporal change in direction of surface normal vector of test surface.

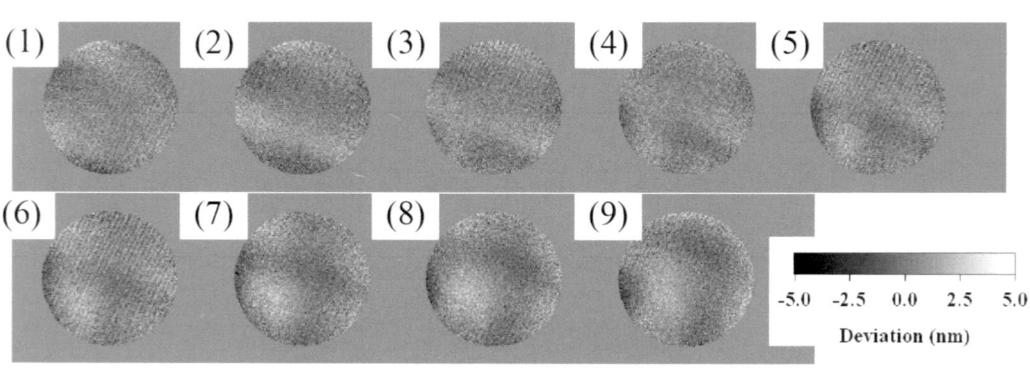


Fig. 54. Differences between ten-measurement average and overall average of 100 measurements. The difference changes slightly due to air current.

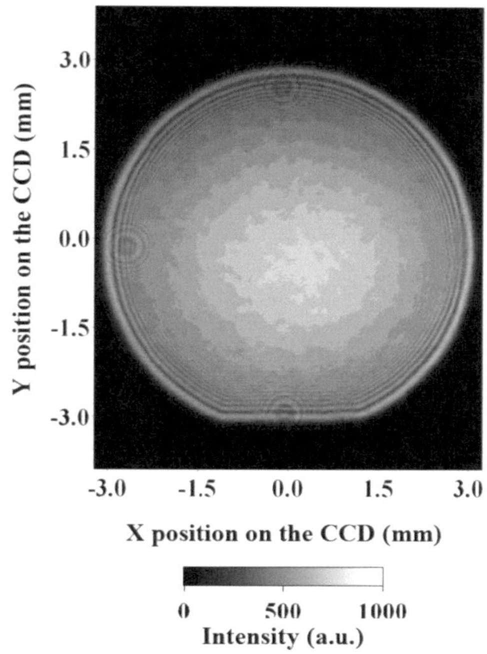


Fig. 55. Light intensity distribution on test mirror surface.

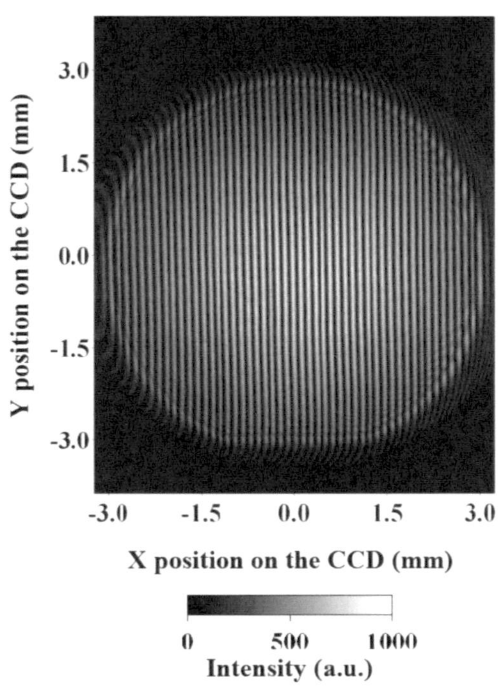


Fig. 56. Typically-observed interference pattern on CCD sensor.

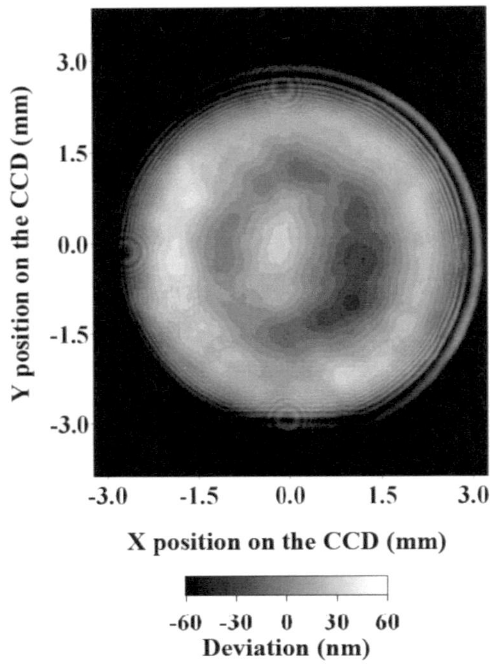


Fig. 57. Deviation of test wavefront from ideal sphere on CCD surface.

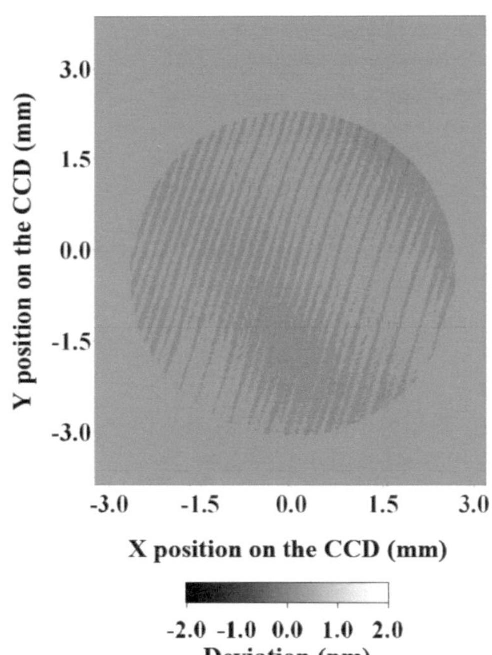


Fig. 58. Difference between averages of first and second 50 measurements.

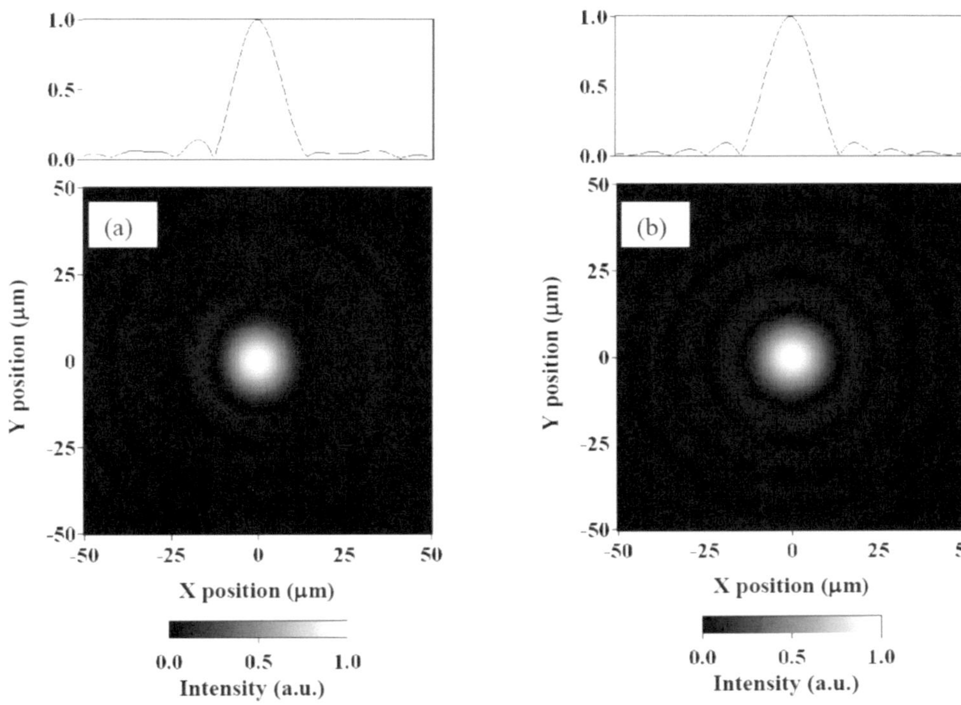


Fig. 59. (a) experimental and (b) ideal intensity map test wave at focal point.

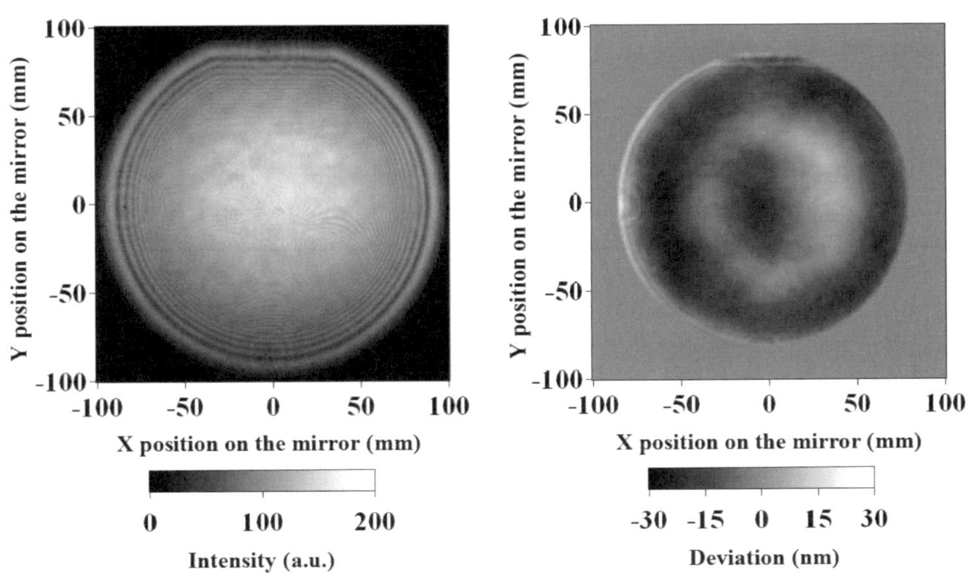


Fig. 60. Reconstructed light intensity map (a) and surface figure (b) on test mirror surface.

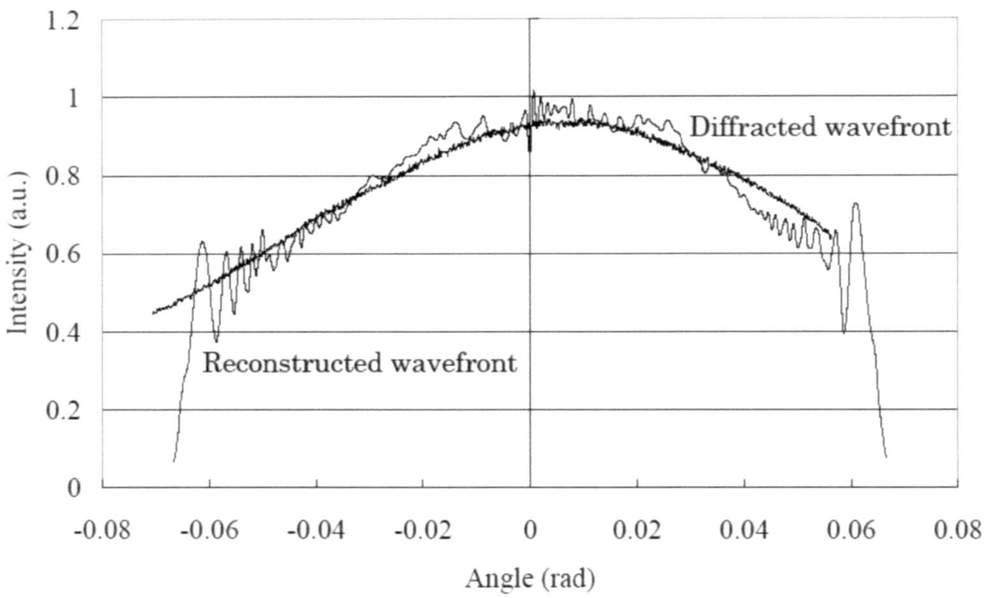


Fig. 61. Comparison of intensity distribution between reconstructed wavefront and fiber-output.

### 6.2 Discussion

Several measurement datasets are taken to evaluate reproducibility of the PS/PDI measurement without altering the experimental apparatus. After a five-hour warm-up period, the first dataset was recorded, and 30 minutes elapsed before the second data was recorded. The third dataset was measured 6 hours after the first dataset. The difference between the first and second datasets is displayed in fig. 62 on the left, and the difference between the first and third datasets is shown on the right. Although air current noise overlapped with the results in fig. 62, the measurement reproducibility between the first and third datasets is 0.5 nm PV and 0.12 nm rms. In fig. 62, geometrically symmetric patterns are observed, which are caused by numerical error in the reconstruction procedure. In the numerical integration, spatially discretized elements such as the CCD pixels affect the result (Born & Wolf, 1959). In addition, the reconstructed wavefront on the test mirror surface is also presented in the image discretization, and causes Moire fringes by coupling the numerical error pattern described above (Freischlad & Koliopoulos, 1990). Improvement of reproducibility in the consecutive measurement requires more warm-up time before the measurement begins. Final evaluation is reproducibility of the measurement results obtained by rotating the test mirror about its optical axis at 90 degrees. The difference between the two results is presented in fig. 63. Here, Moire fringes also degrade the structural coincidence, and the measurement reproducibility is 2.0 nm PV and 0.85 nm rms. A more accurate procedure for numerical integration will produce much better results in the future.

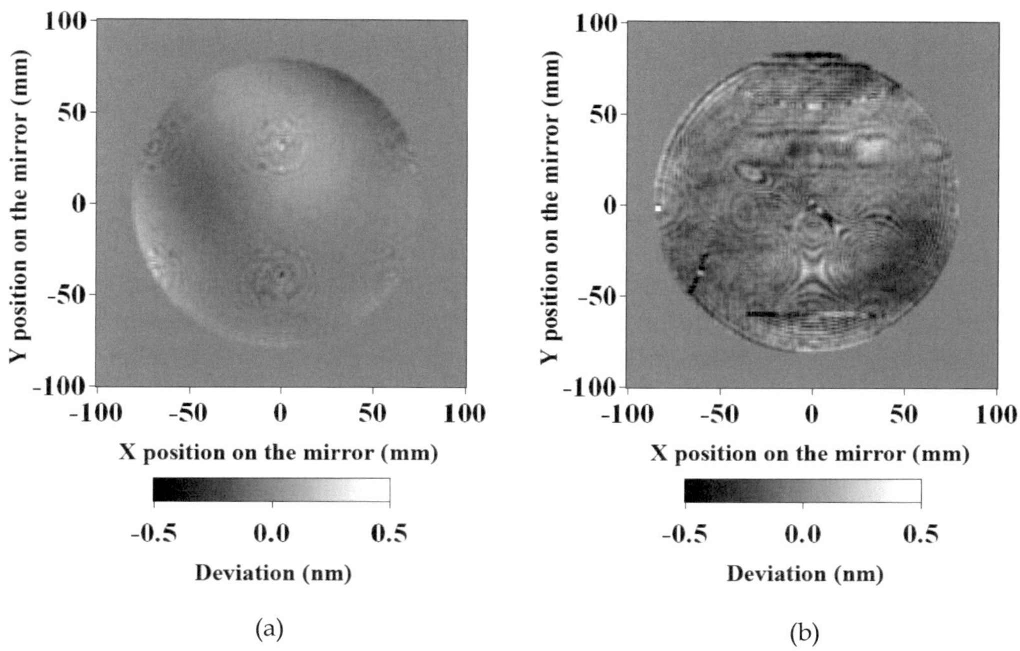


Fig. 62. Measurement reproducibility between two datasets taken with interval of 30 minutes (a) and 6 hours (b), respectively.

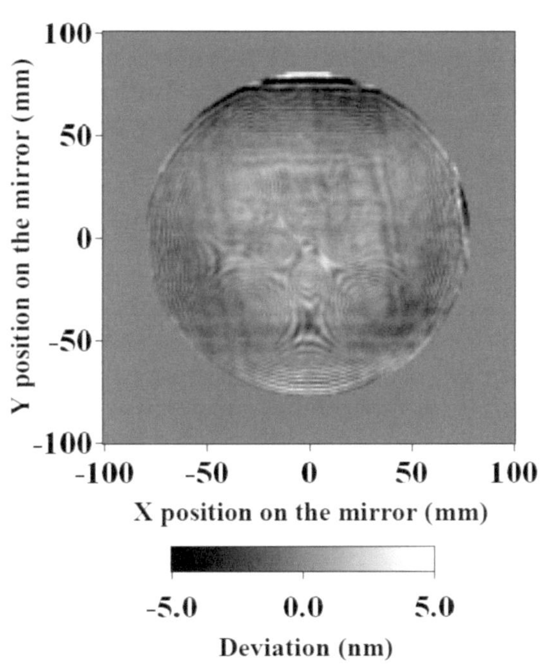


Fig. 63. Difference between before and after rotation of mirror.

## 7. Measurement of aspherical mirrors

### 7.1 Spherical concave mirror with high asphericity

A highly accurate surface figure is needed not only for a spherical surface but also paraboloidal, ellipsoidal, and flat surfaces. In an ordinary PS/PDI system, a highly aspherical test surface emanates condensed interference fringes on the CCD, and correct recording and analysis of the fringes are not possible. In our PDI system however, location of the two point sources is very flexible, as well as the position of projection plane. Therefore, an appropriate fringe pattern can be obtained for the aspherical surfaces and large-aperture optics. In this last section, trial experiments are conducted to measure an aspherical concave mirror, an off-axis paraboloidal mirror, and a mirror-polished silicon wafer. Fig. 64 shows a photo and schematic of a sample concave mirror having a design value of 1000 mm ROC. In the beginning, the optical apparatus shown in fig. 6 is used for the measurement. However, the test surface is highly aspherical, and the reflected wave did not converge well. The intensity distribution after reflection on the test surface is shown in fig. 65, and fig. 66 shows the interference pattern between the reference wavefront and the wavefront shown in fig. 65. These two images are ellipsoidal due to the aspherical test surface. In the ellipsoidal image, the ROC value in the vertical direction is larger than in the horizontal direction. Spatial density of the fringes shown in fig. 66 is an acceptable range for the CCD sensor, and the PS/PDI measurement was implemented. The intensity distribution at the focal point of the test wave is reconstructed in the same manner as described in section 6 and shown in fig. 67. The point might not be the focal point, leading to the conclusion that the spherical test wave assumption could be incorrect. Another method for determining the location data of the reference fiber end and CCD is necessary. The procedure described in section 4 after the mirror measurement was introduced. In this procedure, only the fiber end for the test wave generation is repositioned to observe the interference pattern caused by two diffraction waves. This data will provide the location data of the fibers and CCD. Here, the low-quality focal pattern shown in fig. 67 could possibly cause degradation of a common path, like the configuration shown in fig. 6. Hence, reconstruction of the test wavefront is simulated numerically with misalignment of the fiber end for the test wave, and the results are displayed in fig. 68. Even if the test surface is spherical, misalignment of the fiber end results in an imperfect phase map as shown in fig. 68. Therefore the location coordinate of the fiber end for the test wave must be determined. For this purpose, the concave mirror measurement in section 6 is applied to this experiment. When the fiber end for the test wave is positioned at the center of curvature of the spherical concave mirror, light reflected by the mirror focus on the fiber end again, and movement of the fiber end and the focal point are symmetrical about the center of curvature. The movement of the fiber end and location of focal point, which is determined after the mirror measurement, location coordinate of the fiber end for the test wave is determined. In fig. 69, the measurement procedures mentioned above are shown. In step (1), the interference pattern caused by two diffraction waves is measured, and locations of both the reference fiber end and CCD sensor are determined and fixed. In step (2), considering the measurement of concave mirror described in section 6, the focal point or reference fiber end is determined. In step (3), measurement of the test mirror shown in fig. 64 is conducted and the data is analyzed with data obtained in (1) and (2). Fig. 70 shows the resultant test wavefront on the CCD. In 100 consecutive measurements, several datasets exhibit an interesting behavior, an example of which is shown in fig. 71. This may be caused by overreaction of the unwrap algorithm for phase calculation, and such a phenomenon is induced by mechanical vibration or considerable change in the interference pattern during measurement. Excluding the invalid data, the averaged wavefront data shown in fig. 70 is calculated from 94 datasets. Next, the surface figure of the mirror is reconstructed using fig. 65 and fig. 70, and the results are shown in fig. 72. The left image of Fig. 72 confirms that the test mirror is a circle, and the diameter is estimated to be 109.2 mm by referring to the three markers on the outer edge. The surface shape appears to be ellipsoidal, which is consistent with fig. 65. Fig. 72(left) is compared with the actual intensity distribution of the diffraction wave, and they agree well as shown in fig. 73. Fig. 74 shows the measurement reproducibility of the sample rotation evaluated using the same method used to produce fig. 63. The reproducibility is 5 nm PV and 1.2 nm rms. The test mirror has asphericity of 10 μm, and our PS/PDI system exhibited a successful measuring capability for such a highly aspherical surface.

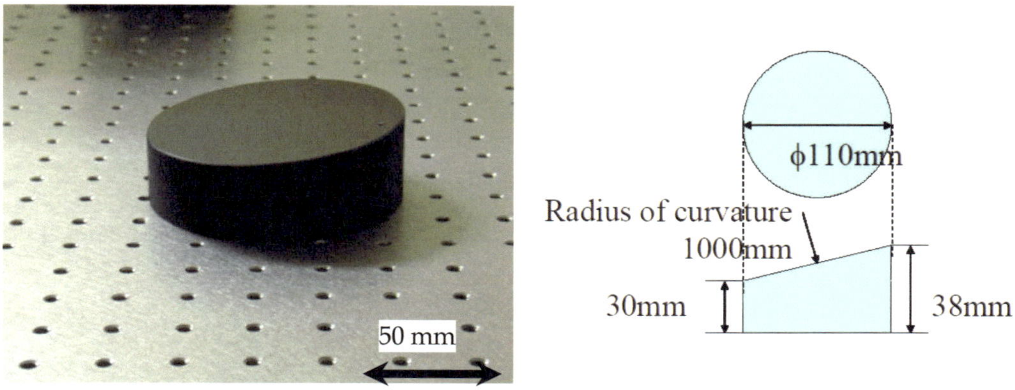


Fig. 64. Photo and drawing of aspherical concave mirror measured in this section.

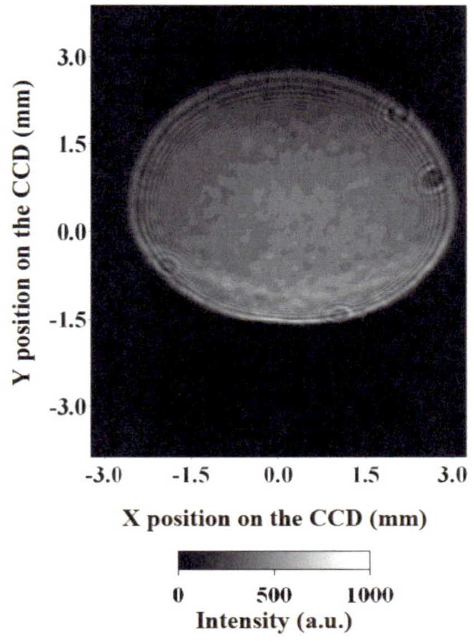


Fig. 65. Intensity distribution of reflected test wavefront on CCD.

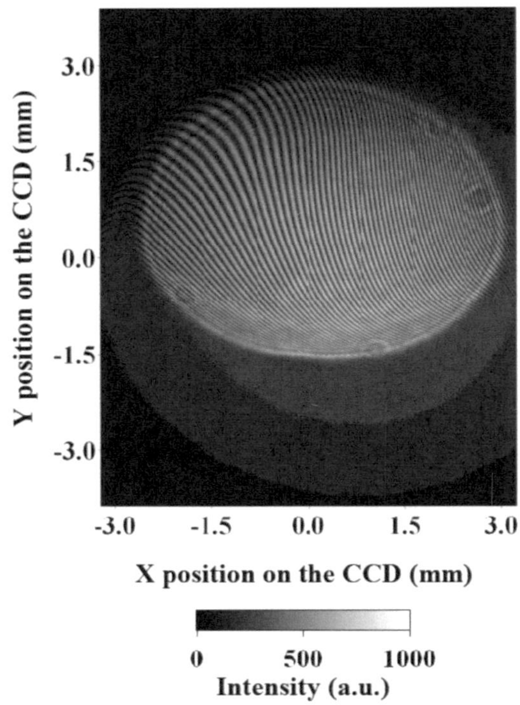


Fig. 66. Typical interference pattern taken at CCD surface.

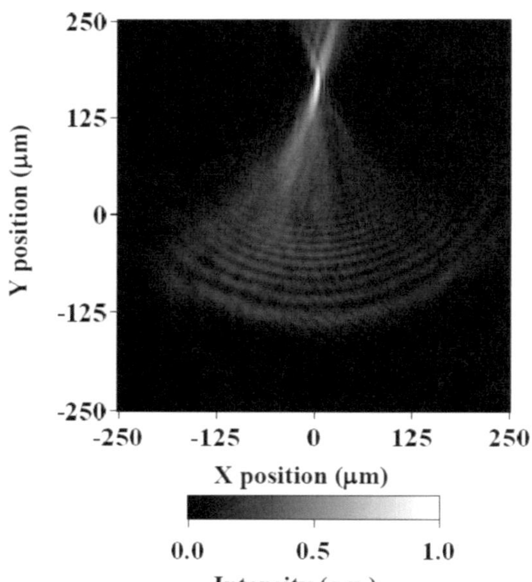


Fig. 67. Reconstructed intensity distribution at focal point of test wave.

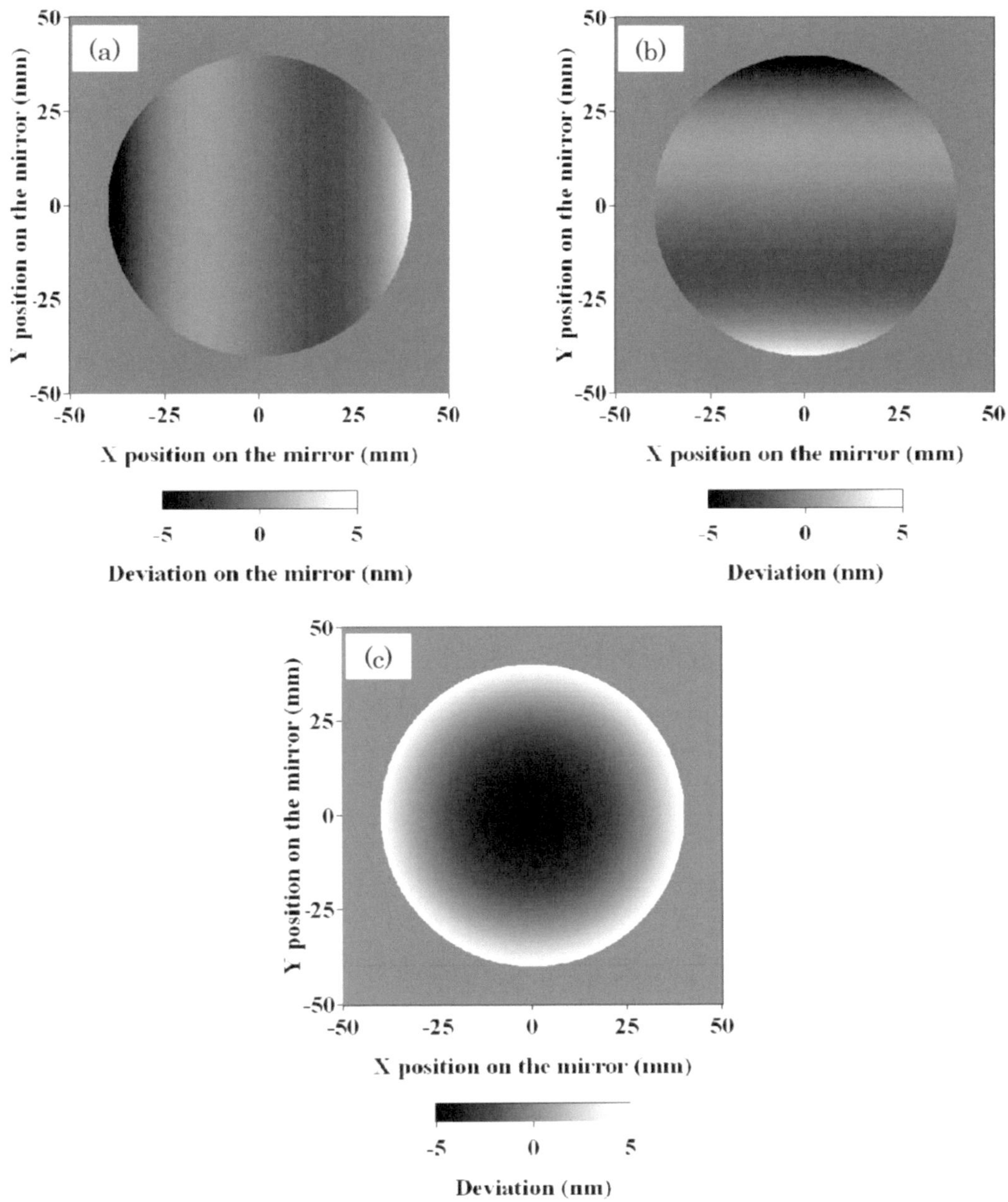


Fig. 68. Numerical simulation of error in phase map caused by misalignment of fiber end at (a) x=1 mm, (b) y=1 mm, and (c) z=50 μm.

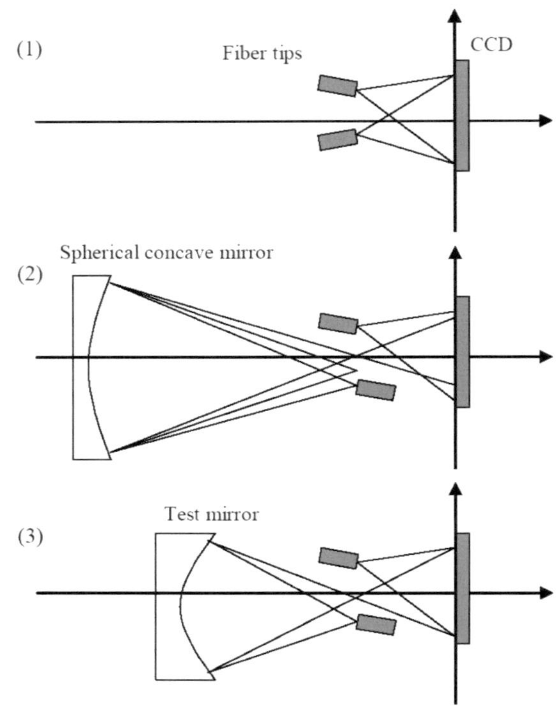


Fig. 69. Three step measurement for highly aspherical concave mirror.

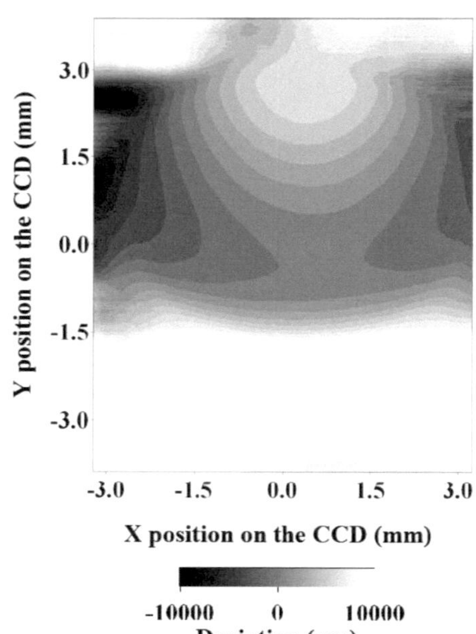


Fig. 70. Wave front deviation of test wavefrom ideal sphere on CCD surface.

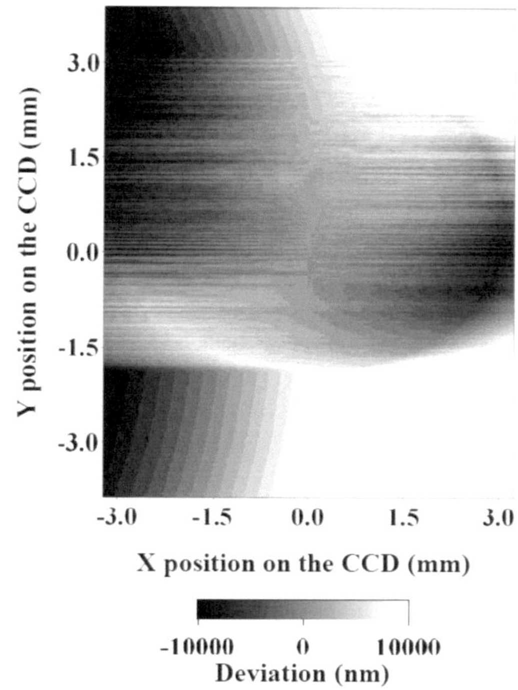


Fig. 71. Example of vibration-induced phase error enhanced by unwrap process.

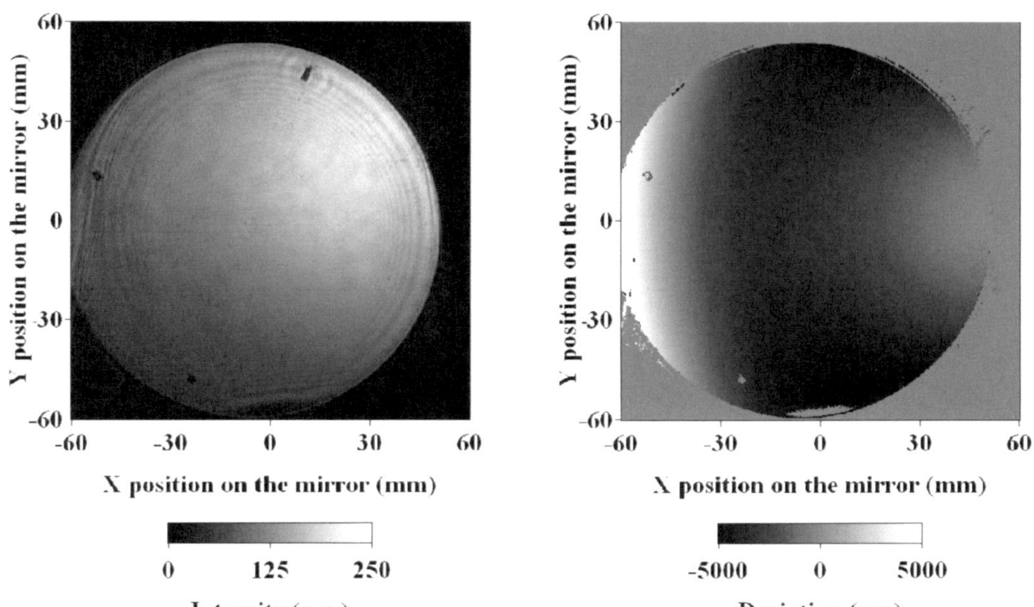


Fig. 72. Reconstructed light intensity distribution (a) and surface figure (b).

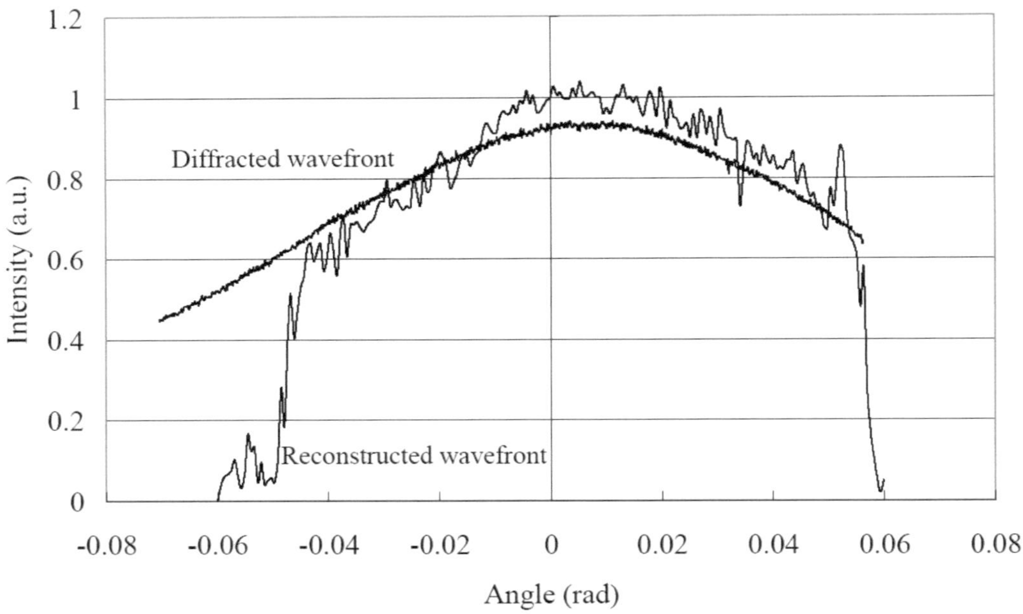


Fig. 73. Comparison of intensity distribution between reconstructed wavefront and fiber-output.

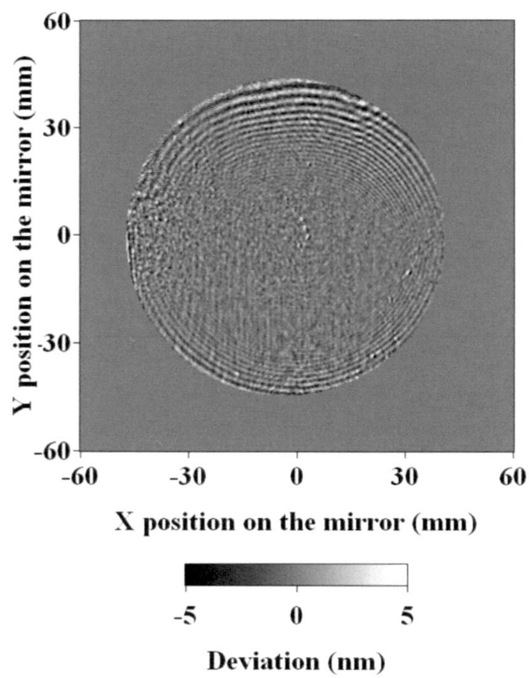


Fig. 74. Difference between before and after rotation of mirror.

### 7.2 Off-axis parabolic mirror

The alternative off-axis parabolic mirror shown in fig. 75 is also measured. The mirror has a 360 mm outer diameter and a 258 mm inner diameter, and the mirror surface has a cross section defined by the equation of $y=x^2/4f$ (f=75 mm, 258 mm < x < 360 mm). This mirror is designed to focus many laser beams onto a single point, and cannot converge the diffraction wave from a fiber end. Hence, the optical configuration shown in fig. 6 is no longer valid, and the mirror measurement is performed with the alternative set up presented in fig. 76. The test wavefront is converted to a plane wave by positioning the fiber end to the focal point of the mirror. Appropriate positioning of the reference fiber end causes generation of a usable interference pattern on the CCD surface. The wavefront of the reference wave converts to a plane wave through a relatively long propagation distance, and the resultant interference pattern is not very dense. The position of the test fiber end is adjusted in accordance with fig. 77. Namely, the fiber end is moved to the actual focal point of the mirror which is illuminated by the plane light wave. Fig. 78 shows the data from the optical power meter, which depends on the three dimensional position of the fiber end. Changes in these curves are used to determine a position of fiber end accurately. The interference pattern recorded by the optical configuration of fig. 78 is shown in fig. 79, where distance between the test mirror and CCD is 1 m. The resultant reconstructed wavefront is shown as the deviation from a flat surface in fig. 80. In the numerical process, locations of the reference fiber end and CCD are determined by relying on the situation of interference between the plane wave and the spherical wave on the CCD surface. Thus, the interference patter is coaxial. In fig. 80, vertical bleary stripes correspond to tool marks scratched during the cutting of the mirror surface. Reproducibility of the data shown in fig. 80 for 30 minutes is evaluated and displayed in fig. 81, and it is 10 nm PV and 1.3 nm rms. Fig. 81 has a vibration-induced coaxial error pattern at double the frequency of the interference pattern due to the vacuum system supporting the mirror. Finally, influence of the misalignment of the test wave fiber is studied. In the measurement shown in fig. 76, the PS/PDI data provides the distribution of the relative optical path differences in the collimated beam from the test mirror. The optical path distribution must be subtracted from data such as the data in fig. 80. In fig. 82, the upper three images display the distribution of relative optical path difference caused by misalignment of the test wave fiber end in three directions, and the lower three images display the consequent deformation of the wavefront from a flat profile. When the fiber end is moved 120 μm in the Y-direction, the result is shown in fig. 83 and is consistent with the lower middle image in fig. 82.

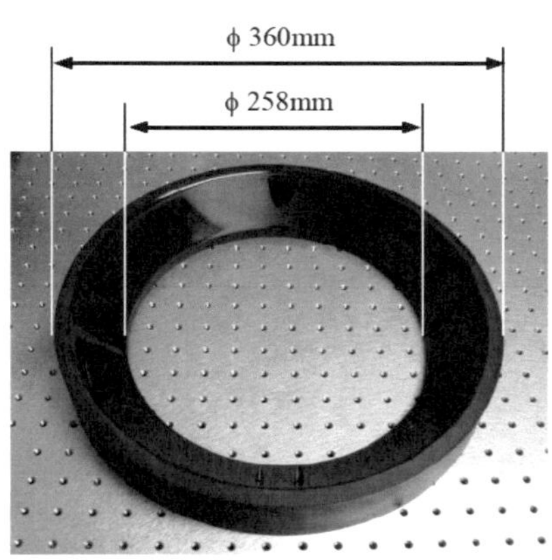


Fig. 75. Photo of aspherical mirror having off-axis paraboloid of revolution.

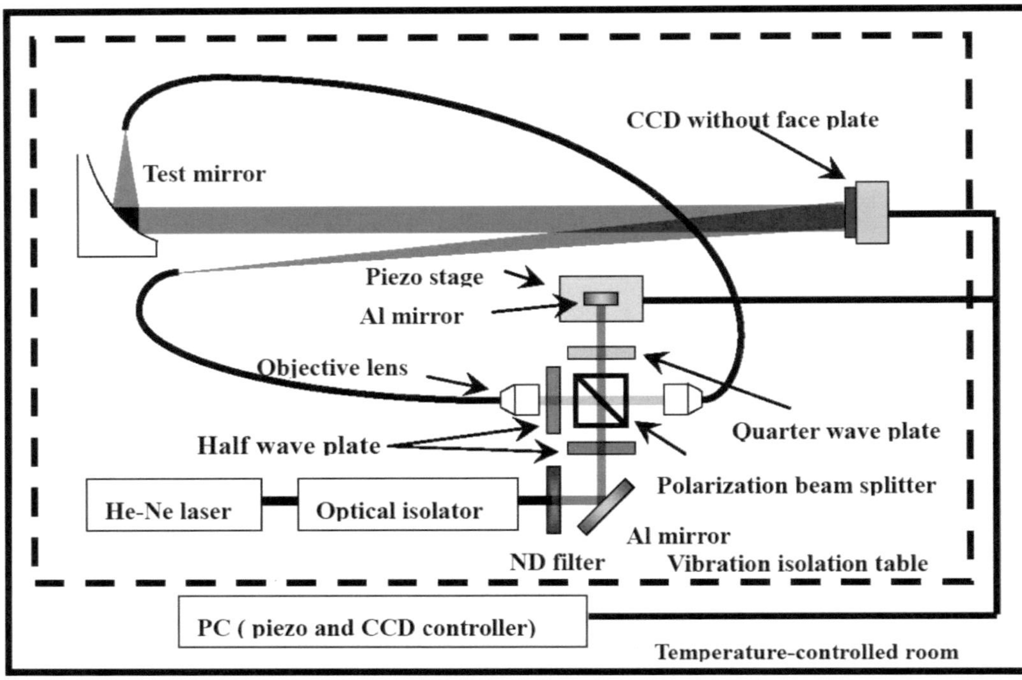


Fig. 76. Optical configuration of the PS/PDI system for off-axis parabolic mirror.

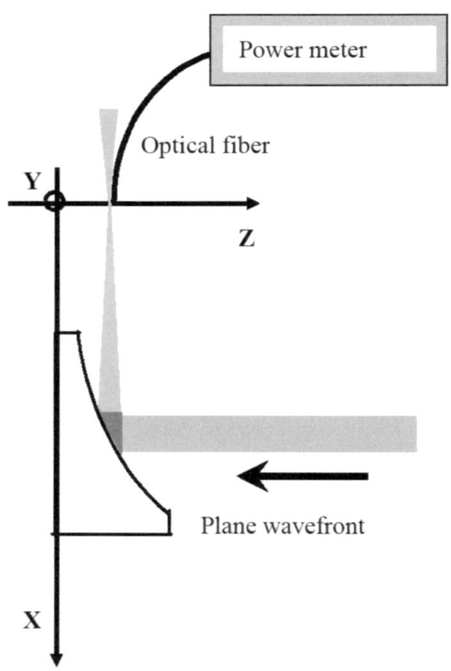


Fig. 77. Installation of optical fiber end to focal point of test mirror.

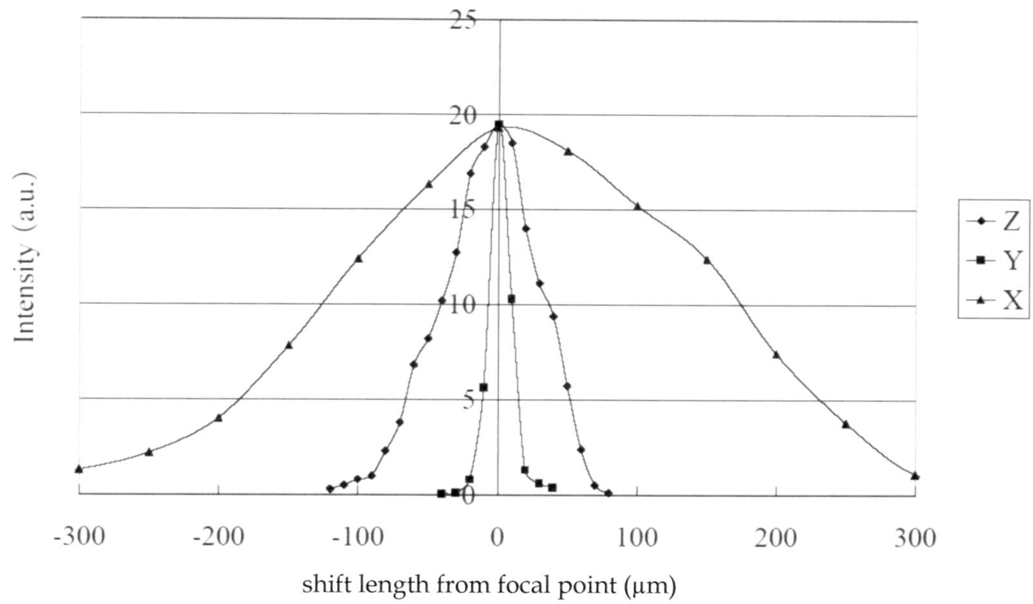


Fig. 78. Output of optical power meter, as indicated in fig. 77, is sensitive enough to locate fiber end for alignment.

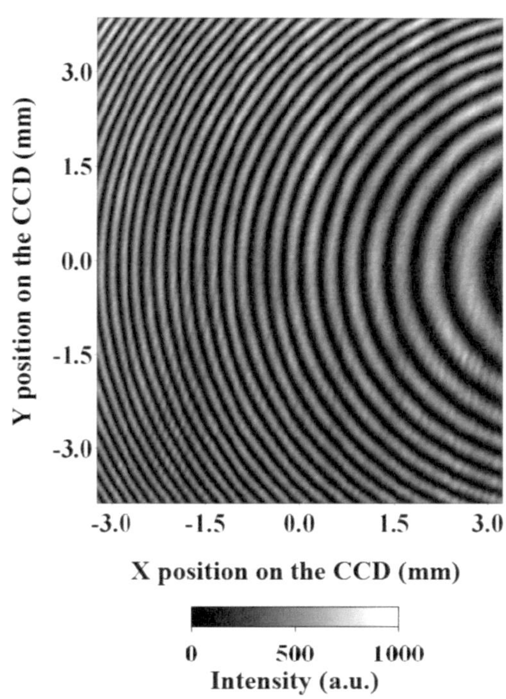


Fig. 79. Typical interference pattern recorded in measurement of the parabolic mirror.

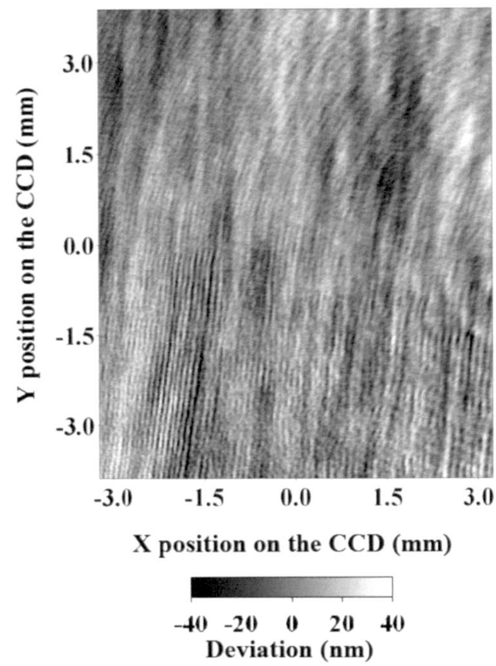


Fig. 80. Deviation of reconstructed surface figure from flat surface.

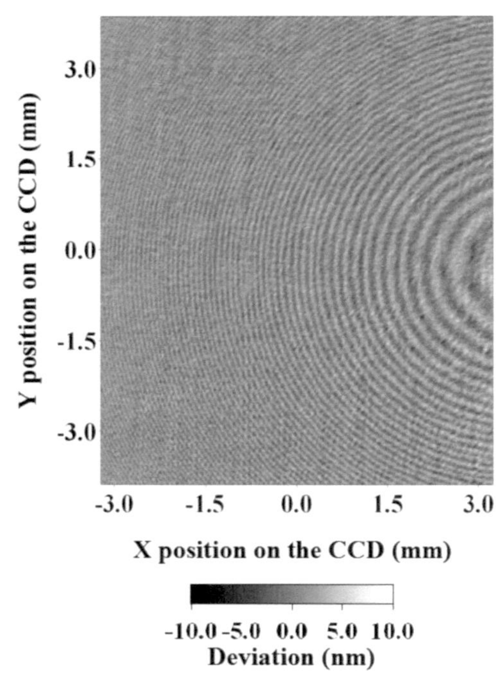


Fig. 81. Reproducibility of the measured wavefront shown in fig. 80.

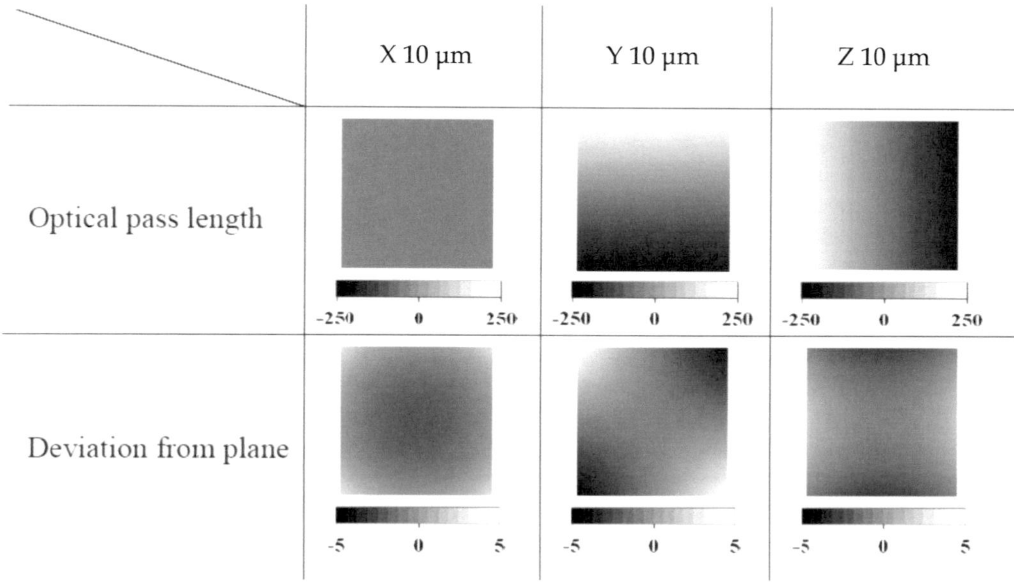


Fig. 82. Numerically simulated measurement error induced by misalignment of fiber end in the parabolic mirror experiment.

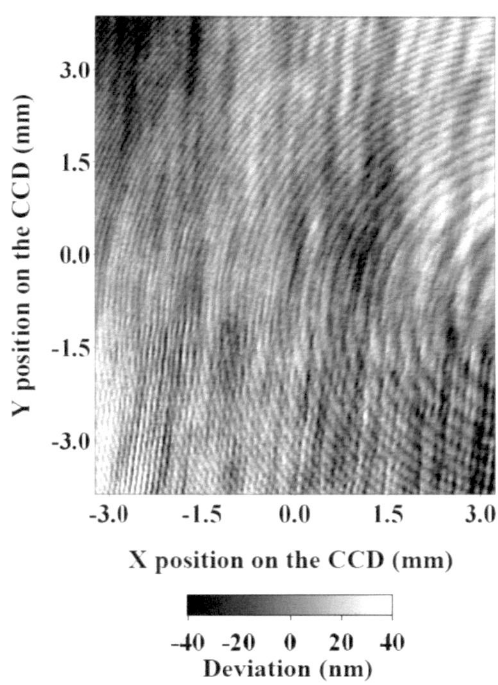


Fig. 83. Deviation of reconstructed wavefront from flat surface in condition of 120 μm misalignment of fiber end for test wave front.

### 7.3 Measurement of silicon wafer as flat surface

Measurement of a flat surface may be possible using our PDI system only. By positioning the two fiber ends properly, a usable interference pattern is recorded by the optical configuration shown in fig. 84. The flat test mirror is placed in front of the CCD obliquely, and the surface normal is at 45 degrees to the surface normal of the CCD. The test wave reflected by the test mirror includes information of the test surface figure, and an interference pattern could be generated by the reference wave as shown in fig. 84. Fig. 85 shows the typical interference pattern, and the reconstructed wavefront, which is the deviation from an ideal sphere, is displayed in fig. 86. The wavefront data is converted to the surface figure of the silicon wafer using the ray trace method and is shown in fig. 87. The surface figure is also measured by a white-light interferometer as shown in fig. 88. The datasets agree well in the low-frequency components.

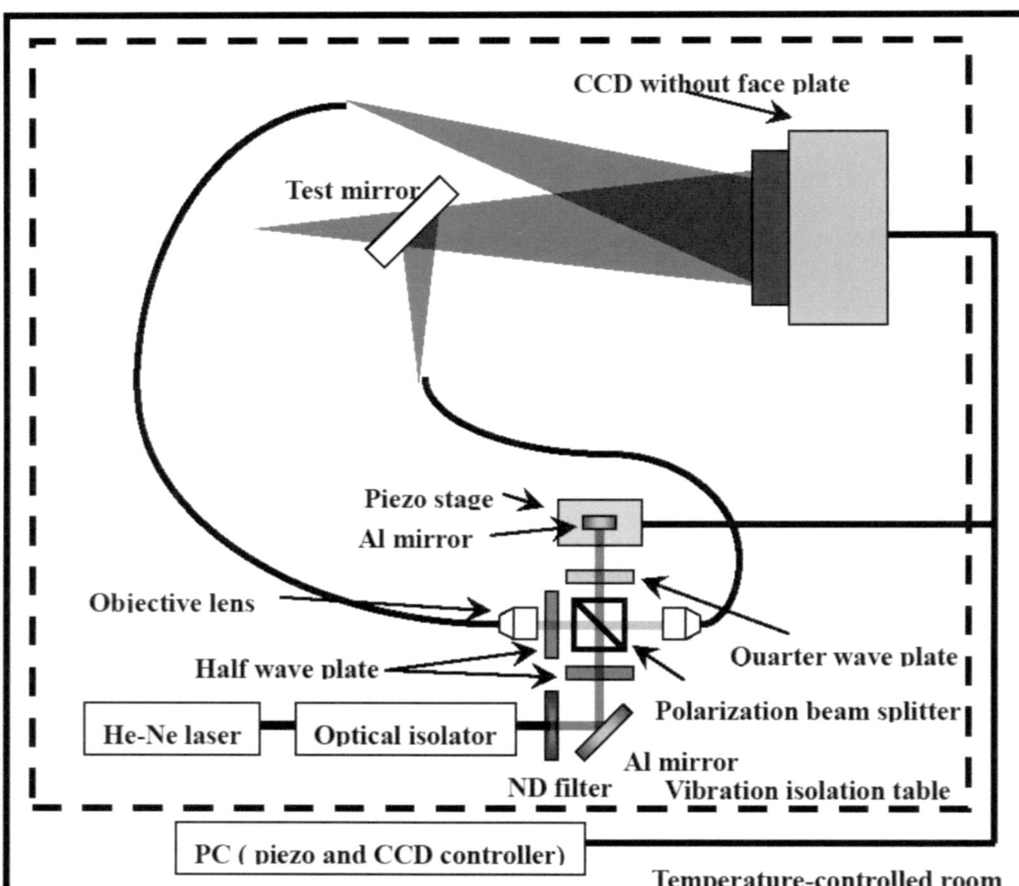


Fig. 84. Optical configuration of the PS/PDI system for measurement of silicon wafer surface.

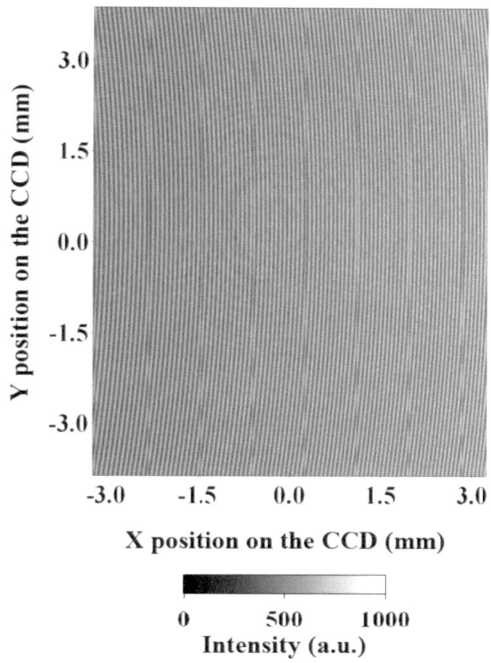


Fig. 85. Typically observed interference pattern in measurement of silicon wafer.

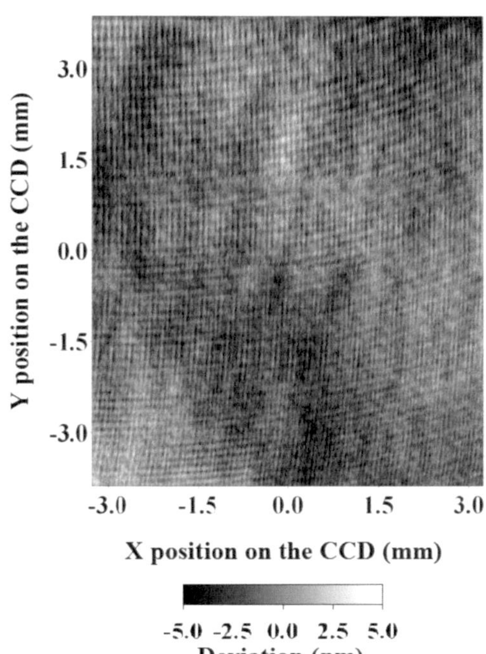


Fig. 86. Deviation of measured wavefront on CCD from ideal sphere.

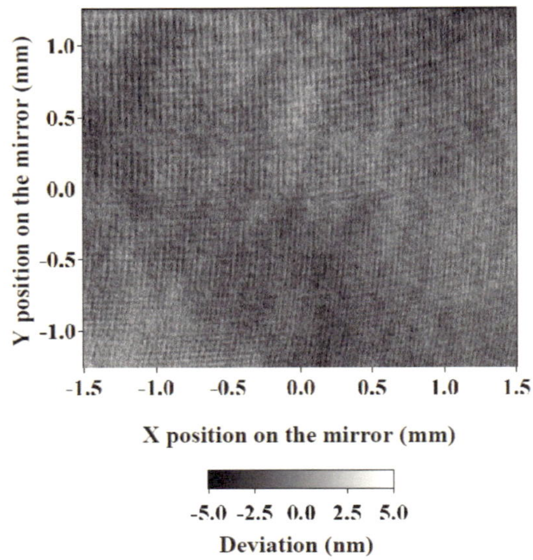


Fig. 87. Numerically derived surface figure of silicon wafer by ray trace.

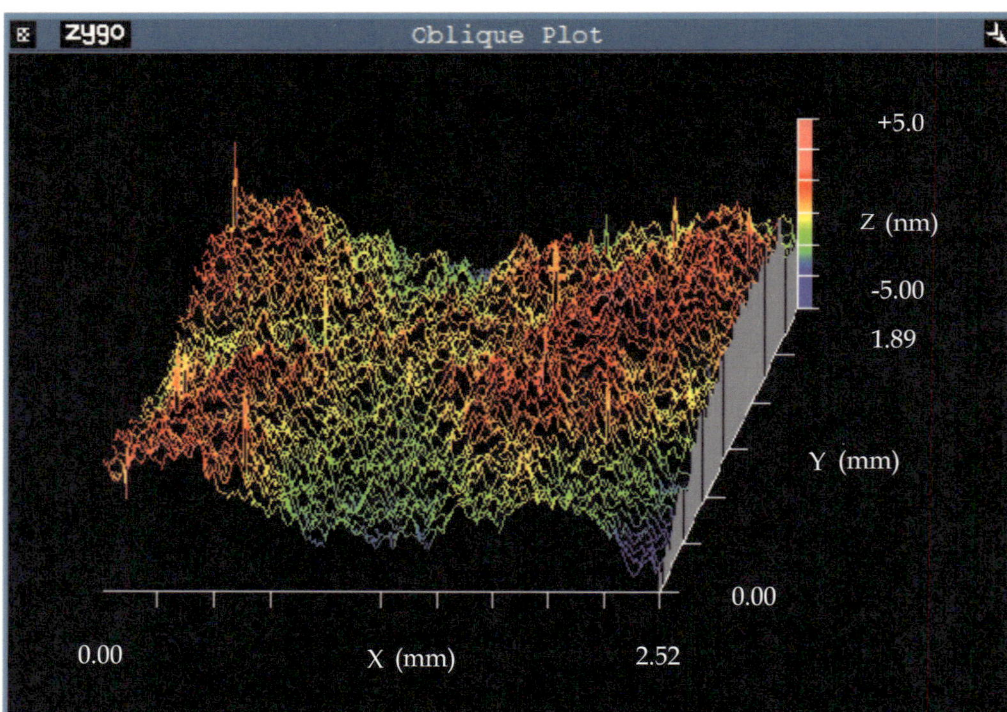


Fig. 88. Surface figure of the silicon wafer measured by commercial white light interferometer.

## 8. Conclusion

In this chapter, known issues in previously studied PDI optics are explained and a novel PDI optical system is proposed for ultra-precise measurement of absolute surface figure for large-aperture optics. The proposed PS/PDI optics system is developed with two single-mode optical fibers. Details of both the apparatus and analysis of interference fringe are explained. In the previous PDI system, a single diffraction wave from a tiny aperture acts as both the test wave and reference wave, and half of the wavefront is used for surface figure measurement. A new configuration was developed for the interferometry using two small apertures as origins of diffraction waves, where the apertures are ends of single-mode optical fibers. These fibers are pointed in opposite directions. Hence, the entire wavefront will be used for surface figure measurement. The two fibers must be close to each other to realize common-path interferometry. Therefore, fiber ends are sharpened by chemical etching and the apexes are polished obliquely for optimization of emission directions. To obtain highly spherical diffraction waves, the polishing angle is carefully determined by numerical simulation based on the Fresnel-Kirchhoff integral. In the measurement process, the CCD camera, which records the interference pattern, is synchronized to a PZT stage, which controls the relative optical phase by changing the length of the optical path, to implement the phase-shift algorithm. The entire control procedure of the interferometer is studied and optimized.

Measurement accuracy of the PS/PDI system is evaluated experimentally by the interference of two diffraction waves. Namely, in the PS/PDI system, two optical fiber ends emit diffraction waves, and these ends are capable of generating an interference pattern when they are oriented in the same direction. The two diffraction waves having highly spherical wavefronts allow estimation of measurement accuracy of the interferometry system. An interference pattern generated by two fiber ends has several tilt fringes due to the optical configuration between the fibers and CCD camera. Such fringes are analyzed numerically by a phase-shift algorithm, and the resultant phase map is unwraped to obtain a continuous initial phase map. In the initial phase map, location information of both fibers and the CCD is included. We described an analytical method to find their positions from the initial phase map. By using the least-squares method, their spatial configuration can be determined accurately due to interference between two ideal spherical waves. In the PS/PDI system, phase-shifting interference patterns must finally be generated on a CCD surface for digital recording. For the purpose of forming the interference pattern on a CCD, an imaging lens is not acceptable due to optical aberrations. A fiber optic window (FOW) serving as the projection plane for the interference pattern is carefully examined. Several types of measurement errors, including the FOW-induced disturbance in the interferogram, are considered. The FOW-induced error is highly reproducible and can be eliminated by subtracting one dataset from another. Random error caused by mechanical vibration is eliminated by averaging many datasets. Temporal fluctuation of laser light is reduced with a long warm-up time for the system. Although the FOW error is suppressed, its residual quantity is not negligible for the sub-nm accuracy of the PS/PDI. The latest PS/PDI system accepts direct projection of an interferogram on a special CCD camera without any projection plane or lens. This particular CCD camera has no protection plate, and the bare

CCD sensor records the interference pattern directly without any unwanted stray light or local interference. As a result, measurement accuracy of the PS/PDI system is estimated to be 0.7 nm PV and 0.07 nm rms for mirror measurement.

This chapter clearly describes the necessity of a wavefront reconstruction process for PS/PDI measurement of aspherical or non-ideal mirrors. The explicit result of PS/PDI or phase map on the CCD sensor must be processed numerically for precise reconstruction of test mirror surface figure. Since wavefront propagation is developed by numerical simulation based on the Kirchhoff-Helmholtz integral theorem, interference between two diffraction waves emitted from optical fiber ends is analyzed. Assuming PS/PDI measurement of an ideal spherical concave mirror, the wavefront analysis returns highly important results. That is to say, a phase map, which can be derived at the CCD surface, depends on light intensity distribution on the test mirror surface, but such a phase map can be transformed to an exact surface figure through the wavefront reconstruction process. In the case of an aspherical mirror having high a surface figure error at a high spatial frequency, the phase map on the CCD surface is very different from the real surface shape. For such a situation, the wavefront reconstruction procedure is quite effective. In the case of an aspherical mirror with low-frequency surface error, the reflected wavefront does not have considerable distortion during its propagation because the propagation distance is not as long. Of course, the wavefront reconstruction process results in much better surface figure data. Finally, the wavefront reconstruction process based on the KH integral will produce an absolute surface figure of the test mirror with the transformation of phase map measured on the CCD sensor serving as a projection plane.

This study discussed the PS/PDI measurement of a nearly ideal spherical concave mirror, and highly precise measurement of the absolute surface figure with sub-nanometer accuracy is confirmed in a series of experimental measurements. In the measurement of the mirror surface, the support method for the mirror is a critical issue. We adopted a three-point support using small steels balls combined with a supporting needle. Although tilt of the mirror surface is induced by thermal expansion of the needle, considerable deformation of the mirror surface is not observed. These mirror fixing techniques are the best options at present. From 100 consecutive measurement results, reproducibility of the results is evaluated with two averaged datasets derived from the first and second 50 datasets. The evaluated value is worse than that estimated in section 4. This is caused mainly by air current between the optical fiber end and the test mirror, and such a considerable influence can be reduced by averaging many datasets. Many tilt fringes exist in the actual mirror measurement even with an ideal optical set up, which must be analyzed to determine location data of the reference fiber end and CCD sensor. Since the test mirror is a nearly ideal spherical, the location data is derived by assuming that the test wave is still highly spherical after reflection. As a result, the surface figure of the test mirror is reconstructed precisely and without issues in accordance with the instruction described in section 4. With two data results recorded at a six hour interval, the measurement reproducibility is estimated to be 0.5 nm PV and 0.12 nm rms. These values are worse than those in section 4 due to both numerical error in the integration process and air current in the experimental atmosphere. Measurement reproducibility with rotation of the mirror is evaluated by rotating the test mirror about its optical axis. The results before and after a 90-degree rotation of the mirror exhibit a reproducibility of 2.0 nm PV and 0.85 nm rms in the two reconstructed datasets.

Trial measurements of several aspherical surfaces using our PS/PDI system are described and discussed. The surfaces studied were a spherical concave mirror having large asphericity, an off-axis parabolic surface, and a silicon wafer as a flat surface. The results increased the availability of our PS/PDI system for aspherical surfaces. The test concave mirror has a large deviation from an ideal sphere, and the focal point is not clear. Therefore, locations of the two fiber ends and CCD are determined by procedures described in sections 4 and 6, and the wavefront reconstruction process is performed using the location data. The mirror surface figure is calculated by averaging efficient data in 100 consecutive measurements. Measurement reproducibility evaluated from two data before and after a mirror rotation is 10 nm PV and 1.4 nm rms. In the measurement of the parabolic mirror, the fiber end for the test wave is set at the focal point of the surface, and the reflected or collimated test wavefront interfered with the reference wave. The resultant wavefront has a reproducibility of 10 nm PV and 1.3 nm rms. Misalignment of the fiber end for the test wave is studied and appearance of low-frequency error is evaluated. In the measurement of the silicon wafer, the diffraction wavefront reflected by the wafer surface is measured using the PS/PDI because the wave includes information of the wafer surface figure. The wavefront deviation data from an ideal sphere is converted to the surface figure of the silicon wafer by the ray trace method. The resultant figure data agrees well with data from a white-light interferometer in the low-frequency components.

## 9. Acknowledgment

A part of this work was supported by a Grants-in-Aid for Scientific Research (KAKENHI) (22560109, 19360066, 16360067) from Japan Society for the Promotion of Science (JSPS). Authors express their sincere gratitude to Dr. Toshiaki Matsuura (Mitsutoyo Corporation) for his intense study in doctoral course. Finally authors are grateful to editors for giving them the chance to contribution to this book.

## 10. References

Ando, M. and 67 co-authors (2001), Stable Operation of a 300-m Laser Interferometer with Sufficient Sensitivity to Detect Gravitational-Wave Events within Our Galaxy, Phys. Rev. Lett. 86, pp. 3950–3954.

Born, M. and Wolf, M. (1959), Principles of Optics. Pergamon Press: London, New York, Paris, Los Angeles.

Chen, S., Li, S., Dai, Y., and Zheng, Z. (2006), Iterative algorithm for subaperture stitching test with spherical interferometers, J. Opt. Soc. Am. A 23, n. 5, pp. 1219-1226.

Creath, K. and Hariharan, P. (1994), Phase-shifting errors in interferometric tests with high-numerical-aperture reference surfaces, APPLIED OPTICS, Vol. 33, No. 1, pp. 24-25.

de Groot, P. (1995), Derivation of algorithms for phase-shifting interferometry using the concept of a data-sampling window, Applied Optics Vol. 34, No. 22, pp. 4723-4730.

de Groot, P. (1995), Phase-shift calibration errors in interferometers with spherical Fizeau cavities, Applied Optics Vol. 34, No. 22, pp. 2856-2863.

de Groot, P. J. (1995), Vibration in phase-shifting interferometry, J. Opt. Soc. Am. A, Vol. 12, No. 2, pp. 354-365.

Ferraro, P., Del Core, C., Miccio, L., Grilli, S., De Nicola, S., Finizio, A. and Coppola, G. (2007), Phase map retrieval in digital holography: avoiding the undersampling effect by a lateral shear approach, Optics Letters, Vol. 32, No. 15, pp. 2233-2235.

Freischlad, K. and Koliopoulos, C. L. (1990), Fourier description of digital phase-measuring interferometry, J. Opt. Soc. Am. A/Vol. 7, No. 4, pp. 542-551.

Goldberg, K. A., Naulleau, P. P., Denham, P. E., Rekawa, S. B., Jackson, K., Anderson, E. H., and Liddle, J. A. (2004), At-wavelength alignment and testing of the 0.3 NA MET optic, J. Vac. Sci. Technol. B 22, pp. 2956-2961.

Greivenkamp, J. E. (1987), Sub-Nyquist interferometry, Applied Optics, vol. 26, n. 24, pp. 5245-5258.

Handa, S., Mimura, H., Yumoto, H., Kimura, T., Matsuyama, S., Sano, Y. and Yamauchi, K. (2008), Highly accurate differential deposition for X-ray reflective optics. Surface and Interface Analysis, 40: pp. 1019–1022.

Harris, F. J. (1978), On the use of windows for harmonic analysis with the discrete Fourier transform, Proceedings of the IEEE, Volume: 66 Issue:1, pp. 51-83.

Higashi, Y., Endo, K., Kume, T., Enami, K., Uchikoshi, J., Ueno, K., Mori, Y. (2007),Surface gradient integrated profiler for x-ray and EUV optics: self calibration method of measured position for an off-axis parabolic mirror (f=150mm) measurement, Proc. of SPIE vol. 6704, 67040D.

Iemmi, C., Moreno, A., and Campos, J. (2005), Digital holography with a point diffraction interferometer, OPTICS EXPRESS, Vol. 13, No. 6, pp. 1885-1891.

Kazuya Ota; Takahiro Yamamoto; Yusuke Fukuda; Katsura Otaki; Iwao Nishiyama; Shinji Okazaki (2001), Advanced point diffraction interferometer for EUV aspherical mirrors (Proceedings Paper), Proceedings of SPIE Vol. 4343, pp.543-550.

Larkin, K. G. and Oreb, B. F. (1992), Design and assessment of symmetrical phase-shifting algorithms, J. Opt. Soc. Am. A, Vol. 9, No. 10, pp. 1740-1748.

Matsuura, T., Okagaki, S., Nakamura, T., Oshikane, Y., Inoue, H., Nakano, M., and Kataoka (2007), T., Measurement Accuracy in Phase-Shifting Point Diffraction Interferometer with Two Optical Fibers, OPTICAL REVIEW Vol. 14, No. 6, pp. 401–405.

Matsuura, T., Okagaki, S., Oshikane, Y., Inoue, H., Nakano, M. and Kataoka, T. (2008), Numerical reconstruction of wavefront in phase-shifting point diffraction interferometer by digital holography. Surface and Interface Analysis, 40: 1028–1032.

Matsuura, T., Udaka, K., Oshikane, Y., Inoue, H., Nakano, M., Yamauchi, K., Kataoka, T. (2010), Spherical concave mirror measurement by phase-shifting point diffraction interferometer with two optical fibers, Nuclear Instruments and Methods in Physics Research A 616, pp. 233–236.

Matsuyama, S., Mimura, H., Katagishi, K., Yumoto, H., Handa, S., Fujii, M., Sano, Y., Shimura, M., Yabashi, M., Nishino, Y., Tamasaku, K., Ishikawa, T. and Yamauchi, K. (2008), Trace element mapping using a high-resolution scanning X-ray fluorescence microscope equipped with a Kirkpatrick-Baez mirror system. Surface and Interface Analysis, 40: pp. 1042–1045.

Medecki, H., Tenjnil, E., Goldberg, K. A., & Bokor, J. (1996), Phase-shifting point diffraction interferometer, Optics Letters, Vol. 21, Iss. 19, pp. 1526–1528.

Millerd, J. E., Brock, N. J., Hayes, J. B., Wyant, J. C. (2004), Instantaneous phase-shift point-diffraction interferometer, Proceedings of SPIE, Vol. 5531, pp. 264-272.

Nomura, T., Murata, S., Nitanai, E., and Numata, T. (2006), Phase-shifting digital holography with a phase difference between orthogonal polarizations, Applied Optics Vol. 45, No. 20, pp. 4873-4877.

Oshikane, Y., Kataoka, T., Endo, K., Noguchi, A., Inoue, H. and Nakano, M. (2003), Development of Phase-Shifting Point Diffraction Interferometry with Fiber Point Sources, J. of the Japan Society of Precision Engineering 69, pp. 678-682.

Otaki, K., Ota, K., Nishiayama, I., Yamamoto, T., Fukuda, & Y., Okazaki, S. (2002). Development of the point diffraction interferometer for extreme ultraviolet lithography: Design, fabrication, and evaluation, J. Vac. Sci. Technol. B 20, pp. 2449-2458 (2002).

Otaki, K., Yamamoto, T., Fukuda, T., Ota, K., Nishiyama, I., and Okazaki, S. (2002), Accuracy evaluation of the point diffraction interferometer for extreme ultraviolet lithography aspheric mirror, J. Vac. Sci. Technol. B 20, pp. 295-300.

Sato, S., Miyoki, S., Ohashi, M., Fujimoto, M., Yamazaki, T., Fukushima, M., Ueda, A., Ueda, K., Watanabe, K., Nakamura, K., Etoh, K., Kitajima, N., Ito, K., and Kataoka, I. (1999), Loss factors of mirrors for a gravitational wave antenna, APPLIED OPTICS, Vol. 38, No. 13, pp. 2880-2885.

Schulz, G. and Schwider, J. (1967), Precise Measurement of Planeness, Applied Optics vol. 6, n. 6, pp. 1077-1084.

Smartt, R. N. and Strong, J. (1972), Point-Diffraction Interferometer, Journal of the Optical Society of America, 62, p. 737.

Smartt, R. N., and Steel, W. H. (1975), Theory and Application of Point-Diffraction Interferometers, Jpn. J. Appl. Phys. 14 Supplement 14-1 pp. 351-356.

Sommargren, G. E. (1996), Phase Shifting Diffraction Interferometry for Measuring Extreme Ultraviolet Optics, OSA TOPS on Extreme Ultraviolet Lithography, Vol. 4,.

Stulen, R.H. & Sweeney, D.W. (1999). Extreme ultraviolet lithography, IEEE J. of Quantum Electronics, vol. 35, pp. 694-699 (1999).

van Wingerden, J., Frankena, H. J., and Smorenburg, C. (1991), Linear approximation for measurement errors in phase shifting interferometry, APPLIED OPTICS, Vol. 30, No. 19, pp. 2718-2729.

Yamaguchi, I., Ida, T., Yokota, M., and Yamashita, K. (2006), Surface shape measurement by phase-shifting digital holography with a wavelength shift, Applied Optics Vol. 45, No. 29, pp. 7610-7616.

Yamamura, K. and Takai, H. (2008), Figuring of elliptical hard X-ray focusing mirror using 1-dimensional numerically controlled local wet etching. Surface and Interface Analysis, 40: pp. 1014–1018.

Yumoto, H., Mimura, H., Kimura, T., Handa, S., Matsuyama, S., Sano, Y. and Yamauchi, K. (2008), Stitching interferometric metrology for steeply curved x-ray mirrors. Surface and Interface Analysis, 40: pp. 1023–1027.

# Part 3

## New Optical Fibers

# 15

# "Crystalline" Plastic Optical Fiber with Excellent Heat-Resistant Property

Atsuhiro Fujimori
*Saitama University*
*Japan*

## 1. Introduction

General "polymer crystals" essentially both crystalline and amorphous regions. It is well known that crystalline polymers construct hierarchical structures ranging from lamellae on the nanometer scale to spherulite on the mesoscopic scale.[1-3] The polymer crystals in these crystalline polymers are generally formed by the folding of the main chain. In many cases, since these folded parts and interspherulite chains form the amorphous region, crystalline polymers are essentially intermingled states of the crystalline and the amorphous regions. Therefore, crystalline polymers are not a suitable candidate for use in plastic optical fibers (POFs) and film-type optical waveguides (FOWs) because of the occurrence of light refraction at the crystalline/amorphous interface. Consequently, amorphous POFs lack heat resistance and dimensional stability.

However, if the construction of extremely homogeneous crystalline POFs is realized, "crystalline" POFs with excellent heat resistance and dimensional stability can be developed. The heat-resistant POFs will efficiently demonstrate their optical ability in a circuit exposed to a high temperature of more than 125 °C; so far there have been no products of heat-resistant POFs that can sustain temperatures higher than 125 °C. If the heat-resistant POFs are realized, light wiring in automobiles will also be achieved; the heat-resistant POFs will not only connect the AV equipment but also connect the control system around the engine. As a result, the overall body of an automobile will become lighter. This future technology is based mainly on "crystalline fluorinated polymers" having a high crystallinity. Generally, polytetrafluoroethylene (PTFE; $-(CF_2\text{-}CF_2)_n-$) and its copolymers easily form rigid helices in order to yield extended-chain crystals. It seems difficult for PTFE to form a lamellae structure because of its rigid molecular chain.[4-8] In addition, since tetrafluoroethylene copolymers obtained by the incorporation of several comonomers exhibit extremely fast crystallization rates,[9] their spherulites generally cannot be observed until they are sufficiently large. Therefore, PTFE exhibits a high degree of crystallinity of over 90%.[10-12]

Poly[tetrafluoroethylene-*co*-(perfluoroalkylvinylether)] (*abbrev.* EFA (alkyl = ethyl) or PFA (alkyl = propyl))[13] has a unique role in the plastics industry due to its inertness, heat resistance, and low coefficient of friction in a wide temperature range. Generally, fluorinated compounds and fluoropolymers have excellent chemical resistance, oil resistance, and oil- and water-shedding resistance.[14-17] They have been used as rubbers at high temperatures and in several lubricating fluorine manufactured products.

However, in the field of fundamental science, structural studies on fluorinated polymers have progressed slowly since the time these polymers were first reported by Bunn and Howells in 1954.[18] We could find very few reports on the systematic structural studies on PTFE or tetrafluoroethylene-based fluorinated copolymer because this compound is difficult to synthesize due to the emission of poisonous gases.[4, 6]

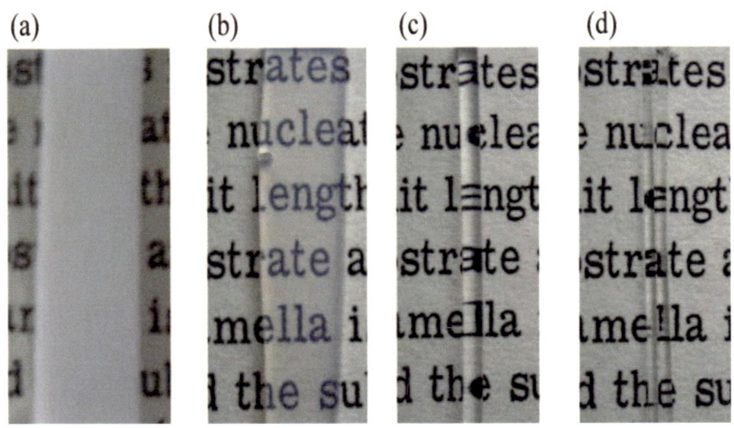


Fig. 1. Changes in transparency of several processed materials of "crystalline" fluorinated copolymers: (a) bulk EFA, (b) pressed processing sheet, (c) crystalline fiber with drawn ratio = 3, (d) crystalline fiber with drawn ratio = 5.

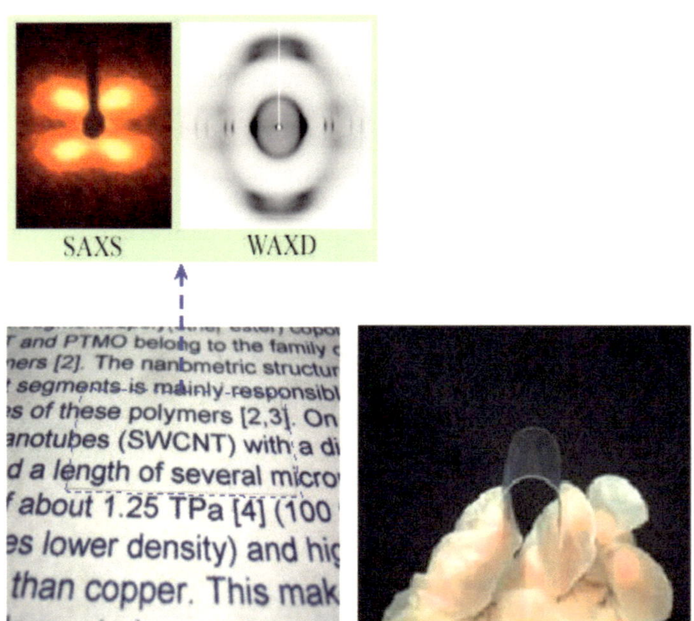


Fig. 2. Photograh of crystalline, transparent, and flexible film made by fluorinated copolymer, and their SAXS and WAXD patterns.

Further, although EFA is a crystalline polymer, processed EFA samples that have a high crystallinity are as transparent as amorphous flexible polymers such as polymethymethacrylate (PMMA)[19] and poly(ethylene terephthalate) (PET), as shown Fig. 1 and Fig. 2. This experimental fact is not well known. Probably, since the transparency of organic materials depends on the existence of differences in electron density between the crystalline and the amorphous regions, it is considered that a high crystallinity of EFA relates closely to the occurrence of transparency. Additionally, processed EFA tubes break into pieces just like glass tubes when an excessive bending force is applied upon them. It is obvious that the enhancement of these unique properties of the processed EFA POFs and FOWs is a result of the changes in the crystal structure and crystalline morphology of EFA fibers that take place during the drawing process. Further, fluorinated polymers do not absorb infrared light because of their stretching vibration and a lack of C-H bonds.[20, 21] Hence, a "crystalline" POF and FOW made by fluorinated polymers transports not only visible light but also infrared light.

In this chapter, the changes in the fine structure and lamella arrangement of the fibers formed by tetrafluoroethylene copolymers upon drawing are investigated by using wide-angle X-ray diffraction (WAXD) and small-angle X-ray scattering (SAXS) methods. We have found very few reports on the studies on the structural changes in fluorinated polymers upon drawing, whereas there are many reports of studies on hydrogenated polymers. Therefore, this study may also be valuable as fundamental research in the field of polymer physics. In addition, we have discussed the relationships between the origins in order to elucidate the occurrence of transparency and structural changes in molecular arrangements.

## 2. Experimental

### 2.1 Materials

#### 2.1.1 Fluorinated copolymer

The fluorinated copolymers used in this study were provided by DuPont-Mitsui Fluorochemicals Co. Ltd. EFA is a random copolymer obtained from the copolymerization of tetrafluoroethylene $-(CF_2\text{-}CF_2)_n-$ and perfluoroethylvinylether $-(CF_2\text{-}CF(OCF_2CF_3))_n-$. The amount of comonomers of these materials was about 3 wt%. The molecular weight of the EFA processed to a crystalline fiber form was about 600,000. This molecular weight was examined by a computer simulation on the basis of the viscoelasticity of the fiber in a molten state because it is difficult to dissolve these polymers in an organic solvent.

#### 2.1.2 Drawing of EFA POFs and FOWs

EFA POFs and FOWs were drawn uniaxially by using a hand-drawing apparatus in an air oven at 280 °C. The surface of the POFs and FOWs specimen was marked at intervals of 2 mm in order to measure the draw ratios. The drawing speed was fixed at 20 mm/min, and the fiber was annealed at 280 °C for 3 min before drawing. Using this method, we obtained fibers with excellent transparency (Figs. 1(c) and 1(d)).

### 2.2 Experimental methods

#### 2.2.1 Small-angle X-ray scattering (SAXS)

The crystalline morphology of the drawn EFA copolymers was characterized with a SAXS instrument (M18XHF, MAC Science Co.) consisting of an 18-kW rotating-anode X-ray generator with a Cu target (wavelength, $\lambda$ = 0.154 nm) operated at 50 kV and 300 mA.[22] This instrument comprised a pyrographite monochromator, pinhole collimation system (ϕ ~0.3, 0.3, and 1.1 mm), vacuum chamber for the scattered beam path, and a two-dimensional imaging plate detector (DIP-220). The sample-to-detector distance was adjusted to 710 mm. The exposure time for each sample was 30 min. For the SAXS measurements, each sample (thickness: approximately 0.5 mm) was placed in a sample holder so that its position remained unchanged. The theoretical detection limit of the SAXS measurement in this study almost corresponded to the value of q = 0.128 $nm^{-1}$ estimated by using the camera distance (from sample to the imaging plate) in the apparatus. However, the actual detection limit examined by counting the pixel numbers of enlarged SAXS patterns on the monitor of an analytical computer was q = 0.170 $nm^{-1}$ (dashed line in the profile of Fig. 3). Hence, the observable maximum value of the long period between the centers of gravities of the lamellae in this study was 36.9 Å.

#### 2.2.2 Wide-angle X-ray diffraction (WAXD)

In order to obtain the WAXD data for the drawn fibers, an R-axis diffractometer (Rigaku Co.) was operated at 45 kV and 200 mA to generate Cu*K*α radiation (λ = 0.1542 nm). WAXD photographs of the samples were taken at room temperature by using a graphite monochromator and a 0.3-mm pinhole collimator. Diffraction data were recorded on a cylindrical imaging plate detector equipped with an interface to a computer system. The camera length was 127.4 mm, and the exposure time was 600 s.

#### 2.2.3 Estimation of thermal properties and transparency

Thermal analyses were carried out by using a Seiko Instruments model DSC200 differential scanning calorimeter (DSC). The DSC measurements were performed at a standard scanning rate of 10.0 °C $min^{-1}$. A sample mass of about 5.00 mg was used for all the DSC measurements. As usual, the scanning of DSC measurements and the heating and cooling cycle were repeated twice in order to examine the difference between the peak position and transition enthalpy in the first and second heating. UV-vis spectra of EFA films were measured using a UV-vis spectrophotometer (V-650, JASCO).

## 3. Results and discussion

### 3.1 Changes in lamellae arrangement of transparent "crystalline" EFA POFs and FOWs

Figure 3 shows the SAXS pattern and normalized one-dimensional SAXS profiles, where q is the scattering vector ($q = 4\pi \sin\theta/\lambda$; θ = Bragg angle), of the undrawn transparent crystalline EFA POWs. A ring-shaped SAXS pattern was observed, which indicated the formation of an isotropic random lamella texture. In the case of PTFE, the SAXS pattern was obscure, and the corresponding profile exhibited extremely low intensity because this polymer almost formed an extended chain and not a lamellae structure.[23] On the contrary, it was found that the tetrafluoroethylene copolymer formed lamellae structures since the undrawn EFA used in this study exhibited isotopic SAXS patterns. The long period of the undrawn sample was estimated to be 27.0 nm. A high-crystallinity EFA sample formed relatively thicker lamellae than the general hydrogenated crystalline polymers.

On the basis of the results of the SAXS measurements of the undrawn EFA fiber, we suggested the following lamella model for tetrafluoroethylene copolymers. According to A. Keller's suggestion,[1] it was assumed that general crystalline polymers form a regular sharp hold. However, the tetrafluoroethylene copolymers used in this study did not form an arrangement of these adjacent reentries because of the existence of a rigid molecular chain and a lack of flexibility. It seemed that the folded parts formed in the ether bond-rich region within the fluorinated main chain. However, so many perfluoroalkylvinylether units could not have contributed to the formation of the folded parts because the ratio of the absolute amounts of the comonomers was extremely low. Hence, we proposed a "switch-board type" lamellae model of these tetrafluoroethylene copolymers, shown in Fig. 4, according to P. J. Flory's suggestion.[23, 24] In this case, it was supposed that there existed a relatively large amorphous region because of the existence of the large long-period structure estimated by SAXS. From the qualitative estimation of the lamella thickness based on the crystallization degree obtained from the DSC measurements, the thickness of the crystalline regions of the EFA lamella form was estimated to vary within a range from 8 to 15 nm (as calculated by using the fusion enthalpy of as-polymerized PTFE, $\Delta H_{endo}$ (58.4 J $g^{-1}$), as the standard fusion enthalpy of EFA, $\Delta H_{endo,\,0}$).[23] The existence of the thick amorphous layer (over 10 nm) also supports the validity of our proposed switch-board type lamella model.

Figure 5 shows the SAXS patterns and corresponding lamella arrangement models for DR1 (draw ratio = 1.0, undrawn), DR3, and DR5 transparent crystalline POFs of EFA. A ring-shaped SAXS pattern was observed for the undrawn DR1 sample (Fig. 5 (a)), while two- or four-point patterns were observed for the DR3 (Fig. 5 (b)) or DR5 (Fig. 5 (c)) fiber samples. The former indicated a random lamellar texture (Fig. 5 (a')), and the latter indicated some lamella structures oriented with respect to the draw direction.

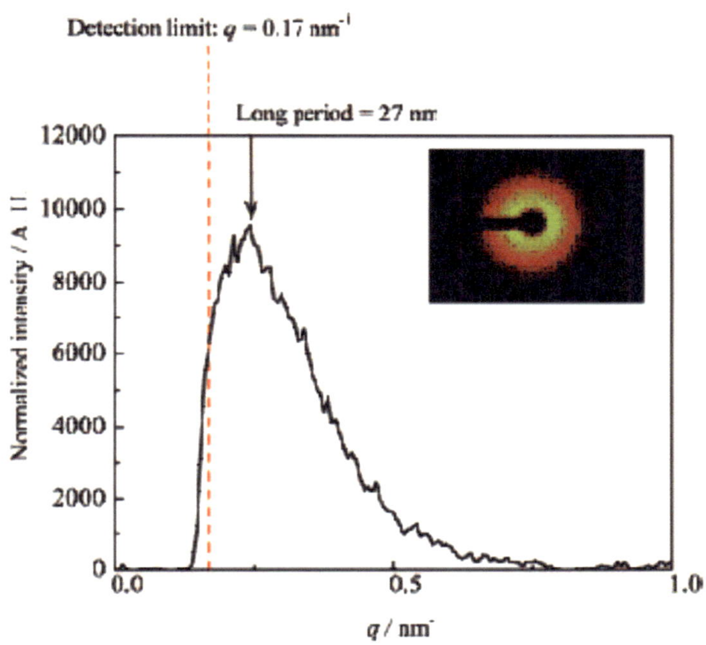


Fig. 3. SAXS pattern and profile of undrawn EFA 'crystalline' POF.

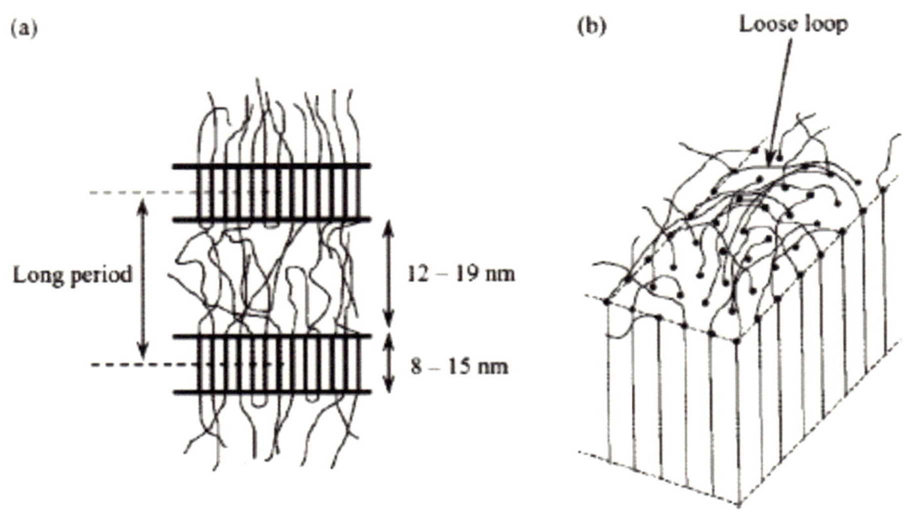


Fig. 4. Schematic illustrations of "switchboard-type" lamella models of fluorinated copolymers like an EFA (a) along the *c*-axis, and (b) in an *a*-*b* plane.

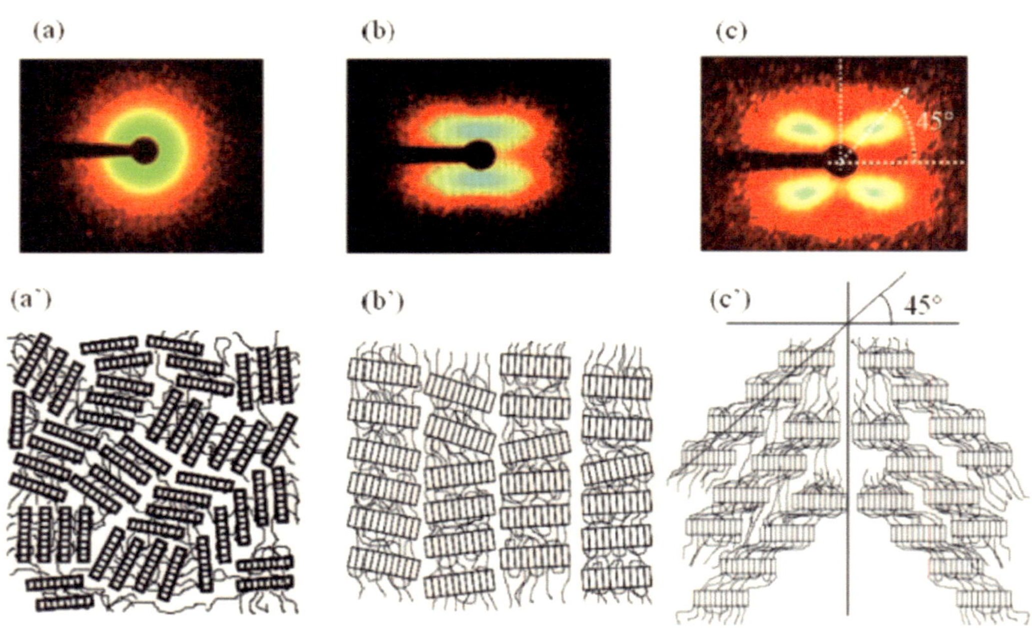


Fig. 5. Changes in SAXS patterns and corresponding lamella arrangement models of EFA transparent 'crystalline' POF with drawing; (a), (a') undrawn, (b), (b') 3 times, and (c), (c') 5 times drawing.

The appearance of the two-point SAXS patterns implied the formation of an arrangement of lamellae parallel to the draw direction (Fig. 5 (b')). As the fiber was drawn further, the interlamella and/or intralamella slips probably occurred, forming the small kink bands in the lamellae. The slip angle of the interlamellae was 45° as calculated by using the position of the strongest spot in the SAXS picture. In accordance with the changes in lamellae, the grain boundaries or amorphous parts between two neighboring lamellae were also distributed regularly towards the draw direction, and they thus resulted in a periodic change in density in the direction normal to them, which accounted for the four-point diffraction pattern. That is, with an increase in the elongation of the EFA sample, a particular kind of layer structure, an alternately tilted lamella arrangement known as the herringbone, was formed inside the fibers (Fig. 5 (c')). Similar results were obtained in the case of drawn polyethylene (PE) fibers previously.[25] The long periods or interplanar spacings were calculated to be 33.9 and 35.3 nm for DR3 and DR5, respectively. These values were larger than the interplanar spacing of the undrawn sample (27.0 nm). This feature of the long periods corresponded well with that of PE, polypropylene (PP), and polyester.[25-30] From the viewpoint of enhancing transparency by using the drawing process, EFA fibers exhibited the elongation of the amorphous region with an increase in density in this region and indicated a resultant increase in the long period upon drawing.

Figure 6 shows the change in SAXS patterns upon drawing. SAXS patterns remained essentially unchanged even upon carrying out the drawing process for five times. However, from the results of the examination of light conductivity in db/km units for EFA fibers using infrared light (at $\lambda$ = 850 nm), most superior abilities were confirmed in the DR5 fibers, and their transmission ability was observed to decrease gradually upon drawing for over six times. Moreover, the drawn EFA fiber broke when the elongation equaled almost nine times the original value. Just before breaking, the color of the drawn EFA fiber became white because of the appearance of many microvoids and/or defects and the light dispersion caused by these voids and/or defects. In order to estimate the changes in lamella thickness and differences in electron density upon drawing, plots of the draw ratio *vs.* long periods and normalized intensity of SAXS profiles are shown in Fig. 7. The values of the long period saturated at about DR3, and the normalized intensity was almost constant from DR4 to DR8. That is, the increase in the lamella thickness containing an amorphous region stopped at DR3 (about 35 nm). After that, although the density of the amorphous region increased gradually upon drawing, a partial appearance of the voids might have occurred simultaneously. As a result, the difference in the overall density between the crystalline and the amorphous regions in the EFA fiber remained unchanged for a draw ratio of more than 4.

### 3.2 WAXD study on crystal structure of tetrafluoroethylene-based polymers

A typical example of the WAXD patterns for the drawn EFA fibers (DR8) is shown in Fig. 8(a). Almost all spots existed on the equator line. Therefore, we have mainly discussed the WAXD profiles integrated along the equatorial direction in this section. Figure 8(b) shows a comparison of the WAXD profiles of the unoriented PTFE and the EFA samples. The lack of an amorphous curve around $2\theta$ = 15° was a peculiarity of the PTFE extended-chain crystal. A halo curve of the EFA appeared due to the existence of an amorphous region in the interlamella parts. However, the crystalline peak positions in both profiles were almost the same since the structure and main-chain arrangement in the crystalline region of EFA

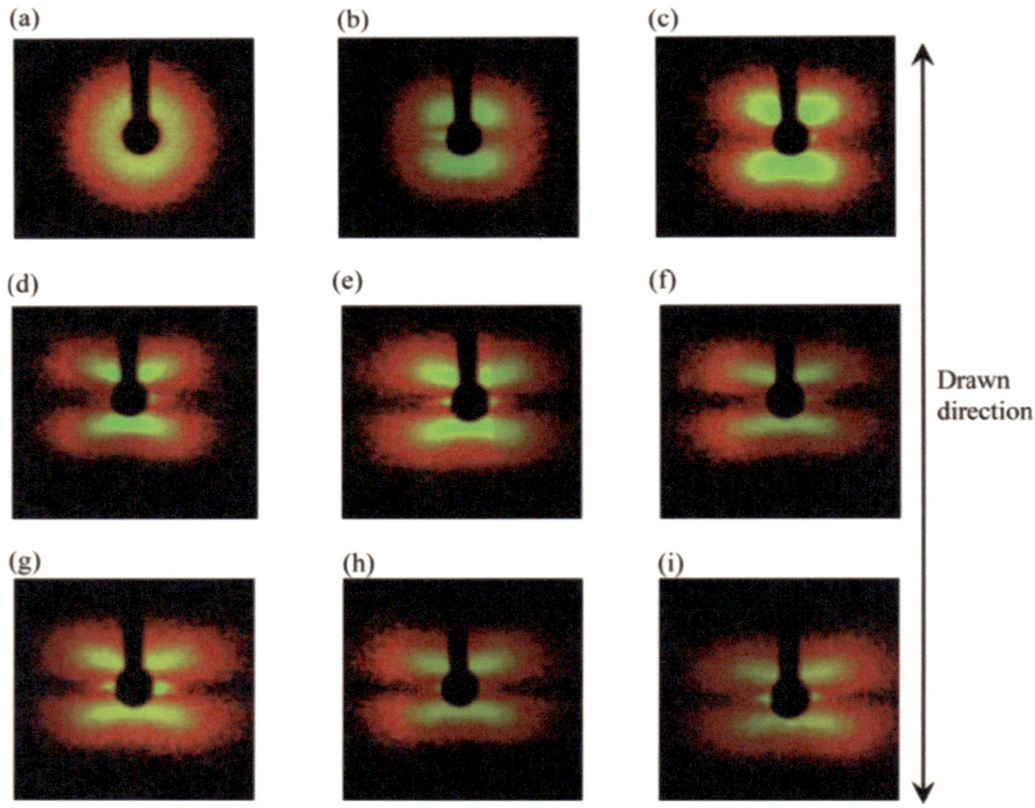


Fig. 6. Changes in SAXS patterns of EFA "crystalline" POFs with drawing at a ratio of (a) 1.0, (b) 1.5, (c) 2.0, (d) 3.0, (e) 4.0, (f) 5.0, (g) 6.0, (h) 7.0, and (i) 8.0.

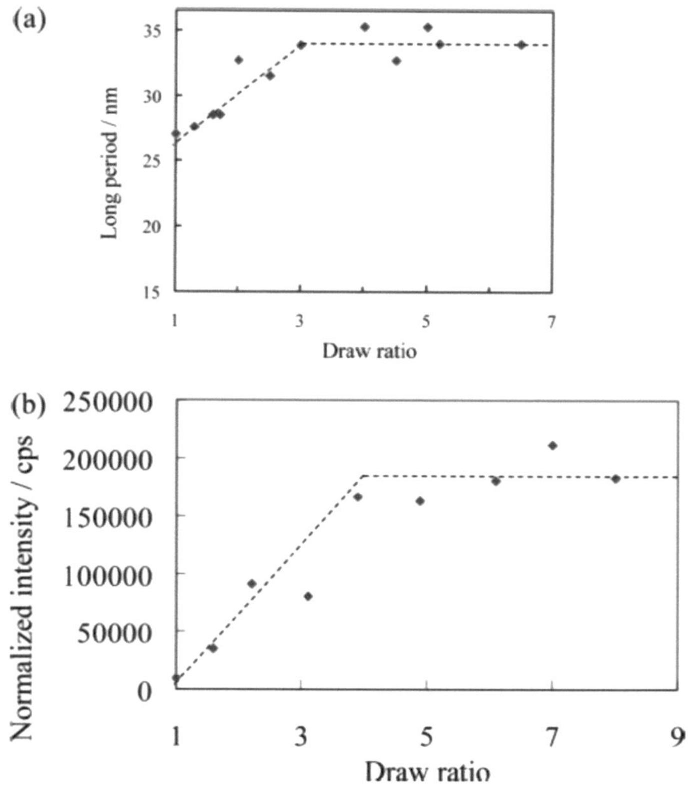


Fig. 7. Plots of draw ratio vs (a) long periods and (b) normalized intensity estimated by SAXS measurements of EFA "crystalline" POFs.

comprised repeating tetrafluoroethylene parts. That is, there was no difference between the structure of the crystalline region of PTFE and that of EFA.

Furthermore, most inner WAXD spots of an EFA fiber (Fig. 8(a); the shadow next to the beam stopper) existed clearly when $2\theta = 9°$. These WAXD results included a very important result with regard to the fluorinated polymer crystal. The peak at around $2\theta = 18.0°$ in the WAXD profiles of tetrafluoroethylene and its copolymers was assigned to the (100) reflection in the quasi-hexagonal system according to the literature documented about 50 years ago.[18, 31-33] Moreover, we could not find any reports related to the inner peak around $2\theta = 9°$. However, in the present WAXD profiles, small peaks at around $2\theta = 9°$ were confirmed and reproduced well by the high-power measurement using an X-ray diffractometer with an imaging plate as the detector. Further, in the WAXD profile of the oriented rod-shaped material processed by isostatic pressing of PTFE, this peak was clearly enhanced (Fig. 8(c)). In addition, Fig. 9 shows the changes in this peak in the WAXD profiles of the transparent crystalline EFA fiber upon drawing and the well-reproduced appearance of this peak in any type of fluorinated copolymers. From the result of Fig. 9(a), it was found that the intensity of this peak around $2\theta = 9.0°$ increased gradually with an increase in the draw ratio. Figure 9(b) shows the WAXD profiles of several fluorinated copolymers such as PTFE, poly[tetrafluoroethylene-*co*-(hexafluoropropylene)] (FEP), PFA, PFA containing PTFE

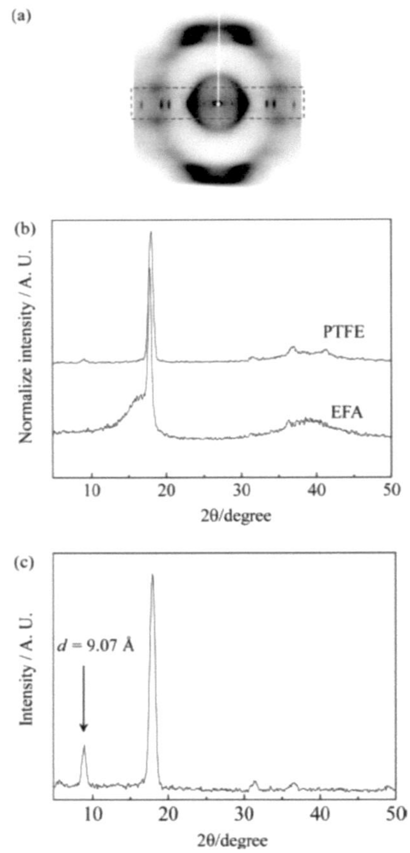


Fig. 8. (a) WAXD patterns of "crystalline" POFs of EFA at draw ratio = 8. (b) Comparison of WAXD profiles of EFA to PTFE. (c) WAXD profile of PTFE orientated rod formed by isostatic extrusion.

particles as nucleators, low molecular weight EFA (250,000), middle molecular weight EFA (300,000), and high molecular weight EFA (600,000) containing PTFE particles. All WAXD profiles of fluoropolymers used in this study contained this small peak at almost the same position. That is, this small peak around $2\theta = 9.0°$ reflected that the genuine crystal structure of fluorinated polymers was always confirmed in the WAXD profiles of tetrafluoroetthylene-based polymers. Furthermore, the intensity of this peak increased upon the formation of an orientated structure due to uniaxial drawing. However, no previous reports that confirm the presence of these small peaks exist, except for the paper we published recent year.[22] It appears that the existence of this diffraction peak has been overlooked for about 50 years. In our previous report, we speculated that the peak at about $2\theta = 9°$ might correspond to the genuine (100) reflection.[23] In the present report, we clearly assert an interpretation of this peak and the crystal structure and partially modify our previous interpretation. In our previous work,[23] we suggested that the previously reported lattice constant needed to be modified and the lengths of the *a*- and *b*-axes be doubled. In addition, the reflection at around $2\theta = 18.0°$ would be attributed to the (200) peak. If this did

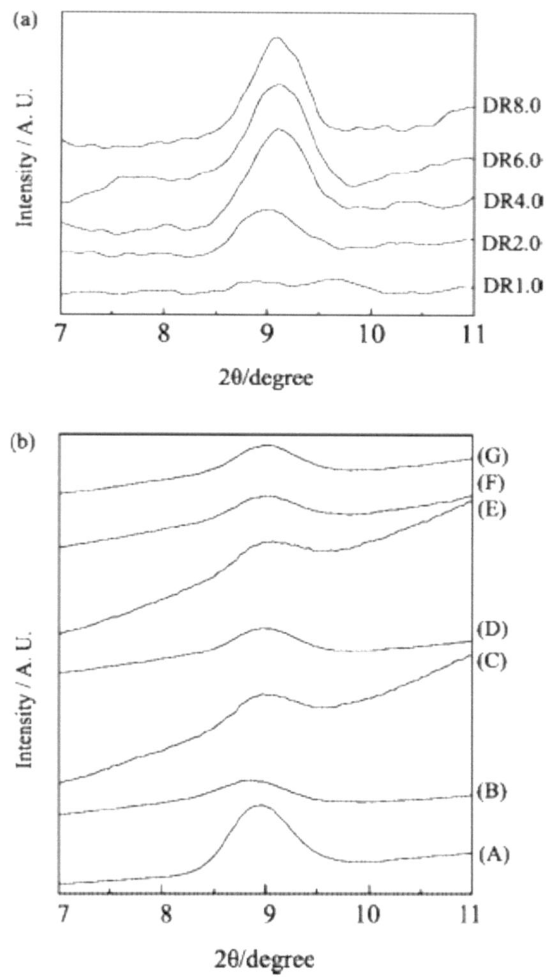


Fig. 9. WAXD patterns of (a) drawn EFA POFs, and (b) several fluorinated polymers in bulk: (A) PTFE, (B) FEP, (C) PFA, (D) PFA containing PTFE particle as nucleator, (E) low molecular weight EFA, (F) high molecular weight EFA, (G) EFA containing PTFE particle as nucleator.

not occur, the reflective indexes of the small peaks at about $2\theta = 9°$ could not be determined. Figures 10(a) and 10(b) show the reciprocal lattice of PTFE and other perfluorinated copolymers observed along the *c*-axis under the suggestion that the parts forming the crystal region had the same structure for tetrafluoroethylene and tetrafluoroethylene copolymers. The proposed lattice constant of PTFE[23] corresponded to $a = b = 11.08$ Å, $c = 16.8$ Å, $\alpha = 90°$, $\beta = 90°$, and $\gamma = 119.3°$ (Fig. 8(b), quasi-hexagonal system) and improved upon the reports by Bunn, et al., Starkweather Jr., et al., Clark, et al.,[18, 31-33] and other investigation groups (Fig. 10(a), $a = b = 5.54$ Å, $c = 16.8$ Å, $\alpha = 90°$, $\beta = 90°$, and $\gamma = 119.3°$ (quasi-hexagonal system)). However, the reciprocal lattice of Fig. 9(b) described a base-centered hexagonal lattice, whereas a base-centered lattice cannot exist in a group of hexagonal lattices. In addition, the reason for the appearance of a (100) reflection (peak at $2\theta = 9°$) weaker than a

(200) one (peak at 2θ = 18°) was not clear. Therefore, we reproposed the necessity of modifying the lattice constant of tetrafluoroethylene and its copolymers in the present work. We reconsidered the packing mode of fluorinated chains from a hexagonal to an orthorhombic system, as shown in Figs. 10(c) and 10(d). In the reciprocal lattice in Fig. 10(c), all WAXD reflection peaks confirmed in this study existed at a point of intersection in reciprocal lattice and all reflective indexes were decided. In this case, the peaks at 2θ = 9° and 18° corresponded to the (100) and (110) reflection peaks, respectively. The lattice constants of this packing system were estimated to be $a$ = 9.58 Å, $b$ = 5.54 Å, and $c$ = 1.69 Å (α = β = γ = 90°). The hexagonal lattice essentially had the structural analogy of an orthorhombic one. In addition, the appearance of peaks at 2θ = 9° and 18° was based on a different plane. Hence, the relation between intensities was not contradictory to an indexing rule. The three-dimensional packing model of the fluorocarbon chain in the crystalline region is shown in Fig. 10(d). The validity of our proposed orthorhombic system of the crystalline fluorinated polymer was also supported by the estimation in a reciprocal lattice along the meridional direction. Figure 11 shows the possibility for applying an orthorhombic lattice to an index WAXD reflection spot along the meridional direction of the drawn EFA fiber at DR5. As mentioned above, we considered the EFA chains as an orthorhombic packing in the crystal region, and the highest diffraction peak in the profile was interpreted as a (110) reflection in this lattice in the following discussion.

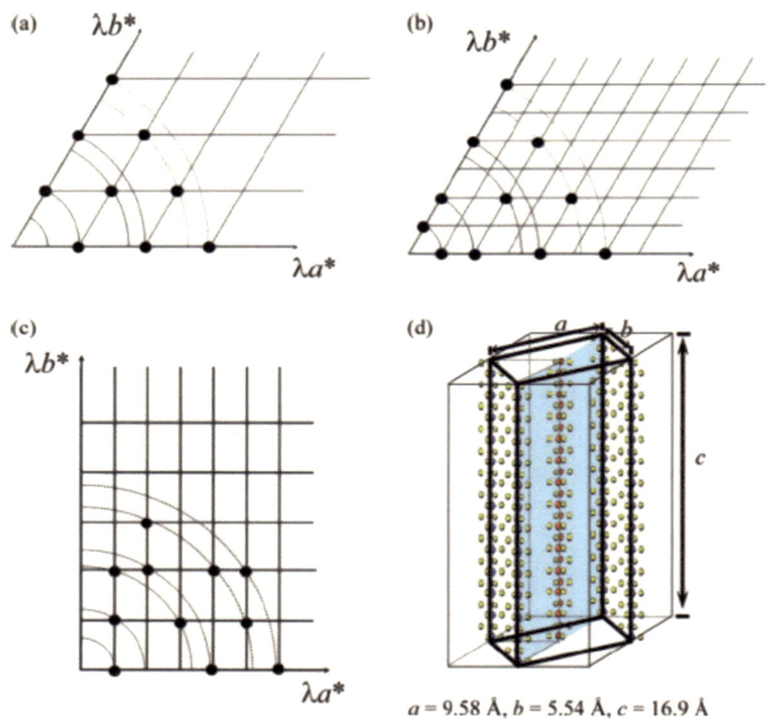


Fig. 10. Reciprocal lattices of crystalline region for several fluorinated polymers (PTFE, EFA, and so on) represented by WAXD data: (a) previously reported quasi-hexagonal lattice, (b) a quasi-hexagonal lattice twice elongated $a$- and $b$-axis, (c) our proposed orthorhombic lattice, and (d) packing model of fluorinated chains in orthorhombic lattice.

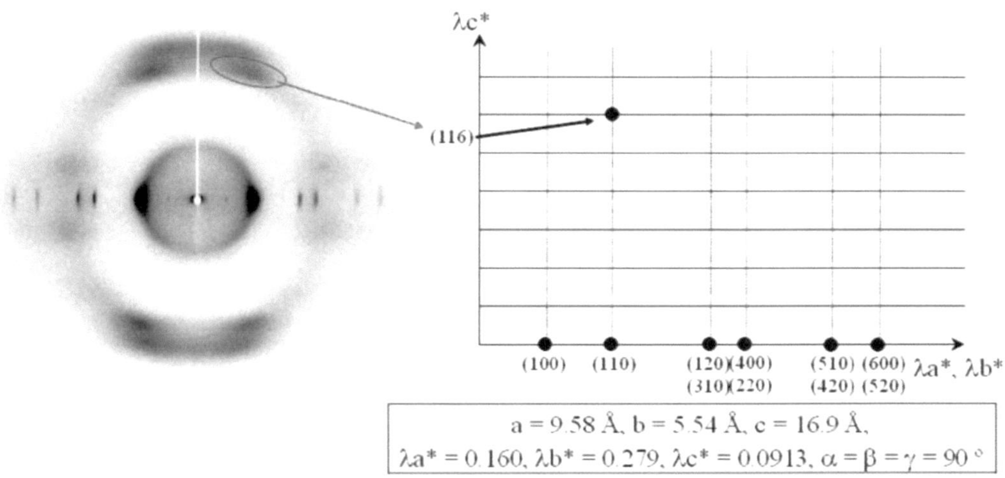


Fig. 11. WAXD patterns and reciprocal lattice in the $\lambda b^*$-$\lambda c^*$ plane of crystalline region for EFA transparent fiber at DR5.

### 3.3 Fine structure estimation of transparent crystalline EFA POFs and FOWs upon drawing at subnanometer scales by WAXD

Figure 12 shows the WAXD patterns of the transparent EFA fiber at several drawing ratios. We can clearly see the gradual enhancement of the WAXD spots along the equator line upon drawing. From the viewpoint of one-dimensional profiles scanned along the equatorial direction, the peak intensity of (110), (120), (220), and (420) reflections in the orthorhombic lattice increased gradually with an increase in draw ratio (Fig. 13(a)). The intensities of (110) peaks normalized by sample size and thickness almost saturated at DR5, as observed from the plot of Fig. 13(b) whereas the sizes of crystallite in the fiber estimated by Schereer's formula[34] are almost constant value all over the draw ratio. That is, it was considered that the increase in the crystallinity of the EFA fiber at the subnanometer scale actually reached a constant value.

In order to evaluate the degree of orientation for the *c*-axis of the EFA crystallites along the draw direction, we calculated the orientation function (*f*) proposed by Hermans and co-workers[35] using the azimuthal WAXD profiles. The function *f* was defined as

$$f\varphi = \frac{1}{2}(3 < \cos^2\varphi > -1).,\quad 0 < f_\phi < 1,$$

where $\varphi$ is the angle between the *c*-axis and the draw direction, and $<\cos^2\varphi>$ is obtained from the (110) and (120) azimuthal profiles by using Wilchinsky's procedure[36] (Fig. 14(a)). Figure 14(b) shows the change in the orientation function of the EFA crystallites ($f\varphi$) as a function of the draw ratio, where $f_\varphi$ increased with the draw ratio up to DR = 2.5, after which it reached a saturation value of around 0.8. These findings suggested that the orientation of an EFA crystallite in the fiber was complete at a draw ratio of 2.5. This value was almost the same as the draw ratio of the saturation value of a long period estimated by SAXS. That is, the orientation of the crystallite and the elongation of lamella reached

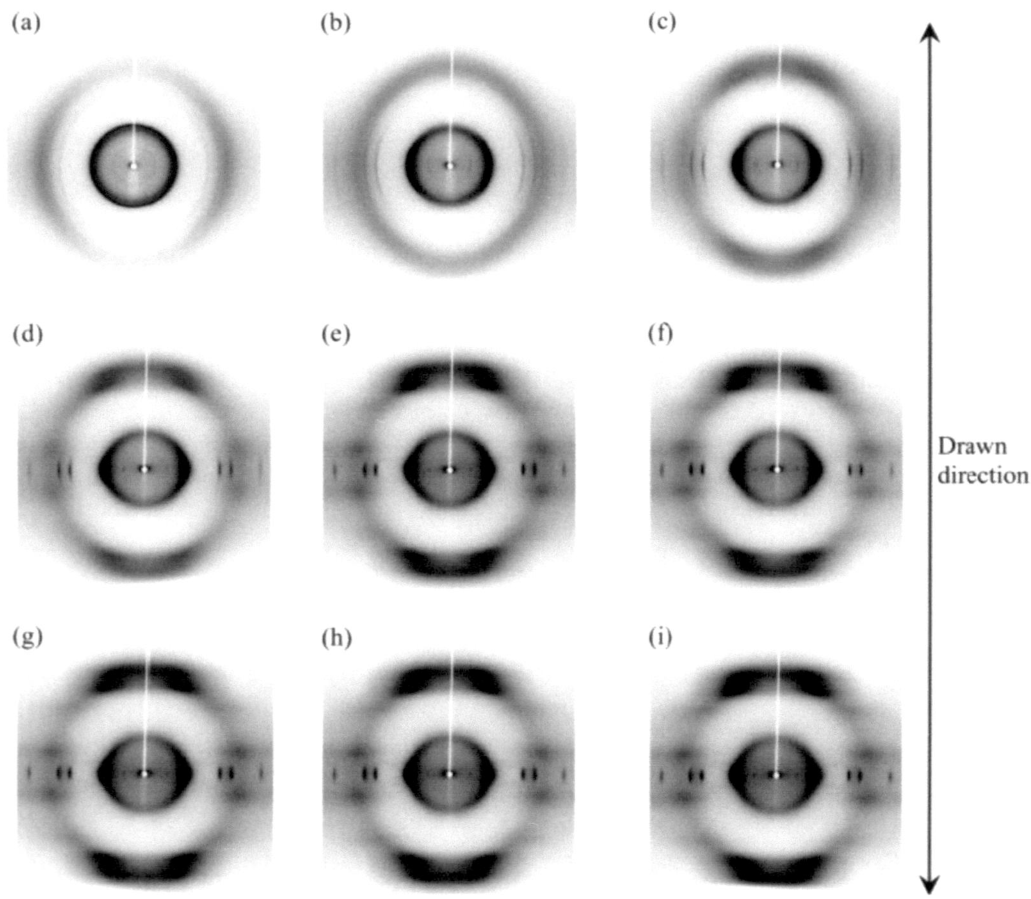


Fig. 12. WAXD patterns of EFA plastic optical fiber at several drawing ratio at room temparture: (a) undrawn, (b) DR ) 1.5, (c) DR ) 2.0, (d) DR ) 3.0, (e) DR ) 4.0, (f) DR ) 5.0, (g) DR ) 6.0, (h) DR ) 7.0, (i) DR ) 8.0.

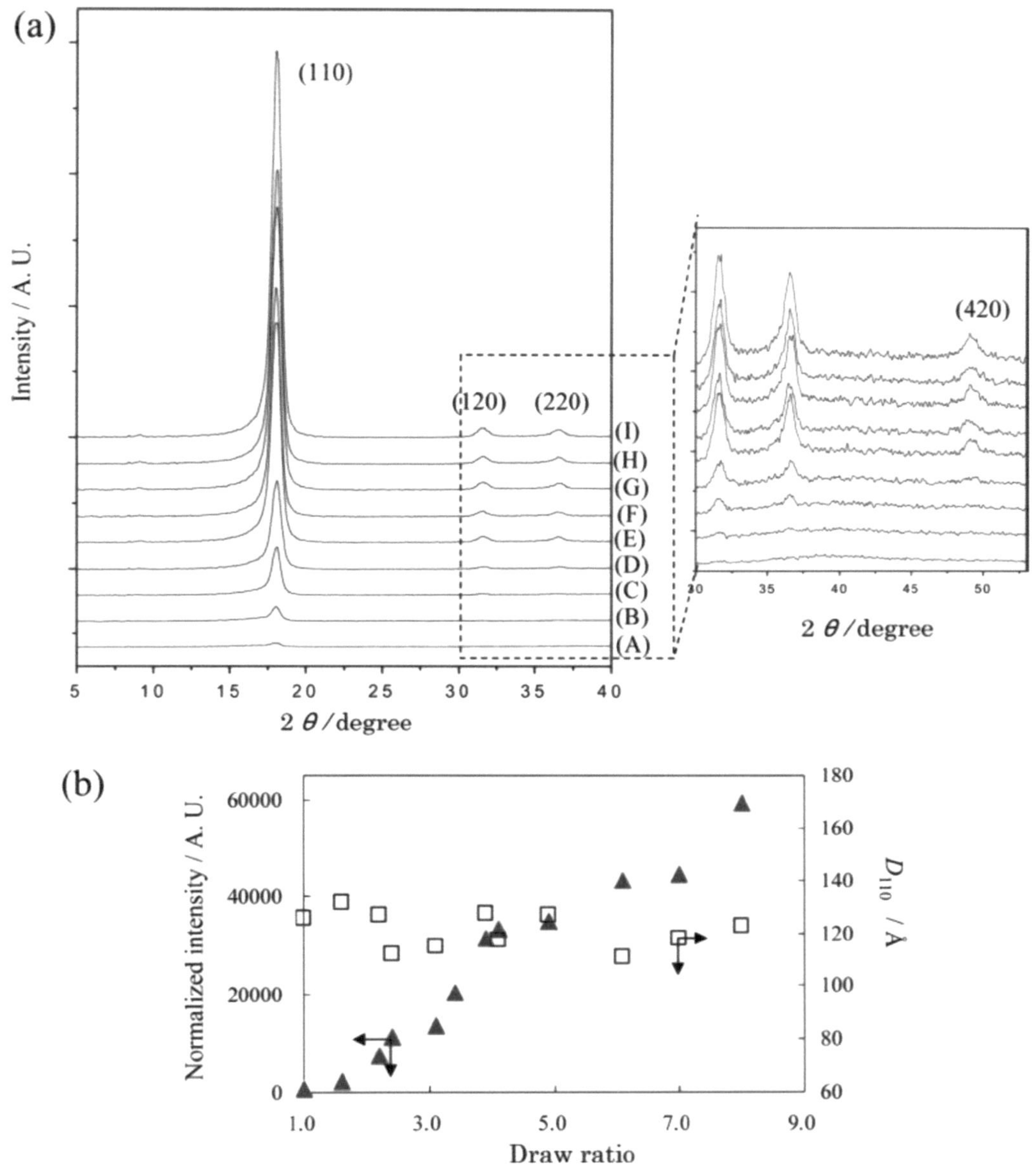


Fig. 13. (a) WAXD profiles of EFA plastic optical fiber with drawing at room temparture: (A) undrawn, (B) DR1.5, (C) DR2.0, (D) DR3.0, (E) DR4.0, (F) DR5.0, (G) DR6.0, (H) DR7.0, (I) DR8.0. (b) Changes in normalized WAXD intensity and crystallite sizes with drawing estimated by Scherrer's formula.

constant values almost simultaneously. Then, the quasi-crystallization process by drawing progressed up to DR5, which was the saturation value of the normalized intensity estimated on the basis of the WAXD patterns. Judging from the draw ratio of the saturation of the SAXS intensity, the increase in the electron density of the amorphous region and the partial appearance of voids might be a simultaneous occurrence upon further drawing. The sample of the crystalline EFA fiber at DR5 was the most transparent and exhibited the highest conductivity of infrared light among all the drawn fibers used in this study. In conclusion, the functionality of light transmittance was closely related to the solid-state structure of the crystalline EFA fiber.

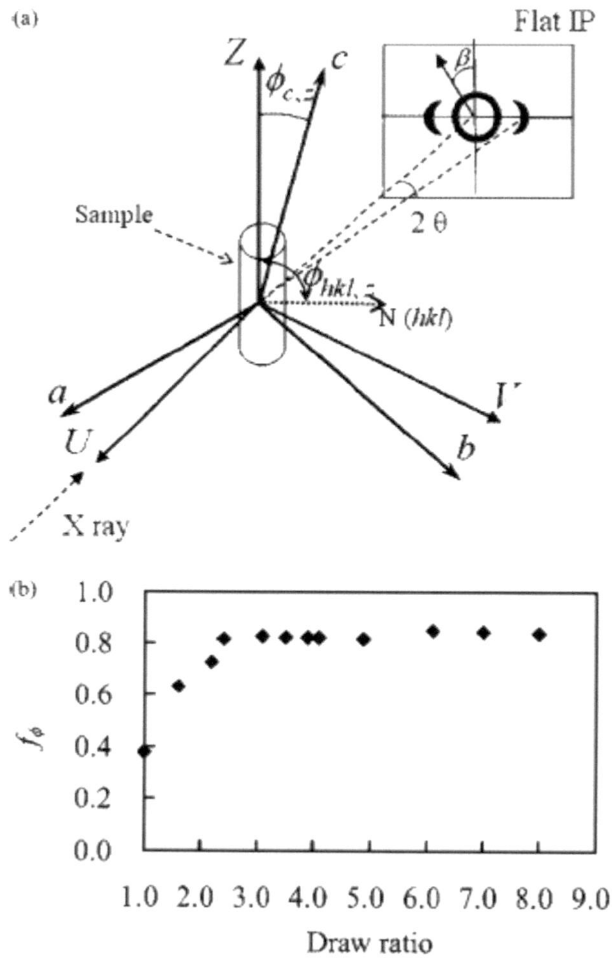


Fig. 14. (a) Schematic representation of Wilchinsky method to estimate orientation coefficient of crystallite. (b) Plot of drawn ratio vs orientation coefficient of crystallite in EFA POF.

Figure 15(a) shows results of DSC measurements of EFA "crystalline" fibers at several drawn ratios in order to estimate crystallization degree and lamella thickness. Areas of melting peaks on thermograms related to fusion enthalpy are gradually increased with drawn ratios. Crystallization degree as calculated by using the fusion enthalpy of as-polymerized PTFE, $\Delta H_{endo}$ (58.4 J g-1), as the standard fusion enthalpy of EFA, $\Delta H_{endo,0}$,[25] are plotted to drawn ratios of EFA fibers (Figure 15(b)). The linearity of changes in crystallinity of drawn fiber wellcorresponds to dependency of WAXD (110) intensity on drawing (Figure 12(b)). Further, from the qualitative estimation of the lamella thickness based on the crystallization degree, the thickness of the crystalline regions of the EFA lamella form was estimated to vary within a range from 6 to 16 nm (Figure 15(c)). In the case of DR5 fiber with most superior transmission ability of infrared light, almost 50% crystallinity and 11 nm lamella thickness are estimated. Therefore, it seems that the enhancement of transmission ability is not caused by increases of crystallinity, but reducing of differences in density between crystal and amorphous region. Probably, a high light transmission rate is not brought about by formation of extreme homogeneous crystalline fiber, but by formation of like a "fringed micelle-type" lamella arrangement which has an indistinct lamella-interface based on the enhancement of density for amorphous parts by drawing. In the case over six times drawing, since transition from amorphous part to crystalline part occurrs in EFA fiber, the density reduction of amorphous region and increases of differences in density between crystal and amorphous parts have developed. As a result, it seems that the transmission ability of infrared light decreases over six times drawing to EFA fibers.

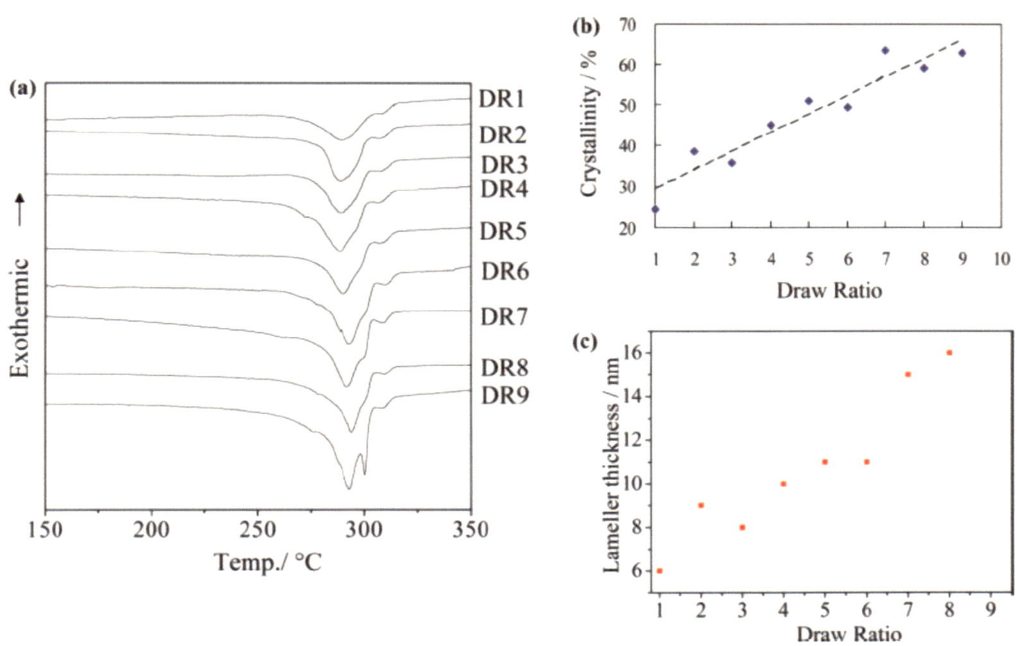


Fig. 15. (a) DSC thermograms of drawn EFA POFs at several ratios (scanning rate, 10 °C min-1). (b) Plot of drawn ratio vs crystallinity of drawn EFA fibers at several ratios. (c) Plot of lamellar thickness vs crystallinity of drawn EFA fibers at several ratios.

Figure 16 shows the schematic illustrations of the hierarchical structures ranging from the lamellae on the nanometer scale to the crystal structure on the subnanometer scale of a transparent EFA fiber.[37] We suggested that the crystal structure of the crystalline fluorinated polymers such as PTFE, EFA, PFA, and so on, form the orthorhombic system. The crystalline fiber of EFA had a herringbone arrangement in lamella when it was drawn over five times. Upon further drawing, the density in the amorphous region increased gradually. However, the overall differences in electron density between the crystalline and the amorphous regions were almost invariable. Probably, the progression of further transparency and the ability of light conductivity were brought about by a reduction in the difference in density. As an ideal type of extremely transparent crystalline fiber, the formation of a fringed micelle-type lamella arrangement may be desirable because of the low differences in densities inside the fibers.

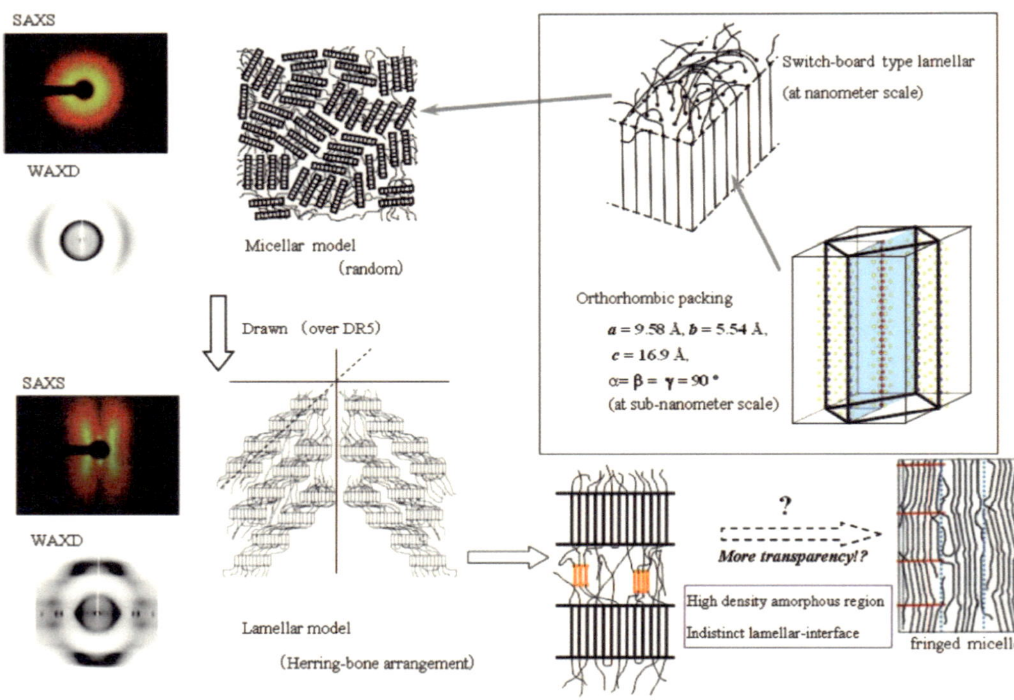


Fig. 16. Schematic illustrations of hierarchical structures from lamellae on the nanometer scale to crystal structure on the subnanometer scale of EFA POF.

Finally, in order to estimate three-dimensional structural formation, SAXS and WAXD measurements from the several incident direction of piled up crystalline EFA FOWs were carried out by using annealed DR=3 sample. Figure 17 shows SAXS and WAXD patterns of EFA FOWs at through, side, and edge direction. At the side-direction, obscure four-point SAXS pattern with void scattering and WAXD fiber pattern were confirmed. In the case of edge-direction, SAXS patterns show only void scattering, and WAXD indicate isotropic Debye ring. From the results of these measurements, schematic illustration of three-dimensional lamella arrangement was shown in Fig. 18. In this case, according to our previous work,[23, 38, 39] "switch-board" type lamella was adopted as structural units. From the view of through and side direction, two-dimensional stacked lamella arrangement forms the "herring-bone" arrangement. However, randomly isotropic structure is observed from edge direction. That is to say, lamella in the drawn EFA films formed uniaxially cylindrical symmetric arrangement. In the case of using this type EFA film as FOWs, it supposes that anisotropy of light conductivity direction occur. Along the through and side direction, visible and infrared light will be efficiently conducted, while edge direction will impede the transmission of lights. Figure 19 shows quantitative data of the transparency of the undrawn EFA film and drawn films by using UV-isible spectrometer. Because a "crystalline" FOWs made by fluorinated polymers efficiently transports infrared light, the λ= 850 nm of wavelength is adopted in this estimation. The film thickness is normalized by 500 μm. The transparency of infrared light in this film linearly increases with drawing ratio in both cases of films with drawing at 200 °C and fixed annealing at 280 °C after drawing. However, transparency of films treated by fixed annealing method is always inferior to that of films drawn at 200 °C only. This result is based on the difference of electron density between crystal and amorphous region. Probably, fixed annealing contributes acceleration of transition from a part of amorphous region to the crystal region. Crystallization of amorphous parts brings about formation of lower density amorphous region. As a result, difference of density between crystal and amorphous region become large and transparency of films decreases.

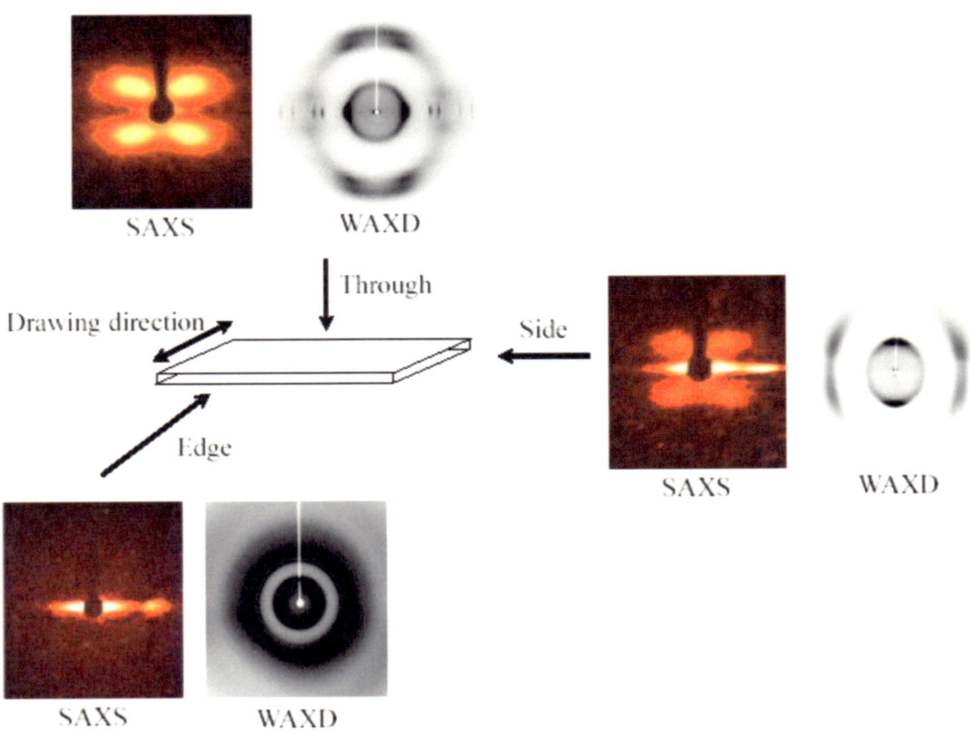


Fig. 17. SAXS and WAXD patterns of drawn EFA FOWs (fixed annealing at 280 °C after drawing at 200 °C) with through, side, and edge direction.

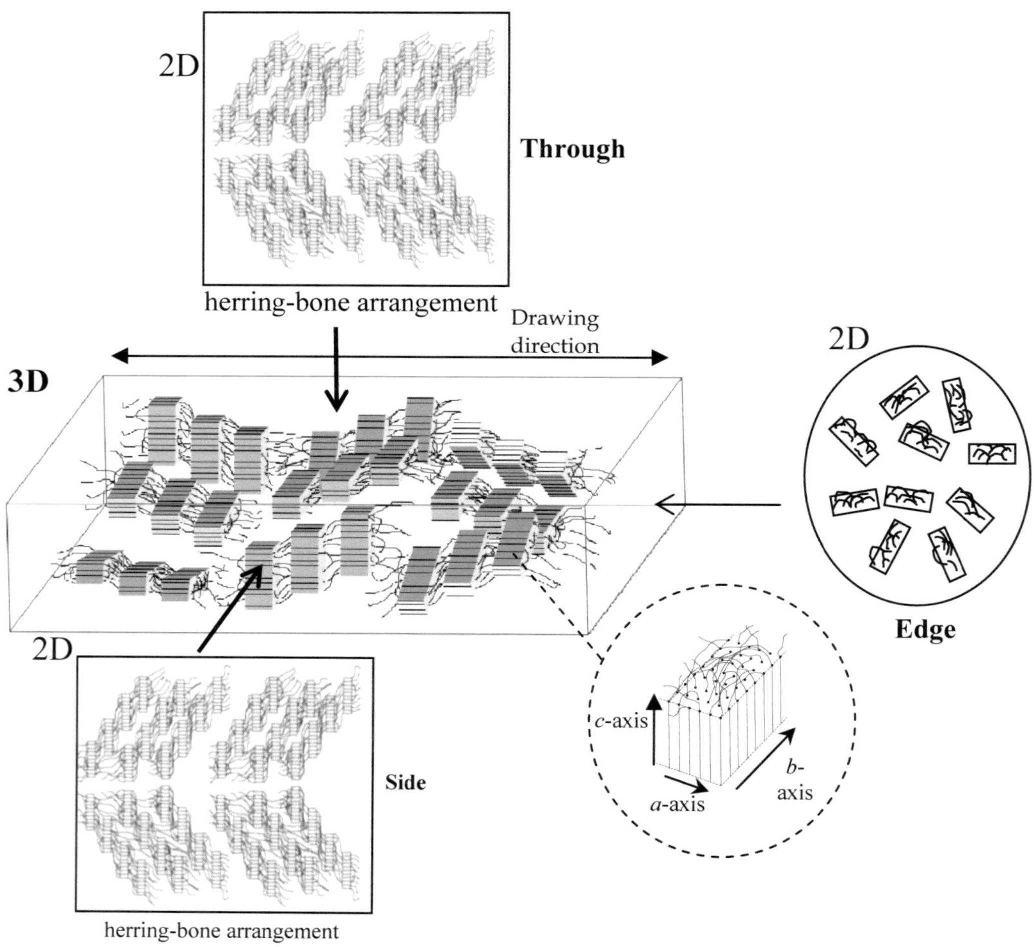


Fig. 18. Illustration of stacked lamellar in drawn EFA FOWs (fixed annealing at 280 °C after drawing at 200 °C).

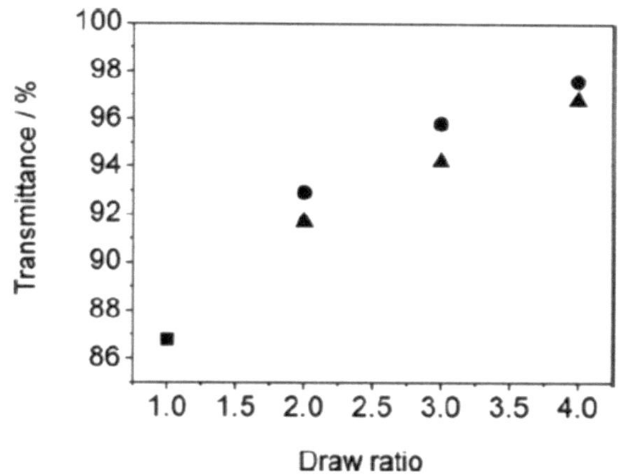


Fig. 19. Plots of drawing ratio *versus* transparency of infrared right (λ = 850 nm) : ■, undrawn ; •, drawn at 200 °C; ▲, fixed annealing at 280 °C after drawing at 200 °C.

## 4. Conclusion

The changes in fine structure upon drawing transparent crystalline EFA fibers and films were investigated by WAXD and SAXS measurements. EFA was crystallized as a lamella crystal in the POFs and FOWs although the polytetrafluoroethylene homopolymer itself usually forms extended-chain crystals. EFA exhibited thicker lamellae (thickness: at least 27 nm) as observed by the SAXS measurement. In this type of crystalline fluorinated copolymers, we considered the formation of a switchboard-type lamellae model according to Flory's suggestion. With an increase in the drawing of the fibers and films, four-point SAXS diagrams developed in the photograph of EFA transparent fibers, which implied that a particular type of layer structure, the alternately tilted lamella arrangement known as the herringbone, was formed. Furthermore, it was found that the previously proposed packing mode of general fluorinated polymers was required to be reconsidered from quasi-hexagonal to orthorhombic in a reciprocal lattice in order to assign all the reflective indexes obtained by using high-resolution WAXD measurements. Furthermore, the orientation of the crystallite and the elongation of lamella of EFA were completed simultaneously in the drawn fibers. The quasi-crystallization process progressed upon further drawing up to five times. After that, an increase in the density of the amorphous region and a partial appearance of voids probably occurred simultaneously. The crystalline EFA fiber at DR5 exhibited excellent transparency and infrared light conductivity. The light transmission property was related closely to the lamella arrangement, crystal structure, and difference in the crystalline/amorphous density of crystalline EFA optical fibers and optical waveguide.

## 5. References

[1] Keller, A., *Phil. Mag.*, 1957, *2*, 1171.
[2] Till, P. H., *J. Polym. Sci.*, 1957, *24*, 301.
[3] Fischer, E. W., *Z. Naturforsch*, 1957, *12*(a), 753.
[4] Burdon, J.; Tatlow J. C.; "*Advances in Fluorine Chemistry*," Vol.1 (eds. M. Stacey, J. C. Tatlow, A. G. Sharp, Academic Press, New York, 1960), pp.129–165.

[5] Patrick, C. R.; Stacey, M.; Tatlow, J. C.; Sharpe, A. G., *"Advances in Fluorine Chemistry,"* Vol. 2, Butterworths Publications Ltd., London, 1961, pp.1–34.
[6] *"Fluoropolymers 2"* in Topics in Applied Chemistry, eds. by Hougham, G., et al. (Kluwer Acad. / Plenum Pub., New York, 1999).
[7] Symons, N. K. J., *J. Polym. Sci., A,* 1963, *1*, 2843.
[8] Rahl, R. J.; Evanco, M. A.; Frendericks, R. J.; Reimschuessel, A. C., *J. Polym. Sci. A-2*, 1972, *1*, 1337.
[9] Ozawa, T., *Bull. Chem. Soc. Jpn.*, 1984, *57*, 952.
[10] Marega, C.; Marigo, A.; Causin, V.; Kapeliouchko, V.; Nicoló, E. D.; Sanguineti, A., *Macromolecules*, 2004, *37*, 5630.
[11] Marega, C.; Marigo, A.; Garbuglio, C.; Fichera, A.; Martorana, A.; Zannetti, R., *Macromol. Chem.*, 1989, *190*, 1425.
[12] Marega, C.; Marigo, A.; Cingano, G.; Zannetti, R.; Paganetto, G., *Polymer*, 1996, *37*(25), 5549.
[13] Lee, J. C.; Namura, S.; Kondo, S.: Abe, A., *Polymer*, 2001, *42*, 8631.
[14] Overney, R. M.; Meyer, E.; Frommer, J.; Brodbeck, D.; Luthi, R.; Howald, L.; Güntherodt, H. J.; Fujihira, M.; Takano, H.; Gotoh, Y., *Nature*, 1992, *359* 133.
[15] Overney, R. M.; Meyer, E.; Frommer, J.; Güntherodt, H. J., *Langmuir*, 1994, *10*, 1281.
[16] Fujimori, A.; Shibasaki, Y.; Araki, T.; Nakahara, H.,*Maclomol. Chem. Phys.*, 2004, *205*, 843.
[17] Fujimori, A.; Araki, T.; Nakahara, H., *Chem. Lett.*, 2000, 898.
[18] Burn, C. W.; Howells, E. R., *Nature*, 1954, *18*, 549.
[19] Koike, Y., Polymer, 1991, *32*, 1737.
[20] Koike, Y.; Naritomi, M.; Japan Patent 3719733, US Patent5783636, EU Patent 0710855, KR Patent, 375581, CN Patent ZL 951903152, TW Patent 090942, 1994.
[21] Ishigure, T.; Kano, M.; Koike, Y., *J. Lightw. Technol.*, 2000, *18*, 178.
[22] Nam, P. H.; Ninomiya, N.; Fujimori, A.; Masuko, T.; *Polym. Eng. Sci.*, 2006, *46*(6), 703.
[23] Fujimori, A.; Hasegawa, M.; Masuko, T., *Polym. Int.*, 2007, *56*, 1281
[24] Flory, P. J., *J. Am. Chem. Soc.*, 1962, *84*, 2857.
[25] Tanaka, K.; Seto, T.; Hara, T.; Tajima, Y., *Rep. Prog. Polym. Phys. Jpn.*, 1964, *7*, 63.
[26] Kaji, K.; Mochizuki, T.; Akiyama, A.; Hosemann, R., *J. Mater. Sci.*, 1978, *13*, 972.
[27] Samuels, R. J., *J. Macromol. Sci.*, 1970, *701*, 241.
[28] Butler, M. F.; Donald, A.N., *J. Appl. Polym. Sci.*, 1998, *67*, 321.
[29] Stribeck, N.; Sapoundjieva, D.; Denchev., Z.; Apostolov, A. A.; Zachmann, H. G.; Stamm, M.; Fakirov. S.; *Macromolecules*, 1997, *30*, 1339.
[30] Hernández, J. J.; Gracía-Gutiérrez, M. C.; Nogals, A.; Rueda, D. R.; Sanz. A.; Sics, I; Hsiao, B. S.; Roslaniec, Z.; Broza, G.; Ezquerra, T. A., Polymer, 2007, *48*, 3286.
[31] Sperati, C. A., Starkweather, H. W. Jr., *Adv. Polym. Sci.*, 1961, 2, 465.
[32] Burn, C. W., Cobbold, A. J., Palmer, R. P., *J. Polym. Sci.*, 1958, 19, 365.
[33] (a) Clark, E. S. Muus, L. T., *Z. Krist.*, 1962, *117*, 119, (b) Clark, E. S., Muus, L. T., *Z. Krist.*, 1962, *117*, 108.
[34] Klug, H. P.; Alexander, L E., *X-ray Diffraction Procedures*, John Wiley and Sons, New York, 1974.
[35] (a) Hermans P. H.; Platzek, P., *Kolloid Z.*, 1939, *88*, 68, (b) Hermans, J. J.; Hermans, P. H.; Vermaas, D.; Weidinger, A., *Rec. Trav. Chim, Pays-Bas*, 1946, *65*, 427.
[36] Wilchinsky, Z. W., *J. Appl. Phys.*, 1959, *30*, 792.
[37] Hayasaka, Y.; Fujimori, A., *Trans. Mater. Res. Soc. Jpn.*, 2008, *33*, 83-86.
[38] Fujimori, A.; Hayasaka, Y,, *Macromolecules*, 2008, *41*, 7606.
[39] Fujimori, A.; Numakura, K.; Hayasaka, Y, *Polym. Eng. Sci.*, 2010, *50*, 1295.

# 16

# Design and Characterization of Single-Mode Microstructured Fibers with Improved Bend Performance

Vladimir Demidov, Konstantin Dukel'skii and Victor Shevandin
*S.I. Vavilov Federal Optical Institute, St. Petersburg*
*Russia*

## 1. Introduction

Over the last few years, clear progress has been made in research and development of single-mode optical fibers with a large core (when core diameter exceeds 10 μm). Such advances were stimulated essentially by growing requirements for means of high power laser radiation transmission. The urgent problem of laser beam delivery lies in the necessity of the primary Gaussian power distribution of light inherent to many laser sources to be maintained without both temporal and spatial distortions. So optical fibers that support only a single transverse mode prove to be the most appropriate technique for efficient light transfer in production areas of complex or compact architecture. But there are still a number of limitations to cope with. For instance, as the power density of generated laser beams increases, the fiber core has to be expanded adequately in order to minimize the impact of undesirable nonlinear effects such as Raman scattering, Brillouin scattering and self-phase modulation. Moreover, fiber material will exhibit irreversible breakdown if the power level equals or exceeds the critical damage threshold.

Conventional single-mode fibers with step-index or graded-index refractive index profile can be acceptably adapted for the realization of large cores. However, the core dimensions enlargement permanently results in the reduction of the refractive index difference between the core and the cladding (Δn). This, in turn, affects adversely the numerical aperture of the fiber (NA), which then has to be reduced twice from its standard values of larger than 0.1 to achieve core diameters of approximately 15 μm at a wavelength around 1 μm (Tunnermann et al., 2005). Such NA lowering weakens considerably the fiber waveguiding so the optical fiber becomes very sensitive to various perturbations, especially to bending effects. Further decrease of NA will require keeping the uniformity of the core refractive index in the vicinity of $10^{-4}$ - $10^{-5}$. It is technologically unattainable when using chemical vapor-phase deposition methods for the fiber preform fabrication.

An alternative flexible approach to solve this challenge is based on exploiting unique wave guiding properties of microstructured optical fibers (MOFs), also known as photonic crystal fibers or holey fibers. MOF design can relatively easily provide extended cores and hence large effective mode areas that nowadays reach values of even thousands of $\mu m^2$. This phenomenon perfectly coordinates with the ability to manage accurately the effective Δn value at a level of as low as 0.0001 or less. Furthermore, MOFs, as opposed to single-mode fibers of a conventional design, can assure robust fundamental mode propagation over a broad wavelength range within the transparency window of silica. The only restriction has to be taken into account while manufacturing microstructures with a large core relates to reasonable control over spectral position of bend-loss edge crucial for the fiber application potential.

Typical silica-based MOF structure is defined by a certain number of air holes arranged in a regular triangular lattice running along the entire length of the fiber (Knight et al., 1996, 1997). One missing central hole filled with a glass introduces a defect in the lattice, acting as the guiding core with the refractive index of undoped fused silica. Holes surrounding the core area serve as the cladding with the effective refractive index lower than that of the core due to presence of the air. Provided that the degree of air content expressed commonly by the k-parameter (i.e. the ratio of the air hole diameter d to the lattice pitch $\Lambda$) does not exceed 0.45 (Mortensen, 2002), the fiber supports a single transverse mode for any wavelength (the endlessly single mode regime). The most suitable manner of core expansion is scaling of cross-sectional fiber dimensions without changes in the given lattice structure. But on the understanding that the fundamental mode acts as the leaky one due to bending in the short-wavelength region, the position of bend-induced leakage boundary moves to longer wavelengths while increasing the core size (Nielsen et al., 2004b). Consequently, the spectral operation range steadily narrows that provokes fibers to be allocated on spools of greater radiuses. This prevents MOFs from being widely exploited in industrial laser or beam delivery applications with standardized curve parameters.

In this work we have concentrated our efforts on finding and implementation of a few novel MOF designs that could effectively combine the large core dimensions and the expanded spectral operation range as compared to classical MOFs. It is obvious that new structures should be actualized by applying principles different from the basic concepts of the standard MOF technology. Here we will focus on two special approaches: 1) competent manipulation of $\Lambda$-parameter in the selected wavelength region; 2) ensuring proper fiber conditions for the establishment of a substantial difference in attenuation coefficients of the fundamental ($LP_{01}$) and the higher order ($LP_{11}$) modes (differential modal attenuation).

## 2. MOFs with a multi-element core

### 2.1 Background on large-core structure design

MOF guides light along the core via the modified total internal reflection mechanism similar to that of conventional single-mode fiber (Knight, 2003). However, in some cases MOF can demonstrate specialty features that have no analogies in conventional waveguide theory.

Our previous investigations (Dukel'skii et al., 2005, 2006) indicate that the lattice pitch $\Lambda$ (in general, the ratio of the wavelength $\lambda$ to the pitch $\Lambda$) is the most significant parameter responsible for such main optical property as the capability of light confinement. For MOFs with air holes assembled in a triangular pattern we revealed the discrete transition between the availability of wave guidance and the lack of it. The exact position of this transition strongly depended on the $\Lambda$ value.

Mentioned effect appeared in three forms: 1) absence of light canalization in short-length samples (Dukel'skii et al., 2005); 2) appreciable increase in attenuation coefficient of multimode samples with $k \geq 0.8$ in the short-wavelength region of the spectra (Dukel'skii et al., 2006); 3) intensive short-wavelength leakage of the fundamental mode power into the outer fiber cladding in single-mode samples (Nielsen et al., 2004b).

For example, we observed straight fiber segments of a length from 2 to 5 centimeters in a microscope under white light launched in each sample at arbitrary angles. When Λ, defined as the hole-to-hole spacing in the first air-hole ring, exceeded ~ 10 μm, samples were characterized by the weakening of waveguiding properties in the visible part of the spectra (Dukel'skii et al., 2005). To be more precise, the core ceased to canalize light and the distribution of light over the fiber cross-section became totally uniform. No changes in the launch fiber conditions could modify that uniformity. Ultimately, the samples with $k \geq 0.8$ supported light propagation, whereas the samples with $k < 0.6$ diverged.

It is a well-known fact that conventional optical fiber made of glass materials with different refractive indexes (as well as corresponding fiber preform) can guide light irrespective of its transverse dimensions. The core diameter enlargement leads only to the increase in amount of excited modes, but not to the absolute lack of waveguiding properties. MOF technology does not imply light guidance of the initial capillary stack (preform) due to its particular structure dissimilar to the resultant MOF design. At the same time, the ability of the fiber drawn from the stack to guide light strongly depends on the transverse microstructure dimensions (d and Λ).

The next interesting phenomenon intrinsic to MOFs with a triangular cladding structure we detected after the long fiber samples (about 100 meters in length) had been investigated concerning the optical loss measurements (Figure 1).

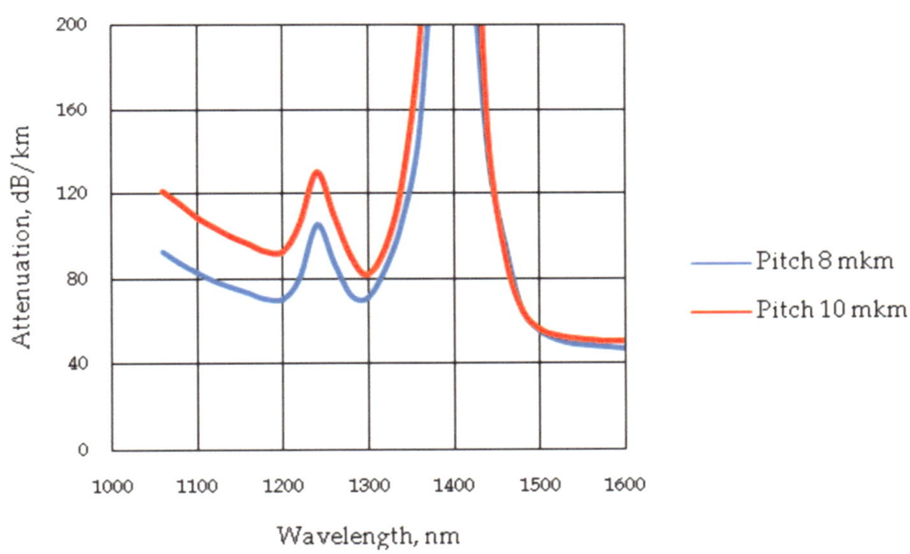


Fig. 1. Spectral attenuation pattern of the MOFs with $k \geq 0.8$ depending on the Λ-parameter (Dukel'skii et al., 2006).

As Figure 1 shows, the increase in optical losses occurred according to the fiber diameter growth, especially in the short-wavelength part of the spectra. Evidently, the Λ-parameter increased in direct proportion to the outer fiber diameter expansion. So we had the same situation as described above: the increase in hole-to-hole spacing definitely impaired wave guidance. The effect was observed while handling only multimode fiber samples with $k \geq 0.8$ and was not connected with macrobending or microbending losses.

The third aspect of Λ manipulation included the fact that abovementioned short-wavelength leakage of the fundamental mode power can dramatically enhance when core diameter rises from 20 to 35 μm (Nielsen et al., 2004b). In practical applications this unfavourable phenomenon forces MOFs to be placed on spools with augmented radiuses, extended, for example, from normal radius of 8 centimeters for communication fiber up to non-standard radius of 16 centimeters. The enhancement of optical power leakage can be formally interpreted by the impact of the Λ-parameter that proportionally increases with the fiber core diameter enlargement. So selecting suitable Λ for a given spectral region, we could control the position of modal leakage boundary.

Thus, we found out that qualified adjustment of Λ-parameter in MOFs characterized by a large core size (up to 35 μm) can be promising for the purpose of deriving a set of special properties. However, aids and concepts for achievement and implementation of these features are not trivial. One of the feasible ways to improve light confinement is to comprise the core of several elements. In this case the core becomes bigger in comparison with the core of a typical MOF structure, although the outer fiber diameter remains permanent. On the basis of geometrical considerations, it is possible to substitute not one, as ordinary, but seven or nineteen central capillaries in the initial triangular array for one solid rod made of the same material as the fiber cladding. By means of the substitution method the values of Λ-parameter can be reduced by two (7 central capillaries) or three (19 central capillaries) times in the resultant MOF as compared to a standard 1-element-core analog of the same core size.

It should be noted that there are several publications (Limpert et al., 2005, 2006) reporting on development of 7- and 19-element-core MOFs for generation of high power laser radiation, but no detailed information about modal consistence or bending performance is provided. Superior theoretical analysis (Saitoh et al., 2005) shows that the endlessly single-mode regime of operation for the 7-element-core MOF can be realized under the higher order mode cut-off condition $k < 0.046$. Apparently, implementation and multiple reproduction of such tiny structure correspond with huge technological difficulties. Furthermore, it seems even more difficult to carry out practically the condition for the single-mode operation of the 19-element-core MOF, which is expected to be extremely low (continuing a row for the phase higher order mode cut-off condition for the 1-element-core structure $k < 0.45$ and for the 7-element-core structure $k < 0.046$).

So taking into account the effect Λ-parameter has on the capability of light confinement, we stated a goal of creating a family of single-mode MOF structures with a large multi-element core that will not be subjected to strong influence of macrobending losses.

### 2.2 MOFs with the core comprised of 7 missing holes

In the first stage of our research we have successfully produced a series of MOFs with the core design presented in Figure 2. The arrangement included the 7-element core area and five air-hole rings organizing the light-reflecting cladding. All experimental samples were drawn from capillary stacks using commercially-available synthetic silica tubes and rods with OH-content in concentration of several ppm. The required value of k-parameter was obtained by the appropriate adjustment of capillary pressure in the high temperature zone of the stack during the drawing process. The outer surfaces of the elements were purified beforehand in order to reduce the influence of mechanical contaminations which content, however, in some cases was quite uncontrollable due to holding the whole technological process in normal laboratory conditions.

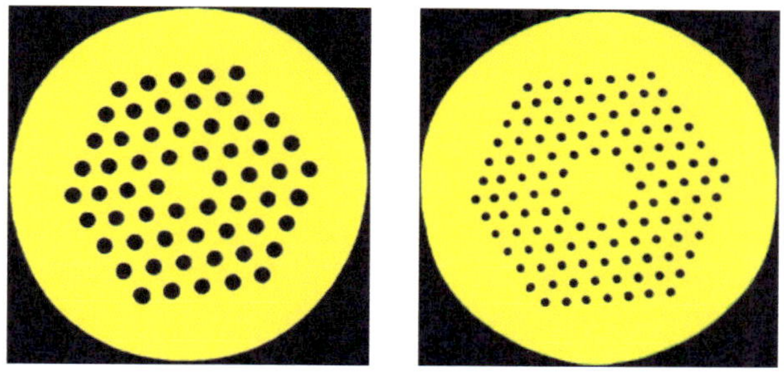

Fig. 2. Microscopic images of 1-element-core (left) and 7-element-core (right) MOF structures.

### 2.2.1 Core diameter of 20 μm

We started with the core size that allows a typical MOF to be single-moded over a relatively wide wavelength range in the near infrared part of the spectra. The comparison of attenuation coefficients between the 7-element-core MOF with k = 0.40 and the standard 1-element-core analog (LMA-20 produced by Crystal Fibre A/S) with k ~ 0.49 is presented in Figure 3. Optical loss measurements were made for a bending diameter of 16 centimeters.

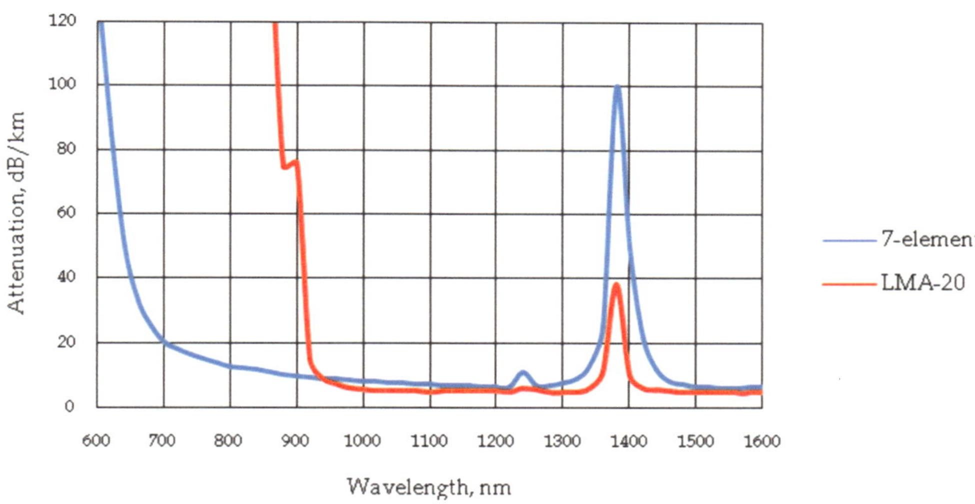


Fig. 3. Spectral attenuation pattern of 7-element-core and 1-element-core (Nielsen et al., 2004b) MOF structures.

Despite the fact that the decrease in the Λ value from 13.2 μm (Nielsen et al., 2004b) to 6 μm in our fiber has a positive effect on the position of bend-loss edge which shifts from 900 to approximately 650 nm, there is no difference in the spectral attenuation behavior between two MOF structures under study. In other words, both curves are smooth and have a low growing tendency while moving to shorter wavelengths. Moreover, they display identically the dramatic increase in attenuation coefficient due to leakage conditioned by the stationary bending radius. Two spectral peaks associated with wavelengths 1250 and 1380 nm apply naturally to hydroxyl groups absorption.

We have determined that no higher order mode cut-off can be identified. In the case of conventional single-mode fiber (for instance, SMF-28) higher order mode cut-off appears as an abrupt power decrease in attenuation curve, so far as $LP_{11}$-mode leaks intensively in the spectral region close to cut-off wavelength. Noted decrease can be described by the exact value of 4.8 dB (Jeunhomme, 1983) determined by either presence or absence of $LP_{11}$-mode radiation at the output end of the fiber. The value of 4.8 dB does not depend on the fiber length and should be observed undoubtedly in the fiber under investigation, since total light attenuation of the measured piece normally does not exceed 10 dB.

A slight decrease in the core size from 20 to 18 μm enables an additional absorption peak due to non-bonding oxygen to be clearly revealed at wavelength λ = 630 nm (Figure 4). One may notice further move of the short-wavelength bend-induced leakage boundary to the ultraviolet part of the spectra. The value of k-parameter can also affect the position of macrobending loss edge owing to a greater or lower contrast between the core and the cladding effective refractive indexes. In any case air-filling fraction needs to be controlled accurately because in some cases it may lead to multimode regime of operation.

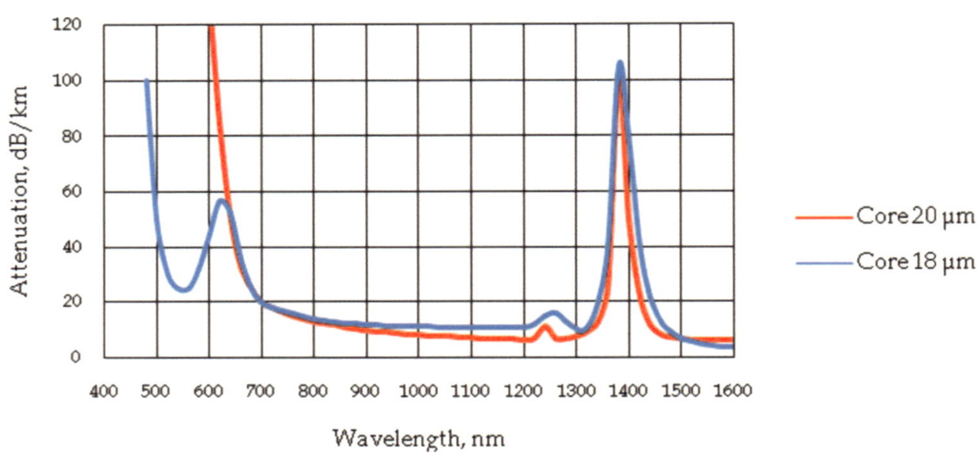


Fig. 4. Dependence of the optical losses on the decrease in the MOF core diameter.

Theoretically, the endlessly single-mode MOF is realized when k < 0.45 (Mortensen, 2002) in case the core is represented by one missing air hole. Referred fiber (Nielsen et al., 2004b) performs the lattice structure with k ~ 0.49 that is close enough to the endlessly single-mode regime condition. The fiber is definitely interpreted to be the single-mode one over the entire spectral range studied from ~ 900 to 1600 nm (Figure 3, red curve). So we suppose that the MOF with the 7-element core of 20 μm in diameter and k-parameter equal to 0.40 is also single-mode in consequence of the absence of power drops in smooth attenuation spectra from 650 to 1600 nm. Nevertheless, we have investigated roughly modal properties of the fabricated fiber. To this effect, we launched radiation from He-Ne laser (λ = 633 nm) into the sample and then observed visually typical Gaussian far-field intensity distribution on the screen placed at the distance of approximately 10 centimeters far from the fiber output. In addition, the rated value of half-divergence angle was equal to several hundredths of a radian that completely corresponded with the fundamental mode operation as well.

Those first raw results inspired us to carry more profound analysis of the 7-element-core MOF structures.

### 2.2.2 Core diameter of 25 μm

To verify the preceding assumption of the single-mode behavior of radiation propagated along the 7-element-core MOF with k = 0.40, extra procedures were carried out approaching the opposite to λ = 633 nm part of the spectral range. Those actions were addressed towards a new set of fibers with the core size of 25 μm in diameter and k-parameter ranging from 0.19 to 0.50.

Figure 5 shows the quasi-single-mode character of the spectral attenuation curve regardless of the k-parameter value. All presented spectra do not contain any noticeable peaks of the higher order mode cut-off. Besides that, there exists pronounced short-wavelength leakage boundary specified by the fiber placement on a spool with a diameter of 16 centimeters. One can also see that the position of this boundary depends directly on the k-parameter: the larger the k value the shorter the wavelength of bend-loss edge. Basic levels of attenuation are defined by various degree of the initial stack purification.

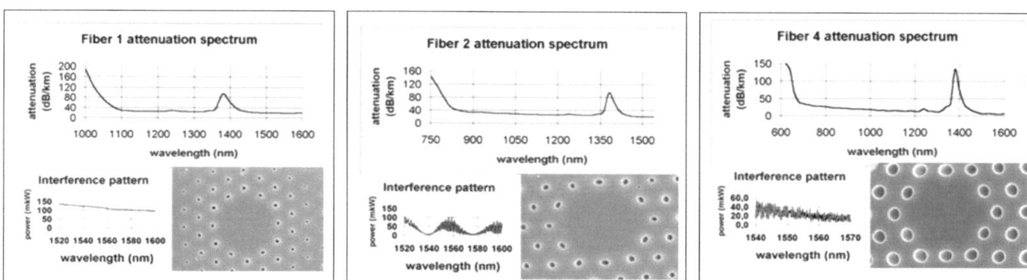


Spectral attenuation and modes beats in the fiber with k = 0.19

Spectral attenuation and modes beats in the fiber with k = 0.27

Spectral attenuation and modes beats in the fiber with k = 0.50

Fig. 5. Dependence of the optical losses on the MOF geometrical parameters.

The fibers, presented in Figure 5, having a length from 20 to 100 meters were investigated by means of the modal beats method. For this purpose, spectrally narrow radiation of tunable semiconductor laser (λ = 1520 - 1580 nm) was launched into a piece of SMF-28 connected with the target MOF sample. The output signal passed through a reciprocal piece of SMF-28 to PT2010 optical power meter, resulting in the modal beats pattern, registered as a change in the power distribution of light. Thus, the output signal could present the beats between two or more guided modes, i.e. intermodal interference (Figure 5, central, right), or the absolute lack of the beats in the single-mode regime (Figure 5, left). We controlled the validity of the modal beats method by comparing two schemes of the experiment in the selected spectral region. The subsidiary technique consisted in visualization of infrared radiation (λ ~ 1550 nm) on a special screen yielding an implicit coincidence with the data given by the main scanning method.

It should be noted that in spite of the smooth character of all spectral attenuation curves, given in Figure 5, and the absence of power drops corresponding to the higher order mode cut-off, the modal beats method has designated clear pattern of modal interaction. Those new results have confirmed the uncertainty of the single-mode behavior of the 7-element-core structures with the core diameter of 20 μm, examined in the previous paragraph.

In the case of excitation of the second order mode the output signal can be described according to the following expression:

$$I = A^2 + B^2 + 2AB\cos(\omega \Delta n_{eff} L / c), \quad (1)$$

where A and B are the amplitudes of $LP_{01}$ and $LP_{11}$ modes respectively, ω is the circular frequency, $\Delta n_{eff}$ is the difference between effective mode indexes, L is the sample length and c is the speed of light in vacuum.
We have applied the discrete Fourier transform and, as a result, have determined the spatial frequencies v corresponding to the peaks of the interference pattern. Then we have calculated the effective index difference between the fundamental and the excited second order mode by applying formula:

$$\Delta n_{eff} = v\lambda^2 / L. \quad (2)$$

The experimental data are summarized in Table 1.

| Fiber | k-parameter | Length, m | v, $nm^{-1}$ | $\Delta n_{eff}$ |
|---|---|---|---|---|
| 1 | 0.19 | 19 | No beats | – |
| 1a | 0.19 | 0.8 | 0.06 | 0.0002 |
| 2 | 0.27 | 2.0 | 0.8 | 0.00096 |
| 3 | 0.34 | 1.6 | 0.7 | 0.00105 |
| 3a | 0.34 | 3.2 | 1.5<br>3.2 | 0.0010<br>0.0024 |
| 4 | 0.50 | 2.8 | 1.3<br>1.8<br>4.0<br>4.8 | 0.0011<br>0.0015<br>0.0034<br>0.0041 |

Table 1. Dependence of the effective mode index difference between the fundamental and the second order mode on the MOF structure (Agruzov et al., 2008).

As the experimental data show, vast majority of the MOF samples are characterized by the propagation of at least two guided modes. The value of the effective mode index difference $\Delta n_{eff}$ ~ 0.001 exists for all presented samples regardless of the k-parameter. Only exception to the general tendency is Fiber 1 with k ~ 0.19, described by the absolute lack of the modal beats even when resolution capacity of registering system is the order of magnitude higher than the common level enough for the clear interference pattern visualization in all other cases. A piece of Fiber 1 having a length of 19 meters is single-moded, while a shorter piece of the same fiber (Fiber 1a) demonstrates the presence of the second order mode at the output end. So the higher order mode can be characterized by the essentially greater attenuation coefficient than the fundamental one.
Further investigations aimed at the determination of the spectral operation width of Fiber 1. But the attempt to obtain the pattern of far-field intensity distribution utilizing the available light source (He-Ne laser) failed: laser radiation intensively leaked away from the core area and filled the entire cladding even in the piece of about 1 meter in length. All other fibers from the list of Table 1 have demonstrated the modal interference pattern at λ = 633 nm: power distribution of light depended on the input fiber geometry and on the conditions of

light propagation along the MOF. A bend or a mechanical stress influenced definitely the far-field pattern so the intensity peak moved from one part of the spot to another.
Reverting to the subject of the absence of $LP_{11}$-mode cut-off in the spectral attenuation patterns (Figures 3, 5), we should state that the higher order mode (or modes) exists simultaneously with the fundamental one in the spectral interval $\lambda$ = 600 - 1600 nm. Normalized frequency V is weakly dependent on wave number. That fact differs MOFs from other types of lightguides. So for the determined wavelength range V-parameter varies negligibly (Mortensen et al., 2003), preserving its magnitude almost invariable (Figure 6). Since V-parameter directly defines the amount of guided modes, a number of them can coexist persistently within a spectral range, specific for each fiber, and at the same time undergo the infinite attenuation by reason of the huge power leakage at the identical wavelength in the blue part of the spectra.

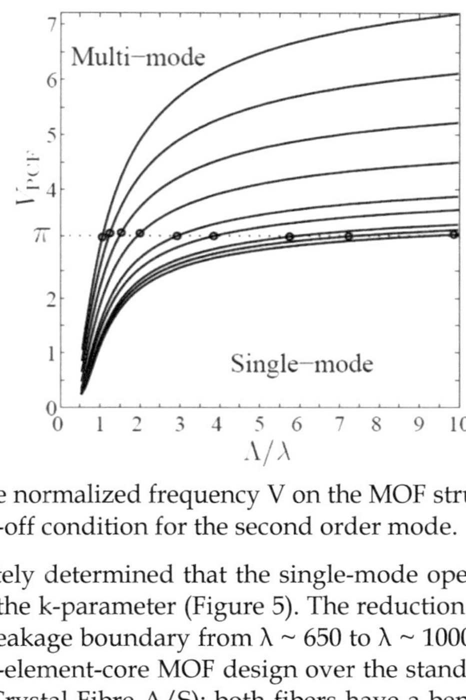


Fig. 6. Dependence of the normalized frequency V on the MOF structure (Mortensen et al., 2003). $V_{PCF} = \pi$ is the cut-off condition for the second order mode.

Finally, we have accurately determined that the single-mode operation can be achieved by the suitable selection of the k-parameter (Figure 5). The reduction of k-parameter causes the move of bend-induced leakage boundary from $\lambda \sim 650$ to $\lambda \sim 1000$ nm. Unfortunately, there is no preference of the 7-element-core MOF design over the standard 1-element-core analog (LMA-25 produced by Crystal Fibre A/S): both fibers have a bend-loss edge located in the wavelength region of 1 μm while being bent on a spool of 16 cm in diameter.

### 2.2.3 Core diameter of 35 μm

The improved situation takes place in case of the further core enlargement approaching 35 μm. It is a well-known fact that such great leap in the core diameter strongly affects the width of the spectral operation range (Nielsen et al., 2004b) and actually transforms the fiber into a 'single-frequency' optical element which is operable exclusively at $\lambda$ = 1550 nm. We have yielded some positive results in development of the bend-resistant MOF design having a core of 35 μm in diameter.
There are three fiber samples, presented in Table 2, made of the same initial capillary stack with the air-filling fraction variable within the range k = 0.2 - 0.4. The experimental data

displays the way how the reduction of k-parameter influenced the modal properties of light propagated. When k was about 0.4 the fiber was the multimode one both in red and infrared parts of the spectra. As the value of k-parameter decreased approximately twice, the fiber turned into the single-mode one, at first, in the visible part of the spectra and then in the infrared part.

| Cross-section structure | 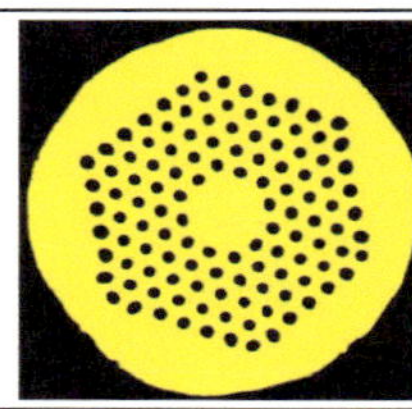 | | 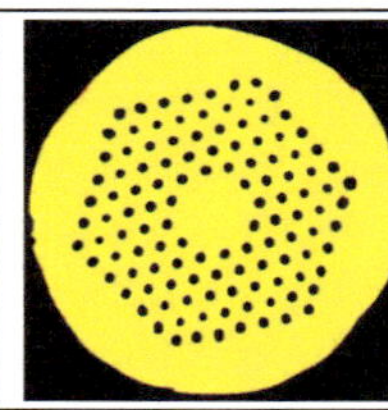 | | 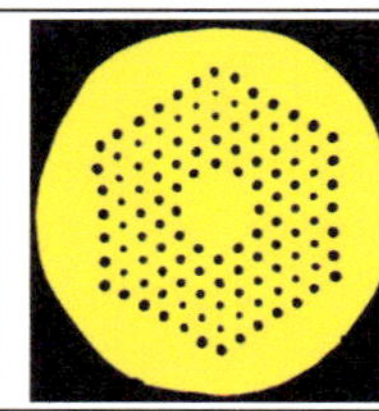 | |
|---|---|---|---|---|---|---|
| k-parameter | 0.4 | | 0.3 | | 0.2 | |
| Wavelength, nm | 633 | 1550 | 633 | 1550 | 633 | 1550 |
| Modal consistence | Multi | Multi | Single | Multi | Leakage | Single |

Table 2. Dependence of the modal properties on the k-parameter reduction.

Additional investigations of the fundamental mode spot size were carried out. For this purpose, conventional single-mode fiber with the core diameter of 8 μm was attached to the MOF. Under mutual end-facet scanning the Gaussian-like power distribution of light was obtained. The MOF was excited by laser radiation at $\lambda$ = 1550 nm. The results have shown that the mode spot size in the 7-element core structure is about 26 μm that corresponds equally to the 1-element core analog (Nielsen et al., 2003) despite the difference both in the amount of air holes, surrounding the core area (12 or 6), and in the k-parameter values (0.19 or ~ 0.49).

It is necessary to note that all presented MOF structures are not strongly single-mode ones if you keep in mind the existence of inherent to MOFs the endlessly single-mode regime of light propagation. For the 7-element-core microstructure this regime is provided at $k < 0.046$ (Saitoh et al., 2005). It is clear that the structure with such low value of k-parameter cannot be correctly realized in practice. Nevertheless, the single-mode operation can be carried out due to the significant difference in attenuation coefficients of the fundamental and the higher order modes.

For the benefit of such point of view, the behavior of the modal properties dependent on the exact k-parameter value testified (Table 2). As it may be seen, for the k = 0.3 in the red part of the spectra there existed only the fundamental mode, though in the infrared part we have observed several modes. This situation cannot be explained otherwise than by the strong attenuation coefficient of the higher order modes. In the case of conventional fiber made of the materials with different refractive indexes we observe the opposite tendency according to the expression (Snyder & Love, 1983):

$$V = 2\pi a\sqrt{n_1^2 - n_2^2}\,/\,\lambda\,, \tag{3}$$

where a is the core radius, $n_1$ and $n_2$ are the core and the cladding refractive indexes respectively. The decrease of normalized frequency V causes the reduction of the amount of excited modes, so in the red part there would be several modes and in the infrared part only the fundamental one. This dependence can also be retained in the case of MOFs with correction for the effective values of the refractive index (Russell, 2006).

Finally, we declare that the higher order mode undergoes considerably strong attenuation in the visible part of the spectra than the fundamental one. This assertion has the corresponding interpretation: the shorter the wavelength the larger divergence due to diffraction and/or leakage of the higher order mode into the gaps between the air holes (Russell, 2006). By means of varying k and Λ parameters it is possible to fit the proper conditions in which the fundamental mode attenuates slightly in comparison with the higher order mode. Here we must say that the latest statement is correct only for the determined wavelength range.

Spectral attenuation pattern (Figure 7) shows the obvious preference of the 7-element-core MOF design over the standard 1-element-core analog (LMA-35 produced by Crystal Fibre A/S): the position of bend-induced leakage boundary moves to the blue part of the spectra for about 100 nm. Comparison is considered for a bending diameter of 32 centimeters.

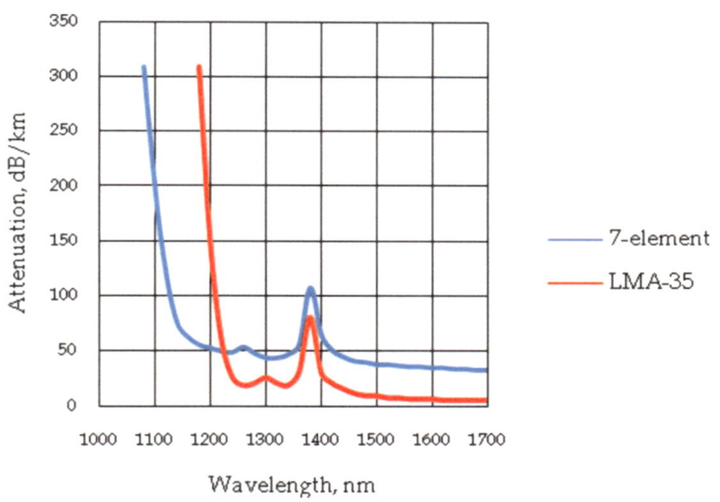


Fig. 7. Spectral attenuation pattern of 7-element-core and 1-element-core (Nielsen et al., 2004b) MOF structures.

### 2.3 MOFs with the core comprised of 19 missing holes

Even more impressive results have been obtained in the development of the MOFs with the core area formed by the initial substitution of 19 central capillaries in the original stack for one solid rod (Figure 8).

We have observed the similar situation for the 19-element-core MOFs with k = 0.3 as for the MOFs discussed in previous paragraph. If look carefully at the family of the spectral attenuation curves, presented in Figure 9, one can note that the increase of the core diameter leads directly to the shift of bend-induced leakage boundary of the fundamental mode to longer wavelengths within the spectral range explored. Also there is no evidence of the higher order mode cut-off in attenuation curves. All the fibers turned out to be the single-mode ones with the appreciable preference in the spectral operation range widening over the 7-element-core MOF structures and, what is more, over the 1-element-core analogs.

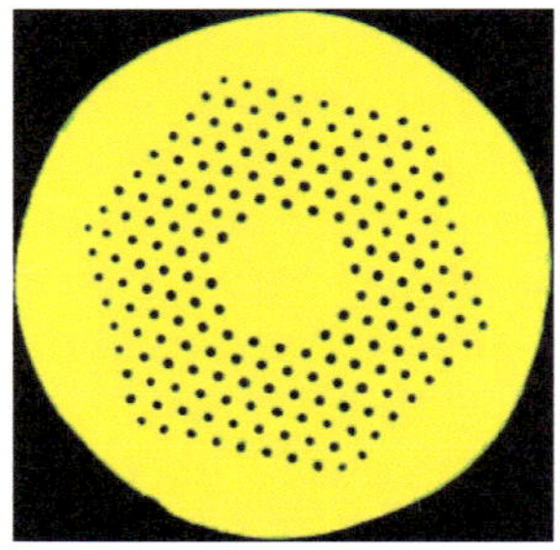

Fig. 8. Microscopic image of the 19-element-core MOF structure.

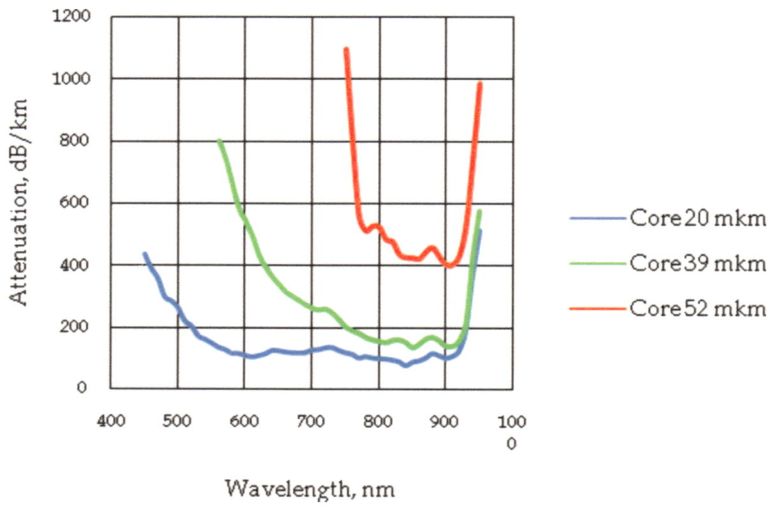


Fig. 9. Spectral attenuation pattern of the 19-element-core MOFs.

At the same time, solid analysis of the modal consistence of the 19-element-core MOFs have shown the complicated behavior of the higher order mode (Table 3).

| Fiber | Core diameter, μm | Short-wavelength leakage boundary, nm (Spool diameter 16 cm) | Length, m | k-parameter | Spool diameter 16 cm | | Spool diameter 32 cm | |
|---|---|---|---|---|---|---|---|---|
| | | | | | 633 nm | 1550 nm | 633 nm | 1550 nm |
| 1 | 45 | 1200 | 6 | ~ 0.2 | Leak. | 2 modes | 2 modes | |
| 2.1 | 27 | 600 | 2.5 | < 0.2 | 2 modes | 2 modes | | |
| 2.2 | 27 | 600 | 10 | < 0.2 | 1 mode | 1 mode | | |
| 2.3 | 27 | - | 25 | < 0.2 | 1 mode | 1 mode | | |

Table 3. Dependence of the modal properties of the 19-element-core structure on the MOF geometrical parameters (Agruzov et al., 2010).

In fact, all presented samples are the multimode ones in case of straight fiber segments so far as the 19-element core guides a few higher-order transverse modes under k > 0.2. However, both the controllable reduction of k-parameter and the fiber placement on a spool decrease the amount of the excited modes. The single-mode behavior of the MOF bent on a spool of 16 centimeters in diameter states stationary at the fiber lengths of more than 10 meters. Then the fiber demonstrates the improved bending resistance properties and hence excellent leakage characteristic. At shorter lengths the presence of the higher order mode radiation at the output end of the fiber is unavoidable.

Particularly, Figure 10 illustrates how successfully the Λ-parameter reduction can be carried out in the 19-element-core MOF (Fiber 2.2).

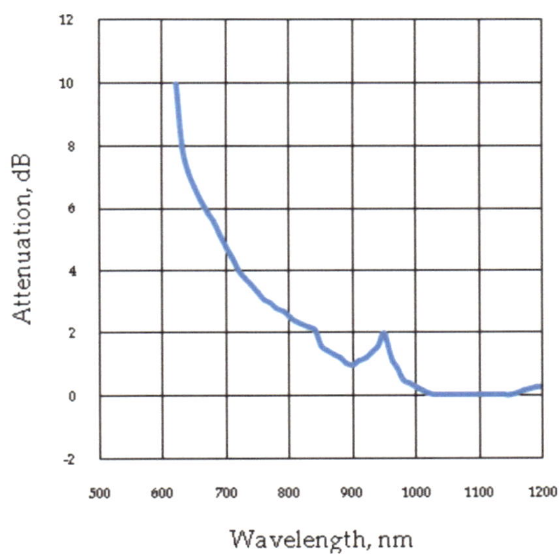


Fig. 10. Spectral attenuation pattern of Fiber 2.2 (k < 0.2).

The position of bend-loss edge is located in the red part of the spectra even when the fiber is placed on a spool of 16 centimeters in diameter. It is rather difficult to compare accurately how far the features of the 19-element-core MOF vary from the standard 1-element-core analog because of the lack of literary data. In any case the leakage characteristic is better than that of the 1-element core of 20 μm in diameter (Nielsen et al., 2004b), where optical losses increase dramatically at the wavelength λ ~ 900 nm. As in the case of the 7-element-core MOF, the single-mode propagation takes place when the k-parameter is equal or less than 0.2. For the larger k values the fiber definitely becomes the multimode one.

Our investigations (not included in Table 3) also have shown that straight pieces of Fiber 2.2 and Fiber 2.3 support several transverse modes at the lengths available in normal laboratory areas (up to 10 meters). This circumstance restricts the application potential of the 19-element-core MOFs, especially if it is necessary to obtain guaranteed single-mode regime. For example, this may take place in high power beam delivery systems, where radiation is passed from the stationary placed laser to the variable operation area. In such situation the modal consistence strongly depends on the fiber configuration. On the other hand, if one needs to transport the energy through the multibend sleeve of the fiber placed in production areas of complicated architecture, the priority of the 19-element-core MOF is evident.

### 2.4 Special attenuation mechanism

While investigating the modal properties of the aforementioned MOF structures with the large 7- or 19-element core, we have defined a specific mechanism when the attenuation coefficient of the higher order mode substantially exceeds the same parameter of the fundamental one. This phenomenon had the most decisive effect on the modal consistence of the MOFs with the core of 35 μm in diameter, allowing the fibers with $k < 0.2$, being bent on a standard spool with a bending diameter of 16 centimeters, to propagate only the fundamental mode within a broad spectral range $\lambda = 600 - 1550$ nm. Draw attention that, theoretically, the higher order mode cut-off condition for this fiber is expected to be technologically unfeasible (since for the 7-element-core fiber k-parameter, in theory, has to be as low as 0.046 to ensure the single-mode operation). So the great difference in attenuation coefficients of $LP_{01}$ and $LP_{11}$ modes enables the implementation of the 19-element-core MOF with the absolutely workable $k \sim 0.2$, operating in the single-mode regime (Fiber 2.2 and Fiber 2.3).

To determine the proportion of the attenuation coefficients, sufficient for the single-mode operation via the differential modal attenuation, we have estimated the optical losses of the higher order mode. For this purpose, we have measured the depth of modulation while registering several patterns of the modal beats. The length of the investigated Fiber 2.2 sample was varied deliberately to achieve distinct patterns (1 meter, 1.5 meters and 2 meters). This procedure has shown the inverse influence of the fiber length on the depth of modulation: less pronounced pattern corresponded to larger sample lengths. Thus, the attenuation coefficient of the higher order mode was evaluated to be ~ 5 dB/m. At the same time, the fundamental mode attenuation coefficient, measured by a cut-back technique, have been rated at a level of tens of dB/km. So now we can state that the fundamental mode propagation via the differential modal can be effectively realized in MOF structures when the higher order mode attenuation coefficient is of at least two orders of magnitude larger than the analogous parameter of the fundamental mode.

The next goal consisted in detailed analysis of the conditions opportune enough for the establishment of the differential modal attenuation. The implementation of microstructures with large cores, irrelevant to 7 or 19 elements, and great Λ-parameter values seemed to be the most applicable means to collect the data.

## 3. MOFs based on the differential modal attenuation mechanism

In the previous part of the work the special attenuation mechanism has been shown. It described the situation when the attenuation coefficient of the fundamental mode may be essentially lower than the same optical parameter of the higher order mode. Starting from that point, we have concentrated our efforts on the extensive research of a few novel MOF designs that could successfully correspond with a set of special requirements: appreciable difference in optical losses of the first two modes ($LP_{01}$ and $LP_{11}$), single-mode operation, high bending resistance and fiber placement on a standard transport spool of 16 centimeters in diameter. Among the others we have tested structures with the circular cladding distribution, with the special $C_{3V}$ cladding symmetry and with the 1-element shifted core. All of these MOF structures seemed to fit with the requirements, especially with the enforcement of the higher order mode to undergo the enhanced attenuation. We can say in advance that the differential modal attenuation mechanism better displays in case of the core diameters of more than 30 μm and the k-parameters as large as 0.60.

### 3.1 Investigation procedures

In order to investigate carefully the modal properties of the MOFs as a function of the transverse fiber dimensions, we used the layout: a set of semiconductor lasers ($\lambda$ = 658, 808, 980 and 1550 nm), objective lenses, micrometer screws, digital CCD-camera (640 x 480 with the pixel size of 7 μm), fiber cutter and personal computer for data handling. We have been registering near-field or far-field patterns of the radiation propagated along the fibers while varying the input coupling conditions. For this purpose, laser radiation was launched into the test fiber under diverse apertures to achieve the most powerful signal on a CCD-camera screen. The fundamental mode specified by the Gaussian power distribution of light was the easiest to excite. If under varying the input coupling geometry we observed only the fundamental mode alteration (the modal spot became larger or smaller uniformly) and the higher order mode did not appear at the output end of the fiber, we explicitly considered the situation to be the single-mode one (Figure 11, left). The amount of modes in tables below was noted as 1. In the other case, when we clearly ascertained the distortion of the Gaussian power distribution or mode superposition with nearly equivalent peak power levels (Figure 11, right), we denoted the amount of the excited modes as 2.

In addition, we have made the evaluation of the mode spot size of the fiber samples with the definite single-mode regime of operation. We have modified the well-known expression for the half-divergence angle (Mortensen et al., 2002) to the form:

$$\omega = \lambda L / \pi W , \tag{4}$$

where W is the mode spot size measured at $1/e^2$ level of peak intensity on the CCD-camera screen and L is the distance between the screen and the fiber end-facet.

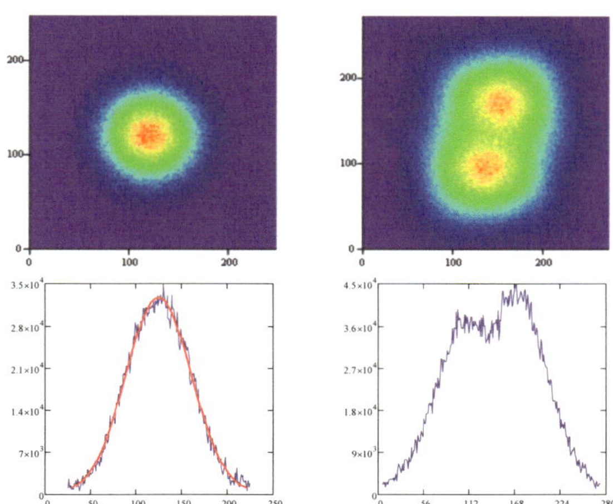

Fig. 11. Far-field patterns of the MOFs: fundamental mode propagation with the Gaussian approximation (left) and mode superposition (right).

### 3.2 MOFs with the circular cladding distribution

We have designed and manufactured a novel type of silica-based MOF containing a solid glass core of 30 μm in diameter and three air-hole rings organizing light-reflecting cladding, as it is shown in Figure 12. The similar structure containing 8 circles was mentioned in (Martelli et al., 2007)

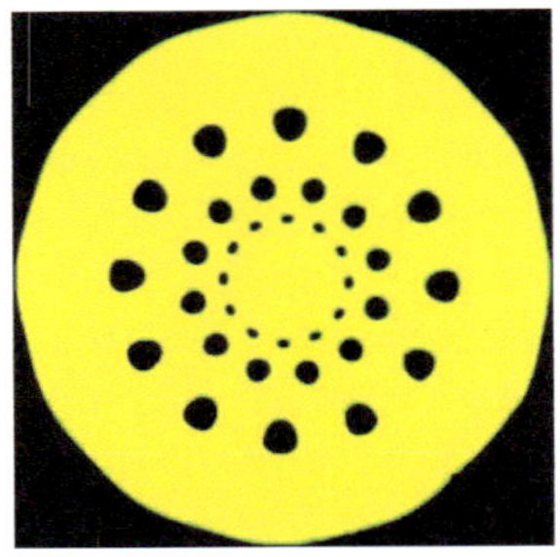

Fig. 12. The MOF structure with the circular cladding distribution.

The principal difference between the standard triangular MOF structure and the circular one is that in the latter case the amount of air-filled channels does not increase within next ring but remains invariable while going from the core area to the outer fiber boundary. In whole, the number of air holes in each ring has to be relatively large to provide a great contrast between the core refractive index and the effective cladding refractive index. In our case the number of air holes in each of the successive rings surrounding the core area was 12, which, in our opinion, is large enough to guarantee satisfactory bending characteristics. On the other hand, a quite large number of air holes will ensure good light confinement in the core area. Another important feature, worthy of attention, is the shape of the fundamental mode spot. So far as the air holes distribution in the first ring replicates the form of nearly a circle, the MOF structure performs the circular-like shape of the modal spot ($C_{12V}$ symmetry), that in some cases may be more preferable as compared to a typical six-fold rotational one ($C_{6V}$ symmetry).
As it is illustrated in Figure 12, at first, we have tended to equalize the k-parameter value in the air-hole rings, because in the triangular lattice that parameter, actually, is approximately constant over the MOF cross-section. In order to fabricate this structure we used three sets of the initial capillaries made of quartz glass. They were characterized by the outer diameters of 1.25, 2.10 and 3.30 millimeters. The ratio of the inner to the outer diameters in all three sets of capillaries was equal to ~ 0.50.
We have managed to achieve the single-mode operation over the entire spectral range studied ($\lambda$ = 658 - 1550 nm) in the derived structure, though the attenuation coefficient was too large in consequence of the tremendous leakage of the fundamental mode power into the fiber curve (standard spool of 16 centimeters in diameter). The weak dependence of the mode spot diameter on the wavelength confirms that process (Figure 13).

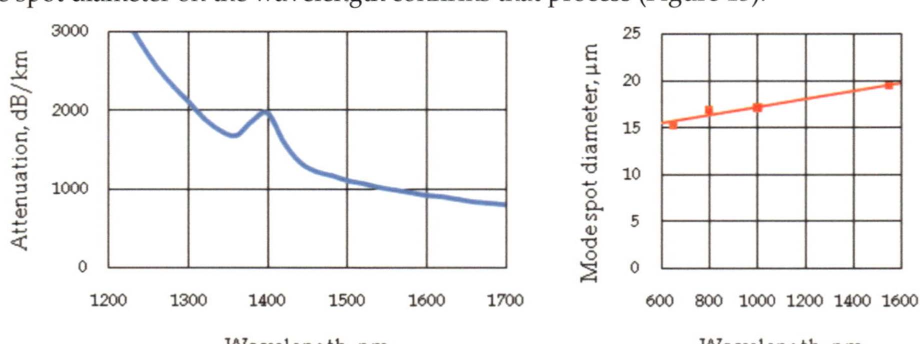


Fig. 13. Spectral attenuation pattern (left) and mode spot diameter (right) of the first series of the MOFs with the circular cladding distribution.

The sensitivity of the propagated radiation to bend may be explained largely by its great leakage through the silica gaps between the air holes in the external ring.
In the second and third series of MOFs with the circular cladding distribution we deliberately used the initial capillaries with the increased ratio of the inner to the outer diameters (Table 4).

| Air-hole ring | Ratio in the second series | Ratio in the third series |
|---|---|---|
| 1 (nearest to the core area) | 0.8 | 0.8 |
| 2 | 0.75 | 0.8 |
| 3 | 0.75 | 0.8 |

Table 4. Inner/outer diameters ratio for the circular-cladding structures.

Using the capillaries with the geometrical parameters presented in Table 4 we have obtained tangible reduction in the optical losses of the fundamental mode. Moreover, there was no evidence of the second order mode radiation both in 3-meter-length and 30-meter-length samples. The leakage losses of the $LP_{11}$ mode have been estimated to be about 10 dB/m. The experimental results are accumulated in Table 5.

| Core diameter, μm | Modal consistence (Fundamental mode spot diameter, μm) | | | | Cross-section |
|---|---|---|---|---|---|
| | $\lambda$ = 658 nm | $\lambda$ = 808 nm | $\lambda$ = 980 nm | $\lambda$ = 1550 nm | |
| 30 | 1<br>(17.1) | 1<br>(18.3) | 1<br>(18.8) | 1<br>(24.0) | 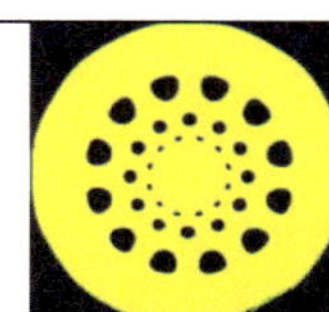 |
| 40 | Leakage | 1<br>(25.7) | 1<br>(27.3) | 1<br>(31.5) | 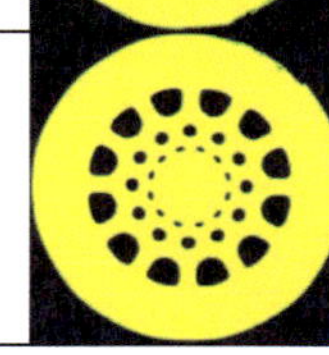 |

Table 5. Modal consistence of the second (30 μm) and the third (40 μm) series of the MOFs with the circular cladding distribution.

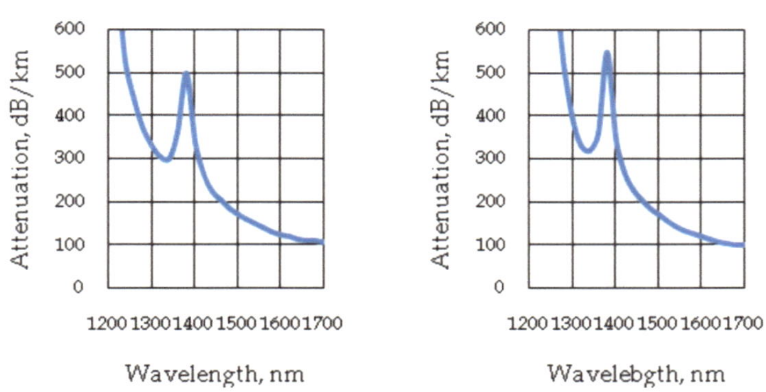


Fig. 14. Spectral attenuation pattern of the MOFs with the circular cladding distribution: the core diameters are 30 μm (left) and 40 μm (right).

As Figure 14 illustrates, the MOF with the core diameter of 30 μm has excessively large optical losses. However, a certain increase in the air-filling fraction in each of three air-hole rings and simultaneous enlargement of the core size (up to 40 μm) give the structure optimal correlation between high attenuation coefficient of the upper modes and reasonable bend performance of the fundamental mode. Table 5 shows that the MOF with the core diameter of 40 μm is characterized by the relatively large mode spot size, ranging from 25.7 to 31.5 μm in the inspected spectral range. Despite the significant attenuation coefficient, about 0.15 dB/m at $\lambda$ = 1550 nm, the MOF is operable for a standard bending diameter of 16 centimeters. In identical conditions, a well-known triangular structure with the 1-element core undergoes the infinite attenuation (Nielsen et al., 2004b).

We assert that the considerable reduction in the attenuation coefficient of the MOFs with the circular cladding distribution can be achieved by both proper variation of the air-hole sizes and appropriate adjustment of the silica gaps between them. Moreover, the presence or the absence of the higher order mode at the output end of the fiber is determined mainly by the air content in the first air-hole ring. At the same time, the air-filling fractions of the second and the third rings, especially the external, define the leakage loss level of the fundamental mode. As additional experiment has shown, the minimal attenuation in the multimode regime (Figure 15) was measured to be ~ 8 dB/km caused mainly by impurities. This MOF has a potential to be used in high-NA applications.

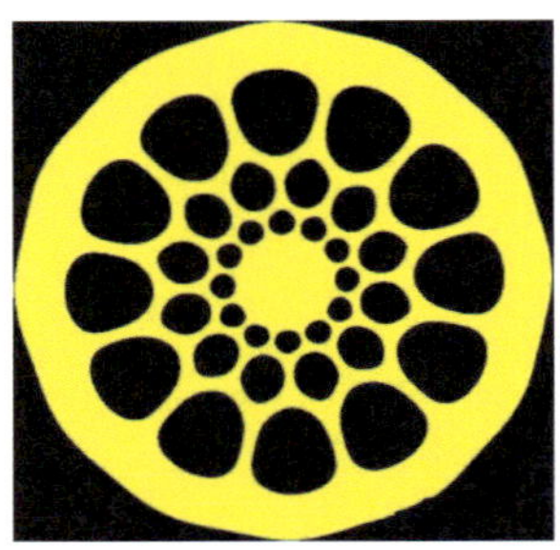

Fig. 15. Microscopic image of the multimode MOF structure with the circular cladding distribution.

### 3.3 MOFs with the $C_{3V}$ cladding symmetry

We have also fabricated a series of the MOF structures that can be treated as an adaptation of standard 1-element-core design with a triangular arrangement of air holes (Russell, 2006). A key variation consisted in the intentional introduction of an alternation of large and small air holes to the MOF design attaining so called $C_{3V}$ transverse symmetry (Figure 16), as opposed to the standard $C_{6V}$ symmetry.

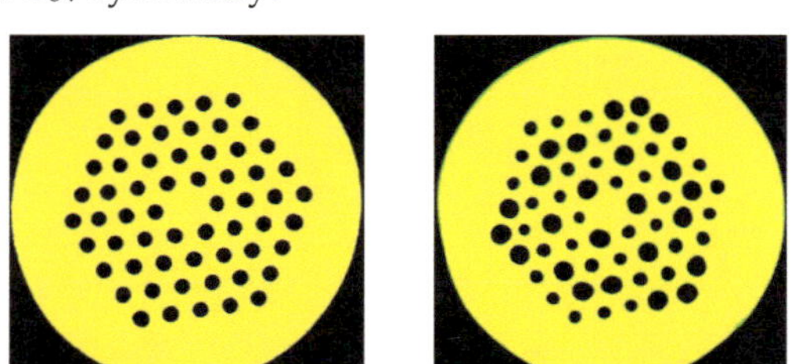

Fig. 16. Microscopic images of the MOFs: typical $C_{6V}$ symmetry (left) and non-standard $C_{3V}$ symmetry.

The alternation of large and small air holes, especially in the air-hole ring closest to the core area, may confidently lead to the enhancement of the higher order mode attenuation. The main mechanism remained the same we had applied earlier for the structure with the circular cladding distribution. We talk about the controllable leakage of the second order mode power through the silica gaps between each pair of neighboring holes.
To achieve the single-mode operation in presented MOF structures we used quartz capillaries of the same outer diameter (1.75 μm), but of the diverse inner diameter. The exact ratios of the inner/outer diameters were 0.6 and 0.8. Additionally, we controlled the air-filling fraction during the fiber drawing process by varying two essential technological parameters we could easily change. We mean the capillary pressure and the drawing temperature. The experimental results are given in Table 6. Optical measurements were made for a bending diameter of 16 centimeters.

| Core diameter, μm | Modal consistence (Fundamental mode spot diameter, μm) | | | | Cross-section |
|---|---|---|---|---|---|
| | λ = 658 nm | λ = 808 nm | λ = 980 nm | λ = 1550 nm | |
| 25 | Leakage | Leakage | 1 (16.8) | 1 (19.6) | 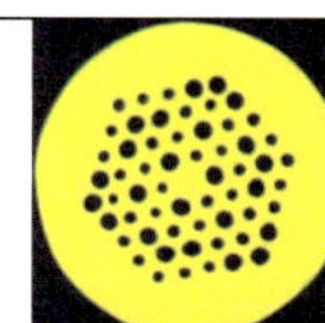 |
| 35 | Leakage | Leakage | Leakage | 1 (24.4) | 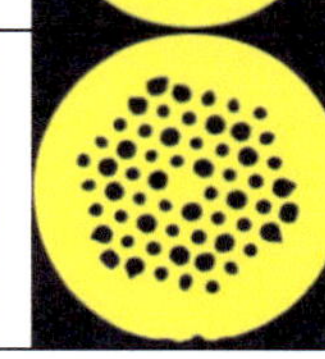 |

Table 6. Modal consistence of the MOFs with the $C_{3V}$ cladding symmetry.

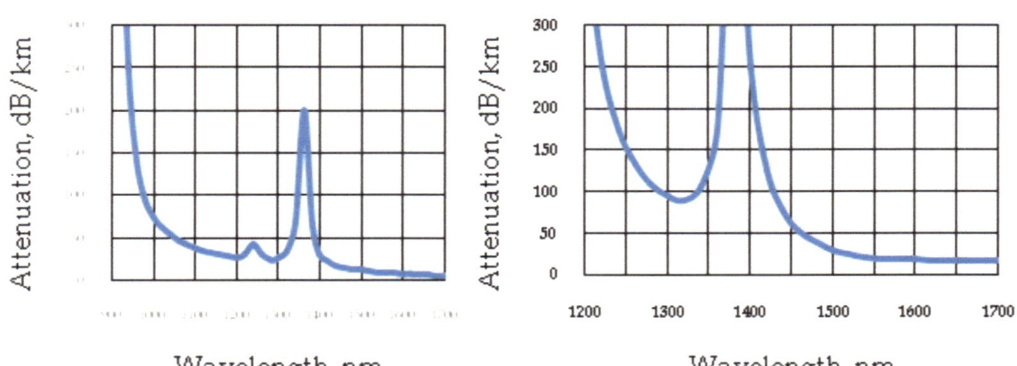


Fig. 17. Spectral attenuation pattern of the MOFs with the $C_{3V}$ cladding symmetry: the core diameters are 25 μm (left) and 35 μm (right).

We have detected some positive features of the MOFs with the $C_{3V}$ cladding symmetry including single-mode regime and high bending resistance properties. The most impressive situation takes place in the fiber with the core diameter of 35 μm, as the spectral operation range can be expanded for about 300 nm to the blue part of the spectra in comparison with the MOF of a typical triangular configuration with the air holes of invariable dimensions. Classical MOFs are still required to be placed on non-standard spools having a diameter of 32 centimeters (Nielsen et al., 2004b), while our fibers are operable for a bending diameter of 16 centimeters. In addition, as opposed to the MOFs with the circular cladding distribution (discussed in previous paragraph), the fibers with the $C_{3V}$ cladding symmetry can guarantee optical losses of less than 20 dB/km at $\lambda = 1550$ nm (Figure 17).

### 3.4 MOFs with the shifted core

This paragraph will cover in detail main optical properties as well as bend characteristics of another MOF structure based on the declared concept of the differential modal attenuation.
Theoretically, the higher order mode of a typical MOF with triangular arrangement of air holes, as opposed to the fundamental mode, may be characterized by a significantly larger degree of the power penetration into the fiber cladding (Tsuchida et al., 2005; Russell, 2006). Applying this specific mode feature, we made an assumption that the MOF structure with the core shifted for the pitch value from its usual location in the center of the lattice structure will exhibit enhanced leakage losses of the higher order mode. At the same time, the leakage losses of the fundamental mode will maintain their value as if the fiber structure is completely retained. So we can implement the special MOF design that enables a relatively simple control of the higher order mode attenuation. A conditioning factor, as in the previous cases we have already studied (MOF structures with the circular cladding distribution and the $C_{3V}$ cladding symmetry), is the reasonable balance towards the dimensions of the air holes, especially in the ring closest to the core area, and the spaces between them filled with a glass material. We have successfully produced by classical stack-and-draw technique the following MOF structure (Figure 18).

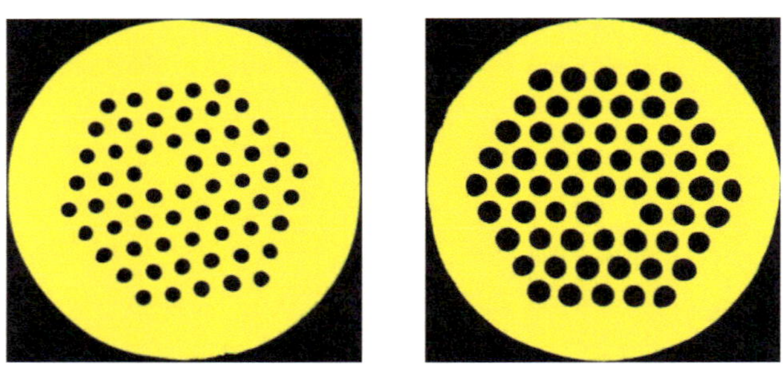

Fig. 18. Microscopic images of the fabricated MOF structures with the shifted 1-element core: k = 0.45 (left), k = 0.75 (right).

When the core is displaced from the center of the lattice, the peripheral part of the effective $LP_{11}$-mode area may intensively leak away into the outer fiber cladding which is then situated nearer than in the MOF of a standard central configuration. Thus, the fundamental mode operation is not strongly defined by the phase cut-off condition, known as $k < 0.45$ for the second order mode (Mortensen, 2002). As we will show further, the endlessly single-mode regime may be carried out at certain lengths of the fiber even when the k-parameter value is for more than 40% larger relative to its theoretical value. In this case the $\Lambda$-parameter and the curve diameter resulting in the bend performance are crucial for the establishment of the differential mode attenuation.

The experimental results are collected in Table 7. It should be noted that we intentionally fabricated several samples with the standard central-core design to compare two different transverse structures. All optical measurements were made for a bending diameter of 16 centimeters.

One can see from Table 7 that the core diameter enlargement from 12.5 to 20 μm under the k-parameter being nearly constant (k ~ 0.50) caused the additive attenuation of the higher order mode (samples $2_{SHIFT}$ and $3_{SHIFT}$).

| Fiber | Core diameter, μm | k | Length, m | Modal consistence at λ (Fundamental mode spot size, μm) | | | |
|---|---|---|---|---|---|---|---|
| | | | | 658 nm | 808 nm | 980 nm | 1550 nm |
| $1_{SHIFT}$ | 12.5 | 0.35 | 10 | 1 (10.4) | 1 (11.8) | 1 (11.8) | 1 |
| $2_{SHIFT}$ | 12.5 | 0.52 | 10 | 2 | 2 | 2 | 2 |
| $3_{SHIFT}$ | 20.0 | 0.51 | 10 | 1 (13.4) | 1 (14.1) | 1 (14.2) | 1 |
| $4_{SHIFT}$ | 20.0 | 0.60 | 10 | 1 (12.3) | 1 (13.0) | 1 (13.4) | 1 |
| $5_{SHIFT}$ | 20.0 | 0.81 | 10 | 2 | 2 | 2 | > 2 |
| $1_{CENT}$ | 20.0 | 0.65 | 20 | Leakage | 1 (15.3) | 1 (15.8) | 2 |
| $2_{CENT}$ | 22.0 | 0.65 | 20 | Leakage | 1 (13.9) | 1 (14.9) | 2 |
| $3_{CENT}$ | 22.0 | 0.67 | 20 | 1 (12.6) | 1 (13.7) | 1 (14.0) | 2 |

Table 7. Modal properties of the MOFs with the shifted and the central cores.

In theory, the core expansion leads to the increase of normalized frequency V, representing exactly the amount of guided modes. This situation can be simply illustrated with the expression (Birks et al., 1997):

$$V = 2\pi\Lambda\sqrt{n_{co}^2 - n_{cl}^2}\,/\lambda\,, \tag{5}$$

where $n_{co}$ and $n_{cl}$ are the refractive index of the core and the effective refractive index of the cladding respectively.

So when V-parameter increases, the existing modes become less sensitive to perturbations, for example, to macro- and microbending, and the higher order mode attenuation coefficient has to be lower. The alternative process in our experiment may be explained by the mechanism of the enhanced modal leakage through the spaces between the air holes, which dimensions extend in the direct proportion to Λ-growth when the whole structure is scaled. Generally, the modal consistence of the MOFs with shifted and central cores approximately coincides if deal with the same core diameters and k-parameters values. The most impressive difference is that in the visible part of the spectra (λ = 658 nm) the fundamental mode radiation of the central-core MOFs intensively leaks away from the core area into the outer fiber cladding, while it good confines in the shifted-core ones. Such high bending resistance properties can be elucidated by the increased air-filling fraction (sample $4_{SHIFT}$) and the mode spot size that is decreased as compared to the spot size in the central-core MOF (samples $4_{SHIFT}$ and $1_{CENT}$). In addition, the mode spot size in MOFs with the shifted core is weakly dependent on the wavelength that is a distinctive feature of MOFs as a class of lightguides with the triangular arrangement of the air holes (Nielsen et al., 2004a). Despite the slight reduction in the fundamental mode spot size, the shifted-core MOFs have an obvious preference over the central-core analog. It consists in the expansion of the spectral operation range.

Thus, we have accurately determined that the position of bend-induced leakage boundary of the MOFs with the shifted core is located in the visible part of the spectra. To identify it numerically we have produced several samples of the length from 50 to 100 meters. The respective curve is presented in Figure 19.

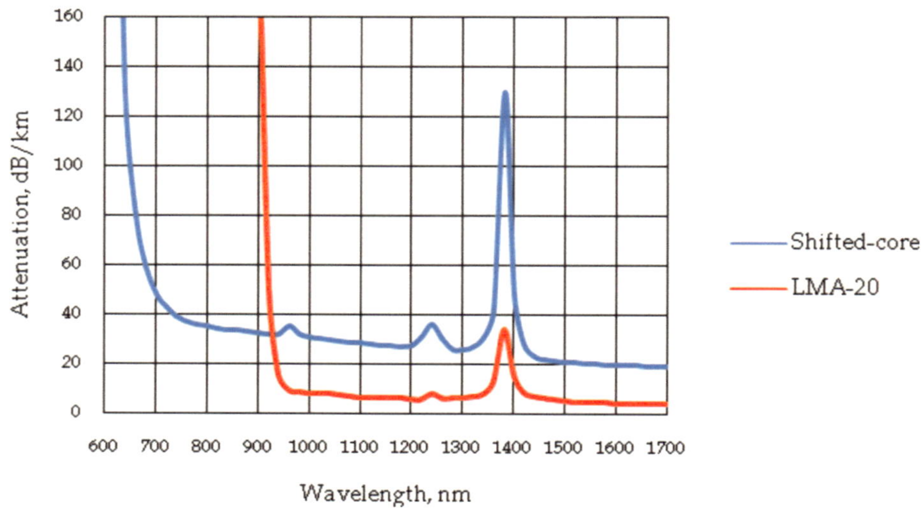


Fig. 19. Spectral attenuation pattern of the MOFs with the core diameter of 20 μm.

As Figure 19 shows, leakage of the fundamental mode power occurs in the wavelength region λ ~ 650 nm. At the same time, in the central-core MOF with the identical core size strong leakage occurs at λ ~ 900 nm (spool diameter of 16 centimeters) and at λ ~ 650 nm (spool diameter of 32 centimeters) (Nielsen et al., 2004b). In other words, the MOF with the shifted core can be characterized by the noticeably improved stationary bending resistance due to the relatively large value of k-parameter which is equal to 0.60. The spectral operation range widening is about 250 nm that is an obvious preference over the typical MOF structure.

Similar processes can be observed in the MOF with the shifted core of 34 μm in diameter (Figure 20).

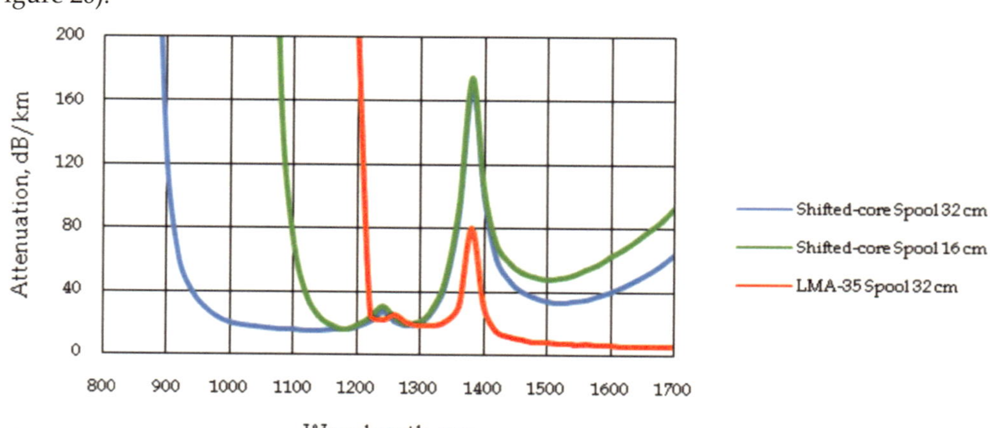


Fig. 20. Spectral attenuation pattern of the MOFs with the core diameter of 34 - 35 μm.

The increase in the k-parameter value up to 0.65 allows the fiber being placed on a standard spool of 16 centimeters in diameter to operate still in the single-mode regime. The mode spot size is approximately 26 μm that coincides with the data for the central-core MOFs (Nielsen et al., 2003).

We assert that further increase in the air-filling fraction (up to 0.70), caused by the natural desire to expand the range of working frequencies as much as possible, leads to the display of the impurity of the higher order mode in the power distribution of light at the output end of the fiber. So we consider the MOF with the shifted core of 34 - 35 μm in diameter and the k-parameter of 0.65 to be the optimal bend-resistant design combining robust fundamental mode propagation and the expanded spectral operation interval.

It should be noted that the investigated MOF with the shifted core of 34 μm in diameter exhibits a certain growth of the attenuation coefficient in the spectral region $\lambda > 1200$ nm (Figure 20). As we have already ascertained mode spot size is weakly dependent on the working wavelength in the fibers with the air holes assembled in the triangular array (Nielsen et al., 2004a) and the fundamental mode leakage occurs in the short-wavelength part of the spectra (Nielsen et al., 2004b). When the fiber core is shifted, the effective mode area largely penetrates into the outer fiber cladding as compared to the central-core fiber, that corresponds adequately with the certain increase in the attenuation coefficient while going to longer wavelengths.

The validity of the ultimate role of differential modal attenuation in achievement of the single-mode operation may be illustrated by the following data (Table 8).

<table>
<tr><th rowspan="2">Length, m</th><th colspan="4">Wavelength, nm</th></tr>
<tr><th>λ = 658 nm</th><th>λ = 808 nm</th><th>λ = 980 nm</th><th>λ = 1550 nm</th></tr>
<tr><td colspan="5">MOFs with the shifted core and k = 0.65</td></tr>
<tr><td>50</td><td>1 mode</td><td>1 mode</td><td colspan="2" rowspan="4">2 modes</td></tr>
<tr><td>5</td><td>1 mode</td><td>1 mode</td></tr>
<tr><td>3</td><td>1 mode</td><td>2 modes</td></tr>
<tr><td>1.5</td><td>2 modes</td><td>2 modes</td></tr>
<tr><td colspan="5">MOFs with the shifted core and k =0.68</td></tr>
<tr><td>40</td><td>1 mode</td><td>1 mode</td><td colspan="2" rowspan="3">2 modes</td></tr>
<tr><td>5</td><td>1 mode</td><td>2 mode</td></tr>
<tr><td>2.5</td><td>2 mode</td><td>2 mode</td></tr>
</table>

Table 8. Dependence of the modal consistence on the fiber length.

Table 8 illustrates the modal consistence of the short-length (from 1.5 to 5 meters) samples of the MOFs with the shifted core of 20 μm in diameter. We have compared these experimental data with those obtained for the long-length samples. During the measurement procedures the fibers were placed on standard spools of 16 centimeters in diameter. As one can see, short-length samples were characterized by the multimode regime of operation which transformed into the single-mode while increasing the fiber length. The higher order mode attenuated almost completely after passing the length of approximately 5 meters (sample with k = 0.65). On the basis of the statement that total attenuation of the higher order mode could be estimated to be of at least 10 dB, we got the bottom boundary for the attenuation coefficient equal to ~ 2 dB/m at $\lambda$ = 658 - 808 nm. That result corresponded with the data we have obtained earlier while working on the MOFs with a multi-element core. Table 8 shows

that further increase in the k-parameter up to 0.68 decreased the attenuation coefficient of the higher order mode that, in turn, limited steadily the spectral operation range.
Finally, we declare that in optimized conditions the expansion of the spectral operation range of the shifted-core MOFs may achieve 300 nm in the near infrared part of the spectra (Figure 20) as compared to the central-core MOFs (Nielsen et al., 2004b). This effectively combines with the large air-filling fraction (k = 0.65) that makes the fiber less susceptible to macrobending or microbending effects.

## 4. Conclusion

In this work, we have reported on the recent results in investigating main optical properties of several series of MOFs. Two of them (7- and 19-element-core designs) have already been discussed earlier, whereas three others (the circular cladding distribution, the $C_{3V}$ cladding symmetry and the shifted-core structure) represent novel design. The modal consistence of radiation propagated along the fiber cores and the spectral attenuation curves have been provided.
We have accurately determined that the basic priorities of novel MOF structures are the expanded spectral operation range (up to 300 nm) and the bend-resistant performance. We have managed to achieve these features by ensuring proper conditions for the high air-filling fraction of the cladding structure. Simultaneously, we have successfully assured the condition for the higher order mode to undergo strong attenuation via considerable power leakage into the outer fiber cladding. That point principally diverges our approach from the classical one, which guarantees the phase cut-off condition for the propagation of the second order mode. The mechanism of the differential modal attenuation has been clearly stated in the case of the shifted-core MOF design when the core area was located close to the outer fiber cladding. The mentioned circumstance has led to the strong attenuation coefficient of the higher order mode that has been estimated to be in the vicinity of 2 - 6 dB/km. In other cases the physical foundation of the differential modal attenuation was not so clear, though it effectively worked allowing the MOFs to support practically only a single transverse mode over a broad spectral range.
The fiber designs discussed in this work can be successfully applied in high power laser technology or in laser beam delivery applications with standardized curve parameters.

## 5. References

Agruzov, P. M.; Kozlov, A. S.; Petrov, M. P.; Dukel'skii, K. V.; Komarov, A. V.; Ter-Nersesyants, E. V.; Khokhlov, A. V. & Shevandin, V. S. (2008). Mode composition of holey fibers with a large seven-element core. *Journal of Optical Technology*, Vol. 75, No. 11, (November 2008), pp. 747-749, ISSN 1070-9762

Agruzov, P. M.; Dukel'skii, K. V.; Komarov, A. V.; Ter-Nersesyants, E. V.; Khokhlov, A. V. & Shevandin, V. S. (2010). Developing microstructured lightguides with a large core, and an investigation of their optical properties. *Journal of Optical Technology*, Vol. 77, No. 1, (January 2010), pp. 59-62, ISSN 1070-9762

Birks, T. A.; Knight, J. C. & Russell, P. St. J. (1997). Endlessly single-mode photonic crystal fiber. *Optics Letters*, Vol. 22, No. 13, (July 1997), pp. 961-963, ISSN 0146-9592

Dukel'skii, K. V.; Komarov, A. V.; Ter-Nersesyants, E. V.; Khokhlov, A. V. & Shevandin, V. S. (2005). Realization of photonic crystal fibers in S.I. Vavilov Federal Optical Institute. *Proceedings of AIS'05 and CAD'2005*, ISBN 5-9221-0621-X, Divnomorsk, September 2005

Dukel'skii, K. V.; Kondrat'ev, Yu. N.; Komarov, A. V.; Ter-Nersesyants, E. V.; Khokhlov, A. V. & Shevandin, V. S. (2006). How the pitch of a holey optical fiber affects its lightguide properties. *Journal of Optical Technology*, Vol. 73, No. 11, (November 2006), pp. 808-811, ISSN 1070-9762

Jeunhomme, L. B. (1983). *Single-Mode Fiber Optics: Principles and applications*. Marcel Dekker, ISBN 978-082-4770-20-4, New York, USA

Knight, J. C.; Birks, T. A.; Russell, P. St. J. & Atkin, D. M. (1996). All-silica single-mode optical fiber with photonic crystal cladding. *Optics Letters*, Vol. 21, No. 19, (October 1996), pp. 1547-1549, ISSN 0146-9592

Knight, J. C.; Birks, T. A.; Russell, P. St. J. & Atkin, D. M. (1997). All-silica single-mode optical fiber with photonic crystal cladding: errata. *Optics Letters*, Vol. 22, No. 7, (April 1997), pp. 484-485, ISSN 0146-9592

Knight, J. C. (2003). Photonic crystal fibres. *Nature*, Vol. 424, No. 6950, (August 2003), pp. 847-851, ISSN 0028-0836

Limpert, J.; Deguil-Robin, N.; Manek-Honninger, I.; Salin, F.; Roser, F.; Liem, A.; Schreiber, T.; Nolte, S.; Zellmer, H.; Tunnermann, A.; Broeng, J.; Petersson, A. & Jakobsen, C. (2005). High-power rod-type photonic crystal fiber laser. *Optics Express*, Vol. 13, No. 4, (February 2005), pp. 1055-1058, ISSN 1094-4087

Limpert, J.; Schmidt, O.; Rothhardt, J.; Roser, F.; Schreiber, T.; Tunnermann, A.; Ermeneux, S.; Yvernault, P. & Salin, F. (2006). Extended single-mode photonic crystal fiber lasers, *Optics Express*, Vol. 14, No. 7, (April 2006), pp. 2715-2720, ISSN 1094-4087

Martelli C.; Canning J.; Gibson B. & Huntington S. *Optics Express*, Vol. 15, No. 26, (December 2007), pp. 17639-17644, ISSN 1094-4087

Mortensen, N. A. (2002). Effective area of photonic crystal fibers. *Optics Express*, Vol. 10, No. 7, (April 2002), pp. 341-348, ISSN 1094-4087

Mortensen, N. A.; Folkenberg, J. R., Skovgaard, P. M. W. & Broeng, J. (2002). Numerical aperture of single-mode photonic crystal fibers. *IEEE Photonic Technology Letters*, Vol. 14, No. 8, (August 2002), pp. 1094-1096, ISSN 1041-1135

Mortensen, N. A.; Folkenberg, J. R.; Nielsen, M. D. & Hansen, K. P. (2003). Modal cut-off and the V-parameter in photonic crystal fibers. *Optics letters*, Vol. 28, No. 20, (October 2003), pp. 1879-1881, ISSN 0146-9592

Nielsen, M. D.; Folkenberg, J. R. & Mortensen, N. A. (2003). Single-mode photonic crystal fiber with an effective area of 600 $\mu m^2$ and low bending losses. *Electronics Letters*, Vol. 39, No. 25, (December 2003), pp. 1802-1803, ISSN 0013-5194

Nielsen, M. D.; Folkenberg, J. R.; Mortensen, N. A. & Bjarklev, A. (2004). Bandwidth comparison of photonic crystal fibers and conventional single-mode fibers. *Optics Express*, Vol. 12, No. 3, (February 2004), pp. 430-435, ISSN 1094-4087

Nielsen, M. D.; Mortensen, N. A.; Albertsen, M.; Folkenberg, J.R.; Bjarklev, A. & Bonacinni, D. (2004). Predicting macrobending loss for large-mode area photonic crystal fibers. *Optics Express*, Vol. 12, No. 8, (April 2004), pp. 1775-1779, ISNN 1094-4087

Russell, P. St. J. (2006). Photonic-Crystal Fibers. *Journal of Lightwave Technology*, Vol. 24, No. 12, (December 2006), pp. 4729-4749, ISSN 0733-8724

Saitoh, K.; Tsuchida, Y.; Koshiba, M. & Mortensen, N. A. (2005). Endlessly single-mode holey fibers: the influence of core design. *Optics Express*, Vol. 13, No. 26, (December 2005), pp. 10833-10839, ISSN 1094-4087

Snyder, A. W. & Love, J. D. (1983). *Optical Waveguide Theory*. Chapman and Hall Ltd, ISBN 978-041-2099-50-2, London, UK

Tsuchida, Y.; Saitoh, K. & Koshiba, M. (2005). Design and characterization of single-mode holey fibers with low bending losses. *Optics Express*, Vol. 13, No. 12, (June 2005), pp. 4770-4779, ISSN 1094-4087

Tunnermann, A.; Schreiber, T.; Roser, F.; Liem, A.; Hofer, S.; Zellmer, H.; Nolte, S. & Limpert, J. (2005). The renaissance and bright future of fibre lasers. *Journal of Physics B: Atomic, Molecular and Optical Physics*, Vol. 38, No. 9, (April 2005), pp. S681-S693, ISSN 0953-4075

# 17

# Fabrication and Applications of Microfiber

K. S. Lim[1], S. W. Harun[1,2], H. Arof[2] and H. Ahmad[1]
[1]*Photonic Research Center, University of Malaya, Kuala Lumpur,*
[2]*Department of Electrical Engineering, Faculty of Engineering, University of Malaya, Kuala Lumpur, Malaysia*

## 1. Introduction

Microfibers have attracted growing interest recently especially in their fabrication methods and applications. This is due to a number of interesting optical properties of these devices, which can be used to develop low-cost, miniaturized and all-fiber based optical devices for various applications (Bilodeau et al., 1988; Birks and Li, 1992 ). For instance, many research efforts have focused on the development of microfiber based optical resonators that can serve as optical filters, which have many potential applications in optical communication and sensors. Of late, many microfiber structures have been reported such as microfiber loop resonator (MLR), microfiber coil resonator (MCR), microfiber knot resonator (MKR), reef knot microfiber resonator as an add/drop filter and etc. These devices are very sensitive to a change in the surrounding refractive index due to the large evanescent field that propagates outside the microfiber and thus they can find many applications in various optical sensors. The nonlinear properties of the micro/nanostructure inside the fiber can also be applied in fiber laser applications. This chapter thoroughly describes on the fabrication of microfibers and its structures such as MLR, MCR and MKR. A variety of applications of these structures will also be presented in this chapter.

## 2. Fabrication of microfiber

### 2.1 Flame brushing technique

Flame brushing technique (Bilodeau et al., 1988 ) is commonly used for the fabrication of fiber couplers and tapered fibers. It is also chosen in this research due to its high flexibility in controlling the flame movement, fiber stretching length and speed. The dimension of the tapered fiber or microfiber can be fabricated with good accuracy and reproducibility. Most importantly, this technique enables fabrication of biconical tapered fibers which both ends of the tapered fiber are connected to single-mode fiber (SMF). These biconical tapered fibers can be used to fabricate low-loss microfiber based devices.

Fig.1 shows a schematic illustration of tapered fiber fabrication based on flame brushing technique. As shown in Fig. 1, coating length of several cm is removed from the SMF prior to the fabrication of tapered fiber. Then the SMF is placed horizontally on the translation stage and held by two fiber holders. During the tapering, the torch moves and heats along the uncoated segment of fiber while it is being stretched. The moving torch provides a uniform heat to the fiber and the tapered fiber is produced with good uniformity along the heat region. To monitor the transmission spectrum of the microfiber during the fabrication, amplified spontaneous emission (ASE) source from an Erbium-doped fiber amplifier (EDFA) is injected into one end of the SMF while the other end is connected to the optical spectrum analyzer (OSA). Fig. 2(a) shows diameter variation of the biconical tapered fiber fabricated using the fiber tapering rig while Fig. 2(b) shows the optical microscope image of the tapered fiber with a waist diameter of 1.7μm. With proper tapering parameters, the taper waist diameter can be narrowed down to ~800nm as shown in Fig. 2(c).

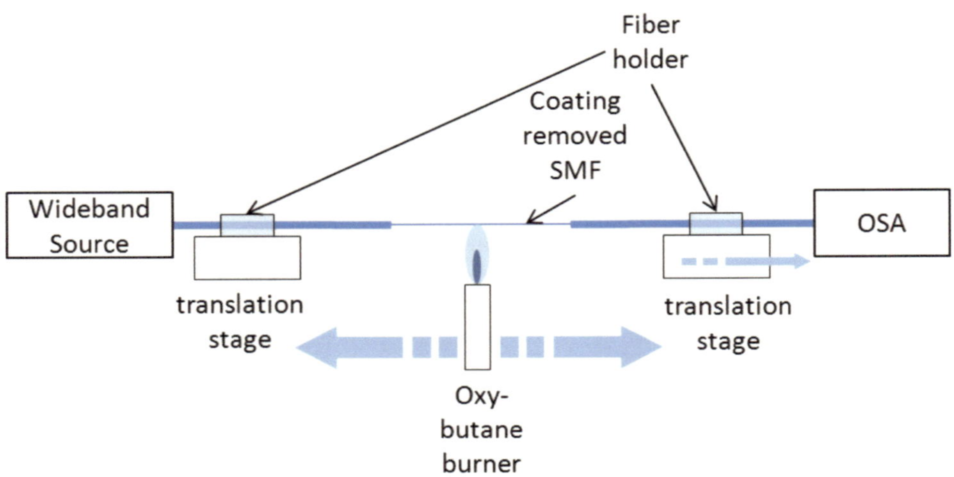


Fig. 1. Tapered fiber fabrication using flame brushing technique.

Adiabaticity is one of the important criteria in fabricating good quality tapered fibers. It is commonly known that some tapered fibers suffer loss of power when the fundamental mode couples to the higher order modes. Some fraction of power from higher order modes that survives propagating through the tapered fiber may recombine and interfere with fundamental mode. This phenomenon can be seen as interference between fundamental mode $HE_{11}$ and its closest higher order mode $HE_{12}$. This results to a transmission spectrum with irregular fringes as shown by the dotted graph in Fig. 3 and the excess loss of the tapered fiber is ~0.6dB (Ding et al., 2010; Orucevic et al., 2007 ). This tapered fiber is not suitable to be used in the ensuing fabrication of microfiber devices. The solid curve in the same figure shows the transmission of a low loss tapered fiber with approximately more than 4mm transition length and the insertion loss lower than 0.3dB. Some analysis suggests that the coupling from fundamental mode to higher order modes can be minimized by optimizing shape of the tapers. In practice, adiabaticity can be easily achieved by using sufficiently slow diameter reduction rate when drawing tapered fibers or in other words manufacture tapered fibers with sufficiently long taper transition length. A detail discussion on the adiabatic criteria and optimal shapes for tapered fiber will be presented in the next section.

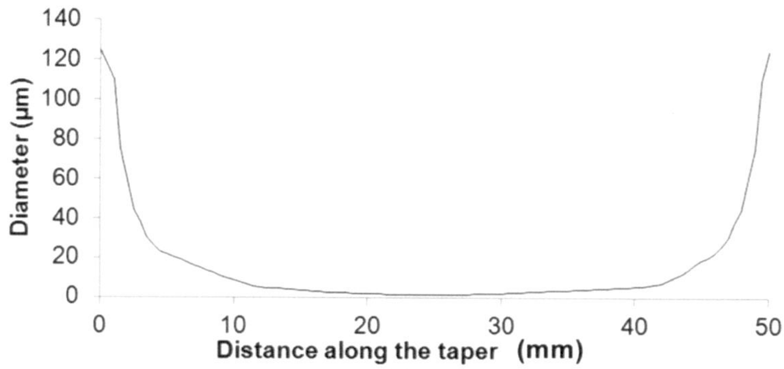


(a)

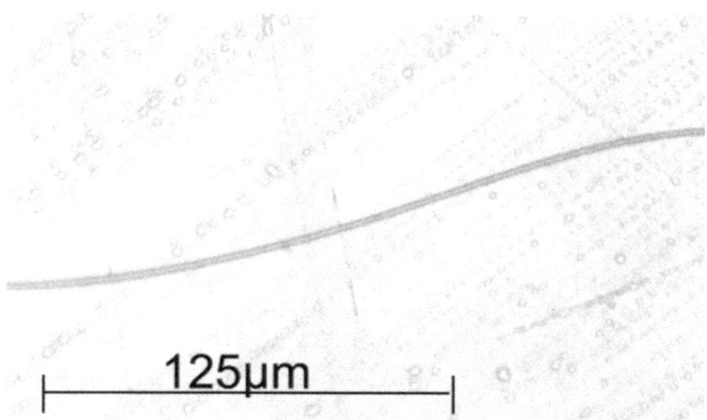


(b)

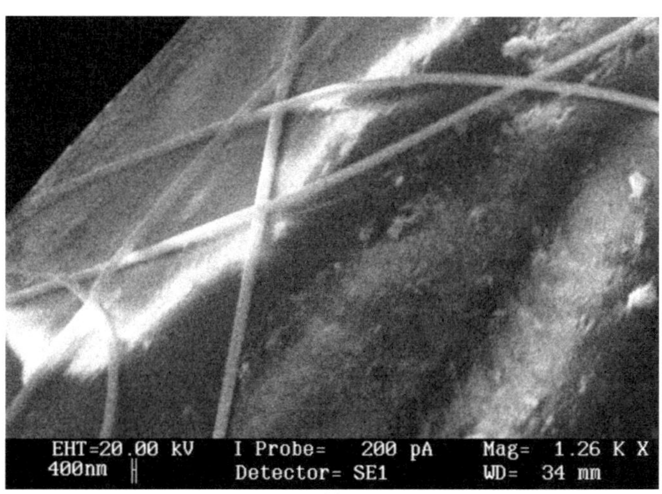


(c)

Fig. 2. (a)The diameter variation of a biconical tapered fiber fabricated in the laboratory (b) Optical microscope image of tapered fiber with a waist diameter of 1.7 µm (c) SEM image of a ~700nm waist diameter tapered fiber.

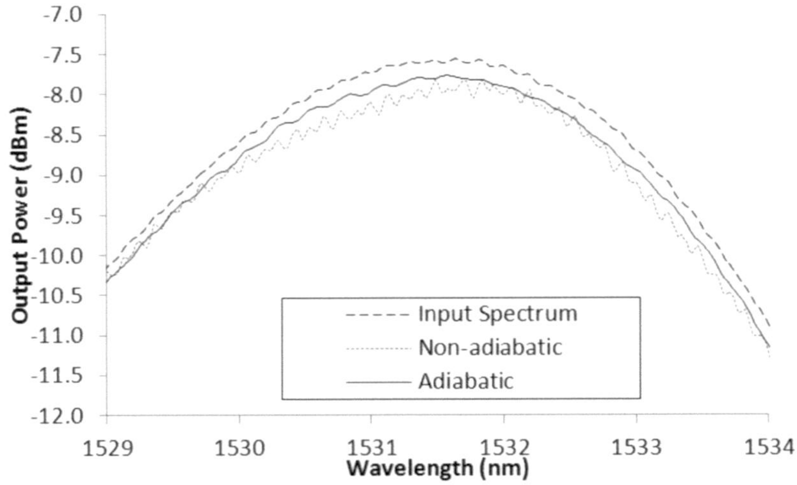


Fig. 3. Output spectra from a microfiber with 10 cm long and ~3μm waist diameter. Input spectrum from EDFA (dashed), adiabatic taper (solid) and non-adiabatic taper (dotted).

### 2.2 Adiabaticity criteria

Tapered fiber is fabricated by stretching a heated conventional single-mode fiber (SMF) to form a structure of reducing core diameter. As shown in Fig. 4, the smallest diameter part of the tapered fiber is called waist. Between the uniform unstretched SMF and waist are the transition regions whose diameters of the cladding and core are decreasing from rated size of SMF down to the order micrometer or even nanometer. As the wave propagate through the transition regions, the field distribution varies with the change of core and cladding diameters along the way. Associated with the rate of diameter change of any local cross section, the propagating wave may experience certain level of energy transfer from the fundamental mode to a closest few higher order modes which are most likely to be lost. The accumulation of this energy transfer along the tapered fiber may result to a substantial loss of throughput. This excess loss can be minimized if the shape of the fabricated tapered fiber follows the adiabaticity criteria everywhere along the tapered fiber (Birks and Li, 1992; Love et al., 1991 ).

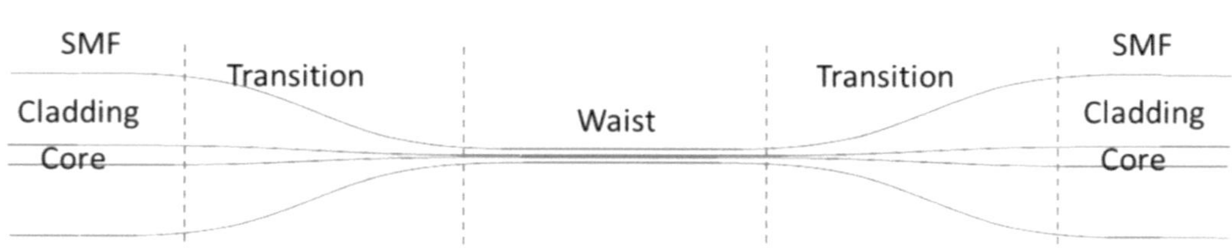


Fig. 4. Typical diameter profile of a tapered fiber.

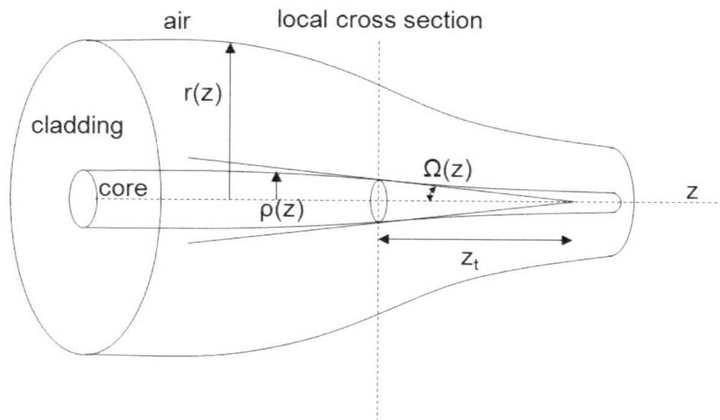


Fig. 5. Illustration of the taper transition.

Fig. 5 gives an illustration of a tapered fiber with decreasing radius where $z$ denotes the position along the tapered fiber. Theoretically, an adiabatic tapered fiber is based on the condition that the beat length between fundamental mode $LP_{01}$ and second local mode is smaller than the local taper length-scale $z_t$.

$$z_b > z_t \tag{1}$$

Referring to illustration in Fig. 5, $z_t$ is given by

$$z_t = \rho / \tan\Omega \tag{2}$$

where $\rho = \rho(z)$ is the local core radius and $\Omega = \Omega(z)$ is the local taper angle.
The beat length between two modes is expressed as

$$z_b = \frac{2\pi}{\beta_1 - \beta_2} \tag{3}$$

where $\beta_1 = \beta_1(r)$ and $\beta_2 = \beta_2(r)$ are the propagation constants of fundamental mode and second local mode respectively. From the above equations, Inequality (1) can be derived to

$$\left|\frac{d\rho}{dz}\right| = \tan\Omega < \frac{\rho(\beta_1 - \beta_2)}{2\pi} \tag{4}$$

where $\frac{d\rho}{dz}$ is the rate of change of local core radius and its magnitude is equivalent to $\tan\Omega$.
For the convenience of usage and analysis, Inequality (4) is rewritten as a function of local cladding radius r=r(z),

$$\left|\frac{dr}{dz}\right| < \frac{r(\beta_1 - \beta_2)}{2\pi} \tag{5}$$

Based on this condition, adiabatic tapered fiber can be acquired by tapering a fiber at a smaller reduction rate in diameter but this will result to a longer transition length. Considering practical limitations in the fabrication of fiber couplers or microfiber based devices, long tapered fiber may aggravate the difficulty in fabrication. For the purpose of miniaturization, short tapered fiber is preferable. To achieve balance between taper length and diameter reduction rate, a factor *f* is introduced to Inequality (5) and yields

$$\frac{dr}{dz} = -\frac{fr(\beta_1 - \beta_2)}{2\pi} \tag{6}$$

where the value of *f* can be chosen between 0 to 1. Optimal profile is achieved when $f = 1$. Practically, tapered fiber with negligibly loss can be achieved with $f = 0.5$ but the transition length of the tapered fiber is 2 times longer than that of the optimal tapered fiber.

### 2.3 Shape of tapered fiber

When a glass element is heated, there is a small increment in the volume under the effect of thermal expansion. However, the change in volume is negligibly small not to mention that the volume expansion wears off immediately after the heat is dissipated from the mass. It is reasonable to assume that the total volume of the heated fiber is conserved throughout the entire tapering process. Based on this explanation, when a heated glass fiber is stretched, the waist diameter of the fiber is reduced. The calculation of varying waist diameter and length of extension can be made based on the idea of 'conservation of volume'. Birks and Li (1992) presented simple mathematical equations to describe the relationship between shapes of tapered fiber, elongation distance and hot-zone length. Any specific shape of tapered fiber can be controlled by manipulating these parameters in the tapering process. The differential equation that describes the shape of the taper is given by

$$\frac{dr}{dx} = -\frac{r}{2L} \tag{7}$$

where *L* denotes the hot-zone length and r denotes the waist diameter.
The function of radius profile is given by the integral

$$r(x) = r_0 \exp\left(-\frac{1}{2}\int \frac{dx}{L}\right) \tag{8}$$

To relate the varying hot-zone length *L* with the elongation distance *x* during the tapering process, *L* can be replaced with any function of *x*. Linear function

$$L(x) = L_o + \alpha x \tag{9}$$

makes a convenient function for the integral in Eqn (8).

$$r(x) = r_0 \left(1 + \frac{\alpha x}{L_0}\right)^{-1/2\alpha} \tag{10}$$

where $r_0$ denotes the initial radius of the fiber. To express the taper profile as a function of *z*, distance along the tapered fiber is given as;

$$r(z) = r_0 \left(1 + \frac{2\alpha z}{(1-\alpha)L_0}\right)^{-1/2\alpha} \tag{11}$$

By manipulating the value of α, several shapes of tapered fiber can be produced such as reciprocal curve, decaying-exponential, linear and concave curve. Several examples of calculated taper shape based on different values of α can be found in the literature of (Birks and Li, 1992 ). Consider the case of tapered fiber with decaying-exponential profile as shown in Fig. 6, the fabrication of such tapered fiber requires a constant hot-zone length (α=0). From the theoretical model presented above, the function for the decaying-exponential profile is given by

$$r(z) = r_0 \exp(-z / L_0) \tag{12}$$

Based on this profile function, narrower taper waist can be achieved by using a small hot-zone length in the fabrication or drawing the taper for a longer elongation distance. Tapered fiber with a short transition length can be achieved from reciprocal curve profile based on positive value of α particularly with $\alpha = 0.5$.

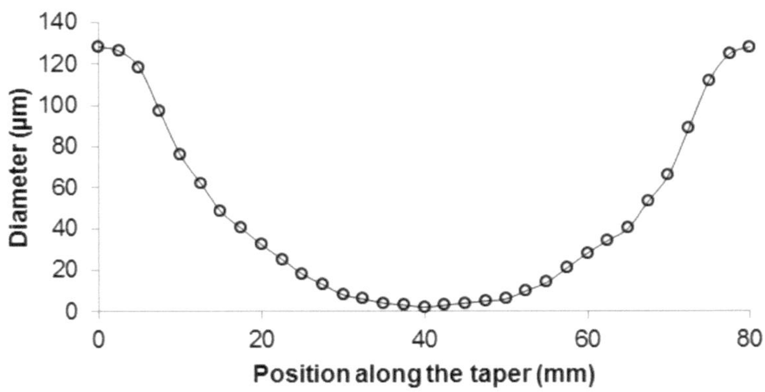


Fig. 6. A tapered fiber with decaying-exponential profile fabricated using a constant hot-zone $L_0$=10mm.

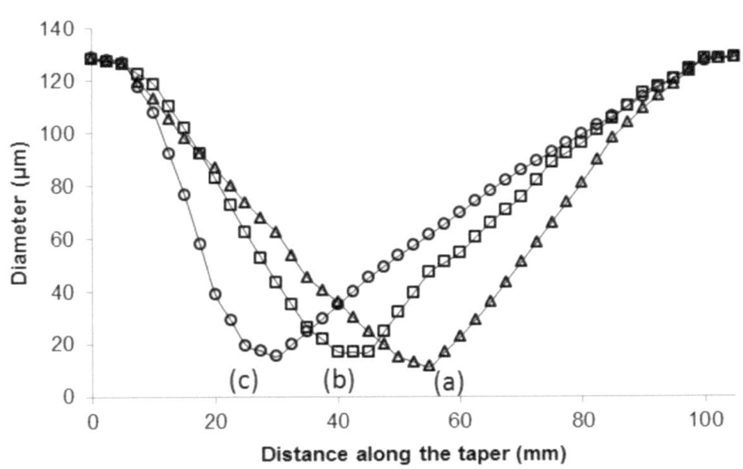


Fig. 7. Three linear taper profiles (a-c) with its smallest waist point at different positions on the tapered fibers. Profile (a) has its smallest waist point at the center of the tapered fiber.

Linear taper profile can be produced using $\alpha$ = -0.5 and the profile function is given by

$$r(z) = r_0\left(1 - \frac{2z}{3L_0}\right) \tag{13}$$

Fig. 7 shows typical examples of linear taper profiles. As shown in the figure, profile (a) has the smallest waist diameter, which located at the center of the tapered fiber. By doing some simple modification on the tapering process, the smallest waist point can be shifted away from the center to one side of the tapered fiber as shown by profiles (b) and (c) in Fig. 7. These profiles are found useful in the fabrication of wideband chirped fiber bragg grating, in which the grating is written on the transition of the tapered fiber. Long linear shape tapers make good candidates for the fabrication of such devices (Frazão and et al., 2005; Mora et al., 2004; Ngo et al., 2003; Zhang et al., 2003 ). On the other hand, linear profile tapers can be used for optical tweezing because of its capability to converge the optical wave to a high intensity at the taper tip (Liu et al., 2006; Xu et al., 2006 ). Microscopic objects are attracted to the high intensity field driven by the large gradient force at the taper tip. Fig. 8 gives a good example of such tapered fiber with 15cm linear taper profile. It was produced by using a long initial hot zone length $L_o$ = 7cm and long elongation distance.

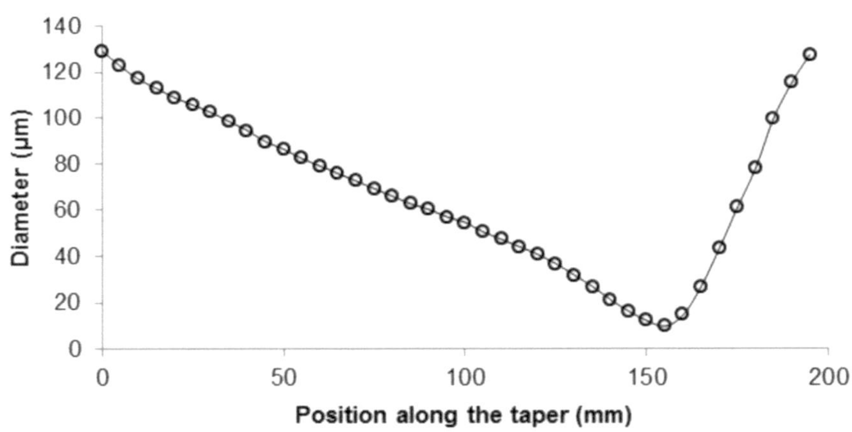


Fig. 8. The diameter of tapered fiber is linearly decreasing from ~128µm to ~10µm along the 15cm transition.

### 2.4 Throughput power of a degrading tapered fiber

In the high humidity environment, the concentration of water molecules in the air is high and very 'hazardous' to tapered fibers/microfiber. The increasing deposition of particles (dust) and water molecules on the microfiber is one of the major factors which causes adsorption and scattering of light that lead to perpetual decay in transmission (Ding et al., 2010 ). In an unprotected environment, freestanding microfibers may sway in the air due to the air turbulence. A small mechanical strength induced can cause cracks in the glass structure which may result to an unrecoverable loss in the microfibers (Brambilla et al., 2006 ).

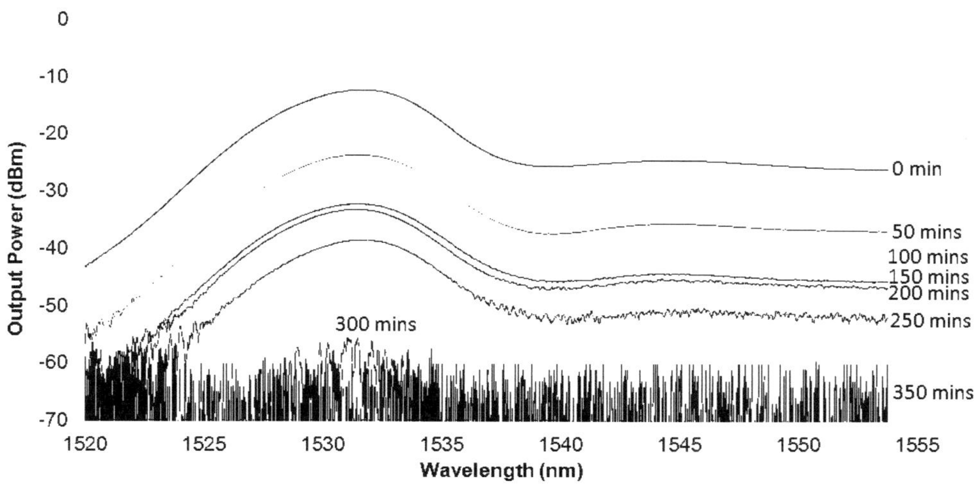


Fig. 9. The throughput power of the 10 cm long and ~3um diameter tapered fiber degrades over time.

Fig. 9 shows the output spectra of a tapered fiber in an unprotected environment. After the tapered fiber was drawn, it was left hanging in an open air and transmission spectrum was scanned and recorded every 50mins as presented in Fig. 9. Over the time, the deposition of dust and water molecules on the taper waist accumulated and the insertion loss of the tapered fiber increased over time. The output power of the tapered fiber dropped monotonically and eventually the power has gone too low beyond detection after 350mins. The throughput of the tapered fiber can be recovered by flame-brushing again as suggested in (Brambilla et al., 2006 ) but there is a risk that the tapered fiber can be broken after several times of flame-brushing and this solution is not practical. Despite the fact that the experiment was carried out in an air-conditioned lab where the humidity was lower (40-60%) but it was still too high for the tapered fibers. Besides, free standing tapered fibers are vulnerable to air turbulence or any sharp objects. New strategies for handling these tapered fibers are crucial for the ensuing research and fabrication of microfiber devices. In order to achieve that, this research team has been motivated to devise a packaging method to address all the problems mentioned earlier which will be discussed in the next section.

## 3. Packaging of microfiber

### 3.1 Embedding microfiber photonic devices in the low-index material

Besides the fast aging of bared microfiber in the air, the portability is another issue encountered when the microfiber is required at a different location. Moving the fabrication rig to the desired location is one way of solving the problem but it is not practical. Without a proper technique, it is risky to remove the tapered fiber from the fiber tapering rig and deliver it intact to another location. Xu and Brambilla (2007) proposed a packaging technique by embedding microfiber coil resonators in a low-index material named Teflon. Microfiber or microfiber device can be coated with or embedded in Teflon by applying some Teflon resin in solution on them and leave the solution to dry for several ten minutes. The resin is solidified after the solvent has finished evaporating from the solution, the optical properties and mechanical properties of the microfiber devices can be well preserved in the material for a very long time (Xu and Brambilla, 2007 ). Jung et al. (2010) had taken slightly different approach by embedding microfiber devices in a low-index UV-curable resin. The resin is solidified by curing it with UV-light. Here, the detail procedure of embedding a microfiber device in the low-index UV-curable resin is demonstrated.

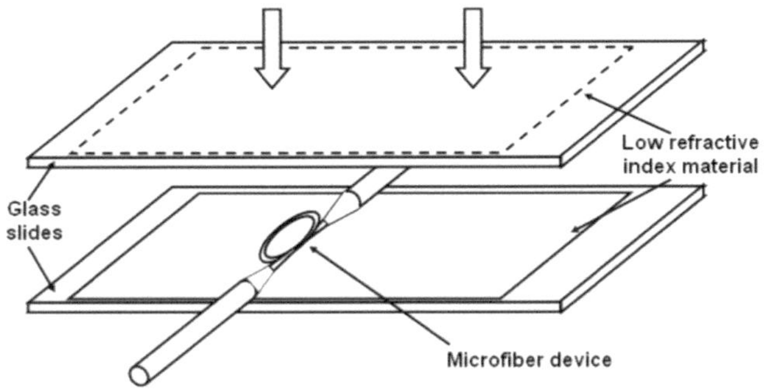


Fig. 10. Illustration of microfiber device embedded in a low refractive index material and sandwiched between two glass plates.

First, the assembled microfiber device is laid on an earlier prepared glass plate with a thin and flat layer of low refractive index material (UV-Opti-clad 1.36RCM from OPTEM Inc.) as shown in Fig. 10. The material has a refractive index of 1.36 at 1550nm. The thickness of the low refractive index material is approximately 0.5mm which is thick enough to prevent leakage of optical power from the microfiber to the glass plate. Some uncured resin is also applied on surrounding the microfiber device before it is sandwiched by another glass plate with the same low refractive index resin layer from the top. It is essentially important to ensure that minimum air bubbles and impurity are trapped around the fiber area between the two plates. This is to prevent refractive index non-uniformity in the surrounding of microfiber that may introduce loss to the system. During the tapering, coiling and coating processes, we monitored both the output spectrum and the insertion loss of the device in real time using the ASE source in conjunction with the OSA. The uncured resin is solidified by the UV light exposure for 3 ~7 minutes and the optical properties of the microfiber device are stabilized. The image of the end product is as shown in Fig. 11.

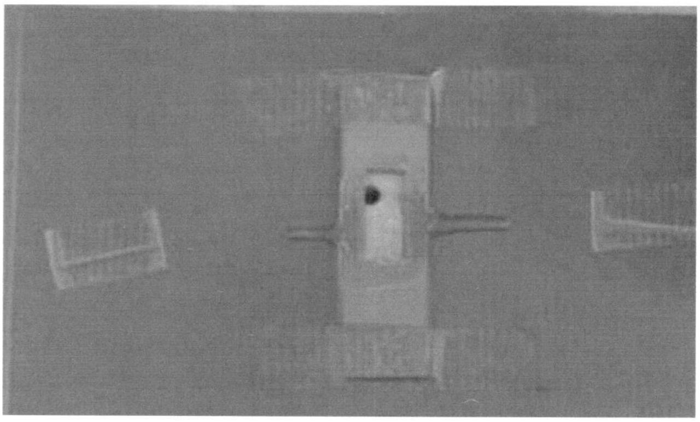

Fig. 11. The image of the end product of an embedded MLR in the low-index resin.

Fig. 12 shows recorded output spectra of the MKR at several intermediate times during the process of embedding it in a low-index UV-curable resin. The first spectrum in Fig. 12(a) was recorded right after an MKR was assembled. The fringes in the spectrum indicate that the resonance condition had been achieved in the MKR but the resonance extinction ratio remains appalling ~3dB. The MKR was benignly laid on an earlier prepared glass slide with thin layer of low refractive index material. After that, some low-index resin (refractive index ~1.36 ) in solution was applied onto the MKR by using a micropipette. Fig. 12(b) shows the stabilized output spectrum of the MKR and the improved resonance extinction ratio ~10dB. This phenomenon can be attributed to the reduction of index contrast; the mode field diameter (MFD) is the microfiber was expanded when it was immersed in the resin and the coupling efficiency of the MKR was altered. The changes of coupling coefficient and round-trip loss of the MKR may have induced critical coupling condition in the MKR and enhanced the resonance extinction ratio. At time = 4 minutes, UV-curing was initiated and a little bit of fluctuation is observed in the output power and extinction ratio (refer Fig. 12(c)) during the curing process. After UV-curing for 6 minutes, the output spectrum became very stable and the resin was finally solidified. Fig. 13 shows the optical microscope image of the embedded MKR in UV-curable resin.

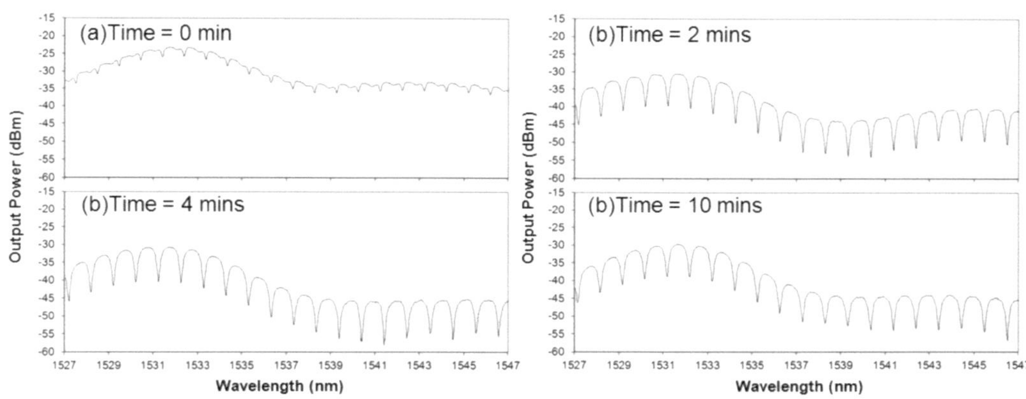


Fig. 12. Embedding an MKR in a low-index material. The time in each graph indicates when the output spectrum of the MKR is recorded. a) MKR is freestanding in the air b) some low-index resin applied on the MKR c) UV curing is initiated and d) resin is solidified.

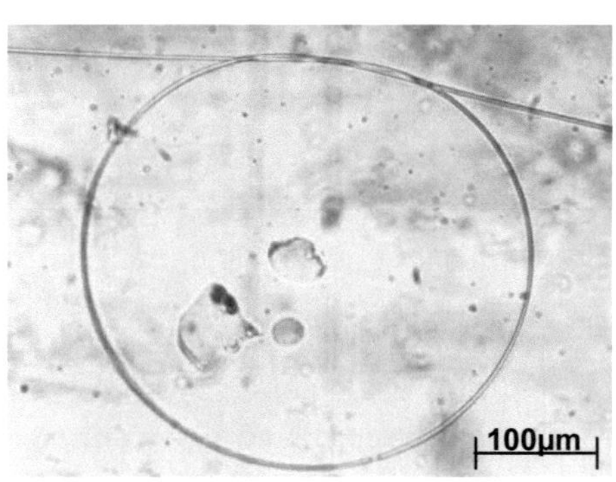


Fig. 13. Optical microscope image of an MKR embedded in UV-curable resin.

### 3.2 Packaging tapered fiber in a perspex case

Tapered fibers are susceptible to the air turbulence and the pollution of dust and moisture when exposed to air. It is very fragile when removed from the fiber tapering rig and maintaining the cleanliness of the tapered fiber in an unprotected condition is difficult. However, another simple packaging method had been devised to address all the difficulties mentioned. For the purpose of long term usage and ease of portability of the tapered fiber, a proper packaging process is essential. In the previous section, microfiber device is embedded in the low-index UV-curable resin to maintain the physical structure and resonance condition of the devices. Although, the refractive indices of the materials are in the range of 1.3~1.4 which is slightly lower than refractive index of silica tapered fiber and the mode can be still be confined within the tapered fiber but some optical properties such as numerical aperture (NA) and MFD will be altered due to the change in refractive index difference between silica microfiber and ambient medium when embedded in the low-index material. In the context of maintaining small confinement mode area and high optical nonlinearity, this method may not be a good idea.

In this section, a new packaging method is proposed where the tapered fiber is kept in a perspex case. The taper waist is kept straight and surrounded by the air without having any physical contact with any substance or object thus maintaining its optical properties in the air. The following part of this section provides detail descriptions of this packaging method. First, an earlier prepared perspex tapered fiber case which was made of several small perspex pieces with a thickness of 2.5mm was used in housing the tapered fiber. The perspex case mainly comprises of a lower part and upper part. Both parts of the perspex case were specially prepared in such a way that the benches at both ends of the perspex case were positioned exactly at the untapered parts of the tapered fiber. After a fresh tapered fiber was drawn, the lower part of the perspex case was carefully placed at the bottom and in parallel with the tapered fiber. That can done with the assistance of an additional translation stage. Then, the perspex case was slowly elevated upward until both benches touch both untapered parts of the tapered fiber as shown in Fig. 14(a). After that, some UV-curable optical adhesive (Norland Product, Inc) was applied to the untapered fibers that laid on the benches before the upper part of the tapered fiber case covered the tapered fiber from the top as shown in Fig. 14(b).

The UV-curable adhesive was used to adhere both the upper part and the lower part of fiber taper case. Despite that the refractive index of the optical adhesive(~1.54) is higher than silica glass (1.44) but the adhesive was only applied to untapered fiber and the light confined within the core of the fiber is unaffected. To cure the UV-curable adhesive, 9W mercury-vapour lamp that emits at ~254nm was used. Depending on the adhesive volume and its distance from the mercury-vapour lamp, the curing time takes for 2-8mins. After the adhesive was solidified and both case parts were strongly adhered to each other (Refer Fig. 14(c)). During the process illustrated in Fig. 14(a)-(c), the fiber taper was held by the two fiber holders in fiber taper rig and this helped to keep the fiber taper straight until the completion of the UV-curing process. After that the fiber taper and its case can be safely removed from the fiber holders. The fiber taper packaged inside the perspex case may remain straight permenantly. In the contrary, the fiber taper may suffer higher insertion loss if the taper fiber was bent during the packaging process. On the other hand, it is essential to prevent any physical contact between the taper waist with human hands or other objects.

The dust or moisture on the fiber taper may introduce loss to the transmission. In the final step, the perspex case was sealed by wrapping it with a piece of plastic wrap. This can minimize the pollution of dust or air moisture in the perspex case for a very long period of time. This fiber taper can be kept in storage for a week and possibly a fortnight without having an increment of loss more than 1.5dB however it is subject to taper dimension and its usage in the experiment.

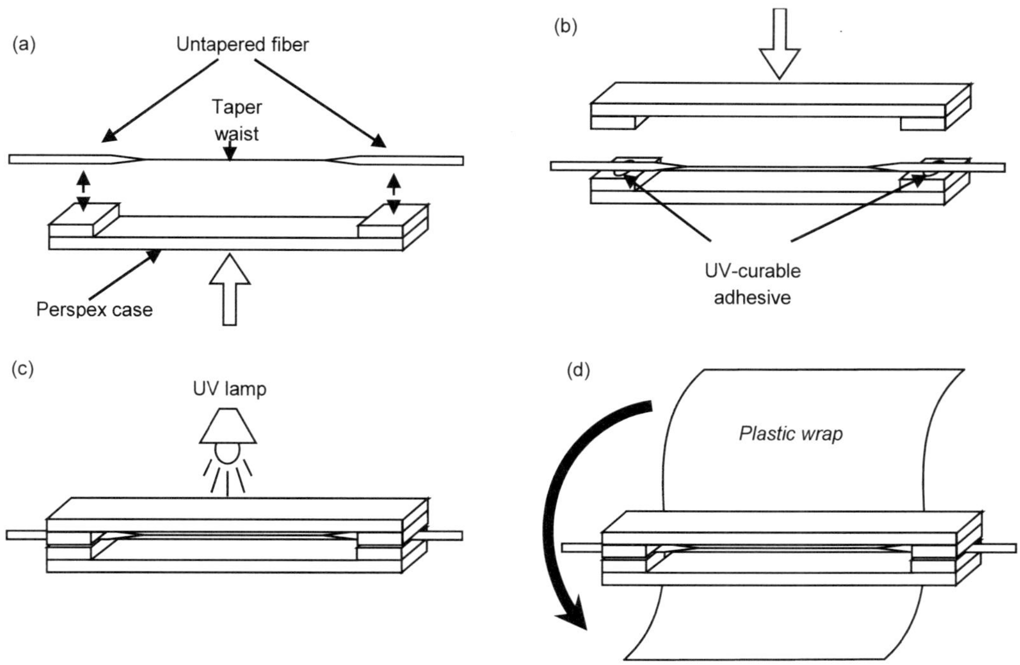


Fig. 14. Schematic illustration for tapered fiber packaging process.

To observe the characteristic of the tapered fiber as well as the reliability of the tapered fiber case over time, an observation on the transmission spectrum was conducted on the packaged tapered fiber for 6 days. Figs. 16(a) and (b) show the 6 days output spectra and output power observation, respectively for the packaged tapered fiber. Unlike the monotonic decrease in throughput power observed in Fig. 9(a), the curve of every transmission spectrum is closely overlaid to each other with a small power variation <1.2dB in the graph. Refer to Fig. 16(b), the variation of the total output power is spontaneous which can be attributed to the fluctuation of power at the ASE source and change of ambient temperature. In comparison with the taper fiber without packaging, obviously Perspex case plays its role well in preserving the tapered fiber to a longer lifespan; it enables portability and allows integration with more complex optical fiber configurations away from the fiber tapering rig.

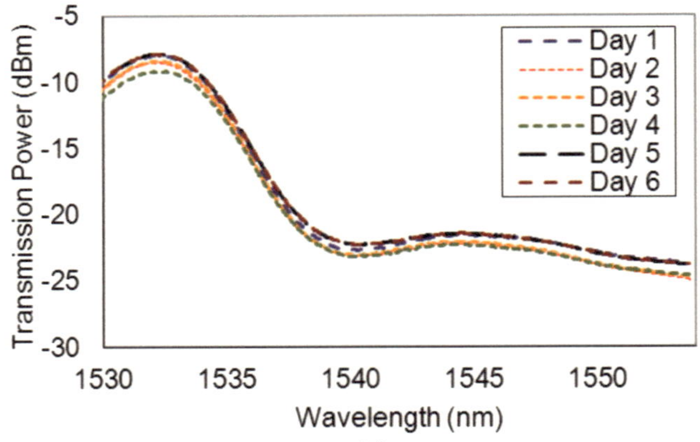


(a)

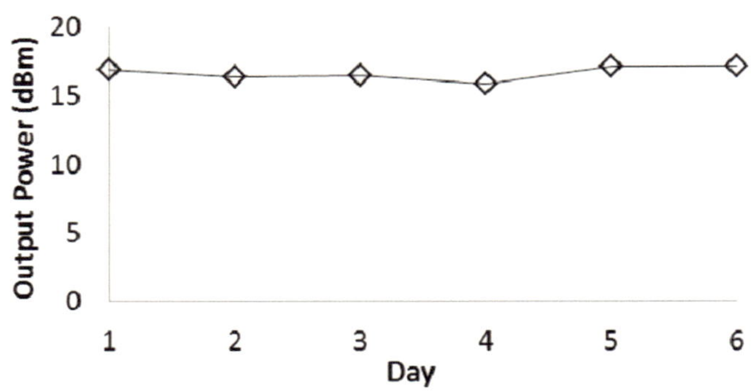


(b)

Fig. 16. The 6 days comparison of (a) output spectrum and (b) output power variation of the 10cm long and ~3um diameter tapered fiber packaged in perspex case.

## 4. Optical microfiber devices

Optical microfiber devices have attracted growing interest recently especially in their simple fabrication methods. This is due to a number of interesting optical properties in this device, which can be used to develop low-cost, miniaturized and all-fiber based optical devices for various applications (Guo et al., 2007 ). For instance, many research efforts have been focused on the development of microfiber/nanofibers based optical resonators that can serve as optical filters, which has many potential applications in optical communication, laser systems (Harun et al., 2010 ), and sensors (Hou et al., 2010; Sumetsky et al., 2006 ). Many photonic devices that are conventionally fabricated into lithographic planar waveguides can also be assembled from microfibers. Recently, there are many microfiber devices have been reported such as MLR (Harun et al., 2010; Sumetsky et al., 2005 ), MCR (Sumetsky, 2008; Sumetsky et al., 2010; Xu and Brambilla, 2007; Xu et al., 2007 ), MKR (Jiang et al., 2006; Lim et al., 2011; Wu et al., 2009 ), reef knot microfiber resonator as an add/drop filter (Vienne et al., 2009 ) , microfiber mach-zehnder interferometer (MMZI) (Chen, 2010; Li and Tong, 2008 ) and etc. These microfiber based devices have the similar functionalities, characteristics and possibly the same miniaturizability with the lithographic planar waveguides. In future, these microfiber based devices may be used as building blocks for the larger and more complex photonic circuits. In this chapter, the transmission spectrum and the corresponding theoretical model of three microfiber based devices are presented, namely MLR, MKR and MMZI. In addition, some of the important optical properties of these devices will be reviewed and discussed.

### 4.1 Microfiber Loop Resonator (MLR)

MLRs are assembled from a single mode microfiber, which is obtained by heating and stretching a single mode fiber. In the past, many MLRs have been demonstrated. For instance and Bachus (1989) assembled a 2mm diameter MLR from an 8.5μm tapered fiber where the coupling efficiency can be compromised by the large thickness of the microfiber. However the deficiency was compensated by embedding the MLR in a silicone rubber which has lower and near to the refractive index of silica microfiber. The transmission spectrum with a frequency spectral range (FSR) of 30GHz is observed from the MLR (Caspar and Bachus, 1989 ). Later on, Sumetsky et al. had demonstrated the fabrication of MLR from a ~1μm diameter waist microfiber which has the highest achieved loaded Q-factor as high as 120,000 (Sumetsky et al., 2006 ). Guo et al. demonstrated wrapping a ~2 μm diameter microfiber loop around copper wire which is a high-loss optical medium. By manipulating the input-output fiber cross angle, the loss induced and the coupling parameter in the resonator can be varied. In the condition when the coupling ratio is equivalent to the round-trip attenuation, the MLR has achieved critical coupling and the transmission of resonance wavelength is minimum. In their work, critical coupling condition have been achieved which resonance extinction ratio as high as 30dB had been demonstrated (Guo et al., 2007; Guo and Tong, 2008 ).

#### 4.1.1 Fabrication of MLR

Fig. 19 shows an example of ~3mm loop diameter MLR assembled from a ~2.0 μm waist diameter microfiber. Similar to other optical ring resonators, MLR has a 'ring' but manufactured from a single mode microfiber. This fabrication can be carried out with the assistance of two 3D translation stages as illustrated in Fig. 20. By aligning the three-axial position of each translation stage and twisting one of the pigtails, the microfiber is coiled into a loop. If the microfiber is sufficiently thin, the van der Waals attraction force between two adjacent microfibers is strong enough to withstand the elastic force from the bending microfiber and maintain the microfiber loop structure. The diameter of the loop can then be reduced by slowly pulling the two SMFs apart using the translation stages. Due the large evanescent field of the microfiber, a coupling region is established at the close contact between the two microfibers and a closed optical path is formed within the microfiber loop. Since the MLR is manufactured from an adiabatically stretched tapered fiber, it has smaller connection loss because microfiber based devices do not have the input-output coupling issue encountered in many lithographic planar waveguides. Despite the difference in the physical structure and fabrication technique between MLR and the conventional optical waveguide ring resonator, they share the same optical characteristics.

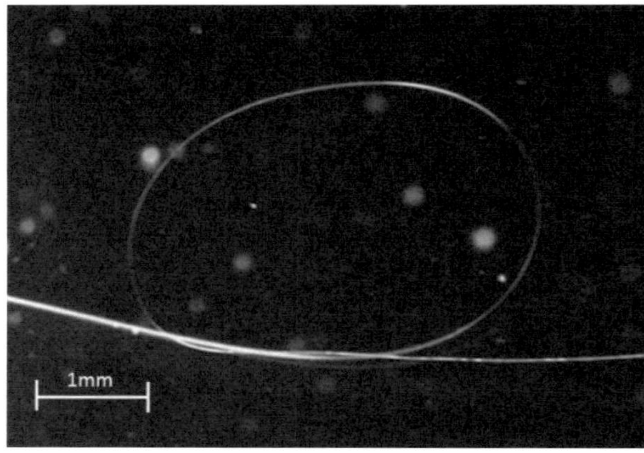


Fig. 19. Optical microscope image of an MLR.

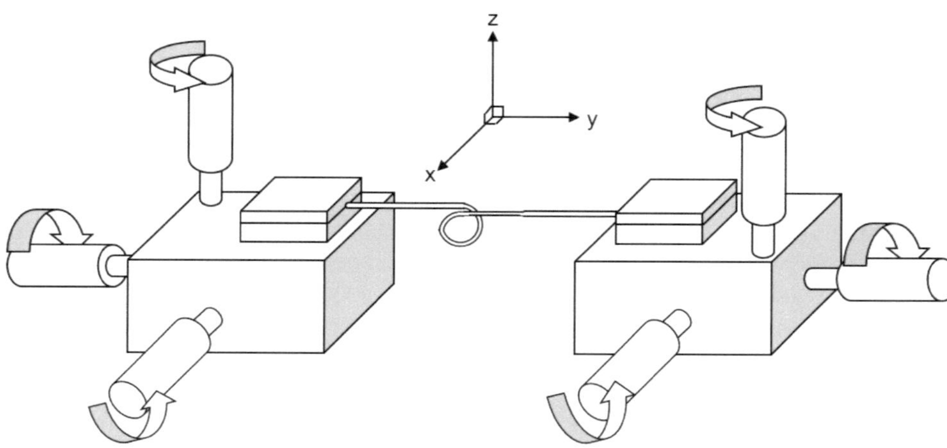


Fig. 20. Manufacture of MLR by using two three-dimensional stages.

### 4.1.2 Theory

The microfiber guides light as a single mode waveguide, with the evanescent field extending outside the microfiber. This evanescent field depends on the wavelength of operation, the diameter of the fiber and the surrounding medium. If the microfiber is coiled onto itself, the modes in the two different sections can overlap and couple to create a resonator. On every round-trip of light in the loop, there are fractions of light energy exchange between the two adjacent microfibers at the coupling region, the input light is allowed to oscillate in the closed loop and the resonance is strongest when a positive interference condition is fulfilled which can be related to this equation.

$$n\lambda_R = L \tag{14}$$

where $L$ is the round-trip length, $\lambda_R$ is the wavelength of the circulating waves and $n$ is an integer. Positive interference occurs to those circulating waves and the wave intensity is building up within the microfiber loop. The relationship in Eqn (14) indicates that each wavelength is uniformly spaced and periodic in frequency, a well-known characteristic of an optical multichannel filter. The amplitude transfer function for MLR is given as (Sumetsky et al., 2006 );

$$T = \frac{\exp(-\alpha L/2)\exp(j\beta L) - \sin K}{1 - \exp(-\alpha L/2)\exp(j\beta L)\sin K} \tag{15}$$

sin$K$ denotes the coupling parameter where $K = \kappa l$, $\kappa$ is the coupling coefficient and $l$ is the coupling length. For every oscillation in the MLR, the circulating wave is experiencing some attenuation in intensity attributed to non-uniformity in microfiber diameter, material loss, impurity in the ambient of microfiber and bending loss along the microfiber loop. However, these losses can be combined and represented by a round-trip attenuation factor, exp($aL/2$) in Eqn (15) while exp($j\beta L$) represents phase increment in a single round-trip in the resonator. The intensity transfer function is obtained by taking the magnitude squared of the amplitude transfer function in Eqn (15).

$$|T|^2 = \frac{\exp(-\alpha L) + \sin^2(K) - 2\exp(-\alpha L/2)\sin(K)\cos(\beta L)}{1 + \exp(-\alpha L)\sin^2(K) - 2\exp(-\alpha L/2)\sin(K)\cos(\beta L)} \tag{16}$$

The resonance condition occurs when

$$\beta L = 2m\pi \tag{17}$$

where $m$ is any integer. The critical coupling occurs when

$$\sin K_c = \exp(-\alpha L/2) \tag{18}$$

The FSR is defined as the spacing between two adjacent resonance wavelengths in the transmission spectrum which is given by

$$\text{FSR}, \ \Delta\lambda \approx \frac{\lambda^2}{n_{eff}L} \tag{19}$$

or

$$\Delta\lambda \approx \frac{\lambda^2}{n_{eff}\pi D} \tag{20}$$

where D is the diameter of the circular loop.
In addition to the characteristic parameters mentioned earlier, Q-factor and finesse F are two important parameters that define the performance of the MLR. The Q-factor is defined as the ratio of resonance wavelength to the bandwidth of the resonance wavelength, the full wave at half maximum, FWHM (Refer Fig. 21). It is given as;

$$Q = \frac{\lambda}{FWHM} \tag{21}$$

The finesse is defined as the FSR of the resonator divided by the FWHM;

$$F = \frac{FSR}{FWHM} \quad (22)$$

Due to the narrow bandwidth at the resonance wavelengths, MLR also functions as a notch filter (Schwelb, 2004 ). The attenuation at the resonance wavelength can be used to filter out/drop the signal from specific channels in the WDM network by suppressing the signal power. In DWDM network, the spacing between two adjacent channels in the network is small therefore notch filter with narrow resonant bandwidth is preferable so that the signals from adjacent channels are unaffected by the attenuation in the drop channel. Based on the relationship in Eqn (22), narrow resonant bandwidth (FWHM) can be found in high finesse filter.

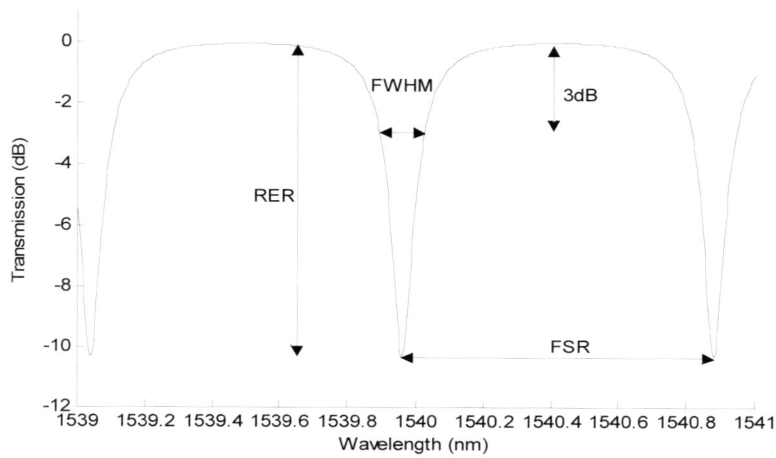


Fig. 21. Typical transmission spectrum of an MLR. The labels in the graph indicates the terminology used in the chapter. RER is an abbreviation for resonance extinction ratio.

### 4.1.3 Transmission spectra of MLRs

The typical transmission spectra of an MLR with different FSRs are shown in Fig. 22. For better clarity of viewing, the transmission spectra with different FSRs are presented in a increasing order from the top to the the bottom in the figure. These transmission spectra were recorded from a freestanding MLR in the air, started from a large loop diameter and the diameter is decreasing in step when the two microfiber arms of the MLR are stretched. Exploiting the van der Waals attraction force between the two microfibers in the coupling region, the resonance condition of the MLR can still be maintained during the stretching of microfiber. In the measurement, the loop diameters are at approximately 1.9mm, 1.4mm, 1.1mm, 0.8mm and 0.6mm which corresponds to FSR values of 0.275nm, 0.373nm, 0.493nm, 0.688nm and 0.925nm, respectively in the C-band region as shown in Fig. 22. These variations of FSR and loop diameter are very consistent with the reciprocal relationship expressed in Eqn (23). The loop diameter of an MLR is restricted by the microfiber elastic force, the smaller is the loop diameter the greater is the elastic force. Thus, it is difficult to keep the microfiber loop in shape when the loop diameter is very small and the MLR loses its resonance condition when the loop opens.

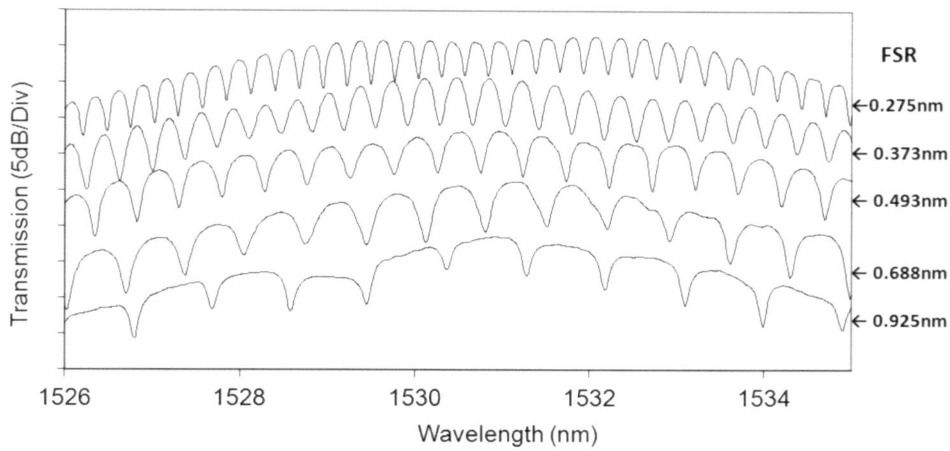


Fig. 22. Transmission Spectra of an MLR with increasing FSR (from top to bottom).

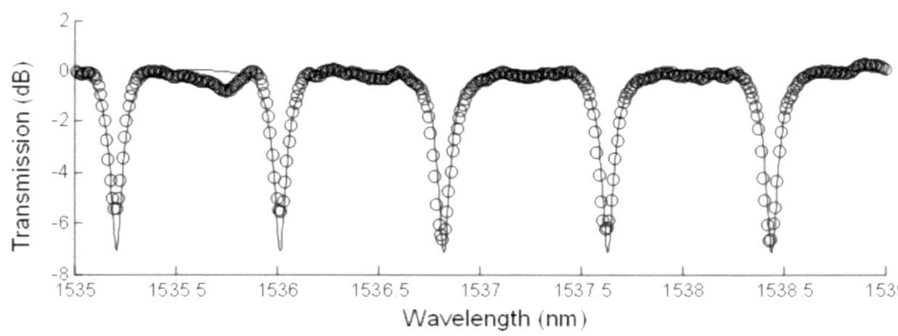


(a)

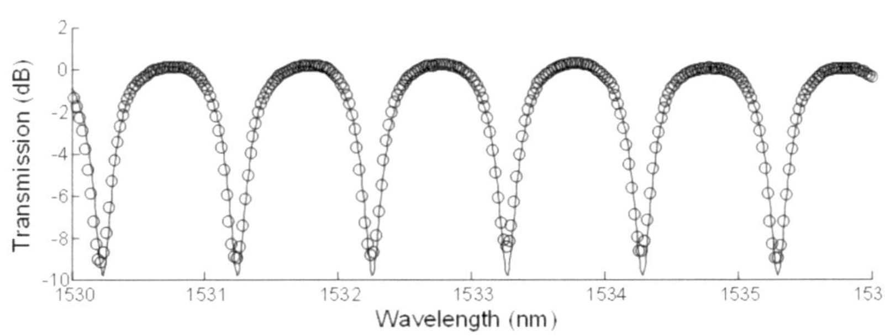


(b)

Fig. 23. The fitting of experimental data (circles) with the characteristic equation (solid line). (a) Q-factor ~18,000 and finesse ~9.5 (b) Q-factor ~5700 and finesse ~3.8

Fig. 23(a) and Fig. 23(b) show the fitting of experimental data with the (intensity) analytical model based on characteristic equation in Eqn (19). The best-fit parameters for the transmission spectrum in Fig. 23(a) are $L$ = 2.03mm, exp $(-aL/2)$ = 0.8853 and sin$K$ = 0.7354. The measured FSR ~0.805nm from transmission spectrum in Fig. 23(a) is in agreement with the calculated FSR ~ 0.807nm. The bandwidth at the resonance wavelength, FWHM is ~0.085nm indicates that the Q-factor and finesse of the MLR are 18,000 and ~9.5. The best fit parameter for Fig. 23(b) are $L$ = 1.61mm, sin $K$ = 0.5133 and exp ($-aL/2$) = 0.6961. The measured FSR and FWHM are ~1.02nm and ~0.27nm respectively which indicate the values of Q-factor and finesse are ~5700 and ~3.8. In the comparison between the two spectra, the finesse provides a good representation in weighting the bandwidth between passband and stopband. The higher is the finesse the narrower is the stop-band compared with pass-band.

### 4.2 Microfiber Knot Resonator (MKR)

MKR is assembled by cutting a long and uniform tapered fiber into two. One tapered fiber is used for the fabrication of microfiber knot while the other one is used to collect the output power of the MKR by coupling the two tapered fiber ends and guides the output light back to an SMF. The fabrication of microfiber knot can be done by using tweezers. The coupling region of the MKR is enclosed by a dashed box in Fig. 24 where the two microfibers intertwisted and overlapped in the resonator. In comparison with MLR, MKR does not rely on van der Waals attraction force to maintain the coupling region yet it can achieve stronger coupling due to the rigid intertwisted microfibers structure at the coupling region. The knot structure can withstand strong elastic force of the microfiber and maintain a rigid resonator structure with a more stable resonance condition. Based on the same microfiber diameter, MKR of smaller knot diameter can be easily manufactured than that of MLR. However, MKR suffers a setback in a high insertion loss due to the cut-coil-couple process where the evanescent coupling between output microfiber and collector microfiber contributes a large fraction in the total insertion loss. The microfiber diameter in the range of 1~3μm is preferable because thinner microfiber is very fragile and it breaks easily in the fabrication of MKRs. Nonetheless, the operating principle of MKR is identical to MLR as it is based on self-touching configuration thus the same characteristic equation can be used to describe the transmission spectrum of MKR.

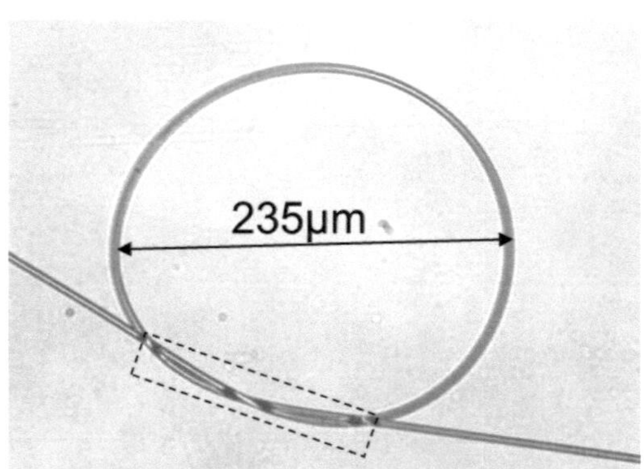


Fig. 24. Optical microscope image of an MKR.

#### 4.2.1 Transmission spectra of MKRs

MKR offers better capability in achieving smaller knot diameter. The knot can withstand the strong elastic force of microfiber and it can achieve a small knot diameter that cannot be achieved in the MLR. Fig. 25 shows transmission spectra of an MKR assembled in the laboratory. The transmission spectra are presented in an increasing order from the top to bottom of the figure and the values are 0.803nm, 1.030nm, 1.383nm, 1.693nm, 2.163nm and 2.660nm within the vicinity of 1530nm. The corresponding knot diameters are approximately 640μm, 500μm, 370μm, 310μm, 240μm and 190μm.

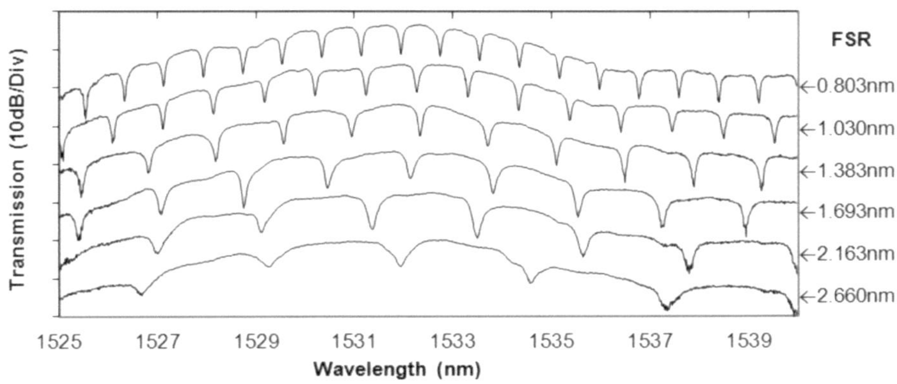


Fig. 25. Transmission Spectra of an MKR with increasing FSR (from top to bottom).

### 4.2.2 Resonance condition of microfiber knot resonator immersed in liquids

Recently, microfiber resonators are suggested in numerous applications particularly in the sensing applications (Lim et al., 2011; Sumetsky et al., 2006 ). The operating principles of these sensors rely on the characteristics of the resonance, the variation of the position of resonance wavelength and the resonance extinction ratio with the sensing parameters, temperature, refractive index and etc. (Wu et al., 2009; Xu et al., 2008 ). The resonance condition of a resonator relies on the index contrast between microfiber and its ambient medium, evanescent field strength and distance between two microfibers in the coupling region. Large evanescent field which can be found in thinner microfibers is one of the solutions to achieving higher coupling in microfiber resonators. The large fraction of light intensity in the evanescent field allows stronger mode interaction between two microfibers and yields high coupling coefficient. Caspar et al. suggest embedding the microfiber resonator into a medium that has a slightly lower refractive index than that of silica. Due to the small index contrast, the microfiber has a larger evanescent field which yields stronger coupling in the resonator (Caspar and Bachus, 1989; Xu and Brambilla, 2007 ). Besides being used as a post-fabrication remedy for improving the resonance condition of the resonator, embedding also offers good protection from the fast aging process and enabling portability for the microfiber devices. Vienne at el. have reported that when a microfiber resonator is embedded in low-index polymer, the optimal resonance wavelength is down-shifted by ~20% (Vienne et al., 2007 ). However, there is very few literatures that provide mathematically analysis on the effect of embedding in low index contrast medium to the resonance condition of the resonator. In order to achieve a better understanding, an experiment on an MKR immersed in liquid solutions was carried out. MKR was used in the experiment due to its rigid knot structure and strong interfiber coupling. The knot structure and resonance condition could be easily maintained during the immersing process.

Unlike MLR that exploits van der Waals attractive force to maintain the structure of the loop, MKR has a more rigid knot structure with interfiber twisted coupling. Nonetheless, both MLR and MKR share the same optical properties, the same transmission equation can be used to describe both structures. Fabrication of an MKR started with fiber tapering using heat and pull technique. After a single-mode biconical tapered fiber was drawn, it was cut at one third part of the waist which the longer section of tapered fiber was used for the fabrication of knot by using tweezers. Then the second section was used as collector fiber by evanescent coupling (Tong et al., 2003 ) with the output port of the MKR. Immediately after that, the transmission spectrum of the freestanding MKR in the air was recorded by an OSA. After that, the MKR was embedded in propan-2-ol solution that has a refractive index (RI) of 1.37. First, the MKR was slowly laid horizontally on an earlier prepared flat platform deposited with a thin layer of propan-2-ol. Using a micropipette, a small volume of propan-2-ol solution was dropped onto the MKR and had it entirely immersed in the solution. The structure of the microfiber knot was intact and the resonance was maintained. This is the crucial part that distinguishes MKR from MLR. It is very difficult to maintain the loop structure and resonance of MLR when immersed in the liquid.

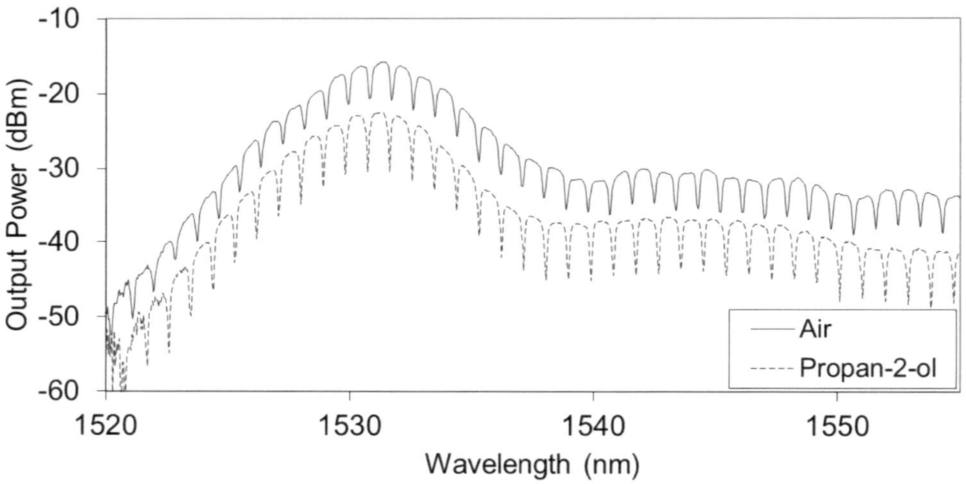


Fig. 26. The transmission spectra of MKR in the air (solid) and propan-2-ol solution (dashed).

Fig. 26 shows the overlaid transmission spectra of the MKR in the air (solid) and solution (dashed). Refering to the peak powers of both spectra, it is easy to determine that the MKR had suffered an additional ~7dB excess loss after it was immersed in the solution. The drop in coupling efficiency at the evanescent coupling between MKR output microfiber and collector microfiber constituted a large fraction in this excess loss. On the other hand, the resonance extinction ratio of the MKR had improved from ~5dB to ~8dB. In the analysis of resonance characteristics, the coupling parameter, $\sin\kappa\ell$ and round-trip attenuation factor of MKR, $\exp(-aL/2)$ can be extracted from the best-fit curves (lines) for the offset experimental data (circles) in Fig. 27(a) and Fig. 27(b) based on the transfer function in (18).

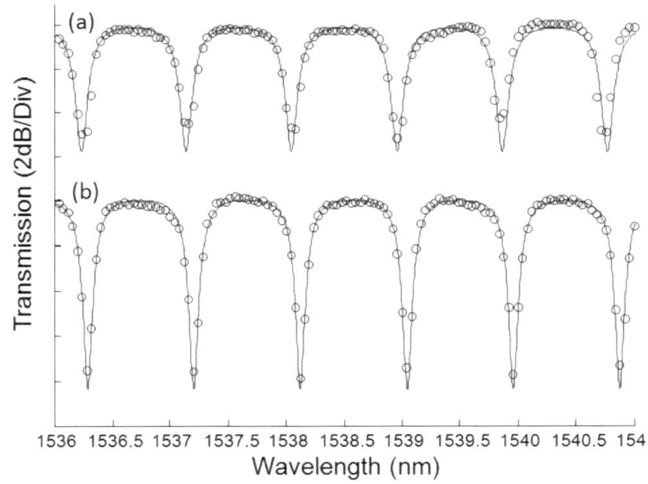


Fig. 27. The offset experimental data (circles) with its best fit curve (solid line) (a) air, RI ~1.00 (b) propan-2-ol, RI ~1.37.

In the air, the best fit parameters for the transmission spectrum in Fig. 27(a) are sin$\kappa\ell$ = 0.6207 and exp(-$aL$/2)= 0.8547. When the MKR was immersed in the propan-2-ol solution, the best fit parameters for Fig. 27(b) are sin$\kappa\ell$ = 0.6762 and exp(-$aL$/2) = 0.8361. The reduction in round-trip attenuation factor can be attributed to the small index contrast between microfiber and ambient medium when immersed in the solution so that the bending loss at the microfiber knot is higher. The output - collector coupling loss is excluded from this analysis as it only affects the total output power (position in the vertical axis) and it can be eliminated in the offset spectrum. Based on Eqn (18), the resonance state of the resonator can be determined from the following expression

$$\delta = \left| \frac{\exp(-\alpha L/2) - \sin\kappa l}{1 - \exp(-\alpha L/2)\sin\kappa l} \right| \tag{23}$$

Smaller value of $\delta$ indicates that the state of resonance is closer to the critical coupling condition and it yields larger resonance extinction ratio. In fact, the resonance extinction ratio can be estimated by

$$RER \sim 20\log_{10}\delta \tag{24}$$

Comparing the two spectra, the spectrum in Fig. 27(b) has higher coupling and smaller round-trip attenuation factor which give smaller value in $\delta$ = 0.3679 if compared with $\delta$ = 0.5060 obtained from the spectr sin$\kappa\ell$ and resulting the lower coupling value. The next experimental data may provide an example for such self-defeating scenario.The transmission spectra of an MKR in the air and low-index UV-curable resin (UV-Opti-clad 1.36RCM from OPTEM Inc.) with an RI of ~1.36 are as shown in Fig. 28(a) and Fig. 28(b) respectively. The coupling parameter, sin$\kappa\ell$ had dropped from 0.7132 to 0.6247 when it was immersed in the water. On the other hand, the round-trip attenuation factor, exp(-$aL$/2) suffers greater fall from 0.9432 to 0.7538. In spite of that, the resonance extinction ratio had increased from ~2dB to ~10dB. This is in agreement with the decreasing value of $\delta$ from 0.7027 to 0.2440 and the state of resonance is closer to critical coupling condition.

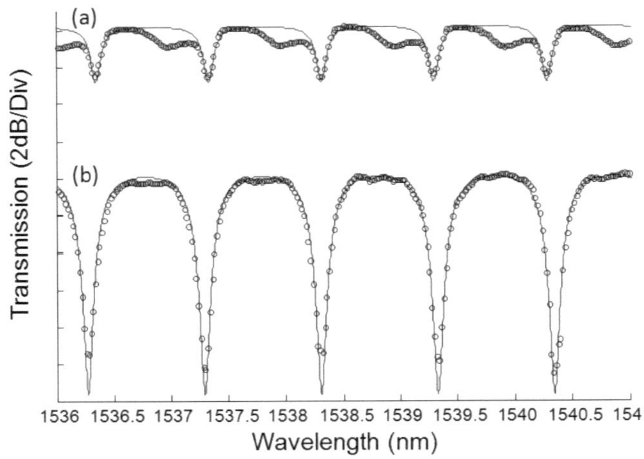


Fig. 28. The transmission spectra of MKR in different ambient mediums (a) air, RI ~1.00 (b) low-index resin, RI ~1.36.

Immersing MKR in a near-index medium do not always promise an improvement in the resonance condition or RER. There is a possibility that the changes in round-trip attenuation factor and coupling parameter yield larger value of $\delta$ and decreases the RER. Fig. 29 provides an example for this scenario. The best-fit parameter for the experimental data in the air (solid) are $\sin\kappa\ell$ = 0.6235 and $\exp(-aL/2)$ = 0.8145 respectively. After the MKR is immersed in the water (dashed), the values have varied to $\sin\kappa\ell$ = 0.7833 and $\exp(-aL/2)$ = 0.9339. In the air, the low value of round-trip attenuation factor can be attributed to the large amount of deposited dust on the microfiber surface which was introduced from the tweezers during the fabrication of microfiber knot. After it was immersed in the water, some portion of the dust might have been 'washed' away and that increases the round-trip attenuation factor. The value of $\delta$ has decreased from 0.3881 to 0.5609 which is an indication of the resonance state deviates from critical coupling.

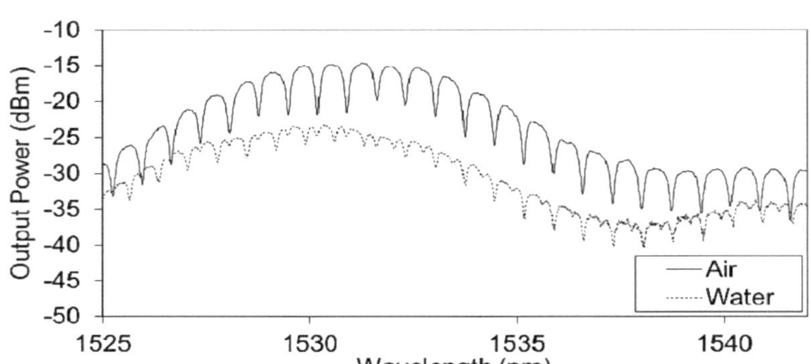


Fig. 29. Example of an MKR with decreased resonance extinction ratio after it is immersed in the water (RI ~1.33).

The purpose of using liquid solutions of different RIs is to investigate the influence of different index-contrasts to the resonance characteristics of the MKR. However, there was no significant indication observed in the experiment showing difference between the solutions. The only obvious changes were observed at the moment when the MKRs were immersed in the solutions. It is believed that the index contrasts induced by these three liquids are within a narrow range from 0.07 to 0.11 and the differences among them are too small to make significant impact on the characteristics of the MKR. On the other hand, we believe that that the microfiber waist diameter and orientation of the microfibers in the coupling region have an important relationship with the resonance of MKR. More investigations pertaining to those parameters are needed.

#### 4.2.3 Polarization dependent characteristic

Microfiber based resonator exhibits strong dependence on its input polarization state. Similar characteristic was reported by Caspar et al. (Caspar and Bachus, 1989 ) where the extinction ratio of the resonator varies with the change of input state of polarization.

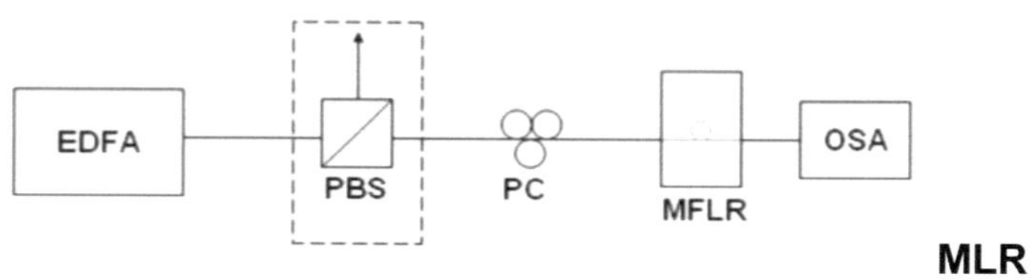


Fig. 30. Experimental set-up to investigate the polarization dependent characteristic of the MLR (Lim et al., 2011 ). Polarized wideband source from EDFA is acquired with the aid of PBS (Dashed box), an unpolarized wideband source can be obtained by removing PBS from the setup.

In the investigation of polarization state dependent characteristics, a simple experimental setup as shown in the Fig. 30 is established. First, the unpolarized ASE source from an EDFA was linearly polarized by a PBS, followed by a PC for controlling the state of polarization (SOP) of the polarized source before it was fed into the MLR. Fig. 31 shows the transmission spectra of the MKR at various resonance conditions of the MLR, which was obtained at different input wave SOP. Both spectra (i) in Fig. 31(a) and Fig. 31(b) show the transmission of the MLR based on an unpolarized input wave (without PBS), which resonance condition was unaffected by the adjustment of the PC. In contrast, the resonance condition for the MLR with polarized input wave was sensitive to the PC adjustment. By carefully adjusting the PC, the resonance extinction ratio could be enhanced or reduced as depicted in spectra (ii) and (iii) of both Fig. 31(a) and Fig. 31(b). Nevertheless, the wavelength of each peak and the FSR remained unchanged regardless of input wave SOP. It is appropriate to attribute this phenomenon to the polarization dependent coupling in the MLR where the interfiber coupling, twisting and alignment of microfiber in the coupling region are accounted for the coupling coefficient difference between two orthogonal polarization states (Bricheno and Baker, 1985; Chen and Burns, 1982; Yang and Chang, 1998 ). Associated with the coupling coefficient, the resonance extinction ratio of a microfiber based resonator can be improved by using an optimized polarized input light. However, more efforts are required for more in-depth study and to explore the possible applications in many fields such as multi-wavelength laser generation and sensing.

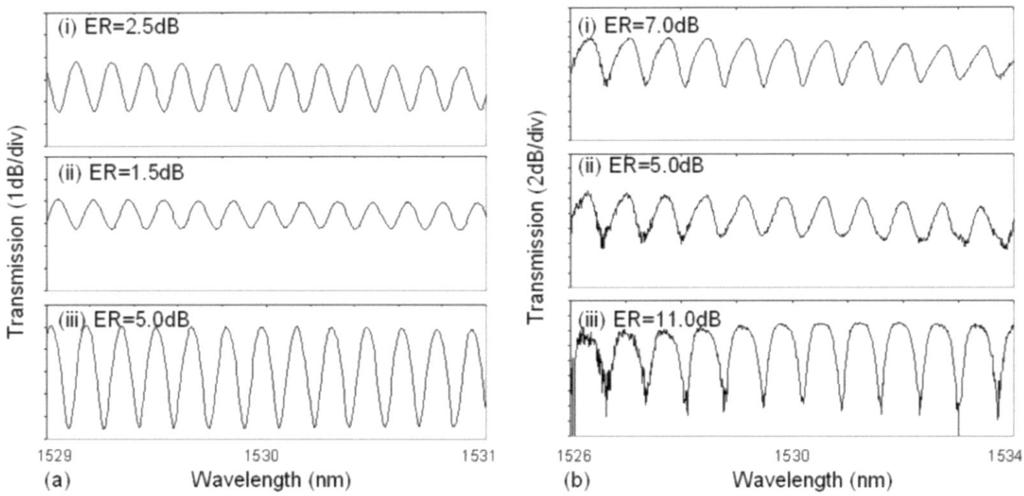


Fig. 31. The spectra of two MLRs for different input wave SOP. (a) FSR=0.162nm at 1530nm (b) FSR=0.71nm at 1530nm.

### 4.2.4 Thermal dependent characteristic

Microfiber based devices are very sensitive to variation of ambient temperature due to the strong dependence of the microfiber dimension and refractive index on temperature. In a thermally unstable environment, these devices may experience thermal drift in the transmission spectrum and fluctuation in the transmission power. However, this problem can be alleviated by the placing the devices in a temperature controlled housing (Dong et al., 2005 ). Sumetsky et al. demonstrated a MLR based ultrafast sensor for measurement of gas temperature. Taking advantage of the close contact between the MLR and air, the change in gas temperature in the ambient of MLR can be determined from the transmission power at the resonance wavelengths within a short response time of several microseconds (Sumetsky et al., 2006 ). Besides, the positions of resonance wavelengths are found to be sensitive to temperature change. The spectral shift of an MLR can be expressed in a linear function of temperature. The property enables temperature measurement based resonance wavelength shift with higher accuracy (Wu et al., 2009; Zeng et al., 2009 ).

MKRs exhibit the similar optical properties with MLRs. The free spectral range of MKR takes in the form of Eqn (22). Based on this equation, the variations in effective index $n_{eff}$ and round-trip length $L$ may lead to transmission spectral shift and their relationship can be expressed as

$$\frac{\Delta\lambda_{res}}{\lambda_{res}} = \left( \frac{\Delta n_{eff}}{n_{eff}} + \frac{\Delta L}{L} \right)_{Temp.} \tag{25}$$

In relation with temperature, both terms on the right hand side of Eqn (25) expresses two linear thermal coefficients; thermal-optic coefficient (TOC) and thermal expansion coefficient (TEC) [13]. With this interpretation, Eqn (25) can be rewritten as

$$\frac{\Delta\lambda_{res}}{\lambda_{res}} = (\alpha_{TOC} + \alpha_{TEC})\Delta T \tag{26}$$

Fig. 32(a) shows the transmission spectra of MKR at temperatures of 30°C, 35°C and 40°C. The spectral shift is approximately 26pm for every temperature increment of 5°C and the linearity between wavelength shift and temperature change can be seen from the linear fitting of experimental data in Fig. 32(b).

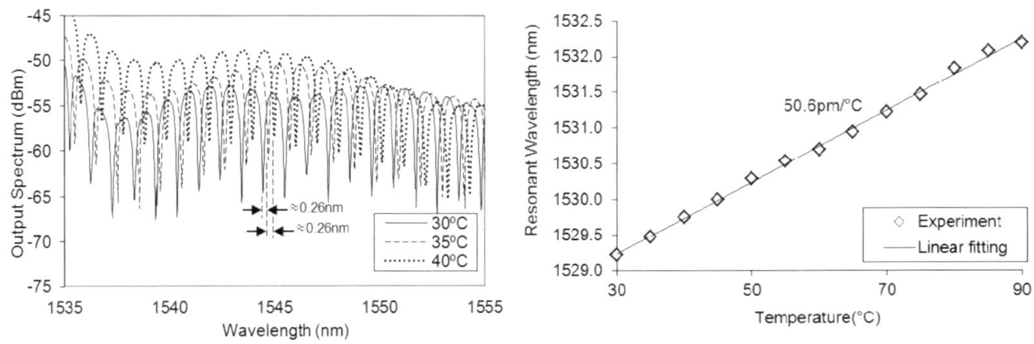


(a) (b)

Fig. 32. (a) The output spectra of an MKR at temperature of 30°C(solid), 35°C(dashed) and 40°C (dotted) (b) The temperature response of the MKR has spectral sensitivity of 50.6pm/°C.

This characteristic has opened up new possibilities for temperature sensing and spectral control based on temperature manipulation. It provides a solution for stabilizing the spectrum of the device which often affected by the thermal drift. Dynamic spectral shift in optical filter can be realized by exploiting this characteristic. In addition, the insusceptibility of fiber-optic components to electrical noise has made these devices very attractive for many industrial sensing applications.

### 4.2.5 Microfiber knot resonator based current sensor

A variety of fiber optic based current sensors have been investigated in recent years using mainly a single mode fiber (SMF) of clad silica. They are typically divided into two categories, where one is based on Faraday Effect and the other is based on thermal effect. The former is capable to remotely measure electrical currents, but the device requires a long fiber due to the extremely small Verdet constant of silica. The latter needs a short length of fiber but requires complex manufacturing techniques to coat fibers with the metals. Recently, a resonant wavelength of the MLR has been experimentally reported to shift with electric current applied to the loop through a copper rod. An acceptable transmission loss is achieved despite the fact that copper is not a good low-index material to support the operation of such structure (Guo et al., 2007 ). This finding has opened up a way to enable dynamic and efficient spectrum control for optical filters by manipulating an electric current dependence spectral shift characteristic of the microfiber based resonator. Microfiber based devices have a strong dependence on temperature due to the thermal expansion characteristic and thermo-optic effect of silica glass. As discussed earlier in Section 12, the transmission spectrum of a microfiber based device shifts as the ambient temperature varies

and the relationship between these two variables is well described by the linear equation in Eqn. (26).
In this section, spectral tunable MKR is demonstrated based on the idea of thermally induced resonant wavelength shift. By manipulating the applying electric current through the microfiber knot wrapped copper wire, the copper wire acts as a heating element and induces temperature change in the MKR. The transmission spectrum of the MKR shifts corresponds to the temperature change. These modified MKRs can be used as low-cost and fast response tunable optical filters which are useful in the applications of optical signal processing, WDM communication and etc. On the other hand, this opto-electrical configuration may operate as a dynamic current sensor with strong immunity to electric noise. In addition, it has a dynamic operational range extending to the regime of extreme high temperature or pressure.

**Fabrication**

First, a ~2μm diameter silica microfiber is fabricated from a SMF using flame-brushing method (Graf et al., 2009 ). Then the microfiber is cut and separated into two unequal parts in which the longer one is used in the knot fabrication and the other one is used as a collector fiber to collect the transmitted light from the MKR (Jiang et al., 2006 ). During the fabrication of the knot, the copper wire is inserted into the knot which diameter is bigger than the diameter of the copper wire (Refer Fig. 33(a)). The light path from the knot resonator is completed by coupling the two microfiber ends. At least ~3 mm of coupling length between two microfibers is required to achieve strong van der Waal attraction force to keep them attached together. The microfiber knot diameter is then reduced and fastened on the copper wire by pulling microfibers from both arms of the microfiber knot as illustrated in Fig. 33(b).

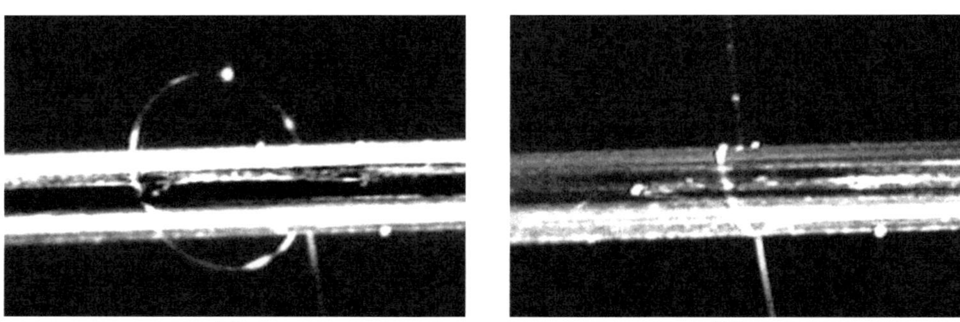

(a) (b)

Fig. 33. Optical microscope image of MKR tied on a copper wire.

The optical characteristics of the resonator are strongly affected by the tensile strain on the microfiber arms of the MKR induced by the pulling on the microfibers arms. It is essential to reduce the tension on the both arms of the MKR by moving the fiber holders a bit closer to the microfiber knot after the knot is fastened. In spite of that, there is a very little change at the knot diameter and the resonance condition of the MKR remains good and stable after the tension is released.

**Theoretical Background**

By wrapping microfiber on a current loaded conductor rod, the conductor rod acts as a heating element. It generates heat and increases the temperature of the microfiber. As discussed earlier in Section 13, the transmission spectrum of a microfiber based device shifts as its ambient temperature varies and the relationship between these two variables is well described by the linear equation in Eqn. (26). Consider the linear relationship between the temperature change and heat energy generated by the conducting current, the relationship between wavelength shift and the conducting current, $I$ can be expressed in the form of

$$\frac{\Delta\lambda_{res}}{\lambda_{res}} \propto \frac{\rho I^2}{A} \tag{27}$$

where $\rho$ and $A$ represent the conductor resistivity and the cross sectional area of the conductor rod, respectively. The term $\rho / A$ in Eqn. (27) is equivalent to the resistance per unit length of the conductor material. The resistivity of the copper rod is $1.68 \times 10^{-8}\ \Omega \cdot m$.

**Current Response**

For optical characterization of the MKR, broadband source from amplified spontaneous emission is first launched into and guide along the SMF and then squeezed into the microfiber through the taper area. The light transmitted out from the MKR is collected by the collection fiber and measured by an OSA. The optical resonance is generated when light traversing the MKR. When an alternating current flows through the copper wire, heat is produced in the wire to change temperature. Because the MKR is in contact with the copper wire, any temperature changes will influence the refractive index and the optical path length of the MKR.

Fig. 34 shows the resonant spectral of the MKR tied on a copper wire with various current loadings. In the experiment, the applying current is uniformly increased from 0 to 2A.

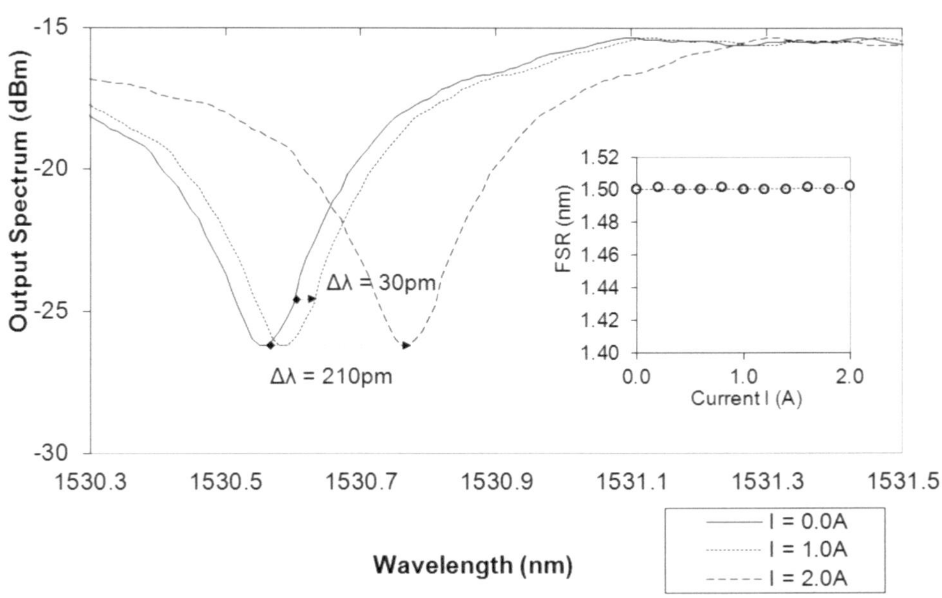


Fig. 34. Resonant wavelength shift of the MKR tied on a copper rod loaded with different current. Inset shows unchanged FSR with the increasing current.

In the spectrum, the resonant wavelength shifts to a longer wavelength the increasing of conducting current in the copper wire. The response time of the wavelength shift is approximately 3s and the spectrum comes to steady condition after 8s. Therefore, each spectrum is recorded at ~10s after the copper wire is loaded with an electric current. At loading current I = 1.0A, the resonant wavelength is shifted by ~30pm from 1530.56nm to 1530.59nm and at I = 2.0A, the resonant wavelength is further shifted to 1530.77nm, 210pm from the original wavelength. Inset of Fig. 34 shows the free spectral range of the transmitted spectrum against the applying currents. As shown in the inset, FSR of the MKR remains unchanged at 1.5 nm with the increasing current. The calculated Q factor and finesse of the MKR are ~4400 and 4.3 respectively. It is also observed that the transmission spectrum always shift towards the longer wavelength direction with increasing current regardless of the current flow direction, and the spectrum returns to original state once current supply is terminated.

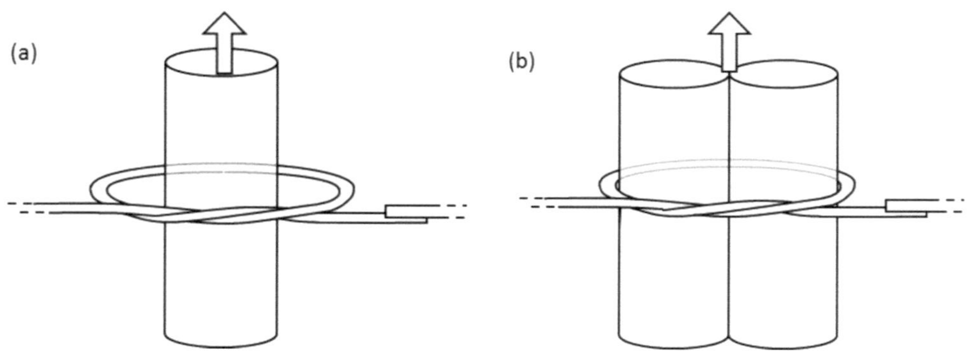


Fig. 35. Schematic illustrations of microfiber knot tied on (a) single copper wire (b) two copper wires with identical wire diameter of ~200μm.

Fig. 35(a) and Fig. 35(b) give schematic illustrations of MKR wrapped on a single copper rod and two copper rods respectively. The measured FSR and knot diameter of the single-rod MKR are 1.7nm and ~317μm respectively while for the two-rod MKR, the measured FSR and knot diameter are 1.46nm and ~370μm. An direct current is applied through the copper wire and the resonant wavelength shift is investigated against the applying current. At small current of < 0.5A, no significant resonant wavelength shift is observed. Beginning at 0.6A, the resonant wavelength shifts gradually toward the longer wavelength. At applying current of 2.0A, a wavelength shift of 0.208nm and 0.09nm are achieved with single and two copper wires configurations of Fig. 35(a) and Fig. 35(b), respectively. Fig. 36 shows the resonant wavelength shift against a square of current ($I^2$) for both configurations. The data set of each configuration can be well fit with a linear regression line with a correlation coefficient value r > 0.95. This justifies the linear relationship stated in Eqn. (27). In comparing the conductor wire cross-sectional area between the two configurations, the two-wire configuration is twice larger than the single-wire configuration. Based on the relation in Eqn. (27), the tuning slope of the wavelength shift with $I^2$ of the two-wire configuration should be a half of the single-wire configuration. The slope of each linear line is 51.3pm/A$^2$ (single rod) and 19.5pm/A$^2$ (two rods) nonetheless it is reasonable to attribute the mismatch between the analysis and experiment to the different orientation and position of the rod(s) in the MKR. The tuning slope of the current sensor can be further increased by using different conductors with higher resistivity such as nichrome, constantan, graphite and etc which are commonly used as heating elements. However, the suitability in integration with microfiber or other opto-dielectric device remains uncertain.

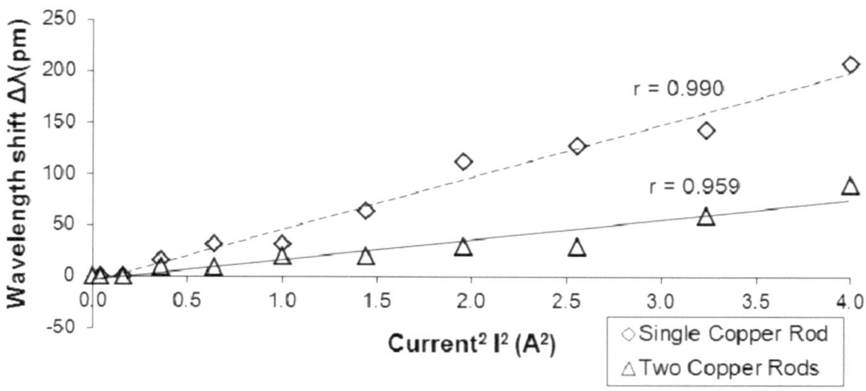


Fig. 36. Current response of MKRs based on single-wire and two-wire configurations. The calculated resistance of the single copper wire and two copper wire are 0.53 Ω m$^{-1}$ and 0.26Ω m$^{-1}$ respectively.

### 4.3. Microfiber Coil Resonator (MCR)

Microfiber coil resonators (MCRs) possess similar functionality with other microfiber resonators. It is useful for the applications of optical filtering, lasers and sensors. Additionally, it can be employed as an optical delay line for the optical communication network with small compactness. It is fabricated by winding a long microfiber on a low-index dielectric rod or a rod coated with low-index material (Sumetsky et al., 2010 ). The helical structure of microfiber coil enables propagation of light along the microfiber, across between the turns of microfiber in the forward and backward directions as illustrated in Fig. 37.

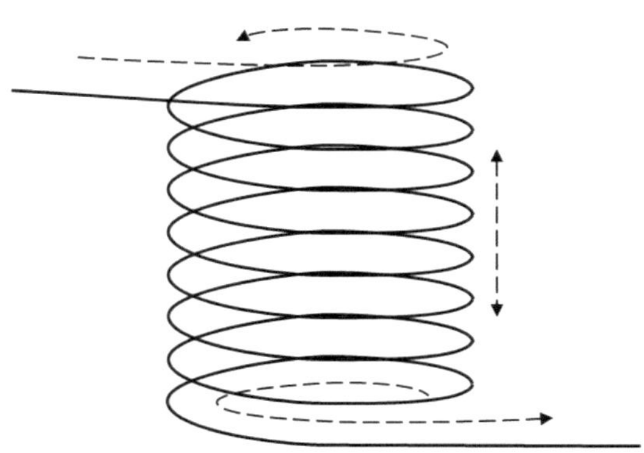

Fig. 37. Helical structure of an MCR and the propagation direction of the light in the resonator.

Fig. 38 shows the transmission spectra of an MCR of increasing number of microfiber turns. The microfiber was wound on a 0.5mm-diameter low-index coated rod. Basically, the optical characteristics a 1-turn MCR is exactly identical to that of MLR, for instance the interference fringes in the MCR transmission spectrum are equally spaced as shown in Fig. 38(a). From the spectrum, the measured FSR is ~0.8nm and the estimated diameter of the coil is ~0.6mm which is slightly larger than the rod diameter. When making additional turns to the coil, it is important to ensure overlapping or touching between turns to establish coupling between them. For every additional turn is made, the transmission spectrum of the MCR is altered. Figs. 38(b)-(d) show the transmission spectra of a 2-turn, 3-turn and 4-turn MCR fabricated in the laboratory. Consider the elastic force from the bent microfiber; it is very difficult to maintain the resonance condition and increasing the number of turns at the same. With the assistance of microscope, the ensuing coiling work can be alleviated. Nonetheless, the reproducibility of MCR was difficult and tedious. Compared with the other microfiber resonators, MCRs have more complicated light propagation properties (Hsu and Huang, 2005 ).

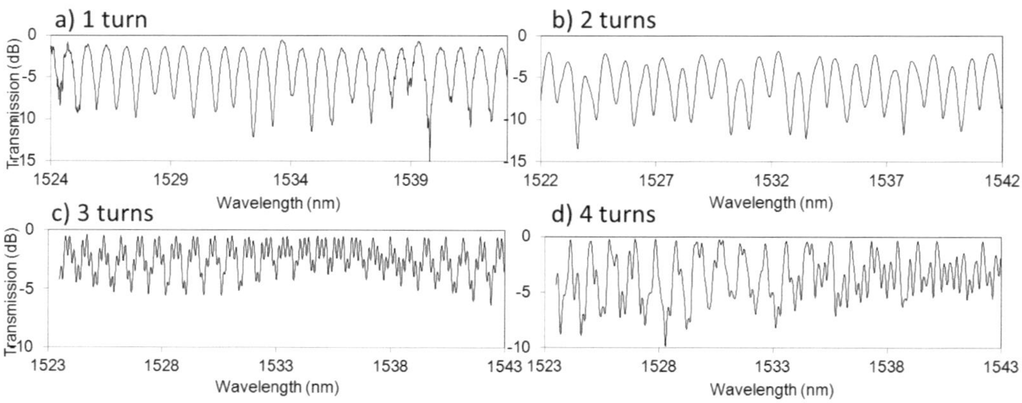


Fig. 38. Transmission spectra of a) 1 turn b) 2 turns c) 3 turns and d) 4 turns MCR.

## 5. Summary

In the past, several fabrication techniques for tapered fibers/microfibers have been suggested. In this chapter, fabrication of microfiber based on flame brushing technique is reviewed. Flame brushing technique is commonly used for the fabrication of fiber couplers and tapered fibers. This technique enables fabrication of biconical tapered fibers which are important components for the manufacture of microfiber based devices. In order to achieve that, a fiber tapering rig was assembled. The heat source comes from an oxy-butane torch with a flame width of 1mm. Two stepper motors are incorporated in the rig to control the movement of the torch and translation stage. A biconical tapered fiber with a waist diameter as small as 700nm can be achieved with the rig. To achieve low loss tapered fibers, the shape of the taper should be fabricated according to adiabaticity criteria. In the laboratory, tapered fibers with linear and decaying-exponential profiles have been demonstrated. To provide protection to the tapered fibers, they are embedded in a low-index material or packaged into a perspex case. These protection measures can prolong the life span and stabilize the temperal performance of these tapered fibers and microfiber based devices.

Three microfiber based devices have been reviewed in this chapter namely MLR, MKR and MMZI. In the first section, the fabrication of MLR is introduced. MLR is assembled from a single mode microfiber by coiling it into a loop. A closed optical path within the loop is established when the two microfibers are put in close contact with each other and form an evanescent coupling. Then, the theoretical model of the self-touching MLR is presented and used for curve-fitting with the experimental data. The important characteristic parameters of the transmission spectra can be extracted from the best-fit curve in the experimental data. In the next section, MKR is presented. Similar to MLR, MKR shares the similar transmission characteristics and the same theoretical model for the MLR can be applied to MKR. However, MKR outperforms MLR in several aspects for instance MKR has more stable and stronger coupling due to its small spacing between the two coupling microfibers in the coupling region. In addition, the structure of MKR is more rigid and robust with the interfiber twisted coupling in the resonator. Nonetheless, MKRs suffer a setback in higher insertion loss if compared with MLRs because of the cut-coil-couple process in the fabrication. The evanescent coupling between output microfiber and collector microfiber contributes a large fraction in the total insertion loss. When MKR is embedded into a medium of different refractive index, the coupling in the MKR varies and it alters the resonance state of the resonator. By curving the experimental data for transmission spectra with the theoretical model, the coupling coefficient and round-trip attenuation factor of the MKR can be extracted from the best-fit curve. The results indicate that the state of resonance of an embedded MKR has been altered and it is closer to critical coupling condition.

Besides, microfiber based resonators exhibit an interesting polarization dependent characteristic in the investigation. The coupling coefficient in the resonator is dependent on input state of polarization and it can be manipulated by using a PBS and a PC. In relation with the coupling, the resonance extinction ratio (RER) of transmission spectrum can be varied by tuning the PC. In the experiment, the variation of RER between 5dB and 11dB was observed. In the past, several literatures have reviewed that the refractive index and dimension of silica microfiber have a strong dependence on temperature. The resonance extinction ratio and resonance wavelength can be expressed in functions of temperature. In order to gain insight into this characteristic, an investigation was conducted to study temperature response of an MKR. Both theoretical analysis and experimental result indicate that the spectral shift in the transmission of MKR is linearly proportional to the increment of temperature. This characteristic can be exploited in many applications particular in sector industrial sensors. By wrapping the microfiber knot on a conductor rod, MKR can perform as a current sensor. When an electric current is loaded through the conductor rod, the rod acts as a heating element and increases the temperature of the MKR. As a result, transmission spectrum of the MKR is shifted. This opto-electric configuration can be used a current sensor and an electric controlled optical filter. In the final section, the fabrication of MCR was demonstrated. An MCR of different number of turns was assembled and tested.

## 6. References

Bilodeau, F., K. O. Hill, S. Faucher, and D. C. Johnson, (1988), Low-loss highly overcoupled fused couplers: Fabrication and sensitivity to external pressure, *Lightwave Technology, Journal of* 6, 1476-1482.

Birks, T. A., and Y. W. Li, (1992), The shape of fiber tapers, *Lightwave Technology, Journal of* 10, 432-438.

Brambilla, G., F. Xu, and X. Feng, (2006), Fabrication of optical fibre nanowires and their optical and mechanical characterisation, *Electronics Letters* 42, 517-519.

Bricheno, T., and V. Baker, (1985), All-fibre polarisation splitter/combiner, *Electronics Letters* 21, 251-252.

Caspar, C., and E. J. Bachus, (1989), Fibre-optic micro-ring-resonator with 2 mm diameter, *Electronics Letters* 25, 1506-1508.

Chen, C.-L., and W. K. Burns, (1982), Polarization characteristics of single-mode fiber couplers, *Microwave Theory and Techniques, IEEE Transactions on* 30, 1577-1588.

Chen, Yi-Huai; Wu, Yu; Rao, Yun-Jiang; Deng, Qiang; Gong, Yuan, (2010), Hybrid mach-zehnder interferometer and knot resonator based on silica microfibers, *Optics Communications* 283, 4.

Ding, Lu, Cherif Belacel, Sara Ducci, Giuseppe Leo, and Ivan Favero, (2010), Ultralow loss single-mode silica tapers manufactured by a microheater, *Appl. Opt.* 49, 2441-2445.

Dong, H., G. Zhu, Q. Wang, H. Sun, N. K. Dutta, J. Jaques, and A. B. Piccirilli, (2005), Multiwavelength fiber ring laser source based on a delayed interferometer, *Photonics Technology Letters, IEEE* 17, 303-305.

Frazão, O., and et al., (2005), Chirped bragg grating fabricated in fused fibre taper for strain–temperature discrimination, *Measurement Science and Technology* 16, 984.

Graf, J. C., S. A. Teston, P. V. de Barba, J. Dallmann, J. A. S. Lima, H. J. Kalinowski, and A. S. Paterno, 2009, Fiber taper rig using a simplified heat source and the flame-brush technique, Microwave and Optoelectronics Conference (IMOC), 2009 SBMO/IEEE MTT-S International.

Guo, Xin, Yuhang Li, Xiaoshun Jiang, and Limin Tong, (2007), Demonstration of critical coupling in microfiber loops wrapped around a copper rod, *Applied Physics Letters* 91, 073512-3.

Guo, Xin, and Limin Tong, (2008), Supported microfiber loops for optical sensing, *Opt. Express* 16, 14429-14434.

Harun, S., K. Lim, A. Jasim, and H. Ahmad, (2010), Fabrication of tapered fiber based ring resonator, *Laser Physics* 20, 1629-1631.

Harun, S. W., K. S. Lim, A. A. Jasim, and H. Ahmad, (2010), Dual wavelength erbium-doped fiber laser using a tapered fiber, *Journal of Modern Optics* 57, 2111 - 2113.

Hou, Changlun, Yu Wu, Xu Zeng, Shuangshuang Zhao, Qiaofen Zhou, and Guoguang Yang, (2010), Novel high sensitivity accelerometer based on a microfiber loop resonator, *Optical Engineering* 49, 014402-6.

Hsu, Shih-Hsin, and Yang-Tung Huang, (2005), Design and analysis of mach?Zehnder interferometer sensors based on dual strip antiresonant reflecting optical waveguide structures, *Opt. Lett.* 30, 2897-2899.

Jiang, Xiaoshun, Limin Tong, Guillaume Vienne, Xin Guo, Albert Tsao, Qing Yang, and Deren Yang, (2006), Demonstration of optical microfiber knot resonators, *Applied Physics Letters* 88, 223501-223501-3.

Jung, Y, G. S. Murugan, G. Brambilla, and D. J. Richardson, (2010), Embedded optical microfiber coil resonator with enhanced high-q, *Photonics Technology Letters, IEEE* 22, 1638-1640.

Li, Yuhang, and Limin Tong, (2008), Mach-zehnder interferometers assembled with optical microfibers or nanofibers, *Opt. Lett.* 33, 303-305.

Lim, K. S., S. W. Harun, S. S. A. Damanhuri, A. A. Jasim, C. K. Tio, and H. Ahmad, (2011), Current sensor based on microfiber knot resonator, *Sensors and Actuators A: Physical* 167, 377–381.

Lim, K. S., S. W. Harun, A. A. Jasim, and H. Ahmad, (2011), Fabrication of microfiber loop resonator-based comb filter, *Microwave and Optical Technology Letters* 53, 1119-1121.

Liu, Zhihai, Chengkai Guo, Jun Yang, and Libo Yuan, (2006), Tapered fiber optical tweezers for microscopic particle trapping: Fabrication and application, *Opt. Express* 14, 12510-12516.

Love, J. D., W. M. Henry, W. J. Stewart, R. J. Black, S. Lacroix, and F. Gonthier, (1991), Tapered single-mode fibres and devices. I. Adiabaticity criteria, *Optoelectronics, IEE Proceedings J* 138, 343-354.

Mora, J., A. Diez, M. V. Andres, P. Y. Fonjallaz, and M. Popov, (2004), Tunable dispersion compensator based on a fiber bragg grating written in a tapered fiber, *Photonics Technology Letters, IEEE* 16, 2631-2633.

Ngo, N. Q., S. Y. Li, R. T. Zheng, S. C. Tjin, and P. Shum, (2003), Electrically tunable dispersion compensator with fixed center wavelength using fiber bragg grating, *Lightwave Technology, Journal of* 21, 1568-1575.

Orucevic, Fedja, Valérie Lefèvre-Seguin, and Jean Hare, (2007), Transmittance and near-field characterization of sub-wavelength tapered optical fibers, *Opt. Express* 15, 13624-13629.

Schwelb, O., (2004), Transmission, group delay, and dispersion in single-ring optical resonators and add/drop filters-a tutorial overview, *Lightwave Technology, Journal of* 22, 1380-1394.

Sumetsky, M., (2008), Basic elements for microfiber photonics: Micro/nanofibers and microfiber coil resonators, *J. Lightwave Technol.* 26, 21-27.

Sumetsky, M., Y. Dulashko, J. M. Fini, and A. Hale, (2005), Optical microfiber loop resonator, *Applied Physics Letters* 86, 161108-161108-3.

Sumetsky, M., Y. Dulashko, J. M. Fini, A. Hale, and D. J. DiGiovanni, (2006), The microfiber loop resonator: Theory, experiment, and application, *Lightwave Technology, Journal of* 24, 242-250.

Sumetsky, M., Y. Dulashko, and S. Ghalmi, (2010), Fabrication of miniature optical fiber and microfiber coils, *Optics and Lasers in Engineering* 48, 272-275.

Tong, Limin, Rafael R. Gattass, Jonathan B. Ashcom, Sailing He, Jingyi Lou, Mengyan Shen, Iva Maxwell, and Eric Mazur, (2003), Subwavelength-diameter silica wires for low-loss optical wave guiding, *Nature* 426, 816-819.

Vienne, G., Li Yuhang, and Tong Limin, (2007), Effect of host polymer on microfiber resonator, *Photonics Technology Letters, IEEE* 19, 1386-1388.

Vienne, Guillaume, Aurélien Coillet, Philippe Grelu, Mohammed El Amraoui, Jean-Charles Jules, Frédéric Smektala, and Limin Tong, (2009), Demonstration of a reef knot microfiber resonator, *Opt. Express* 17, 6224-6229.

Wu, Yu, Yun-Jiang Rao, Yi-huai Chen, and Yuan Gong, (2009), Miniature fiber-optic temperature sensors based on silica/polymer microfiber knot resonators, *Opt. Express* 17, 18142-18147.

Xu, Fei, and Gilberto Brambilla, (2007), Embedding optical microfiber coil resonators in teflon, *Opt. Lett.* 32, 2164-2166.

Xu, Fei, Gilberto Brambilla, and David J. Richardson, 2006, Adiabatic snom tips for optical tweezers.

Xu, Fei, Peter Horak, and Gilberto Brambilla, (2007), Optical microfiber coil resonator refractometric sensor, *Opt. Express* 15, 7888-7893.

Xu, Fei, Valerio Pruneri, Vittoria Finazzi, and Gilberto Brambilla, (2008), An embedded optical nanowire loop resonator refractometric sensor, *Opt. Express* 16, 1062-1067.

Yang, Szu-Wen, and Hung-Chun Chang, (1998), Numerical modeling of weakly fused fiber-optic polarization beamsplitters. I. Accurate calculation of coupling coefficients and form birefringence, *Lightwave Technology, Journal of* 16, 685-690.

Zeng, Xu, Yu Wu, Changlun Hou, Jian Bai, and Guoguang Yang, (2009), A temperature sensor based on optical microfiber knot resonator, *Optics Communications* 282, 3817-3819.

Zhang, J., P. Shum, X. P. Cheng, N. Q. Ngo, and S. Y. Li, (2003), Analysis of linearly tapered fiber bragg grating for dispersion slope compensation, *Photonics Technology Letters, IEEE* 15, 1389-1391.

Zhang, Rui, Jörn Teipel, Xinping Zhang, Dietmar Nau, and Harald Giessen, (2004), Group velocity dispersion of tapered fibers immersed in different liquids, *Opt. Express* 12, 1700-1707.

# 18

# Influence of Current Pulse Shape on Directly Modulated Systems Using Positive and Negative Dispersion Fibers

Paloma R. Horche[1] and Carmina del Río Campos[2]
[1]ETSI Telecomunicación, Universidad Politécnica de Madrid,
[2]Escuela Politécnica Superior, Universidad San Pablo CEU,
Spain

## 1. Introduction

The proliferation of high-bandwidth applications has given rise to a growing interest, between network providers, on upgrading networks to deliver broadband services to homes and small businesses. There has to be a good balance between the total cost of the infrastructures and the services that can be offered to the end users, as they are very sensitive to equipment outlay, requiring the use of low-cost optical components. Coarse Wavelength Division Multiplexing (CWDM) is an ideal solution to the tradeoff between cost and capacity (Thiele, 2007). This technology uses all or part of the 1270 to 1610 nm wavelength fiber range with an optical channel separation of about 20 nm. This channel separation allows the use of low-cost, uncooled, Directly Modulated Lasers (DML). The main advantages of these devices associated with uncooled operation are (Nebeling, 2007); No integration of TEC and cooler required, less complexity for control electronics, reduced power consumption, only laser diode current required, lower device cost. Otherwise, the direct modulation of the laser current leads to a modulation of the carrier density giving rise to a *chirp frequency*. This results in a broadened linewidth and a laser wavelength drift (Henry, 1982; Linke, 1985). Since wavelength chirp was recognized several years ago, (Koch, 1988; Hinton, 1993) many papers have addressed the causes and implications of chirp on the optical system performance (Cartledge, 1989; Hakki, 1992; Horche, 2008). However, a large part of the research has focused on fiber transmission properties. The idea of these studies is, generally, based on considerations regarding the interaction of the laser frequency chirp with the fiber dispersion. It is known that if the chirp parameter ($\alpha$-factor) is positive, as is always the case for directly modulated lasers, then the frequency components of the leading edge of the pulse will be blue-shifted and the trailing edge red-shifted. If the pulses are transmitted over a negative dispersion fiber where the blue wavelength is slower than the red, to some extent, pulse compression and significant transmission performance improvement is expected (Morgado & Cartaxo, 2001; Tomkos et al., 2001).
To counteract the effect of chromatic dispersion, the system can be moved to a non-zero negative dispersion-shifted fiber, but this method has some problems that are difficult to resolve when it comes to a CWDM metropolitan or access network; The traffic must be interrupted. This method may only be used for upgrading long wavelength channels where

this fiber has much lower dispersion values. And, cost-effectiveness, changing the fiber is expensive both to buy and to lay.

In DMLs, the frequency chirp for large-signal modulation can be determined from the shape of the electric pulse applied to the optical source (modulation current) (Coldren & Corzini, 1995). Since the DML chirp parameter is tunable through the modulation current applied to the diode laser, optimum operating-conditions in terms of the chirp/dispersion interactions can be set for fibers having different amounts and signs of dispersion.

It is the aim of this Chapter to discuss and compare how the shape of the modulated signal (e.g, exponential-wave, sine-wave, Gaussian, etc.) can improve the system performance when using both positive and negative dispersion fibers. With this method it is possible to improve each of the WDM system channels individually, offering a low-cost solution since it only involves changes in the transmitters and avoiding the replacement of the fiber.

The aim of this chapter is also to present analytical and simulation results pertaining to the transmission of chirped optical signals in a dispersive fiber.

This Chapter is organized as follows: Section 2 deals with the theoretical background of optical fiber dispersion; In Section 3 we analyze the power and chirp waveforms for different DMLs both theoretically and by simulation; Sections 4 and 5 focus on the study of the interplay between the laser chirp and the dispersion of both Gaussian pulses and pulses of an arbitrary shape, upon which the subject of this work is based; Section 6 details the system modeling together with the presentation of the parameters used in our simulation for the different fibers, DMLs and modulation current; Section 7 discusses the simulation results for the determination of the optimum modulation current under different scenarios; Section 8 studies the accumulated phase along the link as a method of improving the system performance. In Section 9 we analysis the transmission performance over different lengths of positive and negative dispersion fibers through computer simulation. Finally, a brief summary and the conclusions are presented in Section 10.

## 2. Factors contributing to dispersion in optical fibers

When a short pulse of light travels through an optical fiber, its power is "dispersed" in time so that the pulse spreads into a wider time interval. The principal sources of dispersion in single optical fibers are material dispersion, waveguide dispersion, polarization-mode dispersion and non-linear dispersion. The combined contributions of these effects to the spread of pulses in time are not necessarily additive (Agrawal, 2010).

The PMD-induced pulse broadening is characterized by the root mean-square (RMS) $\sigma_T \equiv D_p\sqrt{L}$, where $D_p$ is the PMD parameter. Measured values of $D_p$ vary from fiber to fiber in the range $D_p$ = 0.01–1 ps/√km. Because of the √L dependence, PMD induced pulse broadening is relatively small compared with the other effects. Indeed, for fiber lengths < 400 km, as is the case of metro-CWDM systems, the PMD can be ignored.

The combined effects of material dispersion and waveguide dispersion which is referred to as chromatic dispersion is, in general, the major source of pulse broadening and it may be determined by including the wavelength dependence of the optical fiber refractive indices, $n_1$ and $n_2$ when determining $d\beta/d\omega$ from the characteristic equation of the fiber.

Since $\beta$ is a slowly varying function of this angular frequency, one can see where various dispersion effects arise by expanding $\beta$ in a Taylor series about a central frequency $\omega_0$ (Keiser, 2010). Expanding $\beta$ in a Taylor series yields

$$\beta(\omega)=\beta_0+\beta_1(\omega-\omega_0)+\frac{1}{2}\beta_2(\omega-\omega_0)^2+\frac{1}{6}\beta_3(\omega-\omega_0)^3+\cdots \quad (1)$$

where $\beta_m$ denotes the $m^{th}$ derivative of $\beta$ with respect to $\omega$ evaluated at $\omega = \omega_0$.
From (1) it can be obtained the velocity at which the energy in a pulse travels along a fiber (*group velocity*) and the *group-velocity dispersion* parameter, known as the GVD, which determines how much an optical pulse would broaden on propagation inside the fiber. They are defined as

$$v_g = \frac{1}{\left.\frac{\partial\beta(\omega)}{\partial\omega}\right|_{\omega=\omega_0}} = \frac{1}{\beta_1} \quad ; \quad GVD = \left.\frac{\partial^2\beta(\omega)}{\partial\omega^2}\right|_{\omega=\omega_0} \equiv \beta_2 \quad (2)$$

In some optical communications systems, the frequency spread $\Delta\omega$ is determined by the range of wavelengths $\Delta\lambda$ emitted by the optical source. It is customary to use $\Delta\lambda$ in place of $\Delta\omega$. By using $\omega = 2\pi\, c/\lambda$, 24) can be written as

$$D = -\frac{2\pi c}{\lambda^2}\beta_2 \quad (3)$$

where $D$ is called the *dispersion parameter* and is expressed in units of ps/(km-nm).
In the fourth term of (1), the factor of $\beta_3$ is known as the *third-order dispersion*. This term is important around the wavelength at which $\beta_2$ is equal to zero (*zero-dispersion wavelength*).
Higher-order dispersive effects are governed by the dispersion slope $S = dD/d\lambda$. The parameter $S$ is also called a *differential-dispersion* parameter. By using (3), it can be written as

$$\beta_3 = \frac{\partial\beta_2}{\partial\omega} = -\frac{\lambda^2}{2\pi c}\frac{\partial\beta_2}{\partial\lambda} = -\frac{\lambda^2}{2\pi c}\frac{\partial}{\partial\lambda}\left[-\frac{\lambda^2}{2\pi c}D\right] \quad (4)$$

$$S = \frac{4\pi c}{\lambda^3}\beta_2 + \left(\frac{2\pi c}{\lambda^2}\right)^2\beta_3 \quad \left[\frac{ps}{(nm)^2\cdot km}\right] \quad (5)$$

The numerical value of the dispersion slope $S$ plays an important role in the design of CWDM systems. Since $S > 0$ for most fibers, different channels have slightly different GVD values. This feature makes it difficult to compensate dispersion for all channels simultaneously.

## 3. Directly modulated laser

DFB lasers are the workhorse in WDM systems, both as cooled and uncooled devices. In order to provide cost-effective solutions for CWDM systems, directly modulated, uncooled distributed feedback (DFB) lasers are preferred (DFB-DML).
As regards cost, CWDM systems are usually single-span (unlike their DWDM counterparts), inasmuch as they do not use any type of in-line optical amplification. Since there is no optical amplification, the output power of the laser needs to be sufficient to sustain the system's loss budget. This involves the transmission fiber, multiplexer loss together with additional splice and connector losses.

A typical N-wavelength CWDM system, as standardized in ITU-T G.695, is shown in Fig. 1.

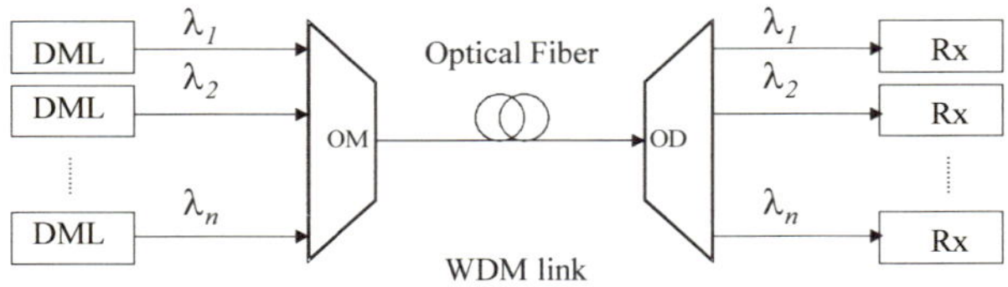


Fig. 1. Proposed optical transmission system.

The transmitter block diagram implemented in our studies consists of a bit random generator, which determines the sequence of bits, $a_k$, that will be sent to an electric pulse generator, in NRZ format, which injects a modulation current, $I(t)$ into the laser diode (DML), uncooled. A block diagram of the directly modulated transmitter to be considered is illustrated in Fig. 2.

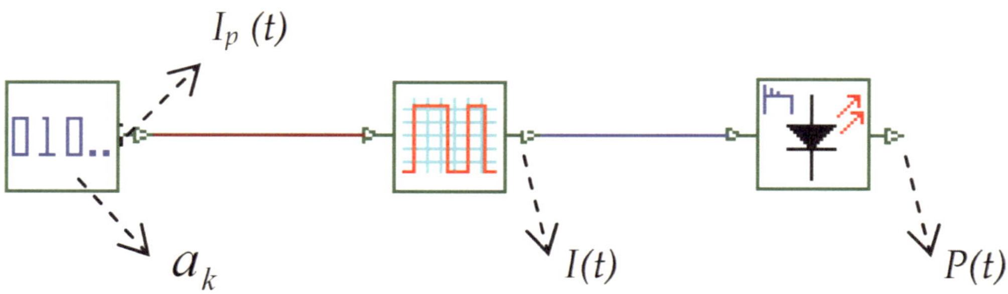


Fig. 2. Directly modulated laser scheme.

The injected laser current is given by the expression (Hinton & Stephens, 1993)

$$I(t) = I_b + \sum_{k=-\infty}^{\infty} a_k I_p(t - kT) \tag{6}$$

where $I_b$ is the bias current, T is the period of the modulation pulse, $a_k$ is the sequence of bits transmitted ($a_k$ = 1 (0) if a binary one (zero) is transmitted during the $k_{th}$ time), and $I_p(t)$ is the applied current pulse.

In a free-chirp source, the optical power output pulse of the laser, $P(t)$, is given by

$$P(t) = \eta_0 . \frac{h\nu}{q} . \sum_{k=-\infty}^{+\infty} a_k I_p(t - kT) \tag{7}$$

where $\eta_0$ is the differential quantum efficiency of the laser, $h\nu$ the photon energy at the optical frequency $\nu$, and $I_p(t)$ the applied current pulse.

However, expression (7) is not applicable in case of directly modulated sources where the injected current that modulates the laser introduces a shift in the emission frequency (chirp frequency). As a consequence, the optical power output pulse is not a linear transformation of the applied current pulse.

### 3.1 Power and chirp waveform from the laser

The optical power and chirping response of the semiconductor laser to the current waveform *I(t)* is determined by means of the large-signal rate equations, which describe the interrelationship between the photon density, carrier density, and optical phase within the laser cavity.

To calculate the optical power at the laser output depending on the value of the modulation current, we take into account the equations for the dynamics of the laser, which are given by the relationships between the density of photons *S(t)*, density of carriers *N(t)* and the phase of the electrical field $\phi(t)$. In this way, the variation in carriers is related to the current waveform injected into the active layer *I(t)* (Cartledge & Srinivasan, 1997; Cartledge et al. 1989; Corvini, 1987), in accordance with the following expression

$$\frac{dN(t)}{dt} = \frac{I(t)}{qV} - G\left[N(t) - N_t\right]S(t) - \frac{N(t)}{\tau_n} \quad (8)$$

and the variation of photons is given by:

$$\frac{dS(t)}{dt} = \Gamma G\left[N(t) - N_t\right]S(t) - \frac{S(t)}{\tau_p} + \frac{\beta \Gamma N(t)}{\tau_n} \quad ; \quad G = g_0 / \left[1 + \varepsilon S(t)\right] \quad (9)$$

where $N_t$ is the carrier density at transparency, $q$ the electron charge, $V$ the volume of the active region, $\tau_n$ the electron lifetime, $\Gamma$ the mode confinement factor in the cavity, $\beta$ is the fraction of spontaneous emission coupled to the laser mode, $\tau_p$ the photon lifetime and G is the gain non-linear coefficient, $\varepsilon$ being the compression factor of gain, $g_0$ the gain slope constant given by $g_0 = a_0 v_g$, where $a_0$ is the active layer gain coefficient and $v_g$ the group velocity.

Thus, direct modulation of the laser current not only imprints the desired transmit data onto the emitted optical intensity, but also leads to a modulation of the carrier density within the laser cavity and in turn to a modulation of the refractive index of the active region. This in turn leads to the modulation of the frequency of the emitted light, a phenomenon that is known as laser chirp (the emission frequency must fulfil the phase condition all the time in order for the laser oscillation to occur). The chirp supposes that, even with singlemode lasers, the value of the instantaneous frequency of the emission is not the same during the optical pulse.

The frequency chirp $\Delta v$ (deviation of the optical frequency at around the unmodulated frequency) is described by the coupled rate equations of the carriers and the optical field in the laser and their analytic solution yields an expression:

$$\Delta v = \frac{1}{2\pi} \frac{d\phi}{dt} \quad (10)$$

$\phi$ being the optical phase, whose variation with the time depends on the linewidth enhancement factor or the Henry factor $\alpha$ (Henry, 1982):

$$\frac{d\phi}{dt} = \frac{1}{2} \alpha \left[ \Gamma v_g a_0 (N(t) - N_t) - \frac{1}{\tau_p} \right] \quad (11)$$

The linewidth enhancement factor $\alpha$ describes the relationship between how the real and imaginary refractive indices are affected by the carrier density. Factor $\alpha$ provides a measure of the coupling strength between the intensity and frequency modulation (FM) of the laser diode. It is defined as (Coldren & Corzine, 1995)

$$\alpha \equiv -\frac{dn/dN}{dg/dN} \tag{12}$$

Here, $n$ is the real part of refractive index of the laser cavity. $dg/dN$ is also known as the differential gain. From (10) and (11), the amount of frequency modulation is proportional to the linewidth enhancement factor which is typically between 4-10 but can be as low as 1 in some active materials.

The output power $P(t)$ is related to the photon density $S(t)$ through the expression

$$P(t) = \frac{V\eta h\nu}{2\Gamma\tau_p} \cdot S(t) \tag{13}$$

where $h$ is the Planck constant and $\eta$ is the differential quantum efficiency.

From the above equations the expression that relates the chirp with the laser output power can be obtained

$$\Delta\nu = \frac{\alpha}{4\pi}\left[\frac{1}{P(t)} \cdot \frac{dP(t)}{dt} + \kappa P(t)\right] \tag{14}$$

$\kappa$ being the adiabatic chirp coefficient, which is directly related to the non-linear gain compression factor

$$\kappa = \frac{2\Gamma}{V\eta h\nu} \cdot \varepsilon \tag{15}$$

The first term in (14), is a structure-independent transient chirp and it produces variations in the pulse temporal width, while the second term is a structure-dependent adiabatic chirp, that produces a different frequency variation between the 1 and 0 bits (if the frequency of "1" is greater than the "0", it means a blue shift) causing a shift in time between the levels corresponding to "1"and "0" when the pulses go through a dispersive media, such as the optical fiber (Koch, 1988; Hakki, 1992).

DMLs can be classified according to their chirp behavior; Transient-chirp dominated DML when the first term of (14) is predominant and adiabatic-chirp dominated DML when the first term of (14) can be ignored compared to the second term of (14) (Hinton & Stephens, 1993). In following sections, we have called a DMLs with a strongly transient chirp dominated behavior DML-T and DML-A when its behavior is strongly adiabatic chirp dominated.

Fig. 3 shows the simulated power (a and c) and chirp (b and d) waveforms for two different DMLs. In this simulation, the modulating current $I(t)$ was made up of 2.5-Gb/s Gaussian current pulses.

In the case of an adiabatic chirp dominated laser (DML-A), 1 and 0 bits have a different frequency proportional to their optical power. The "0" frequency is smaller than the "1"' frequency [see Fig. 3 (b)]. While, in the case of an adiabatic chirp dominated laser (DML-T), the power waveform shows a large power overshoot on the 1's [see Fig. 3 (c)], and the chirp has a significant value during transition states. Both behaviors are consistent with (14)

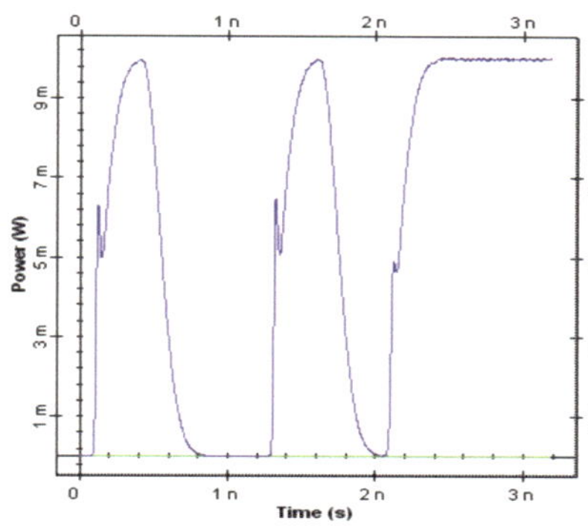


(a) Power waveform from DML-A

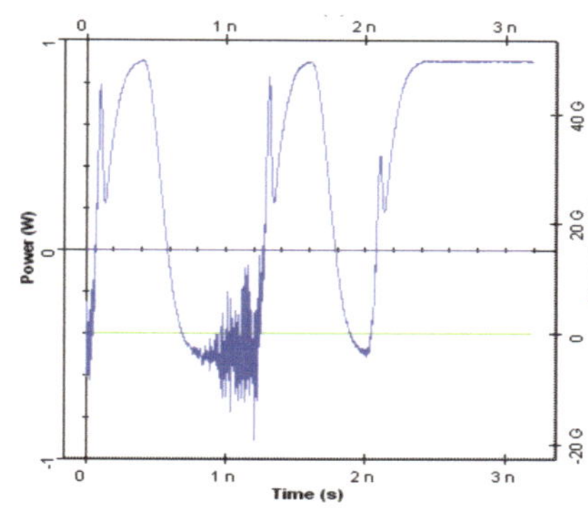


(b) Chirp waveform DML-A

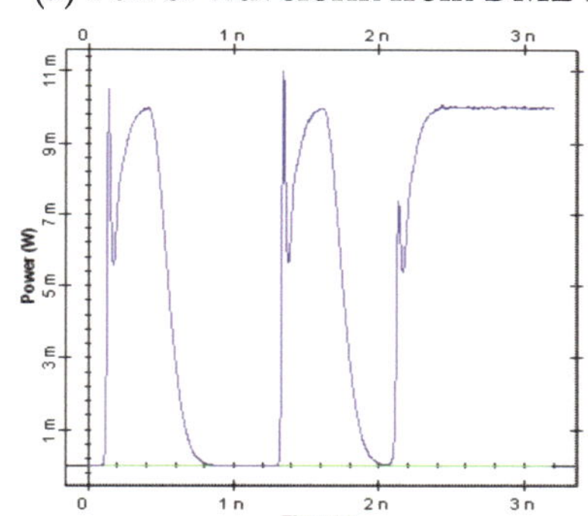


(c ) Power waveform from DML-T

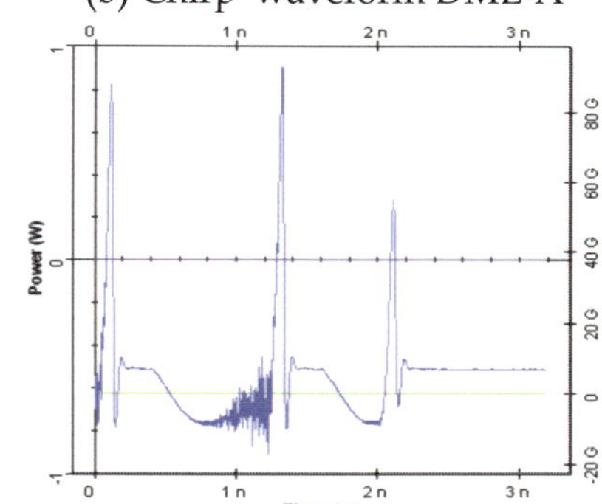


(d) Chirp waveformp DML-T

Fig. 3. Simulated power (left-hand column) and chirp (right-hand column) waveforms for 2.5-Gb/s DML-A (a and b) and DML-T (c and d).

It is important to highlight that through (6-14) the frequency chirp for large-signal modulation can be determined directly from the shape of the modulated signal (e.g, square-wave, sine-wave, Gaussian, etc.). This simple expression is useful when discussing chirping phenomena because it clearly shows the interdependence of intensity and frequency modulation. In particular, the relative phases of frequency and intensity excursions are clearly displayed by (14), and these are of utmost importance in understanding the behavior of chirped signal transmission in dispersive media.

## 4. Transmission characteristics

When an optical pulse is transmitted through a dispersive media, such as an optical fiber, the intensity and shape of the optical signal at the output of the optical fiber, due to the waveform *I(t)*, are related with the dispersive characteristics of the optical fiber. This occurs because the spectral components that constitute the pulse are attenuated and/or phase shifted by different amounts. The effect of dispersion is more dramatic for ultrashort pulses since they have greater spectral widths.

The basic propagation equation that governs pulse evolution inside a single-mode fiber, considering expanding $\beta$ to the third order of (1), is given by (Agrawal, 2010)

$$\frac{\partial A}{\partial z}+\beta_1\frac{\partial A}{\partial t}+j\frac{\beta_2}{2}\frac{\partial^2 A}{\partial t^2}-\frac{\beta_3}{6}\frac{\partial^3 A}{\partial t^3}=0 \tag{16}$$

Upon propagation in a linear lossless dispersive medium, a monochromatic plane wave of frequency $\nu$ traveling a distance $z$ in the $z$ direction undergoes a phase shift and the pulse will be broadened. Figure 4 shows the propagation of an initially unchirped Gaussian pulse through a fiber with anomalous dispersion ($\beta_2$ = GVD < 0). The pulse remains Gaussian, but its width expands and it becomes chirped with a decreasing chirp parameter (down-chirped). In this case the pulse has a decreasing instantaneous frequency.

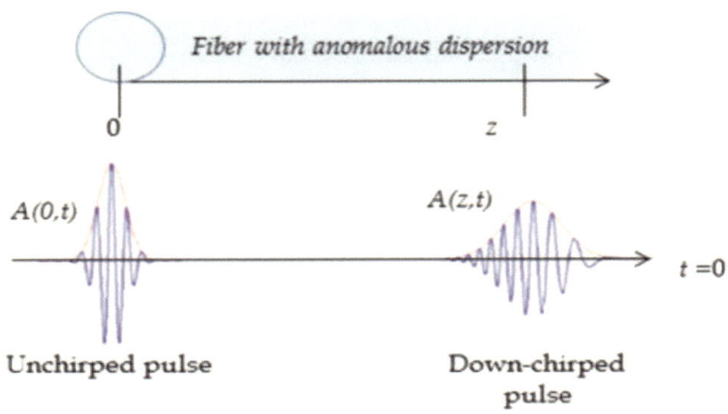


Fig. 4. Transmission of an optical pulse through a dispersive medium.

Thus, a fiber with a GVD parameter converts an unchirped Gaussian pulse of width $\tau_1$ into a chirped Gaussian pulse of width $\tau_2$ and chirp parameter $a$. Therefore, it follows that the process of pulse propagation through an optical fiber is equivalent to a combination of a time delay (second term of (16), group velocity) and a phase filter with chirp parameter $b$ (third and fourth terms of (16) proportional to the distance $z$. By ignoring the third-order dispersion, the chirp parameter can be written in the form (Saleh & Teich, 2007)

$$b=2\beta_2 z=\frac{GVD}{\pi}z \tag{17}$$

A fiber with $\beta_2$ >0 ($b$ > 0) is said to have normal dispersion (negative dispersion coefficient, $D$) and it functions as an up-chirping filter which increase its instantaneous frequency. If $\beta_2$ < 0 ($b$ < 0) is said to have anomalous dispersion (positive dispersion coefficient, $D$) and it corresponds to a down-chirping filter which decreasing its instantaneous frequency.

In normal dispersion, a blue wavelength is slower than red, so the leading edge of the pulse is red-shifted and the trailing edge is blue-shifted. The opposite occurs for anomalous dispersion.

Based on (16) and considering both the case in which the carrier wavelength is far away from the zero-dispersion wavelength (contribution of the $\beta_3$ term is negligible) and the non-linear effects in the fiber to be negligible, the fiber can be seen as a phase filter with a chirp parameter $b$ according to (17), thus:

- For an optical fiber with normal dispersion ($b > 0$) the filter is up-chirping.
- For an optical fiber with anomalous dispersion ($b < 0$) the filter is down-chirping.

A standard single mode fiber (SSMF), optimized for the transmission in the *O-band* (~1310 nm), is a typical fiber with anomalous dispersion whose transmission characteristics are described in *Recommendation ITU-T G.652*. The negative *non-zero dispersion shifted fiber* (N-ZDSF), whose chromatic dispersion coefficient is negative in the *C-band* (and above), is an example of a normal dispersion fiber (*Recommendation ITU-T G.655).*

## 5. The relationship between laser chirp and dispersion

The pulse shape at a distance $z$ inside the fiber is a function of different factors: the initial pulse shape, the chirp associated with this pulse from the DML and the dispersive characteristics of the optical fiber.
As already mentioned, for wavelengths far away from the zero-dispersion wavelength and considering non-linear effects in the fiber as negligible, the fiber transmission can be regarded as a phase filter with a chirp parameter $b$ in accordance with (17) and with a transfer function given by (Saleh & Teich, 2007)

$$H_e(f) = \exp\left[-jb(\pi f)^2\right] \tag{18}$$

In the following sections the influence of the interplay between chirp and dispersion on a propagated pulse is examined by regarding the process of propagation as a phase filter with a transfer function governed by (19) and considering that, in accordance with (14) and (16), the optical output power from a DML is blue shifted during the pulse rise time (increasing frequency) and red shifted during the pulse fall time (decreasing frequency), that is, the pulse is a down-chirped pulse.
Firstly, we are going to consider the propagation of a chirped-Gaussian. Although the Gaussian pulse has an ideal shape that is not encountered exactly in practice, it is a useful approximation that lends itself to analytical studies. In section 5.2 a general formula is derived that can be used to evaluate the waveform of a chirped pulse of arbitrary shape.

### 5.1 Propagation of chirped-Gaussian pulses in an optical fiber

A general Gaussian pulse has a complex envelope $A(z,t)$

$$A(z,t) = A_0 \exp\left[-\frac{1}{2}(1 - ia_1)\frac{t^2}{T_0^2}\right] \tag{19}$$

where $A_0$ is the peak amplitude, $T_0$ represents the half-width of the pulse at the 1/e power point. The parameter $a_1$ in (19) takes into account the frequency chirp of the DML at z = 0 (ignoring the adiabatic chirp, $\kappa = 0$).
When a chirped Gaussian pulse, defined by (19), is transmitted through a phase filter, with a transfer function given by (18), the outcome is also a chirped Gaussian pulse. Thus, the shape waveform of the filtered pulse, after a distance z, is given by the equation governing the evolution of the complex amplitude of the field envelope along the fiber, $A\ (z,\ t)$

$$A(z,t) = \frac{A_0}{\sqrt{1 - ja_2 \cdot \frac{T_1}{T_0}}} \exp\left[ -\frac{(1 - ja_2)}{2T_1^2} t^2 \right] \tag{20}$$

where

$$\frac{2T_1^2}{1 - ja_2} = \frac{2T_0^2}{1 - ja_1} + jb \tag{21}$$

$T_1$ *is* the half-width similar to $T_0$.
Equating the real and imaging parts of (21) leads the width $T_1$ and the chirp parameter $a_2$ associated with the pulse after a distance z

$$b_f(z) = \frac{T_1}{T_0} = \left[ \left(1 + \frac{a_1 b}{2T_0^2}\right)^2 + \left(\frac{b}{2T_0^2}\right)^2 \right]^{1/2} = \left[ \left(1 + \frac{a_1 \beta_2 z}{T_0^2}\right)^2 + \left(\frac{\beta_2 z}{T_0^2}\right)^2 \right]^{1/2} \tag{22}$$

$$a_2 = a_1 + \left(1 + a_1^2\right) \frac{b}{2T_0^2} \tag{23}$$

Therefore, when a Gaussian pulse with an initial chirp $a_1$, travels a distance z in a dispersive media, the outcome is also a chirped Gaussian pulse with altered parameters. The chirp parameter will acquire a new value given by $a_2$ and the pulse will be broadened (or compressed) by a factor $b_f$.
In summary, as expected, the shape of the signal at the output of the optical fiber is affected by the sign of both $\beta_2$ (fiber parameter) and $a$ (transmitter parameter). Compression can occur if an up-chirped pulse travels through a down-chirping (anomalous) medium, or if a down-chirped pulse travels through an up-chirping (normal) medium.
Taking into account the relationship between $\beta_2$ and $D$ given by (5), Fig. 5 shows the pulse broadening factor $b_f$ as a function of the accumulated dispersion, $DL$ for an initial chirped Gaussian pulse from a DML with $T_0$ = 30 ps (equivalent to the bit period for a 10-Gb/s signal), dynamic chirp only, and a wavelength $\lambda$ = 1550 nm and different chirp values (0, ±2 y ±4). In this case, the chirp parameter $a_1$ in (23) is the linewidth enhancement factor $\alpha$ given by (12). In order to be congruent with the notation used here, $a_1$ is taken as $-\alpha$.
When comparing the result for the chirped pulses with $\alpha$ = 2 or 4 (DML) with the result for the unchirped pulse ($\alpha$ = 0), it is observed that for $D > 0$ (SMF fiber), the presence of chirp leads to significantly increased dispersion-induced pulse broadening, which increases in line with the increased chirp. However, for up to a certain amount of negative dispersion (N-NZDSF fiber), the chirp is actually beneficial and leads to pulse compression so that one can enhance the transmission system behavior.
From Fig. 5, it is observed that there is an optimum value for the accumulated dispersion that leads to minimum-width pulse. To determine the value $b_{min}$ at which the pulse has its minimum width $T_1^{\min}$, the derivative of $T_1$ in (22) with respect to $b$ is equated to zero. The result is (Saleh & Teich, 2007).

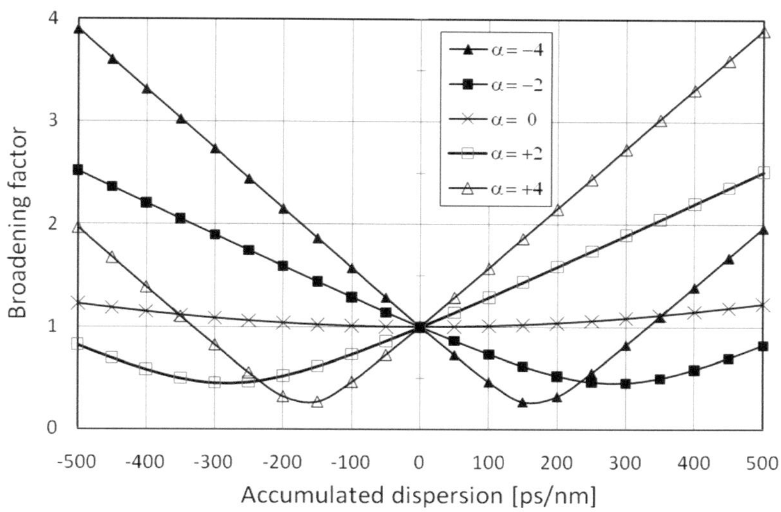


Fig. 5. Calculated pulse broadening factor $b_f$ as a function of the accumulated dispersion: DL.

$$T_1^{\min} = \frac{T_0}{\sqrt{1+a_1^2}}$$
$$b_{\min} = -a_1\left(T_1^{\min}\right)^2 = -2\frac{a_1}{1+a_1^2}T_0^2 \tag{24}$$

From (245) we can calculate the fiber length at which the width $T_1^{\min}$ is reached

$$z_{\min} = -\frac{a_1}{\beta_2\left(1+a_1^2\right)}T_0^2 \tag{25}$$

Based on (25) $z_{min}$ only exists for $a_1\beta_2 < 0$.
As an example, Fig. 6 (a) shows the simulated waveform of a positive chirped Gaussian pulse together with the chirp pulse when it is propagated through a negative dispersion fiber for three fiber lengths. Fig. 6 (b) represents the chirped Gaussian pulse and the phase throughout the pulse ($L_1 < L_2 < L_3$). The minimum width is reached at a length $L_2$ in accordance with (24). It is important to highlight that the minimum width $T_1^{\min}$ is obtained when the chirp value is zero (total compensation between DML-chirp and dispersion) and the accumulated phase is constant. Based on this, we will demonstrate the enhancement performance of an optical system by researching a modulation current waveform so that a constant accumulated phase, along the link, is achieved.

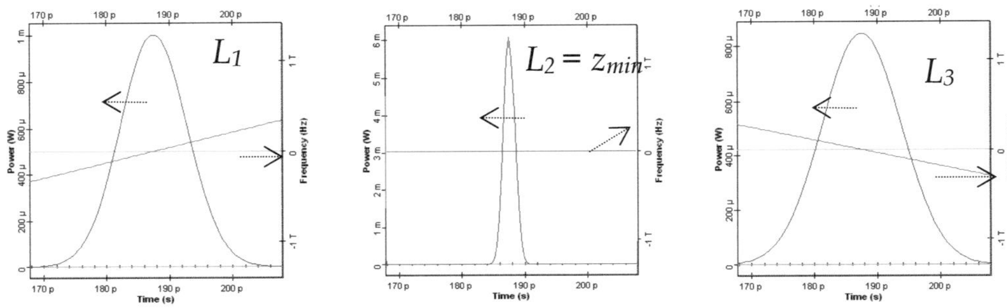


(a) Gaussian pulse and chirp.

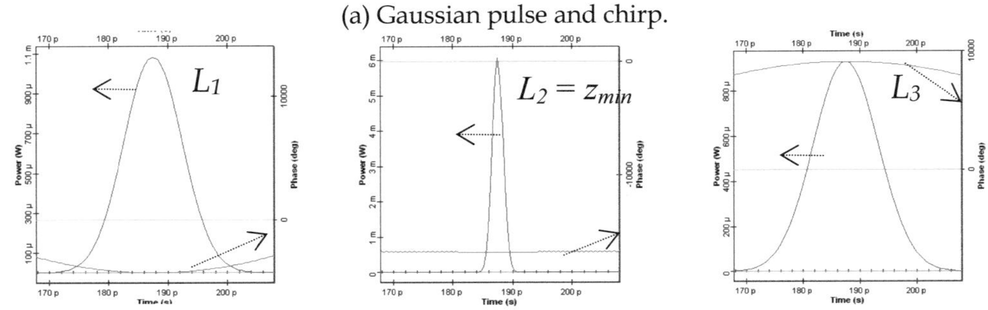


(b) Gaussian pulse and optical phase shift.

Fig. 6. Simulated waveform of a positive chirped Gaussian pulse and chirp (a) when the pulse is propagated through a negative dispersion fiber for three fiber lengths. Chirped Gaussian pulse and phase throughout the pulse (b).

### 5.2 Chirped pulses of arbitrary shape

Equation (20) only describes the propagation of a Gaussian pulse and includes dispersive effects only up to the second order ($\beta_2$). In practice, the pulses cannot be considered Gaussian, mainly for two reasons; as is demonstrated in Section 3.1. The pulse waveform of a DML exhibits an overshoot and ringing in its output power whilst, on the other hand, if a higher-order dispersion governed by β3 is included in (20), the pulse no longer remains Gaussian on propagation because it develops a tail with an oscillatory structure (Agrawal, 2010).

For pulses of an arbitrary shape, a measurement of the pulse width is provided by the *root mean square* (RMS) width of the pulse defined as:

$$\sigma_p = \left( \left\langle t^2 \right\rangle - \langle t \rangle^2 \right)^{1/2} \quad ; \quad \left\langle t^m \right\rangle = \frac{\int_{-\infty}^{\infty} t^m |A(z,t)|^2 \, dt}{\int_{-\infty}^{\infty} |A(z,t)|^2 \, dt} \tag{26}$$

Then, the broadening factor is given by

$$\frac{\sigma^2}{\sigma_0^2} = \left(1 + \frac{a\beta_2 z}{2\sigma_0^2}\right)^2 + \left(\frac{\beta_2 z}{2\sigma_0^2}\right)^2 + \left(\frac{\beta_3 z\left(1 + a^2\right)}{4\sqrt{2}\sigma_0^3}\right)^2 \tag{27}$$

where $\sigma_0$ is the input pulse ($\sigma_0=T_0/\sqrt{2}$). The last term in (27) is the third-order dispersion contribution to the pulse broadening.
If the complex envelope of the optical field at the laser output is given by (Agrawal, 2010)

$$A(0,t) = \sqrt{P(t)}\exp[j\phi(t)] \tag{28}$$

where $P(t)$ is the laser output power and $\phi(t)$ is obtained by the integration of (11).
The output signal complex envelope is obtained by the convolution of the input envelope (28) with the fiber response $h_e(z,t)$ [Fourier transform of (18)]. Thus, the output power at a z point is calculated as

$$P_z(t) = \left|A(z,t) * h_e(z,t)\right|^2 \tag{29}$$

Starting from (29) the shape of a chirped pulse can be determined when it is propagated through fibers with both positive and negative dispersion values.
In general, (29) cannot be easy to solve in an analytic form. In this case, computer simulations are useful to predict the exact influence of the interplay of chirp and dispersion on transmission performance.
The purpose of the following section is to present simulation results pertaining to the transmission of chirped optical signals in a dispersive fiber. In the following, it will be shown that transient-chirp dominated lasers perform better over negative and positive dispersion fibers than adiabatic-chirp dominated lasers. It will be also shown that the transient component of the chirp improves the transmission performance significantly over positive dispersion fibers if the modulation current waveform is chosen appropriately.

## 6. System model

Good modulation performance at high bit-rates, to avoid excessive back-to-back penalties, is one of the main requirements for CWDM transmitters. Since the $a$ parameter (including the $\alpha$ factor and $\kappa$ coefficient) is tunable by means of the modulating current $I(t)$. The optimum operating conditions in terms of the chirp/dispersion interactions can be set for fibers having different amounts and signs of dispersion. It is the aim of this section to discuss and compare how the shape of the modulated signal (e.g. square-wave, sine-wave, Gaussian, etc.) can improve the system performance when using both positive and negative dispersion fibers.
As already mentioned, numerical simulation will be required to model the exact influence of the interplay between chirp and dispersion on transmission performance. The decision on the choice of the transmission fiber characteristics (i.e., absolute dispersion value and its sign) and the DML characteristics (adiabatic or transient) for Metro applications should first be determined through simulations. The parameters of the components involved in the simulation should be sufficiently accurate and representative of the majority of commercially available components, so that useful conclusions on the design and performance of the real system can be obtained. Our simulations are based on commercial software for optical transmission systems and the component parameters used in it have been extracted from the datasheet of trademarks.
Fig. 1 shows the basic CWDM link implemented in our simulations. In this work we are specifically interested in how an optical system is impacted by the pulse shape of

modulation current. But our final purpose is to optimize a CWDM system for use in a metropolitan area optical network. Due to the chirping frequency, when a directly modulated laser is used, the system performance is very sensitive to the presence of spectrally selective components (such as multiplexers, demultiplexers and filters) in the link. As has already been shown analytically in previous sections, the shape of the pulse and its accumulated phase change as a result of their transmission through these components. For this reason, we have simulated the WDM link as shown in Fig. 1, including the optical multiplexer (OM) and demultiplexer (OD) components. Nevertheless, for the sake of clarity in analyzing the results, only the data related to one channel (i.e. the channel allocated at 1551 nm) are presented here. A summary of the most representative parameters used in our simulation is detailed below.

### 6.1 Optical fibers

As already mentioned, the chromatic dispersion coefficient, both in absolute terms and in sign, is one of the most influential fiber parameters in transmission performance when a DML is used as a transmitter. For this reason, in this work, we have simulated two fibers with opposite signs in their dispersion coefficient; the already laid and widely deployed, single-mode ITU-T G.652 fiber (SMF) and the ITUT-T G.655 fiber (N-NZDSF). It is well known that the SMF fiber dispersion coefficient is positive in the telecommunication band spectrum, from the *O-band* to the *L-band*, and the dispersion coefficient sign of the N-NZDSF fiber is negative around C-band.

In this work, spectral attenuation coefficient, dispersion slope, effective area and the non-linear index of refraction are compliant with SMF-28e and MetroCor Corning® commercial fibers. These fibers are in compliance with ITU-T G.652D and ITUT-T G.655, respectively.

Table I summarizes the values of the extracted parameters for the two fibers at 1551 nm wavelength.

| | SMF | N-NZDSF |
|---|---|---|
| Attenuation Coefficient (dB/km) | 0.20 | 0.20 |
| Dispersion [ps/(nm·km)] | 18 | -5.6 |
| Zero dispersion wavelength (nm) | 1313 | 1605 |
| Effective Group Index of Refraction | 1.4682 | 1.469 |

Table I. Parameters for the two fibers at 1551 nm wavelength.

### 6.2 Transmitter

Distortion in the dispersion-induced waveform is the greatest hindrance that the designer of metro area networks has to contemplate. This distortion in the dispersion-induced waveform can have serious negative effects on the signal transmission, even at very short distances depending on the optical transmitter chosen and the characteristics of its frequency chirp.

The transmitter block diagram implemented in our simulations is shown in Fig. 2. The pulse string was encrypted with a $2^{15}$-1 pseudo-random bit sequence, OC-48 system, at 2.5 -Gb/s. For our purpose, the leading and tailing edges of the applied current pulse $I_p(t)$ have been matched to different forms; exponential, sine and Gaussian (see Fig. 7) .

Exponential: Gaussian: Sine:

$$I_p(t)=\begin{cases}1-e^{-(t/c_r)}\rightarrow 0\le t\le t_1\\ 1\rightarrow t_1\le t\le t_2\\ e^{-(t/c_f)}\rightarrow t_2\le T_b\end{cases}\quad I_p(t)=\begin{cases}e^{-(t/c_r)2}-1\rightarrow 0\le t\le t_1\\ 1\rightarrow t_1\le t\le t_2\\ e^{-(t/c_f)^2}\rightarrow t_2\le T_b\end{cases}\quad I_p(t)=\begin{cases}sin(\pi t/c_r)\rightarrow 0\le t\le t_1\\ 1\rightarrow t_1\le t\le t_2\\ sin(\pi t/c_f)\rightarrow t_2\le T_b\end{cases}\qquad (30)$$

where, $c_r$ and $c_f$ are the rise and fall time coefficients, respectively. $t_1$ and $t_2$, together with $c_r$ and $c_f$, are numerically determined to generate pulses with the exact values of the rise time and fall time parameters, and $T_b$ is the bit period. In all simulated cases, fall and rise time are fixed at the half bit period.

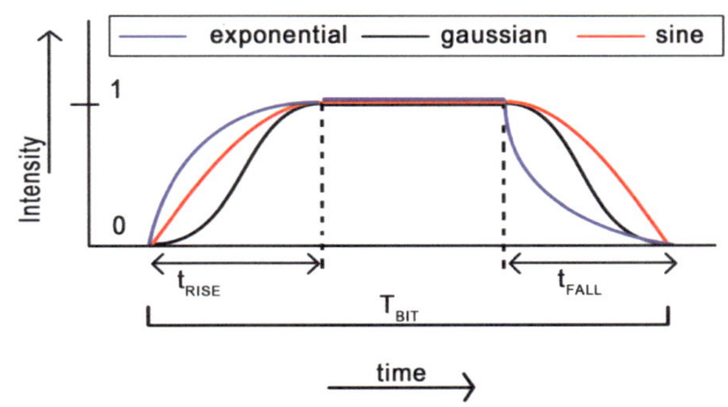


Fig. 7. Exponential, Gaussian and sine pulse shape for the laser modulation current.

To simulate the DMLs, the Laser Rate Equations introduced in (Cartledge & Srinivasan, 1997; Tomkos et al., 2000) has been used as model.

| | *DML-A* | *DML-T* |
|---|---|---|
| *Factor $\alpha$* | 2.2 | 5.6 |
| *Volume active layer (cm $^3$), V* | $5.8\ 10^{-12}$ | $1.1095\ 10^{-10}$ |
| *Gain compression coefficient (cm$^3$), $\varepsilon$* | $5\ 10^{-18}$ | |
| *Quantum efficiency, $\eta$* | 0.19 | |
| *adiabatic coefficient, $\kappa$ (Hz/W)* | $28.7\ 10^{12}$ | $1.5\ 10^{12}$ |

Table II. Simulation parameters for directly modulated lasers DML-A and DML-T.

In the simulation, we have chosen the values of commercial lasers to generate two types of chirp behavior: DML-A with a dominant adiabatic chirp and DML-T with a dominant transient chirp. For this, we work with two typical values for the Henry coefficient, ($\alpha$) (Yu & Guido, 2004; Villafranca et al, 2008; del Rio & Horche, 2008) and the adiabatic coefficient ($\kappa$) have been determined from the parameters$\varepsilon$, $\eta$, and $\Gamma$, in accordance with (17). These values are detailed in Table II.

### 6.3 Other components

The optical receiver was modeled as a *pin* photodiode with a *Responsivity* of 1 A/W and with a thermal noise of 1.84375 *$10^{-22}$ W/Hz, followed by a low-pass Bessel electric filter to eliminate high-frequency noise.

The WDM multiplexer and demultiplexer components are configured as optical Bessel filters (2nd order) with a bandwidth of 10 GHz. They are fixed in all simulations.

### 6.4 Evaluation of system performance

Four different systems have been analyzed in the computer simulations, all based on the optical link shown in Fig. 1; using a DML-A in conjunction with SMF and N-NZDSF fibers, and the same scheme by substituting the DML-A by the DML-T. The performance of the systems is evaluated and the results are compared.

The transmission system performance is often characterized by the *Bit Error Rate* (BER), which is required to be less than $10^{-15}$ for systems with a bit rate greater than 2.5-Gb/s. Experimental characterization of these systems is difficult since the direct measurement of the BER involves considerable time and cost at such a low BER value. Another way of estimating the BER is by calculating the $Q$-factor of the system, which can be modeled more easily than the BER. The BER, with the optimum setting for the decision threshold, depends on the *Q-factor* as (Alexander, 1997)

$$BER = \frac{1}{2} erfc\left(\frac{Q}{\sqrt{2}}\right) \approx \frac{1}{\sqrt{2\pi}} \frac{e^{-\frac{Q^2}{2}}}{Q}; \qquad Q = \frac{|I_1 - I_0|}{\sigma_1 + \sigma_0} \tag{31}$$

where, $I_i$ and $\sigma_i$ are average and variance values, respectively, of the 1 and 0 bits.

On the other hand, the $Q$-factor can be determined from *eye diagram*. The eye diagram provides a visual way of monitoring the system performance: The "*eye opening*" is affected by the dispersive effects accumulated along inside the link and a closing of the eye is an indication that the system is not performing properly. If the effect of laser chirp is small, the eye closure $\Delta$ can be approximated by (Keiser, 2010)

$$\Delta = \left(\frac{4}{3}\pi^2 - 8\right) t_{chirp} DLB^2 \delta\lambda \left[1 + \frac{2}{3}\left(DL\delta\lambda - t_{chirp}\right)\right] \tag{32}$$

where $t_{chirp}$ is the duration of the chirp, $B$ is the bit rate, $D$ the fiber chromatic dispersion, L is the fiber length and $\delta\lambda$ is the maximum wavelength excursion induced by the chirp.

From (32), one approach to minimize the chirp effect is to work at the fiber zero dispersion wavelength. However, in WDM systems each channel has its own dispersion coefficient and it is not possible to implement this solution. Another way of minimizing the chirp effects, whatever the working wavelength, has to be found.

In the following section, the performance of the simulated systems is evaluated in terms of the $Q$-factor.

## 7. Simulations and results

For the simulation of the proposed system (Fig. 1), 100 km of optical fiber was used and, as already mentioned, only the results for the channel allocated at a wavelength of 1551 nm are presented. In all simulated cases the laser was polarized with a bias current slightly below the threshold; which means that no impact on the extinction ratio is considered.

According to (14), the DML chirp characteristics are dependent on their output optical power. In our previous works (Horche & del Rio, 2008, del Rio et al., 2011), we

demonstrated the existence of an optimum optical power that leads to a maximum *Q-factor* ($Q_{max}$). This is why the study includes an analysis of the systems at different output optical powers of the DML. In commercial uncooled DFB-DMLs the average modulated output power can reach 6 dBm or more at room temperature. In our simulations, a peak power for the "1" bit between 0.1 mW to 10 mW (average ~ 6 dBm) has been considered.
For each of these cases the system performance (*Q*-factor) has been analyzed taking into account the three different modulations current types given by (30).

### 7.1 Transmitter with dominant adiabatic chirp, DML-A

In a DML, the output optical power waveform is a function of the modulation current, as a result of using different currents, different optical pulse shapes are obtained. Fig. 8(a) shows the output optical power pulse shape of a DML-A when the modulation current takes exponential, Gaussian and sine forms, respectively.
The three pulses have the typical characteristic shape of the pulses generated by an adiabatic chirp-dominated laser. As can be seen in Fig. 8(a), for the current Gaussian modulation shape, the pulse leading edge is delayed in relation to the exponential and sine forms. This is the result of a higher *turn-on delay* (turn-on delay is the time required for the carrier density to build up to the threshold value before light is emitted). As seen in the following sections, this is one of the reasons why the system with a Gaussian current pulse presents the worst performance.
In the case of an adiabatic chirp-dominated laser, where the transient chirp has been completely "masked" by the adiabatic term, there is a shift between the frequency of the 1 and the 0 bits, the frequency of the "1" being greater than the frequency of the "0" (blue shift). Therefore, the result of the dispersion with the specific chirp characteristics is a high-intensity spike at the front of the pulses and a trailing tail-end for transmission over a positive dispersion fiber (SMF), see Fig. 8 (b). Exactly the opposite effects take place for transmission over a negative dispersion fiber (N-NZDSF), see Fig. 8 (c) [Hakki, 1992]. In the case of DML-A the main role in the transmission performance is played by the absolute value of the dispersion coefficient (rather than its sign) [Tomkos et al., 2001]. Both figures, 8(b) and 8(c), take into account the simulation of a point-to-point optical link without any selective component in wavelength, that is, only the fiber introduces changes into the chirp of the transmitted pulse.
In Fig. 9 the maximum *Q*-factor, minimum BER, of the system in Fig. 1 is shown. Variations in the DML optical power from 0.1 to 10 mW for the three types of modulation current are analyzed for an SMF, Fig. 9(a), fiber and an N-NZDSF fiber, Fig. 9 (b). The graphs show that, regardless of the type of fiber used, the system has a maximum performance for a given value of power ($P_{optimun}$) which agrees with the results obtained in previous works (Horche & del Rio, 2008; Suzuki, 1993). For low levels of channel power ($P_{ch}$), below the optimum power, the *Q*-factor increases with $P_{ch}(t)$, mainly because of a larger amount of power reaches the detector and the performance of the system is improved. For $P_{optimun}$ the interplay of the chirp with the dispersion leads to the best compensation between them. On the other hand, for $P_{ch}$ higher than $P_{optimun}$, the chirp increases with the power which gives rise to a greater frequency shift between the "1" and "0" levels and linewidth broadening producing a greater optical pulse deformation.
Looking at Fig. 9(a) and (b) we can see that the system has a similar behavior with the sine and exponential models and they are far from the Gaussian model. These variations are a result of the similarities/differences that exist in the shape of the pulses at the output of the laser when the different models are considered [see Fig. 8(a)].

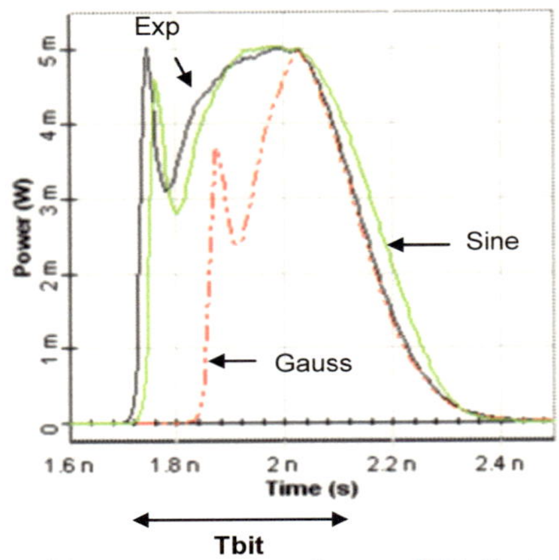


(a) Output pulse shape of DML-A.

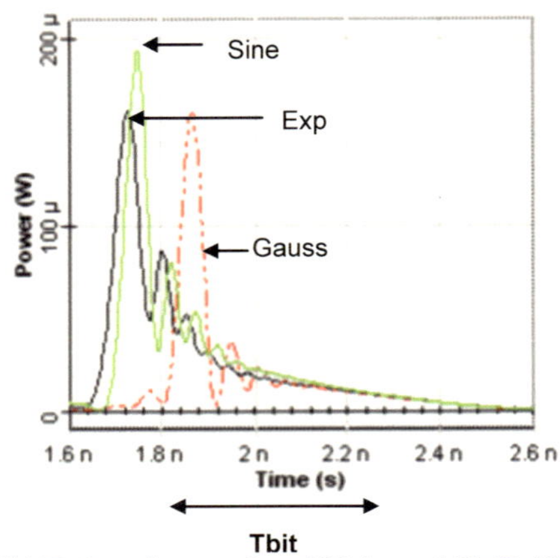


(b) Pulse shape after 100 km of SMF fiber

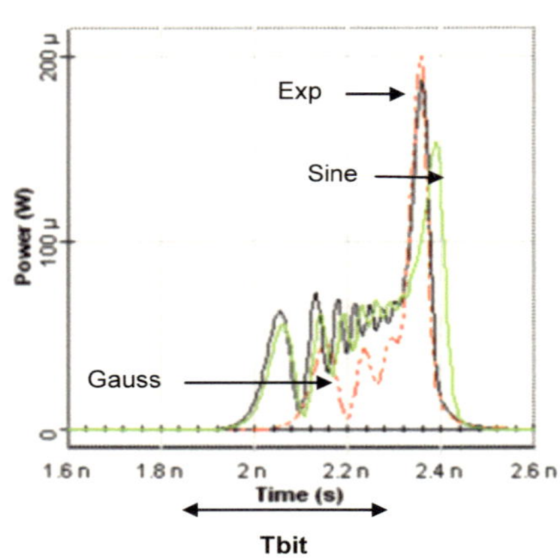


(c) Pulse shape after 100 km of N-NZDSF fiber

Fig. 8. Pulse shape for a DML-A with three types of modulation current: exponential, (thick dotted black line), Gaussian (thin dotted red line) and sine (continuous green line).

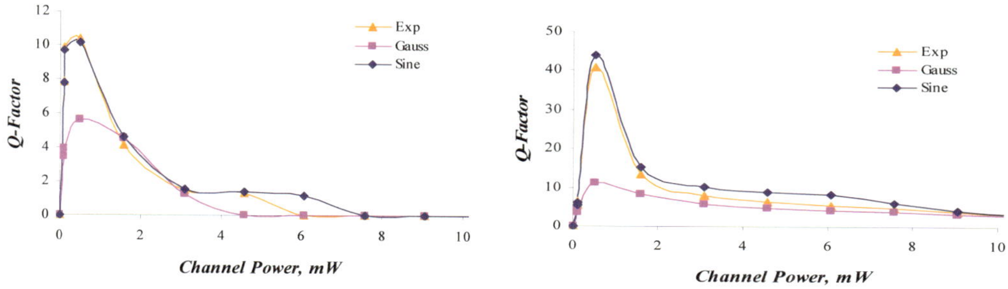


(a) SMF (b) N-NZDSF

Fig. 9. Q-factor as a function of the DML-A output power on fibers with (a) a positive dispersion coefficient and (b) a negative dispersion coefficient.

Comparing the results from the SMF and N-NZDSF, Fig. 9(a) and (b) respectively, the N-NZDSF fiber outperforms the SMF fiber in all three cases of sine, exponential and Gaussian because of the reduced absolute dispersion value.

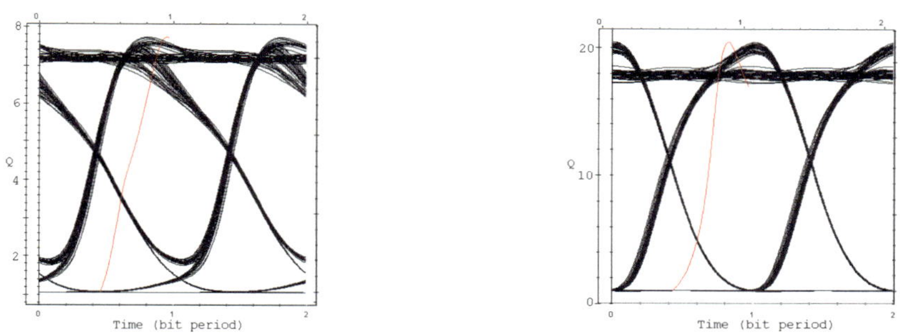


(a) SMF (b) N-DSF

Fig. 10. Eye diagram for a DML-A laser with an optical output of 1 mW after 100 km of SMF (a) and N-NZDSF (b). A sine pulse is considered as the modulation current.

In Fig. 10, the eye diagrams are shown for the case of an adiabatic chirp dominated transmitter after transmission over 100 km of SMF (a) and N-NZDSF (b), respectively. In both case $P_{ch}$ was 1 mW. The different dispersion sign will only affect the asymmetry of the eye diagram as is obvious from the results of Fig. 10.

From the simulations carried out with a DML-A transmitter, it can be concluded that with this type of laser the major factor is the absolute value of the fiber dispersion, and the waveform of the current modulation is not significant

### 7.2 Transmitter with a dominant transient chirp (DML-T)

Transient chirp-dominated laser diodes exhibit significantly more overshoot and ringing in output power and frequency deviations (see the power and chirp waveforms in Fig. 11). In case of a DML-T, the leading edge of the pulse is blue shifted relative to the main portion of the pulse, while the tailing edge is red shifted. The blue-shift chirped portion advances relative to the main portion of the pulses in the case of positive dispersion fibers. This effect produces a pulse spreading and consequently, an inter-symbol interference can occur.

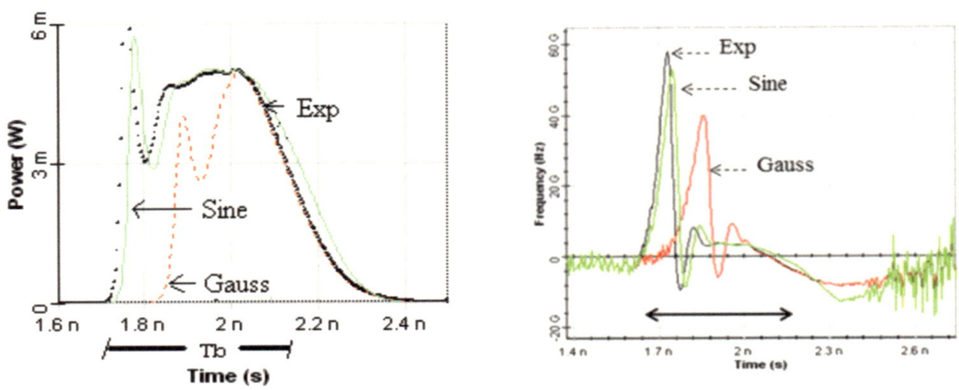


(a) (b)

Fig. 11. Power (a) and chirp (b) waveforms for a DML-T transmitter for three different types of current modulation: exponential, Gaussian and sine.

Fig. 12 shows the *Q*-factor as a function of power for the three models studied. Fig. 12(a) corresponds to the propagation over an SMF fiber and Fig. 12 (b) is dealing with N-NZDSF. Clearly, from the results summarized in Fig. 12 some conclusions can be extracted:

- Again, the minimum quality standard is achieved for the Gaussian model
- In both SMF and N-NZDSF the existence of an optimum optical power for a maximum *Q*-factor is demonstrated. However, with the DML-T transmitter, the dispersion characteristics of the fiber used in the link determine the modulation current pulse shape for which the transmission performance is improved.

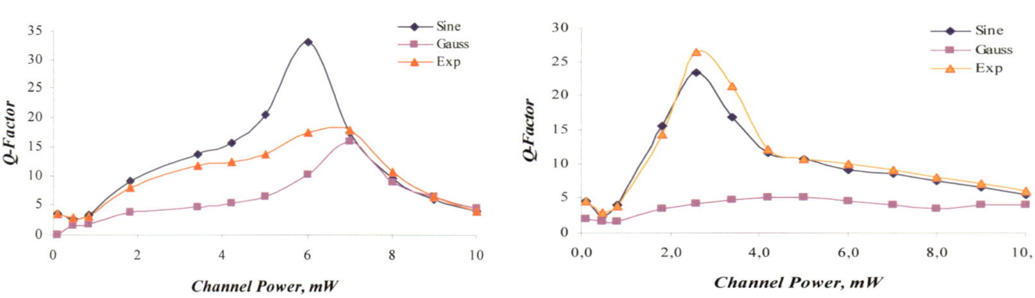


(a) SMF. (b) NZ-NDF.

Fig. 12. Q-factor as a function of the DML-T output power on fiber with (a) a positive dispersion coefficient and (b) a negative dispersion coefficient.

Therefore, by using a negative dispersion fiber (Fig. 12(b)), the exponential current waveform is the best choice together with an output power of 2.6 mW. In this case, the reduction in quality that the sine and Gaussian pulse models have with respect to the exponential is approximately 11% and 84%, respectively.

When using an SMF fiber with a positive dispersion coefficient (see Fig. 12(a)), the modulation current with a sine shape has a better performance, particularly where the power of the laser is about 6 mW.

This improvement in performance observed with sine modulation current, can be understood in terms of the interaction that is produced between the DML-T chirp and pulse shape with the fiber dispersion; as is shown later, the accumulated phase in the received optical bit patterns is practically constant all of the time. When the chirped pulse travels along a fiber, it interacts with the fiber dispersion characteristics, causing a phase shift in the pulse, which is a function of the sign and the absolute value of the dispersion coefficient.
Fig. 13 shows the eye diagram after 100 km of SMF (a) and N-NZDSF (b) using a DML-T transmitter and when the modulation current has a sine shape. The eye corresponding to a transmission over an N-NZDSF fiber is more distorted than that corresponding to transmission over an SMF fiber. This is surprising because of the larger absolute value of the dispersion of SMF fiber compared with N-NZDSF fiber.

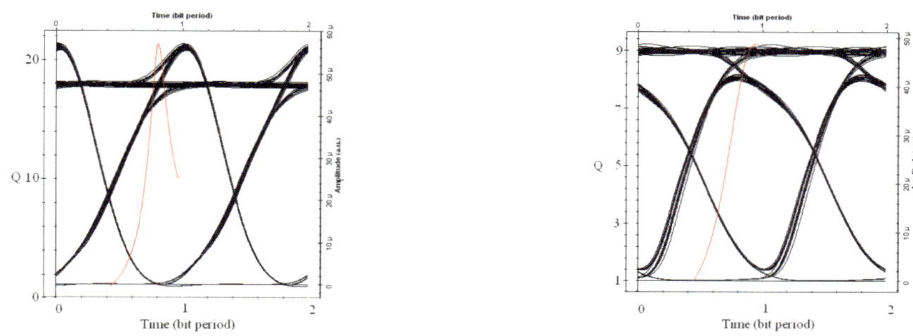


(a) SMF (b) N-NZDSF

Fig. 13. Eye diagram for a DML-T laser with an optical output of 6 mW after 100 km of fiber. A sine pulse is considered as modulation current.

Through previous simulations, it has been demonstrated that directly modulated systems using SMF fibers can reach a similar (or better) $Q$-factor than those using N-NZDSF fibers, through the use of a correct modulation current. It means that through a suitable choice of emitted optical power in the DML together with the shape of the modulation current, the tolerance of the system at chirp can be optimized and its performance improved. This is an important conclusion since the SMF fiber is currently the most commonly installed, even when operating at 1550 nm.

## 8. Study of accumulated phase

All the features of the received waveforms analyzed in the previous section can be easily explained by considering the interaction of the laser chirp and pulse shape with the fiber dispersion. Then, the different $Q$-factors obtained with the sine, exponential and Gaussian models, when a DML-T is used together with an SMF fiber, can be justified simply by studying the accumulated phase in the bit sequence.
In the analysis of the accumulated phase, we must be very careful with the choice of the bit sequence length as the accumulated phase at reception depends on the number of transitions between the 0's and 1's in the bit pattern.
As discussed in section 3.1, the DML chirp depends on the emitted average power. Thus sending a short sequence of alternating 1 and 0 bits, the chirp fluctuates between a maximum value (when sending a "1") and a minimum (when sending a "0") and the

accumulated phase will take a constant average value. Fig. 14(a) shows the accumulated phase for a bit sequence of 8 bits (01010101). However, if the bit sequence has several identical symbols, the results can be very different. For example, Fig. 14 (b) and (c) show the accumulated phase for the "11110000" and "00001111" sequences, respectively.

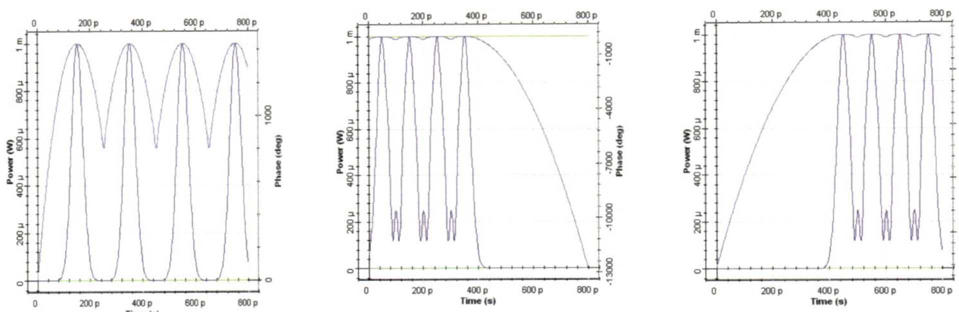


(a) (b) (c)

Fig. 14 The accumulated phase for a bit sequence of 8 bits (a) 01010101 (b) 1111000 and (c) 00001111.

Therefore, to obtain a result independent of the number of symbol transitions, we have to choose a large enough length sequence. On the other hand, this has the disadvantage of requiring excessively long processing times. In our simulation, we used a sequence of 512 bits, which maintains a good balance between accuracy of results and the calculation time. Fig. 15 represents the accumulated phase variation over a sequence of 512 bits, for different DML-T output power for the sine model after 100 km of SMF fiber. As can be seen, there is a single value of power (6 mW) that produces a constant phase variation over time that would be the best choice for the transmission. For this power value the phase variation is constant in time with the accumulated chirp being practically zero.

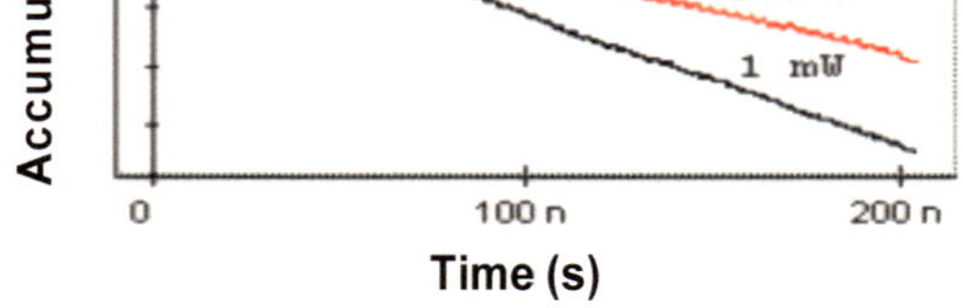


Fig. 15. Phase shifted (chirp) produced in a 512 bit sequence, for the sine model after 100 km of SMF fiber.

The accumulated phase is now represented for a sequence of 512 bits at the output of the DML-T laser with 6mW [Fig. 16(a)] and before the bit stream reaches the receiver [Fig. 16(b)], for the exponential, sine and Gaussian models. It can be seen that the bit stream goes through different accumulated phase variations when going along the SMF fiber and the filter components in the link (multiplexers, demultiplexers). The accumulated phase due to the sine pulse shape is practically constant, unlike that produced for exponential and Gaussian pulses.

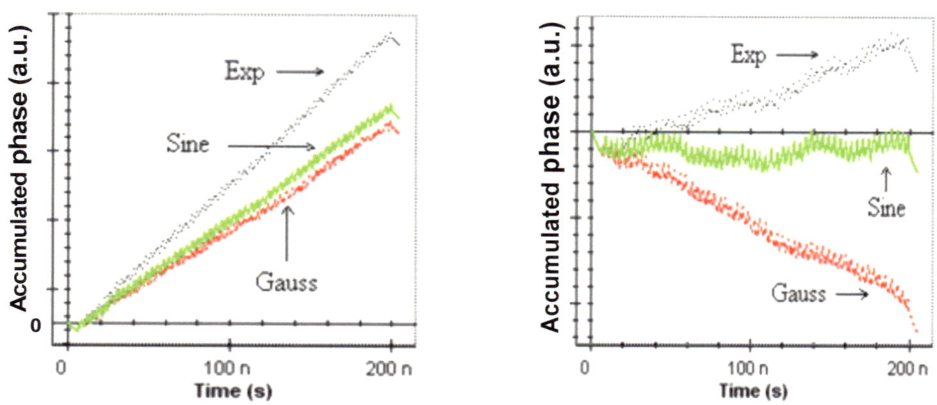


(a) At the output of the DML-T laser (b) At the output of the demultiplexer.

Fig. 16. Accumulated phase shifted (chirp) for a DML-T transmitter with 6 mW of optical power, 512-bit sequence, both before and after 100 km of SMF fiber.

The high dispersion tolerance is mainly because of the phase-correlative modulation between the adjacent bits via a precise control of the frequency chirp in the DML modulation. Therefore, based on (1-8), we can design the parameters of the DMLs (including the drive current (DC) bias) to offer a suitable chirp response for the generation of the phase correlation.

## 9. Analysis of transmission performance through different SMF and N-NZDSF fiber lengths

In order to get a general idea of the behavior of the systems detailed in Fig. 1 with the link length, a set of simulations was carried out by varying the fiber length, $L$. As in the former cases, only the results for a channel centered at 1551 nm are shown. The simulated transmitter was a DML-T similar to that used in the former sections. The output power is $P_{ch}$= 6 mW in case of a SMF fiber [see Fig. 12 (a)] and $P_{ch}$= 2.6 mW when a NZ-NDF fiber is used [see Fig. 12 (b)]. These power values have been selected to obtain the $Q_{max}$.

Fig. 17 represents the $Q$-factor versus fiber length when an SMF fiber (a) and a NZ-NDF fiber (b) are used for sine, Gaussian and exponential modulation current waveforms, respectively. As expected, the $Q_{max}$ is reached for a fiber length of 100 km and a sine modulation current in the case of an SMF fiber. This is the length at which the interaction between the accumulated dispersion in the fiber and the chirp generated in the DML is optimal. Nevertheless, for other fiber lengths, another choice can provide a better performance.

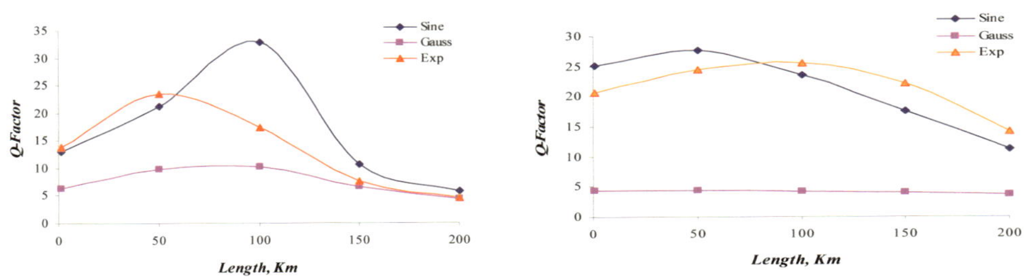


(a) SMF (b) NZ-NDF

Fig. 17. Q-factor as a function of the link distance through fibers with (a) positive dispersion coefficient and (b) negative dispersion coefficient (Pch = 6 mW).

In the case of an N- NZDSF fiber, in Fig. 17 (b), with a $P_{ch}$= 2.6 mW, the $Q_{max}$ is obtained for lengths different from 100 km, i.e. for a sine modulation current, $L$ = 50 km [not exponential as in Fig. 12 (b)]. Therefore, we can conclude that the maximum of the $Q$ factor depends on the modulation current shape, the length of the link and the optical power at the DML output.

## 10. Conclusion

In this chapter we have determined the dependence of the system performance of a directly-modulated optical system as regards the shape of the optical pulse transmitted through the fiber. It has been demonstrated that, if the laser has adiabatic dominant chirp behavior (DML-A), the absolute value of the fiber dispersion coefficient will be a determinant of the system performance but the influence of the current pulse shape is not significant.

We have also demonstrated that, if the laser has transient dominant chirp behavior, it is possible to improve the performance of an already installed system by modifying the shape of the electric current that modulates the laser. As the high dispersion tolerance is mainly because of the phase-correlative modulation between the adjacent bits, then, we can design the parameters of the DMLs, including the modulating current shape, to offer a suitable chirp response for the generation of the phase correlation. For SMF fibers (positive dispersion coefficient), this shape is sinusoidal and for NZ-NDF fibers (negative dispersion coefficient), the shape is exponential, when long-haul inter-office transport systems, corresponding to link distances of about 40 - 100 km, are considered.

In summary, through the appropriate combinations of DML transmitters (adiabatic or transient) and shape current, optical fiber systems can be achieved which are optimized in terms of dispersion and cost. With this method it is possible to improve each of the WDM system channels individually, offering a low-cost solution since it only involves changes in the transmitters and avoiding the replacement of the fiber.

## 11. Acknowledgements

The authors gratefully acknowledge the support of the MICINN (Spain) through project TEC2010-18540 (ROADtoPON).

## 12. References

Agrawal, P. (2010) Fiber-Optic Communication System. *John Wiley & Sons*. ISBN 978-0-470-50511-3.

Alexander S. B. (1997). Optical Communication Receiver Design. SPIE Press/ IEE.

Cartledge, J.C. & Burley, G.S., (1989). The effects of laser chirping on lightwave system performance, *J. Lightwave Technol.*, Vol. 7, No. 3, pp. 568–573, 1989. ISSN: 0733-8724

Cartledge, J.C. & Srinivasan, R.C. (1997). Extraction of DFB Laser Rate Equation Parameters for System Simulation Purposes. *J. of Lightwave Technol.*, Vol. 15, No. 5, May 1997, pp. 852-860. ISSN: 0733-8724.

Coldren L.A., & Corzine, S.W. (1995) *Diode Lasers and Photonic Integrated Circuits*, Wiley Series in Microwave and Optical Engineering.

Corvini, P. & Koch, T. (1987). Computer Simulation of High-Bit-Rate Optical Fiber Transmission Using Single-Frequency Lasers. *J. Lightwave Technol.*, Vol. 5, No. 11, pp. 1591 - 1595. November 1987. ISSN: 0733-8724.

del Río, C. & Horche, P.R. (2008). Directly modulated laser intrinsic parameters Optimization for WDM Systems. *International Conference on Advances in Electronics and Micro-electronics*, ENICS 2008.

del Río, C., Horche, P.R., & Martín, A. (2011), Interaction of semiconductor laser chirp with fiber dispersion: Impact on WDM directly modulated system performance". *Proc. of The Fourth International Conference on Advances in Circuits, Electronics and Micro-electronics*. CENICS 2011, 22-27 August, Nize, France.

Hakki, B.W. (1992). Evaluation of transmission characteristics of chirped DFB lasers in dispersive optical fiber. *J. of Lightwave Technology*, Vol. 10, No. 7, pp. 964 – 970, Jul 1992. ISSN: 0733-8724

Henry, C. H. "Theory of the linewidth of Semiconductor Lasers". *IEEE J. of Quantum Electronics*, Vol. QE-18, pp. 259 - 264, Feb 1982. ISSN: 0018-9197.

Hinton, K.; Stephens, T.; Modeling high-speed optical transmission systems. *IEEE J. Selected Areas in Communications*, Vol. 11, Issue 3, pp. 380 - 392.

Horche, P.R. & del Río, C.. "Enhanced Performance of WDM Systems using Directly Modulated Lasers on Positive Dispersion Fibers". *Optical Fiber Technology*. Vol. 14, No. 2, pp. 102-108, April 2008.

Keiser, G., (200). *Optical Fiber Communications*. McGraw-Hill. ISBN: 978-007-108808.

Koch, T.L.; Corvini, P.J., (1988). Semiconductor laser chirping-induced dispersive distortion in high-bit-rate optical fiber communications systems. *IEEE International Conference on Communications*, ICC '88. Vol. 2, pp. 584 - 587.

Nebeling, M. (2007). CWDM Transceivers, In: *Coarse Wavelength Division Multiplexing; Technologies and Applications*. Edited by Hans-Jörg Thiele & Marcus Nebeling, pp. (57-90), CRC Press Taylor & Francis Group, ISBN-10: 0-8493-3533-7, USA.

Saleh, B.E.A., Teich, M.C., Fundamentals of Photonics. (2007). Wiley Series in Pure an Applied Optics. ISBN: 978-0-471-35832-9, USA.

Suzuki, N., (1993). Simultaneous Compensation of Laser Chirp, Kerr Effect, and Dispersion in 10 Gp/s Long-Haul Transmission Systems. *J. of Lightwave Technology*, Vol. 11, No. 9, Sep 1993. ISSN: 0733-8724

Thiele, H-J & Nebeling, M (2007) *Coarse Wavelength Division Multiplexing; Technologies and Applications*. CRC Press Taylor & Francis Group, ISBN-10: 0-8493-3533-7, USA.

Tomkos I.; Roudas, I.; Boskovic, A.; Antoniades, N.; Hesse, R. & Vodhanel, R. (2000). Measurements of Laser Rate Equation Parameters for Simulating the Performance of Directly Modulated 2.5 Gb/s Metro Area Transmission Systems and Networks. *IEEE Lasers and Electro-Optics Society, LEOS 2000,* Vol 2, pp. 692 – 693, 2000.

Tomkos, I., Chowdhury, D., Conradi, J., Culverhouse, D., Ennser, K., Giroux, C., Hallock, B., Kennedy, T., Kruse, A., Kumar, S., Lascar, N., Roudas, I., Sharma, M., Vodhanel, R. S., & Wang, C.-C. (2001). Demonstration of Negative Dispersion Fibers for DWDM Metropolitan Area Networks, *IEEE J. on selected Topis in Quantum Electronics*. Vol. 7, No. 3, May/Jun 2001, pp. 439-460.

Villafranca, A., Lasobras, Escorihuela, J. R., Alonso, R. & Garcés, I. "Time-resolved Chirp Measurements using complex Spectrum analysis based on stimulated Brillouin Scattering" *Proceedings of the OFC*, Paper OWD4, San Diego (USA), Feb 2008.

Yu, Y., & Giuliani, G. (2004). Measurement of the Linewidth Enhancement Factor of Semiconductor Lasers Based on the Optical Feedback Self-Mixing Effect. *IEEE Photonics Technology Letters*, Vol. 16, No. 4, pp. 990 - 992. April 2004. ISSN:1041-1135

Linke, R.A. (1985). Modulation Induced Transient Chirping in Single Frequency Lasers. *IEEE J. Quantum Electronics*, Vol. QE-21, No. 6, (June 1985), pp. 593-597, ISBN

Morgado, J.A.P & Cartaxo, A.V.T. (2001). Dispersion supported transmission technique: comparation of performance in anomalous and normal propagation regimes. *IEE Proc-Optoeelctron.*, Vol. 148, No. 2, (April 2001), pp. 107-116, ISBN

# Part 4

## Sensors

# 19

# Mechanical Properties of Optical Fibers

Paulo Antunes, Fátima Domingues, Marco Granada and Paulo André
*Instituto de Telecomunicações and Departamento de Física, Universidade de Aveiro*
*Portugal*

## 1. Introduction

Nowadays, optical communications are the most requested and preferred telecommunication technology, due to its large bandwidth and low propagation attenuation, when compared with the electric transmission lines. Besides these advantages, the use of optical fibers often represents for the telecom operators a low implementation and operation cost.

Moreover, the applications of optical fibers goes beyond the optical communications topic. The use of optical fiber in sensors applications is growing, driven by the large research done in this area in recent years and taking the advantages of the optical technology when compared with the electronic solutions. However, the implementation of optical networks and sensing systems in seashore areas requires a novel study on the reliability of the optical fiber in such harsh environment, where moisture, $Na^+$ and $Cl^-$ ions are predominant.

In this work we characterize the mechanical properties, like the elastic constant, the Young modulus and the mean strain limit for commercial optical fibers. The fiber mean rupture limit in standard and Boron co-doped photosensitive optical fibers, usually used in fiber Bragg grating based sensors, is also quantify. Finally, we studied the effect of seawater in the zero stress aging of coated optical fibers. Such values are extremely relevant, providing useful experimental values to be used in the design and modeling of optical sensors, and on the aging performance and mechanical reliability studies for optical fiber cables.

## 2. Mechanical properties

The optical fibers are mainly used as the transmission medium in optical communications systems, nevertheless its applications in sensing technology is growing. Although the optical fiber mechanical properties are important for its use in optical communications (bending radius) is on the sensing applications that these properties are more relevant. In sensing technology the physical properties of the optical fiber are essential for the sensors characterization. Most of the optical fiber based sensors rely on the deformation of the optical fiber to determine the external parameter of interest. As an example, fiber Bragg grating (FBG) are one of the most promising technologies in sensing applications due to it numerous advantages, like small size, reduced weight, low attenuations, immunity to electromagnetic interference and electrical isolation (Antunes et al., 2011). The FBG concept as sensor relies on the mechanical deformation of the optical fiber to measure static or dynamic parameters like deformation, temperature or acceleration, therefore it is crucial to know the mechanical properties of the optical fiber (Antunes et al., 2008).

If an optical fiber is perturbed mechanically, it will suffer a deformation proportional to the amplitude of the perturbation force. This approach is valid for perturbations values lower than the elastic limit of the optical fiber, where the mechanical perturbations are reversible.
The Hooke's law expresses the relation between the perturbation force and the produced deformation, the proportionality is given by the material elastic constant. The Hooke's law is given by the following expression, along the longitudinal axis of the fiber:

$$K = \frac{|F|}{|\Delta L|} \tag{1}$$

where K is the elastic constant and $\Delta l$ is the relative deformation imposed by the action of the perturbation force, *F*.
The fiber Young modulus, $E_G$, is the proportionality constant between the perturbation force per area and the relative deformation:

$$F = E_G A \frac{\Delta l}{l} \tag{2}$$

In expression (2), *A* is area and *l* is length the optical fiber under perturbation. Considering expressions (1) and (2), the elastic constant is given by:

$$|K| = \frac{E_G A}{l} \tag{3}$$

According to expression (2), the slope of the linear region (elastic region) of the perturbation force as a function of the relative deformation represents the product $E_G \times A$. This product can be used in expression (3) to obtain the elastic constant of the optical fiber, knowing its length.
In this work we tested standard optical communications fiber SMF-28e from Corning, which according to the manufacturer specifications, have an uncoated diameter of 125.0±0.3 μm and 245±5 μm with the protective coating and photosensitive optical fiber PS1250/1500 co-doped fiber, from FiberCore, with 125±1 μm diameter without coating.
The strain measurements were made using a Shimadzu AGS-5kND mechanical test machine. The tested optical fibers were: ten samples of standard optical fiber with the acrilate protective coating; nine samples of standard optical fiber without the acrilate protective coating and ten samples of photosensitive optical fiber without the acrilate protective coating. Each sample had a length of 20.0 cm and was glued, with Cyanoacrylate glue, in each extremity to an aluminum plate to make it possible the fixation to the mechanical test machine.
Figure 1 shows photography of one optical fiber samples.

Fig. 1. Photography of a tested optical fiber with the glued aluminum plates on the extremities.

The experimental data collected with the mechanical test machine allows the representation of the force versus strain curves, for each sample of fiber.
Figure 2 shows the force versus strain graphic, representing the force applied along the fiber longitudinal axis, for each of the ten samples of standard optical fiber with the acrilate protective coating.

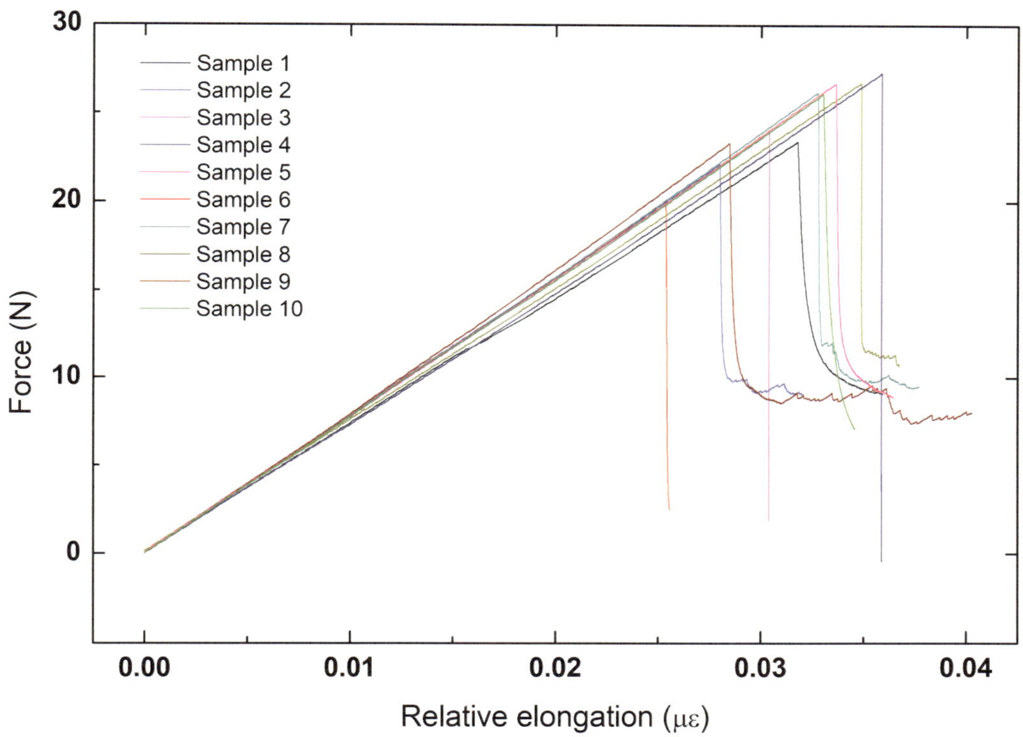


Fig. 2. Force versus strain experimental data for each sample of the standard optical fiber with the acrilate protective coating.

Considering the ten fiber samples of standard optical fiber with the acrilate protective coating, the average value for the product $E_G \times A$ is 780.87±24.99 N and the average rupture force value for the fibers fracture is 24.60±2.38 N.
The force versus strain curves for the standard fiber without the protective coating samples are presented in figure 3.
The average value obtained for the $E_G \times A$ product of the standard optical fiber without the coating was 849.42±6.88 N and the average rupture force was 4.35±1.45 N.
The force versus strain curve for the photosensitive optical fiber without the protective coating is presented in figure 4.

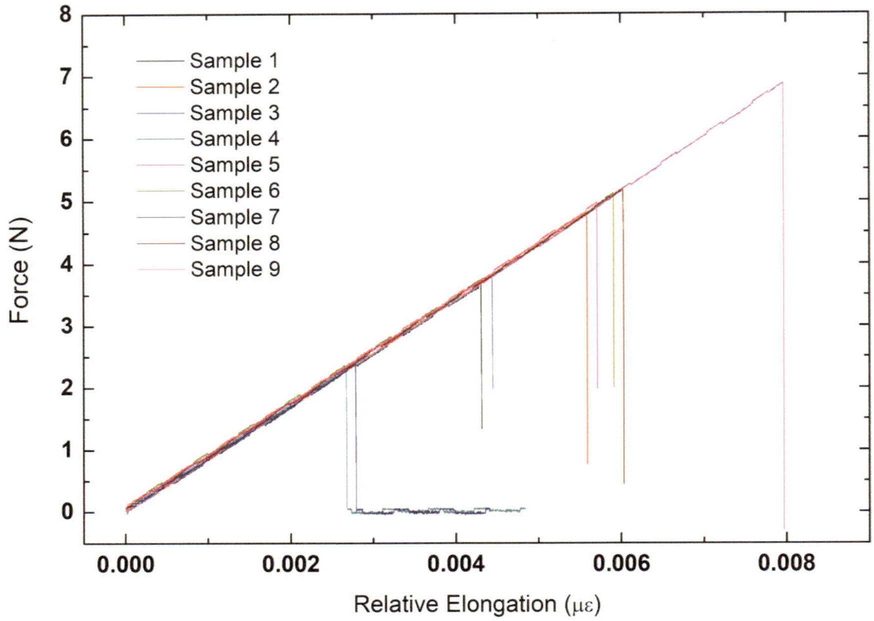


Fig. 3. Force versus strain experimental data for each sample of standard optical fiber without the acrilate protective coating.

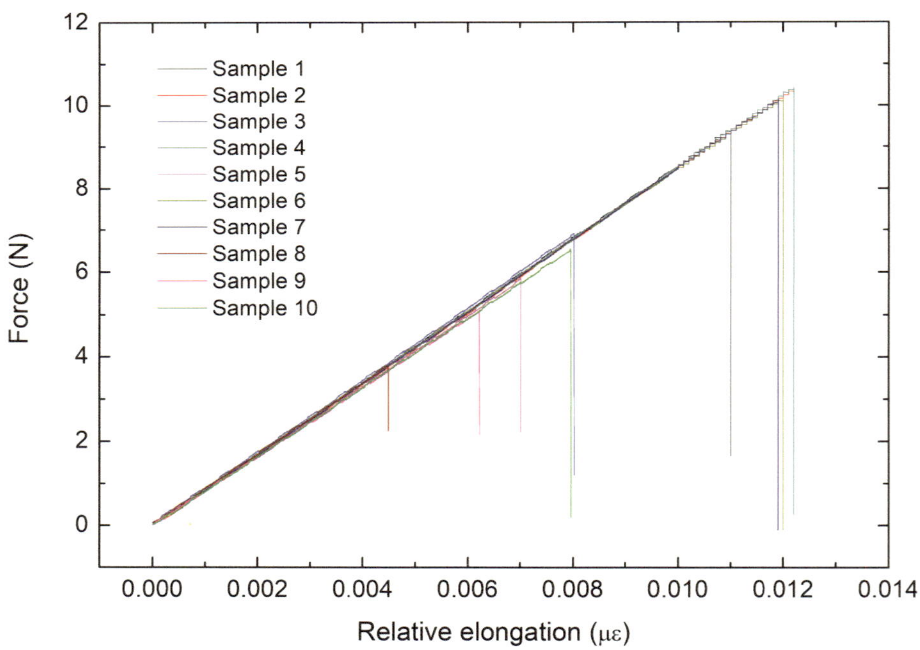


Fig. 4. Force versus strain experimental curve for the ten samples of photosensitive optical fiber without the acrilate protective coating.

For the photosensitive optical fiber without the coating the average value obtained for the $E_G \times A$ product was 841.30±15.06 N and the average rupture limit was 7.57±2.51 N.
Considering the area of each optical fiber type, the Young modulus can be obtained through the $E_G \times A$ product. The Young modulus for each type of optical fiber is presented in table 1, considering the average value of the $E_G \times A$ product.

| *Fiber* | $E_G$ *(GPa)* |
|---|---|
| Standard fiber with coating | 16.56±0.39 |
| Standard fiber without coating | 69.22±0.42 |
| Photosensitive fiber without coating | 68.56±1.47 |

Table 1. Young modulus for different types of optical fiber.

The Young modulus values found in literature (Pigeon et al., 1992; Mita et al., 2000; Antunes et al., 2008) for silica and silica fibers are consistent with the ones we measure. The obtained values can be used in the design and modeling of optical fiber sensors were the fiber can be under some kind of stress, like FBG based sensors.

## 3. Optical fiber degradation behavior

The application of optical fiber in aggressive environments may lead to the degradation of its physic reliability and therefore the performance of the systems in which it is applied (El Abdi et al., 2010). Therefore, the conservation of the optical fiber physical characteristics in harsh environments, where the fiber is exposed to abrasion and moisture, is a key point to assure its good performance.
The optical fiber coatings can provide a robust protection from the extrinsic factors that may decrease its strength and performance. Nevertheless, and in spite of the protection provided by the coatings, the fiber is still permeable to moisture. There are reports of its ability to retain the hydroxyl groups from the water molecules,(Berger et al., 2003; Méndez et al., 2007; El Abdi et al., 2010; André et al., 2011), but also of the diffusion of other ions in addition to the ones from water. Such ions diffusion can be responsible for the optical fiber degradation and the decrease of its strength (Thirtha et al., 2002; Lindholm et al., 2004). Therefore, the behavior of the fiber strength with aging is not only dependent on the moisture present in the environment but also on the diffusion rate of the ions across the coating (Armstrong et al., 1999; Domingues et al., 2010).
The implementation of optical fiber systems (as for example in sensing structures or optical networks) (Ferreira et al., 2009), is dependent on the optical fiber reliability and its lifetime degradation in abrasive environments, where ions like $Na^+$, $Cl^-$, or even moisture are present.

### 3.1 Effect of maritime environment in the zero stress aging of optical fiber

To study the effect of maritime environment in the aging of optical fiber, several samples of a standard single mode fiber (SMF-28eR) fiber, manufactured by Corning, were left in a Sodium Chloride (NaCl) aqueous solution. The fiber under test had a diameter of 125 μm, $d_f$, and a total diameter of 250 μm with the coating, $d_c$. The effect of different NaCl concentrations were studied, namely, 0 (pure deionized water), 35, 100 and 250 g/L. The 35 g/L solution matches the average sea water concentration of NaCl, while the 250 g/L

corresponds to the highest Sodium Chloride concentration found in the Earth, namely in the Dead Sea.
During the tests, samples were removed from the NaCl solution and its fracture stress was measured. For that procedure it was used an experimental setup like the one represented in figure 5.

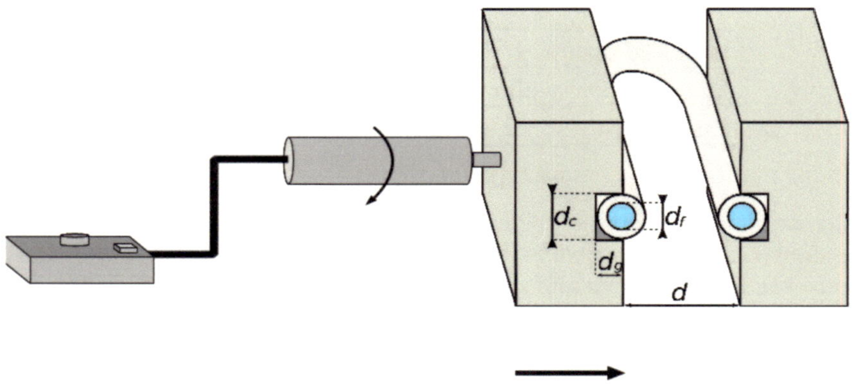


Fig. 5. Illustration of the experimental setup used to measure the fiber fracture stress.

The experimental setup uses a fixe PTFE plate and a moveable PTFE plate with grooves for the fiber fixing. The moveable plate is controlled by an electric translation stage (Newport, model 861). Initially, the plates have a distance between them of 20 mm. This distance is reduced at a speed of 0.55 mm/s. After the fiber break, the distance between plates is measured and related with the fiber fracture stress, which is dependent of the distance between the two plates. The stress applied in the fiber can be calculated through the equation (4)(El Abdi et al., 2010):

$$\sigma = E_0 \varepsilon \left( 1 + \frac{1}{2} \alpha \varepsilon \right) \tag{4}$$

where σ is the stress in the fiber, $E_0$ the fiber young modulus, ε the strain in the fiber and α is a non linearity elastic parameter. The strain ε is given by (5):

$$\varepsilon = 1.198 \left[ \frac{\mathrm{d_f}}{\mathrm{d} - \mathrm{d_c} + 2\mathrm{d_g}} \right] \tag{5}$$

where $d_f$ is fiber diameter without the coating, d the distance between plates, $d_c$ the fiber diameter with coating and $d_g$ is the total depth of the two grooves. The non linearity elastic parameter, α, is given by:

$$\alpha = \frac{3}{4} \alpha' + \frac{1}{4} \tag{6}$$

For an optical fiber α'=6 (El Abdi et al., 2010). By applying equation (4) to the measured data, we can determinate the applied stress at which the fiber fractures. This procedure was executed for the NaCl concentrations previously referred and for different aging periods.

For every sample the average stress values and error were calculated for five identical samples. In figure 6 is presented the results obtained along the aging time for the three concentrations under study.

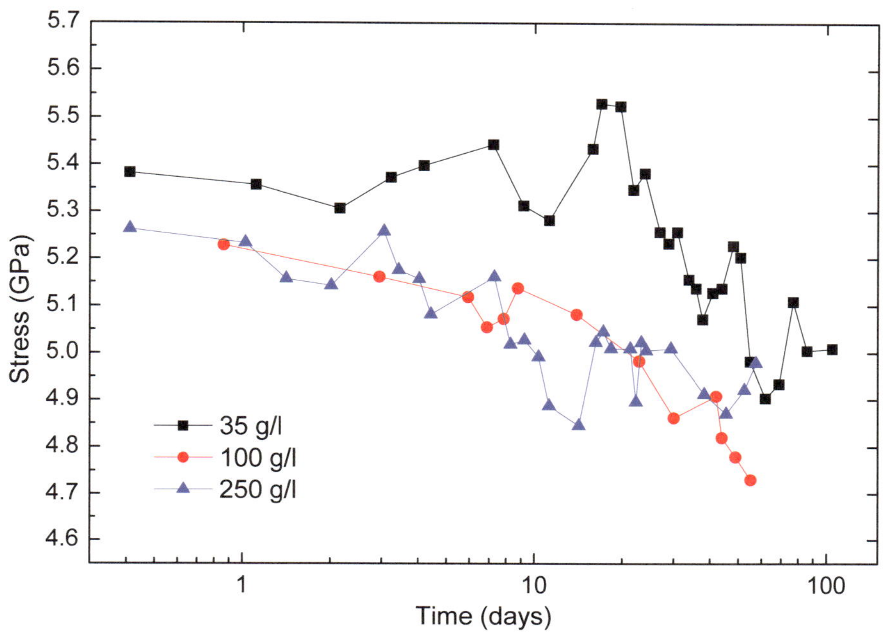


Fig. 6. Fracture stress values along degradation time for the fibers aged in the 35, 100 and 250 g/L NaCl solutions.

Several studies have reported that when the optical fiber is submitted to harsh environments its strength drastically drops after a certain time, showing a fatigue transition generally called as "knee". This behavior represents the sudden decrease of the optical fiber strength and the transition of its strength to a degradation regime (Armstrong et al., 1999; El Abdi et al., 2010). From the analyses of data collected is possible to observe that, for the fibers aged in the solutions with higher NaCl concentrations the strength of the fiber decreases and the "Knee" appears sooner in time.

In order to understand the stress behavior of aged fibers, the fracture stress values obtained for the three concentrations under study were fitted to a Boltzmann function, given by equation (7):

$$y = \frac{(A_1 - A_2)}{1 + \exp((x - x_0)/d_x)} + A_2 \qquad (7)$$

The parameters A1 and A2 are the upper and lower limit of the fiber stress, respectively, $x_0$ is the activation parameter. The $d_x$ parameter represents the function higher slope and assumes, in that point, a value of $(A2-A1)/(4d_x)$.

In figure 7 is presented the behavior of the fracture stress of the fiber as function of the exposition time in a logarithmic scale to the 35, 100 and 250 g/L NaCl concentrated solutions and the fit to the Boltzmann function:

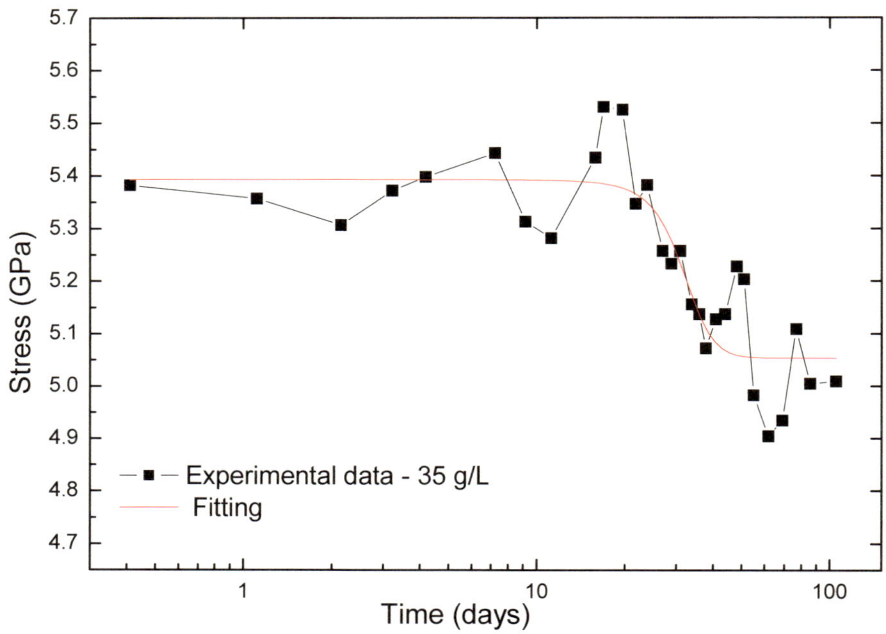
5.7
5.6
5.5
5.4
5.3
5.2
5.1
5.0
4.9
4.8
4.7
Stress (GPa)
Experimental data - 35 g/L
Fitting
1
10
100
Time (days)

a)

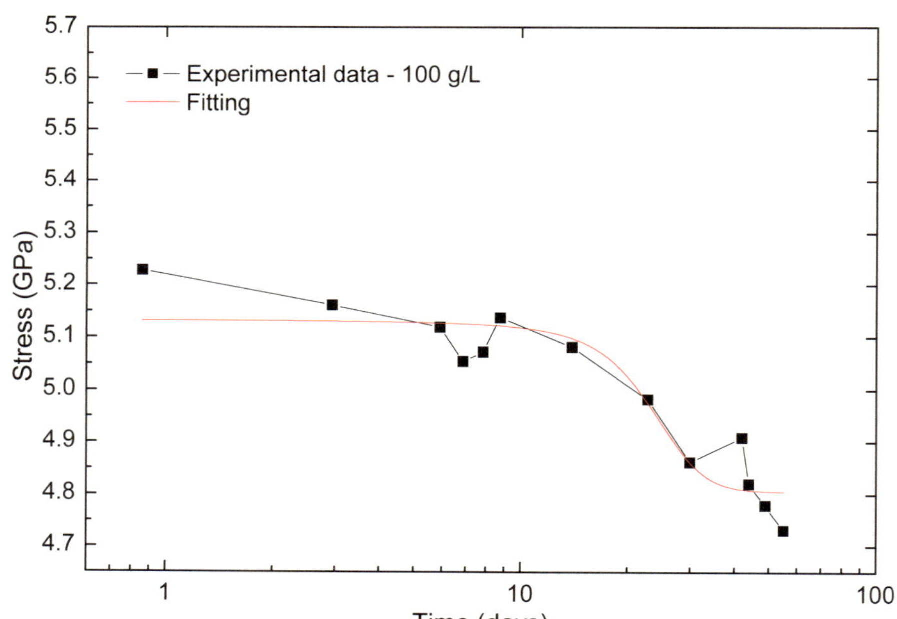
5.7
5.6
5.5
5.4
5.3
5.2
5.1
5.0
4.9
4.8
4.7
Stress (GPa)
Experimental data - 100 g/L
Fitting
1
10
100
Time (days)

b)

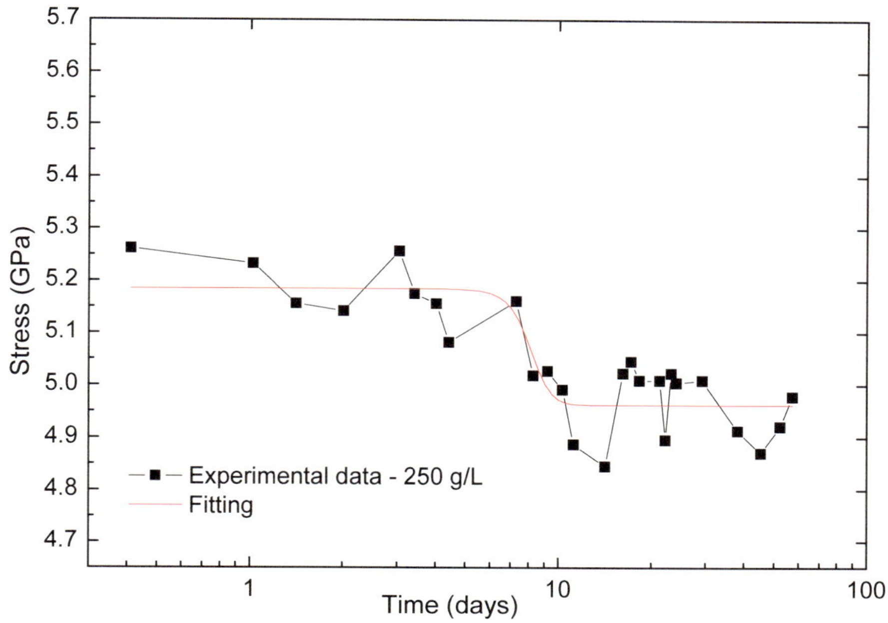


c)

Fig. 7. Fracture stress along time for the fibers aged in the a) 35, b) 100 and c) 250 g/L NaCl solutions and fitting to the Boltzmann function.

The table 2 displays the Boltzmann fit parameters for the three different concentrations.

| Concentration (g/L) | A1 (GPa) | A2 (GPa) | x0 (days) | dx (days$^{-1}$) |
|---|---|---|---|---|
| 35 | 5.39 | 5.05 | 31.34 | 3.97 |
| 100 | 5.13 | 4.80 | 23.13 | 4.30 |
| 250 | 5.18 | 4.96 | 8.05 | 0.65 |

Table 2. Boltzmann function fitting parameters.

From the values in table 2 we can see that, the time at which the strength transition occur ($x_0$) decreases with the increase of the NaCl concentration. If we establish a relation between the x0 value and the time at which the "knee" appears, we can affirm that the "Knee" will show up earlier for higher concentrations of NaCl.

This connection between the NaCl concentration and the time at which the strength transition occurs, is related to the ability of the ions in solution to infiltrate the fiber coating, react with the fiber glass surface, and remove the products through the coating (Thirtha et al., 2002). So, we can say that the diffusion through the coating of the species in solution or present in the environment in which the fiber is placed, has a major role on the aging behavior of the fiber, once such diffusion implies the decrease of the fiber strength and as consequence the decrease of its lifetime. In the analysis of the strength transition parameter, figure 8, we verify that the strength transition period has a degradation rate of 0.1174 days/[NaCl].

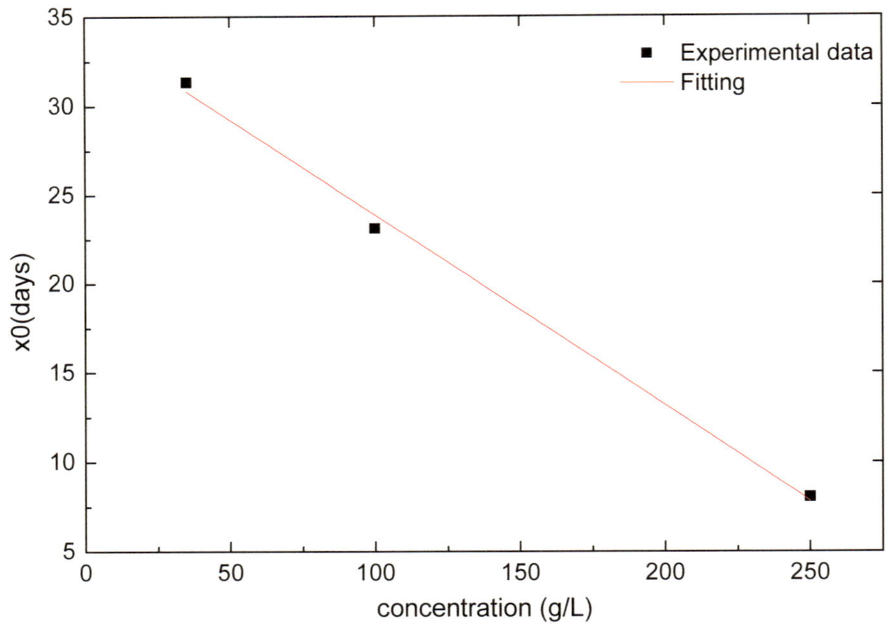


Fig. 8. Degradation rate of the aged optical fiber.

According to the authors (Danzer et al., 2007), the probability of fracture in a material increases with the number of flaws and with its dimension. Based on such assumption, the study of the probability to failure of the different aged samples will give us the information regarding the number of flaws in the samples.

To implement this study it was used the statistical Weibull's law given by:

$$\ln\left[\ln\left\{\frac{1}{1-F}\right\}\right]=m\left[\ln(\sigma)-\ln(\sigma_0)\right] \tag{8}$$

This law establishes the relation between the probability of fiber failure, F, with the applied stress σ. The first term represents the cumulative failure probability, and its evolution with the increase of the failure stress, ln(σ). The parameters $\sigma_0$ and m are constant, being m also referred to as the Weibull slope.

In the figure 9 it is represented the cumulative failure probability for fibers with no degradation and the ones aged in a 35 g/L solution of NaCl for 20 and 86 days.

It is possible to observe that the higher the aging period, the lower is the stress necessary to fracture the fiber.

Also for the different concentrations used and for the same aging periods, figure 10, it is clear that the higher the concentration of NaCl, the lower is the stress need for fracture to occur.

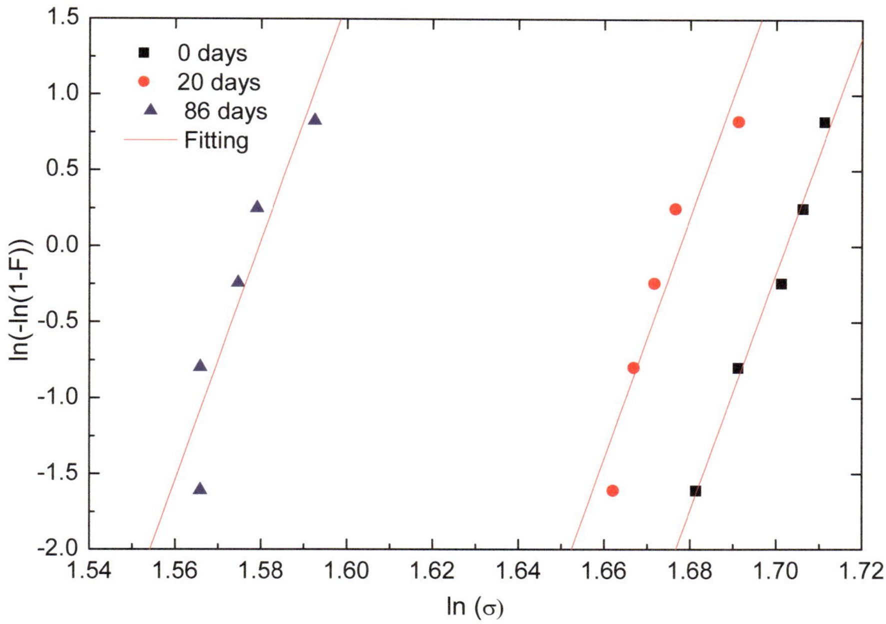


Fig. 9. Cumulative failure probability for fibers aged in the same NaCl concentration for different periods of time.

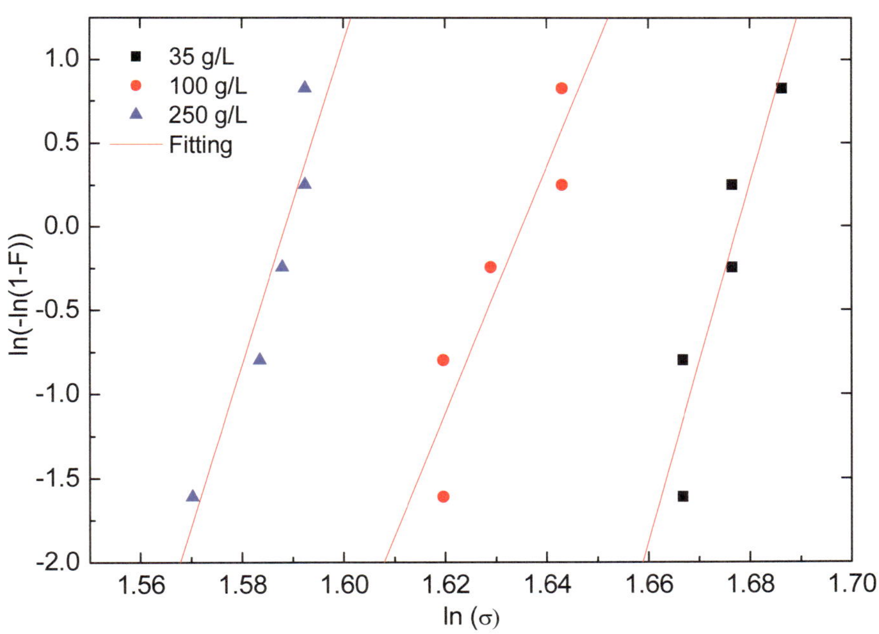


Fig. 10. Cumulative failure probability for fibers aged in different NaCl concentrations for the same period of time.

Through these latest analysis we may assume that the number of flaws in the fiber increases with the aging time and with the concentration of NaCl ions in solution.

### 3.2 Microscopic analysis of the aged optical fiber

In addition to the analytical study of the fibers, also its microscopic condition was analyzed through optical microscopy and SEM images.
In figure 11 we present the optical microscopy images, taken with an Olympus BH2 and the digital camera Sony DKC-CM30, for fibers degraded on a 35 g/l NaCl aqueous solution a) during 31 days and b) 105 days (Domingues et al., 2010).

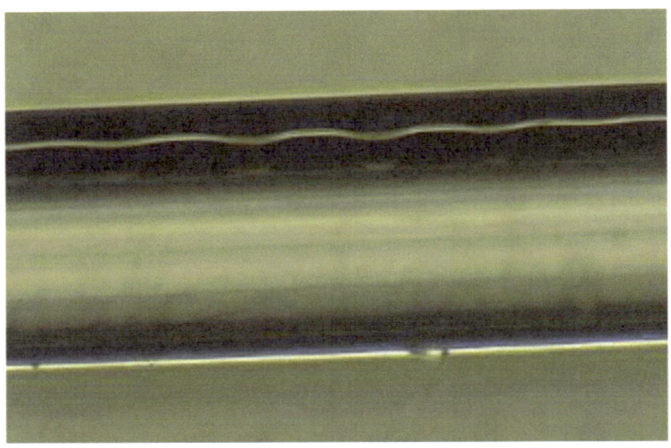

a)

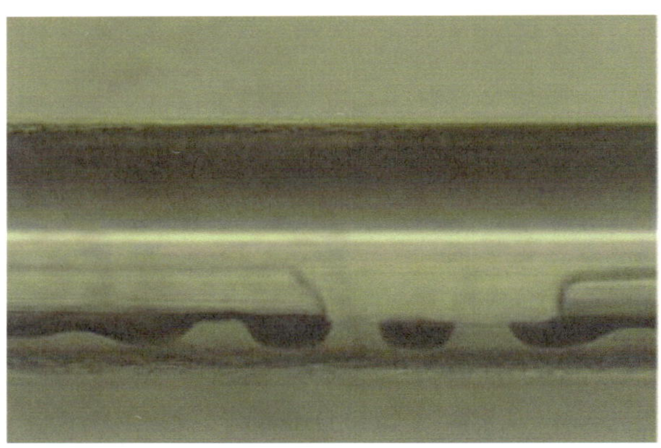

b)

Fig. 11. Microscopy images from fibers aged in a 35 g/l NaCl aqueous solution a) during 31 days and b) 105 days.

In these images we can see the difference in the protective polymer as consequence of its degradation.

The SEM images were obtained with an Hitachi SU-70 apparatus after carbon evaporation. In figure 12, are represented some of the images collected, for the three concentrations under study.

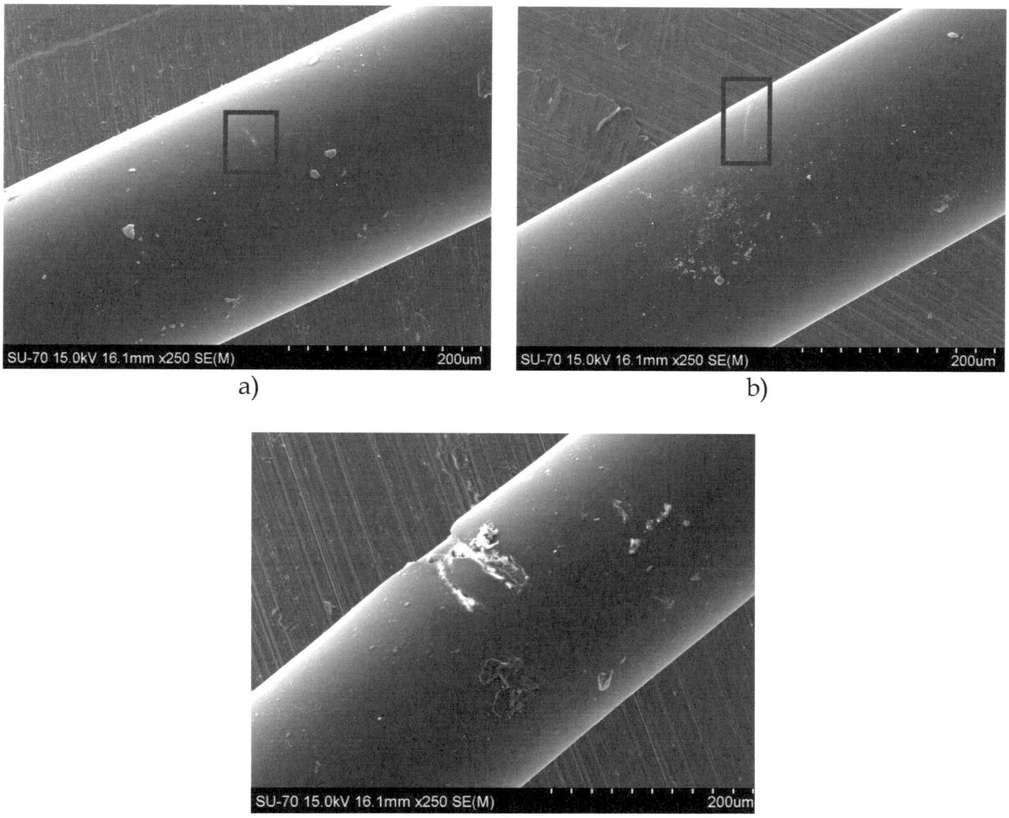


a) b) c)

Fig. 12. SEM images from fibers aged in a) 35g/L NaCl solution for 105 days, b) 100g/L NaCl solution for 69 days and c) 250g/L NaCl solution for 64 days.

In these samples, it is possible to identify the damage induced in the coating in the three samples, being the most relevant, the one presented by the sample aged in a 250 g/L solution. But also in addition to the degradation, we can identify some NaCl crystal deposited in the fiber's surface.

## 4. Conclusions

We characterized the mechanical properties for commercial optical fibers. The Young modulus obtained has the value of 69.2 GPa for the standard optical fiber without the external acrilate protective coating. The effect of seawater in the zero stress aging of coated optical fiber, shows its increasing degradation with the Sodium Chloride concentration. The degradation rate has a value of 0.1174 days/[NaCl].

These results can be useful for the design and modeling of optical sensors, and on the aging performance of optical fiber deployed in telecommunication networks.

## 5. References

André, P. S., F. Domingues and M. Granada (2011). Impact of the Maritime Environment on the Aging of Optical Fibers, *Proceedings of CLEO2011: Laser Science to Photonic Applications*, Baltimore, USA.

Antunes, P., H. Lima, N. Alberto, L. Bilro, P. Pinto, A. Costa, H. Rodrigues, J. L. Pinto, R. Nogueira, H. Varum and P. S. André (2011). Optical sensors based on FBG for structural helth monitoring, in: *New Developments in Sensing Technology for Structural Health Monitoring*, S. C. Mukhopadhyay, Springer-Verlag.

Antunes, P., H. Lima, J. Monteiro and P. S. André (2008). Elastic constant measurement for standard and photosensitive single mode optical fibres, *Microwave and Optical Technology Letters*, Vol. 50, No. 9, pp. 2467-2469, ISSN: 1098-2760.

Armstrong, J. L., M. J. Matthewson, M. G. Juarez and C. Y. Chou (1999). The effect of diffusion rates in optical fiber polymer coatings on aging, *Optical Fiber Reliability and Testing*, Vol. 3848, pp. 62-69, ISSN: 0277-786X.

Berger, S. and M. Tomozawa (2003). Water diffusion into a silica glass optical fiber, *Journal of Non-Crystalline Solids*, Vol. 324, No. 3, pp. 256-263, ISSN: 0022-3093.

Danzer, R., P. Supancle, J. Pascual and T. Lube (2007). Fracture statistics of ceramics - Weibull statistics and deviations from Weibull statistics, *Engineering Fracture Mechanics*, Vol. 74, No. 18, pp. 2919-2932, ISSN: 0013-7944.

Domingues, F., P. André and M. Granada (2010). Optical Fibres Coating Aging induced by the Maritime Environment, *Proceedings of MOMAG2010*, Brasil.

El Abdi, R., A. D. Rujinski, M. Poulain and I. Severin (2010). Damage of Optical Fibers Under Wet Environments, *Experimental Mechanics*, Vol. 50, No. 8, pp. 1225-1234, ISSN: 0014-4851.

Ferreira, L. F., P. F. C. Antunes, F. Domingues, P. A. Silva, R. N. Nogueira, J. L. Pinto, P. S. Andre and J. Fortes (2009). Monitorization of Sea Sand Transport in Coastal Areas Using Optical Fiber Sensors, *2009 Ieee Sensors, Vols 1-3*, pp. 146-150

Lindholm, E. A., J. Li, A. Hokansson, B. Slyman and D. Burgess (2004). Aging behavior of optical fibers in aqueous environments, *Reliability of Optical Fiber Components, Devices, Systems, and Networks Ii*, Vol. 5465, pp. 25-32, ISSN: 0277-786X.

Méndez, A. and T. F. Morse (2007). *Specialty Optical Fibers Handbook*, Elsevier Inc.,

Mita, A. and I. Yokoi (2000). Fiber Bragg Grating Accelerometer for Structural Health Monitoring. Fifth International Conference on Motion and Vibration Control (MOVIC 2000). Sydney, Australia.

Pigeon, F., S. Pelissier, A. Mure-Ravaud, H. Gagnaire and C. Veillas (1992). Optical Fibre Young Modulus Measurement Using an Optical Method, *Electronics Letters*, Vol. 28, No. 11, pp. 1034-1035,

Thirtha, V. M., M. J. Matthewson, C. R. Kurkjian, K. C. Yoon, J. S. Yoon and C. Y. Moon (2002). Effect of secondary coating on the fatigue and aging of fused silica fibers, *Optical Fiber and Fiber Component Mechanical Reliability and Testing Ii*, Vol. 4639, pp. 75-81, ISSN: 0277-786X.

# 20

# Fiber Fuse Propagation Behavior

Shin-ichi Todoroki
*National Institute for Materials Science*
*Japan*

## 1. Introduction

A fiber fuse is the continuous self-destruction of optical fiber induced and fed by propagating light. It is triggered by the local heating of a waveguide structure through which a high power beam is being delivered. A typical example is seen along a single mode silica glass optical fiber delivering a few watts of light (see Fig. 1). Once heat-induced high density plasma (or an optical discharge) is captured in the core region, it travels along the fiber toward the light source, consuming the light energy and leaving a hollow damage train. Thus, this phenomenon imposes an inevitable limit on the light power per fiber for current optical communication systems, namely, the service be unable to meet the growing demands of communication traffic.

The first oral presentation describing the fiber fuse phenomenon was given by Kashyap (1988), followed by the first publication by Kashyap & Blow (1988). These appeared just after certain

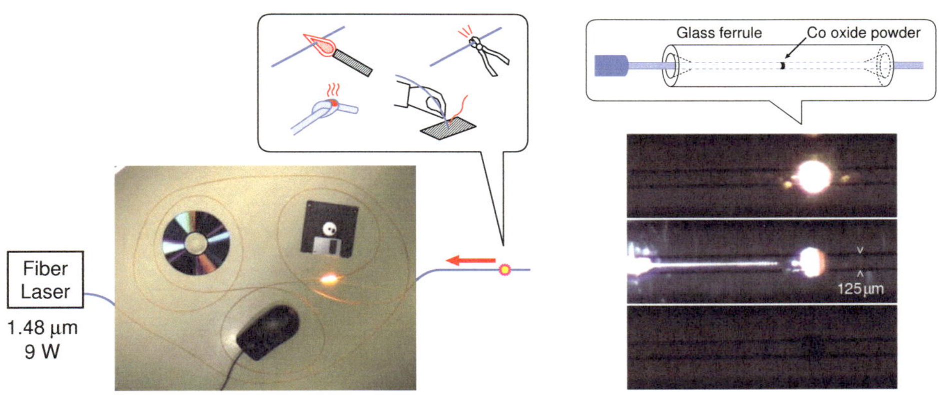


(a) Macroscopic view of fiber fuse propagation. The speed is about 1.2 m/s.

(b) Successive images of fiber fuse ignition (30 frames per second, Todoroki (2005d)).

Fig. 1. Captured video images of fiber fuse ignition and propagation along a single mode silica glass optical fiber pumped by a 9.0 W and 1.48 $\mu$m laser light. See the original video at http://www.youtube.com/watch?v=yjX5dU1EkTk (See also Table 3)

See http://pubman.mpdl.mpg.de/pubman/item/escidoc:1058545 and 1058546 for errata and a Japanese translation, respectively.

important milestones in optical communication, including the ultimate loss reduction of silica glass optical fibers by Kanamori et al. (1986) and the invention of Er-doped fiber amplifiers by Mears et al. (1987). Although this sensational phenomenon must have attracted some interest since its discovery, the number of related papers increased only after the turn of the century as a result of the rapid development of high power light sources (see Fig. 2).

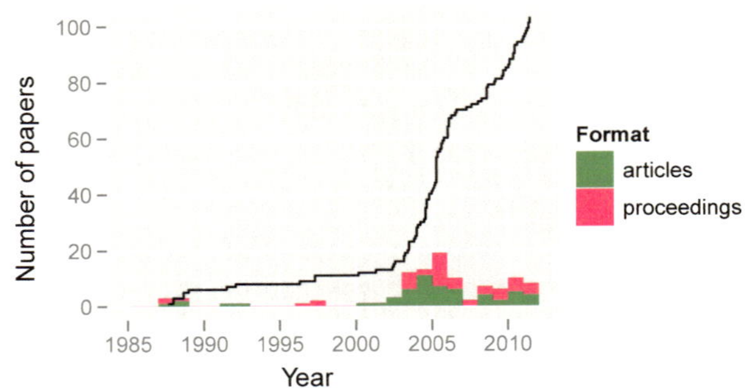


Fig. 2. Presentations and publications on the fiber fuse phenomenon. Total number (line) and yearly increments (bars). Source: http://www.geocities.jp/tokyo_1406/node3.html (See also Table 4)

Most of these researches were conducted from practical standpoints and have included techniques for the prevention and termination of a fiber fuse. In addition, studies from theoretical and/or microscopic viewpoints have been undertaken since 2003 (see Fig. 3). This paper briefly summarizes recent studies of macroscopic fiber fuse propagation where the fuse is regarded as a point without internal structures (Section 2) and the microscopic behavior of traveling plasma in relation to periodic void formation (Section 3). All the descriptions relate to silica-based fibers unless otherwise specified.

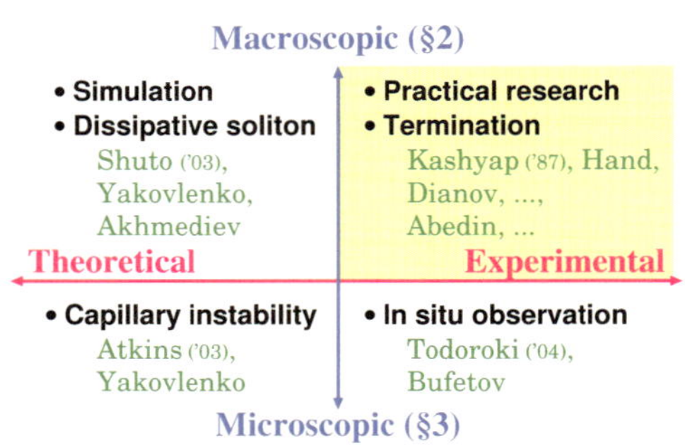


Fig. 3. Classification of fiber fuse studies with some key words and researchers.

## 2. Macroscopic behavior

### 2.1 Fiber fuse initiation, propagation and damage

Optical fibers were primarily developed to be transparent so that they would transmit light as far as possible. Therefore, at first it seems strange that transmitting light can actually destroy an optical fiber. However, it becomes reasonable when we consider that the absorption of the waveguide materials increases at elevated temperatures. Heat is the key to fiber fuse initiation.

Kashyap (1988) reported a steep increase in absorption over 1050 °C through a one-meter-long Ge-doped single-mode silica glass fiber (see Fig. 4). Shuto et al. (2004a) suggested three factors as possible origins of this absorption: (i) point-defect (Ge $E'$ center) formation (see Eqs. (1) and (2)), (ii) electronic conductivity due to the thermal ionization of a Ge-doped silica core (Eq. (2)), and (iii) the thermochemical production of SiO in silica glass (Eq. (3)). Davis et al. (1996) reported that the threshold power density required for fiber fuse initiation was of the order of 3 $\mathrm{MW/cm^{-1}}$ among typical single mode fibers operated at 1.06 $\mu$m regardless of fiber type or core composition.

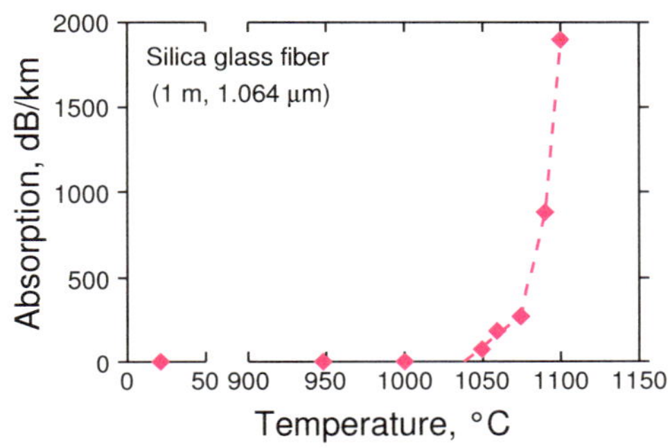


Fig. 4. Temperature dependence of the absorption in a single-mode Ge-doped silica glass fiber reported by Kashyap (1988).

$$\begin{array}{c}\backslash\\ -\mathrm{Ge}\\ /\end{array}\!-\mathrm{O}-\!\begin{array}{c}/\\ \mathrm{Si}-\\ \backslash\end{array} \longrightarrow \begin{array}{c}\backslash\\ -\mathrm{Ge}\\ /\end{array}\!-\!\begin{array}{c}/\\ \mathrm{Si}-\\ \backslash\end{array} + \frac{1}{2}\mathrm{O}_2 \uparrow \tag{1}$$

$$\begin{array}{c}\backslash\\ -\mathrm{Ge}\\ /\end{array}\!-\!\begin{array}{c}/\\ \mathrm{Si}-\\ \backslash\end{array} \longrightarrow \begin{array}{c}\backslash\\ -\mathrm{Ge}^{\bullet}\\ /\end{array}\;\begin{array}{c}/\\ \mathrm{Si}^{+}-\\ \backslash\end{array} + e^{-} \tag{2}$$

$$\mathrm{SiO_2} \longrightarrow \mathrm{SiO} + \frac{1}{2}\mathrm{O_2} \uparrow \tag{3}$$

$$\mathrm{SiO} \longrightarrow \mathrm{Si} + \mathrm{O} \longrightarrow \mathrm{Si^{+}} + \mathrm{O^{+}} + 2e^{-} \tag{4}$$

Another seemingly strange characteristic of a fiber fuse is the propagation of a bright spot toward the light source. However, this is intuitively understandable if we compare it with a grassfire (see Fig. 5). A solitary wave of fire or plasma persists as an irreversible reaction region being fueled from the front, emitting light and heat all around, and leaving cinders or damage behind it. Such a solitary wave can be considered a "dissipative soliton" as discussed later.

In the fiber fuse reaction region, ionized gas plasma is enclosed by a molten glass layer, and this situation is maintained by the pump laser energy. Shuto (2010) mentioned that most of the plasma is generated through the reaction described by Eq. (4). The temperature is estimated to be more than a few thousand degrees Kelvin by Hand & St. J. Russell (1988) and Dianov, Fortov, Bufetov, Efremov, Rakitin, Melkumov, Kulish & Frolov (2006).

(a) Grassfire at Mt. Wakakusayama, a UNESCO World Heritage site, before the onset of spring in Nara, Japan. ©Yoshitaka Inoue

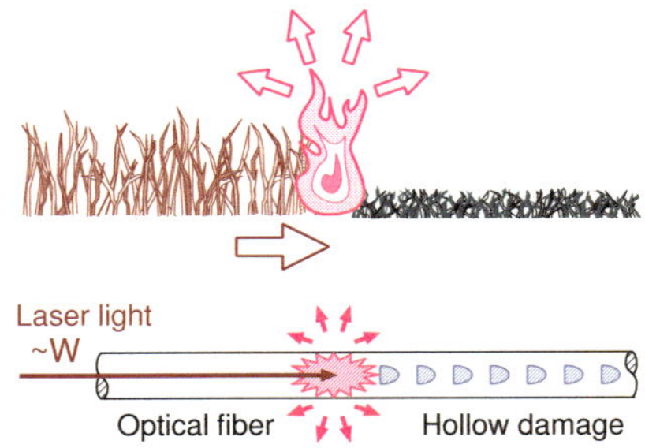


(b) Energy flows into/from a propagating reaction zone indicated with arrows.

Fig. 5. Comparison between grassfire and fiber fuse.

The propagation speed and pump power dependence of a fiber fuse have been investigated by many researchers. The speed is generally of the order of meters per second and increases almost linearly with pump laser power. However, Davis et al. (1997) pointed out that the slope vs. power density varies for different types of fibers.
As for the threshold power needed for fuse propagation, $P_{th}$, Dianov, Bufetov, Frolov, Plotnichenko, Mashinskii, Churbanov & Snopatin (2002) pointed out that mode field diameter (MFD), $2r_s$, is the dominating factor, i. e., the threshold power density ($I_{th} = P_{th}/\pi r_s^2$) varies approximately inversely with the MFD for several types of fibers pumped at 1.06–1.48 $\mu$m (see also Bufetov & Dianov (2005)). On the other hand, Seo et al. (2003) reported a linear relation between $P_{th}$ and MFD, which is parallel to the Dianov's relation. Table 1 lists the $P_{th}$ values for some standard silica based fibers.

| Fiber | $P_{th}$/W | $\lambda/\mu$m | $P_{th}$/W | $\lambda/\mu$m | |
|---|---|---|---|---|---|
| SMF | 1.0 | 1.064 | 1.4 | 1.467 | Seo et al. (2003) |
| | | | ~1.2 | 1.48 | Todoroki (2005c) |
| | | | 1.39 | 1.55 | Abedin & Morioka (2009) |
| DSF | 1.2 | 1.064 | 0.65 | 1.467 | Seo et al. (2003) |
| | | | ~1.1 | 1.55 | Abedin (2009) |
| DCF | | | ~0.7 | 1.55 | Abedin (2009) |

DSF: Dispersion Shifted Fiber, DCF: Dispersion Compensating Fiber

Table 1. Threshold power of fiber fuse propagation, $P_{th}$, for various fibers. See also Fig. 6 in Takenaga, Omori, Goto, Tanigawa, Matsuo & Himeno (2008).

After the fiber fuse has passed, a hollow damage train is left behind, and the fiber is no longer able to guide light. Figure 6 shows an example of fuse damage; a fiber fuse propagated from right to left and terminated at the position shown by the arrow due to a gradual reduction in the pump power. The pair of horizontal lines surrounding these voids is the border of the region modified by the passage of the hot plasma. Its diameter is larger than the original core size seen on the left of the arrow and increases with the pump power.
Kashyap (1988) detected $O_2$ gas inside the voids using Raman microscopy. The adjacent glass layer is expected to be densified. Dianov et al. (1992) provided supporting evidence, namely that refractive index around the void increased after the passage of the fiber fuse and subsequently decreased after fiber annealing for several seconds at about 1000 °C.

Fig. 6. Fiber fuse damage train in a standard single mode fiber (Corning SMF-28). After the ignition of the fiber fuse, the pump power of the laser (1.48 $\mu$m) decreased from 7 W (c), 3.5 W (b) to ~1.2 W (a) until its self-extinction at the position shown by the arrow.

The top of the void train provides a strong scattering point if the light is launched again. Yamada et al. (2011) warned that this scattered light heats and burns the surrounding coating and nylon jacket.

### 2.2 Dissipative soliton and termination technology

If we wish to avoid a fiber fuse, we should consider the energy flow into and from the traveling plasma. A "dissipative soliton" is a useful concept for this purpose and is defined by Akhmediev & Ankiewicz (2005) as follows:

> A dissipative soliton is a localized structure which exists for an extended period of time, even though parts of the structure experience gain and loss of energy and/or mass. ··· These solitons exist in "open" systems which are far from equilibrium.

Figure 7 compares a dissipative soliton and a conventional soliton that transfers energy and/or mass.

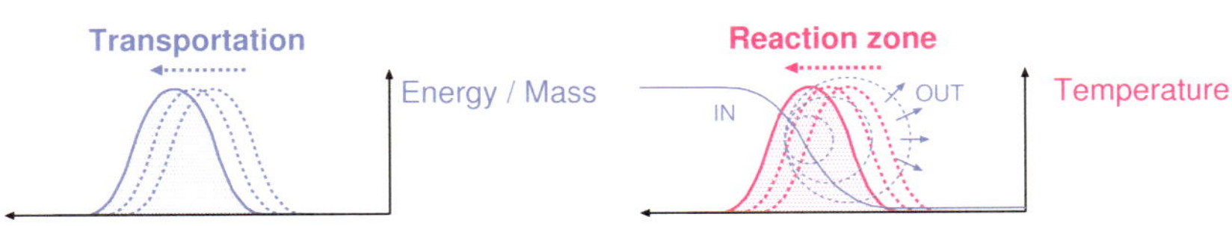


Fig. 7. Comparison of conventional and dissipative solitons.

On the basis of this energy flow, several researchers performed computer simulations of fiber fuse propagation and succeeded in reproducing the experimental propagation speed values. Their equations describing the energy flow are listed in Table 2. For the absorption coefficient, Rocha et al. (2009) used the Arrhenius form, $\alpha = \alpha_0 \exp(-T'/T)$, whereas other researchers used a custom function reproducing the steep temperature dependence shown in Fig. 4.

Shuto et al. (2004b) performed precise calculations using polar coordinates in which radiation heat loss appeared in the boundary condition. They also discussed the wavelength and power density dependence of the propagation speed. Akhmediev et al. (2008) described fiber fuse propagation as a "dissipative soliton" on the basis of a simple one dimensional calculation. Rocha et al. (2009) introduced a radiation loss term from the fiber surface and simulated three different types of fibers using MFD and $\alpha_0$ as parameters (see also Rocha et al. (2010)).

The dissipative soliton disappears when the supply of energy or matter is reduced, or the radiant heat and/or light is increased so that the system parameters move outside the range in which the soliton can exist. With a fiber fuse, it stops when the pumping power falls below the threshold power, $P_{\text{th}}$, or when the energy removal is increased. This is the design principal behind "fiber fuse terminators", which have been proposed by many researchers.

| | | Light-induced heat | Diffusion | Radiation |
|---|---|---|---|---|
| Shuto et al. (2004b), Golyatina et al. (2004) : | $\rho C_{\mathrm{p}} \frac{\partial T}{\partial t} =$ | $\alpha(T) I$ | $+\kappa \left( \frac{\partial^2 T}{\partial z^2} + \frac{\partial^2 T}{\partial r^2} + \frac{1}{r} \frac{\partial T}{\partial r} \right)$ | |
| Akhmediev et al. (2008) : | $\frac{\partial T}{\partial t} =$ | $\chi \alpha(T) I$ | $+D \frac{\partial^2 T}{\partial z^2}$ | $-k(T - T_0)$ |
| Rocha et al. (2009) : | $\rho C_{\mathrm{p}} \frac{\partial T}{\partial t} =$ | $\alpha_{\mathrm{Arr}} I$ | $+\kappa \frac{\partial^2 T}{\partial z^2}$ | $-\sigma_{\mathrm{s}} \epsilon_{\mathrm{e}} (T^4 - T_0^4)$ |

Table 2. Equations used for simulating fiber fuse propagation. $\rho$: density, $C_{\mathrm{p}}$: specific heat, $T$: plasma temperature, $T_0$: environment temperature, $\alpha(T)$ and $\alpha_{\mathrm{Arr}}$: absorption coefficient (see text), $\kappa$: thermal conductivity, $\sigma_{\mathrm{s}}$: Stefan-Boltzmann constant, $\epsilon_{\mathrm{e}}$: surface emissivity.

Hand & Birks (1989) demonstrated a fiber fuse termination at a special segment where the waveguide structure was modified to expand the pump laser beam (see Fig. 8 (a)). This is because a reduced power density in the core region exhausts the feed for the plasma. Later, Yanagi et al. (2003) developed a detachable device for practical use. However, we should note that this device cannot capture the fuse if the pump power exceeds the assumed limit.

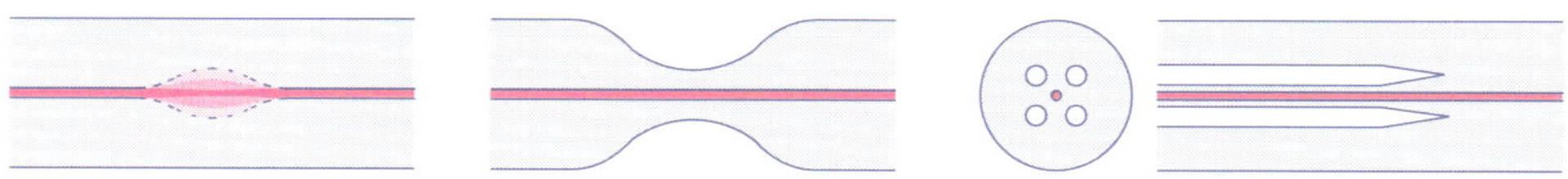

(a) Mode field expansion. (b) Waist formation by etching. (c) Hole-assisted fiber spliced with conventional fiber.

Fig. 8. Waveguide structures for in-line fiber fuse terminators.

Dianov, Bufetov & Frolov (2004) proposed a different structure in which the thickness of the cladding layer is reduced by chemical etching as illustrated in Fig. 8 (b). The fiber fuse is arrested here because its internal pressure cannot be maintained at this weakened segment and this causes deformation. Takenaga, Tanigawa, Matsuo, Fujimaki & Tsuchiya (2008) discovered that fiber fuse propagation is not as easy in hole-assisted fibers (HAFs) as in conventional fibers. The termination occurs near the splice point of these fibers (see Fig. 8 (c)) and the penetration length into the HAF side increases as the pump power decreases. Thus, it is reasonable to consider that the holes release the internal pressure of the plasma to destabilize it. This behavior was directly observed by Hanzawa et al. (2010) with an ultra-high speed video camera. Recent proposals for devices using holey fibers are discussed in section 2.3.

Another approach to fiber fuse termination is to interrupt the light source after detecting a propagating fiber fuse. Abedin et al. (2009) proposed a remote detection method that employs a backreflected light from the void train (see Fig. 9 (a)). They found that characteristic signals appeared in optical coherence-domain reflectometry after the ignition of the fiber fuse.

Rocha et al. (2011) demonstrated a local detection technique that uses a fiber Bragg grating as a temperature sensor placed in thermal contact with the propagation line (see Fig. 9 (b)). A heat pulse of a few degrees centigrade was detected as an increase in the Bragg wavelength monitored with an optical interrogator.

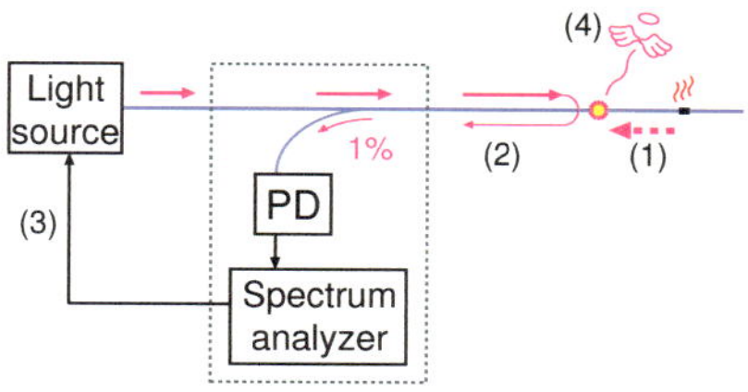


(a) Backreflected light detection.

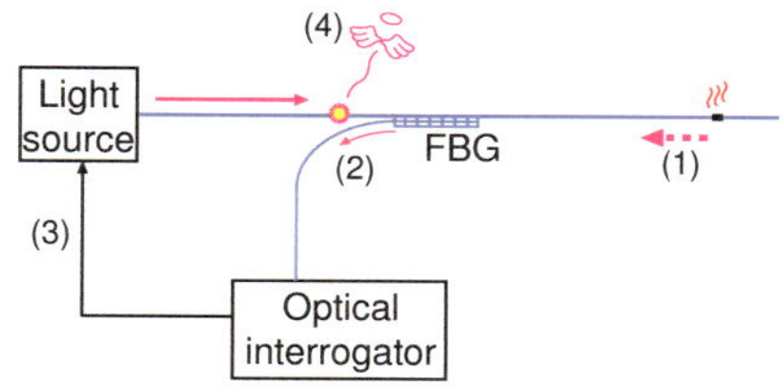


(b) Monitoring temperature increase.

Fig. 9. Schematic diagram of fiber fuse terminator. (1) propagation after ignition, (2) signal detection, (3) interruption of the light source and (4) termination.

### 2.3 Fiber fuse under special conditions

A fiber fuse can be initiated not only with a CW laser but also with a pulsed laser. Kashyap (1988) reported that mode locking has little effect on the velocity of damage propagation, but does alter the shape of the periodic voids. Dianov, Fortov, Bufetov, Efremov, Frolov, Schelev & Lozovoi (2006) demonstrated a detonation-like propagation mode with velocities up to 3 km/s that appeared when a Q-switched laser was used for pumping (wavelength: 1.064 $\mu$m, pulse repetition frequency: 5 kHz, pulse duration: 250 ns, pulse energy: up to 0.6 mJ, and maximum pulse power: 3.0 kW (3.8 W average)). Traveling plasma appeared only for a pulse duration and the damage included large cracks with a diameter up to 120 $\mu$m.

There have been several reports on non-conventional fibers. Lee et al. (2006) investigated fuse propagation over polarization-maintaining fibers and found that the power threshold for fast axis alignment is larger than that for slow axis alignment. Wang et al. (2008) showed some damage photographs of crack propagation in double-clad fiber for high power use.

Photonic crystal fibers (PCFs) also allow fiber fuse propagation. However, Dianov, Bufetov, Frolov, Chamorovsky, Ivanov & Vorobjev (2004) demonstrated that their threshold power is approximately ten times higher than that for conventional fibers. The reason is apparently the same as that discussed for HAFs above; the air holes reduce the plasma density. Thus, it is possible to use this fiber to make a fiber fuse terminator. Kurokawa & Hanzawa (2011) observed a fuse termination in situ at a splice point between PCF and conventional fiber. Ha et al. (2011) proposed another terminator using a hollow optical fiber.

Dianov, Bufetov, Frolov, Mashinskii, Plotnichenko, Churbanov & Snopatin (2002) reported the destruction of chalcogenide and fluoride glass fibers pumped at less than 1 W. They observed a distraction wave without plasma that thermally decomposed the entire cross section of the fiber. This is because these materials decompose at a much lower temperature than silica glass.

## 3. Microscopic behavior

Another characteristic behavior of a fiber fuse is the formation of periodic bullet-shaped voids. These voids are formed just after the passage of running plasma, and this has been experimentally confirmed by ultra-high speed photography (Bufetov & Dianov (2005); Bufetov et al. (2005), or see Fig. 10).

Two ideas have been proposed for the driving force behind this void formation. Atkins et al. (2003) suggested that Rayleigh instability, or positive surface tension, minimizes the interface area between plasma and molten glass following the analogy of falling water droplets and an air jet in a fluid (see Fig. 11). On the other hand, Yakovlenko (2004) pointed out that this idea has certain inadequacies, that is, it overlooks the high viscosity of molten silica and the

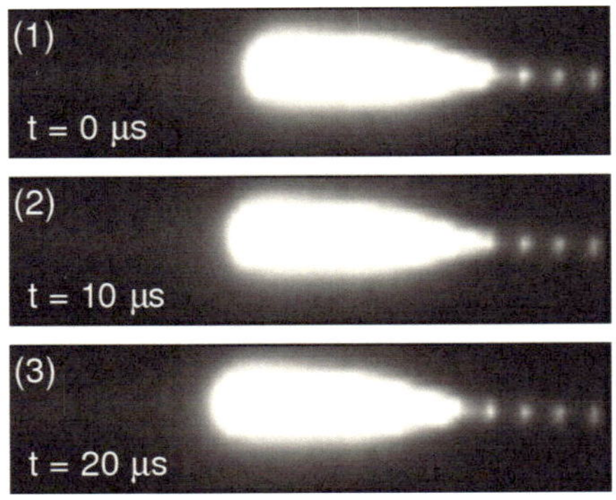


Fig. 10. Ultra-high speed photographs of fiber fuse propagation pumped with 9 W 1.48 $\mu$m light (Todoroki (2005b)). Discrete scattering points from the periodic voids (22 $\mu$m-interval) are clearly seen after the running plasma.

disappearance of surface tension at the elevated temperatures. Instead, he mentioned the electrostatic repulsion between negative charge layers induced at the plasma-liquid interface (see also Yakovlenko (2006a)). However, neither idea can explain why the voids look like bullets. Although Yakovlenko (2006b) pointed out that a bullet shape appeared in his simulated temperature profile of a running plasma, there is no description of its periodic appearance in the time domain.

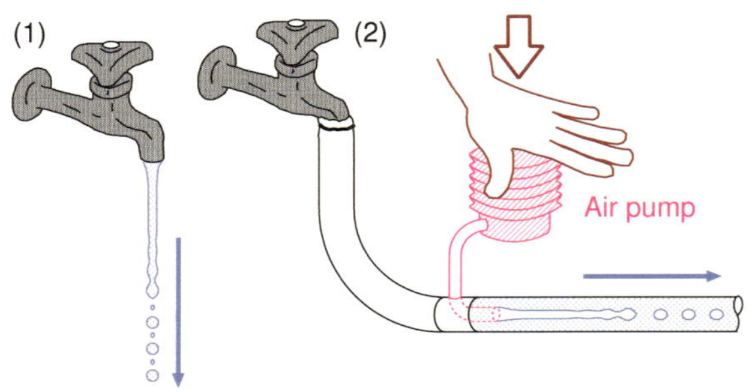


Fig. 11. Examples of jet breakdown due to Rayleigh instability, (1) A jet of water broken up into droplets (see Isenberg (1992) p. 131) and (2) a bubble train in a water flow (see Chandrasekhar (1981) p. 540).

Meanwhile, I have been investigating the mechanism of periodic void formation through an experiment-based approach, namely, a morphological analysis of void shapes and ultra-high speed photography. Section 3.1 discusses the requirements for periodic void formation. Then, the quench-induced deformation of the hollow melt is clarified by a statistical analysis of damage photographs (Section 3.2). Finally, the mechanism of bullet-like void formation is proposed (Section 3.3). All the discussion relates to silica-based step-indexed single-mode optical fibers unless otherwise specified.

### 3.1 Pump power dependence of hollow damage morphology

We should note that not all the fused damages has the appearance of periodic bullets. Figure 12 shows the front part of a fused damage train that remained in Corning SMF-28 fibers after the light sources had been turned off (1.2–9 W, 1.48 $\mu$m, CW; Todoroki (2005c)). Bullet-shaped periodic voids appear when the pump power exceeds 2 W (see (a)–(d)) or ∼1.3 W (g) near the propagation threshold. The latter mode is very unsteady with respect to fluctuations in laser power and/or waveguide structure, i. e. the periodicity disappears easily and sometimes the fuse vanishes[1]. For the former stable mode, the interval of the periodic voids increases with the pump power and a long and narrow void is left at the top of the damage train.

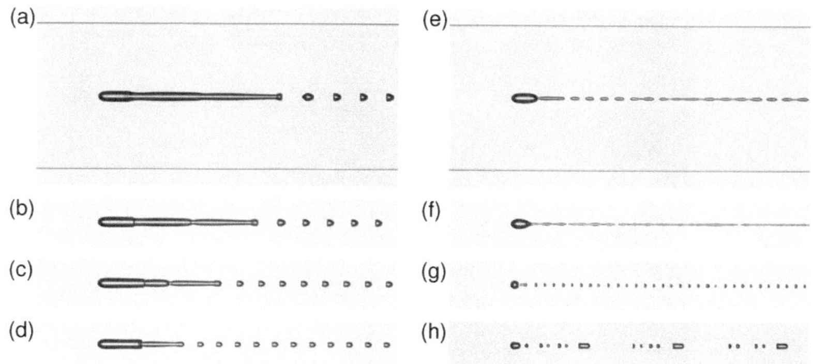


Fig. 12. Optical micrographs showing the front part of the fiber fuse damage generated in single-mode silica glass fibers (Corning SMF-28). The powers of the pump laser (1.48 $\mu$m) are (a) 9.0 W, (b) 7.0 W, (c) 5.0 W, (d) 3.5 W, (e) 2.0 W, (f) 1.5 W, (g) ∼1.3 W, and (h) ∼1.2 W. The thin two lines at the top and bottom of (a) and (e) are the edges of the 125 $\mu$m diameter fiber. (Todoroki (2005c))

These top voids are valuable evidence for exploring the state of traveling plasma. In fact, these photographs show good agreement with the in-situ image of fiber fuse propagation shown in Fig. 13. In both cases, the shape along the axial direction is asymmetric when the pump power exceeds 2 W. Thus, this asymmetric shape is expected to be the origin of the stable periodic void formation. However, these in-situ images provide no further information about the void formation process owing to their poor resolution. Thus, the shape of the damage sites are analyzed instead in the next subsection.

### 3.2 Deformation of hollow melt during quenching

The period of one void formation is estimated to be a few tens of microseconds based on the propagation speed and void interval. For example, the periodic voids shown in Fig. 12 (a)–(d) were generated every 18.7 $\mu$s (9 W) – 25.4 $\mu$s (3.5 W). During this period, the plasma was found to propagate at a constant speed (Todoroki (2005a)). Since the damage structure varies according to the moment at which the light source is turned off during this cycle, at least 40 samples were prepared in order to collect a variety of damage patterns.

In addition, to maximize the quenching rate of the melt, a fiber fuse was terminated in a segment where the colored nylon jacket (0.9 mm$\phi$) over the cladding had been removed beforehand (about 20 cm at the maximum). This is because color pigments in the nylon jacket scatters the visible radiation from the inside and the backscattered light is re-absorbed by the melt with absorptive species shown in Eqs. (1)–(3) to generate heat. This light-heat conversion

[1] For example, see the video in Fig. 2 of Todoroki (2005d) in which the fuse stopped after it passed through a splicing point in a jacket-free segment.

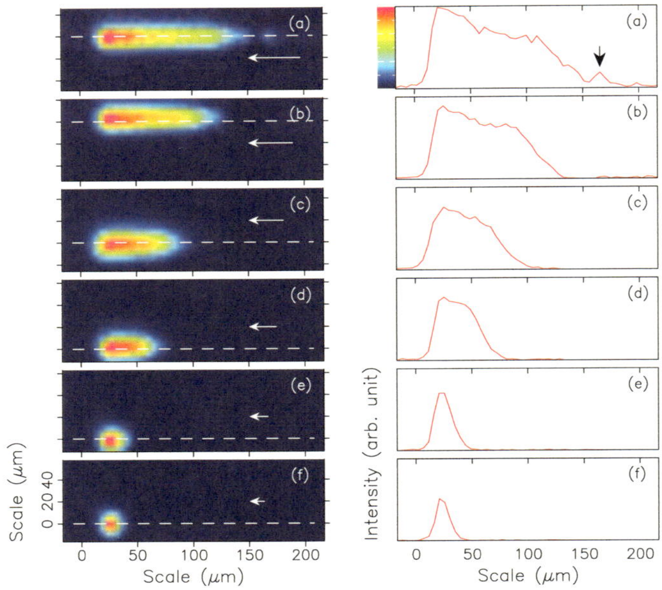


Fig. 13. Ultrahigh-speed photograph of visible light emission from a fiber fuse (left, gray-scale image is converted to colored-scale image, exposure time: 1 $\mu$s) and intensity profiles along the dashed line in each image (right). The powers of the pump laser (1.48 $\mu$m) are (a) 9.0 W, (b) 7.0 W, (c) 5.0 W, (d) 3.5 W, (e) 2.0 W, and (f) 1.5 W. Each horizontal arrow indicates the distance that the plasma moves in 40 $\mu$s. (Todoroki (2005c))

occurs throughout the fiber fuse propagation but is absent in the bare fiber segment just before the fiber fuse termination[2].

To compare the damage sites precisely, the void size should be properly normalized because it varies sensitively with the pump power. For example, the interval of the periodic voids, $\Lambda$, among the 81 samples pumped with 9 W 1.48 $\mu$m light ranged from 21.7 $\mu$m to 22.7 $\mu$m. This must be due to the fluctuation of pump laser power and the loss of the fiber between the fuse and the light source. Thus, the following two parameters are defined on the basis of $\Lambda$,

$$x_1' = x_1/\Lambda, \qquad l_2' = l_2/\Lambda \tag{5}$$

where $x_1$ is a parameter describing the top position of the first void in a virtual scale graduated on $\Lambda$ (see the vertical lines in Fig. 14) and $l_2$ is the length of the second void.

The left column in Fig. 15 ((1) –(6)) shows selected photographs of the samples pumped with 9 W 1.48 $\mu$m light. They are sorted in order of increasing $x_1'$, with the intention of rearranging them in chronological order within the void formation cycle (Todoroki (2005a;c)). The sorted photographs suggest that $l_2'$ has a tendency to decrease with increasing $x_1'$ (see the inset table). This correlation among all the 40 samples is shown on the left in Fig. 16.

The sequence of the photographs seems to capture the moment at which the long top void is divided in two by a melt bridge (Todoroki (2005a;c)). However, we should not overlook a possibility that these structures are the result of the modification that occurred during the fiber

[2] The plasma decay time in a bare fiber is reported to be less than 7 $\mu$s (Todoroki (2005a)). Detailed description of this light-heat conversion will be published soon (Todoroki (2011)).

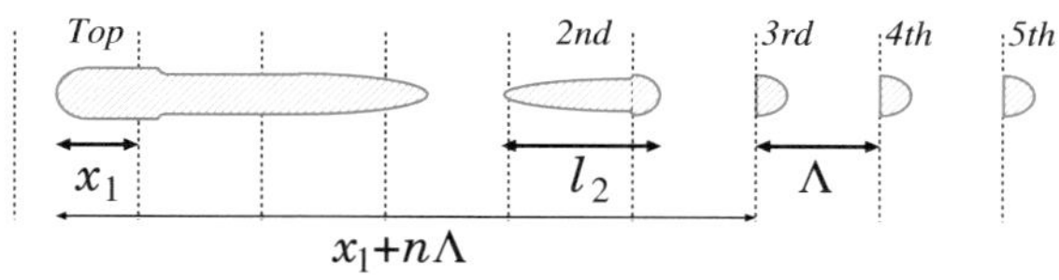


Fig. 14. Definition of the parameters used in Eq. (5). The vertical lines are placed at the bottom of bullet-like voids at intervals of $\Lambda$. Then, the distance between the top of the first void and periodic voids is written as $x_1 + n\Lambda$ ($n$: natural number).

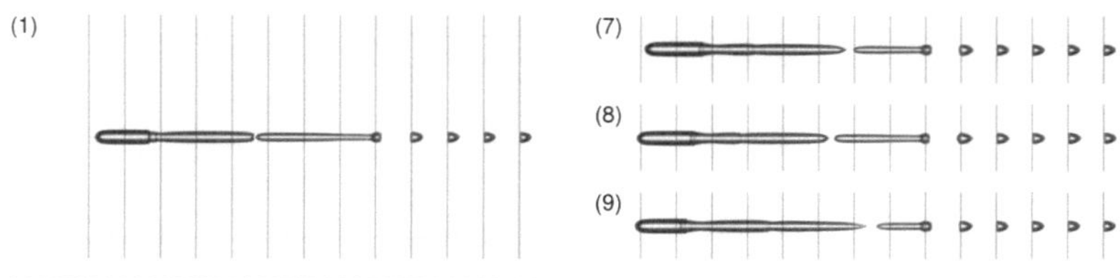


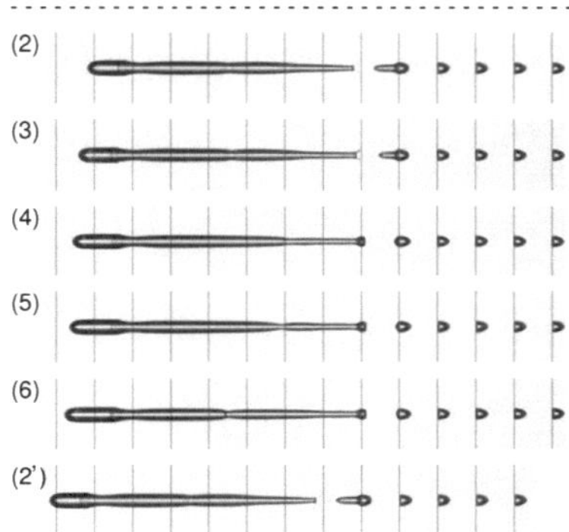


| | quickly quenched $x_1'$ | quickly quenched $l_2'$ | | slowly quenched $x_1'$ | slowly quenched $l_2'$ |
|---|---|---|---|---|---|
| (1) | −0.16 (=0.84) | 3.52 | (7) | −0.10 (=0.90) | 2.20 |
| | | | (8) | 0.13 | 2.68 |
| | | | (9) | 0.18 | 1.54 |
| (2) | 0.19 | 0.93 | | | |
| (3) | 0.43 | 0.82 | | | |
| (4) | 0.57 | 0.44 | | | |
| (5) | 0.66 | 0.42 | | | |
| (6) | 0.79 | 0.40 | | | |
| | $\Lambda$ = 22.1 $\mu$m | | | $\Lambda$ = 21.9 $\mu$m | |

Fig. 15. Optical micrographs of the damage sites pumped with 9 W 1.48 $\mu$m light. (1) – (6): samples whose quenching point has no plastic coating, and (7) – (9): with coating (see text). The photograph at the bottom (2′) is the same as that at the top (2), shifted 22.1 $\mu$m to the left. The inset table lists the size parameters defined in Eq. (5), that are plotted as red points in Figs. 16 and Fig. 17.

fuse quenching period. Fortunately, the histogram of $l_2'$ shown in the right of Fig. 16 provides a clue to this problem. That is, the $l_2'$ distribution is strongly biased below 1.0. In other words, void structures like (1) shown in Fig. 15 ($l_2' > 1.0$) appeared less frequently than (2) – (6).

If these void structures are stable during quenching, the $l_2'$ distribution is expected to be independent on the quenching rate. Thus, another set of the samples was prepared in which a fiber fuse was quenched in a colored nylon jacketed segment with an outer diameter of 0.9 mm. Hereafter, they are referred to as 'slowly quenched' whereas the previous samples were referred to as 'quickly quenched'. The result shown in Fig. 17 is clearly different from that shown in Fig. 16; the correlation between $x_1'$ and $l_2'$ became considerably weaker and the $l_2'$ histogram became relatively flat.

This behavior is well explained by an assumption that the melt surrounding the plasma tends to form a bridge inside the cavity after the plasma has been extinguished and before the melt is frozen (see the lower white arrow in Fig. 18 between (3-a) and (c)). In other words, after the quench begins, a melt with a long hollow space ($l_2' < 1$) becomes unstable and forms a bridge inside the cavity ($l_2' > 1$). In this case, the bridge position is influenced very little by

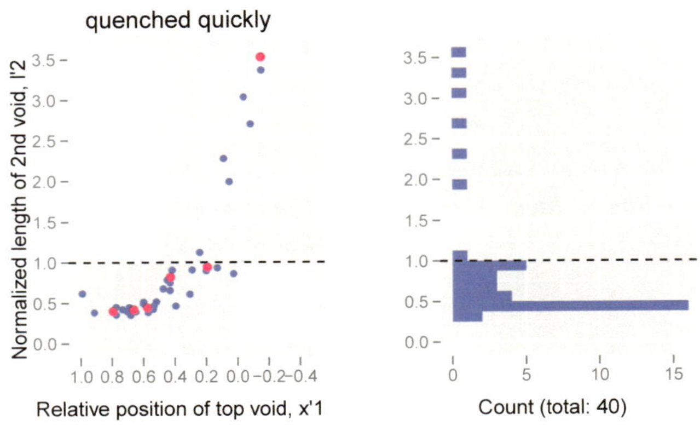


Fig. 16. Correlation between $x_1'$ and $l_2'$ (left) and histogram of $l_2'$ (right) for the samples pumped with 9W 1.48 $\mu$m light in which the fiber fuse was extinguished in a coating-free segment (see text). The red points are the results for samples (1) – (6) shown in Fig. 15.

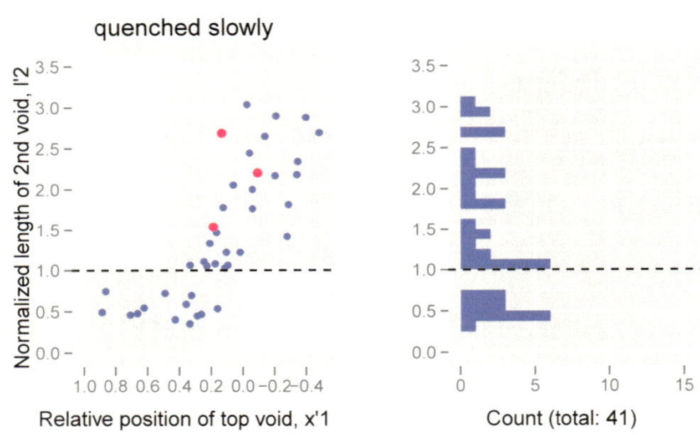


Fig. 17. Correlation between $x_1'$ and $l_2'$ (left) and histogram of $l_2'$ (right) for the samples pumped with 9W 1.48 $\mu$m light in which the fiber fuse was extinguished in a coated segment (see text). The red points are the results for samples for (7) – (9) shown in Fig. 15.

the moment at which the power is turned off, and there is little correlation between $x_1'$ and $l_2'$ in the resulting structures. In addition, the bridge width in the slowly quenched samples is larger than that of the quickly quenched samples (compare Fig. 15 (7) – (9) and (1)). This suggests that bridge growth was promoted by slow quenching.

Possible driving forces for this bridge formation are, firstly, the sudden pressure/temperature decrease in the hollow cavity that occurs as a result of the laser power being switched off and the subsequent deposition of quenched gas, and secondly, the negative surface tension of glass melt (Yakovlenko (2006b)). The temperature dependence of the surface tension is given by Eötvös formula, $\gamma(T) = k(\rho/M)^{2/3}(T_{\rm cr} - T)$, where $k$ is a constant, $\rho$ density, $M$ molar weight, and $T_{\rm cr}$ critical temperature. Although the experimental data for the silica melt are not available, Yakovlenko (2006b) mentioned that the tension is expected to be negative at elevated temperatures (more than $\sim$3000 K) and surface creation promoted.

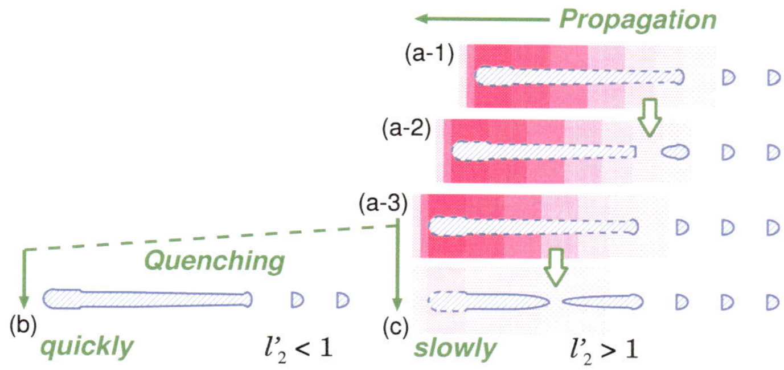


Fig. 18. Model of void formation during propagation and quenching of fiber fuse. Bridge formation is shown by white arrows.

The bridge that remained in the quickly quenched samples with $l_2' > 1$ as in Fig. 15 (1) were probably formed after the power had been turned off. The low probability of the samples with $l_2' > 1$ shown on the right in Fig. 16 supports this hypothesis. Moreover, this is confirmed by the recent result that I obtained by the in-situ videography of fiber fuse propagation shown in the left of Fig. 19. If there is a bridge before quenching, i. e. during propagation, no light emission is expected from the bridge. However, the centers of the intensity profiles have no dark regions.

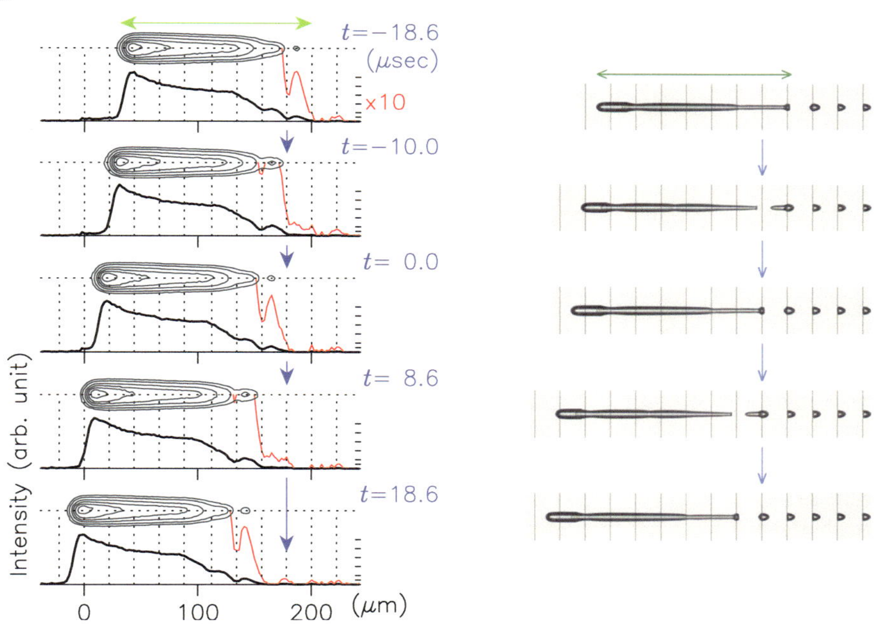


Fig. 19. Intensity profiles of visible light emission from a fiber fuse propagating through a single-mode optical fiber pumped by 9 W 1.48 $\mu$m light (left, exposure time: 0.37 $\mu$s) and photographs of hollow damage for comparison, generated under the same condition (right). The vertical line interval is 22 $\mu$m. The vertical dashed lines are placed on the scattering points from the periodic voids. The photographs are the same as those shown in Fig. 15 (2) and (4).

This is more clearly recognized when we compare these profiles with the void photographs as shown on the right. These photographs are arranged to coincide with the in situ images with respect to the following two points; (1) the full width of the profile matches the length of the top void (see the green arrows on the top) and (2) the weak scattering points on the vertical dashed line match the top of the periodic voids on the vertical solid line. The blue arrows are reference points for this comparison.

Consequently, the melt surrounding the traveling plasma tends to form a bridge inside the cavity after the light source has been turned off. However, this action is suppressed by fast quenching (see Fig. 18 (b)). Therefore, the damage photographs of quickly quenched samples without such a bridge (see Fig. 15 (2) – (6)) constitute useful data for discussing the periodic process of fiber fuse propagation in the next subsection.

### 3.3 Bridge formation during fiber fuse propagation

According to the sequence in Fig. 15 (2) – (6), the bridge formation process occurs between (6) and (2′). The hottest region of the plasma passed this position $\sim$120 $\mu$s earlier and the temperature is decreasing. Its quenching rate is much slower than that induced by suddenly turning off the light source because the hot plasma is still alive nearby and is moving away at about 1 m/s (= 1 $\mu$m/$\mu$s). Thus, this situation, illustrated in Fig. 18 (a-1) and (a-2), is almost the same as the slow quenching case (see the lower white arrow) except that the bridge is compressed by the hot plasma.

This compression explains why the periodic voids look like bullets. This process is frozen in the photographs shown in Fig. 15 (2) – (6). After a bridge appears, the detached void begins to shrink due to the pressure from the hot plasma until the surrounding melt solidifies. Since the rear side of the new void solidifies earlier than the front, the rear shape remains round whereas the front becomes flatter.

It is interesting to find that an early sign of bridge formation is recorded in the in-situ observation result shown on the left in Fig. 19. A weak modulation appeared on the tail of the light emission profile and its interval is the same as that of the periodic voids. However, it appeared only when the pump power was 9 W (see the black arrow in Fig. 13). This modulation may suggest the instability of the plasma or the surrounding melt as proposed by Atkins et al. (2003) (Rayleigh instability) and Yakovlenko (2004) (induced electrostatic repulsion) but further study is needed.

The void formation sequence can be modified by controlling certain external conditions. Bufetov et al. (2008) observed a large-scale periodic void train in an optical fiber that allows the interference of the $LP_{01}$ and $LP_{02}$ modes. Figure 20 shows the void train surrounded by a region of modified refractive index. Its interval was found to coincide with that of the interference pattern.

Todoroki (2008) reported the breakage of the periodic void pattern over hetero-core splice points as shown in Fig. 21. Traveling plasma temporarily stopped forming voids when the core was expanded from HI 1600 to SMF-28e, whereas it left some long voids when the core size was reduced. It is interesting to find a similar tendency in Bufetov's case at the inflection points of the interference. That is, the void train disappears at the segment where the mode field is increasing (see dashed lines b in Fig. 20), and a long void appears where the mode field is reduced (see dashed lines a). Moreover, some of the voids have a flat area whose direction is opposite to that of the regular bullet-like voids (see arrows in Figs. 20 and 21). This behavior must be due to the modulated internal pressure of the plasma. To clarify the mechanism,

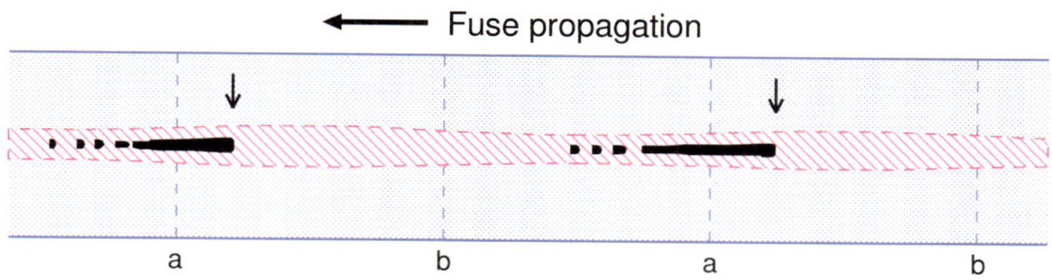


Fig. 20. Illustration of large-scale periodic void train left in an optical fiber allowing the interference of two propagation modes of 1.07 $\mu$m light. This is reproduced from a photograph in Bufetov et al. (2007). The diameter of the fiber cladding is 125 $\mu$m. The vertical dashed lines are on the inflection points of the modified refractive-index area shown as a red hatched region.

further investigation is needed including the in-situ observation of traveling plasma and a statistical analysis of quickly quenched void samples.

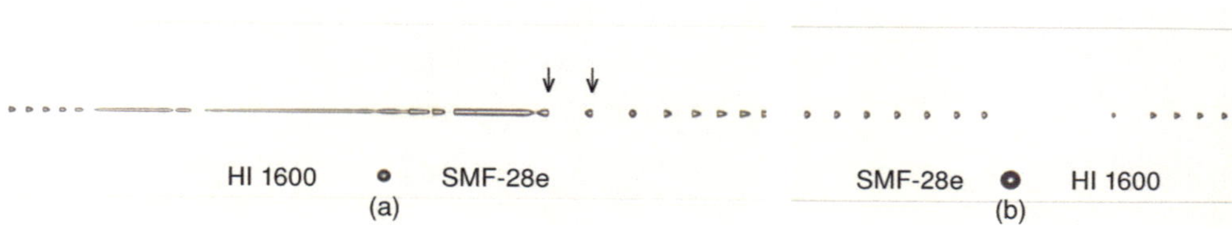


Fig. 21. Photographs of fiber fuse damage over the splicing point between HI 1060 and SMF-28e (Corning). The hollow spheres in the cladding are located at the border between two fibers and were accidentally captured during splicing. The fuse was initiated with 9W 1.07 $\mu$m light that propagates in a multi-mode through SMF-28e.

## 4. Summary

Knowledge accumulated about fiber fuse propagation since 1988 is briefly summarized. From a macroscopic viewpoint, the dissipative soliton concept and an analogy with grassfire help us understand this strange phenomenon. The strong heat-induced absorption of silica glass and the highly confined supply of laser energy cause captured plasma to shift to the light source along the fiber leaving catastrophic damage behind it. From a microscopic viewpoint, the periodic void formation process was unveiled by the statistical analysis of void shapes and ultra-fast videography. The bullet-like shape of the damage train results from the formation of an intrinsic bridge inside the hollow silica melt behind the traveling plasma and the successive compression of detached voids under a steep temperature gradient along the fiber.

## 5. Acknowledgements

I am grateful to Mr. Keisuke Aizawa, Mr. Arata Mihara and Mr. Yousuke Suzuki (Photron Ltd.) for helping with the ultrahigh-speed videography experiment.

| | Todoroki (2005c) | Todoroki (2005d) | Todoroki (2008) |
|---|---|---|---|
| Ignition: | — | Fig. 3* | — |
| Propagation: | Fig. 1, Fig. 3* | — | Movie S1* & S2* |
| Self-termination: | — | Fig. 2, Fig. 7* | — |
| Void formation model:† | Fig. 7 | — | — |

Table 3. Fiber fuse video clips available on the net. *: Ultra-high speed video, †: outdated because all the samples were slowly quenched (see Section 3.2).

| | *Russia* | *Japan* |
|---|---|---|
| Kashyap (1988) | | |
| Kashyap & Blow (1988) | | |
| Hand & St. J. Russell (1988) | | |
| Hand & Birks (1989) | | |
| Davis et al. (1996) | Dianov et al. (1992) | |
| Davis et al. (1997) | | |
| | Dianov, Bufetov, Frolov, Plotnichenko, Mashinskii, Churbanov & Snopatin (2002) | |
| *Atkins et al. (2003)* | Dianov, Bufetov, Frolov, Mashinskii, Plotnichenko, Churbanov & Snopatin (2002) | Yanagi et al. (2003) |
| | | Seo et al. (2003) |
| | Dianov, Bufetov, Frolov, Chamorovsky, Ivanov & Vorobjev (2004) | |
| | Dianov, Bufetov & Frolov (2004) | *Shuto et al. (2004a)* |
| | *Yakovlenko (2004)* | *Shuto et al. (2004b)* |
| | *Golyatina et al. (2004)* | |
| | Bufetov & Dianov (2005) | Todoroki (2005b) |
| | Bufetov et al. (2005) | Todoroki (2005c) |
| | | Todoroki (2005a) |
| | | Todoroki (2005d) |
| Lee et al. (2006) | Dianov, Fortov, Bufetov, Efremov, Frolov, Schelev & Lozovoi (2006) | |
| | Dianov, Fortov, Bufetov, Efremov, Rakitin, Melkumov, Kulish & Frolov (2006) | |
| | *Yakovlenko (2006a)* | |
| | *Yakovlenko (2006b)* | |
| | Bufetov et al. (2007) | |
| | Bufetov et al. (2008) | Takenaga, Omori, Goto, Tanigawa, Matsuo & Himeno (2008) |
| *Akhmediev et al. (2008)* | | Takenaga, Tanigawa, Matsuo, Fujimaki & Tsuchiya (2008) |
| Wang et al. (2008) | | Todoroki (2008) |
| | | Abedin & Morioka (2009) |
| | | Abedin et al. (2009) |
| *Rocha et al. (2009)* | | Abedin (2009) |
| | | *Shuto (2010)* |
| *Rocha et al. (2010)* | | Hanzawa et al. (2010) |
| Ha et al. (2011) | | Kurokawa & Hanzawa (2011) |
| Rocha et al. (2011) | | Yamada et al. (2011) |
| | | Todoroki (2011) |

Table 4. Chronological table of fiber fuse research cited in this paper. Theoretical works are given in italics.

## 6. References


Abedin, K. S. (2009). Remote sensing of fiber fuse propagation using RF detection, *The Technical Report of The Proceedings of The Institute of Electronics, Information and Communication Engineers, OPE, Optoelectronics* 109(159): 43–46.

Abedin, K. S. & Morioka, T. (2009). Remote detection of fiber fuse propagating in optical fibers, *Proceedings of Optical Fiber Communication/National Fiber Optic Engineers Conference*. (OThD5).

Abedin, K. S., Nakazawa, M. & Miyazaki, T. (2009). Backreflected radiation due to a propagating fiber fuse, *Optics Express* 17(8): 6525–6531.

Akhmediev, N. & Ankiewicz, A. (2005). Dissipative solitons in the complex Ginzburg-Landau and Swift-Hohenberg equations, *Dissipative Solitons*, Vol. 661 of *Lecture Notes in Physics*, Springer-Verlag, Berlin, pp. 1–17.

Akhmediev, N., St. J. Russell, P., Taki, M. & Soto-Crespo, J. M. (2008). Heat dissipative solitons in optical fibers, *Physics Letters A* 372(9): 1531–1534.

Atkins, R. M., Simpkins, P. G. & Yablon, A. D. (2003). Track of a fiber fuse: a Rayleigh instability in optical waveguides, *Opt. Lett.* 28(12): 974–976.

Bufetov, I. A. & Dianov, E. M. (2005). Optical discharge in optical fibers, *Physics-Uspekhi* 48(1): 91–94.

Bufetov, I. A., Frolov, A. A., Dianov, E. M., Fortov, V. E. & Efremov, V. P. (2005). Dynamics of fiber fuse propagation, *Optical Fiber Communication Conference, 2005. Technical Digest. OFC/NFOEC*, Vol. 4, Anaheim, CA. (OThQ7).

Bufetov, I. A., Frolov, A. A., Shubin, A. V., Likhachev, M. E., Lavrishchev, C. V. & Dianov, E. M. (2007). Fiber fuse effect: New results on the fiber damage structure, *Proceedings of the 33rd European Conference on Optical Communication*, Vol. 1, IEE's Photonics Professional Network, Berlin, Germany, pp. 79–80. (Mon 1.5.2).

Bufetov, I. A., Frolov, A. A., Shubin, A. V., Likhachev, M. E., Lavrishchev, S. V. & Dianov, E. M. (2008). Propagation of an optical discharge through optical fibres upon interference of modes, *Quantum Electronics* 38(5): 441–444.

Chandrasekhar, S. (1981). *Hydrodynamic and Hydromagnetic Stability*, International Series of Monographs on Physics (Oxford, England), Dover Publications.

Davis, D. D., Mettler, S. C. & DiGiovani, D. J. (1996). Experimental data on the fiber fuse, *in* H. E. Bennett, A. H. Guenther, M. R. Kozlowski, B. E. Newnam & M. J. Soileau (eds), *27th Annual Boulder Damage Symposium: Laser-Induced Damage in Optical Materials: 1995*, Vol. 2714 of *SPIE Proceedings*, SPIE, pp. 202–210. (Boulder, CO, USA, 30 Oct. 1995).

Davis, D. D., Mettler, S. C. & DiGiovani, D. J. (1997). A comparative evaluation of fiber fuse models, *in* H. E. Bennett, A. H. Guenther, M. R. Kozlowski, B. E. Newnam & M. J. Soileau (eds), *Laser-Induced Damage in Optical Materials: 1996*, Vol. 2966 of *SPIE Proceedings*, SPIE, pp. 592–606. (Boulder, CO, USA, 7 Oct 1996).

Dianov, E. M., Bufetov, I. A. & Frolov, A. A. (2004). Destruction of silica fiber cladding by the fuse effect, *Opt. Lett.* 29(16): 1852–1854.

Dianov, E. M., Bufetov, I. A., Frolov, A. A., Chamorovsky, Y. K., Ivanov, G. A. & Vorobjev, I. L. (2004). Fiber fuse effect in microstructured fibers, *IEEE Photon. Technol. Lett.* 16(1): 180–181.

Dianov, E. M., Bufetov, I. A., Frolov, A. A., Mashinskii, V. M., Plotnichenko, V. G., Churbanov, M. F. & Snopatin, G. E. (2002). Catastrophic destruction of fluoride and chalcogenide optical fibers, *Electron. Letters* 38(15): 783–784.

Dianov, E. M., Bufetov, I. A., Frolov, A. A., Plotnichenko, V. G., Mashinskii, V. M., Churbanov, M. F. & Snopatin, G. E. (2002). Catastrophic destruction of optical fibres of various composition caused by laser radiation, *Quantum Electron.* 32(6): 476–478.

Dianov, E. M., Fortov, V. E., Bufetov, I. A., Efremov, V. P., Frolov, A. A., Schelev, M. Y. & Lozovoi, V. I. (2006). Detonation-like mode of the destruction of optical fibers under intense laser radiation, *J. Exp. Theo. Phys. Lett.* 83(2): 75–78.

Dianov, E. M., Fortov, V. E., Bufetov, I. A., Efremov, V. P., Rakitin, A. E., Melkumov, M. A., Kulish, M. I. & Frolov, A. A. (2006). High-speed photography, spectra, and temperature of optical discharge in silica-based fibers, *IEEE Photon. Technol. Lett.* 18(6): 752–754.

Dianov, E. M., Mashinskii, V. M., Myzina, V. A., Sidorin, Y. S., Streltsov, A. M. & Chickolini, A. V. (1992). Change of refractive index profile in the process of laser-induced fiber damage, *Sov. Lightwave Commun.* 2: 293–299.

Golyatina, R. I., Tkachev, A. N. & Yakovlenko, S. I. (2004). Calculation of velocity and threshold for a thermal wave of laser radiation absorption in a fiber optic waveguide based on the two-dimensional nonstationary heat conduction equation, *Laser Physics* 14(11): 1429–1433.

Ha, W., Jeong, Y. & Oh, K. (2011). Fiber fuse effect in hollow optical fibers, *Opt. Lett.* 36(9): 1536–1538.

Hand, D. P. & Birks, T. A. (1989). Single-mode tapers as 'fibre fuse' damage circuit-breakers, *Electron. Lett.* 25(1): 33–34.

Hand, D. P. & St. J. Russell, P. (1988). Solitary thermal shock waves and optical damage in optical fibers: the fiber fuse, *Opt. Lett.* 13(9): 767–769.

Hanzawa, N., Kurokawa, K., Tsujikawa, K., Matsui, T., Nakajima, K., Tomita, S. & Tsubokawa, M. (2010). Suppression of fiber fuse propagation in hole assisted fiber and photonic crystal fiber, *J. Lightwave Technology* 28(15): 2115–2120.

Isenberg, C. (1992). *The Science of Soap Films and Soap Bubbles*, new edn, Dover Publications.

Kanamori, H., Yokota, H., Tanaka, G., Watanabe, M., Ishiguro, Y., Yoshida, I., Kakii, T., Itoh, S., Asano, Y. & Tanaka, S. (1986). Transmission characteristics and reliability of pure-silica-core single-mode fibers, *J. Lightwave Technology* 4(8): 1144–1150.

Kashyap, R. (1988). Self-propelled self-focusing damage in optical fibres, *Lasers '87; Proc. the Tenth Int. Conf. Lasers and Applications*, STS Press, McLean, VA, pp. 859–866. (Lake Tahoe, NV, Dec. 7-11, 1987).

Kashyap, R. & Blow, K. J. (1988). Observation of catastrophic self-propelled self-focusing in optical fibres, *Electron. Lett.* 24: 47–49.

Kurokawa, K. & Hanzawa, N. (2011). Fiber fuse propagation and its suppression in hole-assisted fibers, *IEICE Transactions on Communications* E94.B(2): 384–391.

Lee, M. M., Roth, J. M., Ulmer, T. G. & Cryan, C. V. (2006). The fiber fuse phenomenon in polarization-maintaining fibers at 1.55$\mu$m, *Proc. of the Conference on Lasers and Electro-Optics (CLEO)*. (JWB66).

Mears, R. J., Reekie, L., Jauncey, I. M. & Payne, D. N. (1987). Low-noise erbium-doped fibre amplifier operating at 1.54 $\mu$m, *Electronics Letters* 23(19): 1026–1028.

Rocha, A. M., Antunes, P. F. C., Domingues, M. F. F., Facão, M. & André, P. S. (2011). Detection of fiber fuse effect using FBG sensors, *IEEE Sensors Journal* 11(6): 1390 –1394.

Rocha, A. M., Facão, M. & André, P. S. (2010). Study of fiber fuse effect on different types of single mode optical fibers, *in* D. Faulkner (ed.), *NOC/OC&I 2010 Proceedings : 15th European Conference on Networks and Optical Communications and 5th Conference on Optical Cabling and Infrastructure (ISBN: 9789729341939)*, Universidade do Algarve, Faro-Algarve, Portugal, pp. 71–75. (Presented on June 10).

Rocha, A. M., Facão, M., Martins, A. & André, P. S. (2009). Simulation of fiber fuse effect propagation, *International Conf. on Transparent Networks – Mediterranean Winter, 2009*, Angers, France. (FrP.12).

Seo, K., Nishimura, N., Shiino, M., Yuguchi, R. & Sasaki, H. (2003). Evaluation of high-power endurance in optical fiber links, *Furukawa Review* (24): 17–22.

Shuto, Y. (2010). Evaluation of high-temperature absorption coefficients of ionized gas plasmas in optical fibers, *IEEE Photon. Technol. Lett.* 22(3): 134–136.

Shuto, Y., Yanagi, S., Asakawa, S., Kobayashi, M. & Nagase, R. (2004a). Evaluation of high-temperature absorption coefficients of optical fibers, *IEEE Photon. Technol. Lett.* 16(4): 1008–1010.

Shuto, Y., Yanagi, S., Asakawa, S., Kobayashi, M. & Nagase, R. (2004b). Fiber fuse phenomenon in step-index single-mode optical fibers, *IEEE J. Quantum Electronics* 40(8): 1113–1121.

Takenaga, K., Omori, S., Goto, R., Tanigawa, S., Matsuo, S. & Himeno, K. (2008). Evaluation of high-power endurance of bend-insensitive fibers, *Proceedings of Optical Fiber Communication/National Fiber Optic Engineers Conference*. (JWA11).

Takenaga, K., Tanigawa, S., Matsuo, S., Fujimaki, M. & Tsuchiya, H. (2008). Fiber fuse phenomenon in hole-assisted fibers, *Proceedings of the 34th European Conference on Optical Communication*, Vol. 5, pp. 27–28. (P.1.14).

Todoroki, S. (2005a). Animation of fiber fuse damage, demonstrating periodic void formation, *Opt. Lett.* 30(19): 2551–2553.

Todoroki, S. (2005b). In-situ observation of fiber-fuse propagation, *Jpn. J. Appl. Phys.* 44(6A): 4022–4024.

Todoroki, S. (2005c). Origin of periodic void formation during fiber fuse, *Optics Express* 13(17): 6381–6389.

Todoroki, S. (2005d). Transient propagation mode of fiber fuse leaving no voids, *Optics Express* 13(23): 9248–9256.

Todoroki, S. (2008). In situ observation of modulated light emission of fiber fuse synchronized with void train over hetero-core splice point, *PLoS ONE* 3(9): e3276.

Todoroki, S. (2011). Threshold power reduction of fiber fuse propagation through a white tight-buffered single-mode optical fiber, IEICE Electronics Express 8 (accepted).

Wang, J., Gray, S., Walton, D. & Zenteno, L. (2008). Fiber fuse in high-power optical fiber, *in* M.-J. Li, P. Shum, I. H. White & X. Wu (eds), *Passive Components and Fiber-based Devices V*, Vol. 7134 of *SPIE Proceedings*, SPIE, pp. 71342E–1–9. (Hangzhou, China).

Yakovlenko, S. I. (2004). Plasma behind the front of a damage wave and the mechanism of laser-induced production of a chain of caverns in an optical fibre, *Quantum Electron.* 34(8): 765–770.

Yakovlenko, S. I. (2006a). Mechanism for the void formation in the bright spot of a fiber fuse, *Laser Physics* 16(3): 474–476.

Yakovlenko, S. I. (2006b). Physical processes upon the optical discharge propagation in optical fiber, *Laser Physics* 16(9): 1273–1290.

Yamada, M., Koyama, O., Katsuyama, Y. & Shibuya, T. (2011). Heating and burning of optical fiber by light scattered from bubble train formed by optical fiber fuse, *Proceedings of Optical Fiber Communication/National Fiber Optic Engineers Conference*. (JThA1).

Yanagi, S., Asakawa, S., Kobayashi, M., Shuto, Y. & Naruse, R. (2003). Fiber fuse terminator, *The 5th Pacific Rim Conference on Lasers and Electro-Optics*, Vol. 1, p. 386. (W4J-(8)-6, Taipei. Taiwan, 22-26 Jul. 2003).

# 21

# Radiation Induced by Charged Particles in Optical Fibers

Xavier Artru and Cédric Ray
*Université de Lyon, Université Lyon 1, CNRS-IN2P3,*
*Institut de Physique Nucléaire de Lyon*
*France*

## 1. Introduction

The electric field of a charged particle passing through or near an optical fiber induces a transient charges and currents in the fibrer medium (1; 2). These charges and current radiates electromagnetic waves, both outside the fiber (free light) and inside (guided light). This chapter is devoted to the guided light, which will be referred to as PIGL, for *Particle Induced Guided Light*.

If the fiber radius is large enough and the particle passes trough it, as in Fig. 1, both PIGL and oustide radiation can be considered as transition radiation and becomes Cherenkov radiation when the particle velocity exceeds that of light in the medium. This is the basis of the quartz fibre particle detectors (3–5). Let us mention two other uses of optical fibers as particle detectors : (i) as dosimeters, through the effect of darkening by irradiation (6); (ii) in scintillating glass fibers for particle tracking.

Here we will consider fibers of radius $a$ comparable to the wavelength, in which case the standard OTR or Cherenkov descriptions are not appropriate. Two types of PIGL have to be considered :

- Type I : The particle passes *near* or *through* a straight or weakly bent part of the fibre, far from an extremity. Translation invariance along the fiber axis is essential.

- Type II : The particle passes near or through an end of the fiber or an added structure (e.g., metallic balls glued on the fibre surface), which is not translation invariant.

## 2. Particle-induced guided light of Type-I

The PIGL intensity will be calculated in the framework of quantized fields used by Glauber (7). We will use relativistic quantum units units familiar to particle physicists : $\hbar = c = \varepsilon_0 = \mu_0 = 1$. $\lambda\!\!\!^{-} \equiv \lambda/2\pi = 1/\omega$. The Gauss law is written $\nabla \cdot \mathbf{E} = \rho$, not $4\pi\rho$. $e^2/(4\pi) = \alpha = 1/137$.

### 2.1 Expansion of the field in proper modes

The fiber is along the $\hat{\mathbf{z}}$ axis. The cylindrical coordinates are $(r, \phi, z)$. $\mathbf{r} = (x, y)$ is the transverse position. $x \pm iy = re^{\pm i\phi}$.

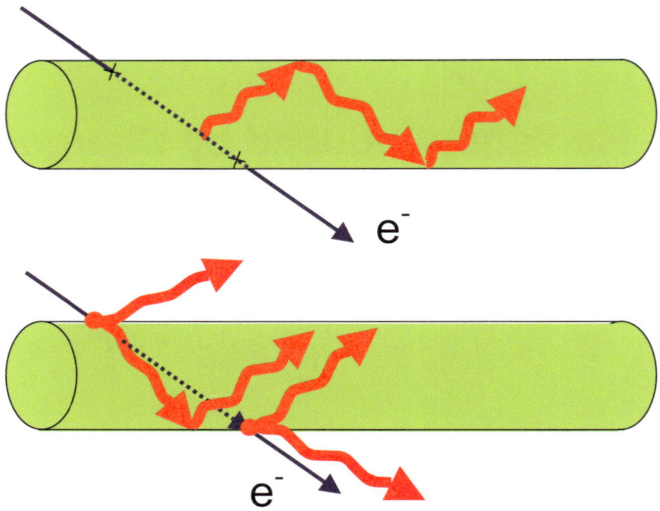


Fig. 1. Standard mechanisms of production of light inside a fiber, by an electron passing through. Top: Cherenkov radiation. Bottom: Transition radiation.

The quantized electromagnetic field $\mathbf{E}^{op}$ in presence of the fiber can be expanded in propagation modes :

$$\mathbf{E}^{\mathrm{op}}(t,\mathbf{X}) = \int_0^\infty \frac{d\omega}{2\pi} \sum_m a_m(\omega)\, \vec{\mathcal{E}}^{(m)}(\omega;\mathbf{X})\, \exp(-i\omega t) + \text{hermit. conj.} \tag{1}$$

The complex-valued field

$$\vec{\mathcal{E}}^{(m)}(\omega;\mathbf{X}) = \mathbf{E}^{(m)}(\omega;\mathbf{r})\, \exp(ipz) \tag{2}$$

is a "photon wave function". $m = \{M,\nu,\sigma\}$ is a collective index which gathers the total angular momentum $M \equiv J_z = L_z + S_z$ of the photon, the radial quantum number $\nu$ and the direction of propagation $\sigma = \mathrm{sign}(p) = \pm 1$. $a_m$ and $a_m^\dagger$ are the destruction and creation operators of a photon in the mode $m$. $\omega$ and $p$ are linked by the dispersion relation,

$$\omega = \omega_m(p) \quad \text{or} \quad p = p_m(\omega)\,. \tag{3}$$

The $\nu$ spectrum has a discrete part for guided modes and a continuous part for free modes. The summation over $m$ in (1) implies that $\nu$ is treated as a fully discrete variable, for simplicity. This is actually the case if we quantize the field inside a cylindrical box.
The quantized magnetic field is expanded like in (1). $a_m$ and $a_m^\dagger$ obey the commutation rules

$$\left[a_{M,\nu,\sigma}(\omega), a^\dagger_{M',\nu',\sigma'}(\omega')\right] = 2\pi\, \delta(\omega-\omega')\, \delta_{MM'}\, \delta_{\nu\nu'}\, \delta_{\sigma\sigma'}\,. \tag{4}$$

For a fixed $\omega$ the modes $m$ are orthonormal in the sense

$$\int d^2\mathbf{r}\, \left[\mathbf{E}^{(m)*}(\omega;\mathbf{r}) \times \mathbf{B}^{(n)}(\omega;\mathbf{r}) + \mathbf{E}^{(n)}(\omega;\mathbf{r}) \times \mathbf{B}^{(m)*}(\omega;\mathbf{r})\right]_z = \omega\, \delta_{mn}\,. \tag{5}$$

For $n = m$, the left-hand side is the power carried by the fiber in the mode $m$, which is $\hbar\omega$ (= one photon) per unit of time.

Equations (1), (4) and (5) correspond to Eqs. (2.29b), (2.25b) and (2.14a) of Ref.(7). The correspondance would be $\mathbf{f_k} \to -i\,(2/\omega)^{1/2}\,\mathbf{E}^{(m)}$, but we use the continuous variable $\omega$ instead of a fully discrete set of quantum numbers. $a_m$ and $\mathbf{E}^{(m)}$ differ from those of Ref.(1) by a factor $(dp/d\omega)^{1/2} = v_g{}^{-1/2}$. The factor 2 in (5) was forgotten in Refs.(1; 2), leading to an overestimation of the photon production yield by a factor 2.

### 2.2 Wave functions of the fiber modes

The propagation modes in optical fibers can be found in several textbooks, e.g. (8). Nevertheless, it is useful to present a short review based on states of definite angular momentum $M$.

We assume that the fiber has an homogeneous refractive index $n = \sqrt{\varepsilon}$ and no clad. For a guided mode the phase velocity $v_{\text{ph}} = \omega/p$ is in the interval $[1/n, 1]$. The photon transverse momentum is $q = \sqrt{\varepsilon\omega^2 - p^2}$ inside the fiber and $i\kappa = i\sqrt{p^2 - \omega^2}$ (evanescent wave) outside the fiber. The longitudinal parts of the fields have $S_z = 0$ therefore their orbital angular momentum $L_z$ is equal to $M$. Using cylindrical coordinates $(r, \phi, z)$ they write

$$E_z(\mathbf{r}) = i\,e^{iM\phi}\,f_z(r), \qquad B_z(\mathbf{r}) = e^{iM\phi}\,h_z(r), \tag{6}$$

Both in medium and in vacuum $f_z$ and $h_z$ obey the same differential equation

$$\left[\partial_r^2 + r^{-1}\partial_r - M^2/r^2 + k_T^2(r)\right]\, f_z \text{ or } h_z = 0 \quad (\text{except for } r = a) \tag{7}$$

where $k_T^2(r) = q^2$ inside the fiber and $k_T^2(r) = -\kappa^2$ outside the fiber.

The piecewise solutions of (7) are Bessel functions $J_M$ or $K_M$. From the fact that $f_z$ and $h_z$ are continuous at $r = 0$ and $r = a$ and decreasing at $r \to \infty$, it follows that $h_z(r)/f_z(r)$ is independent on $r$. We write

$$f_z(r) = c_E\,\psi(r), \qquad h_z(r) = c_B\,\psi(r), \tag{8}$$

$$\psi(r) = J_M(qr) \text{ inside}, \quad \psi(r) = c_K\,K_M(\kappa r) \text{ outside}, \quad c_K = \frac{J_M(qa)}{K_M(\kappa a)}.$$

The transverse components $\mathbf{E}_T$ and $\mathbf{B}_T$ can be expressed either in terms of the radial and azimuthal basic vectors, $\hat{\mathbf{e}}^r = \mathbf{r}/r$ and $\hat{\mathbf{e}}^\phi = \hat{\mathbf{z}} \times \hat{\mathbf{e}}^r$,

$$\begin{aligned} \mathbf{E}_T &= e^{iM\phi}\,\left(f_r(r)\,\hat{\mathbf{e}}^r + f_\phi(r)\,\hat{\mathbf{e}}^\phi\right) \\ \mathbf{B}_T &= e^{iM\phi}\,\left(h_r(r)\,\hat{\mathbf{e}}^r + h_\phi(r)\,\hat{\mathbf{e}}^\phi\right), \end{aligned} \tag{9}$$

or in terms of the $S_z = \pm 1$ eigenvectors $\hat{\mathbf{e}}^\pm = (\hat{\mathbf{x}} \pm i\hat{\mathbf{y}})/2$ :

$$\begin{aligned} \mathbf{E}_T &= e^{i(M-1)\phi}\,f_-(r)\,\hat{\mathbf{e}}^+ + e^{i(M+1)\phi}\,f_+(r)\,\hat{\mathbf{e}}^- \\ i\,\mathbf{B}_T &= e^{i(M-1)\phi}\,h_-(r)\,\hat{\mathbf{e}}^+ + e^{i(M+1)\phi}\,h_+(r)\,\hat{\mathbf{e}}^- \end{aligned} \tag{10}$$

with $f_\pm = f_r \pm i f_\phi$ and $-i h_\pm = h_r \pm i h_\phi$. The $\hat{\mathbf{e}}^+$ and $\hat{\mathbf{e}}^-$ parts of the fields have orbital momenta $L_z = M \mp 1$, therefore their radial dependence are Bessel functions of order $M \mp 1$ :

$$\begin{aligned} f_\pm(r) &= c^\pm_{fJ}\, J_{M\pm1}(qr)\ (r \le a), \qquad c^\pm_{fK}\, K_{M\pm1}(\kappa r)\ (r > a), \\ h_\pm(r) &= c^\pm_{hJ}\, J_{M\pm1}(qr)\ (r \le a), \qquad c^\pm_{hK}\, K_{M\pm1}(\kappa r)\ (r > a). \end{aligned} \tag{11}$$

The Maxwell equations relate the transverse fields to the longitudinal ones. The formula in the $\{\hat{\mathbf{e}}^r, \hat{\mathbf{e}}^\phi\}$ basis can be found in (8). Translated in the $\{\hat{\mathbf{e}}^+, \hat{\mathbf{e}}^-\}$ basis they give

$$c^\pm_{fJ} = (\pm p\, c_E - \omega\, c_B)/q, \qquad c^\pm_{fK} = \mp(q\, c_K/\kappa)\, c^\pm_{fJ},$$

$$c^\pm_{hJ} = (\pm p\, c_B - \omega\varepsilon\, c_E)/q, \qquad c^\pm_{hK} = (-p\, c_B \pm \omega\, c_E)\, c_K/\kappa.$$

The continuity of $h_z, h_r, h_\phi, f_z, f_\phi$ and $\epsilon(r) f_r$ at $r = a$ leads to

$$\frac{c_B}{c_E} = -MQ\left[\frac{J'_M(u)}{uJ_M(u)} + \frac{K'_M(w)}{wK_M(w)}\right]^{-1} = -\frac{1}{MQ}\left[\frac{\varepsilon J'_M(u)}{uJ_M(u)} + \frac{K'_M(w)}{wK_M(w)}\right] \tag{12}$$

where $u \equiv qa,\ w \equiv \kappa a$ and

$$Q = \left(u^{-2} + w^{-2}\right) p/\omega = \left(\varepsilon\, u^{-2} + w^{-2}\right) \omega/p.$$

From the two expressions of $c_B/c_E$ in (12) one obtains

$$\left[\frac{J'_M(u)}{uJ_M(u)} + \frac{K'_M(w)}{wK_M(w)}\right] \cdot \left[\frac{\varepsilon J'_M(u)}{uJ_M(u)} + \frac{K'_M(w)}{wK_M(w)}\right] = M^2 \left(\frac{1}{u^2} + \frac{1}{w^2}\right) \cdot \left(\frac{\varepsilon}{u^2} + \frac{1}{w^2}\right), \tag{13}$$

which, together with $u^2 = (\varepsilon\omega^2 - p^2)a^2$ and $w^2 = (p^2 - \omega^2)a^2$, determines the dispersion relation (3).

#### 2.2.1 Normalization of the mode wave functions

The $z$-component of the Pointing vector of the complex field is

$$\mathcal{P}^{(m)}(\mathbf{r}) = 2\,\mathrm{Re}\left\{\mathbf{E}^{(m)*} \times \mathbf{B}(m)\right\}_z = \mathrm{Re}\left\{f^*_-(r)\, h_-(r) - f^*_+(r) h_+(r)\right\}.$$

Using (11) and integrating over $\mathbf{r}$ gives the mode power

$$\begin{aligned} P^{(m)} = P^{(m)}_{\mathrm{int}} + P^{(m)}_{\mathrm{ext}} &= \int_0^a 2\pi r\, dr \left\{c^-_{fJ}\, c^-_{hJ}\, J^2_{M-1}(qr) - c^+_{fJ}\, c^+_{hJ}\, J^2_{M+1}(qr)\right\} \\ &+ \int_a^\infty 2\pi r\, dr \left\{c^-_{fK}\, c^-_{hK}\, K^2_{M-1}(\kappa r) - c^+_{fK}\, c^+_{hK}\, K^2_{M+1}(\kappa r)\right\}. \end{aligned} \tag{14}$$

The coefficient $c_E$ has to be adjusted to get the normalization (5).

Fig. 2 shows the phase velocity $v_{\mathrm{ph}} = \omega/p$ of the lowest mode ($M = \pm1, \nu = 1$) called $HE_{11}$ and the external fraction of the mode power, as a function of $\omega$. The index of refraction is $n = 1.41$ (fused silica).

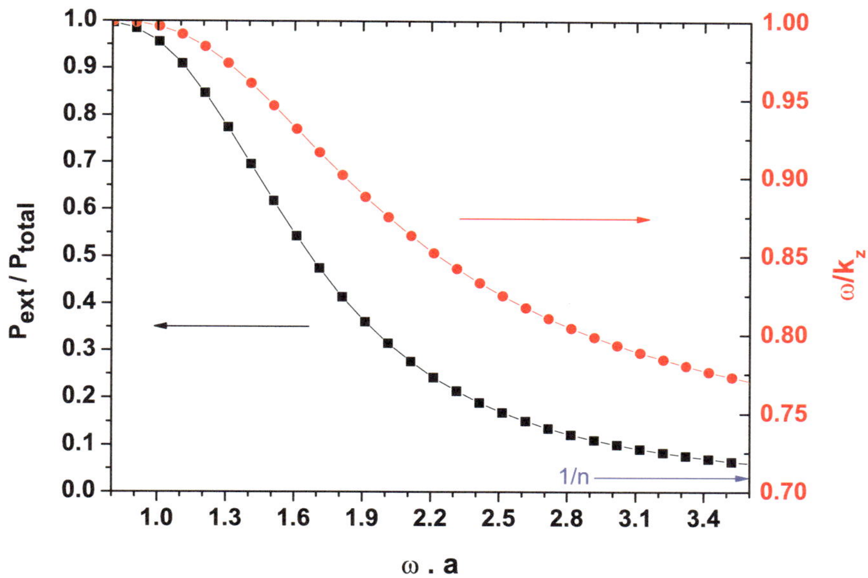


Fig. 2. Phase velocity $v_{\rm ph} = \omega / p$ (balls, right scale) and external fraction of the power (squares, left scale) for the $HE_{11}$ mode.

### 2.2.2 Linearly polarized modes

When changing $M$ into $-M$, the above defined field modes change as follows :

$$\begin{aligned}\{\mathbf{E}_T, E_z, \mathbf{B}_T, B_z\}^{(-M)} &= (-1)^M \;\Pi(0^\circ)\;\{\mathbf{E}_T, E_z, \mathbf{B}_T, B_z\}^{(M)} \\ &= \Pi(90^\circ)\;\{\mathbf{E}_T, E_z, \mathbf{B}_T, B_z\}^{(M)} \\ &= (-1)^M\;\{\mathbf{E}^*_T, -E^*_z, \mathbf{B}^*_T, -B^*_z\}^{(M)} . \end{aligned} \tag{15}$$

$\Pi(\alpha)$ is the operator of mirror reflection about the plane $\phi = \alpha$, for instance

$$\Pi(0^\circ)\{E_x, E_y, E_z\}(x, y, z) = \{E_x, -E_y, E_z\}(x, -y, z)$$

and a similar formula for $\mathbf{B}$, with an extra $(-)$ sign since it is a pseudovector. The linear combination

$$\{\mathbf{E}, \mathbf{B}\}^{(M,0^\circ)} = \left[\{\mathbf{E}, \mathbf{B}\}^{(M)} + (-1)^M \{\mathbf{E}, \mathbf{B}\}^{(-M)}\right] / \sqrt{2} \tag{16}$$

is even under $\Pi(0^\circ)$ and has real $\mathbf{E}_T$. For $M = 1$,

$$\mathbf{E}_T{}^{(1,0^\circ)} = [f_-(r)\,\hat{\mathbf{x}} + f_+(r)\,(\cos 2\phi\,\hat{\mathbf{x}} + \sin 2\phi\,\hat{\mathbf{y}})] / \sqrt{2} \tag{17}$$

is the state whose dominant $(f_-)$ part is linearly polarized parallel to $\hat{\mathbf{x}}$.

### 2.3 Bent fiber

Bending the fiber has several effects :

- a) small break-down of the degeneracy (*i.e.*, slightly different dispersion relations) between the polarized states $(M, 0^\circ)$ and $(M, 90^\circ)$, where $0^\circ$ is the azimuth of the bending plane),

- b) co-rotation of the transverse wave function $\vec{\mathcal{E}}^{(m)}(\omega;\mathbf{X})$ with the unit vector $\hat{\mathbf{s}}$ tangent to the local fiber axis.
- c) escape of light by tunneling through a centrifugal barrier.

For large enough bending radius, effects a) and c) can be ignored. Effect b) is non-trivial when the bending is skew (not planar). Instead of (2), we have

$$\vec{\mathcal{E}}^{(m)}(\omega;\mathbf{X}) = \mathcal{R}_{\mathrm{f}}(s)\,\mathbf{E}^{(m)}\left(\omega;\,\mathcal{R}_{\mathrm{f}}^{-1}(s)\,\mathbf{r}\right)\,\exp(ips)\,, \tag{18}$$

where $\mathbf{X}_{\mathrm{f}}(s)$ is the point of the fiber axis nearest to $\mathbf{X}$, $s$ its curvilinear abscissa and $\mathbf{r} = \mathbf{X} - \mathbf{X}_{\mathrm{f}}(s)$ (see Fig. 3 left). $\mathcal{R}_{\mathrm{f}}(s)$ is a finite rotation matrix resulting from a succession of infinitesimal rotations $\mathcal{R}(\hat{\mathbf{s}} \to \hat{\mathbf{s}} + d\hat{\mathbf{s}})$ :

$$\mathcal{R}_{\mathrm{f}}(s+ds) = \mathcal{R}(\hat{\mathbf{s}} \to \hat{\mathbf{s}} + d\hat{\mathbf{s}}) \circ \mathcal{R}_{\mathrm{f}}(s)\,, \quad \mathcal{R}_{\mathrm{f}}(0) = \mathrm{I}\,, \tag{19}$$

$\mathcal{R}(\hat{\mathbf{s}} \to \hat{\mathbf{s}}')$ denoting the rotation along $\hat{\mathbf{s}} \times \hat{\mathbf{s}}'$ which transforms $\hat{\mathbf{s}}$ into $\hat{\mathbf{s}}'$. Taking into account the non-commutativity of the rotations, we have

$$\mathcal{R}_{\mathrm{f}}(s) = \mathcal{R}\left(\hat{\mathbf{s}}, \Omega(s)\right) \circ \mathcal{R}(\hat{\mathbf{z}} \to \hat{\mathbf{s}})\,. \tag{20}$$

where $\hat{\mathbf{z}}$ is the orientation of the beginning of the fiber, $\mathcal{R}\left(\hat{\mathbf{s}}, \alpha\right)$ stands for a rotation of angle $\alpha$ about $\hat{\mathbf{s}}$ and $\Omega(s)$ is the dark area on the unit sphere in Fig. 3 (right). For a state of given angular momentum $M$ in (18) one can replace $\mathcal{R}_{\mathrm{f}}(s)$ by $\mathcal{R}(\hat{\mathbf{z}} \to \hat{\mathbf{s}})$ and take into account the first factor of (20) by the *Berry phase* factor $\exp[-iM\Omega(s)]$. If the fiber is bent in a plane, $\Omega(s) = 0$.

### 2.4 Mode excitation by a charged particle

When a particle of charge $Ze$ passes *trough* or *near* the fiber, it can create one or several photons by spontaneous or stimulated emission. Neglecting its loss of energy and momentum, the particle acts like a cassical current and the excitation of the quantum field is a *coherent state* (7). The spontaneous photon emission amplitude in the mode $m$, corresponding to Eqs. (7.11) and (7.16) of (7), is

$$R^{(m)}(\omega) = \frac{Ze}{\omega} \int d\mathbf{X}(t) \cdot \vec{\mathcal{E}}^{(m)*}(\omega;\mathbf{X})\;\exp(i\omega t) \tag{21}$$

for a mode normalized according to (5). The photon spectrum of spontaneous emission in the mode $m$ reads

$$\frac{d\mathcal{N}^{(m)}_{\mathrm{phot}}}{d\omega} = \frac{\omega}{2\pi P^{(m)}(\omega)}\,\left|R^{(m)}(\omega)\right|^2. \tag{22}$$

Thanks to the factor $P^{(m)}(\omega)$ given by (14) in the denominator, this expression is invariant under a change of the normalisation of the mode fields.

### 2.5 Straight fiber and particle in rectilinear uniform motion

For a particle following the straight trajectory

$$\mathbf{X} = \mathbf{b} + \mathbf{v}t\,, \quad \mathbf{b} = (b,0,0)\,, \quad \mathbf{v} = (0,v_T,v_L)\,, \tag{23}$$

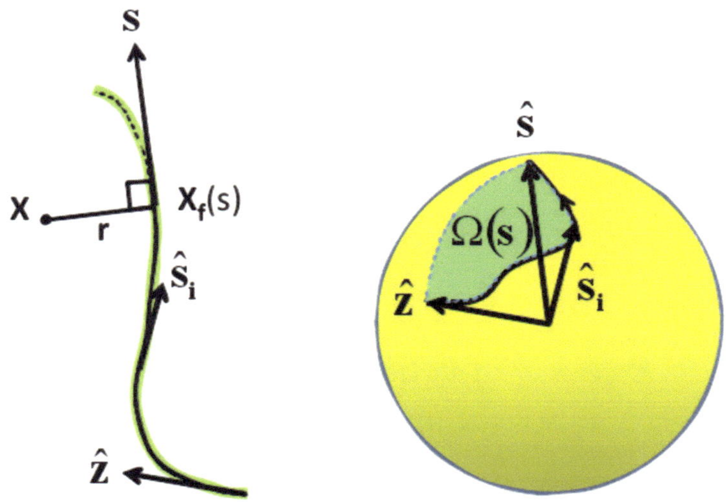


Fig. 3. Left: bent fiber and definition of $\mathbf{X}_f(s)$ and $\mathbf{r}$. Right: curve drawn by the extremities of successive tangent vectors ($\hat{\mathbf{z}}$, $\hat{\mathbf{s}}_i$, $\hat{\mathbf{s}}$, ...) on the unit sphere and definition of the solid angle $\Omega(s)$. The dotted arc of circle represents the "most direct" rotation, $\mathcal{R}(\hat{\mathbf{z}} \to \hat{\mathbf{s}})$, transforming $\hat{\mathbf{z}}$ int $\hat{\mathbf{s}}$.

Eqs.(21) and (2) give

$$R^{(m)}(\omega) = \frac{Ze}{\omega} \int_{-\infty}^{\infty} dy \left[ E_y^{(m)}(x,y) + \frac{v_L}{v_T} E_z^{(m)}(x,y) \right]^* \exp\left( iy \frac{\omega - v_L\, p}{v_T} \right) \tag{24}$$

Using (6-11) one arrives at the pure imaginary expression

$$R^{(m)}(\omega) = \frac{-iZe}{\omega} \int_0^{\infty} dy \, \{\cos[\eta y + (M-1)\phi]\, f_-(r) \\ - \cos[\eta y + (M+1)\phi]\, f_+(r) + 2(v_L/v_T)\, \cos(\eta y + M\phi)\, f_z(r)\} \tag{25}$$

with $r = \sqrt{b^2 + y^2}$, $\phi = \tan^{-1}(y/b)$ and

$$\eta = (v_L\, p - \omega)/v_T = (v_L - v_{\text{ph}})\, p/v_T\,.$$

### 2.6 Limit of small crossing angle

For small crossing angle $\theta = \tan^{-1}(v_T/v_L)$ the integrand of (24) becomes large due to the $v_L/v_T$ factor of the third term, although $f_z$ is generally small. On the other hand, unless $|v_L - v_{\text{ph}}| \lesssim v_T/(pa)$, the integrand oscillates fast in the region $|y| \lesssim a$ where the field is important and the amplitude is strongly reduced. One therefore expects an almost monochromatic peak at $\omega = \omega_C(v)$ fixed by the "fiber Cherenkov condition"

$$v_{\text{ph}}(\omega_C) \equiv \omega_C/p(\omega_C) = v \tag{26}$$

and the dispersion relation (3). The case $\theta = 0$ (electron runnig parallel to the fiber), where $\omega \equiv \omega_C$, has been studied in Refs.(9; 10).

### 2.7 Slightly bent fiber or particle trajectory

Local curvatures of the trajectory or of the fiber can be neglected and formula (25) is accurate enough when the crossing angle $\theta$ is large. Let us consider the case where the particle trajectory, the fiber or both are slightly curved, but at angles not far from the $\hat{\mathbf{z}}$ direction. Then we have to use (18) instead of (2) in (21). However we can omit the rotation matrix $\mathcal{R}_{\mathrm{f}}(s)$ and make the approximation

$$d\mathbf{X}(t)\cdot\vec{\mathcal{E}}^{(m)*}(\omega;\,\mathbf{X})\simeq v\,dt\,E_z^{(m)}(\omega;\,\mathbf{r})\,\exp(ips)\,. \tag{27}$$

Thus we can rewrite (21) as

$$R^m(\omega)=\frac{Zev}{\omega}\int dt\,E_z^*[(\omega;\,\mathbf{r}(t)]\,\exp[i\omega t-ip_m(\omega)s]\,,$$
$$\mathbf{r}(t)=\mathbf{X}_{\mathrm{p}}(t)-\mathbf{X}_{\mathrm{f}}(s)\,,\quad s=\int v\,dt\,\cos\theta(t)\,. \tag{28}$$

Here again the integrand oscillates too fast - and the amplitude is too small - when $\omega$ is not close to $\omega_C$. The total photon number in the mode $m$ is

$$\mathcal{N}_{\mathrm{phot}}^{(m)}=2Z^2\,\alpha\,v^2\int\frac{d\omega}{\omega\,P^{(m)}(\omega)}\int dt'\,E_z^*[(\omega;\,\mathbf{r}(t')]\int dt''\,E_z[(\omega;\,\mathbf{r}(t'')] \tag{29}$$
$$\exp\{i\omega(t'-t'')-ip(\omega)\,(s'-s'')\}\,. \tag{30}$$

To first order in $\omega-\omega_C$ the exponential can be written as

$$\exp\{i\omega\,(T-S/v_g)+i[\omega_C/v_g-p_m(\omega_C)]\,S\} \tag{31}$$

where $t'-t''=T,\ s'-s''=S$ and $v_g=d\omega/dp$ is the group velocity at $\omega=\omega_C$. Neglecting the variations of the other factors with $\omega$, the integration over $\omega$ yields a factor $2\pi\delta(T-S/v_g)$. From the second line of (28), we have $S/T\simeq v\cos\theta(t)\simeq v$ at small $S$ and $T$, therefore $\delta(T-S/v_g)=\delta(T)/[1-v/v_g]$. One finally obtains

$$\mathcal{N}_{\mathrm{phot}}^{(m)}=\frac{4\pi Z^2\alpha\,v^2}{\omega_C\,P^{(m)}(\omega_C)}\,\frac{1}{|1-v/v_g(\omega_C)|}\int dt\,|E_z[\omega_C;\,\mathbf{r}(t)]|^2\,. \tag{32}$$

The energy of the light pulse is obtained by multiplying by $\omega_C$. This formula applies in particular to the limit of small crossing angles considered above. The photon number increases linearly with the path length over which the particle travels inside or close to the fiber.

### 2.8 Numerical results for straight electron trajectory and straight fiber

The dimensionless photon spectrum $\omega d\mathcal{N}_{\mathrm{phot}}/d\omega$ in the fundamental mode $HE_{11}$ of a fused silica fiber is plotted in Fig. 4 for three impact parameters, $b=0.2\,a$ (penetrating trajectory), $b=a$ (tangent trajectory) and $b=1.5\,a$ (fully external trajectory), and two particle velocity vectors, $(v_L,v_T)=(0.88,0.1)$ and $(v_L,v_T)=(0.85,0.5)$, corresponding to large and moderate angle respectively. We took the sign of $M$ to be the same as the $J_z$ of the particle.

The spectra are harder for penetrating trajectories, due to (i) the discontinuity of the fields at the fiber surface, (ii) the lower importance of the evanescent field at high frequency.

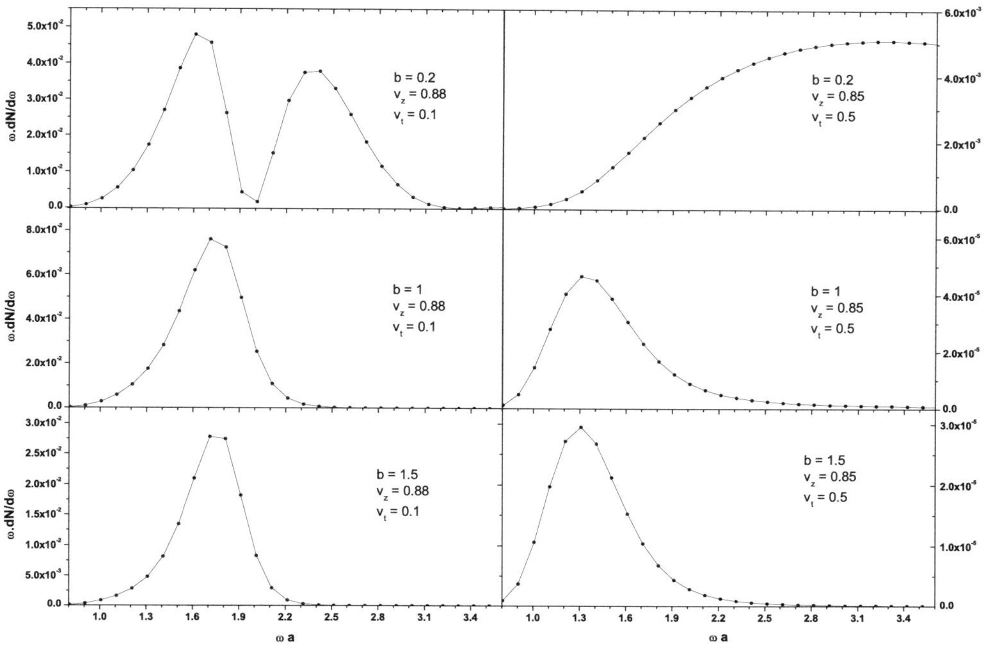


Fig. 4. Dimensionless photon spectrum $\omega d\mathcal{N}_{\rm phot}/d\omega$ as a function of $\omega a$ in the $HE_{11}$ mode for six types of the particle trajectory and $M = +1 = \text{sign}(v_T)$.

In the large angle - penetrating case, the dimensionless yield is of the order of $\alpha = e^2/(4\pi) = 1/137$. In the tangent case it is much smaller. Note the peak at a relatively small frequency, where the wave travels mainly outside the fiber (see Fig. 2). At still smaller frequency, the wave function of the mode becomes too much diluted, which explains the vanishing yields at small $\omega$ in the six curves.

In the $b = 0.2\,a$ and $v_T = 0.1$ case, we have a dip at $\omega a = 2$ instead of an expected Cherenkov peak fixed by Eq.(26). This is a peculiarity of the odd $M$ modes when $b$ is small : if $b = 0$, then $\phi$ in (25) is either $-\pi/2$ or $+\pi/2$ and, at the Cherenkov point ($\eta = 0$), $\cos(\eta y + M\phi)$ is zero in the whole integration range.

A separate figure (Fig. 5) at small crossing angle ($v_T/v_L = 0.03/0.95$) shows the narrow peak of "fiber Cherenkov light" at the position $\omega a \simeq 1.4$ predicted by (26) and Fig. 2. The half-width at half maximum, 0.06, corresponds roughly to the condition $|v_L - v_{\rm ph}| \lesssim v_T/(pa)$ mentioned in Paragraph 2.6.

## 2.9 Polarisation

If $b = 0$, the $HE_{11}$ guided light is linearly polarized in the particle incidence plane. If $b \neq 0$, some circular polarization is expected. One could naively expect that the favored photon angular momentum $M$ has the sign of the azimutal speed of the particle, i.e. the sign of $v_T$ in (23), but this is not always true. What matters in fact is not the sign of $M$ but the sense of rotation of the electric field of the mode in the *moving plane* $z = vt$. In this plane the azimuth of the field varies like $M(\omega t - pz) = (v_{\rm ph} - v_L)\, M\omega t/v_{\rm ph}$. If the moving plane is faster than

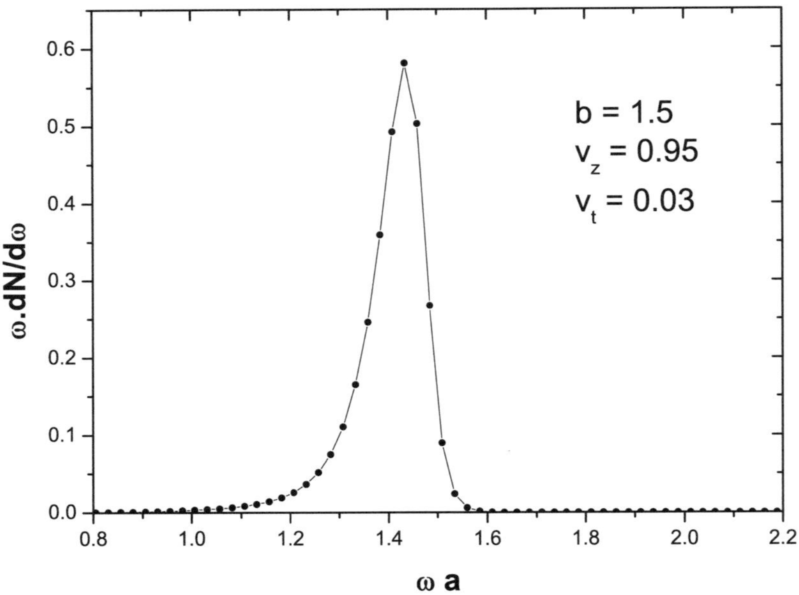


Fig. 5. Photon yields in the $HE_{11}$ mode with $M = +1$ for a small crossing angle : $(v_L, v_T) = (0.95, 0.03)$ ; $b = 1.5a$.

the wave, the field rotates in the opposite way. Thus the favored sign of $M$ is the sign of $(v_{\text{ph}} - v_L)\, v_T$. This can be seen from (25) : if $M$ and $\eta$ have the same sign, the integrand oscillate faster and the amplitude is reduced.

In Figs. 4 and 5, $M$ has the sign of $v_T$. This circular polarization is favored at $v_{\text{ph}} > v_L$, whence $\omega < \omega_C(v_L)$, and unfavored at $v_{\text{ph}} < v_L$, whence $\omega > \omega_C(v_L)$. This partly explains the asymmetric shape of the fiber Cherenkov peak in Fig. 5. Changing the sign either of $M$ or of $v_T$ should result in a harder spectrum.

### 2.10 Interferences with periodically bent trajectory or bent fiber

With an undulated trajectory, as in Fig. 6a or an undulated fiber as in Fig. 6b, one can have several meeting points, the PIGL amplitude of which, given by (25) or (28), add coherently. Let $L_f$ and $L_p$ be the lengths of the fiber and of the particle trajectory between two meeting points. Two successive fiber-particle interactions are separated in time by $\Delta t = L_f / v$ and their phase difference is

$$\Delta\Phi = p\, L_f - \omega\, \Delta t = \omega\, \left( L_f / v_{\text{ph}} - L_p / v \right) . \tag{33}$$

If $N$ equivalent meeting points are spaced periodically, the frequency spectrum is

$$\left( \frac{d\mathcal{N}^{(m)}}{d\omega} \right)_{N\,\text{meeting}} = \left( \frac{d\mathcal{N}^{(m)}}{d\omega} \right)_{1\,\text{meeting}} \times \frac{\sin^2(N\Delta\Phi/2)}{\sin^2(\Delta\Phi/2)} . \tag{34}$$

The last fraction is the usual interference factor in periodical systems, e.g. in undulator radiation. For large $N$ it gathers the photon spectrum in quasi-monochromatic lines fixed by

$$\omega\left(L_f/v_{\rm ph} - L_p/v\right) = 2k\pi \quad (k \text{ integer})\,. \tag{35}$$

If the fiber bending is *not planar*, but for instance helicoidal (Fig. 6c), the left- and right circular polarisations have different phase velocities. Their propagation amplitudes acquire an additional phase $\phi_B = -M\Omega$, called the *Berry phase*, where $\Omega$ is the solid angle of the cone drawn by the local axis of the fiber (11) (as if $\hat{\mathbf{s}}$ coincides with $\hat{\mathbf{z}}$ in Fig. 3). The preceding condition becomes

$$\omega\left(L_f/v_{\rm ph} - L_p/v\right) = 2k\pi - \phi_B\,. \tag{36}$$

The interferences disappear when the velocity spread of the charged particle beam is such that the variation of $\omega\, L_p/v$ is more than, say, $2\pi$.

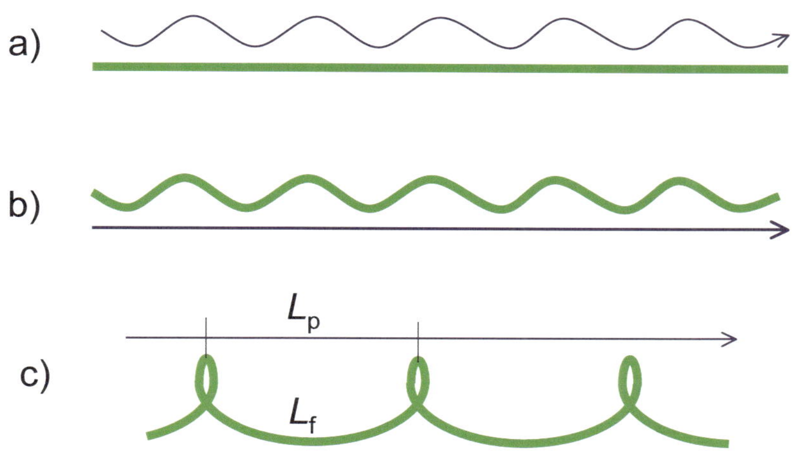


Fig. 6. Periodically bent particle trajectory (a), planar bent fiber (b) and helical bent fiber (c). $L_p$ and $L_f$ are the lengths of the curved or straight periods, for the particle and the fiber respectively.

### 2.11 Application of type-I PIGL to beam diagnostics

PIGL in a monomode fiber is intense enough not for single particle detection, but for beam diagnostics.

The "fiber Cherenkov radiation" can be used to measure the velocity of a semi-relativistic particle beam, using the dependence of $v_{\rm ph}$ on $\omega$ shown in Fig. 2.

In a periodically bent fiber, the interference can test the velocity spread of the beam.

At large crossing angle, a fiber can measure the transverse profile of the beam with a resolution of the order of the diameter $2a$. No background is made by real photons coming from distant sources (for instance synchrotron radiation from upstream bending magnets). Indeed, such photons are in the continuum spectrum of the radial number $\nu$, therefore they are not captured by the fiber, but only scattered. This is an advantage over beam diagnostic tools like optical transition radiation (OTR) and optical diffraction radiation (ODR). The translation invariance along the fiber axis, which guarantees the conservation of $\nu$, is essential for this property.

The resolution power of PIGL is also not degraded by the large transverse size $\sim \gamma\lambdabar$ of the virtual photon cloud at high Lorentz factor $\gamma = (1-v^2)^{-1/2}$. Indeed, the virtual photons at transverse distance $\gg \lambdabar$ are almost real, therefore are not captured by the fiber.

## 3. Particle-induced guided light of Type-II

The second type of PIGL is produced at a place where the fiber is not translation invariant. We consider two examples : 1) PIGL from the cross section of a cut fiber, 2) PIGL assisted by metallic balls glued to the fiber. These devices are represented in Fig. 7.

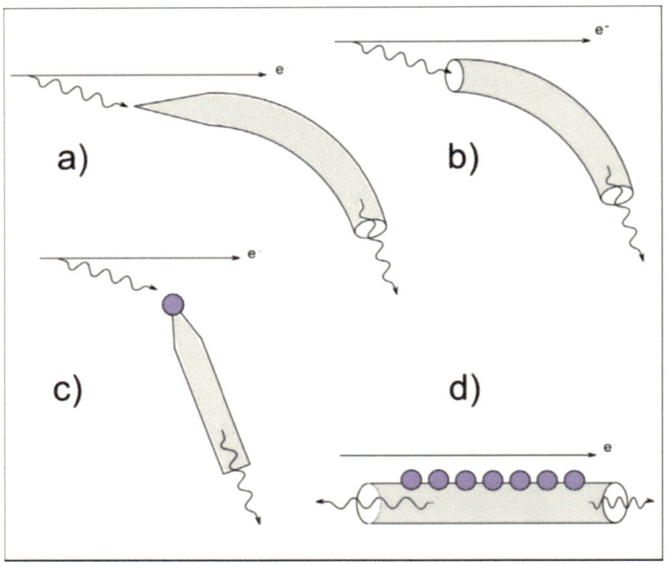


Fig. 7. Part of fiber which can capture virtual photons for Type-II PIGL : a) conical end ; b) sharp-cut end ; c) metallic ball glued on one end ; d) regularly spaced metallic balls glued along the fiber.

### 3.1 PIGL from the cross section of a cut fiber

The entrance section of a sharp-cut fiber can catch free real photons and convert them into guided photons. Assuming that the photons are incident at small angle with the fiber axis, the energy spectrum captured by the fiber in the mode $m = \{M, \nu\}$ is given by

$$\frac{dW^{(m)}}{d\omega} = \frac{1}{2\pi P^{(m)}(\omega)} \times \left| \int d^2\mathbf{r} \left[ T_B(\mathbf{r})\, \mathbf{E}_T{}^{(m)*}(\omega;\mathbf{r}) \times \mathbf{B}_T{}^{\text{in}}(\omega;\mathbf{r}) \right.\right.$$
$$\left.\left. + T_E(\mathbf{r})\, \mathbf{E}_T{}^{\text{in}}(\omega;\mathbf{r}) \times \mathbf{B}_T{}^{(m)*}(\omega;\mathbf{r}) \right] \right|^2 . \qquad (37)$$

where $\{\mathbf{E}^{in}, \mathbf{B}^{in}\}$ is the incoming field on the cutting plane. $T_E(\mathbf{r})$, $T_B(\mathbf{r})$ are the Fresnel refraction coefficients at normal incidence, given by

$$T_E(\mathbf{r}) = 2/(1+\sqrt{\varepsilon(\mathbf{r})}), \quad T_B(\mathbf{r}) = \sqrt{\varepsilon(\mathbf{r})}\, T_E(\mathbf{r}) . \qquad (38)$$

$\varepsilon(\mathbf{r})$ is the local permittivity of the fiber. Outside the fiber, $T_E(\mathbf{r}) = T_B(\mathbf{r}) = 1$. Equation (37) is deduced from the orthonormalization relation (5).
With some caution (37) can be applied to the capture of virtual photons from the Coulomb field of a relativistic particle passing near the entrance face (see Fig. 7b). The transverse component of this field is given by (12; 13)

$$\mathbf{E}_T{}^{\text{in}}(\omega;\mathbf{r}) = \frac{Ze\,\omega}{2\pi\gamma v^2 b}\, K_1\left(\frac{\omega b}{\gamma v}\right)\mathbf{b}, \quad \mathbf{B}_T{}^{\text{in}}(\omega;\mathbf{r}) = \mathbf{v}\times\mathbf{E}_T{}^{\text{in}}(\omega;\mathbf{r})\,. \tag{39}$$

Here $\mathbf{b} = \mathbf{r} - \mathbf{r}_{\text{particle}}$ is the impact parameter relative to the particle. It must be large enough compared to $\lambda\!\!\!^{-}$, otherwise the incoming photon is too different from a real one.

### 3.2 PIGL from a conical end of fiber

The sharp-cut fiber has a wide angular acceptance but is not optimized for capturing the virtual photon cloud accompagning an ultrarelativistic particle, which has an angular divergence $\sim 1/\gamma$. A more efficient capture is possible with a narrow conical end (Fig. 7a), at the price of a smaller acceptance. The wave function of a parallel photon may be quasi-adiabatically transformed into a guided mode without too much loss. This should be true for the photons of the Coulomb field in the impact parameter range $\lambda\!\!\!^{-} \ll b \lesssim \gamma\lambda\!\!\!^{-}$, which are quasi-real and have a small transverse momentum $k_T \sim 1/b$.

### 3.3 PIGL from metallic balls

It is also possible to capture a virtual photon with a metallic ball glued to the fiber, either at the extremity (Fig. 7c) (14; 15), or on the side as in Fig. 7d. Then a plasmon is created (16; 18), which has some probability $p_f$ to be evacuated as guided light in the fiber.
A rough estimate of the capture efficiency can be obtained when the impact parameter of the particle is large compared to the ball radius $R$ and the time scale $\Delta t \sim b/(\gamma v)$ of the transient field is short compared to the reduced period $1/\omega = \lambda\!\!\!^{-}$ of the plasmon : the particle field boosts each electron of the ball with a momentum $\mathbf{q} \simeq 2Z\alpha\,\mathbf{b}/(vb^2)$. It results in a collective dipole excitation of the electron cloud, of energy

$$W(b) \simeq \frac{4\pi R^3 n_e}{3}\left(\frac{2Z\alpha}{vb}\right)^2 \frac{1}{2m_e} = \frac{2Z^2\alpha}{3v^2}\,\frac{\omega_P^2 R^3}{b^2} \qquad (R \ll b \ll \gamma v\lambda\!\!\!^{-})\,, \tag{40}$$

where $\omega_P = (4\pi\alpha\, n_e/m_e)^{1/2}$ is the plasma frequency of the infinite medium. For a spherical ball the dipole plasmon frequency is simply given by $\omega = \omega_P/\sqrt{3}$, assuming the Drude formula $\varepsilon = 1 - \omega_P^2/\omega^2$ and neglecting the retardation effects (case $R \lesssim \lambda\!\!\!^{-}$). The number of stored quanta is then

$$\mathcal{N}(b) = \frac{W(b)}{\omega} \simeq \frac{2Z^2\alpha}{v^2}\cdot\frac{R^3}{\lambda\!\!\!^{-}\, b^2}\,. \tag{41}$$

Taking $b_{min} = R$ and $b_{max} = \gamma v\lambda\!\!\!^{-}$, the cross section for this process is

$$\sigma = \int_{b_{min}}^{b_{max}} 2\pi\, b\, db\, \mathcal{N}(b) \simeq \frac{4Z^2\alpha}{v^2}\cdot\frac{R^3}{\lambda\!\!\!^{-}}\cdot\ln\frac{\gamma v\lambda\!\!\!^{-}}{R}\,. \tag{42}$$

More precise values of the plasmon frequencies are used in (16–18) in the context of Smith-Purcell radiation. Retardation effects and other mutipoles are taken into account in (17; 18). A typical order of the cross section, $\sigma \sim 10^{-2}\lambdabar^{2}$ is obtained with $R \sim \lambdabar$, $Z = 1$, $\gamma v \sim 1$. The plasmon wavelength is typically $\lambdabar \sim 10^2$ nm. Larger cross section can be realized by increasing $R$, but higher multipoles will dominate, unless $\gamma$ is increased simultaneously. Discussions and experimental results about this point are given in (18).
The efficiency of the ball scheme depends on the ball-to-fiber transmission probability $p_f$, which is less than unity because the plasmon may also be radiated in vacuum or decay by absorption in the metal.

### 3.3.1 Interferences between several balls

If several metallic balls are glued at equal spacing $l$ on one side of the fiber (Fig. 7d), constructive interferences (resonance peaks) are obtained when

$$\omega/v \mp p \equiv (1/v \mp 1/v_{\rm ph})\,\omega = 2k\pi/l \quad (k\,\text{integer})\,, \tag{43}$$

$\omega$ and $p$ being linked by (3). The $-$ and $+$ signs correspond respectively to lights propagating forward and backward in the fiber. The forward light has the highest frequency. This process is in competition with the Smith-Purcell radiation from the balls, where $\mp 1/v_{\rm ph}$ is replaced by $-\cos\theta_{\rm rad}$. We can call it "guided Smith-Purcell" radiation. It is advantageous to choose $l$ such that $\omega$ lies on a plasmon resonance of the ball.

### 3.3.2 Shadowing

The guided Smith-Purcell spectrum for $N$ balls can be written as

$$\left(\frac{d\mathcal{N}^{(m)}}{d\omega}\right)_{N\,\text{balls}} \simeq \left(\frac{d\mathcal{N}^{(m)}}{d\omega}\right)_{1\,\text{ball}} \times \frac{\sin^2(N\Delta\Phi/2)}{\sin^2(\Delta\Phi/2)} \times \text{shadow factor}\,. \tag{44}$$

This is similar to (34) except for a *shadow factor* which is less than unity. Indeed, each ball intercepts part of the virtual photon flux, thus makes a shadow on the following balls. The shadow of one ball has a longitudinal extension $l_f \sim v\lambda/(1-v) \sim \gamma^2 v\lambda$. Beyond this region, called *formation zone*, the cloud of virtual photons of wavelength $\lambda$ is practically restored if there is no other piece of matter in the formation zone.
The shadow effect has been directly observed in diffraction radiation (19). In the case of mettalic balls it is included in the rescattering effects studied by García et al (20).

### 3.4 Application of Type-II PIGL to beam diagnostics

Type-II PIGL captures real as well as virtual photons : it acts both as a *near field* and a *far field* detector. Type-II PIGL can therefore be used for beam monitoring, but, like OTR and ODR, it is sensitive to backgrounds from distant radiation sources.
If the particle beam is ultrarelativistic, the quasi-real photons of the Coulomb field at impact parameter up to $b_{max} \sim \gamma\lambdabar$ can be captured. They give the logarithmic increase of (42) with $\gamma$ and a similar one in (37). They can degrade somewhat the resolution power of Type-II PIGL in transverse beam size measurements, but experience with OTR monitors shows that this effect is not drastic (21–24).

## 4. Conclusion

This chapter shows the various possibilities of optical fibers in charged particle beam physics. The phenomenon of light production by a particle passing *near* the fiber, which has some theoretical interest, has not been tested experimentally up to now.
The flexibility of a fiber is an advantage over the delicate optics of OTR and ODR. A narrow fiber has less effects on the beam emittance than the metallic targets used in OTR and ODR.
Much work remains to be done before using the Type-I and Type-II PIGL : find the most convenient wavelength domain (infra-red, visible or ultraviolet) and fiber diameter ; determine the ball-to-fiber transmission coefficients $p_f$, etc.
The fiber has to be monomode if one wants to emphasize the interference effects. However it would be interesting to make simulations and experiments of the excitations of modes higher than $HE_{11}$. In particular the $M = 0$ TM mode has a significant $E_z$ component, therefore may be excited at small crossing angle as much as the $HE_{11}$ mode.

## 5. References


[1] X. Artru and C. Ray, Ed. S.B. Dabagov, Proc. SPIE, Vol. 6634 (2007) ; arXiv: hep-ph/0610129.
[2] X. Artru and C. Ray, Nucl. Inst. Meth. in Phys. Research B 266 (2008) 3725.
[3] P. Coyle et al, Nucl. Instr. Methods in Phys. Research A 343 (1994) 292.
[4] A. Contin, R. De Salvo, P. Gorodetzky, J.M. Helleboid, K.F. Johnson, P. Juillot, D. Lazic, M. Lundin, Nucl. Instr. Methods in Phys. Research A 367 (1994) 271.
[5] E. Janata, Nucl. Inst. Meth. in Phys. Research A493 (2002) 1.
[6] H. Henschel, M. Körfer, J. Kuhnhenn, U. Weinand and F. Wulf, Nucl. Instr. Methods in Phys. Research A 526 (2004) 537.
[7] R.J. Glauber and M. Lewenstein, Phys. Rev. A 43, (1991) 467.
[8] T. Okoshi, *Optical Fibers*, Academic Press, 1982
[9] L.S. Bogdankevich, B.M. Bolotovskii, J. Exp. Theoret. Phys. 32, 1421 [Sov. Phys. JETP 5, 1157] (1957).
[10] N.K. Zhevago, V.I. Glebov, Nucl. Instr. Methods A 331 (1993) 592; Zh. Exp. Teor. Fiz. 111 (1997) 466.
[11] A. Tomita and R. Chiao, Phys. Rev. Lett. 57 (1986) 937.
[12] J.D. Jackson, *Classical Electrodynamics*, John Wiley & Sons, Inc. (1962).
[13] W.K.H. Panofsky and M. Phillips, *Classical Electricity and Magnetism*, Addison-Wesley, Inc. (1962).
[14] T. Kalkbrenner, M. Ramstein, J. Mlynek and V. Sandoghdar, J. Microsc. 202 (2001) 72.
[15] P. Anger, P. Bharadwaj and L. Novotny, Phys. Rev. Lett. 96 (2006) 113002.
[16] N. K. Zhevago, Europhys. Lett. 15 (1991) 277.
[17] F.J. García de Abajo and A. Howie, Phys. Rev. Lett. 80 (1998) 5180.
[18] N. Yamamoto, K. Araya and F.J. García de Abajo, Phys. Rev. B 64 (2001) 205419.
[19] G. Naumenko, X. Artru, A. Potylitsyn, Y. Popov, L. Sukhikh and M. Shevelev, J. Phys.: Conf. Ser. 236 (2010) 012004.
[20] F.J. García de Abajo, Phys. Rev. Lett. 82 (1999) 2776; F.J. García de Abajo, Phys. Rev. E 61 (2000) 5743.
[21] D.W. Rule and R.B. Fiorito, AIP Conference Proceedings 229 (1991) 315.
[22] V.A. Lebedev, Nucl. Instr. Methods A 372 (1996) 344.

[23] J.-C. Denard, P. Piot, K. Capek, E. Feldl, Proc. of the 1997 Particle Accelerator Conference.
[24] X. Artru, R. Chehab, K. Honkavaara, A. Variola, Nucl. Instr. Methods B 145 (1998) 160.

22

# Non Linear Optic in Fiber Bragg Grating

Toto Saktioto[1,2] and Jalil Ali[1]
[1]*Advanced Photonics and Science Institute, Faculty of Science, Universiti Teknologi Malaysia, Skudai, Johor;*
[2]*University of Riau, Pekanbaru, Riau;*
[1]*Malaysia*
[2]*Indonesia*

## 1. Introduction

A Fiber Bragg Grating (FBG) is a periodic variation of the refractive index of the core in the fiber optic along the length of the fiber. The principal property of FBGs is that they reflect light in a narrow bandwidth that is centered about the Bragg wavelength, $\lambda_B$ (A. Orthonos and K. Kalli, 1999). FBGs are simple intrinsic devices that are made in the fibre core by imaging an interference pattern through the side of the fibre. They are used as flexible and low cost in-line components to manipulate any part of the optical transmission and reflection spectrum. FBG is formed by the periodic variations of the refractive index in the fiber core. Several techniques have been established to inscribe them with UV-lasers. However, these technologies are limited to photosensitive fiber core material, which are unsuitable for high power applications. Only recently modifications have been demonstrated in a non photosensitive fiber but at the expense of longer exposure times (K. W. Chow *et al.*, 2008). FBGs have all the advantages of an optical fibre, such as electrically passive operation, lightweight, high sensitivity with also unique features for self-referencing and multiplexing capabilities. This gives them a distinct edge over conventional devices (Nahar Singh *et. al*, 2006, Govind P. Agrawal 2002). Therefore, FBGs in optical fibers have a wide range of applications, such as for sensors, dispersion compensators, optical fibre filters, and all-optical switching and routing (T. Sun *et. al,*2002). An UV laser source is used to form FBG's in fiber optics either through internal writing or external writing technique (A. Orthonos *et al*, 1995). The novel idea of using soliton is introduced for FBG.

Solitons are particle-like waves that propagate in dispersive or absorptive media without changing their pulse shapes and can survive after collisions. Various types of optical soliton phenomenon have been studied extensively in the area of nonlinear optical physics. These include the nonlinear Schröedinger solitons in dispersive optical fibers, spatial and vortex solitons in photorefractive material, waveguides and cavity solitons in resonators (Y. S. Kivshar and G. P. Agrawal, 2003).

The principal objective of this topic is to investigate the soliton in FBG showing potential energy. The theory involved in the modelling of soliton is based on the coupled-mode theory including the Kerr nonlinearity, group velocity dispersion (GVD) and self phase modulation (SPM) The motion of a particle moving in FBG represents the pulse propagation in the grating structure of fiber optics exhibiting the existence of optical fiber. In order to describe the photon motion, the function of potential energy is depicted.

## 2. Properties of Fiber Bragg Grating

A simple form of Fiber Bragg Grating (FBG) in Figure 1 consists of a periodic modulation of the refractive index in the core of a single-mode optical fiber (Phing, H.S.*et al*, 2008) . These types of uniform fiber gratings, where the phase fronts are perpendicular to the fiber longitudinal axis with grating planes have a constant grating period, Λ.

$$\lambda_B = 2N_{eff}\Lambda \tag{1}$$

where Λ is the spatial period (or pitch) of the periodic variation and $N_{eff}$ is the effective index for light propagating in a single mode fiber.
The Bragg condition is a manifestation of both energy and momentum conservation. Energy conservation requires that the frequency of the incident radiation and the reflected radiation is the same, means

$$\hbar\omega_f = \hbar\omega_i \tag{2}$$

Momentum conservation requires that the incident wave vector, $k_i$, plus the grating wave vector, $K$, equal the wave vector of the scattered radiation, $k_f$. This leads to an equation in which,

$$k_i + K = k_f \tag{3}$$

where the grating wave vector, K, has a direction normal to the grating planes with a magnitude $\frac{2\pi}{\Lambda}$. The diffracted wave vector is equal in magnitude, but opposite in direction to the incident wave vector.

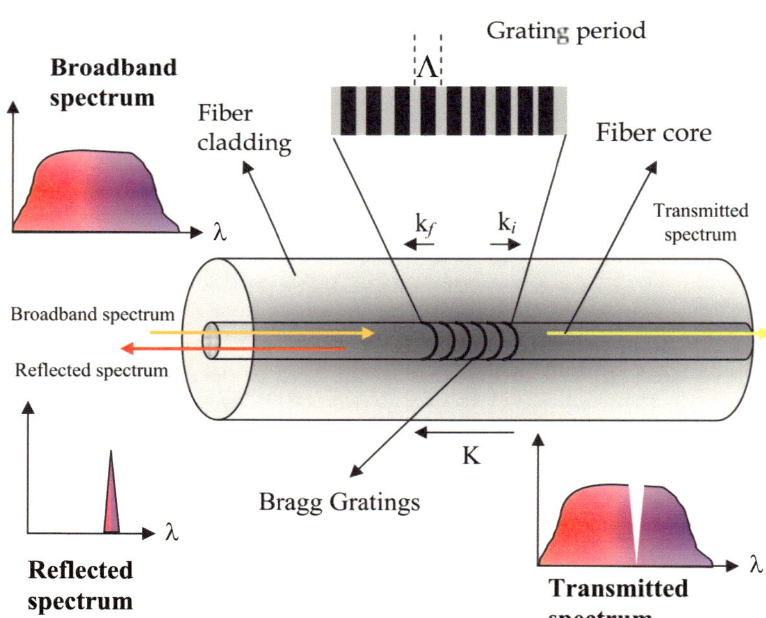


Fig. 1. A basic diagram of Fiber Bragg Grating (A. Orthonos and K. Kalli, 1999).

Hence, the momentum conservation becomes

$$2\left(\frac{2\pi n_{eff}}{\lambda_B}\right)=\frac{2\pi}{\Lambda} \tag{4}$$

Equation (4) simplifies to the first-order Bragg condition

$$\lambda_B = 2n_{eff}\Lambda \tag{5}$$

$\lambda_B$ is the Bragg wavelength. This is the free space center wavelength of the input light that will be back-reflected from the Bragg grating region). $n_{eff}$ is the effective refractive index of the fiber core at free space center wavelength.

## 3. Optical soliton in FBG

The existence of optical solitons in lossless fiber was theoretically demonstrated first by Hasegawa and Tappert in 1973. Bright and dark solitons appear in anomalous and normal dispersion regime respectively. The existence of an optical soliton in fibers is made by deriving the evolution equation for the complex light wave envelope via the slowly varying Fourier amplitude by retaining the lowest order of the group dispersion. This lower order is taken from the variation of the group velocity as a function of light frequency and the nonlinearity. For a glass fiber it is cubic and originates from the Kerr effect (K. Porsezian, 2007). The one soliton solution of the nonlinear Schrödinger equation is given by a sech $T$ function which is characterized by four parameters, the amplitude, the pulsewidth, the frequency, time position and the phase. In particular, the soliton speed is a parameter independent of the amplitude unlike the case of Kortweg de Vries (KdV) soliton. This is important fact in the use of optical soliton as a digital signal. Originally in 1980, L. F. Mollenauer and his colleagues at Bell Laboratories succeeded in observing optical soliton in fiber. During the 1990's, many other kinds of optical soliton were discovered such as spatiotemporal solitons and quadratic solitons (Y. S. Kivshar and G. P. Agrawal, 2003).

Soliton in fibers is formed after the exact balancing of group velocity dispersion (GVD) arising as a combination of material and waveguide dispersion with that of the self-phase modulation (SPM) due to the Kerr nonlinearity. Due to this, a similar soliton-type pulse formation in Fiber Bragg Grating where the strong grating-induced dispersion is exactly counterbalanced by the Kerr nonlinearity through the SPM and cross-phase modulation (CPM) effects. As a result, there is a formation of slowly travelling localized envelope in FBG structures known as Bragg grating solitons. They are often referred to as gap solitons if their spectra lies well within the frequency of the photonic bandgap if the frequency of incident pulse matches the Bragg frequency. Thus based on the pulse spectrum with respect to the photonic bandgap, solitons in FBG can be classified into two categories as either Bragg grating solitons or gap solitons.

There are basically two conditions that one can determine the formation of solitons in FBG. First is based on high intensity pulse propagation in which the refractive index modulation is weak in FBG where nonlinear coupled-mode (NCM) equations are used to describe a coupling between forward and backward propagating modes. The other conditions deals with the low intensity pulse propagation in FBG where the peak intensity of the pulse is assumed to be small enough so that the nonlinear index change, $n_2I$ is much smaller than the maximum value of $\delta n$. Under the low intensity limit, the NCM equations can be reduced to the nonlinear Schrödinger equation by using multiple scale analysis.

## 4. Coupled-mode theory for FBG

Several methods have been adopted to study and analyze the reflection and transmission properties of FBG (R. Kasyhap, 2004, M. Liu and P. Shum, 2006). The pulse propagation in FBG and its effect on Bragg grating affect the wave propagation in optical fibers can be examined using the coupled-mode theory (CMT) and Bloch wave technique. However, in this chapter we take CMT only into consideration.

One of the standard methods of analysis of FBG is using the coupled-mode theory (K. Thyagarajan and A. Ghatak and, 2007). According to this theory, the total field at any value of $z$ can be written as a superposition of the two interacting modes and the coupling process results in a $z$-dependent amplitude of the two coupled modes. It is assumed that any point along the grating within the single-mode fiber has a forward propagating mode and a backward propagating mode. Thus the total field within the core of the fiber is given by

$$\Psi(x,y,z,t) = A(z)\psi(x,y)e^{i(\omega t-\beta z)} + B(z)\psi(x,y)e^{i(\omega t+\beta z)} \tag{6}$$

where $x$, $y$, $z$ refers to space while $t$ refers to variation of time, $A(z)$ and $B(z)$ represents the amplitudes of the forward and backward propagating modes (assumed to be the same order mode), $\psi(x,y)$ represents the transverse modal field distribution, $\omega$ refers to frequency and $\beta$ is the propagation constant of the mode. The total field given by Equation (6) has to satisfy the wave equation given by

$$\nabla^2\Psi + k_0^2 n_g^2(x,y,z)\Psi = 0 \tag{7}$$

where $n_g^2(x,y,z)$ represents the refractive index variation along the fiber. For an FBG it is given by

$$n_g^2(x,y,z) = n^2(x,y) + \Delta n^2(x,y)\sin(Kz) \tag{8}$$

where $K = 2\pi / \Lambda$ represents the spatial frequency of the grating and $\Delta n^2$ represents the index modulation of the grating. For a uniform grating $K$ is independent of z; when $K$ depends on z, such gratings are referred to as chirped gratings. However, now we further focused on uniform gratings.

Substituting Equation (6) and Equation (7) into Equation (8) and making some simplifying approximations, we can obtain the following coupled-mode equations:

$$\frac{dA}{dz} = \kappa B e^{i\Gamma z} \quad \text{and} \quad \frac{dB}{dz} = \kappa A e^{-i\Gamma z} \tag{9}$$

where $\Gamma = 2\beta - K$ and $\kappa$ represents the coupling coefficient given by

$$\kappa = \frac{\omega\varepsilon_0}{8}\iint \psi^* \Delta n^2(x,y)\psi\, dx dy \tag{10}$$

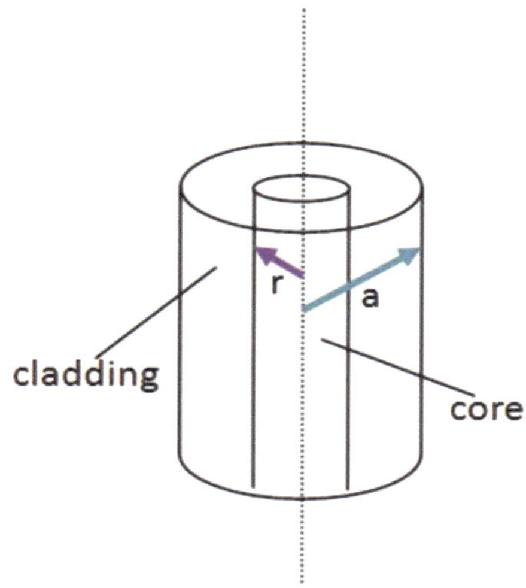


Fig. 2. Cross-section of an optical fiber with the corresponding refractive index profile (R. Kasyhap, 1999).

If the perturbation in the refractive index shown in Figure 2 is constant and finite only within the core of the fiber, then

$$\Delta n^2(x,y) = \Delta n^2, \quad r < a \qquad (11)$$

$$= 0, \quad r > a$$

and we obtain

$$\kappa \approx \frac{\pi \Delta n}{\lambda_B} l \qquad (12)$$

where $\lambda_B$ is the Bragg wavelength and

$$l = \frac{\int_0^a \psi^2 r dr}{\int_0^\infty \psi^2 r dr} \qquad (13)$$

The coupled-mode Equations (9) can be solved using the boundary conditions of

$$A\,(z = 0) = 1 \quad \text{and} \quad B\,(z = L) = 0 \qquad (14)$$

where $L$ is the length of the grating. Equation (14) implies that the incident wave has unit amplitude at z = 0 and the amplitude of the reflected wave at $z = L$ is zero because there is no reflected wave beyond $z = L$. We defined the reflectivity of the FBG by the ratio of the reflected power at z = 0 to the incident power at $z = 0$. Solving the coupled-mode equations and using the boundary conditions we obtain the reflectivity of the grating as follows:

$$R = \frac{\kappa^2 \sinh^2(\Omega L)}{\Omega^2 \cosh^2(\Omega L) + \frac{\Gamma^2}{4}\sinh^2(\Omega L)} \tag{15}$$

where

$$\Omega^2 = \kappa^2 - \frac{\Gamma^2}{4}$$

## 5. Pulse propagation in FBG

Wave propagation in a linear periodic medium has been studied extensively using coupled-mode theory. In the case of a dispersive nonlinear medium, the refractive index is given as

$$\bar{n}(\omega, z, I) = \bar{n}(\omega) + n_2 I + \delta n_g(z) \tag{16}$$

where $n_2$ is the Kerr coefficient and $\delta n_g(z)$ accounts for periodic index variation inside the grating. The coupled-mode theory can be generalized to include the nonlinear effects if the nonlinear index change, $n_2 I$ in Equation (2.11) is so small that it can be treated as a perturbation.
The starting point consists of solving Maxwell's equations with the refractive index given in Equation (16). When the nonlinear effects are relatively weak, we can work in the frequency domain and solve the Helmholtz equation,

$$\nabla^2 \tilde{E} + \tilde{n}^2(\omega, z)\omega^2 / c^2 \tilde{E} = 0 \tag{17}$$

The forward and backward propagating modes in FBG due to Bragg reflection can be described using CMT as been explained by Yariv in the distributed feedback structure (K. Senthilnathan, 2003). As usual, the governing equations for the pulse propagation in FBG are derived using Maxwell's equation. In this study the focus is on the frequency domain as the nonlinear effects are assumed to be relatively weak. It can easily be shown that Maxwell's equation are reduced to the following wave equation in the form

$$\frac{\partial^2 \vec{E}}{dz^2} - \frac{\varepsilon(z)}{c^2}\frac{\partial^2 E}{\partial t^2} = 0 \tag{18}$$

where perturbed permittivity, $\varepsilon(z) = \bar{n}^2 + \tilde{\varepsilon}(z)$, $\bar{n}^2$ is the spatial average of $\tilde{\varepsilon}(z)$, and $\bar{n}$ is the average refractive index of the medium. We consider the term $\tilde{\varepsilon}(z)$ with a period $\Lambda$ and define $k_0 = \pi / \Lambda$. Using the Fourier series, $\tilde{\varepsilon}(z)$ can be written as

$$\tilde{\varepsilon}(z) = 2\tilde{\varepsilon}\cos(2k_0 z) \tag{19}$$

This electric field inside the grating can be written as

$$\vec{E}(z,t) = \vec{E}_f(z,t)e^{+i(k_b z - \omega_d t)} + \vec{E}_b(z,t)e^{-i(k_a z - \omega_a t)} + \ldots \tag{20}$$

where $E_{f,b}(z,t)$ represents the forward and backward propagating waves, respectively, inside the FBG structure. Now, inserting Equation (19) and (20) into Equation (18) and considering that the fields $E_{f,b}(z,t)$ are varying slowly with respect to $\omega_0^{-1}$ in time and $k_0^{-1}$ in space, the resulting frequency domain coupled mode equations can be written as

$$i\frac{\partial \vec{E}_f}{\partial z} + i\frac{\bar{n}}{c}\frac{\partial \vec{E}_f}{\partial t} + \kappa \vec{E}_b = 0 \tag{21}$$

$$-i\frac{\partial \vec{E}_b}{\partial z} + i\frac{n}{c}\frac{\partial \vec{E}_b}{\partial t} + \kappa \vec{E}_f = 0$$

The value of κ represents the coupling between the forward and backward propagating waves in the FBG. The set of Equations (21) are called linear coupled-mode (LCM) equations in which the non-phase-matched terms have been neglected. The LCM equations assume slowly varying amplitudes rather than the electric field itself. Note that CMT is an approximate description that is valid for shallow gratings and for wavelength close to the Bragg resonance.

## 6. Potential energy distribution in FBG

In the presence of Kerr nonlinearity, using CMT, the NLCM equations can be written as

$$i\frac{\partial \vec{E}_f}{\partial z} + i\frac{\bar{n}}{c}\frac{\partial \vec{E}_f}{\partial t} + \kappa \vec{E}_b + \left(\Gamma_s\left|\vec{E}_f^2\right| + 2\Gamma_x\left|\vec{E}_b^2\right|\right)\vec{E}_f = 0$$

$$-i\frac{\partial \vec{E}_b}{\partial z} + i\frac{\bar{n}}{c}\frac{\partial \vec{E}_b}{\partial t} + \kappa \vec{E}_f + \left(\Gamma_s\left|\vec{E}_b^2\right| + 2\Gamma_x\left|\vec{E}_f^2\right|\right)\vec{E}_b = 0 \tag{22}$$

where $E_f$ and $E_b$ are the slowly varying amplitudes of forward and backward propagating waves, $\bar{n}$ is the average refractive index, and $\Gamma_s$ and $\Gamma_x$ are SPM and Cross-Phase modulation terms. In Equation (22) the material and waveguide dispersive effects are not included due to the dispersion arising from the periodic structures dominates the rest near Bragg resonance condition. Noted that the above NLCM equations are valid only for wavelengths close to the Bragg wavelength.

Now, by substituting the stationary solution to the above coupled-mode equations is by assuming

$$E_{(f,b)}(z,t) = e_{(f,b)}(z)e^{-i\hat{\delta}ct/\bar{n}} \tag{23}$$

where $\hat{\delta}$ is the detuning parameter. Using the stationary solution in Equation (22), we obtain

$$i\frac{de_f}{dz} + \hat{\delta}e_f + \kappa e_b + \left(\Gamma_s\left|e_f\right|^2 + 2\Gamma_x\left|e_b\right|^2\right)e_f = 0$$

$$-i\frac{de_b}{dz} + \hat{\delta}e_b + \kappa e_f + \left(\Gamma_s\left|e_b\right|^2 + 2\Gamma_x\left|e_f\right|^2\right)e_b = 0 \tag{24}$$

Equation (24) represents the time-independent light transmission through the grating structure, where $e_f$ and $e_b$ are the forward and backward propagating modes κ represents $n_{0k}$, $\left(n_{0k} = \dfrac{n_{01} - n_{02}}{\pi}\right)$ where $n_{01}$ is the core refractive index and $n_{02}$ is the cladding refractive respectively, $\Gamma_s$ represents Self Phase Modulation and $\Gamma_x$ represents Cross-phase modulation effects. This has been extensively investigated by many researchers. The NLCM equations are non-integrable in general. But in a few cases, NLCM equations have exact analytical solutions representing the solitary wave solutions. However, Christoudolides and Joseph have obtained the soliton solution to the NLCM equation, known as slow Bragg soliton, under the integrable massive Thirring model where the SPM and detuning parameter is set to zero. After using suitable transformation, it is used in nonlinear optics as a simple model to explain the self-induced transparency effect. Using the Stokes parameters they derived the relation of energy density for the stationary solution for the NLCM equation in terms of the Jacobi elliptic function.

There are some possible interesting soliton-like solutions apart from these stationary solutions. In the fiber Bragg grating, these soliton-like solution for the NLCM equations carry a lot of practical importance.

## 7. Solution of optical soliton using NLCM

Wave propagation in optical fibers is analyzed by solving Maxwell's Equation with appropriate boundary conditions. In the presence of Kerr nonlinearity, using the coupled-mode theory, the nonlinear coupled mode equation is defined under the absence of material and waveguide dispersive effects. The dispersion arising from the periodic structure dominates near Bragg resonance conditions and it is valid only for wavelengths close to the Bragg wavelength.

In order to explain the formation of Bragg soliton, consider the Stokes parameter since it will provide useful information about the total energy and energy difference between the forward and backward propagating modes. In this study, the following Stokes parameter are considered where

$$\begin{aligned} A_0 &= \left|e_f\right|^2 + \left|e_b\right|^2, \\ A_1 &= e_f e_b^* + e_f^* e_b, \\ A_2 &= i\left(e_f e_b^* - e_f^* e_b\right) \\ A_3 &= \left|e_f\right|^2 - \left|e_b\right|^2 \end{aligned} \quad \text{and} \tag{25}$$

with the constraint $A_0^2$ equals to the sum of $A_1^2 + A_2^2 + A_3^2$. In the FBG theory, the nonlinear coupled-mode (NLCM) equation requires that the total power $P_0 = A_3 = \left|e_f\right|^2 - \left|e_b\right|^2$ inside the grating is constant along the grating structures. Rewriting the NLCM equations in terms of Stokes parameter gives

$$\frac{dA_0}{dz} = -2\kappa A_2, \quad \frac{dA_1}{dz} = 2\hat{\delta} A_2 + 3\Gamma A_0 A_2{}'$$

$$\frac{dA_2}{dz} = -2\hat{\delta}A_1 - 2\kappa A_0 - 3\Gamma A_0 A_1, \quad \frac{dA_3}{dz} = 0 \tag{26}$$

In Equation (26), we drop the distinction between the SPM and cross modulation effects. Hence Equation (26) becomes $3\Gamma = 2\Gamma_x + \Gamma_s$ . It can be clearly shown that the total power, $P_0$ ($=A_3$) inside the grating and is found to be constant meaning it is conserved along the grating structure. In the derivation of the anharmonic oscillator type equation, it is necessary to use the conserved quantity. This is obtained in the form $\hat{\delta}A_0 + \frac{3}{4}\Gamma A_0^2 + \kappa A_1 = C$, where $C$ is the constant of integration and $\hat{\delta}$ is the detuning parameter. Equation (27) can further be simplified to (Yupapin, P.P. *et al*, 2010),

$$\frac{d^2 A_0}{dz^2} - \alpha A_0 + \beta A_0^2 + \gamma A_0^3 = 4\hat{\delta}C \tag{27}$$

where $\alpha = 2\left[2\hat{\delta}^2 - 2\kappa^2 - 3\Gamma C\right]$ , $\beta = 9\Gamma\hat{\delta}$ and $\gamma = \frac{9}{4}\Gamma^2$. Equation (27) contains all the physical parameter of the NLCM equation. Physically, $a$ represents the function of detuning parameters, and phase modulation factors (SPM and CPM). $\beta$ represents the function of phase modulation factors (SPM and CPM) and the detuning parameters. Lastly, $\gamma$ represents the phase modulation factor (SPM and CPM). In general, $a$, $\beta$ and $\gamma$ are the oscillation factors.

## 8. Potential energy distribution in FBG structures

In order to describe the motion of a particle moving within a classical anharmonic potential, we have the solution of Equation 28 in the form of

$$V(A_0) = -\alpha\frac{A_0^2}{2} + \beta\frac{A_0^3}{3} + \gamma\frac{A_0^4}{4} \tag{28}$$

It represents the potential energy distribution in a FBG structures while the light is propagating through the grating structures.

In Equation (28), $\beta$ is not considered due to power conservation along the propagating of this FBG structure. The qualitative aspects of the potential well will change if the nonlinearity parameter of the wave equation is varied.

Figure 3 depicts the double-well potential under Bragg resonance condition where $\beta = 0$, $\gamma = 0.23$ and $\alpha$ is varied from 0.1 to 1.0. Photon with power of less than the total power, $P_0$ will only travel inside the well unless their energy exceeds the energy level. This would allow the photon to move outside the well.

Figure 4 explains the optimized point for various values of $\alpha$. The graph clearly shows that the optimized points decreased exponentially when values of $\alpha$ are increased. However, when $\alpha >>1$, the trend of the curve is no longer valid since it turns into an almost linear relationship.

Figure 5 shows the motion of photon in double well potential under different values of gamma for the Bragg resonance condition of $\gamma$ from 0.13 to 0.53. Note that the increment of gamma which is between $0.53<\gamma<1$ will reduce the double well potential to a single well potential.

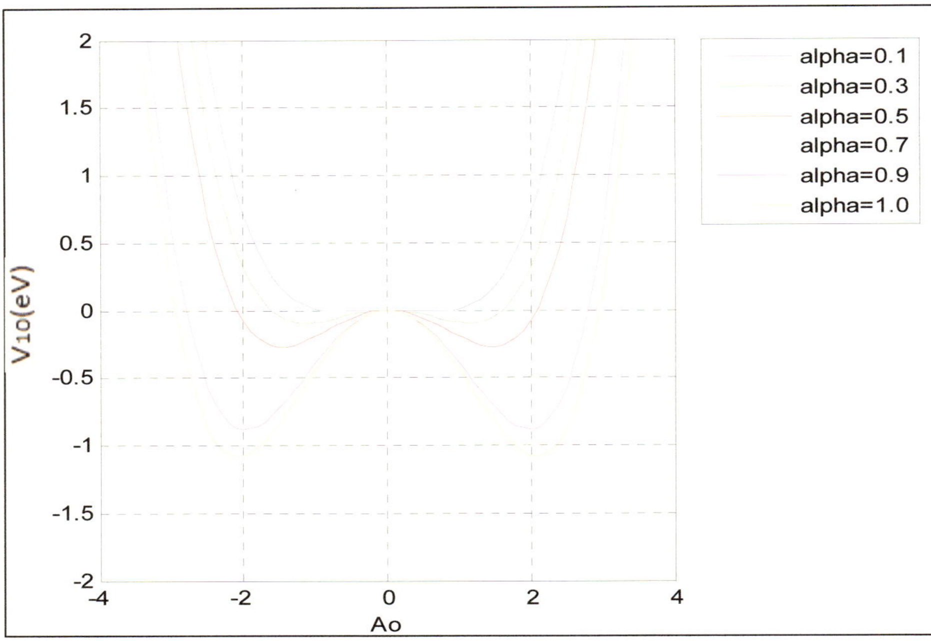


Fig. 3. The motion of photon in double well for different values of α.

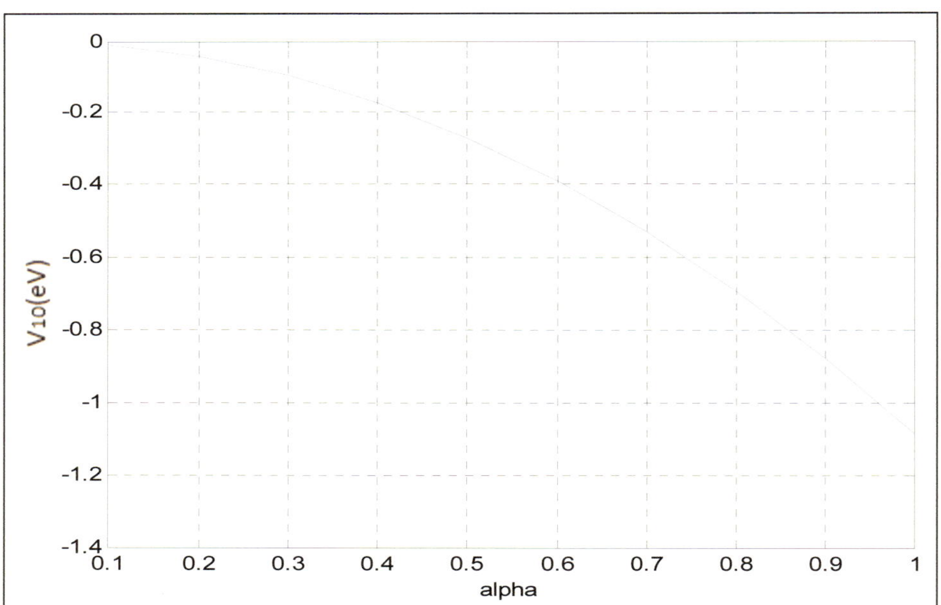


Fig. 4. The optimized point of the double well potential for different values of α.

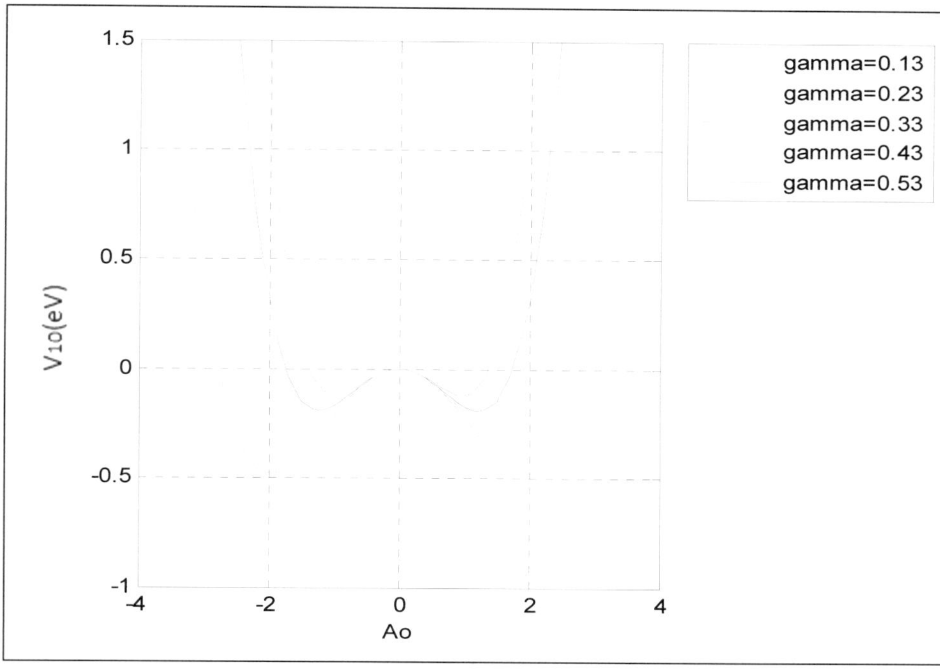


Fig. 5. Under Bragg resonance condition the system possesses double well potential for $\gamma$ 0.13 to 0.53.

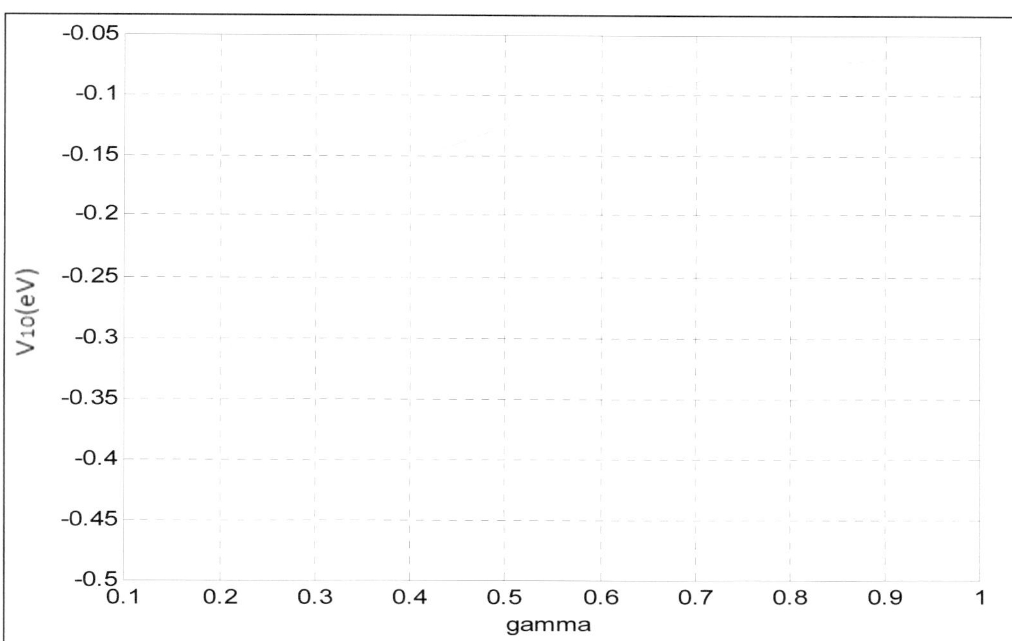


Fig. 6. The optimized point of the double well potential when $\gamma$ = 0.1 to 1.0.

Figure 6 describes the optimized point for varies of gamma, $\gamma$. Parametric variation of gamma produces a potential energy function which increases exponentially. However, when $\gamma >> 1$, a plateau is observed. This shows that it is not valid if $\gamma \to \infty$.

## 9. External disturbance of potential energy photon in FBG

By using Equation (28) where another term is considered, then we can have

$$\frac{d^2 A_0}{dz^2} - \alpha A_0 + \beta A_0^2 + \gamma A_0^3 + \theta A_0^4 = 4\hat{\delta}C \tag{29}$$

where $\alpha = 2\left[2\hat{\delta}^2 - 2\kappa^2 - 3\Gamma C\right]$ , $\beta = 9\Gamma\hat{\delta}$ , $\gamma = \frac{9}{4}\Gamma^2$ and $\theta = f(\theta)$. To simplify Equation (29), it is assumed the parameters of $\alpha$, $\beta$ and $\gamma$ is independent with respect to parameter $\theta$. Equation (29) contains all the physical parameter of the NLCM equation.

In order to describe the motion of a particle moving with the classic anharmonic potential, where the external disturbance is involved then we have the solution as follows,

$$V(A_0) = -\alpha\frac{A_0^2}{2} + \beta\frac{A_0^3}{3} + \gamma\frac{A_0^4}{4} + \theta\frac{A_0^5}{5} \tag{30}$$

It represents the potential energy distribution in the Fiber Bragg Grating structures.

Figure 7 depicts the motion of photon in a potential well which changes when few nonlinear parameters are taken into account as shown is Equation (30). Photon is trapped by the $\alpha$ parameter which is depicted by legend V. When $\alpha$ is too large, the potential well produces $A_0$ increases and have a wider of double well. The $\gamma$ parameter is shown by X legend. When $\gamma$ is large, the potential well produces $A_0$ increases. Suppose that the source is imposed to FBG than initial power is used to generate the particles. It shows that double well potential well is not symmetrical and the potential energy will decrease at certain region and is shown in Figure 7 in legend Y. The other effect is the disturbance at potential energy by legend Z where photon cannot be trapped symmetrically. It will tend to equilibrium but it is not stable where the photon leaves the potential curve as a losses.

In terms of parametric function, we can describe it as follows. The change in $\alpha$ will affect the dip of the potential well. If $\alpha$ is approximately too small, the shape of the potential well develop into a single potential well. The occurrence of $\beta$ effect in the motion of photon gives an effect to the negative region which means $A_0 < 0$. The effect of $\gamma$ also shows that the width of potential well will decrease if we increase the value of $\gamma$. Therefore if we increase the value of gamma, we can assume that the photon will be localized and can be trapped. In addition, another nonlinear factor $\theta$, it will change the shape of potential well rapidly. We could say that if we include the existence of $\theta$, the shape of potential well becomes chaotic. The photon does not only move within a certain region that is known as the potential well and moving freely.

Figure 8 shows the effect of external disturbance, $\theta$. It shows that by increasing the value of $\theta$, it will also affect the change in $\gamma$. In other words, the negative part of $Ao$ will be influenced it potential energy. The different values of $\gamma$ will produce different profiles. By simulating, we assumed that the increased of $\gamma$ value from 0.3 to 0.9, the curve will be positioned within the region **C**. The peak of $V$ for each $\gamma$ from 0.3 to 0.9 describes $\theta$ increases linearly and large gradient compare to the initial V. This represents that potential energy cannot maintain photon to be trapped and equilibrium state if $\gamma$ is relatively small.

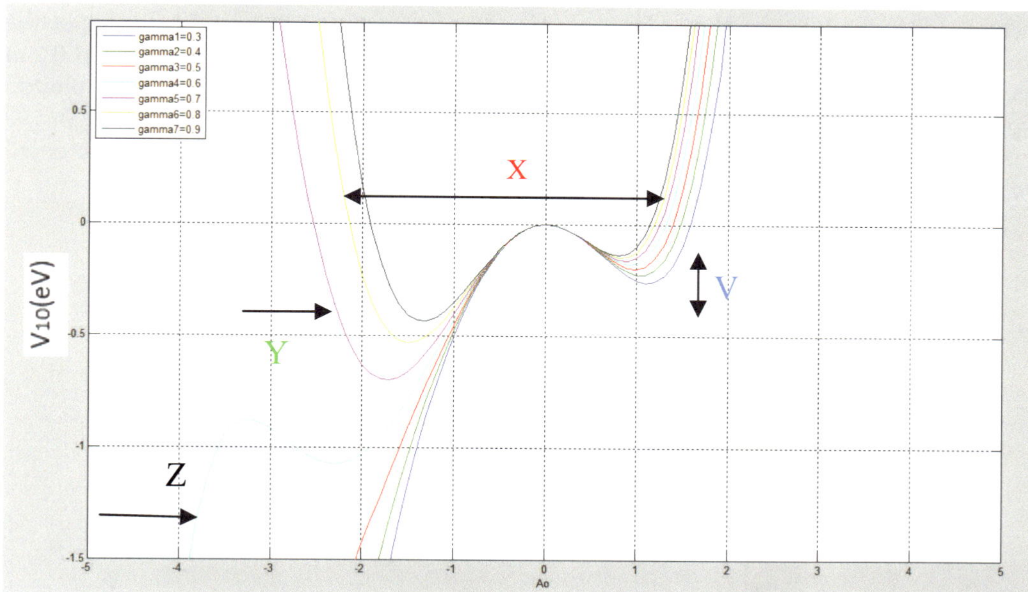


Fig. 7. The motion of photon in potential well for $a$ = 0.9, β = 0.3, $\theta$ = 0.09 and $\gamma$ is varies from 0.3 to 0.9.

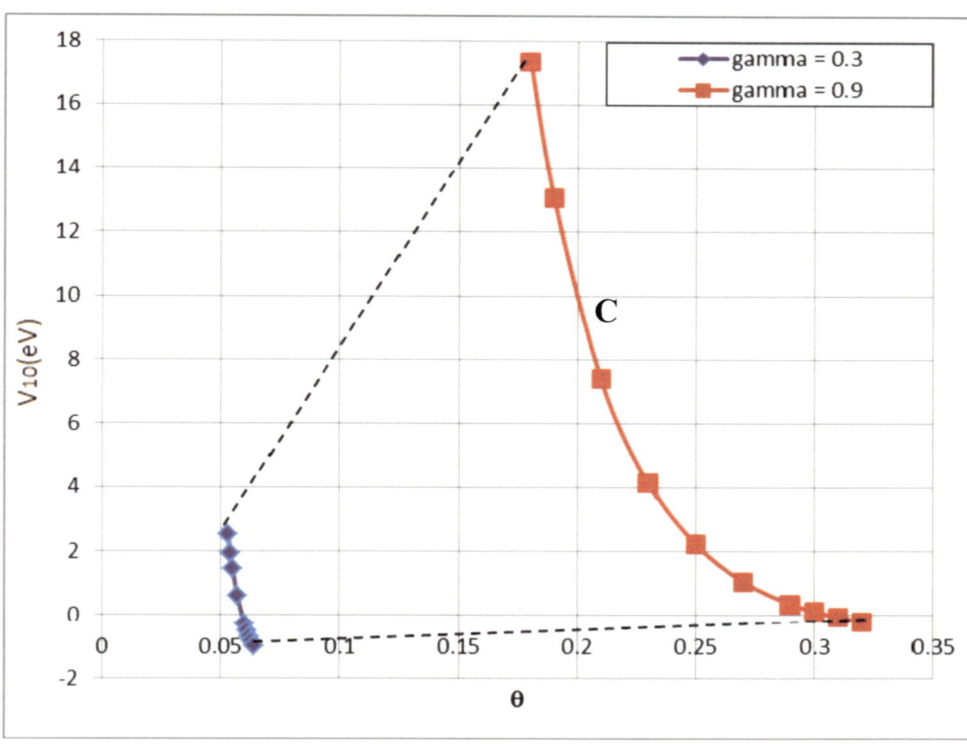


Fig. 8. The effect of theta, $\theta$ to γ and shape of the potential well of the photon.

In Figure 9 it can be shown that by increasing the value of $\beta$ the potential energy of the potential well will be reduced. The highest potential drop occurs within the range of $\beta$, 0.2 to 0.3. If the disturbance is large, it requires a high potential energy to maintain the photon especially for $\gamma$ = 0.7. In other words, increasing the $\gamma$ value will affect the shape of the potential well in terms of the potential energy. It will affect the equilibrium of the potential well and therefore the trapped photons are no longer being trapped or localized.

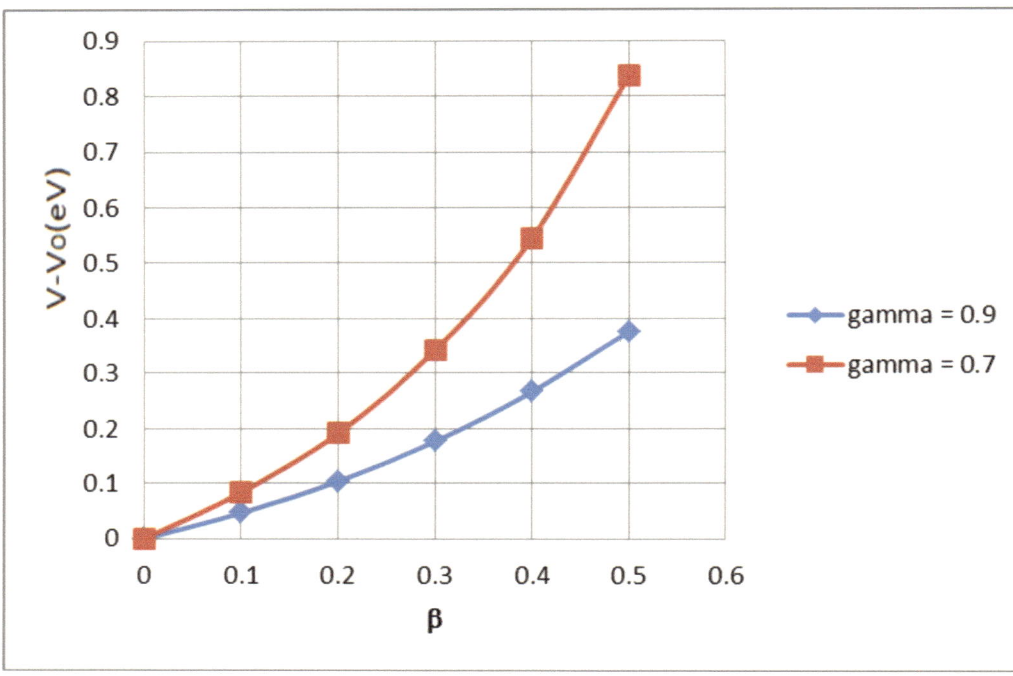


Fig. 9. The disturbance to the potential energy by $\beta$ factor.

## 10. Photon due to external energy perturbation in potential well

Figure 10 depicts the motion of photon in potential well which changes when few nonlinear parameters is take into account as described by Equation (30). There are theoretically some comments in this figure. Photon is trapped by $\alpha$ parameter which is depicted by legend V. When $\alpha$ is too large, the potential well produces $A_0$ increases and have a wider double well. The value of $\gamma$ parameter is shown by X legend. When $\gamma$ is large, the potential well produces

$A_0$ increases. Suppose that the source is imposed to FBG than initial power is used to generate the particles. It shows that double well potential well is not symmetric and potential energy will decrease at the certain region in legend Y. The other effect is the perturbation of potential energy by legend Z where photon cannot be trapped symmetrically. It will tend to equilibrium but it is not stable where it can go for losses.
The change in the parametric function can be easily described in terms of $\alpha$, $\beta$ and $\gamma$. The dip in the potential well will transform with a single potential well when $\alpha$ is extremely small. $\beta$ affects the photon motion which in turn will effect to the negative region of the potential well when $A_0 < 0$.The effect of $\gamma$ shows that the width of potential well will decrease if we increased the value of $\gamma$. The photon will be trapped when $\gamma$ is increased. The shape of the potential well can be controlled by a nonlinear factor $\theta$. The changes in the value of $\theta$ lead to a chaotic behaviour of the potential well. Under these conditions the photon can either move within certain specific regions or act as a free particle. Thus, Figure 10 illustrates the single perturbation as described by the nonlinear parameter, $\theta$.

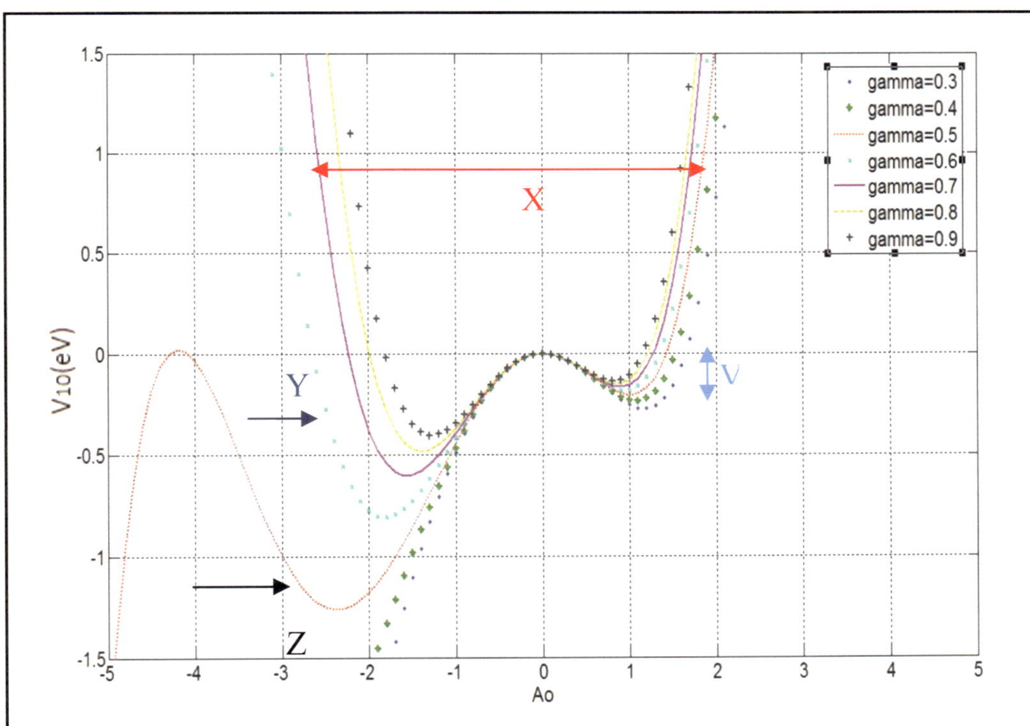


Fig. 10. The motion of photon in potential well for $\alpha = 0.9$, $\beta = 0.3$, $\theta = 0.09$ and $\gamma$ is varies from 0.3 to 0.9.

Now consider the case in Equation (27) with a set of constraints which is governed by $\phi_{(e)} = \sum_{n=0}^{\infty} A_0{}^n$ . The perturbation factor then is

$$\frac{d^2 A_0}{dz^2} = \phi_e{}'' \big|_{n=0} \tag{31}$$

If Equation (30) is accumulated using the external perturbation then

$$\phi'' \big|_{n=0} + \sum_{\substack{n=0 \\ m=1}}^{\infty} C_m A_0{}^n = \psi$$

where $\psi$ is a function of $f(\delta, C, C_{m,})$ and $C_m = [\alpha, \beta, \gamma, ...]$

The value of $m = 2n$ for $n = 1, 2, 3, ..., m = 2n + 1$ for $n = 0, 1, 2, ...$

$C$ is constant and $C = (C_1, C_2, C_3, ..., C_m)$. The value of $C$ is linear to $A_0$ but not to $V$. Equation (31) can then be modified by

$$V(A_0) = \sum_{\substack{m=1 \\ n=0}}^{\infty} C_m A_0^n \tag{32}$$

Equation (32) represents the complete potential energy distribution in the Fiber Bragg Grating structure. We believe at this juncture, the potential function is modified from Conti and Mills (C. Conti and S. Trillo, 2001). Using well-known Duffing oscillator type equation, analogically it is written as

$$\phi_e{}'' + \sum_{\substack{m=1 \\ n=0}}^{\infty} C_m A_0{}^n = 0 \tag{33}$$

For multi perturbation of nonlinear parameters, two major shapes will be simplified in the series term. The coupled mode equations are solved under different conditions when soliton is used for FBG writing. The cases examined are (i) when there is no energy disturbance (ii) the effect of potential energy disturbance factor (iii) potential energy with the highest disturbance factor.

When multi perturbations are considered then the photon will be trapped and untrapped for various conditions. As depicted in Figure 11, it explains the extrapolation of the graph if more factors of perturbation added into Equation (32). The addition of parametric factors by the higher odd number, Figure 11 (a) will allow the photon to move in a well, and Figure 11 (b) will lead the photon to be untrapped and higher even number. It is clearly shown in the graphs that as $n >> \infty$ , the value of $|A_0|$ will remain constant in the range of $-2 < A_0 < 2$. However, when the value of $V_{(0)}$ is equal to zero, there are many possibilities of $A_0$, meaning the exact value of intentsity, $A_0$ to trap the photon is difficult to determine in this condition. If the parametric factor considered is too large then we may conclude that the photon is in indifferent state part of the equlibrium.

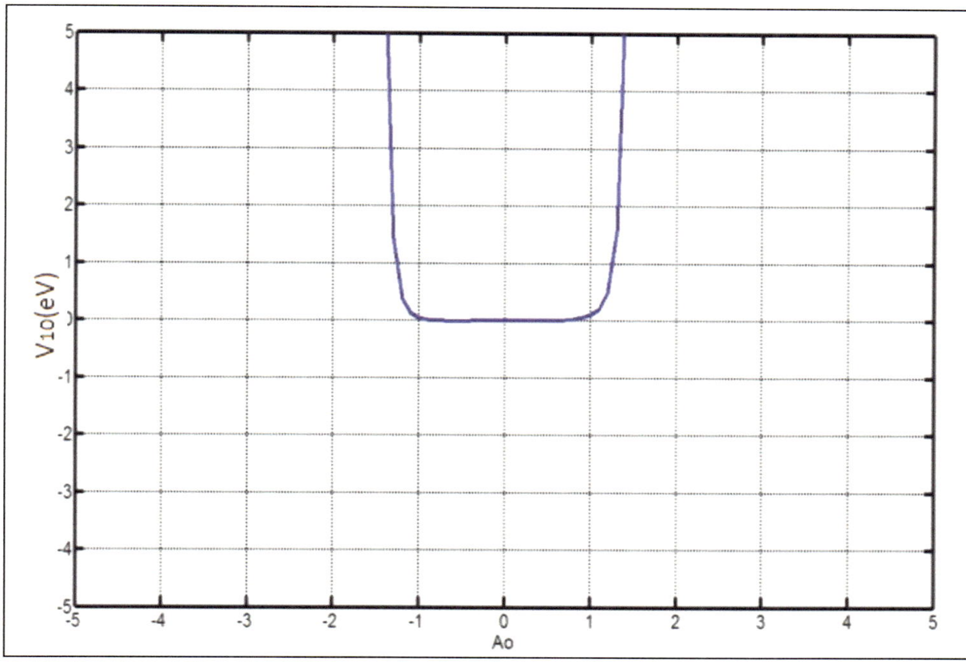


(a)

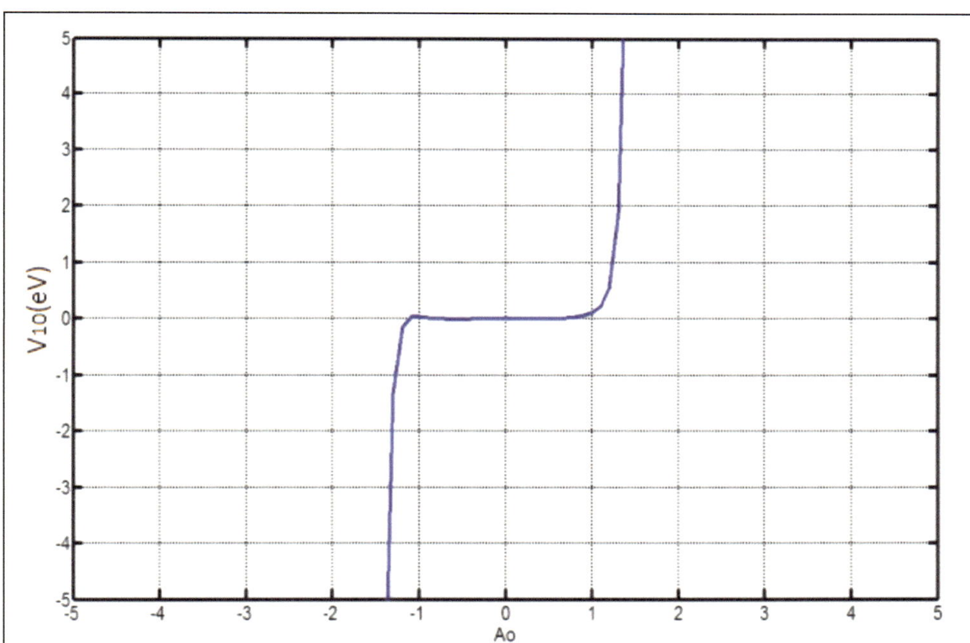


(b)

Fig. 11. The disturbance factor that affect the shape of the potential well of the motion of photon.

The stationary solutions of Equation (32) are applied neither for bright nor dark soliton solution since the dominant parametric factor in contributing $A_0$ is unknown. However, from Equation (32) we have

$$A_0 = f(C_m, z) \tag{34}$$

Under these conditions, the frequencies with photonic band gap keep forming an envelope after the exact balancing at grating-induced dispersion with nonlinearity. Either decay or increase, the forward and backward waves are transferred by Bragg reflection process. The total energy of the system, potential energy function is equal to zero having multi perturbation which is $-1<A_0<1$ and if $V \rightarrow \infty$, $A_0 = 2$.

## 11. Conclusion

The novel idea of using soliton in Fiber Bragg Gratings (FBG) shows that the motion of a particle moving in FBG represents the pulse propagation in the grating structure of FBG. This indicates the existence of optical soliton. It is described in terms of the photon motion and as a function of potential energy. Results obtained show that photon can be trapped by nonlinear parameters of potential energy which are identified as $\alpha$, $\beta$, $\gamma$ and $\theta$.

In the first simulation results of nonlinear parametric studies of photon in a FBG, we have successfully shown that the changes of nonlinearity parameter will affect the motion in the potential well. This will influence the existence of Bragg soliton in a fiber Bragg grating. In the second simulation results, we have added new nonlinear parameter which is known as $\theta$. We have preset the value of $\theta$, $\alpha$ and $\beta$ and vary the value of $\gamma$ over certain range. From the results, it is depicted that the factor $\theta$ will affect the shapes of potential well. If the existence of $\theta$ is taken into account, the potential well profile becomes chaotic.

The simulation data are then expanded on the multi perturbation of potential energy photon in FBG. It shows that the change of $a$ affect the dip of the potential well. The occurrence of $\beta$ effect in the motion will affect the soliton propagation in the region for $A_0<0$. The effect of $\gamma$ shows that the width of potential well will decreased if the value of $\gamma$ is increased. However, another nonlinear factor, $\theta$ will turns the shape of potential well rapidly which necessities the multi perturbation studies. When multi perturbations are considered, the photon will be trapped and entrapped under various conditions. From this, we may conclude the addition of nonlinear parametric factors by the higher odd number will allow the photon to move in a well instead to be entrapped with the higher even number. It is found from this study that the potential well under Bragg resonance condition is not symmetrical and conserved. The higher perturbation series representing the potential well is much indifferent of the equilibrium in both odd and even nonlinear parametric factor of $n$.

As a conclusion, these studies have successfully shown that it is plausible to use soliton for FBG writing and the soliton can be controlled by manipulating the parametric effects which are $\alpha$, $\beta$, $\gamma$ and $\theta$. The model developed in this topic can be further extended by optimizing the nonlinear parameters in terms of the potential energy, soliton trapping and its applications as optical tweezers. The model can be tested by developing compact miniature FBG inscribing system using laser diode.

## 12. Acknowledgment

We would like to thank the Institute of Advanced Photonic Science, Faculty of Science, Universiti Teknologi Malaysia (UTM) and Physics Department, Faculty of Math and Natural Sciences, University of Riau, Pekanbaru, Indoneseia for generous support in this research.

## 13. References

Andreas Orthonos and Kyriacos Kalli. (1999)."Fiber Bragg Gratings - Telecommunications and Sensing". Artech House Boston, London.

Andreas Orthonos and Xavier Lee. "Novel and Improved Methods of Writing Bragg Gratings with Phase Masks". IEEE Photonics Technology Letters, Vol. 7, No. 10, October 1995.

C. Conti and S. Trillo, "Bifurcation of gap solitons through catastrophe theory". Phys. Rev. E 64 (2001), 036617.

Govind P. Agrawal. (2002). "Fiber Optic Communication System". Wiley-Interscience, U.S.A.

K. Senthilnathan et al. "Grating solitons near the photonic bandgap of a fiber Bragg grating," ScienceDirect, Chaos, Solitons and Fractals 33 (2007) 523-531.

K. Thyagarajan and Ajoy Ghatak. (2007). "Fiber Optic Essentials". Wiley-Interscience, Canada.

K. W. Chow, Ilya M. Merhasin, Boris A. Malomed, K. Nakkeeran, K. Senthilnathan and P. K. A. Wai. "Periodic waves in fiber Bragg gratings". *Physical Review E*, 77, 026602 (2008).

M. Liu and P. Shum. "Simulation of soliton propagation in a directional coupler," Springer, Optical and Quantum Electronics (2006) 38:1159-1165.

Nahar Singh, Subhash C. Jain, Vandana Mishra, G. C. Poddar, Palvinder Kaur, Himani Singla, A. K. Aggrawal and Pawan Kapur. "Fibre Bragg grating - based sensing device for petrol leak detection". *Current Science*. Vol. 9, No. 2, January 2006.

Phing, H.S., Ali, J., Rahman, R.A., Saktioto. "Growth dynamics and characteristics of fabricated Fiber Bragg Grating using phase mask method". *Microelectronics Journal*. (2008).

R. Kashyap, *Fiber Bragg Gratings* (Academic Press, San Diego, 1999)

T. Sun, S. Pal, J. Mandal, K. T. V. Grattan. "Fibre Bragg grating fabrication uysing fluoride excimer laser for sensing and communication applications". *Central Laser Facility Annual Report 2001/2002*.

Yuri S. Kivshar and Govind P. Agrawal. (2003). "Optical Solitons: From Fiber to Photonics Crystal". Academic Press, U.S.A.

Yupapin, P.P., Saktioto, T., Ali, J., Photon trapping model within a fiber bragg grating for dynamic optical tweezers use, *Microwave and Optical Technology Letters* 52 (4), pp. 959-961.(2010)

# 23

# Optical Fibers and Optical Fiber Sensors Used in Radiation Monitoring

Dan Sporea[1], Adelina Sporea[1],
Sinead O'Keeffe[2], Denis McCarthy[2] and Elfed Lewis[2]
[1]*National Institute for Laser, Plasma and Radiation Physics, Laser Metrology Laboratory,*
[2]*Optical Fibre Sensors Research Centre, University of Limerick*
[1]*Romania,*
[2]*Ireland*

## 1. Introduction

By their very nature, optical fibers and, by extension, intrinsic and extrinsic optical fiber-based sensors are promising devices to be used in very different and complex environments considering their characteristics such as: capabilities to work under strong electromagnetic fields; possibility to carry multiplexed signals (time, wavelength multiplexing); small size and low mass; ability to handle multi-parameter measurements in distributed configuration; possibility to monitor sites far away from the controller; their availability to be incorporated into the monitored structure; wide bandwidth for communication applications. In the case of the optical fibers, the possibility to be incorporated into various types of sensors and actuators, free of additional hazards (i.e. fire, explosion), made them promising candidates to operate in special or adverse conditions as those required by space or terrestrial applications (spacecraft on board instrumentation, nuclear facilities, future fusion installations, medical treatment and diagnostics premises, medical equipment sterilization). Major advantages to be considered in using optical fibers/optical fiber sensors for radiation detection and monitoring refer to: real-time interrogation capabilities, possibility to design spatially resolved solutions (the capability to build array detectors), in-vivo investigations (i.e. inside the body measurements).

As information on the behavior and operation of optical fibers/optical fiber sensors under irradiation conditions are scattered over a great variety of journal papers and conference contributions dealing with many different fields (nuclear science and engineering; measurement science; material science; radioprotection; nuclear medicine and radiology; sensor design; radiation dosimetry; fusion installations concepts; particle accelerators; astrophysics and space science; defense and security; lasers, optics, optical fibers and optoelectronics; physics and applied physics; scientific instrumentation; radiation effects) we decided to design this book chapter as a comprehensive review on the subject. The chapter opens with some general considerations on the radiation–matter interaction, and continues with a review of irradiation effects on different types of optical fibers (silica optical fibers, plastic optical fibers, special optical fibers), effects which can be considered when radiation sensors are developed. The next issue addressed refers to environments where optical fibers/optical fiber sensors are employed for radiation monitoring/

dosimetry. The main part of the chapter is dedicated to the presentation of major proposed designs for intrinsic and extrinsic optical fiber sensors for radiation measurements. The experimental set-ups and irradiation conditions we used for the assessment of irradiation effects on various optical fibers are introduced and samples from our results illustrate the possible use of such optical fibers in radiation monitoring/dosimetry. The Romanian team focused its work on tests of silica and sapphire optical fibers subjected to different irradiation conditions (alpha particles, beta, gamma and X rays, neutron), while the Limerick group investigated radiation detectors based on plastic optical fibers, under gamma and X-rays irradiation.

Radiation induced effects in materials and devices are evaluated based on the energy losses resulting upon the interaction between highly energetic radiation and matter (Wrobel, 2005). Losses associated to these interactions can imply atoms ionization or can induce non-ionizing effects, such as changes of the vibrational/rotational states of molecules, atoms vibrations or atoms displacement. Generically, the term highly energetic radiation covers: a) uncharged particles - photons (gamma-rays, X-rays); b) charged light (electrons, positrons) and heavy (different types of ions) particles; c) nucleus constituents (neutrons, protons). Accelerated alpha particles (two protons and two neutrons bounded) or electrons, as well as high energy neutrons or photons can produce an atom's ionization. Depending on their energy, charge particles and nucleons can lead to either elastic or non-elastic interactions. On the other hand, lower energy photons can contribute to phenomena such as Rayleigh scattering (elastic scattering of photons by atoms or molecules), Compton scattering (the decrease of a X-ray or gamma-ray energy as it undergoes a non-elastic interaction with matter, the lost energy being transferred to a scattering electron as part of a ionizing process), photo-electric effect (the emission of electrons by a solid, liquid or gas upon the absorption of photons in that material) or pair production (the generation of an electron-positron pair as photons, having sufficient energy, interact with a nucleus). The occurrence of a particular phenomenon depends on the energy spectrum of the incident photons and the interaction cross sections (depending on the material involved). Light charged particles (i.e. electrons) when travelling inside a material are deflected by the surrounding atoms and lose energy, which is converted to photon energy, so, a continuum spectrum electromagnetic radiation is generated (Bremsstrahlung or deceleration radiation). Cerenkov radiation (a continuum spectrum optical radiation having a higher intensity at the UV wavelength end of the optical spectral range) is produced when a charged particle (i.e. electron) propagates within a dielectric medium with a speed higher than the phase velocity of the light in that medium. The molecules of the medium are polarized by the travelling particles and, returning to the ground state, emit visible optical radiation.

## 2. Optical fibers performances under irradiation

Exposed to ionizing radiation, silica optical fibers exhibit effects such as: radiation induced absorption (RIA), radiation induced luminescence (RIL), increase of the optical radiation scattering as it propagates over the fiber length, thermoluminescence, change of the waveguide refractive index. RIA and RIL effects contribute, generally speaking, to the degradation of the signal-to-noise ratio of the signal transmitted over the optical fiber guide. Attempts were made to reformulate the problem and to use these effects as a measure of the dose rate/total dose of the radiation to which the optical fiber is exposed.

In assessing the behavior of silica optical fibers under irradiation one has to consider several premises which have a major influence on the irradiation induced absorption, and hence on a would-be application of such optical fibers to radiation dosimetry. The RIA is affected by:

1. the manufacturing conditions, the parameters related to the technology used in producing the optical fiber: the deposition conditions, the draw process characteristics - draw speed, fiber drawing tension, the preform deposition temperature, oxygen-to-reagent ratio (02/R) used during core and clad deposition (Friebele, 1991; Girard et al., 2006; Hanafusa et al., 1986);
2. the existence, prior to the irradiation, of some precursors (Miniscalco et al., 1986);
3. the dopants present in the optical fiber core or cladding (pure silica, or doped with Ce, Er, Ge, F, N, P, Yb, high-OH, low-OH, high-Cl, low-Cl, H2-loading), (Arvidsson et al., 2009; Berghmans, 2006; Berghmans et al., 2008; Bisutti et al., 2007; Brichard & Fernandez Fernandez, 2005; Friebele, 1991; Girard et al., 2004a; Girard et al., 2004b; Girard et al., 2008; Griscom et al., 1996; Henschel et al., 1992; Kuyt et al., 2006; Lu et al., 1999; Mady et al., 2010; Paul et al., 2009; Regnier et al., 2007; Vedda et al., 2004; Wijnands et al., 2007), in some situations such ingredients contribute to the radiation hardening (Brichard et al., 2004; Brichard & Fernandez Fernandez, 2005; Girard et al., 2009);
4. the residual substances remaining after the manufacturing process, as for example chlorine having an associated color center in the UV spectral range, which can extend into the visible (Girard et al., 2008; Girard & Marcandella, 2010);
5. the type of radiation to which the optical fiber is subjected (Arvidsson et al., 2009; Bisutti et al., 2007; Brichard et al., 2001; Calderón et al., 2006; Girard et al., 2004a; Girard et al., 2004b; Girard et al., 2006; Girard et al., 2008; Merlo & Cankoçak, 2006; Toh et al., 2004; Yaakob et al., 2011);
6. the irradiation conditions: total dose, dose-rate, steady-state, pulsed, cyclic (Arvidsson et al., 2009; Berghmans, 2006; Berghmans et al., 2008; Kuhnhenn, 2005; Lu et al., 1999; Paul et al., 2009 ; Thériault, 2006; Wijnands et al., 2007);
7. the annealing effect and the temperature stress applied to the optical fiber during or post irradiation (Okamoto et al., 2004; Thériault, 2006; Toh et al., 2004; Tsuchiya et al., 2011);
8. the optical spectral band considered for the RIA evaluation (Regnier et al., 2007);
9. the optical fiber coating (Brichard, 2002; Gusarov et al., 2008b);
10. the photobleaching process and the level of the optical power injected into the optical fiber for measurement and its wavelength (Arvidsson et al., 2009; Miniscalco et al., 1986; Thériault, 2006; Treadaway et al.,1975).

The complexity of changes induced by highly energetic radiation in the optical transmission of silica optical fibers is governed by the optical fiber design, i.e. core-cladding dopants and diameter ratio (Deparis et al., 1997; Friebele, 1991; Girard et al., 2004a; Girard et al., 2004b; Kuhnhenn, 2005; Paul et al., 2009), and by the multitude of color centers activated under the irradiation process, such as (Brichard & Fernandez Fernandez, 2005; Origlio, 2009):

1. Ge centers: GeE', Ge(1), Ge(2), GEC, GeX, Ge-NBOH (Non-Bridging Oxygen Hole) (Alessi et al., 2010 ; Bisutti et al., 2007; Girard et al., 2006 ; Girard et al., 2008; Lu et al., 1999; Wijnands et al., 2007) ;
2. P centers: P1, P2, P4, Phosphorus-Oxygen-Hole - POHC (Bisutti et al., 2007; Girard et al., 2006; Girard et al., 2011 ; Paul et al., 2009; Wijnands et al., 2007) ;
3. silica related paramagnetic centers (Non-Bridging Oxygen Hole - NBOHC, PerOxy-Radical - POR, SiE', Self-Trapped Hole - STH 1/2) or diamagnetic centres (Oxygen

Deficient Centre - ODC; Peroxy Linkage - POL), (Berghmans et al., 2008; Girard et al., 2008).

In order to distinguish the contribution of various color centers to the irradiation induced absorption in silica optical fibers, complementary investigations, apart from the off-line/on-line optical transmission measurements, were carried-out: Electron Paramagnetic Resonance - EPR (Girard et al., 2008; Radiation effects, 2007; Sporea et al., 2010a; Weeks & Sonder, 1963), luminescence (Girard et al., 2005; Girard et al., 2006; Sporea et al., 2010a; Miniscalco et al., 1986), thermoluminescence (Espinosa et al., 2006; Hashim et al., 2008; Mady et al., 2010; Sporea et al., 2010b; Yaakob et al., 2011), confocal microscopy luminescence and Raman analysis (Girard et al., 2008; Origlio, 2009), time resolved photoluminescence (Cannas et al., 2008). Based on the results derived from different sources, a decomposition of the optical absorption spectra by using Gaussian bands associated to known color centers can be performed (Girard et al., 2008; Girard et al., 2011; Sporea et al., 2010a).

Apart from the aspects previously described (increase of the optical absorption in irradiated optical fibers, post irradiation luminescence and thermoluminescence effects) other phenomena have to be considered as prospective candidates for the use of optical fibers in radiation monitoring:

1. generation of Cerenkov radiation when optical fibers are subjected to a charged particle flux;
2. radiation induced luminescence (RIL) in optical fibers under irradiation (Skuja et al., 1996).

Traveling in a dielectric, transparent media, a charged particle (i.e. electron), having a velocity greater than the optical radiation phase velocity in that medium, interact with medium molecules by polarizing them. As the molecules return very fast to their ground state an optical radiation is emitted (Cerenkov, 1958). The optical radiation has a continuum distribution over the UV-visible spectral range.

As the emitted radiation spectrum varies with $1/\lambda^3$, the spectrum of the Cerenkov radiation is predominantly in UV-blue. Investigations on the generation of Cerenkov radiation were carried out for both silica and plastic optical fibers, such as polymethylmethacrylate - PMMA (Intermite et al., 2009; Jang et al., 2010a). The efficiency of coupling to and detecting the Cerenkov radiation with an optical fiber is a function of the type of optical fiber (its refractive index), the particles energy, the irradiation geometry: angle of incidence of the particle beam on the optical fiber axis, the optical fiber NA, the distance between the particle trajectory and the optical fiber axis (Goettmann et al., 2005; Jang et al., 2010a; Jang et al., 2011a).

In the case of a PMMA fiber with $n$ = 1.49, for an electron beam energy of 6 MeV the maximum value for $\theta$ is $47.7^0$ (Jang et al., 2010a), while for a silica optical fiber ($n$ = 1.46) and a energy of the electron beam of 175 keV the Cerenkov radiation angle was evaluated to be about $46^0$ (Intermite et al., 2009).

Gamma-rays can lead to the generation of Cerenkov radiation only if a Compton converter target is placed between the gamma source and the optical fiber. Photons incident on the converter produce electrons through the Compton Effect, with electrons further used to generate Cerenkov radiation. The efficiency of the conversion from gamma-ray to optical radiation depends on (Pruett et al., 1984): the target material (the best one proved to be beryllium), its thickness, the distance between the target and the optical fiber, the type of optical fiber, the angle between the gamma-ray trajectory (as a collimating scheme was employed) and the optical fiber axis, the energy of the gamma-ray (investigations were done with a $^{60}Co$ source - the 1.17 and 1.33 MeV lines, and a $^{24}Na$ with its 1.38 and 2.76 MeV lines).

Optical materials, as per se or included in optical fibers, exhibit radioluminescence (radiation induced luminescence – RIL): visible radiation is emitted as long as the materials are exposed to ionizing ration. This signal is generally superposed over the Cerenkov spectrum, but can be distinguished by its narrow band emission spectra. The radioluminescent peaks observed, function of the material investigated, its dopants and the type of the irradiation, are located from UV - 185 nm, 250 nm, 285 nm, to visible - 420 nm, 450 nm, 696 nm (Toh et al., 2002; Treadaway et al., 1975; Yoshida et al., 2007). The superposition of the two spectra makes the discrimination of the luminescence signal difficult, and so, several methods to improve the S/N in detecting the radioluminescent signal against the Cerenkov radiation were developed: the gated detection of the luminescence (Justus, 2006; Plazas, 2005), different time frames or the different angular distribution of the two signals are used to detect the RIL (Akchurin et al., 2005); the subtraction method by employing a dummy optical fiber to record the Cerenkov radiation separately; an optical filter or a set-up based on a spectrometer (Archambault, 2005; Lee et al., 2007b; Jang et al., 2010a; Jang et al., 2011a).

An alternative to the generation of radioluminescence in optical materials is represented by the use of commercially available scintillating optical fibers as those delivered by Saint-Gobain Crystals and Detectors – France (Saint-Gobain, 2005) - core material: Polystyrene, core refractive index: 1.6, cladding material: acrylic, cladding refractive index: 1.49, NA: 0.58, scintillation efficiency: 2.4%, trapping efficiency: 3.4%, by Kuraray Co. (Japan) or specially designed ones, as Ce doped optical fibers (Chiodini et al., 2009).

Additional effects produced in silica optical fibers under irradiation refer to:

1. the change of the silica density observed either during gamma irradiation tests on distributed optical fiber sensors based on Brillouin scattering: a non-linear increase with the total dose of the measured frequency and Brillouin FWHM line width (Alasia et al., 2006), or under soft X-ray or reactor irradiation (Primak et al., 1964);
2. the modification of the refractive index of silica subjected to gamma-ray (Fernandez Fernandez et al., 2003).

In concluding this section on silica optical fibers, it is worth mentioning the behaviour of fiber Bragg gratings (FBGs) manufactured in silica optical fibers as they are exposed to irradiation, considering some authors suggestion to use such sensors as radiation detectors (Krebber et al., 2006). Several tests were run to evaluate the FBGs radiation hardness under neutron and gamma-ray exposure, in order to assess their possible use for temperature or stress monitoring in nuclear environments. Various types of FBGs as it concerns the optical fiber material, and the fabrication technology were used: Ge-doped (10 mol % $GeO_2$, Accutether AT120) with and without $H_2$ or $D_2$ loading (Gusarov et al., 2002; Gusarov et al., 2008a), B/Ge co-doped photo sensitive fiber (Maier, 2005), low-P content and N-doped optical fibers (Gusarov et al., 1999a), grating written by classical UV methods or using fs UV laser radiation (Gusarov et al., 2010). Gamma-ray pre-irradiation treatment or UV exposure of the sample before the nuclear reactor in core irradiation was also carried out in order to increase the immunity to irradiation effects (Fernandez Fernandez et al., 2001; Gusarov et al., 2008a). The Bragg gratings were investigated either in transmission or reflection, the parameters tested under irradiation were: the grating peak wavelength shift, the device transmission/reflectivity, the Full-Width Half-Maximum (FWHM), the temperature sensitivity of the device. Studies revealed that the FBGs coating can play an important role in the radiation sensitivity of the sensor (Gusarov et al., 2008). Generally, under gamma irradiation, the FWHM parameter was not affected by the irradiation, while a shift of the peak wavelength was noticed, as being dependent of the total irradiation dose and the dose rate - the wavelength shift is directed towards longer (Caponero et al., 2007; Fernandez Fernandez et al., 2002; Maier et al., 2005a) or shorter wavelengths (Maier et al., 2005), accompanied by a saturating effect and the decrease of the device reflectivity. Depending on the fabrication process and on the optical fiber composition, highly doped Ge optical fibers proved to be less sensitive to gamma irradiation, while hydrogen loaded optical fibers show a higher sensitivity (Gusarov et al., 1999; Gusarov et al., 2002). In some situations, under mixed gamma-neutron irradiation, the FWHM of the FBGs had also increased, suggesting the partial erasing of the written grating (Fernandez Fernandez et al., 2001). The temperature coefficient of the FBG is not affected by gamma irradiation (Gusarov et al., 1999b).

Investigations into the radiation degradation of Poly(methyl methacrylate) (PMMA) (Yoshida & Ichikawa, 1995) show that the effects of ionizing radiation can be divided into two sections, namely main-chain scission (degradation) and crosslinking. In many polymers both processes take place in parallel with one another, however in certain cases the scission predominates the crosslinking, and such polymers are known as degrading polymers. PMMA, shown in Fig. 1, is one such polymer.

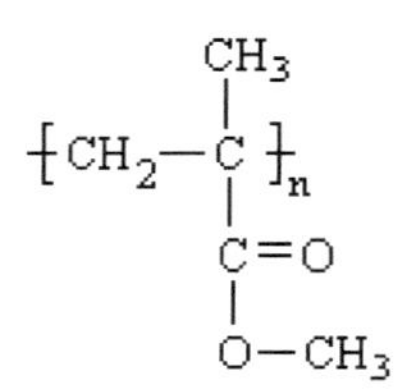

Fig. 1. PMMA polymer molecule.

Yoshida and Ichiwkawa (Yoshida & Ichikawa, 1995) report that the side-chain is initially affected by the gamma irradiation and the radical formed is a precursor for the main chain scission. When PMMA is irradiated with ionizing radiation, such as gamma radiation, a free radical is generated on the ester side-chain, $-COO\dot{C}H_2$ .

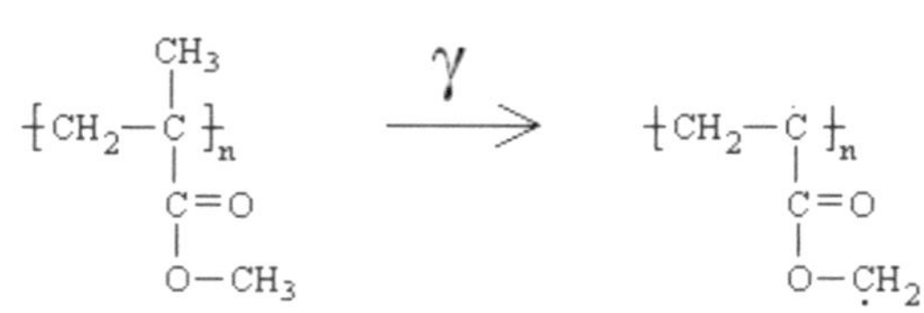

(1)

This side chain radical may be generated in a number of ways (Ichikawa, 1995), by direct action of ionizing radiation:

$$-COOCH_3 + \gamma \rightarrow -COO\dot{C}H_2 + H \quad (2)$$

by proton transfer of the side-chain cation:

$$-COOCH_3 + \gamma \rightarrow -COOCH_3^{\cdot +} + e^-$$
$$-COOCH_3^{\cdot +} \rightarrow -COO\dot{C}H_2 + H^+ \tag{3}$$

or by hydrogen abstraction:

$$-COOCH_3 + H \rightarrow -COO\dot{C}H_2 + H_2 \tag{4}$$

The side-chain radical remains stable at temperatures below 200 K. Above 210 K, the side-chain radical coverts to a scission type radical due to the detachment of the side chain. This is followed by the β-scission of the main-chain radical and is shown in equations 5 and 6.

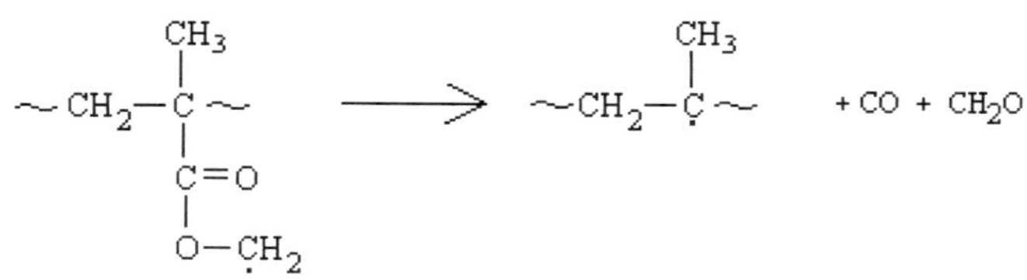


(5)

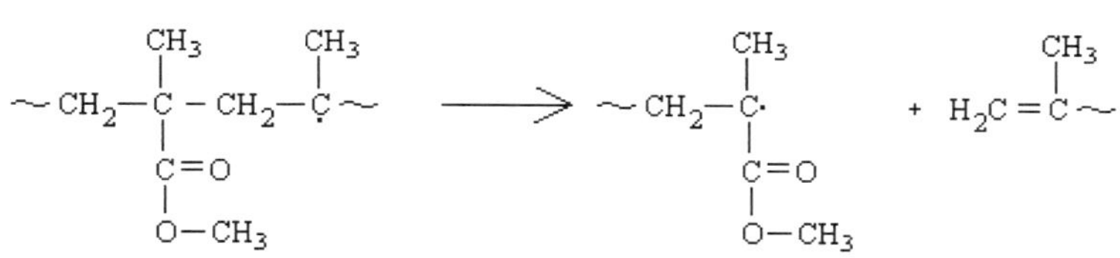


(6)

The Beer-Lambert law can be used to determine the radiation dose absorbed. The absorption co-efficient, α, is given by equation 7 (O'Keeffe et al, 2007).

$$\alpha = -\frac{1}{L_0}\log\left\{\frac{P(\lambda)}{P^0(\lambda)}\right\}, \tag{7}$$

where $\frac{P(\lambda)}{P^0(\lambda)}$ is the ratio of the spectral radiant power of light transmitted through the irradiated dosimeter to that of the light transmitted in the absence of the dosimeter.

The Beer-Lambert law is used to determine the radiation-induced attenuation. The radiation-induced attenuation is given by equation 8.

$$RIA(dB) = -\frac{10}{L_0}\log\left\{\frac{P_T(\lambda,t)}{P_T^0(\lambda)}\right\}, \tag{8}$$

here $L_0$ is the irradiated length of fiber, $P_T(\lambda,t)$ is the measured optical power in the irradiated fiber and $P_T{}^0(\lambda)$ is the optical power of the reference fiber.

## 3. Environments for use of optical fiber sensing in radiation dosimetry

The possible applications of optical fibers and the optical fiber sensors in radiation monitoring and dosimetry refer to:

1. measurement of the absorbed dose in radiotherapy (Andersen et al., 2002; Jang et al., 2009; Justus et al., 2006) and brachytherapy (Suchowerska et al., 2007);
2. spatial dose distribution in the case of linear electron and proton accelerators for medical treatment (Bartesaghi et al., 2007; Jang et al., 2010a; Lee et al., 2007a);
3. evaluation of beam losses (dose rate, total dose, location) in particle accelerators (Henschel et al., 2004; Intermite et al., 2009; Wulf & Körfer, 2009), beam profiling (Wulf & Körfer, 2009), and the operating conditions of an electron storage ring (Bahrdt et al., 2009; Rüdiger et al., 2008);
4. synchrotron radiation beam profile diagnostics (Byrd et al., 2007 ; Chen et al., 1996);
5. neutron or mixed gamma-ray neutron dosimetry (Bartesaghi et al., 2007a; Jang et al., 2010b);
6. the investigation of isotopic composition of cosmic rays (Connell et al., 1990);
7. radiation dosimetry in computed tomography (Jones & Hintenlang, 2008; Moloney, 2008);
8. distributed radiation dosimetry for beta & gamma rays, and neutrons (Naka et al., 2001);
9. beam profile in the case of free electron lasers (Goettmann et al., 2007) or proton beams (Benoit et al., 2007);
10. remote monitoring of ground water or soil for radioactive contamination (Jones et al., 1993);
11. radiation protection and monitoring of nuclear installations (Magne et al., 2008);
12. monitoring of radioactive waste (Nishiura & Izumi, 2001);
13. reconstruction of the charge particle tracks (Adinolfi et al., 1991; Angelini et al., 1989; Atkinson et al., 1988; Mommaert, 1992; Nakajima et al., 2009 ; Yukihara et al., 2006);
14. the use as transfer detectors for dosimetric calibrations (Tremblay et al., 2010);
15. space dosimetry (Yukihara et al., 2006);
16. tritium detection (Jang et al., 2010c).

In addition, optical fiber-based dosimeters can find in the near future their way in more common environments. One of the main application areas of gamma radiation is in sterilisation, in particular, the sterilisation of medical products. The ability of radiation to kill pathogenic micro-organisms is the basis on which the sterilisation of medical products depends. It is used in a wide variety of products such as hypodermic needles, surgical sutures, blood handling equipment, implant substances and tissues, surgical gloves and utensils, catheters, dental supplies, etc. It is still unknown how exactly the radiation kills micro-organisms, however it is thought to be associated with the damage caused by the radiation to the deoxyribonucleic acid (DNA) of the micro-organism (McLaughlin et al., 1989). The radiation sensitivity of micro-organisms depends on the amount of DNA in the nucleus, but it also depends on a number of other factors, such as the environment they are in. Micro-organisms irradiated in an aerobic environment are more sensitive than those in an anaerobic environment. Also, those in water are more sensitive than those in the dry state. The exact dose of gamma radiation used in the sterilisation of medical products varies for different countries but is usually between 25 kGy and 35 kGy (McLaughlin et al., 1989).

Gamma radiation can be used for a number of different applications within the food industry. Food can be treated to prevent sprouting in onions, garlic and potatoes, to extend the shelf-life of mushrooms, cherries and strawberries, to eradicate insects in grain and fruit, to kill pathogenic microorganisms in fish and meat, to pasteurize dried herbs and spices, and to delay ripening of fruit and vegetables. Radiation can also be used to prevent the spoilage process, which commonly leads to the rotting of food and wastage. This is due to radiation slowing down the physiological, chemical and biochemical changes occurring in the food and killing micro-organisms and insects. Although radiation does not prevent the drying out of fresh food, its application allows the food to be treated in a sealed package, which prevents both the desiccation process and the microbial re-contamination. Food irradiation treatment depends on the ability of radiation to kill cells and alter the enzyme activities of the food. The killing action of radiation, which inhibits cell division, is used to prevent sprouting, reduce the number of viable micro-organisms on the food, prevent the hatching of insect eggs and larvae, and kill or sterilize the insects in the food. A radiation treatment of between 1 kGy to 7 kGy, depending on the product, can significantly reduce the number of viable pathogenic micro-organisms, e.g. *Salmonella, Escherichia coli* and *Lysteria,* which can contaminate food causing serious food poisoning (McLaughlin et al., 1989, O'Keeffe & Lewis, 2009).

A number of other applications for radiation are also being investigated, such as in the treatment of sludge, to reduce the amount of bacteria and infectious micro-organisms, before it may be used as a fertiliser. It is also used in the preservation of ancient objects having historic or artistic value by killing the micro-organisms that can destroy the organic material that the objects are made of or have components of (McLaughlin et al., 1989).

## 4. Experimental set-up and irradiation conditions

Based on the previous considerations, the best assessment of the color centers' dynamics (the equilibrium between color center generation and the recovery process through photobleaching or heating), for different irradiation conditions (dose-rate, temperature, type of radiation) can be achieved during on-line investigations, by optical absorption measurements. Within such an experimental frame, a broadband spectral analysis of the optical transmission characteristics of the optical fiber is recommended in order to catch the complexity of the phenomena involved. In order to achieve this goal we used a set-up composed of combinations of the equipments listed below:

1. broadband light sources, Analytical Instrument Systems DT 1000 CE stabilized deuterium-tungsten lamp for the UV-visible range or a StellarNet SL1 Tungsten Halogen lamp for spectral measurements in the visible-IR domain;
2. a scientific-grade QE65000, TEC-cooled optical fiber spectrometer from Ocean Optics (1024 x 58 pixels; spectral sensitivity from 200 nm to 1100 nm; 0.065 counts/e- sensitivity; 65% quantum efficiency at 250 nm; 1000:1 S/N ratio; 25000:1 dynamic range/single acquisition; 16 bit resolution; from 8 ms to 15 min. integration time; 3 RMS counts dark noise);
3. a NIR InGaAs TEC-cooled mini spectrometer (1024 pixels; spectral sensitivity from 900 nm to 1700 nm; 4000:1 dynamic range, 12 bit resolution; 4000:1 S/N ration; from 1 ms to 30 s integration time) from StellarNet;

a. NIRX-SR InGaAs, 2 TEC stages, mini spectrometer (1024 pixels, spectral sensitivity from 1600 nm to 2300 nm; 4000:1 dynamic range, 16 bit resolution; from 1 ms to 200 ms integration time) from StellarNet;

b. Avantes optical fiber multiplexers for UV-visible of visible-IR (2 x 8 channels; > 60% optical throughput; > 99% optical repeatability; < 60 ms switching time between adjacent positions).

Several versions of the basic set-up (Sporea & Sporea, 2005) were developed to accommodate various experiments. The overall system is integrated under the National Instruments graphical programming environment and one such experimental implementation is illustrated in Fig. 2 *a*. The experimental instrumentation was used for on-line investigations related to the irradiation effects on silica optical fibers (mostly in the UV range where numerous color centers are developed) and sapphire optical fibers (Sporea & Sporea, 2007; Sporea et al., 2010a; Sporea et al., 2010b). By connecting only one end of the tested optical fiber to different spectrometers, according to the case with or without the use of a multiplexer, the radioluminescent signal was acquired during different types of irradiations. As the temperature plays an important role in recovery of the radiation induced optical absorption some on-line experiments were run by simultaneously irradiating and heating the optical fibers. For gamma-ray irradiation in a cylindrical geometry (the optical fiber, 2 to 5 m long, is coiled and the gamma source is placed in the middle of the circle formed by the fiber) a two sections set-up was used: on the lower level the investigated optical fiber is kept at room temperature, while the optical fiber placed on the higher stage can be electrically heated during the irradiation. A sketch of this mechanical design was presented in Fig. 1 of a previous paper (Sporea et al., 2010a). In that case, the two stages are thermally isolated. For the situation when optical fibers are irradiated by charge particles (electrons or protons) or neutrons, an electrically controlled heater was developed, to rise only locally the temperature of one of the optical fibers, as two short (10 to 15 mm long), similar samples are simultaneously irradiated (Fig. 2 *b*). In this case, each sample is placed into a separate quartz tube, one of these tubes being thermally isolated and coated on the inner side with a heat reflecting layer. In this tube, along with the optical fiber a thermally resistant wire is located parallel to the fiber. By applying a current to this wire the thermally isolated tube is heated. A thermocouple was used to monitor and record in real time the temperature inside the tube. In this way, the optical absorption of both samples is registered in real time, with one of the optical fibers also being subjected to temperature stress.

The set-ups previously detailed were utilized to evaluate the quality of optical fibers suitable for intrinsic or extrinsic radiation monitors (based on optical absorption or radioluminescence) operating at atmospheric pressure. In evaluating materials for alpha particles dosimetry another set-up was designed (Fig. 2 *c*), where the radioluminescence signal is picked-up through a multimode, high UV responsivity optical fiber (Sporea et al., 2010b).

The general conditions for gamma irradiation were described in another paper (Sporea et al., 2010a). In the case of the beta ray irradiation, run at the Linear Accelerator Laboratory of the National Institute for Laser, Plasma and Radiation Physics, the following operating conditions apply: mean electron energy: 6 MeV, the electron beam current: 1 μA, the pulse repetition rate - 100 Hz; the pulse duration - 3,5 μs; the beam diameter - 10 cm; the spot uniformity - +/- 5 %. Proton irradiation was performed at the Tandem irradiation facility of the "Horia Hulubei" National Institute of R&D for Physics and Nuclear Engineering–IFIN-HH, with a beam current of 1 nA, mean dose rate of 200 Gy/s, a total dose of 2.8 MGy, the proton energy of 14 MeV, and the beam diameter of about 3 mm. In some cases, off-line measurements were carried out in order to evaluate the optical absorption recovery either at room temperature or after heating with a laboratory programmable Memmert oven, controlled over a RS-232 serial interface.

a b c

Fig. 2. Set-ups for the on-line evaluation of irradiation induced optical absorption and radioluminescence: a – the data acquisition and processing system running under LabLIEW graphical programming environment; b – detail of the device used for temperature control of the optical fibers during the irradiation; c – detail of the multimode optical fiber acting as a guiding path of radioluminescence signal generated under alpha particles irradiation (Sporea et al., 2010b).

The two teams jointly investigated, under X-ray irradiation, the characteristics of several extrinsic optical fiber sensors developed at the University of Limerick. Tests in Ireland were done with the ORBITA microfocus X-ray inspection system, while the investigations in Romania were performed with the X-ray micro-tomographic unit developed by Dr. Ion Tiseanu at the National Institute for Laser, Plasma and Radiation Physics. The ORBITA equipment operates within the maximum limits of 0-160 kV, 0- 500 μA. In Romania the sensors were tested with a focused, quasi-monochromatic energy of 17.5 KeV (McCarthy et al., 2011).

## 5. Optical fiber sensors classification and design

As mentioned in the "Introduction", the use of optical fibers in sensor applications has many advantages over conventional sensors, such as electrochemical and semiconductor sensors. Optical fibers are made of a dielectric material and as such are chemically inert. This makes them very suitable in chemical sensing or in chemically harsh environments. They also provide immunity from electromagnetic interferences and their high electrical isolation makes them suitable for use in lightning protection, high-voltage and medical applications. Optical fibers are also capable of withstanding high temperatures, up to 400 60°C (CeramOptec). These characteristics allow optical fiber sensors to be used in environments not permitted by electrical sensors. Through the use of optical fibers it is possible to have great distances to the measuring point allowing for remote sensing in hostile environments, such as in high-radiation-level areas in the vicinity of a nuclear reactor. The possibility of small, simple interfaces along with lightweight fiber technology results in very appealing sensors. Optical fiber sensors have also been shown to have high sensitivity, large dynamic range along with high resolutions (O'Keeffe, 2008).

The use of plastic optical fibers within the sensor industry is a relatively new technology when compared to glass optical fibers. Plastic optical fibers have been shown to exhibit good transmission in the visible region of the electromagnetic spectrum. Due to the large fiber cross-section of plastic optical fibers, generally a 1 mm core, connecting to the light source and detector is non-problematic. This means that, in contrast to glass optical fibers, no expensive precision components are required for centring the fibers. The large fiber core diameter also means that more light can be transmitted and subsequently incident on the detector, resulting in a higher sensitivity of the sensor. Minor contamination, e.g. dust on the fiber end face, does not result in the complete failure of the sensor system due to its large core diameter. Consequently, fibers can be connected on site in industrial environments with relative ease and without affecting the system. PMMA is also easy to cut, grind and melt and so an uncomplicated process, requiring relatively little time, for processing the end faces is necessary to achieve a clean and smooth surface. Plastic optical fibers are also considered easier to handle when compared to glass optical fibers. Glass fibers tend to break when bent around a small radius, which does not occur with plastic fibers. These fibers are extremely low in cost when compared with glass optical fibers. The properties of PMMA plastic optical fibers result in relatively economical connectors for the system, which further contribute towards a low cost solution (Weinert, 1999).

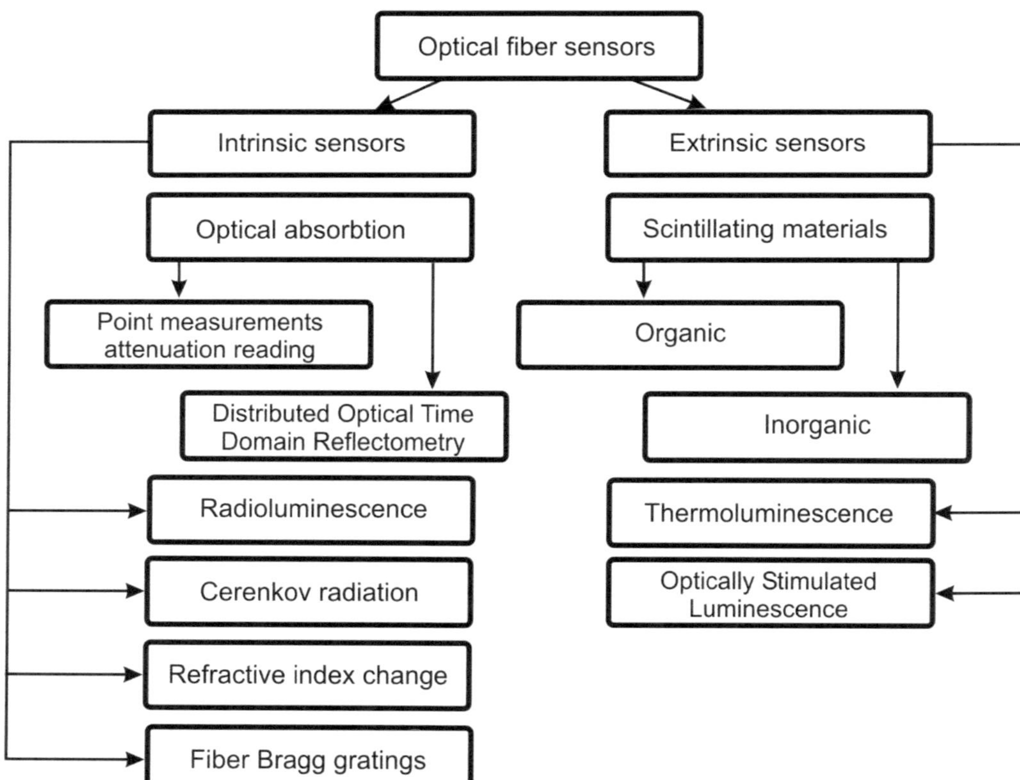


Fig. 3. Classification of the optical fiber sensors for radiation detection according to the operating principle.

Optical fiber sensors can be characterised under two main types: *intrinsic sensors*, where the interaction occurs within the optical fiber itself and *extrinsic sensors*, where the optical fibers are used to guide the light to and from the region where the light interacts with the measurand. Fig. 3 synthesizes the classification of optical fiber-based radiation sensors. Evanescent wave sensors function by causing the light guided in the fiber to couple with the variable to be measured via the evanescent field and have features of both intrinsic and extrinsic devices. It is possible to further classify optical fiber sensors according to the type of modulation to be used (Lopez-Higuera, 1998):

a. Amplitude or intensity sensors detect the amount of light that is a function of the influencing environment. Their ability to use either incoherent or coherent light sources together with simple optical components makes for a low-cost system.
b. Phase or interferometric sensors detect a modulation in phase caused by the measurand. This type of source requires coherent light sources making them a relatively costly system. However, they can provide a very high level of sensitivity.
c. Polarimetric sensors detect a modulation in the polarisation of the light due to the variable to be measured.
d. Spectroscopic sensors detect a modulation in the spectrum of optical radiation due to the influencing variable.

Radiation dosimetry is fundamental to the wide range of radiation processes and as such is the focus of several research endeavours. A number of physical and chemical sensors, which can be subdivided into liquid, solid and gaseous systems, are available to measure ionising radiation and researchers are continually looking for ways to improve the systems, be it increasing the sensitivity, providing real-time measurements or significantly reducing the costs. All these factors are important for providing the optimum radiation dosimeter. In selecting an optical fiber sensor for radiation measurement (dose rate, total dose, point sensor, 1D or 2D sensors) several parameters have to be considered (O'Keeffe et al, 2008):

a. Material sensitivity: The ability for radiation to interact with the material is the primary characteristic for any radiation dosimeter. This interaction varies greatly among the different types of radiation dosimeters. The materials utilised in gamma dosimetry must have a high sensitivity to gamma radiation within the dose ranges required for the specific application. Especially for in-vivo medical measurements the sensor has to be sensitive to small dose rate/ total doses, comparable to those involved in radiotherapy.
b. Linearity: The dependency of the sensor out-put signal as function of the input dose/ dose rate. Some applications require a good linearity (no signal saturation) over several decades.
c. The residual signal: It is required that the output signal after its reading (for the case of re-usable sensors such as those based on thermoluminescence or Optically Stimulated Luminescence) to be as small as possible, no cumulative signal from irradiation to irradiation can be accepted or at least this un-read signal has to be reproducible.
d. Post-irradiation fading: The material used as a radiation dosimeter must also show minimum post-irradiation fading. This post-irradiation fading is due to the repair of the physical damage caused to the material during irradiation. Information regarding the irradiation dose is lost if the dosimeter material begins to recover immediately after irradiation.

e. After glow or phosphorescence: in the case of sensors based on radiation induced luminescence a "tail" of the luminescence signal can impinge on the temporal response of the sensor.
f. Time dependence: Many materials also exhibit a time dependence of specific absorbance. This means that it is often required to wait up to 3 h before obtaining an accurate reading. For real time medical applications it is usually important that the dosimeter is fast to measure.
g. Stability and reliability: Stability and reliability of the dosimeter, as with any sensor, are also important. In order to ensure that this can be achieved the dosimeter must be immune to a number of environmental conditions. The effect of humidity and temperature on the dosimeter must be investigated and accounted for. The influence of dose rate on the dosimeter material should also be considered. Immunity to other disturbances, such as those found in electromagnetically harsh environments, is also advantageous for gamma dosimeters as they are often employed in such conditions.
h. Ease of use: For medical applications it is important that the dosimeters are easy to use. For this to be achieved, the system must be easily installed in the area of application without the risk of affecting the measurements. Maintenance should also be minimal. The readout from the dosimeter should be clear and easy to understand. It is also important that the system is low in cost. The availability of clear, easy to implement calibration procedures is an additional advantage.

### 5.1 Intrinsic optical fiber sensors

Radiation dosimetry based on RIA in gradient index P-doped optical fibers was investigated for $^{60}Co$ gamma-ray irradiation (dose rates from 0.01 mGy/s to 1.9 Gy/s) and pulsed electrons (total doses of 30 Gy, 100 Gy and 1 kGy). The measurements were carried out at several fixed wavelengths (670 nm, 850 run, 1300 run, and 1550 nm) or spectrally with a spectrum analyser, over the 500 nm to 1700 nm spectral range. In order to be a reliable radiation detector these optical fibers were checked to exhibit high reproducibility, good linearity (over six decades), reduced dose rate dependence (independent over five decades), low fading (a drop of the irradiation induced attenuation of 10 % after two hours and of 14 % after five hours) and small temperature dependence (above 70 $^0$C the annealing process is significant) and sensitivity to photobleaching (at 10 µW and respectively 10 nW optical power of the laser used to measure the optical absorption no noticeable decrease of the attenuation was observed) (Henschel et al., 1992).
The irradiation induced absorption in doped optical fibers ($GeO_2$ or $P_2O_5$, from 12 to 16 mol/%, in the core), under gamma-ray irradiation at low dose rate (1, 0.1 and 0.01 Gy/h, up to a total dose of 1 Gy) was done to evaluate the possible use of such optical fibers as radiation dosimeters. The absorption was monitored during the irradiation ($^{60}Co$-gamma radiation source of energy 1.25 MeV and Cs-137 radiation source of energy 0.662 MeV) over the wavelength interval 350 nm - 1100 nm (Paul et al., 2009).
As compared to the published data, our investigations focused mostly on the estimation of irradiation effects on special optical fibers, by checking their response in the UV-visible spectral range. We worked with optical fibers such as: solarization resistant; UV-enhanced response; $H_2$-loaded; or operating in the deep-UV, having a core diameter from 200 µm to 1 mm. We evaluated, using the set-up described under headline *4*, optical fibers with different jacket materials: Tefzel, Polyimide or Al (Sporea et al., 2010b). Tests were run under gamma or beta rays, and neutron irradiations. Fig. 4 illustrates some of our results recorded over the

UV-visible spectral range (from 200 nm to 950 nm) for optical fibers irradiated at room temperature.

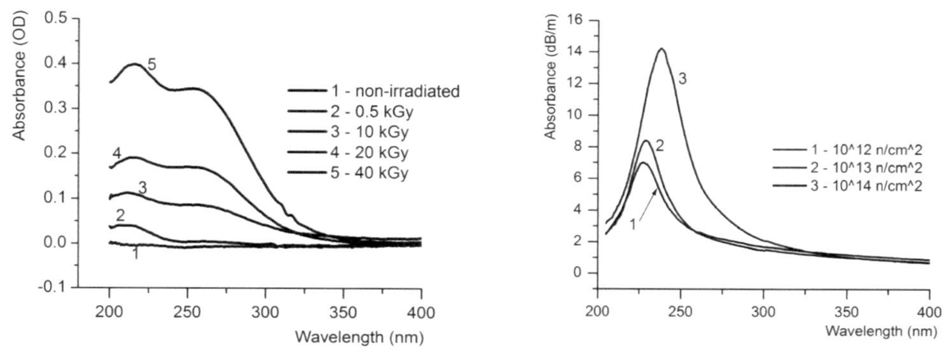


a b

Fig. 4. The changes of the UV optical absorption for: a - 600 µm core diameter, 12 cm long, UV-enhanced, Polyimide jacket optical fiber, irradiated by beta radiation; b - 400 µm core diameter, 10 cm long, solarization resistant, Tefzel jacket optical fiber, subjected to neutron irradiation at three values of the neutron fluences (Sporea et al., 2010b).

The annealing can play an important role in the recovering process of the radiation induced optical absorption. For this reason we run some on-line experiments to evaluate the impact of the ambient temperature during the irradiation process (Fig. 5). The graphs indicate the modifications of the optical attenuation during on-line measurements for a gamma-ray irradiation (dose rate of 36 Gy/min) 400 µm core diameter, UV-enhanced response, $H_2$-loaded, solarization resistant optical fiber. The difference in the peak maxima results from the coupling losses along the connecting fibers. The change of the optical attenuation is higher for the sample subjected to heating (240 $^0$C) during the irradiation. Such behaviour can help to select appropriate optical fibers and irradiation conditions in relation to radiation monitoring.

a b

Fig. 5. The influence on the irradiation induced absorption for a 12 cm long, solarization resistant optical fiber: a – heating to 240 $^0$C is applied to the optical fiber during the irradiation; b – the irradiation is performed at room temperature (Sporea et al., 2010b).

Experimental results (Fig. 4) as well as the corresponding decomposition of the spectral characteristics into Gaussian components (Sporea et al., 2010a) indicate the development of various color centers as the optical fiber is irradiated. In order to estimate the possibility to use UV optical fibers in radiation dosimetry, we measured the change of the optical absorption at two specific wavelengths (Fig. 6 and 7), for different optical fibers, at room temperature, under gamma and beta rays irradiation (Sporea et al., 2010b).
Two wavelengths were monitored in order to optimize the optical fiber as it concerns the radiation sensitivity and linearity of its response with the total dose. The two samples in Fig. 6 *a* exhibit a very good linearity with the total dose if the color center developed at λ = 240 nm is considered. A higher sensitivity can be achieved (in the detriment of the dynamic range) with the same optical fibers, if the measurements are done for λ = 215 nm. The second set of optical fibers (Fig. 6 *b*), being designed for UV applications, are by far less sensitive to gamma irradiation, but they have also a good linear response with the total dose. The results in Fig. 6 indicate a good dynamic range for the investigated optical fibers. Some of the optical fibers tested under gamma-rays were also evaluated under electron irradiation. They showed to have low radiation sensitivity, and a poorer linearity under beta radiation (Fig. 7). The reading at λ = 280 nm for the deep-UV optical fiber designed for laser beam delivery remains almost unchanged with the total dose. The color center produced at λ = 215 nm for the same fiber shows a higher sensitivity and an acceptable linearity up to 40 kGy.

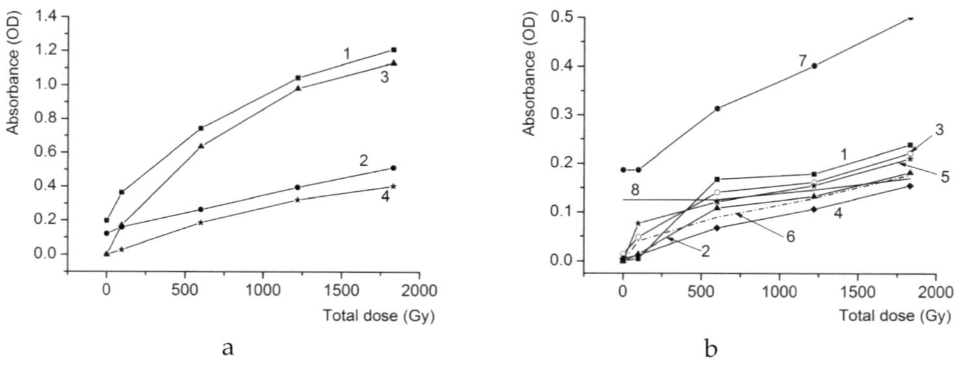


Fig. 6. The optical absorbance as function of the total dose, gamma-ray irradiation (700 Gy/h) for: a – 1 mm core diameter, multimode silica optical fiber, at λ = 215 nm (1) and λ = 240 nm (2); 400 μm core diameter, solarization resistant, Tefzel jacket optical fiber, at λ = 215 nm (3) and λ = 240 nm (4); b – 600 μm core diameter, UV-enhanced, Polyimide jacket optical fiber, at λ = 215 nm (1) and λ = 240 nm (2); 600 μm core diameter, $H_2$-loaded, Polyimide jacket optical fiber, λ = 215 nm (3) and λ = 240 nm (4); 600 μm core diameter, deep UV-enhanced, Polyimide jacket optical fiber, at λ = 215 nm (5) and λ = 240 nm (6); 400 μm core diameter, deep UV improved for laser delivery optical fiber, at λ = 215 nm (7) and λ = 240 nm (8) (Sporea et al., 2010b).

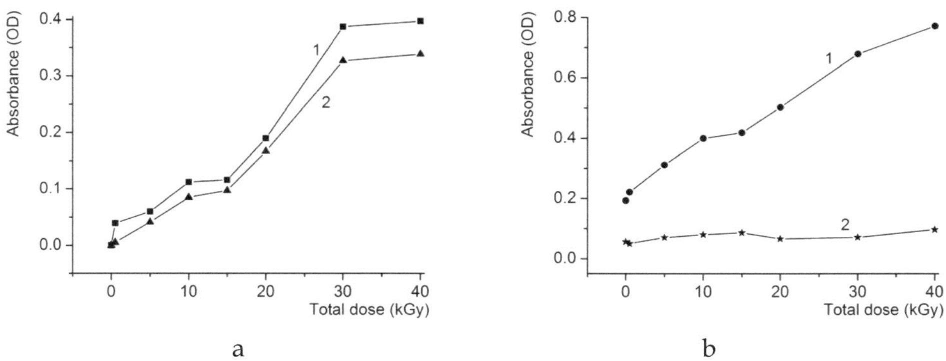


Fig. 7. The optical absorbance as function of the total dose for beta-ray irradiation (6.1 kGy/min) for: a - 600 µm core diameter, deep UV-enhanced, Polyimide jacket optical fiber, at $\lambda = 215$ nm (1) and $\lambda = 280$ nm (2); b - 400 µm core diameter, deep UV improved for laser delivery optical fiber, at $\lambda = 215$ nm (1) and $\lambda = 280$ nm (2) (Sporea et al., 2010b).

The RIA induced by X-ray irradiation in multimode P-doped optical fibers (7 wt.% phosphorus concentration in the optical fiber core) was investigated as preliminary studies on the use of such optical fibers to evaluate the total dose (Girard et al., 2011). Measurements were carried out for a total dose up to 3 kGy, at different temperatures (5 $^0$C; 15 $^0$C; 25 $^0$C, 35 $^0$C and 50 $^0$C), for dose rates of 1 Gy/s; 10 Gy/s and 50 Gy/ s. The optical absorption was monitored over the 350 nm - 900 nm spectral range. No bleaching of the irradiated optical fibers was noticed and the temperature change under the investigation limits plays a minor role. The highest sensitivity for the irradiation induced absorption was observed in the UV part of the spectrum.

The Irish team explored the possible use of change in the optical absorption in POF under irradiation. Fig. 8 shows the transmission spectrum before the fiber was irradiated and after it was irradiated to 3 kGy of gamma radiation. It is immediately clear that, after exposure to gamma radiation, the transmission is attenuated over the entire visible spectrum. The amount of attenuation depends on the wavelength, with lower wavelengths having a significantly higher attenuation than higher wavelengths (O'Keeffe et al., 2007; O'Keeffe & Lewis, 2009; O'Keeffe et al., 2009a).

The possible use of POFs as radiation monitors is exemplified in reference to Fig. 9, when the peak wavelengths were selected for further analysis. It can be seen that as the wavelengths increase, the rate at which the intensity decreases is reduced. 525 nm shows a rapid decrease in intensity and becomes fully attenuated rapidly. In contrast to this, at 650 nm the intensity, while already low to begin with due to the intrinsically high attenuation in PMMA in this region, decreases at a much slower rate and does not become completely attenuated over the course of the experiment.

The sensitivity of the PMMA fiber to ionizing radiation is directly related to the wavelength and thus by selecting the correct wavelength the fiber can be used to monitor over a wide dose range, selecting 525 nm for low doses with high sensitivity and selecting 650 nm for high doses with lower sensitivity.

Radiation-induced attenuation (RIA) calculations, based on equation 8, were performed on the individual wavelengths to determine the exact sensitivity of the fiber to gamma

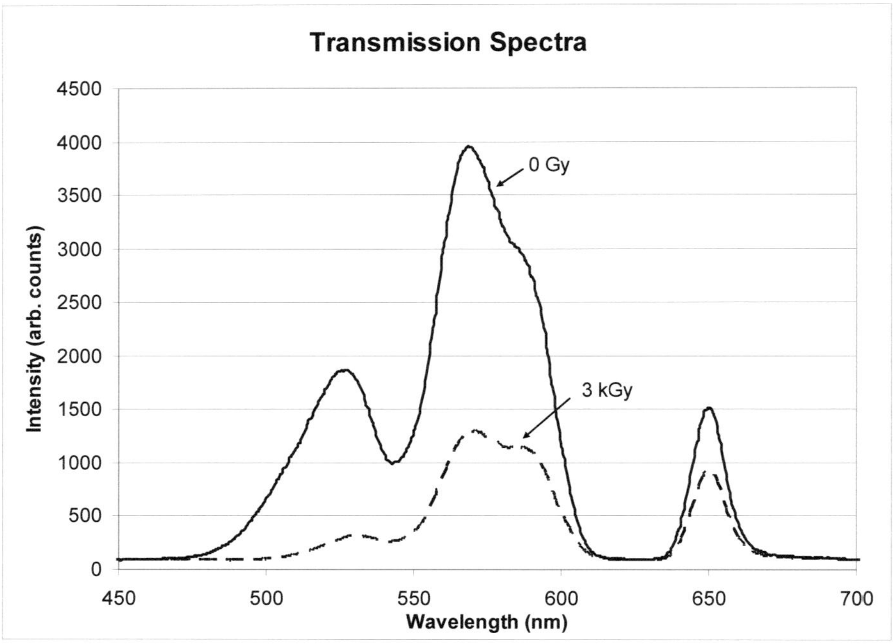


Fig. 8. The transmission of a POF spectra before irradiation and after 3 kGy (O'Keeffe et al, 2009a).

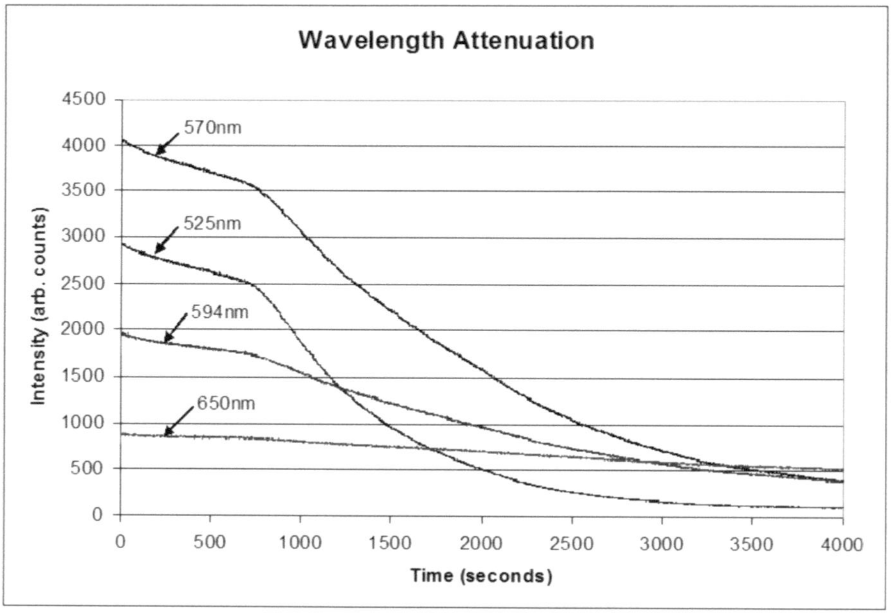


Fig. 9. Intensity of the peak wavelengths over time, as POFs are exposed to gamma radiation (O'Keeffe et al, 2009a).

radiation and the results are presented in Fig. 10 *a*. The wavelength dependence on the RIA sensitivity is evident as the rate of attenuation decreases for increasing wavelength. All the wavelengths exhibit identical characteristics in the RIA with increasing dose. Initially, up to 1 kGy, the RIA increases slowly, followed by significantly higher rate of increase. After this the fiber reaches saturation, whereby the intensity has become completely at that wavelength. The saturation can be seen in Fig. 10 *b*, and indicates the dose range over which the individual wavelengths can monitor. As the wavelength increases the upper-detection limit for that individual wavelength also increases, 4 kGy at 525 nm and 45 kGy at 650 nm (O'Keeffe et al., 2007; O'Keeffe & Lewis, 2009; O'Keeffe et al., 2009a).

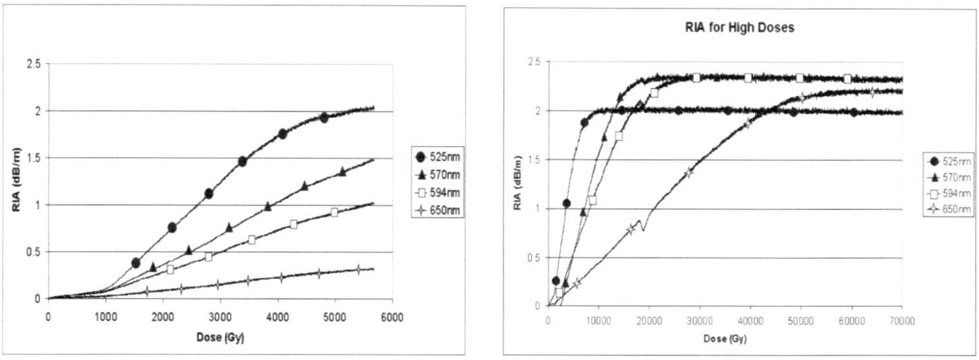


a b

Fig. 10. Radiation-induced attenuation for gamma irradiation; a – total dose up to 5.7 kGy; b – total dose up to 70 kGy (O'Keeffe et al, 2009).

Fig. 11 shows the radiation-induced attenuation at lower radiation doses up to 50 Gy for the two lower peak wavelengths, 525 nm and 570 nm. The determination of lower detection limit of the fiber was limited by the experimental set-up and the high dose rate of the gamma source used, however it can be seen that there is a distinct and immediate increase in the radiation-induced attenuation at these wavelengths. An averaging filter is applied to the data and indicates that beyond 30 Gy there is a steady, quantifiable increase in the radiation-induced attenuation (O'Keeffe et al., 2007; O'Keeffe & Lewis, 2009; O'Keeffe et al., 2009a).

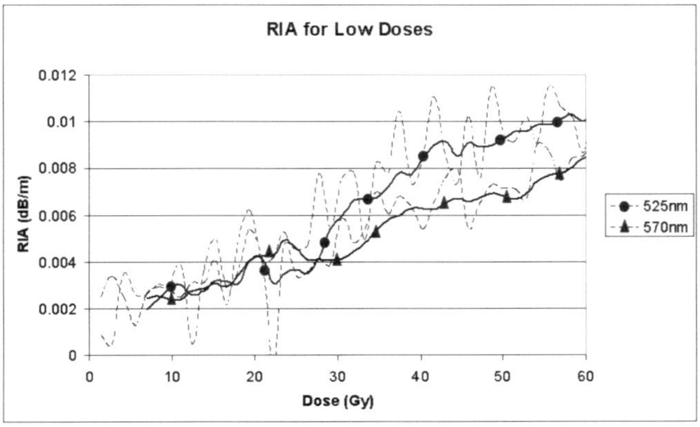


Fig. 11. RIA for low doses of gamma radiation (O'Keeffe et al, 2009a).

Table 1 gives a summary of the sensitivity and dose range of the PMMA fiber for individual wavelengths. It shows that the sensitivity decreases with increasing wavelengths, from 0.6 dBm-1/kGy at 525 nm to 0.06 dBm-1/kGy at 650 nm. The dose range is also dependant on the wavelength as those wavelengths with high sensitivity becoming completely attenuated relatively quickly compared with the higher wavelengths.

| **Wavelength (nm)** | **Sensitivity ($dBm^{-1}/kGy$)** | **Dosimetry range (kGy)** |
|---|---|---|
| 525 | 0. 6 | 0.03 - 4 |
| 570 | 0. 3 | 0.03 - 10 |
| 594 | 0. 2 | 0.5 - 12 |
| 650 | 0.06 | 1 - 45 |

Table 1. Summary of wavelength dependant sensitivity and dose range.

The radiation induced attenuation of PMMA optical fibers exposed to X-ray radiation, using 6 MV photon energy was reported (O'Keeffe et al, 2009b). The transmission spectra were monitored online and were recorded every two seconds for analysis of attenuation changes as the radiation dose was increased. The results of the RIA calculations are shown in Fig. 12 for two peak wavelengths, 565 nm and 594 nm. There is an immediate and distinct increase in the RIA as the PMMA optical fiber is exposed to the X-ray radiation. The initial results indicate a sensitivity of the fibers to radiation of 0.001 dBm-1/Gy.

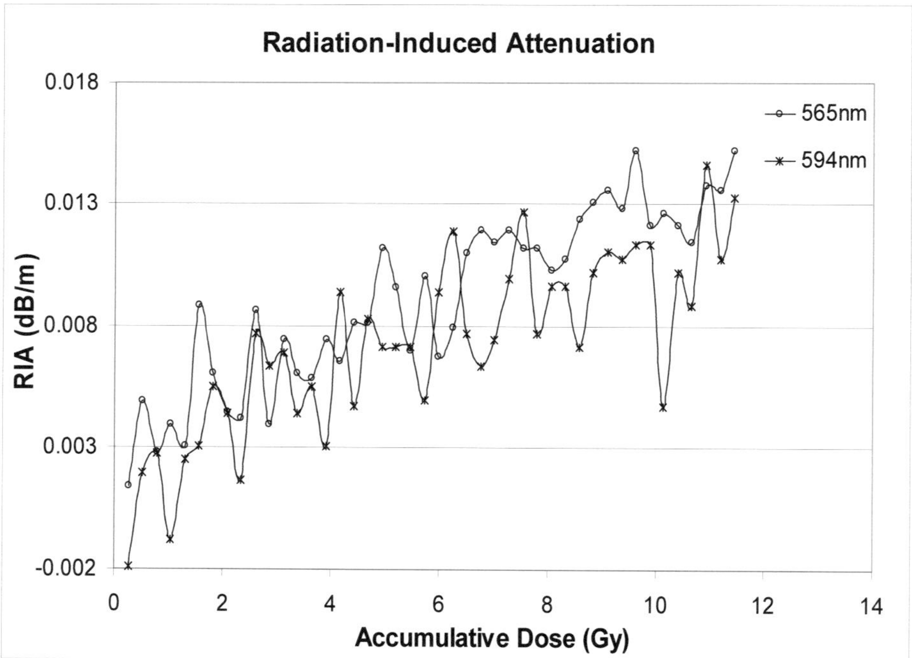


Fig. 12. Radiation Induced Attenuation for two wavelengths, 565nm and 594nm, for POFs under X-ray irradiation (O'Keeffe at al, 2009b).

As the beam losses in particle accelerators have to be checked over long distances a monitoring system, using the principle of Optical Time Domain Reflectrometry (OTDR), was designed for DESY TESLA accelerator (Henschel et al., 2004). The operation of this distributed dosimeter is based on the attenuation induced in optical fibers as they are exposed to radiation associated with machine operation (X-rays, electromagnetic showers). When a short laser pulse is launched into the fiber, some Rayleigh back-scattered optical radiation coming from the fiber sections where color centers are generated by the highly energetic radiation is collected by a detector. From the time of flight and the level of the returned optical radiation pulse a distributed dose mapping can be done, as the pulsed propagation time can be linked to the distance from the laser source of the perturbing non-homogeneities. Within some limits the amount of the detected light can reflect the dose to which the optical fiber segments were exposed. The main limitation of the system is related to the more or less permanent degradation of the optical link transmission during the irradiation. Some solutions to this problem were tested, with not so encouraging results, as for example photobleaching with laser radiation at 830 nm (100 mW CW operation), 670 nm (200 mW/CW), 532 nm (3 W/ CW and 5.5 kW/pulsed).

In the case of P-doped optical fibers, an alternative to extend the operation lifetime of the system can be the thermal annealing (Kuhnhenn et al., 2004). The radiation sensitive optical fiber was subjected to several $^{60}$Co irradiation (dose rates between 2.05 to 3.2 Gy/min, total dose of 1000 Gy) - annealing cycles (heating to 400 $^{0}$C). A cumulated (after several cycles) residual optical attenuation was observed (about 0.25 dB/m), while the radiation sensitivity of the optical fiber changes very little.

Among the advantages, at least theoretically, of such a system reside in: on-line, distributed dose monitoring; permanent access to difficult to reach locations in complex installations; the possibility to "tune" the sensitivity of the system for specific dose and dose rates, based on the selection of the optical fiber and the interrogation wavelengths used. In order to solve the problem related to the limits imposed by the optical fiber length and the increase in time of the fiber attenuation, an alternative set-up was suggested when the same OTDR units interrogates the optical link from both ends. Even multiple distributed sensors can be operated by the same control instrument (Henschel et al., 2004). Nevertheless, there are some difficulties as it concerns the system performances (the optical fiber radiation sensitivity, the needed information on the exposed optical fiber length, the saturation of the attenuation with total dose). For location where the dose has to be more strictly monitored, a set of local optical fiber dosimeters were designed. In this approach, the attenuation changes with the irradiation dose were evaluated with a LED ($\lambda$ = 660 nm, for a higher sensitivity) as an interrogation light source, and a set of optical power meters coupled to the LED though connecting optical fibers and segments of radiation sensitive optical fibers, the last ones located in radiation exposed areas.

Apart from standard silica and plastic optical fibers, microstructured optical fibers (MOF) such as photonic crystal fibers (PCF) and random hole optical fiber (RHOF) were tested under irradiation, to assess their radiation hardness or possible use in radiation dosimetry (Wu et al., 2003). In RHOFs the silica core is surrounded by thousands longitudinal channels, having random positions and dimensions. By coating the holes of a RHOF with a phosphorescent material acting as a converter of X-ray or gamma-ray energy to optical radiation, a dose rate monitor can be built, the optical fiber as a whole acting also as a light guide (Alfeeli et al., 2007).

A sensor based on the detection of the Cerenkov radiation produced in a bunch of radiation resistant, high OH-content optical fibers was used to optimize the injection efficiency of the transfer channel between the synchrotron and the storage ring at DELTA. The set-up makes possible a 2D mapping as the time of flight is used to evaluate the location of the Cerenkov radiation produced along the optical fiber, with a longitudinal spatial resolution of 0.2 m corresponding to a delay of about 1 ns between two Cerenkov flashes, while the spatial positioning of the detecting optical fibers assures the localization along the transversal direction (Rüdiger et al., 2008; Wittenburg, 2008). Higher spatial resolution of the Cerenkov radiation-based sensor can be achieved by employing two parallel positioned optical fibers, one used as reference (it is optimized to generate and guide the maximum amount of Cerenkov radiation) and the other one is structured as a sequence of alternating segments of optical fibers similar to the reference fiber and optical fibers having a much higher attenuation. Both optical fibers are coupled to silicon photomultipliers. In this way, if the segments composing the second optical fiber are short enough, by measuring the attenuated signal as compared to the optical signal picked-up by the reference optical fiber, the location of the generated Cerenkov radiation can be identified (Intermite et al., 2009).
A set of four radiation resistant, high-OH content optical fibers, coupled to photomultiplier tubes to detect the Cerenkov radiation, were installed in equidistant radial arrangement along the undulator section of a free electron laser. The position of a wire scanner crossing the electron beam is correlated with the detected Cerenkov radiation, resulting in beam loss, and so, the profile of the beam can be reconstructed by appropriate data processing of the acquired optical signal (Goettmann et al., 2007).
A distributed radiation dosimeter for beta rays, gamma rays and neutrons was designed by using a plastic optical fiber (POF) subjected to irradiation and by detecting the fluorescence and Cerenkov radiation optical signals at both ends of the optical fiber. The time-of-fly technique applied to pulses detected by photomultiplier tubes combined with time-to-amplitude converters and a multichannel analyzer was used to locate the place where the irradiation occurs (Naka et al., 2001). The operating parameters of the equipment (responsivity, resolution) are function of the energy spectrum of the detected radiation, the optical fiber attenuation and degradation, the emission spectrum of the generated optical radiation.
We explored the possible use of sapphire optical fibers in spectroscopic applications in radiation environments and in radiation dosimetry. 25 cm long 425 µm diameter sapphire optical fibers were measured in relation to the radiation induced attenuation (Fig. 13) and radioluminescence, as they were subjected to gamma-ray, proton and neutron irradiation (Sporea & Sporea, 2007). By analyzing the experimental data we noticed the radiation resistance of these optical fibers and the three peaks spectrum of the RL emission under proton irradiation. As the sapphire fibers proved to be immune to radiation (tests were run for gamma-ray up to 530 kGy total dose, for proton up to 2.8 MGy total dose, and for neutron for fluences up to $3.15 \times 10^{13}$ n/cm$^2$) we can think to have these optical fibers as real time dose rate monitors, by measuring the RL signal (Fig. 14). Because a small increase of the optical attenuation occurs for $\lambda$ = 350 nm to 550 nm after irradiation, dosimetric evaluation based on RL can be performed more accurately if the amplitudes of both emission peaks ($\lambda$ = 440 nm and $\lambda$ = 690 nm) are considered to correct the green emission measurements. The noisy signal in UV is due to a poor transmission of sapphire under 350 nm. The set-up described in section 4 has the capability to acquire the time dependence of data at selected wavelengths, hence there is the possibility to monitor in real the radioluminescence associated to different centers and to have information on the beam energy.

Another intrinsic type of optical fiber sensor, suggested to be used in particle accelerators for evaluation of very high doses, is based on a Mach-Zehnder interferometer. One of the interferometer arms represents the reference, while the other arm is exposed to irradiation. This second arm is formed by a radiation sensitive optical fiber exhibiting a refractive index change with the radiation dose, and, by combining the coherent radiation traveling along the two arms, a variation of the refractive index produced by the irradiation can be detected (Henschel et al., 2000).

Exploring the radiation induced changes in the refractive index of silica optical fibers a FBG-based dosimeter was suggested, applicable to high radiation doses ($10^3$ Gy up to $10^5$ Gy with an accuracy of at least 10%). Measurements were carried out, at different dose rates, under gamma ray irradiation, assuring a strict control and measurement of the temperature during the irradiation (Krebber et al., 2006). The sensors can be optimized for radiation dosimetry by a careful selection of the optical fiber type and the operating wavelength. It was demonstrated theoretically that FBGs written in photonic crystal fibers (PCF) can achieve a high sensitivity to radiation (at least one order of magnitude) as compared to FBGs produced in bulk material (Florous et al., 2007).

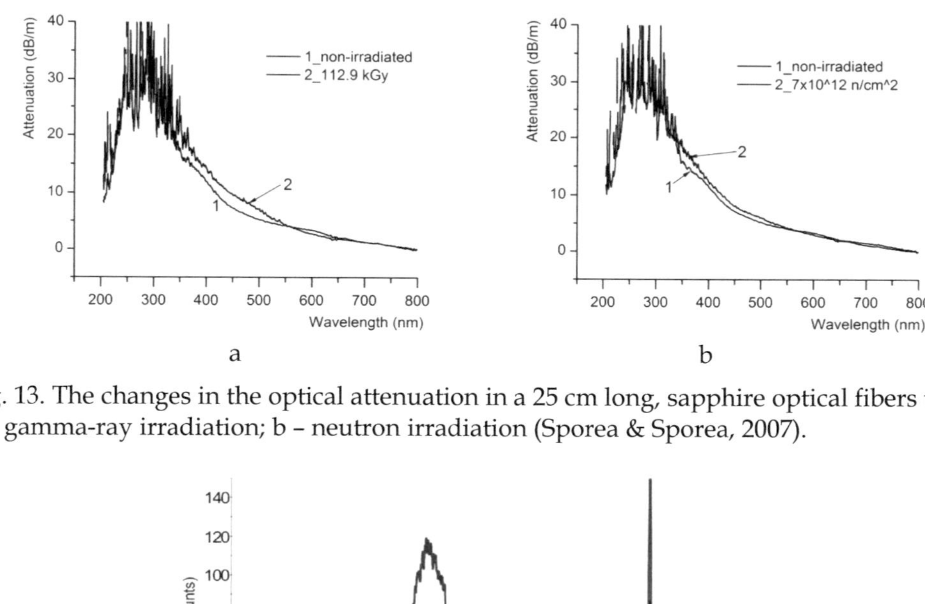


Fig. 13. The changes in the optical attenuation in a 25 cm long, sapphire optical fibers under: a - gamma-ray irradiation; b - neutron irradiation (Sporea & Sporea, 2007).

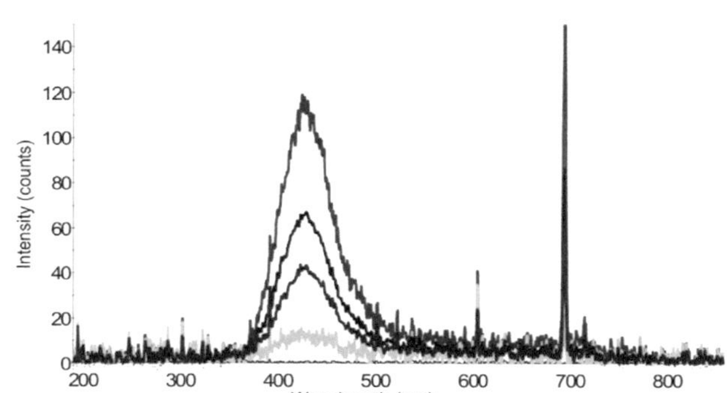


Fig. 14. The radioluminescence emission from a sapphire optical fiber, for proton beam irradiation. The amplitude changes of the RL signal reflects the modifications of the beam current at different irradiation stages (Sporea & Sporea, 2007).

## 5.2 Extrinsic optical fiber sensors

Point measurements of the absorbed dose in radiotherapy were done by coupling a 10 mm long, 1 mm diameter piece of organic scintillator (Saint Gobain Crystals BCF-20) to a POF acting as light guide for the generated optical radiation. A higher sensitivity of the sensor was obtained by outside coating the scintillating tip by a reflecting $TiO_2$ layer. The read-out was performed with a photodiode-amplifier system. The device was tested under proton irradiation at a Varian CLINAC 2100C/D accelerator for a 6 MeV energy beam. The amount of the collected optical radiation varied linearly with the dose rate up to 900 cGy/min. The system, acting as a real time instrument, provides within a small volume, high resolution, water equivalent readings, without the need of temperature, humidity or pressure correction (Jang et al., 2009). A similar detecting optical set-up was used to build a 1D array of sensors for high energy proton beam profiling, with scintillating tips spaced 1 cm and respectively 0.5 cm apart, and the light detection with a 10 channel photodiode array (Lee et al., 2007a).

An alternative approach uses a Bicron BC400 scintillating material glued at the end of a PMMA optical fiber and coated with an Al layer, for improved optical radiation collection (Suchowerska et al., 2007). The dimensions of the sensor make possible the limitation of the background optical noise, fluorescence and Cerenkov emission in the optical fiber (less than 0.1 % of the useful signal).

Another optical fiber dosimeter for use in radiotherapy is composed of fused quartz glass doped with $Cu^{1+}$ ions fused to multimode silica optical fiber, which permits the connection to a photomultiplier and a data collection system. Prior to exposure to 6 MV X-rays from a linear accelerator, the radiation detection tip was pre-sensitised under soft X-ray radiation in order to stabilize its responsivity (Justus et al., 2006). To remove the background noise, composed of the native fluorescence in the coupling optical fibers (having a decay time of several ns) and the Cerenkov radiation (emitted in the ps range), and to keep for the dose measurement only the scintillation signal from the $Cu^{+}$-doped fused quartz optical fiber (lasting several hundreds ms) a gated data acquisition is performed by temporarily selecting the signal to be processed. The measurements carried out indicated the independence of the system response on the dose rate, energy and dose per pulse. Some limits are associated to the long-term stability, degradation under irradiation, directionality of the response (Tanyi et al., 2011).

An array of eight pieces of scintillating optical fibers (BCF-10 type), interlaced with eight pieces of clear optical fibers (BCF-98), both types having a length of 5 mm, and a diameter of 1 mm, was coupled to 16 photomultiplier tubes through a bunch of connecting optical fibers, to form a low resolution, radiation imaging device. The role of the short, clear optical fibers is to collect the perturbing Cerenkov radiation generated into the detecting system during the exposure to charge particle beams. In this way, the Cerenkov "noise" can be subtracted from the signal delivered by the scintillating optical fibers, improving in this way the detecting S/N. The set-up was used to evaluate the depth dose curves in a PMMA (polylmethylmetracrylate) phantom, over a water depth equivalent of 14 cm for proton, and respectively 3 cm for electron irradiation, the spatial resolution being given by the optical fibers spacing of 1 cm (Bartesaghi et al., 2007a). The tissue equivalent response of the detection scheme is given by the polystyrene core of the scintillating optical fibers. A 2D detection scheme based on the same principle was developed (Bartesaghi et al., 2007b). Attempts were made to embed a clear optical fiber into a boron ($B_2O_3$) doped scintillating tip in order to detect neutrons.

A 1D dosimeter composed of 25 individual sensors, each one fabricated by optically coupling a 4 mm/ 400 µm diameter piece of fused-quartz glass doped with $Cu^{1+}$ ions to a POF, which make possible the transfer of the luminescence signal from the radiation detector tips to a CCD read-out system, was used for mapping the radiation filed in computed tomography - CT (Jones & Hintenlang, 2008).

Another approach to compensate the detected signal collected from the scintillating optical fiber (BCF-60) for the Cerenkov radiation generated in the guiding optical fibers uses a CCD readout scheme in connection to two dichroic beamsplitters. The guiding optical fibers output both the fluorescence signal from the scintillator and the Cerenkov radiation. The scintillating material was selected so that it emits optical radiation in the 500 nm to 600 nm spectral band, while the Cerenkov radiation is predominant in the UV-blue spectral range. One of the beamsplitters reflects towards the CCD array the optical signals with wavelengths lower than 500 nm (mostly the Cerenkov radiation), while the other directs to the detector array the optical radiation between 500 nm and 600 nm. Using a suitable calibration procedure the amount of Cerenkov radiation is subtracted from the two optical signals detected by the CCD and the irradiation dose can then be computed (Frelin & Hintenlang, 2006).

In a recent implementation, the sensing tip consisted of a piece of BGO (Bismuth Germanate - $Bi_4Ge_3O_{12}$) attached to a POF, while the detection was done using a photomultiplier (Seo et al., 2011). The BGO was selected as being a very efficient gamma-ray to light converter.

An alternative design used a polystyrene-based organic scintillator (BCF-60, Saint-Gobain) coupled to a POF in order to separate the Cerenkov signal from the scintillating signal by wavelength discrimination. The optical signal was detected by a CCD camera (Jang et al., 2010a).

An extrinsic, off-line radiation dosimeter for beta radiation based on thermoluminescence in Ge-doped and Al-doped silica optical fibers was suggested (Yaakob et al., 2011). Samples of such optical fibers were exposed to 6, 9, and 12 MeV electron beams, together with classical TLD-100 detectors in order to evaluate the newly proposed detecting materials sensitivity, responsivity, linearity, stability and reproducibility. The optical signal realised upon annealing (done at 400 $^0$C for 1 h) was read by a Harshaw 4500 TL instrument. Ge-doped optical fibers proved much more sensitive that the Al-doped ones, and exhibit a quite good linearity. The tested optical fibers make repeated usage possible, as no major degradation appeared.

To avoid the optical noise introduced by the Cerenkov radiation in the UV-visible optical spectral range a new type of optical fiber extrinsic dosimeter was proposed. In this approach, a $Gd_2O_2S$-based scintillator was developed by rare earth ions doping ($Pr^{3+}$, $Yb^{3+}$ and $Nd^{3+}$). The scintillating material was optically coupled to a set of optical fibers in order to guide the light to a CCD detector (Takada et al., 1999). In front of the CCD camera the optical signal, guided by the optical fiber bunch, is filtered with an optical high-pass filter. The system has the advantage that the scintillator emits in this case in the IR, hence, the Cerenkov radiation can be cut-down, improving the S/N ratio on the detecting side. To compensate for the radiation induced degradation of the optical fibers transmission the system was complemented with an Optical Time Domain Reflectometer (OTDR). This instrument monitors and localises the increase of the absorption along an optical fiber, of the same type as those used as light guide fibers, which was placed parallel to the connecting optical fibers. In this way, corrections are introduced to compensate for the degradation of the optical links.

Extrinsic optical fiber radiation sensors, proposed in the mid-60s, rely on the Optically Stimulated Luminescence (OSL). The operation of such a detector is based on semiconductor or insulator materials having, in their wide band gap, two additional energy levels. By exposing the material to ionizing radiation, the energy level close to the conduction band is populated, with the number of the trapped electrons being, more or less, proportional to the irradiation dose. At the end of the irradiation process, the material is exposed to an optical radiation at a specific wavelength and the trapped charges are transferred to the second energy level (located also in the band gap, closer to the valence band). By the recombination of these charges, an optical radiation, representing the reading signal of the detector, is emitted (Aznar[a], 2005; Beddar, n.d.; Edmund, 2007). The most common OSL materials used are: MgS:Ce,Sm; SrS:Eu,Sm; NaCl:Cu; KCI:Eu; KBr:In; RbBr:Ti; BbI:In; A-AI2O3; CaS:Ce,Sm; SrS:Ce,Sm (Liu et al., 2008); KBr:Eu (Klein & McKeever, 2008). Such a sensor can exhibit a very low detecting threshold (down to 10 μGy) and a dynamic range over 6 decades (Benoit et al., 2007). Depending on the materials used, the reading optical radiation is in the visible to near-IR spectral range, while the detected optical radiation is mostly in the UV-visible domain. As the two radiations are spectrally separated the same optical fiber can be employed to carry-out both the stimulation signal (from a light source to the ionising radiation detector) and the luminescence one (from the ionising radiation detector towards the optical detector (photomultiplier, CCD, photodiode)), the luminescence material being optically coupled to one end of the signal guiding optical fiber.

An on-line, OSL dosimeter for the evaluation of both dose rate and total dose was developed by optically coupling a solid state dosimeter of carbon doped aluminium oxide (Al2O3:C) to an optical fiber (Andersen et al., 2002). During radiation exposure, some electrons are trapped for a short time in lattice defects of the solid state material, and, as they recombine with holes, a radioluminescence (RL) signal is emitted. In the mean time, additional charges continue to be trapped, as previously described, until an optical radiation stimulus, with an appropriate wavelength laser beam (in this case, $\lambda$ = 532 nm, 20 mW), realise them and, through recombination with holes, a visible radiation is generated through OSL (Aznar, 2005). Both the excitation light and the detected optical signals (RL & OSL) are guided by the same optical fiber from the laser source and towards the detection/ processing unit. A dichroic beamsplitter is used to couple the stimulation laser radiation to the optical fiber and to separate it from the detected signal, during the detection step. The radioluminescence (RL) signal can be correlated to the irradiation dose rate, while the post-irradiation optically stimulated (OSL) signal can be a measure of the total dose, if appropriate calibrations are performed. That is: the RL signal is increasing with the total dose received, and does not depend on the dose rate, acquired dose history or the radiation quantity. If the measurements are done over well defined time intervals, the change of the RL signal corresponding to each time interval is proportional to the dose rate (Andersen, et al., 2009). The equipment was tested with good results (linearity, sensitivity to Cerenkov radiation, reproducibility, angular dependence) under gamma ray, X rays and electron beam irradiation. For X-ray related applications the best reproducibility values achieved were 1.9 % (RL) and 2.7 % (OSL) at 1 $\sigma$, with a good linearity up to 30 mGy air kerma (Aznar et al., 2005).

According to the previously described detecting scheme (RL measurement during the irradiation and OSL data acquisition after the irradiation was stopped), both the dose rate and the total dose can be calculated from the RL signal. Unfortunately, these results are affected by the Cerenkov radiation. A correction can be made as the total dose computed from the RL signal is evaluated against the total dose obtained form the OSL signal, when no Cerenkov radiation is present. An improvement for this method was proposed when, during the irradiation interval, OSL reading is simultaneously run with the RL signal detection (which in turn is of course accompanied by Cerenkov radiation). In this case, equilibrium has to be established between the filling rate of the trapping levels due to irradiation and their depletion rate produced by optical stimulation. From the acquired data the dose rate and total dose can be derived. The method permits the suppression of the background noise and avoids the saturation of the OSL detector as trapped charges are periodically removed (Gaza et al., 2004). The "dynamic depletion" mechanism was applied also in the case of other OSL materials such as KBr:Eu (Gaza & McKeever, 2006).

The principle of operation discussed above was extended to an array of 15 multiplexed OSL and RL sensors (Magne et al., 2008). In designing such a system several characteristics have to be considered for the validation of the sensor: response linearity, the temperature dependence of the OSL and RL signals, reproducibility of one channel readings and from channel-to-channel, fading (decrease of the read signal occurring between the irradiation and the reading moment).

In designing an OSL sensor aspects related to the dependence of the RL and OSL signals on parameters such as temperature (during the irradiation and optical stimulation) and linear energy transfer - LET (in the case of charge particles: electrons, protons or heavy charged particles - HCP) have to be considered (Edmund, 2007).

A more complex extrinsic radiation dosimeter, capable of evaluating separately gamma-ray and thermal neutrons under a mixed irradiation, was devised (Jang et al., 2010b). It consists of two detectors based on the BCF-20, Saint Gobin scintillating optical fiber coupled through POFs to photomultipliers for the optical radiation detection. In addition, one of the scintillating detectors has a $^{6}Li$ cap used for the conversion of incoming thermal neutrons energy to alpha particles. The sensor was tested against pure gamma-ray and thermal neutrons irradiation by exposing it, separately, to a $^{60}Co$ gamma source and $^{252}Cf$ source of fast neutrons. In the last situation, thermal neutrons were obtained by interposing a 5-cm-thick polyethylene block between the source and the detector. During the gamma irradiation both detectors (with and without the $^{6}Li$ converter) generated the same amount of optical radiation. In the second experiment, when they were exposed only to a mixed gamma - neutron field, the signal from the detector, which includes also the converting material, provides information on both gamma and neutron irradiation, while the scintillating optical fiber generates optical radiation induced only by the gamma irradiation. By subtracting the second signal from the first one the contribution of thermal neutrons can be deduced. By adding a third, dummy optical fiber, which does not include any scintillating material, the background Cerenkov radiation can be taken into account and the measurement results can be compensated for it (Jang et al., 2011b).

Different types of plastic scintillators are available for gamma, alpha, beta and neutron detection, with some of them having short scintillating constants, or operating under high temperatures (Beddar, Bayramov & Sardarly, 2008).

Detection based on radiation induced luminescence was evaluated for neutron in ZnS-Ag (peak $\lambda$ = 450 nm), ZnS-Cu (peak $\lambda$ = 570 nm), $SrAl_20_4$-EuDy (peak $\lambda$ = 500 nm), and gamma-ray in $Al_20_3$-Cr, $Eu0_2$, and $GdO_2$, when such materials were attached at the end of a radiation resistant optical fiber. Such a fiber only has the role in guiding the optical signal to the optical detection part of the sensor (Shikama et al., 2004). Other implementations used LTB:Mn; LTB:Cu and LTB:Cu,Ag,P (Santiago et al., 2009); GaN (Pittet et al., 2009).

Three types of inorganic scintillating material ($Gd_2O_2S$:Tb, $Y_3Al_5O_{12}$:Ce, and CsI:Tl) fixed with an optical epoxy to the end of a bundle of multimode POFs, used to guide the scintillating radiation towards a photomultiplier, were tested as a tritium sensors (Jang et al., 2010c). Radioactive tritium emits beta rays, which incident on the scintillating material produce optical radiation in the 455 nm - 550 nm spectral range. As the energy of the beta radiation used is very low, the detecting material thickness was designed to be 0.1 mm.

When using scintillating materials it is of interest to assess the stability and the degradation of their characteristics under irradiation. Tests were carried out on PWO (Lead Tungstate - $PbWO_4$) LSO ($Lu_2$-$SiO_5$:Ce), and LYSO ($Lu_{2(1-x)}Y_{2x}SiO_5$:Ce crystals (Mao et al., 2009). Upon irradiation, such scintillating crystals can exhibit an increase of the optical absorption, a decrease of the sensitivity, post irradiation phosphorescence (afterglow), a change of the spatial uniformity of their response to radiation. Similar investigations were done on other inorganic scintillators such as: BGO - $Bi_4Ge_3O_{12}$, GSO - $Gd_2SiO_5$, CWO - $CdWO_5$ (Seo et al., 2009).

A comparison between an inorganic scintillating material ($Al_2O_3$:C) and an organic scitillator (BCF-12) reveals their advantages and drawbacks (Beierholm et al., 2009):

a. phosphorescence following the irradiation is present in Al2O3:C;
b. as charge carriers are trapped during the irradiation, Al2O3:C exhibits a memory effect making it suitable for OSL sensors production;
c. Al2O3:C has a higher sensitivity than the organic scintillator, but this parameter changes increases with the total dose;
d. Al2O3:C is sensitive to temperature;
e. the temporal response constitute an advantage for BCF-12 material (3.2 ns luminescence life time) as compared to 35 ms for $Al_2O_3$:C.

Another design of an extrinsic optical fiber radiation detector is based on a phosphorescent tip fixed at the end of a POF. The fiber is prepared by carefully paring back the outer cable insulation of the PMMA fiber using a special tool, exposing a specific length of the PMMA fiber, as shown in Fig. 15 *a*. The exposed fiber is inserted and secured into a plastic cylinder of known uniform diameter. A plastic sleeve is inserted over the fiber cable to ensure relative centering of the fiber core in the plastic molding housing. Fig. 15 *b* shows the PMMA fiber inserted into the plastic cylinder. Once secured, the next step is to prepare the scintillating phosphor material and epoxy mix. The epoxy is formed by adding a known quantity of hardener to another known quantity of resin. While the epoxy is in a liquidated state, a known amount of scintillating phosphor material is combined with the liquidised epoxy. The next step is to inject the epoxy/scintillating phosphor mix into the sealed cylinder mold and let the mixture harden. After a set period of time, the scintillating coated fiber is removed from its mold. Fig. 15 *c* shows the completed fabrication. The fabrication process is described in more detail elsewhere (McCarthy et al., 2010).

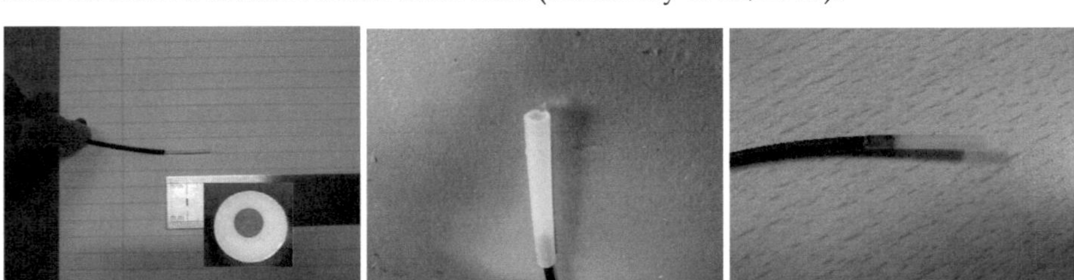

a b c

Fig. 15. The preparation sequences of the phosphorous based extrinsic optical fiber radiation sensor (McCarthy et al., 2010)

The resultant fiber optic sensor described above has been tested and found to respond to excitation from low X-ray energies. The fiber optic sensor connects to a computer controlled fluorescence spectrometer which interprets low levels of light received from the sensor upon excitation from an X-ray source. The fiber optic sensor was exposed to 90 keV X-ray energy and the resultant spectra can be seen below in Fig. 16. The peak wavelength response from the sensor was found to be 544 nm. Sensors using different phosphorus were evaluated under X-ray irradiation in Ireland (broadband X-ray source, energy from 20 keV to 140 keV) and Romania (X-ray quasi-monochromatic 17.5 keV source, I = 800 mA, V = 40 kV), and the emission spectra were recorded with the QE65000 scientific grade spectrometer.

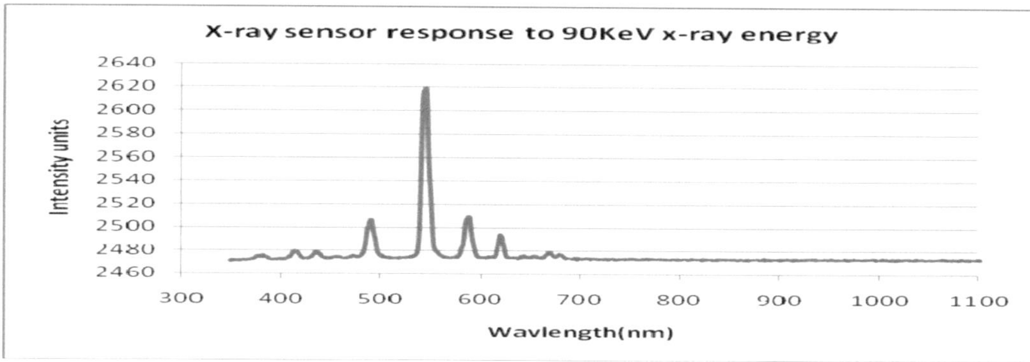


Fig. 16. The spectral response of the X-ray sensor exposed to a 90 keV radiation (McCarthy et al., 2011).

Investigations on the responsivity along the sensor length were carried out by scanning the detector tip along a line parallel to the optical fiber axis (Fig. 17), with a focused (25 μm) X-ray beam (McCarthy et al., 2011).

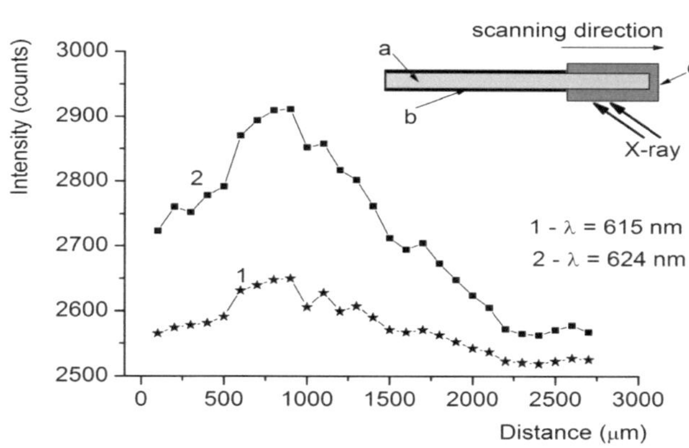


Fig. 17. The spatial responsivity of an extrinsic X-ray sensor for the two peak wavelengths. Insert: a – the POF; b – lightproof jacket; c - phosphorus tip. The focused X-ray beam is scanned along the length of the detecting tip and the profile of the X-ray generated luminescence, as collected by the POSF, is plotted against the distance (McCarthy et al., 2011).

Ge-doped silica optical fibers were investigated as possible thermoluminescence dosimetry (TLD) materials (Hashim et al., 2008). Pieces of 0.8 cm of such optical fibers, carefully measured for their weight (in order to normalize the result to the detector mass) were exposed to X-ray (2 Gy and 10 Gy doses) along with the classical TLD-700 (LiF:Mg,Ti), for comparative evaluation. The signal readout for both materials was done at 300 $^0$C for 25 s, at a heating cycle rate of 25 $^0$C/s, with pre-heating at 160 $^0$C for 10 s. Upon the reading the optical fiber samples were annealed to 400 $^0$C. The Ge-doped optical fibers have a linear response with the total dose up to 10 Gy. As in the case of other TLD materials, the tested optical fibers exhibit fading (drop of the thermoluminescent signal under lightproof conditions): a decay of 19.7 % after six days storage, as compared to the signal readout after the first 24 h from the irradiation, for the 10 Gy total dose, as compared to the 36.1 % drop for the TLD-700 for the same testing conditions.
Previous studies on commercially available optical fibers (Nokia Cable) used 5 mm long fiber samples irradiated by $^{60}$Co gamma radiation at low (from 0.01 Gy to 3 Gy) and high doses (from 4 Gy to 5 kGy). The erase of previously trapped carriers was performed by annealing the samples at temperature from 390 $^0$C to 490 $^0$C. The sensor response was compared to a standard TLD-100 dosimeter, as they were subjected to the same irradiation conditions and readings. The peak of the thermoluminescent signal from the optical fiber appeared at about 230 $^0$C, both for gamma and beta irradiation. Over the entire temperature range, the optical fiber thermally released optical radiation has two peaks, at 400 nm and 575 nm (Espinosa et al., 2006). This investigation did not indicate a significant fading over 15 months of periodic check. The $SiO_2$ optical fiber presented a higher sensitivity than the TLD-100, by 1.3 times.
In another design, 8 mm long pieces of Ge-doped silica optical fibers were investigated under gamma-ray ($^{192}$Ir, mean energy 397 keV) and beta (12 MeV energy) irradiation followed by thermoluminescence (TL) reading (Ong et al., 2009). The reported reproducibility is better that 6 %, the sensor having a linear response between 0.2 Gy and 12 Gy for both gamma and beta irradiation. The thermoluminescent signal does not depend on the dose rate and it is angular independent. The fading effect limits the practical use of these optical fibers as thermoluminescent sensors to two weeks.
The same type of optical fibers (Nokia Cable) was evaluated for their possible utilization in OSL dosimetry, as they were exposed to gamma-rays (Espinosa et al., 2011). This optical fiber has a high reproducibility, a low residual signal after several irradiation-reading cycles and, it is estimated from their fading rate, that they can be employed for about six days after the irradiation.
Post irradiation luminescence and the thermoluminescence signals obtained by off-line investigations, from $^{60}$Co gamma irradiated COTS optical fibers, are illustrated in Fig. 18 (Sporea et al., 2010b). Depending on the characteristics of the generated optical signal (linearity, reproducibility, sensitivity, etc.) commercially available optical fibers can be incorporated into on-line/off-line radiation detecting schemes, by monitoring the peak value of the emission at specific wavelengths.
Another approach for the design of an extrinsic optical fiber radiation monitor can be developed around Ge-doped silica glass. Figure 19 illustrates the change of the optical absorption and the RL signal as such a glass is irradiated by alpha particles. The optical absorbance measurements were done off-line, while the RL emission was detected on-line using the system described in section 4. In considering the inclusion of such a glass into a real time dose rate meter the same corrections have to be made on the RL signal to

compensate for the decrease of the optical transmission of the glass. Such a detector is a single-use type as far as no recovery of the radiation induced color centers was observed (Sporea et al., 2010b). The optical absorption is linked to the total deposed charge, while the RL signal is a measure of the instant beam current.

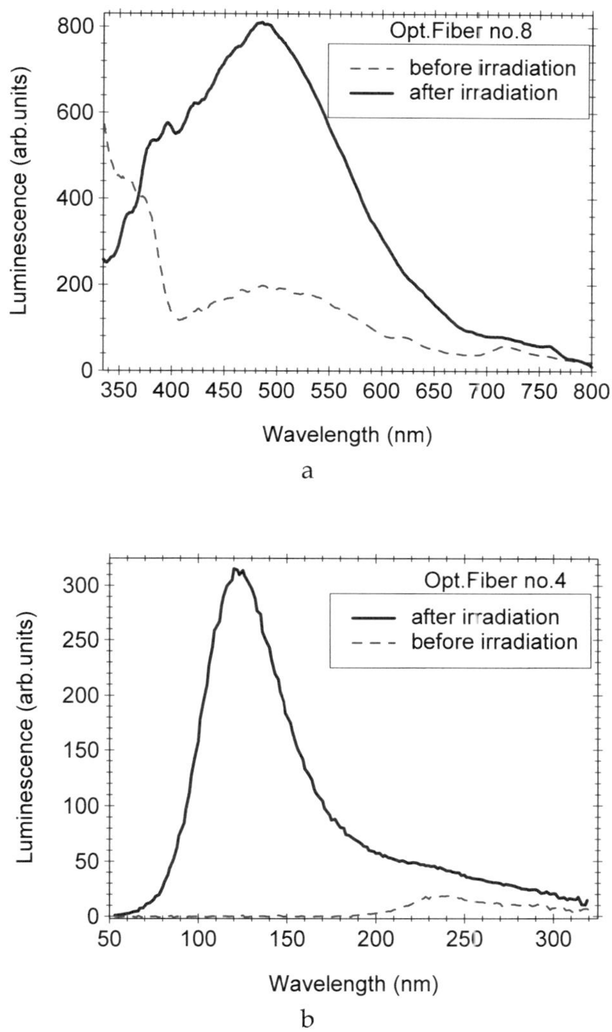


Fig. 18. The luminescence (a) and the (b) thermoluminescence signals after $^{60}Co$ gamma irradiation (total dose 300 Gy, dose rate 158 Gy/h) for: a multimode step-index 400 µm silica core, visible/near-IR optical fiber, with acrylate coating (a); a 600 µm core diameter, deep UV-enhanced, Polyimide jacket optical fiber (b) (Sporea et al.. 2010b).

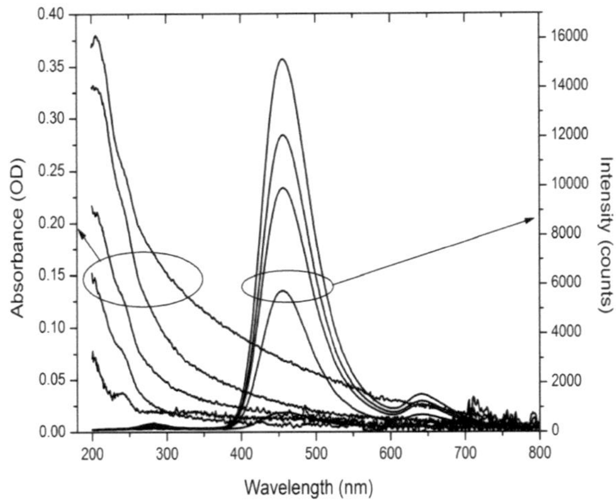


Fig. 19. The changes induced by alpha particle irradiation in a Ge-doped silica glass: a – incease of the optical absorption (left site curves), for various total deposed charge; b – the radioluminescence signal (right site curves), for different instant values of the beam current (Sporea et al., 2010b)

An optical fiber-based system for on-line dosimetry was designed around an optically transparent thermoluminescent glass material containing ZnS nanocrystals and $Cu^{11}$ ions (Huston et al., 1996). In this approach, the thermoluminescent material is optically coupled to a multimode optical fiber. $Nd^{3+}$ ions were additionally included into the ZnS nanocrystals to increase its absorption for the radiation from a semiconductor laser diode emitting 1 W at 807 nm, which is directed through this optical fiber towards the ionizing radiation (the irradiation was done with a $^{60}Co$ source) detecting material. The laser radiation is used in this case to heat the thermoluminescent material, which generates an optical readout signal. This signal is picked-up by the same optical fiber and is directed towards a photomultiplier by a dichroic beamsplitter combined with some color filters. The luminescence signal has in this application two TL peaks at 160 $^0$C and 220 $^0$C. The sensor presents a good linearity with the total dose of doses of medical interest (up to 6 Gy).
Earlier, a multiplexed version was proposed in which the output of a laser diode ($\lambda$ = 840 nm) coupled to an optical fiber is switched between several "probe" optical fibers. Each such "probe" has fixed at its other end an ionizing radiation detection element. Two such detecting elements were suggested: one formed by a phosphorescent material covered by an absorber and glutted directly to the optical fiber end, or, alternatively, a piece of phosphor, covered by an absorber and mounted to the optical fiber, but isolated with a gap of about 0.1 mm (Jones et al., 1993). As the laser radiation strikes the absorber, its energy is transferred to the phosphor and the thermoluminescent effect appears (at about 400 mW of optical power, reading temperature of 220 $^0$C – 240 $^0$C) if the radiation detector was previously irradiated. The generated optical signal is collected by the same system of optical fiber probes and is directed through a wavelength selective beamsplitter towards a photomultiplier. In the first approach, the optical coupling for the thermally generated signal is a better one, but part of

the energy of the laser reading beam is also transferred to the coupling optical fiber. A better sensitivity, from the thermoluminescence point of view, is achieved with the second solution, which has the drawback of a poorer optical coupling for the reading signal. The spectrum of the emitted optical radiation, as well as the temperature of the peak emission, depend on the phosphorus used (in the reported case they are: $CaSO_2$:Mn; $CaF_2$:Tm; $CaF_2$:Dy; $CaSO_4$:Dy). The system proved a good linearity over the 1 mGy to 100 Gy range, under $^{60}$Co irradiation. In this application it is worth underlining the opportunity to operate with one reading laser and one detector a multitude of multiplexed miniature dosimeters.

## 6. Future work

Considering the intrinsic characteristics of optical fiber systems: capabilities to work under strong electromagnetic fields; possibility to carry multiplexed signals (time, wavelength multiplexing); small size and low mass; ability to handle multi-parameter measurements in distributed configuration; possibility to monitor sites far away from the controller, their availability to be incorporated into the monitored structure, numerous groups around the world investigated in the past 20 years the possibility to include optical fiber-based devices into radiation monitoring systems. Most of the implementations relay on radiation induced effects such as: increase of the optical absorption on the optical fiber, generation of optical radiation during the irradiation conditions or post irradiation. The chapter evaluates the main solutions promoted for radiation detectors using the optical fiber as a detecting medium or as a light guide for optical signal transmission. The results obtained up to now are encouraging, but the investigative efforts have to continue as it concerns some basic parameters of these devices: sensitivity, linearity, immunity to dose rate and temperature, dynamic range, response time, etc. Besides the main types of optical fibers mentioned in this review, additional research has to be carried out on novel species of optical waveguides (i.e. microstructured optical fibers, photonic optical fibers, mid-IR optical fibers, new scintillating optical fibers). Tests were also carried out on some plastic optical fibers to evaluate their use as total dose monitors considering the increase of their optical absorption at specific wavelengths.

Further investigations have to be done to check for the reproducibility and the stability of the irradiation induced effects for the proposed optical fibers, as well as the possibility to develop re-usable sensors. The level of the fading effect in the case of sensing materials interrogated by thermoluminescnece has to be verified.

For the radioluminescence detection with sapphire optical fibers and doped silica glasses we intent to proceed to the evaluation of the linearity of the emitted signal as function of the beam parameters (beam current, focusing system current, etc.).

The proposed extrinsic POF sensor has to be improved as it concerns the uniformity of the luminescence signals along the sensing tip, as well as the optical coupling of the luminescence emission to the guiding optical fiber. Additionally, we intend to check its linearity for X-rays irradiation, at higher energies. An extension of the research done on this sensor will be the assessment of beta-ray induced luminescence, in conjunction with the evaluation of effects produced by those irradiations (X-ray and beta-ray) on the optical transmission of the coupling optical fibers.

## 7. Conclusions

An extended review of optical fiber radiation sensors was introduced, with these type of sensors finding applications in various fields, from nuclear industry to radiotherapy, from

space science to particle accelerators. Radiation effects in optical fibers (silica glass, plastic and sapphire) were described along with the classification of the detectors based on intrinsic and extrinsic designs. Examples of authors' experimental results were also included in relation to the tests carried out on silica, plastic and sapphire optical fibers. The evaluation of such optical fibers for the development of radiation monitoring sensors was described.
The chapter emphasis the advantages optical fibers can have in radiation dosimetry: real time measurements (under specific design); small size (able to be used as inside the body or implanted sensors); remote investigations (useful for radiation therapy, particle accelerators, nuclear sites); able to be included into multiplexed or distributed sensor arrays (particle beam diagnostics, monitoring of contaminated sites); susceptible for passive (luminescence, Cerenkov radiation detectors) or active (thermoluminescence, Optically Stimulated Luminescence - OSL) sensing, some of them being water equivalent (an advantage for medical applications).

## 8. Acknowledgements

The work was done in the frame of the bilateral collaboration existing between the National Institute for Laser, Plasma and Radiation Physics in Bucharest, and the University of Limerick, as part of the COST Action TD1001: *Novel and Reliable Optical Fibre Sensor Systems for Future Security and Safety Applications (OFSeSa)*. The Romanian authors acknowledge the financial support of the Romanian Ministry for Education, Research, Sport and Youth in the frame of the research grant 12084/2008. They also want to thank to their colleagues: Dr. S. Agnello, for providing the Ge-doped silica glass sample, Ms. R. Georgescu and Mr. D. Negut for the gamma irradiation, Dr. C. Oproiu for performing the beta-ray irradiation, Dr. M. Secu for running the luminescence and thermoluminescence measurements, Dr. I. Tiseanu for assisting with X-ray irradiation, Dr. I. Vata for help in doing alpha particles and neutron irradiations. The Irish team wishes to acknowledge the support of the European Commission under the 7th Framework Programme through the 'Marie Curie Re-integration' action of the 'Peoples' Programme, (PERG04-2008-239207).

## 9. References

Adinolfi, M.; Angelini, C.; Antinori, F.; Beusch, W.; Cardini, A.; Crennell, D.J.; De Vincenzi, M.; Da Vi, C.; Di Paolo, M.; Di Vita, G. ; Duane, A.; Fabre, J.-P.; Flaminio, V.; Frenkel, A.; Gys, T.; Harrison, K.; Lamanna, E. ; Lucchesi¸ D.; Martelloti, G.; McEwen, J.G.; Morrison, D.R.O.; Penso, G.; Petrera, S.; Roda, C.M.A.; Sciubba, A.; Villalobos-Baillie, O. & Websdale, D.M. (1991). Application of a scintillating-fibre detector to the study of short-lived particle, *Nucl. Instr. Met. Phys. Res.*, Vol. A310, (1991), (485-489)

Akchurin, N.; Atramentov, O.; Carrell, K.; Gümüş, K.Z.; Hauptman, J.; Kim, H.; Paar, H.P.; Penzo, A. & Wigmansa, R. (2005). Separation ofscintillation and Cherenkov light in an optical calorimeter, *Nucl. Instr. Met. Phys. Res.*, Vol. A550, (2005), (185-200), doi:10.1016/j.nima.2005.03.175

Alasia, D.; Fernandez Fernandez, A.; Abrardi1, L.; Brichard, B. & Thévenaz, L. (2006). The effects of gamma-radiation on the properties of Brillouin scattering in standard Ge-doped optical fibres, *Meas. Sci. Technol.*, Vol. 17, (2006), (1091–1094), doi:10.1088/0957-0233/17/5/S25

Alessi, A.; Agnello, S.; Sporea, D.G.; Oproiu, C.; Brichard, B. & Gelardi, F.M. (2010). Formation of optically active oxygen deficient centers in Ge-doped $SiO_2$ by γ- and β-ray irradiation, *J. Non-Cryst. Solids*, Vol. 356, (2010), (275–280), doi:10.1016/j.jnoncrysol.2009.11.016

Alfeeli, B.; Pickrell, G.; Garland, M.A. & Wang, A. (2007). Behavior of random hole optical fibers under gamma ray irradiation and its potential use in radiation sensing applications, *Sensors*, Vol. 7, (2007), (676-688)

Andersen, C.E.; Aznar, M.C.; Bøtter-Jensen, L.; Bäck, S.Å.J.; Mattsson, S. & Medin, J. (2002), Development of optical fibre luminescence techniques for real time in vivo dosimetry in radiotherapy, *Proceedings of the the International Symposium "Standards and codes of practice in medical radiation dosimetry"*, Vol. 2, pp. 353-360, Vienna, November 2002

Andersen, C.E.; Nielsen, S.K.; Greilich, S.: Helt-Hansen, J.; Lindegaard, J.C. & Tanderup, K. (2009). Characterization of a fiber-coupled $Al_2O_3$:C luminescence dosimetry system for online *in vivo* dose verification during $^{192}$Ir brachytherapy, *Med. Phys.*, Vol. 36 No. 3, (March 2009), (708-718), DOI: 10.1118/1.3063006

Angelini, C.; Beusch, W.; Bloodworth, I.J.; Carney, J.N.; Crennell, D.J.; De Vincenzi, M.; Duane, A.; Fabre, J.P.; Fisher, C.M.; Flaminio, V.; Frenkel, A.; Harrison, K.; Hughes, P.; Kinson, J.B.; Lamanna, E.; Leutz, H.; Martellotti, G.; McEwen, J.G.; Morrison, D.R.O.; Penso, G.; Petrera, S.; Quercigh, E.; Roda, C.; Sciubba, A.; Villalobos-Baillie, O.; Votruba, M.F. & Websdale, D.M. (1989). High-resolution tracking with scintillating fibers, *Nucl. Instr. Methods Phys. Res.*, Vol. A277, (1989), (132-137)

Archambault, L. (2005). *Elaboration d'un dosimeter à fibres scintillantes*, PhD Thesis, Faculté des études supérieures de l'Université Laval

Arvidsson, B.; Dunn, K.; Issever, C.; Huffman, B.T.; Jones, M.; Kierstead, J.; Kuyt, G.; Liu, T.; Povey, A.; Regnier, E.; Weidberg, A.R.; Xiang A. & Yef, J. (2009). The radiation tolerance of specific optical fibres exposed to 650 kGy(Si) of ionizing radiation, *JINST 4 P07010*, (2009), doi:10.1088/1748-0221/4/07/P07010

Atkinson, M.N. ; Crennell, D.J.; Fisher, C.M. ; Hughes, P.T.; Kirkby, J.; Fent, J.; Freund, P.; Osthoff, A.; & Pretzl, K. (1988). A high resolution scintillating fiber (SCIFI) tracking device with CCD readout, *Nucl. Instr. Met. Phys. Res.*, Vol. A263, (1988), (333-342)

Aznar, M.C. (2005), *Real-time in vivo luminescence dosimetry in radiotherapy and mammography using $Al_2O_3$:C*, PhD Thesis, University of Copenhagen, 2005

Aznar, M.C.; Hemdal, B.; Medin, J.; Marckmann, C.J.; Andersen, C.E.; Bötter-Jensen, L.; Andersson, I & Mattsson, S. (2005). In vivo absorbed dose measurements in mammography using a new real-time luminescence technique, *Brit. J.Radiol.*, Vol. 78, (2005), (328–334), doi: 10.1259/bjr/22554286

Bahrdt, J.; Feikes, J.; Frentrup, W.; Gaupp, A.; v. Hartrott, M.; Scheer, M. & Wüstefeld, G. (2009). Cherenkov fibers for beam diagnostics at the metrology light source, *Proceedings of the 23rd Particle Accelerator Conference,* paper TU5RFP029, Vancouver, May 2009

Bartesaghi[a], G.; Conti, V.; Prest, M.; Mascagna, V.; Scazzi, S.; Cappelletti, P.; Frigerio, M.; Gelosa, S.; Monti, A.; Ostinelli, A.; Mozzanica, A.; Bevilacqua, R.; Giannini, G.; Totaro, P. & Vallazza E. (2007). A real time scintillating fiber dosimeter for gamma and neutron monitoring on radiotherapy accelerators, *Nucl. Instr. Methods Phys. Res., A*, Vol. 572, (2007), (228-230), doi:10.1016/j.nima.2006.10.323

Bartesaghi[b], G.; Conti, V.; Bolognini, D.; Grigioni, S.; Mascagna, V.; Prest, M.; Scazzi, S.; Mozzanica, A.; Cappelletti, P.; Frigerio, M.; Gelosa, S.; Monti, A.; Ostinelli, A.; Giannini, G. & Vallazza E. (2007). A scintillating fiber dosimeter for radiotherapy, *Nucl. Instr. Methods Phys. Res.*, Vol. A581, (2007), (80–83), doi:10.1016/j.nima.2007.07.032

Bayramov, A.A. & Sardarly, R.M. (2008). Applications of plastic sensors, *Fizika*, Vol. CILD XIV, No. 3, (2008), (149-153)

Beddar, S. (n.d). Scintillation dosimetry: Review, new innovations and applications, http://www.aapm.org/meetings/09SS/documents/32Beddar-PlasticDosimeters.pdf

Beierholm, A.R.; Andersen, C.E.; Lindvold, L.R.; Kjær-Kristoffersenc, F. & Medinc, J., A comparison of BCF-12 organic scintillators and $Al_2O_3$:C crystals for real-time medical dosimetry, *Radiat. Meas.*, Vol. 43, (2008), (898–903), doi:10.1016/j.radmeas.2007.12.032

Benoit, D.; Vaillé, J-R.; Ravotti, F.; Garcia, P. & Dusseau, L. (2007). Optically Stimulated Luminescence Dosimetry, *1st Workshop on Instrumentation for Charged Particle Therapy*, London, May 2007

Berghmans, F. (2006). Ionizing radiation effects on optical components, *NATO Advanced Study Institute, Optical Waveguide Sensing & Imaging in Medicine, Environment, Security & Defence*, Gatineau, October 2006

Berghmans, F.; Brichard, B.; Fernandez Fernandez, A.; Gusarov, A.; Van Uffelen. M. & Girard, S. (2008). An Introduction to Radiation Effects on Optical Components and Fiber Optic Sensors, *Optical waveguide sensing and imaging*, Bock, W.J.; Gannot, I. & Tanev, S., pp. 127-166, Springer Series B: Physics and Biophysics, Dordrecht, The Netherland

Bisutti, J.; Girard, S. & Baggio, J. (2007). Radiation effects of 14 MeV neutrons on germanosilicate and phosphorus-doped multimode optical fibers. *J. Non-Cryst. Solids*, Vol. 353, (2007), (461–465), doi:10.1016/j.jnoncrysol.2006.10.013

Brichard, B.; Borgermans, P.; Fernandez Fernandez, A.; Lammens, K. & Decréton, M. (2001). Radiation effect in silica optical fiber exposed to intense mixed neutron–gamma radiation field, *IEEE T. Nucl. Sci.*, Vol. 48, No. 6, (December 2001), (2069-2073)

Brichard, B.; Fernandez Fernandez, A.; Ooms, H.; Berghmans, F.; Decréton, M.; Tomashuk, A.; Klyamkin, S.; Zabezhailov, M.; Nikolin, I.; Bogatyrjov, V.; Hodgson, E.; Kakuta, T.; Shikama, T.; Nishitani, T.; Costley, A. & Vayakis, G. (2004). Radiation-hardening techniques of dedicated optical fibres used in plasma diagnostic systems in ITER, *J. Nucl. Mater.*, Vol. 329–333, (2004), (1456–1460), doi:10.1016/j.jnucmat.2004.04.159

Brichard, B. & Fernandez Fernandez, A. (2005). Radiation effects in silica glass optical fibers, In: *Short Course Notebook, New Challenges for Radiation Tolerance Assessment*, pp. 95-138, Cap d'Agde, September 2005

Byrd, J.M.; De Santis, S. & Yin, Y. (2007). Fiberoptics-based instrumentation for storage ring beam diagnostics, *Proceedings of 8th European Workshop on Beam Diagnostics and Instrumentation for Particle Accelerator*, pp. 325-327, Venice, May 2007

Calderón, A.; Martínez-Rivero, C.; Matorras, F.; Rodrigo, T.; Sobrón, M.; Vila, I.; Virto, A.L.; Alberdi, J.; Arce, P.; Barcala, J.M.; Calvo, E.; Ferrando, A.; Josa, M.I.; Luque, J.M.; Molinero, A.; Navarrete, J.; Oller, J.C.; Valdivieso, P.; Yuste, C.; Fenyvesi, A. & Molnár, J. (2006). Effects of $\gamma$ and neutron irradiation on the optical absorption of pure silica core single-mode optical fibres from Nufern, *Nucl. Instr. Met. Phys. Res.*, Vol. A 565, (2006), (599–602), doi:10.1016/j.nima.2006.05.228

Cannas, M.; Lavinia, V. & Roberto, B. (2008). Time resolved photoluminescence associated with non-bridging oxygen hole centers in irradiated silica, *Nucl. Instr. Met. Phys. Res.*, Vol. B, No. 266, (2008), (2945–2948), doi:10.1016/j.nimb.2008.03.144

Caponero, M.A.; Baccaro, Stefania; Donisi, D.; Fabbri, F. & Pillon, M. (2007). Characterisation of FBG sensors under ionizing radiation for high energy physics and space physics, In: *Proceedings of the 10th Conference Astroparticles, Particles and Space Physics, Detectors and Medical Physics Applications*, pp. 533-539, Como, October 2007, doi: 10.1142/9789812819093_0092

Cerenkov, P.A., (1958). Radiation of particles moving at a velocity exceeding that of light, and some of the possibilities for their use in experimental physics, *Nobel Lecture*, December, 1958

Chiodini, N.; Vedda, A.; Fasoli, M.; Moretti F.; Lauria, A.; Cantone, M.-C.; Veronese, I.; Tosi, G.; Brambilla, M.; Cannillo, B.; Mones, E.; Brambilla, G. & Petrovich, M. (2009). Ce doped $SiO_2$ optical fibers for remote radiation sensing and measurement, *Proceeidngs of Fiber Optic Sensors and Applications VI. SPIE Defense and Security*, Udd, E.; Du, H.H. & Wang, A., Vol. 7316, doi.org/10.1117/12.818507

Connell, J.J.; Binns, W.R.; Dowkontt, P.F.; Epstein, J.W.; Israel, M.H.; Klann, J.; Webber, W.R. & Kish, J.C. (1990). The scintillating isotope experiment: BEVALAC calibrations of the test models, *Nucl. Instr. Methods in Phys. Res.*, Vol. 294, (1990), (335-350)

Deparis, O.; Griscom, D.L.; Mégret, P.; Decréton, M. & Blondel, M. (1997). Influence of the cladding thickness on the evolution of the NBOHC band in optical fibers exposed to gamma radiations, *J. Non-Cryst. Solids*, Vol. 216, (1997), (124-128)

Edmund, J.M. (2007). *Effects of temperature and ionization density in medical luminescence dosimetry using $Al_2O_3$:C*, PhD Thesis, University of Copenhagen, 2007

Espinosa, G.; Golzarri, J.I.; Bogard, J. & García-Macedo, J. (2006). Commercial optical fiber as TLD material, *Radiat. Prot. Dosim.*, Vol. 119, No. 1-4, (2006), (197-200), doi: 10.1093/rpd/nci564

Espinosa, G. (2011). A study and characterization of the optically stimulated luminescence response of commercial SiO2 optical fiber to gamma radiation, *Rev. Mex. Fís.*,Vol. 57, No. 1, (2011), (30–33)

Fernandez Fernandez, A.; Berghman, F.; Brichard, B.; Decréton, M.; Gusarov, A.I.; Deparis, O.; Mégret, P.; Blondel, M. & Delchambre, A. (2001). Multiplexed fiber Bragg grating sensors for in-core thermometry in nuclear reactors, *Fiber Optic Sensor Technology II, Proc. SPIE*, Vol. 4204, pp. 40-49 , March 2001, doi:10.1117/12.417427

Fernandez Fernandez, A.; Brichard, B., Berghmans, F. & Decréton, M. (2002). Dose-rate dependencies in gamma-irradiated fiber Bragg grating filters, *IEEE T. Nucl. Sci.*, Vol. 49, No. 6, (December 2002), (2874-2878), doi: 10.1109/TNS.2002.805985

Fernandez Fernandez, A.; Brichard, A. & Berghmans, F. (2003). In situ measurement of refractive index changes induced by gamma radiation in germanosilicate fibers,

*IEEE Photonic Tech. L.*, Vol. 15, No. 10, (October 2003), (1428-1430), doi: 10.1109/LPT.2003.818247

Florous, N.J.; Saitoh, K.; Murao, T. & Koshiba, M. (2007). Radiation dose enhancement in photonic crystal fiber bragg gratings: towards photo-ionization monitoring of irradiation sources in harsh nuclear power reactors, *Quantum Electronics and Laser Science Conference, QELS '07*, Baltimore, 2007, doi: 10.1109/QELS.2007.4431290

Friebele, E.J. (1991). Correlation of single mode fiber fabrication factors and radiation response, *Naval Research Laboratory _NRL_ Final Report No. NRL/MR/6505_92-6939*, November 1991

Frelin, A.-M.; Fontbonne, J.-M.; Ban, G.; Batalla, A.; Colin, J.; Isambert, A.; Labalme, M.; Leroux, T. & Vela, A. (2006). A new scintillating fiber dosimeter using a single optical fiber and a CCD camera, *IEEE T. Nucl. Sci.*, Vol. 53, No. 2, (June 2006), (1113-1117), doi:10.1109/TNS.2006.874931

Gaza; R.; McKeever, S.W.S.; Akselrod, M.S.; Akselrod, A.; Underwood, T.; Yoder, C.; Andersen, C.E.; Aznar, M.C.; Marckmann, C.J. & Bötter-Jensen, L. (2004). A fiber-dosimetry method based on OSL from $Al_2O_3$: C for radiotherapy applications, *Radiat. Meas.*, Vol. 38, (2004), (809–812), doi:10.1016/j.radmeas.2003.12.004

Gaza, R. & McKeever, S.W.S. (2006). A real-time, high-resolution optical fiber dosimeter based on optically stimulated luminescence (OSL) of KBr:Eu, for potential use during the radiotherapy of cancer, *Radiat. Prot. Dosim.* (2006), Vol. 120, No. 1–4, (14–19), doi:10.1093/rpd/nci603

Girard[a], S.; Keurinck, J.; Boukenter, A.; Meunier, J.-P.; Ouerdane, Y.; Azaïs, B.; Charre, P. & Vié, P. (2004). Gamma-rays and pulsed X-ray radiation responses of nitrogen-, germanium-doped and pure silica core optical fibers, *Nucl. Instr. Met. Phys. Res.*, Vol. B215, (2004), (187–195), doi:10.1016/j.nimb.2003.08.028

Girard[b], S.; Keurinck, J.; Ouerdane, Y.; Meunier, J.-P. & Boukenter, A. (2004). γ-rays and pulsed X-ray radiation responses of germanosilicate single-mode optical fibers: influence of cladding codopants, *J. Lightwave Technol.*, Vol. 22, No. 8, (2004), (1915 - 1922), doi: 10.1109/JLT.2004.832435

Girard, S.; Vincent, B.; Ouerdane, Y.; Boukenter, A.; Meunier, J.-P. & Boudrioua, A. (2005). Luminescence spectroscopy of point defects in silica-based optical fibers, J. Non-Cryst. Solids, Vol. 351, (2005), (1830–1834), doi:10.1016/j.jnoncrysol.2005.04.043

Girard, S.; Ouerdane, Y.; Boukenter, A. & Meunier, J.-P. (2006). Transient radiation responses of silica-based optical fibers: influence of modified chemical-vapor deposition process parameters, J. Appl. Phys., Vol. 99, (2006), doi: 10.1063/1.2161826

Girard, S.; Ouerdane, Y.; Origlio, G.; Marcandella, C.; Boukenter, A.; Richard, *N.*; Baggio, *J.*; Paillet,P.; Cannas, M.; Bisutti, J.; Meunier,J.-P. & Boscaino, R. (2008). Radiation effects on silica-based preforms and optical fibers—I: experimental study with canonical samples, *IEEE T. Nucl. Sci.*, Vol. 55, No. 6, (December 2008), (3473-3482), doi: 10.1109/TNS.2008.2007297

Girard, S.; Marcandella, C.; Origlio, G.; Ouerdane, Y.; Boukenter, A. & Meunier, J.-P. (2009). Radiation-induced defects in fluorine-doped silica-based optical fibers: Influence of a pre-loading with $H_2$, *J. Non-Cryst. Solids*, Vol. 355, (2009), (1089–1091), doi:10.1016/j.jnoncrysol.2008.11.035

Girard, S. & Marcandella, C. (2010). Transient and Steady State Radiation Responses of Solarization-Resistant Optical Fibers, *IEEE T. Nucl. Sci.*, Vol. 57, No. 4, (2049 - 2055), doi: 10.1109/TNS.2010.2042615

Girard, S.; Ouerdane, Y.; Marcandella, C.; Boukenter, A.; Quenard, S. & Authier, N. (2011). Feasibility of radiation dosimetry with phosphorus-doped optical fibers in the ultraviolet and visible domain, *J. Non-Cryst. Solids*, Vol. 357, (2011), (1871–1874), doi:10.1016/j.jnoncrysol.2010.11.113

Goettmann, W.; Wulf, F.; Körfer, M. & Kuhnhenn, J. (2005). Beam loss position monitor using Cerenkov radiation in optical fibers, *Proceedings of the 7th European Workshop on Beam Diagnostics and Instrumentation for Particle Accelerators*, pp. 301-303, Lyon, June 2005

Goettmann, W.; Körfer, M. & Wulf, F. (2007). Beam profile measurement with optical fiber sensors at FLASH, *Proceedings of 8th European Workshop on Beam Diagnostics and Instrumentation for Particle Accelerator*, pp. 123-125, Venice, May 2007

Griscom, D.L.; Golant, K.M.; Tomashuk, A.L.; Pavlov, D.V. & Tarabrin, Yu.A. (1996). γ-radiation resistance of aluminum-coated all-silica optical fibers fabricated using different types of silica in the core, *Appl. Phys. Lett.*, Vol. 69, (1996), (322-324)

Gusarov[a], A.I.; Starodubov, D.S.; Berghmans, F.; Deparis, O.; Defosse, Y.; Fernandez Fernandez, A.; Decréton, M.; Mégret, P. & Blondel, M. (1999). Comparative study of MGy dose level g-radiation effect on FBGs written in different fibres, presented at *OFS-13*, Kyongju, April 1999

Gusarov[b], A. I.; Berghmans, F.; Deparis, O.; Fernandez Fernandez, A.; Defosse, Y.; Mégret, P.; Décreton, M. & Blondel, M. (1999), High total dose radiation effects on temperature sensing fiber bragg gratings, *IEEE Photon. Tech. L.*, Vol. 11, No. 9, (September 1999), (1159-1161)

Gusarov, A.; Fernandez Fernandez, A.; Vasiliev, S.; Medvedkov, O.; Blondel, M. & Berghmans, F. (2002). Effect of gamma-neutron nuclear reactor radiation on the properties of Bragg gratings written in photosensitive Ge-doped optical fiber, *Nucl. Instr. Met. Phys. Res.*, Vol. B187, (2002), (79-86)

Gusarov[a], A.; Vasiliev, S.; Medvedkov, O.; Mckenzie, I. & Berghmans, F. (2008). Stabilization of fiber Bragg gratings against gamma radiation, *IEEE T. Nucl. Sci.*, Vol. 55, No. 4, (2008), 2205 - 2212, doi: 10.1109/TNS.2008.2001038

Gusarov[b], A.; Chojetzki, C.; Mckenzie, I.; Thienpont, H. & Berghmans, F. (2008). Effect of the fiber coating on the radiation sensitivity of type I FBGs, *IEEE Photon. Tech. L.*, Vol. 20, No. 21, (November, 2008), (1802-1804), doi: 10.1109/LPT.2008.2004699

Gusarov, A.; Brichard, B. & Nikogosyan, D.N. (2010). Gamma-radiation effects on Bragg gratings written by femtosecond UV laser in Ge-doped fibers, *IEEE T. Nucl. Sci.*, Vol. 57, No. 4, (2010), (2024 - 2028), doi: 10.1109/TNS.2009.2039494

Hanafusa, H.; Hibino, Y. & Yamamoto, F. (1986). Drawing condition dependence of radiation-induced loss in optical fibres, *Electron. Lett.*, Vol. 22, No. 2, (1986), (106-108).

Hashim, S.; Ramli, A.T.; Bradley, D.A. & Wagiran,, H. (2008). The thermoluminescence response of Ge-doped optical fibers to X-ray photon irradiation, *J. Fiz. UTM.*, Vol. 3, (2008), (31-37)

Henschel, H.; Köhn, O. & Schmidt, H.U. (1992). Optical fibres as radiation dosimeters, *Nucl. Instr. Met. Phys. Res.*, Vol. B69, (1992), (307-314)

Henschel, H.; Körfer, M.; Wittenburg, K. & Wulf, F. (2000). Fiber optic radiation sensing systems for TESLA, *TESLA Report No. 2000-26*, September 2000

Henschel, H.; Körfer, M.; Kuhnhenn, J.; Weinand, U. & Wulf, F. (2004). Fibre optic radiation sensor systems for particle accelerators, *Nucl. Instr. Met. Phys. Res.*, Vol. A526, (2004), (537–550), doi:10.1016/j.nima.2004.02.030

Huston, A.L.; Justus, B.L. & Johnson, T.L. (1996). Fiber-optic-coupled, laser heated thermoluminescence dosimeter for remote radiation sensing, *Appl. Phys. Lett.*, Vol. 68, No. 24, (June 1996), (3377-3379)

Ichikawa, T., Mechanism of radiation-induced degradation of poly(methyl methacrylate) -- temperature effect, *Nucl. Instr. Met. Phys. Res. Section B: Beam Interactions with Materials and Atoms*, Vol. 105, No. 1-4, (1995), (150-153), doi:10.1016/0168-583X(95)00534-X

Intermite, A.; Putignano, M. & Welsch, C. P. (2009). Feasibility study of an optical fiber sensor for beam loss detection based on a SPAD array, *Proceedings of 9th European Workshop on Beam Diagnostics and Instrumentation for Particle Accelerators*, pp. 228-230, Basel, May 2009

Jang, K.W.; Cho, D.H.; Shin, S.H.; Yoo, W.J.; Seo, J.K.; Lee, B.; Kim, S.; Moon, J.H.; Cho, Y.-H. & Park, B.G. (2009). Characterization of a Scintillating Fiber-optic Dosimeter for Photon Beam Therapy, *Opt. Rev.*, Vol. 16, No. 3, (2009), (383–386)

Jang[a], K.W.; Cho, D.H.; Yoo, W.J.; Seo, J.K.; Heo, J.Y.; Lee, B.; Cho, Y.-H.; Park, B.G.; Moon, J.H. & Kim, S. (2010). Measurement of Cerenkov light in a fiber-optic radiation sensor by using high-energy photon and electron beams, *J. Korean Phys. Soc.*, Vol. 56, No. 3, (March 2010), (765-769), doi: 10.3938/jkps.56.765

Jang[b], K.W.; Yoo, W.J.; Park, J. & Lee, B. (2010). Development of scintillation-fiber sensors for measurements of thermal neutrons in mixed Neutron-gamma fields, *J. Korean Phys. Soc.*, Vol. 56, No. 6, (June 2010), (1777-1780), DOI: 10.3938/jkps.56.1777

Jang[c], K.W.; Cho, D.H.; Yoo, W.J.; Seo, J.K.; Heo, J.Y.; Park, J.-Y. & Lee, B. (2010). Fiber-optic radiation sensor for detection of tritium, *Nucl. Instr. Methods Phys. Res.*, doi:10.1016/j.nima.2010.09.060

Jang[a], K.W.; Yoo, W.J.; Seo, J.K.; Heo, J.Y.; Moon, J.; Park, J.-Y.; Kim, S.; Park, B.G. & Lee, B. (2011). Measurements and Removal of Cerenkov light generated in scintillating fiber-optic sensor induced by high-energy electron beams using a spectrometer, *Opt. Rev.*, Vol. 18, No. 1, (2011), (176–179)

Jang[b], K.W.; Lee, B. & Moon, J.H. (2011). Development and characterization of the integrated fiber-optic radiation sensor for the simultaneous detection of neutrons and gamma rays, *Appl. Radiat. Isotopes*, Vol. 69, (2011) (711–715), doi:10.1016/j.apradiso.2011.01.009

Jones, A.K. & Hintenlang, D. (2008). Potential clinical utility of a fiber optic-coupled dosemeter for dose measurements in diagnostic radiology, *Radiat. Prot. Dosim.*, Vol. 132, No. 1, (2008), (80–87), doi:10.1093/rpd/ncn252

Jones, S.C.; Hegland, J.E.; Hoffman, J.M. & Braunlich, P.F. (1990). A fibre-optic TLD microprobe for remote in-vivo radiotherapy dosimetry, *Radiat. Prot. Dosim.*, Vol. 34, No. 1/4, (1990), (279-282)

Jones, S.C.; Sweet, J.A.; Braunlich, P.; Hoffman, J.M. & Hegland, J.E. (1993). A remote fibre optic laser TLD system, *Radiat. Prot. Dosim.*, Vol. 47, No. 1/4, (1993), (525-528)

Justus, B.L.; Falkenstein, P.; Huston, A.L.; Plazas, M.C.; Ning, H. & Miller, R.W. (2006). Elimination of Cerenkov interference in a fibre-optic-coupled radiation dosimeter, *Radiat. Prot. Dosim.*, Vol. 120, No. 1–4, (2006), (20–23), doi:10.1093/rpd/nci525

Klein, D.M. & McKeever, S.W.S. (2008). Optically stimulated luminescence from KBr:Eu as a near-real-time dosimetry system, *Radiat. Meas.*, Vol. 43, (2008), (883 - 887), doi:10.1016/j.radmeas.2008.01.015

Krebber, K., Henschel, H. & Weinand, U. (2006). Fibre Bragg gratings as high dose radiation sensors?, *Meas. Sci. Technol.*, Vol. 17, (2006), (1095–1102), doi:10.1088/0957-0233/17/5/S26

Kuhnhenn, J.; Henschel, H.; Köhn, O. & Weinand, U. (2004). Thermal annealing of radiation dosimetry fibres, *Proceedings RADECS 2004*, pp. 39-42, Madrid, September 2004

Kuhnhenn, J. (2005). Radiation tolerant fibres for LHC controls and communications, *5th LHC Radiation Workshop*, CERN, November 2005

Kuyt, G.; Regnier, E. & Gilberti, R. (2006). Optical fiber behavior in radioactive environments, presented at *IEEE ICC meeting*, St-Peterburg, 2006

Lee[a], Bongsoo; Jang, Kyoung Won; Cho, Dong Hyun; Yoo, Wook Jae; Kim, Hyung Shik; Chung, Soon-Cheol & Yi, Jeong Han (2007). Development of one-dimensional fiber-optic radiation sensor for measuring dose distributions of high energy photon beams, *Opt. Rev.*, Vol. 14, No. 5 (2007) (351–354)

Lee[b], Bongsoo; Jang, Kyoung Won; Cho, Kyoung Won; Yoo, Wook Jae; Tack, Gye-Rae; Chung, Soon-Cheol; Kim, Sin & Cho, Hyosung. (2007). Measurements and elimination of Cherenkov light in fiber-optic scintillating detector for electron beam therapy dosimetry, *Nucl. Instr. Methods Phys. Res.* Vol. A579, (2007), (344-348), doi:10.1016/j.nima.2007.04.074

Liu, Y.-P.; Chen, Z.-Y.; Ba, W.-Z.; Fan, Y-W.; Du, Y.-Z.; Pan, S.-L. & Guo, Q. (2008). A study on the real-time radiation dosimetry measurement system based on optically stimulated luminescence, *Chinese Phys. C (HEP & NP)*, Vol. 32, No. 5, (May 2008), (381-384)

Lopez-Higuera, J. M. (1998). *Optical Sensors*, Universidad de Cantabria

Lu, P.; Bao, X.; Kulkarni, N. & Brown, K. (1999). Gamma ray radiation induced visible light absorption in P-doped silica fibers at low dose levels, *Radiat. Meas.*, Vol. 30, (1999), (725-733)

Mady, F.; Benabdesselam, M., Mebrouk, Y. & Dussardier, B. (2010). Radiation effects in ytterbium-doped silica optical fibers: traps and color centers related to the radiation-induced optical losses, *RADECS 2010 Proceedings*, Paper LN2, http://hal.archives-ouvertes.fr/hal-00559422/en/

Magne, S.; De Carlan,L.-L.; Sorel, S.; Bordy, J.-M.; Isambert, A. & Bridier, A. (2008). MAESTRO Project: evelopment of a multi-Channel OSL dosimetricsystemfor clinical use, *Workshop on dosimetric issues in the medical use of ionizing radiation EURADOS AM 2008*, Paris, January 2008

Maier, R. R.J.; MacPherson, W.N.; Barton, J.S.; Jones, J.D.C.; McCulloch, S., Fernandez-Fernandez, A.; Zhang, L. & Chen, X. (2005). Fibre Bragg gratings of type I in SMF-28 and B/Ge fibre and type IIA B/Ge fibre under gamma radiation up to 0.54 MGy,

*Proceedeidngs of the 17th International Conference on Optical Fibre Sensors*, Voet, M., Willsch, R., Ecke, W., Jones, J. & Culshaw, B. , SPIE Vol. 5855, pp. 511-514, 2005, doi: 10.1117/12.624037

Mao, R.; Zhang, L. & Zhu, R-Y. (2009). Gamma ray induced radiation damage in PWO and LSO/LYSO crystals, *Proceedings of the 2009 IEEE Nuclear Science Symposium Conference Record*, (2009), (2045-2049), http://authors.library.caltech.edu/19580/1/Mao2009p111452008_Ieee_Nuclear_Science_Symposium_And_Medical_Imaging_Conference_(2008_NssMic)_Vols_1-9.pdf

McCarthy, D.; O'Keeffe, S. & Lewis, E. (2010). Optical fibre radiation dosimetry for low dose applications, presented at *IEEE Sensors Conference 2010*, October 2010, Hawaii

McCarthy, D.; O'Keeffe, S.; Sporea, D.; Sporea, A.; Tiseanu, I. & Lewis, E. (2011). Optical Fibre X-Ray Radiation Dosimeter Sensor for Low Dose Applications, presented at *IEEE Sensors Conference 2011*, October 2011, Limerick

McLaughlin, W.L.; Miller, A .; Boyd, A.W.; McDonald, J. C. & Chadwick, K.H. (1989). *Dosimetry for Radiation Processing*, Taylor & Francis, 1989, ISBN-13: 9780850667400

Merlo, J.P & Cankoçak, K. (2006). Radiation-hardness studies of high OH−content quartz fibers irradiated with 24 GeV protons, *CMS Conference Report*, January 2006, http://cdsweb.cern.ch/record/926539/files/CR2006_005.pdf

Miniscalco, W.J.; Wei, T. & Onorato, P.K. (1986). Radiation hardened silica-based optical fibers, *RADC-TR-88-279 - Final Technical Report December 1986*, Rome Air Development Center, Air Force Systems Command, GTE Laboratories

Moloney, W.E. (2008). *A fiber-optic coupled point dosimetry system for the characterization of multi-detector computer tomography*, Master of Science Thesis, University of Florida, 2008

Mommaert, C. (1992). Optoelectronic readout for scintillating fiber trackers, *Nucl. Instr. Met. Phys. Res.*, Vol. A323, (1992), (477-484)

Naka, R.; Watanabe, K.; Kawarabayashi, J.; Uritani, A.; Iguchi, T.; Hayashi, N.; Kojima, N.; Yoshida, T.; Kaneko, J.; Takeuchi, H. & Kakuta, T. (2001). Radiation distribution sensing with normal optical fiber, *IEEE T. Nucl. Scie.*, Vol. 48, No. 6, (December 2001), (2348-2351)

Nakajima, D.; Őzel-Tashenov, B.; Bianchin, S.; Borodina, O.; Bozkurt, V.; Göküzüm, B.; Kavatsyuk, M.; Minami, S.; Rappold, C.; Saito, T.R.; Achenbach, P.; Ajimura, S.; Ayerbe, C.; Fukuda, T.; Hayashi, Y.; Hiraiwa, T.; Hoffmann, J.; Koch, K.; Kurz, N.; Lepyoshkina, O.; Maas, F.; Mizoi, Y.; Mochizuki, T.; Moritsu, M.; Nagae, T.; Nungesser, L.; Okamura, A.; Ott, W.; Pochodzalla, J.; Sakaguchi, A.; Sako, M.; Schmidt, C.J.; Sugimura, H.; Tanida, K.; Träger, M.; Trautmann, W. & Voltz, S. (2009). Scintillating fiber detectors for the HypHI project at GSI, *Nucl. Instr. Met. Phys. Res.*, Vol. A608, (2009), (287–290), doi:10.1016/j.nima.2009.06.068

Nishiura, R. & Izumi, N. (2001). Radiation sensing system using an optical fiber, *Mitsubishi Electric ADVANCE*, September 2001, pp. 25-28

Okamoto, K.; Toh, K.; Nagata, S.; Tsuchiya, B.; Suzuki, T.; Shamoto, N. & Shikama, T. (2004). Temperature dependence of radiation induced optical transmission loss in fused silica core optical fiber, *J. Nucl. Mater.*, Vol. 329–333, (2004), (1503–1506), doi:10.1016/j.jnucmat.2004.04.243

O'Keeffe, S.; Fernandez Fernandez, A.; Fitzpatrick, C.; Brichard, B. & Lewis, E. (2007). Real-time gamma dosimetry using PMMA optical fibres for applications in the sterilization industry, *Meas. Sci. Technol.*, Vol. 18, No. 10, (2007), (3171-3176)

O'Keeffe, S.; Fitzpatrick, C.; Lewis, E. & Al-Shamma'a, A.I. (2008). A review of optical fibre radiation dosimeters, *Sensor Rev.*, Vol. 28, No. 2, (2008), (136–142), doi 10.1108/02602280810856705

O'Keeffe, S. & Lewis, E. (2009). Polymer optical fibre for in situ monitoring of gamma radiation processes, *Intl. J. Smart Sensing and Intelligent Systems,* VOL. 2, No. 3, (September 2009), (490-502)

O'Keeffe[a], S.; Lewis, E.; Santhanam, A. & Rolland, J.P. (2009), Variable Sensitivity Online Optical Fibre Radiation Dosimeter, presented at *The Eighth IEEE Conference on Sensors (IEEE Sensors Conference)*, October 2009, Christchurch

O'Keeffe[b], S.; Lewis E.; Santhanam, A.; Winningham, A. & Rolland, J.P. (2009). Low dose plastic optical fibre radiation dosimeter for clinical dosimetry applications, presented at *The Eighth IEEE Conference on Sensors (IEEE Sensors Conference)*, October 2009, Christchurch

Ong, C.L.; Kandaiya, S.; Kho, H.T. & Chong, M.T. (2009). Segments of a commercial Ge-doped optical fiber as a thermoluminescent dosimeter in radiotherapy, *Radiat. Meas.*, Vol. 44, (2009), (158–162), doi:10.1016/j.radmeas.2009.01.011

Origlio, G. (2009). *Properties and radiation response of optical fibers: role of dopants*, PhD Thesis, http://portale.unipa.it/export/sites/www/dipartimenti/fisica/home/attachments/tesi_dottXXI/Origlio_PhD.pdf

Paul, M.C.; Bohra, D.; Dhar, A.; Sen, R.; Bhatnagar, P.K. & Dasgupta, K. (2009). Radiation response behavior of high phosphorous doped step-index multimode optical fibers under low dose gamma irradiation, *J. Non-Cryst. Solids,* Vol. 355, (2009), (1496-1507), doi:10.1016/j.jnoncrysol.2009.05.017

Pittet, P.; Lua, Guo-Neng; Galvana, J.-M.; Loisya, J.-Y.; Ismaild, A.; Giraudd, J.-Y. & Balossod, J. (2009). Implantable real-time dosimetric probe using GaN as scintillation material, *Sensors Actuators A: Physical,* Vol. 151, (2009), (29–34), doi:10.1016/j.sna.2009.02.018

Plazas, M. C.; Justus, B. L.; Falkenstein, P.; Huston, A. L.; Ning, H. & Miller, R. (2005). Optical fiber detectors as in-vivo dosimetry method of quality assurance in radiation therapy, *Revista Colombiana de Física,* Vol. 37, No. 1, (2005), (307-313)

Primak, W.; Edwards, E.; Keiffer, D. & Szymansk, H. (1964). Ionization expansion of compacted silica and the theory of radiation-induced dilatations in vitreous silica, *Phys. Rev.*, Vol. 133, No. 2A, (January 1964), (A531-A535)

Pruett, B. L.; Peterson, R. T.; Smith, D. E.; Looney, L. D. & Shelton, Jr., R. N. (1984). Gamma-ray to Cerenkov-light conversion efficiency for pure-silica-core optical fibers, *SPIE Proceedings*, Vol. 506, pp. 10-16, San Diego, August 1984

Radiation effects, The 10 th Europhysical Conference on Defects in Insulation Materials, *Phys. Status Solidi*, Vol. 4, No. 3 , (2007), (1060-1175)

Regnier, E.; Flammer, I.; Girard, S.; Gooijer, F.; Achten, F. & Kuyt, G. (2007). Low-dose radiation-induced attenuation at Infrared wavelengths for P-doped, Ge-doped and pure silica-core optical fibres, *IEEE T. Nucl. Sci.*, Vol. 54, No. 4, (August 2007), (1115-1119), doi: 10.1109/TNS.2007.894180, http://indico.cern.ch/getFile.py/access?contribId=1&resId=1&materialId=0&confId=37455

Rüdiger, F.; Körfer, M.; Göttmann, W.; Schmidt, G. & Wille, K. (2008). Beam loss position monitoring with optical fibres at DELTA, *Proceedings of EPAC 2008*, pp. 1032-1034, Genoa, June 2008

Saint-Gobain Crystals and Detectors (2005). *Scintillating Optical Fibers*, http://www.detectors.saint-gobain.com/uploadedFiles/SGdetectors/Documents/Brochures/Scintillating-Optical-Fibers-Brochure.pdf

Santiago, M.; Prokic, M.; Molina, P.; Marcazzó, J. & Caselli, E. (2009). A tissue-equivalent radioluminescent fiberoptic probe for in-vivo dosimetry based on Mn-doped lithium tetraborate, *WC 2009, IFMBE Proceedings 25/III*, Dössel, O. & Schlegel, W C. (Eds.), pp. 367–370, 2009

Seo, M.W; Kim, J.K. & Park, J.W. (2009). Gamma-ray induced radiation M. W damage in a small size inorganic scintillator, *The Tenth International Conference on Inorganic Scintillators and their Applications SCINT 2009*, Jeju, June 2009

Seo, M.W; Kim, J.K. & Park, J.W. (2011). Test of a fiber optic scintillation dosimeter with BGO tip in a $^{60}Co$ irradiation chamber, *Progress in Nucl. Scie. and Techn.*, Vol. 1, (2011), (186-189)

Shikama, T.; Toh, K.; Nagata, S.; Tsuchiya, B.; Yamauchi, M.; Nishitani, T.; Suzuki, T.; Okamoto, K.; Kubo, N.; Ishihara, M. & Kakuta, T. (2004). Optical fast neutron and gamma-ray detection by radioluminescence, *JAERI-Review*, (2004-017), (101-105)

Skuja, L; Tanimura, K & Itoh, N. (1996). Correlation between the radiation-induced intrinsic 4.8 eV optical absorption and 1.9 eV photoluminescence bands in glassy $SiO_2$, *J. Appl. Phys.*, Vol. 80, No. 6, (Septemer 1996), (3518 - 3525), doi: 10.1063/1.363224

Sporea, D. & Sporea, R. (2005). Setup for the in situ monitoring of the irradiation-induced effects in optical fibers in the ultraviolet-visible optical range, *Rev. Sci. Instrum.*, Vol. 76, No. 11, (2005), doi:10.1063/1.2130932

Sporea, D. & Sporea, A. (2007). Radiation effects in sapphire optical fibers, Phys. Status Solidi (c), Vol. 4, No. 3, (2007), (1356–1359), doi: 10.1002/pssc.200673709

Sporea[a], D., Agnello, S. & Gelardi, F.M. (2010). Irradiation Effects in Optical Fibers, In: *Frontiers in Guided Wave Optics and Optoelectronics*, Pal, B., pp. 49-66, Intech, ISBN 978-953-7619-82-4, Vienna, Austria

Sporea[b], D.; Sporea, A.; Oproiu, C.; Vata, I.; Negut, D.; Secu, M & Grecu, M. (2010). Annual Report, Contract 12084/2008

Suchowerska, N.; Lambert, J.; Nakano, T.; Law, S.; Elsey, J. & McKenzie, D.R. (2007). Afibre optic dosimeter customised for brachytherapy, *Radiat. Meas.*, Vol. 42, (2007), (929–932), doi:10.1016/j.radmeas.2007.02.042

Takada, E.; Kimura, A.; Hosono, Y.; Takahashi, H. & Nakazawa, M. (1999). Radiation distribution sensor with optical fibers for high radiation fields, *J. Nucl. Sci. Techn.*, Vol. 36, No. 8, (August 1999), (641-645)

Tanyi, J. A.; Nitzling, K.D.; Lodwick, C.J.; Huston, A.L. & Justus, B.L. (2011). Characterization of a gated fiber-optic-coupled detector for application in clinical electron beam dosimetry, *Med. Phys.*, Vol. 38, No. 2, (February 2011), (961-967), doi: 10.1118/1.3539737

Thériault, S. (2006). Radiation effects on COTS laser-optimized multimode fibers exposed to an intense gamma radiation field, presented at *Photonics North*, Quebec City, June 2006

Toh, K.; Shikama, T.; Kakuta, T.; Nagata, S.; Tsuchiya, B.; Mori, C.; Hori, J. & Nishitani, T. (2002). Studies of radioluminescence in fused silica core optical fibers, *Proc. SPIE*, Vol. 4786, (2002), doi:10.1117/12.451745

Toh, K.; Shikama, T.; Nagata, S.; Tsuchiya, B.; Suzuki, T.; Okamoto, K.; Shamoto, N.; Yamauchi, M. & Nishitani, T. (2004). Optical characteristics of aluminum coated fused silica core fibers under 14 MeV fusion neutron irradiation, *J. Nucl. Mater.*, Vol. 329–333, (2004), (1495–1498), doi:10.1016/j.jnucmat.2004.04.245

Treadaway, M.J.; Passenheim, B.C. & Kitterer, B.D. (1975). Luminescence and absorption of electron-irradiated common optical glasses, sapphire, and quartz, IEEE *T. Nucl. Sci.*, Vol.22, No. 6, (1975), (2253-2258), doi: 10.1109/TNS.1975.4328115

Tremblay, N.M.; Hubert-Tremblay, V.; Rachel Bujold, Rachel; Beaulieu, L. & Lepage, M. (2010). Improvement in the accuracy of polymer gel dosimeters using scintillating fibers, *J. Phys. Conf. Ser.*, Vol. 250, (2010), (1-4), doi:10.1088/1742-6596/250/1/012076

Tsuchiya, B.; Kondo, S.; Tsurui, T.; Toh, K.; Nagata, S. & Shikama, T. (2011). Correlation between radiation-induced defects, and optical properties of pure fused silica-core optical fiber, under gamma-ray irradiation in air at 1273 K, *J. Nucl. Mater.*, (2011), in press, doi:10.1016/j.jnucmat.2010.12.164

Vedda, A.; Chiodini, N.; Di Martino, D.; Fasoli, M.; Keffer, S.; Lauria, A.; Martini, M.; Moretti, F.; Spinolo, G.; Nikl, M.; Solovieva, N. & Brambilla, G. (2004). Ce3+-doped fibers for remote radiation dosimetry, *Appl. Phys. Lett.*, Vol. 85, No. 26, (December 2004), (6356–6358), doi: 10.1063/1.1840127

Weeks, R. A. & Sonder, E. (1963). The relation between the magnetic susceptibility, electron spin resonance, and the optical absorption of the E1'center in fused silica, In: *Paramagnetic Resonance II*, W. Low, pp. 869-879, Academic Press, LCCN 63-21409, New York

Weinert, A. (1999) *Plastic Optical Fibers*, Publicis MCD Verlag

Wijnands, T.; De Jonge, L.K.; Kuhnhenn, J.; Hoeffgen, S. K. & Weinand, U. (2007). Radiation tolerant optical fibres for LHC beam instrumentation, presented at the *6th LHC Radiation Workshop*, CERN, September 2007

Wittenburg, K. (2008). Beam loss monitors, http://cdsweb.cern.ch/record/1213279/files/p249.pdf

Wrobel, F. (2005). Fundamentals on Radiation-Matter Interaction, *New Challenges for Radiation Tolerance Assessment, RADECS 2005 Short Course Notebook*, Cap d'Agde, September 2005

Wu, F.; Ikram, K. & Albin, S. (2003). Photonic Crystal Fiber for Radiation Sensors, presented at *Workshop on Innovative Fiber Sensors*, Newport News, May 2003

Wulf, F. & Körfer, M. (2009). Beam loss and beam monitoring with optical fibers, *Proceedings of 9th European Workshop on Beam Diagnostics and Instrumentation for Particle Accelerators*, pp. 411-417, Basel, May 2009

Yaakob, N.H.; Wagiran, H.; Hossain, I.; Ramli, A.T.; Bradley, D.A.; Hashim, S.; Ali, H. (2011). Electron irradiation response on Ge and Al-doped $SiO_2$ optical fibres, *Nucl. Instr. Met. Phys. Res.*, Vol. A637, (2011), (185–189), doi:10.1016/j.nima.2011.02.041

Yoshida, H. & Ichikawa, T. (1995). Temperature effect on the radiation-degradation of poly(methyl methacrylate), *Radiat. Phys. Chem.*, Vol. 46, No. 4-6, Part 1, (1995), (921-924), doi:10.1016/0969-806X(95)00293-7

Yoshida, Y.; Muto, S. & Tanabe, T. (2007). Measurement of soft X-ray excited optical luminescence of a silica glass, In: *CP882, X-ray Absorption Fine Structure – XAFS73*, Hedman, B. & Pianetta, P., 2007, American Institute of Physics

Yukihara, E.G.; Sawakuchi, G.O.; Guduru, S.; McKeever, S.W.S.; Gaza, R.; Benton, E.R.; Yasuda, N.; Uchihori, Y. & Kitamura, H. (2006). Application of the optically stimulated luminescence (OSL) technique in space dosimetry, *Radiat. Meas.*, Vol. 41, (2006), (1126 - 1135), doi:10.1016/j.radmeas.2006.05.027

# 24

# Nanoparticles On A String – Fiber Probes as "Invisible" Positioners for Nanostructures

Phillip Olk
*NTNU Trondheim*
*Norway*

## 1. Introduction

Optical fibers have a long and successful history in telecommunications (Howes & Morgan, 1980) and had an enormous global impact on technology and cultural interaction, far beyond just taking telegraphy further (Stephenson, 1996). Aside from transporting data from A to B, optical fibers allowed the exploration of a whole field in photonics research, from fiber-based sensors (Udd, 1990) via various photonic crystal fibers (Zolla et al., 2005) up to quantum optical experiments (Philbin et al., 2008) - current research is presented in this book.
In this chapter, optical fibers are used just as a pointed probe which is basically transparent for light. So most of the originally engineered waveguide properties are ignored, such as low dispersion and absorption for long range applications. Contrary, using optical fibers as scanning probe tips exploits optical and mechanical effects on the nano-scale, i.e., of very close proximity, even less than 5 nm. This is a well-established "abuse" of telecommunication fibers in the Scanning Nearfield Optical Microscope[1], an opto-mechanical tool that is of high benefit in the field of nano-optics (Dunn, 1999). These scanning probe optical microscopes exploit one property of optical fibers that is often overlooked, as it's taken as granted: optical fibers are *transparent*.

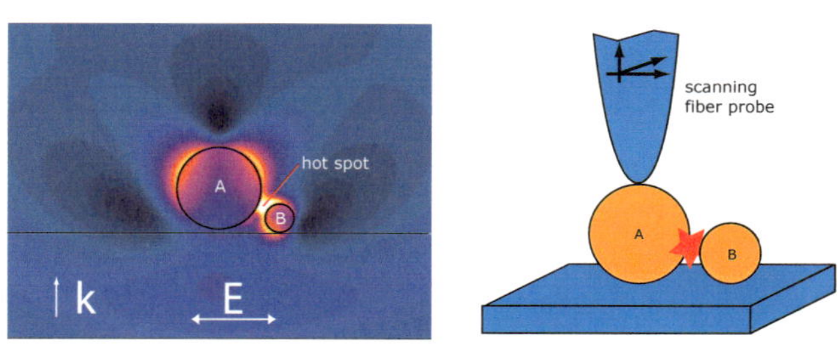


Fig. 1. *Left.* Numerical calculations of the intensity in the vicinity of two metal spheres A and B (Au, 80 nm and 30 nm in diameter, respectively), exposed to a plane wave ($\lambda$ =532 nm in vacuum). The hot spot has a 20x relative intensity w/respect to the incident wave. *Right.* Suggested experimental implementation: particle A is glued to a 3D-positionable (potentially rotatable) fiber probe, whereas particle B is resting on the substrate.

[1] Abbreviated SNOM, or NSOM, depending on Your current position on the globe.

In the field of nano-optics, the effects of near-fields are applied in order to manipulate light on a scale distinctively smaller than the wavelength of that light. One handy effect is the so-called near-field enhancement at metal structures: a tiny metal rod , i.e. less than $\lambda/4$ in length and only a few dozens of nm in diameter, or a metal sphere can serve as dipole antenna when exposed to a propagating electromagnetic field. The electrons just follow the external electrical field which is oscillating with the frequency of light, and if metal geometry, light frequency and electron mobility/velocity are well trimmed, the particle may exhibit a plasma resonance. If one examines the electric field in the very vicinity of the dipole one will encounter local intensities which are higher than the incident field. In other words: a nanometer-scaled lump of metal may serve as an *optical antenna* and provides enhanced local intensities. As the volume of enhanced intensity is rather small, such a construction can be used as a highly localized light source. E.g., this is applied in tip-enhanced Raman spectroscopy (TERS), where "intensity concentration" is crucial in order to increase the Raman response, which tends to have a very small cross section, but scales with $I^2$.
Fig.1 depicts a basic application idea of such a nanoantenna: as calculations hint, Fig.1 *left*, a local field enhancement may exist between two nanoparticles, which in turn might be used for experiments such as fluorescence excitation, Raman scattering and similar. In Fig.1 *right*, a sketch is given for a real experimental implementation of such a bi-particle nano-antenna. One particle is attached to a scanning probe, while the second is resting on a substrate. As the actual local field enhancement is highly dependent on the inter-particle distance and the relative orientation to the polarisation, the scanning fiber tip is used as a manipulator in order to optimise the system. Note that no optical properties of the tip are used.
The following sections will provide a detailed insight into the methods of mounting a such single metal nanoparticle to a scanning probe. The reader might keep in mind that the concepts here are not restricted to optical antennas on fiber tips, but can be transferred to different materials for both probe (glass, Si, metals...) and nanoparticle (metal, luminescent nanodiamonds, micelles, cells, molecules etc.), and thus into other wavelength regimes (Wenzel et al., 2008). For the sake of simplicity, only metal nanoparticles are discussed here in detail.
While the idea of, e.g., particle-enhanced Raman spectroscopy is older than the expression "optical antenna" (Wessel, 1985), it turned out to be rather difficult to determine an appropriate nano-particle and subsequently attach it to a scanning probe fulfilling these conditions:

- Exactly the pre-selected particle shall be attached, not just some random item.
- Only this single one shall be attached, not "two, maybe three".
- The particle (especially its optical properties) shouldn't be altered by the pick up process.
- The position and orientation of the attached particle on the probe tip must be known exactly afterwards.

A scientist who is only limited by "technical feasibility" might think of manufacturing techniques such as electron beam lithography: evaporate metal on a glass needle, cut away the surplus metal, done. This certainly is a valid approach, but not everybody has, wants, likes to use, or can afford such an e-beam tool. The technique, as largely introduced by (Kalkbrenner et al., 2001) and continued here, is relatively cheap: if one can afford a SNOM and a usual optical microscope, particle-decorated tips are well within reach.
Most certainly, alternative experiments can generate very similar information about the system of interest. For example, distance-dependent spectroscopy of particle pairs can be realized by producing vast arrays of particle pairs by some lithography technique (Rechberger

et al., 2003), or by pushing single particles into positions by means of a scanning probe (Bek et al., 2008). These techniques are well suitable for two-dimensional positioning, but an experimentalist may encounter a situation where she or he wants to have full control of both position and orientation of a three-dimensional particle in a three-dimensional space – this can be achieved by moving and rotating the particle-decorated scanning probe (Kalkbrenner et al., 2004). As the fiber, especially if immersed in index matching fluid, is hardly visible, the comparison to marionette string puppets is self-evident.

## 2. Manufacturing particle-decorated fiber tips

This section describes a workflow for attaching a particle to a scanning probe tip for further experiments. Starting from common basics, a few details were optimised, but the procedure may need adaptation for given instruments and circumstances.

### 2.1 The basic probe design

As a start, an "empty" scanning fiber probe is needed. Depending on time, budget and capabilities, one may either buy a ready-made one or build it from scratch using a raw optical fiber, a quartz tuning fork, and glue. This fiber tip is part of a scanning probe system. Depending on the actual setup and experiment, various mechanisms and realisations are imaginable, Fig. 2, and most of them were successfully applied for particle decoration experiments (Kalkbrenner et al., 2001; Olk et al., 2007; Uhlig et al., 2007). For mid-IR and THz applications, even silicon is more or less transparent, so the SNOM tip may be replaced by a slightly modified AFM cantilever (Wenzel et al., 2008).

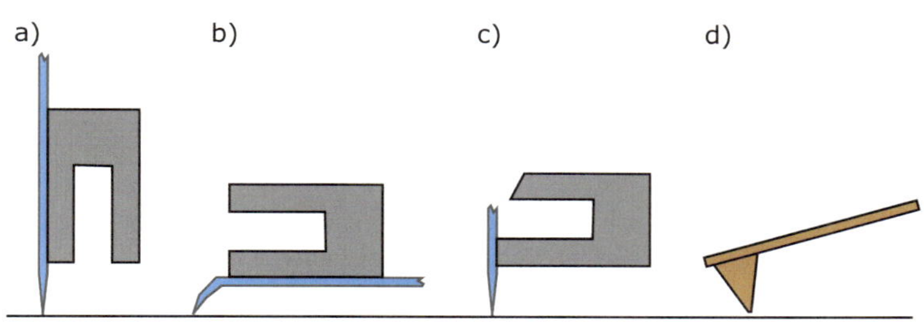


Fig. 2. Variations of scanning probe realizations. a)-c) are based on fiber tips, glued to common quartz tuning forks. d) is a usual AFM tip (Si), which may be sufficiently transparent for IR and THz experiments.

Independent on how the scanning probe is obtained, it is actually desired to have a probe tip *not* as pointed as possible: the tip should not be smaller than the particle that is to be picked. This "trick" allows much bigger tolerances when aligning and positioning the scanning probe with respect to the nanoparticle, Fig.3. In addition, this facilitates the manufacturing process of the pointed fiber itself, as it is not mandatory to use HF acid in order to etch the glass: a heat-and-pull technique, as used in micropipette pullers, can be sufficient.

### 2.2 Particle decoration

This section provides an all-optical method to pre-select, attach, and check the successful attachment of a desired particle. Depending on the actual experiment, some steps can be

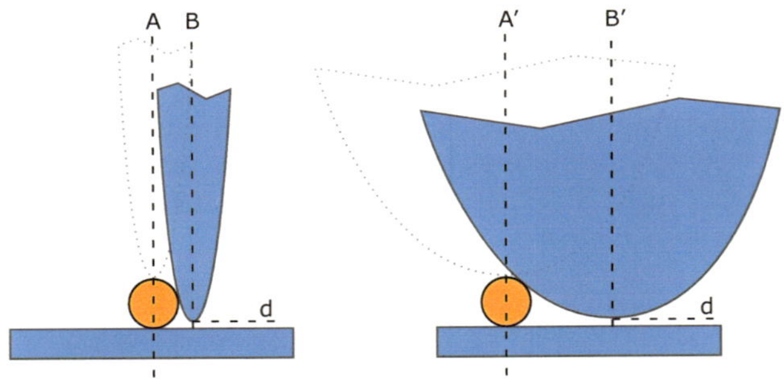


Fig. 3. Tolerances when picking a given particle with a pointed (left) or blunt tip (right). The ideal tip position is hinted by the dotted outlines, and is symmetric to the line $A$ or $A'$, respectively. For a given acceptable minimum vertical tip-sample-distance $d$, the lateral tolerance is small for pointed tips, $\overline{AB}$, and significantly bigger for blunt tips, $\overline{A'B'}$.

omitted (e.g., the orientation check for spherical particles) or might need to be added (e.g. time-correlated photon counting to check the number of emitters in a nanocrystal).

### 2.2.1 Particle preparation

Nanoparticles are usually provided as powder, or as suspension in a liquid. For the latter case, a single droplet is applied onto a glass cover slide, as common in microscopy. Depending on the particle density in the suspension liquid, this droplet may just dry (for very low densities), or be sent across the substrate for tens of seconds by a hand-operated rubber blower. As a result, single nanoparticle "candidates" should be distributed over the slide with a next-neighbour-distance big enough so a) single particles can be analysed optically and b) the more or less blunt fiber tip will pick up only one particle.

### 2.2.2 Optical microscopy of single nanoparticles – the role of homogeneous immersion

For standard microscopy, the Rayleigh criterion is still a valid limit for the resolution. Nonetheless, standard optics are capable to "see" nanoscopic particles:
According to Mie's theory of scattering and absorption of light (plane wave, wavelength $\lambda$) at a small (radius $R < \lambda$ ) conductive sphere, the scattering cross section is proportional to $C_{\mathrm{scat}} \propto R^6$, or, as volume $V = 4/3\pi R^3$:

$$C_{\mathrm{scat}} \propto V^2 \,. \tag{1}$$

So even a very small particle is visible in the eyepiece of a microscope. It can't be *resolved* if close to an other one, but it will produce an Airy disk according to the point spread function of the microscope.[2]
For absorption, Mie provides a similar expression $C_{\mathrm{abs}}$, but this is just proportional to the particle volume:

$$C_{\mathrm{abs}} \propto V \,. \tag{2}$$

In a sketch of these two expressions, Fig.4 *left*, one realises that there exists a crossover size $V_\times$ for the particle, when the lucid scatterer ($C_{\mathrm{scat}} > C_{\mathrm{abs}}$ for $V > V_\times$) turns into an absorber ($V < V_\times$). For a real microscope situation as in Fig.5, this means that the observer faces a

[2] One can still tell single particles from doublets by both intensity and scattering spectrum, as will be explained later.

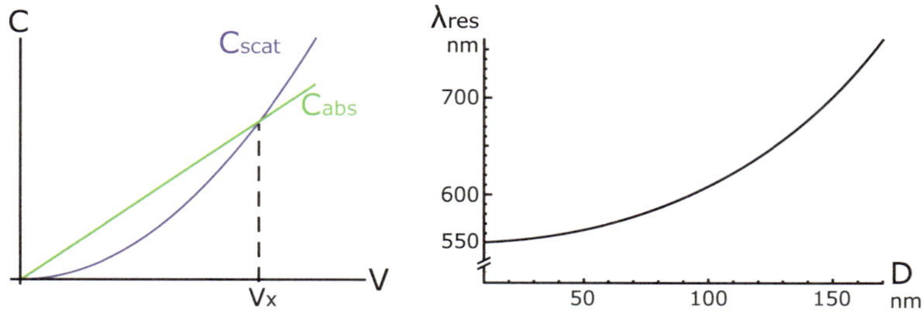


Fig. 4. Optical properties of conductive spheres. *Left*. Qualitative sketch of Mie's scattering cross section $C_{scat} \propto V^2$ and absorption cross section $C_{abs} \propto V$. They cross over at $V_\times$. *Right*. Spectral position $\lambda_{res}$ of the plasma resonance of a particle of diameter $D$. The numerical values are valid for Au spheres, embedded in an effective medium of refraction $n = 1.52$, and a diameter range from 20 to 160 nm. $\lambda_{res}$ is a vacuum wavelength, as measured by a (non-immersed) spectrometer.

contrast problem if the particle size is too small: big particles look bright, but in a transient size region, the residual reflection from the cover slide is about as bright as the scattering from the particle – "optical cloaking" without any cape involved. If the particle is even smaller, and the residual reflection is not too intense, the particles may be visible again, but this time, they are dark-on-bright ("bright" being the residual reflection), Fig.5b).

This situation is tightened for "modern standard" microscopy: immersion lenses easily provide numerical apertures NA > 1. This means that larger portions of the illumination light are reflected totally, as the maximum angle of incidence, $\varphi_{NA}$ is easily bigger than the critical angel of incidence $\varphi_{crit}$, Fig.5b). As a result, the background intensity $I_{bg}$ is elevated, so even a lucid scatterer is "swallowed" in contrast to the intense reflection. The solution to this is the homogeneous immersion, Fig.5c): by suppressing any reflection, only the scattering is observed. The idea is rather old (Abbe, 1873), but this is an example of lack of interdisciplinary exchange: scanning probe people prefer a dry environment for their probes, as this facilitates everything (Q factors, controlling etc.). On the other hand, life scientists, who deal with large NA objective lenses since many decades, never were aware of this problem, as they generally *always* use immersion fluids (mostly water).

Depending on the actual nanoparticle, other techniques might be applied in order to locate and identify single particles: absorption-mediated interference microscopy successfully imaged particles smaller than 2 nm (Boyer et al., 2002), and luminescent particles, such as fluorescent PS spheres or N-V centres in nanodiamonds, provide straight forward access to their location. Alternatively, if one succeeds in (self-)arranging nanoparticles in some fixed grid, only relative positioning of the sample is necessary.

A slightly different train of thoughts is valid for the scanning probe tip, which may be considered as a dielectric sphere. Here, absorption is not as important, but the index of refraction: the closer the indexes of the glass tip and the immersion medium are, the less light is "disturbed". Consequently, the tip can't be located by the bare eye. Indeed, the fiber core itself provides a variation of the index of refraction, an essential property for guiding optical modes, but this is too small to give a significant visual hint on the whereabouts of the fiber.

If there was no immersion medium on the cover slide, the nanoparticle is hardly seen, if the immersion medium matches the slide and the fiber probe too well, the probe vanishes optically. As the experimentalist needs to see both probe tip and particle during the picking

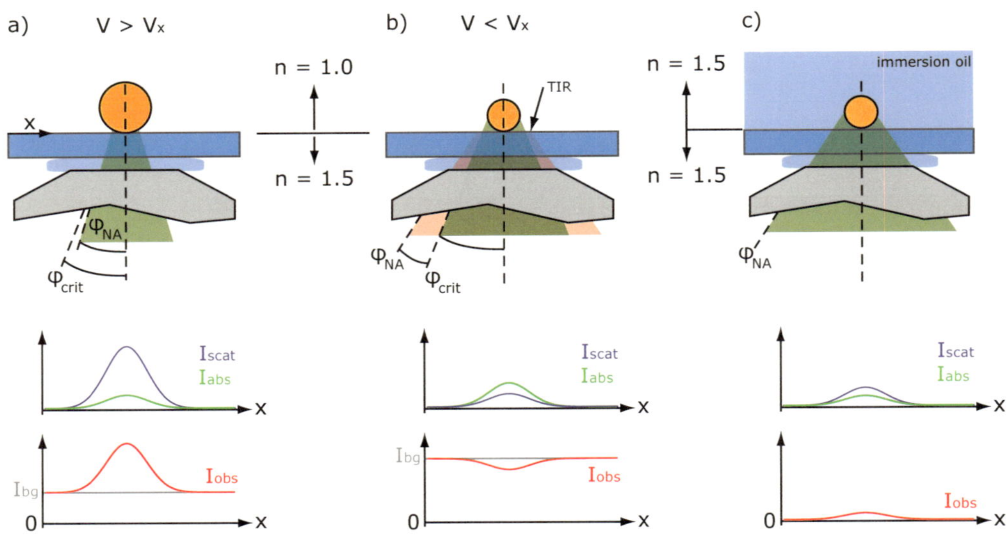


Fig. 5. Illumination schemes (top row) and scattered, absorbed, background and observable intensities for different configurations. a) A sphere of volume $V > V_{\times}$ scatters more than it absorbs, hence $I_{obs} = I_{scat} - I_{abs} + I_{bg}$ shows a positive bump – the contrast of the particle is bright on dark. b) For particle volumes $V < V_{\times}$ the bump is negative, as scattering dominates (dark on bright). In addition, a lens with a larger NA is applied, which increases the background $I_{bg}$ due to total internal reflection (TIR): the red sectors of the incident light are fully reflected. c) Immersion oil on the cover slide removes the refractive index jump and hence any reflection: $I_{bg} = 0$. The ratio $V/\lambda$ becomes $nV/\lambda$, so the effective wavelength in the medium is smaller. This, in turn, pushes the position of $V_{\times}$ towards smaller particle volumes.

procedure, a compromise is needed. Instead of a regular immersion oil, which matches common BK7 cover slides and optical fibers too well, a liquid of a lower refractive index is desired. Glycol[3] and glycerin[4] turned out to be suitable, as they are non-volatile, viscous, not too toxic, their index of refraction can be tuned down by mixing with water, and they are easily obtainable. Keep in mind that the immersion liquid between cover slide and microscope lens, i.e. *under* the slide, must be chosen according to the lens manufacturer's specifications.

### 2.2.3 How to select the right nanoparticle

Selecting an appropriate nanoparticle *before* it is attached to the probe tip is important, as this is the key ability of the procedure provided here. In the case of metal nanoparticles, Mie's theory helps: a single gold sphere, will back-scatter a characteristic spectrum. The peak position $\lambda_0$ is related to the diameter $D$, e.g., a gold sphere embedded in a liquid of a refraction index $n = 1.5$ has its peak position at

$$\lambda_0(D) = \frac{1.15 \cdot 10^9}{2.09 * 10^6 - 20D^2}\,. \tag{3}$$

[3] Ethylene glycol, [107-21-1], $n = 1.43$
[4] Glycerol, [56-81-5], $n = 1.473$

This rule of thumb, derived from (Sönnichsen, 2001), is valid for sphere diameters between 20 and 160 nm, and a powerful handle for size control (Härtling et al., 2008). A plot can be seen in Fig.4 *right*. If the particle of interest was not symmetric, its "effective diameter" would change for different orientations – the peak position is affected according to rotating either the particle or the incident polarisation (Kalkbrenner et al., 2004). As a benefit of this behaviour, the orientation of metal nanowires/ellipsoids/optical antennas can be determined (Olk et al., 2010). In addition, single nanoparticles and multiplets, despite of they can't be resolved, will produce an Airy disk of a red-shifted colour. For a well-known, reproducible species of nanoparticles, the step of spectral analysis may be omitted, as according to equation 1 an intensity analysis might be sufficient in order to identify "healthy", single nanoparticles.
For nanoparticles differing from metal scatterers, alternative identification methods must be provided, according to the actual requirements of the intended experiment.

### 2.2.4 Choice of glue

The adhesive needs to provide these properties:

- The nanoparticle must be fastened securely and reliably.
- The glue shouldn't alter the mechanics of the scanning probe.
- The optical properties of the glue must not interfere with the actual experiment.

For affixing metal nanoparticles to glass, APTMS[5], APTES, or PEI[6] are fine, but other material combinations may demand other specific glues - large plastic spheres can actually be pronged. Detailed procedure descriptions exist in order to obtain densely packed monolayers of these materials. In this application here, such a perfect monolayer isn't mandatory: it is sufficient to dip the probe tip into a 5-10% aqueous solution of APTMS for some tens of seconds. This allows the silane ends of the molecules to bind to the glass surface. Unbound, excess molecules can be rinsed off by a spill of purified water. As a result, a holey monolayer of amino groups is ready to bind to gold or silver.
The last requirement for the glue, being optically neutral, can turn out to be tricky. For larger particles, a patchy monolayer of molecules isn't affecting the effective medium strongly. But in a single-molecule Raman spectroscopy experiment, the Raman signal from the glue must not outshine the response from the molecule of interest.

### 2.2.5 Alignment and attaching

The previous sections provide a scanning probe and a candidate particle. Both tip and particle are very small, so their images are (Airy) disks, probably of different size, colour, and intensity. When setting up a regular scanning probe, the tip is positioned manually some microns over the sample surface. Now the experimentalist aligns the two image disks, frequently jumping forth and back between the two focal planes. A scaled fine drive screw for the microscope will be helpful, as the focal plane of a high-NA lens is very thin. When aligning, two (Airy) disks are centred on each other - the Rayleigh criterion plays no role here. In addition, using a "blunt" tip adds lateral tolerance, as in Fig. 1. In total, the centring tolerance is in the range of the tip radius, which in turn can be close to the (irrelevant) Rayleigh criterion.
Once tip and particle are aligned on the vertical axis, and separated vertically by just one or two micrometers, the actual picking can take place. For most setups, a regular approach procedure at moderate oscillation amplitudes is sufficient: once the tip is "parked" on the

[5] 3-Aminopropyl-trimethoxysilane, [13822-56-5]
[6] Polyethylenemin, [29320-38-5]

nanoparticle for a few seconds, the tip may be retracted. If everything went well, the tip *and the nanoparticle* disappeared from the focal plane - moving the microscope up will bring both back. Moving the tip laterally will quick-check if the particle is actually affixed thoroughly enough. Fig.6 is the corresponding view through the eyepiece of a microscope.

(a) (b) (c)

Fig. 6. Black-and-white optical images of particle-decorated fiber probes. The fiber ($n \approx 1.5$) is immersed in glycol ($n = 1.42$). *Left* The tip produces a V-shaped "shadow" to an external lamp pointing to the upper left of the image. The bright concentric rings (cones actually) stem from fiber-transmitted laser light that is coupled into the other end of the fiber. The white spot in the middle is the back-scattered light from a gold sphere of 80 nm diameter. The illumination scheme corresponds to Fig.5c). *Middle* The laser is switched off, the pointing direction of the external lamp is different now. *Right* The external lamp is switched off, the microscope's illumination is the only remaining light source. The excellent contrast demonstrates the concept of immersed fibers.

To illustrate the possible results, Fig.7 provides some electron microscopy images of particle-decorated tips. Note that an electron image is not necessary for day-to-day quality control, as the procedure, as given here, provides a rigorous check whether the particle is attached or not.

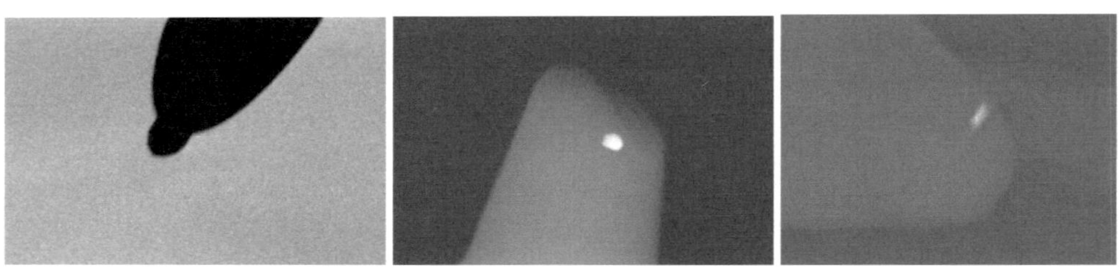

(a) (b) (c)

Fig. 7. SEM images of various particle-decorated tips. a) a 100 nm Au sphere, carbon coated for improved scanning electron imaging. b) Material contrast image of a 80 nm Au sphere. Despite of the broken glass tip, the decoration process worked and produced a usable particle probe. c) Material contrast image of a Ag nanorod, 125 nm in length and 55 nm in diameter. Note the odd spatial orientation of the nanorod.

If the particle refuses to follow the tip, the binding process couldn't take place for several reasons. The nature of these reasons is somewhat unclear, but pressure, time, density of the glue molecules and oscillation amplitude may play a role. Therefore, increasing the normal force by tuning the setpoint may help, as might a reduction of the oscillation amplitude. If this didn't lead to success yet, harsher methods may be appropriate: a deliberate disturbance of the distance controller may lead to the desired result. An audible clearing of the throat, coughing, clapping, and, as the last resort, tapping on the optical table (just a soft finger tip), can induce a gentle, but non-ignorable disturbance. Be aware that due to the small area of a nanoparticle, the forces on it may reach easily the magnitude of GPa, which in turn deforms the spherical particle to a heavily patted lump. While this opens the opportunity of nano-minting[7], the change of the shape leads to a visible(!) redshift of the plasma resonance.

### 2.2.6 Optical antennas - multiple particles

For optical antenna applications, e.g., using a metal nanostructure as a concentrator of the incident field in order to produce a high intensity, say, for Raman scattering experiments, multiple particles turn out to be advantageous, (Fleischer et al., 2008; Jiang et al., 2003; Li et al., 2003; Nie & Emory, 1997; Olk et al., 2007). For this reason, it may seem attractive to obtain particle decorated probes consisting of two (or even more) particles.

Alternatively, one may try to attach the second particle *directly* to the first one, using a specific glue for this purpose: in this case, Au is to be attached to Au, so alkane dithiols (Brust et al., 1995) or more rigid linking molecules (Pramod & Thomas, 2008) are appropriate, if a fume hood is available. A rather flexible link can be achieved by dipping the tip (which is already the first nanoparticle) for 10 minutes into undiluted decanethiol[8]. The residual thiol can be rinsed off by isopropanol. Especially if the second particle is of similar or bigger size than the first one it is recommended to passivate the residual, but unused molecule on the glass tip: for APTMS, exposure to a mild organic acid, mixed into the immersion liquid, turned out to be beneficial. In a straightforward attempt, one may repeat the procedure above and

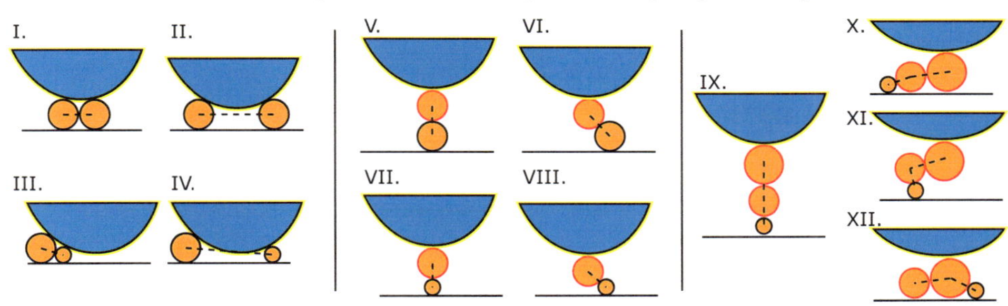


Fig. 8. Possible alignments (only two-dimensional, for the sake of simplicity) for two particles on a tip. The dotted lines connect the particle centres and depict the principal axes. *I.-IV.* If both particles are attached directly to the glass tip. *V.-VIII.* The first particle is affixed to the glass tip by a specific glue (yellow coating), while the second particle is mounted on the first one by a different, specific glue (red coating). *IX.-XII.* Illustration of some mentionable situations for three-particle chains. For small and many particles, quality control by electron microscopy is advised.

affix an additional particle to the glass tip. This will produce particle-decorations as in Fig.8 *left*. Note that in this configuration, the final orientation of the principal axis[9] is dependent on

[7] For comparison: a typical AFM cantilever, spring constant 40 N/m, bent by 50 nm, applies a force of 2 $\mu$N. This force, distributed over the projection of a sphere of 25 nm radius, exceeds 1 GPa, more than the yield strength of many metals.

[8] 1,10-Decanedithiol, [1191-67-9].

[9] The line connecting the two centres of the particles.

the particle size, on the distance between the particles, and even on the order of attaching the particles[10]. Generally, the principle axis will be rather horizontal than vertical.
In terms of accuracy, the concepts of Fig.3 can be transferred to this two-particle system, but note that now, the relative position, and consequently, the orientation of the principal axis, is of fundamental importance for the optical antenna effects, Fig.8 *middle*.
The situation "worsens" for particle chains of three or more particles, Fig.8 *right*. If a straight, vertical chain is desired, one has to mind the ways of how the manufacturing process could go wrong. Of course, such "faulty" assemblies may open opportunities for other usage: a folded chain might be considered as a collection of two-particle antennas, each sensitive for a different polarization orientation and/or wavelength.

### 2.3 Applied scanning particle probes

Two well-established application fields are Scanning Particle-enhanced *Raman Spectroscopy*, and the examination of plasmons on metal nanostructures, e.g. for fluorescence enhancement by means of *optical antennas*, and shall be discussed in detail.
Of course, the concept of a "particle on a string" is not limited to plasmonics and metal particles alone. The reader be reminded of N-V centres in nanodiamonds (Balasubramanian et al., 2008), or quantum dots Aigouy et al. (2006). Future experiments may use single nanowire lasers as a photon source (Johnson et al., 2001), or magnetic nanoparticles as a sensor (Härtling et al., 2010), not to mention antibodies or functional enzymes – the possibilities are virtually endless.

#### Plasmonics - optical antennas

One important field where the plasmonic properties of metal nanostructures on a tip are exploited, is the field of *plasmonics*. This denotes the discipline of engineering and manipulating metal structures and their optical properties.
An obvious field of research is the examination of the interplay among several metal nanostructures. Thanks to near-field optical coupling, the spectral properties of the metal nanoparticle on the tip is manipulated just by bringing it near to other metal structures - and vice versa. The observed effects vary from plasmon resonance de-tuning (Kalkbrenner et al., 2001; Olk et al., 2008) to four-wave-mixing (Danckwerts & Novotny, 2007).
For single particles, the enhanced near-fields are their most important property. Increasing the local intensity on a restricted volume allows for the *selective* excitation of single fluorescent molecules (Bharadwadj et al., 2007; Frey et al., 2004; Sandoghdar & Mlynek, 1999).

#### Raman spectroscopy

All progress in the field of enhanced near-field probes quickly found its way into the Raman spectroscopy, probably the most established application of particle-decorated tips. Due to the very small cross sections in Raman scattering experiments, the Raman community depends on high local fields, generated by Surface-Enhanced Raman Spectroscopy (SERS) and Tip-Enhanced Raman spectroscopy (TERS), using massive but pointed wire tips. The concept of optical antennas for local field enhancement was quickly embraced by the community, combined with SERS and TERS concepts, and quickly driven towards near-field enhancing metal nanostructures (Nie & Emory, 1997; Plieth et al., 1999; Stöckle et al., 2000).
The concept of using just one metal particle as an optical antenna, similar to the proposal

[10] Attaching a larger particle first and well on the symmetry axis of the tip will prevent a second, smaller particle from touching the glass tip.

of (Wessel, 1985), was realised early (Anderson, 2000), and has evolved towards RNA sequencing (Bailo & Deckert, 2008). Note that for scanning particle-enhanced Raman microscopy, mostly AFM-based tips with a "thin" metal film (i.e. sputtered metal coagulated to islands, which we consider as nanoparticles here) are used. Aside from the simple fact that attaching a pre-selected particle, as demonstrated here, came later into being (Olk et al., 2007), each experimentalist needs to consider the advantages and disadvantages of the two approaches: price, availability of tools, reproducibility, expected and desired local enhancement, and experimental skills play roles.

## 3. Demonstration - Fluorescence in the vicinity of two particles

In order to put the concepts of section 2 to a test, two different two-particle systems are analysed by means of fluorescence spectroscopy:
Fig.9 *left* illustrates the basic setup (Olk, 2008): collimated excitation light, controlled in both intensity and polarisation, is focused on the particle system, which is immersed in glycerine dyed with Nile Blue[11]. The linear polarisation of the laser source can either be rotated by a $\lambda/2$ plate or converted to so-called radially polarised light, which produces a focal polarization in the direction of $z$ (Olk et al., 2010). Any fluorescence collected by the focusing lens is guided to a spectrometer. A typical spectrum of pure Nile Blue is in Fig.9 *right*. In a first step, the

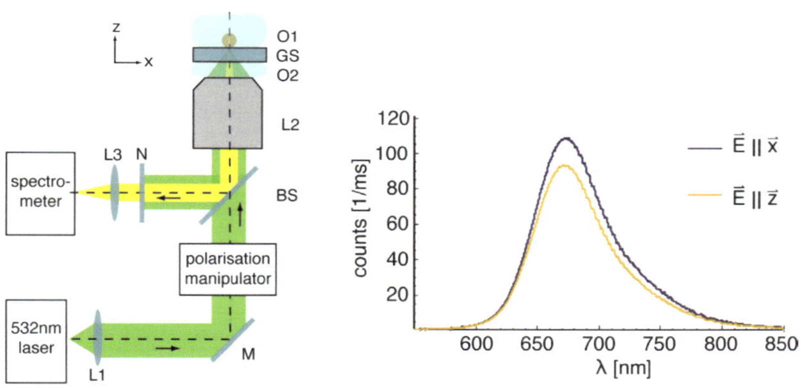


Fig. 9. *Left*. Experimental setup. L2 is a microscope lens requiring immersion oil O2 and a specific glass slip GS in order to perform optimal. O1 is the immersion medium dyed with Nile Blue. *Right*. Typical fluorescence spectra of Nile Blue for two given focal polarizations.

focal plane is moved 30 $\mu$m into the Nile Blue glycol, so the focal volume is fully in material that can fluoresce. Then, a fiber tip carrying one single Au particle of 80 nm diameter is inserted into the focus, and the fluorescence intensities are recorded for radially and linearly polarised light, Fig.10 *left*. In comparison to Fig.9 *right*, the fluorescence signal is reduced, approximately by a factor of two. This is owed to the fact that the fiber tip itself displaces the dye, so a good part of the focal volume consists of non-fluorescent glass. This is an effect that is hardly quantifiable, as the indexes of refraction are close. Usually, this is a desired advantage, but here, the similarity of the indices lets the tip vanish for the human eye. This renders it impossible to position the *bare* tip in the focus in order to quantify its displacement.

[11] [3625-57-8]

A different issue that can be learned from Fig.10 *left* is that for a focal z polarisation, the detected fluorescence intensity is much lower than for horizontally polarised excitation light. This can be explained by the inset in that figure: the region of high nearfields at the particle consists of two lobes capping opposite poles. For x-polarised illumination, both lobes extend into the dye, whereas for z polarisation, one lobe extends into the nonfluorescent fiber tip.

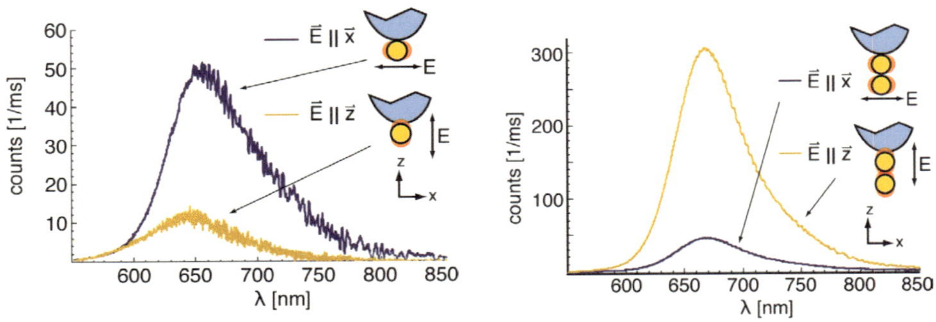


Fig. 10. Fluorescent spectra of Nile Blue for two focal polarisations. *Left* A single Au 80 nm sphere. *Right*. Two Au 80 nm spheres, the second attached to the first.

In a second step, a second Au sphere of 80 nm diameter is attached to the first one, as described in section 2.2.6. In a direct comparison, Fig.10 *right*, the fluorescence intensity is higher for both polarizations. While the signal for horizontal polarisation remained about the same, the signal for vertical polarisation increased enormously - this can be assigned to the enhanced near-field intensity between the two spheres.
A rotation check, i.e. rotating the fiber probe and the polarisation with respect to each other, provides an intensity variation of only 10%, Fig.11 hints that the two particles are close to the symmetric orientation V. in Fig.8.

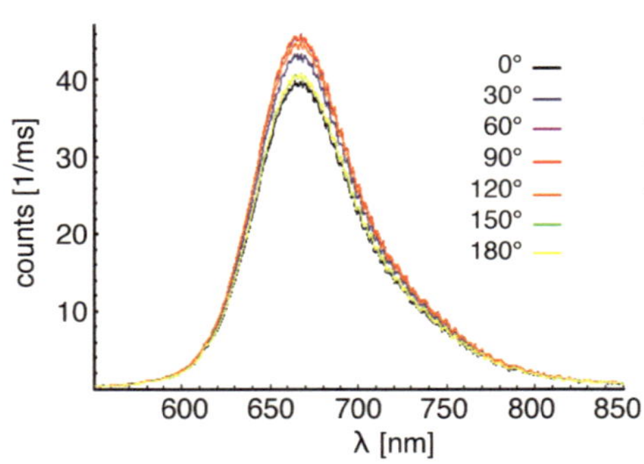


Fig. 11. Orientation-dependent fluorescence in the vicinity of two 80 nm Au spheres. The horizontal polarisation and the probe are rotated around the z-axis. The peak intensity varies by about 10%.

In a third step, the experiment is repeated with a 80 nm particle which carries a smaller particle of 30 nm diameter (both Au). A basic check with excitation light polarised along x,y,z, Fig.12 *left*, shows that the z-direction is not as pronounced anymore as in Fig.10 *right*. The rotation

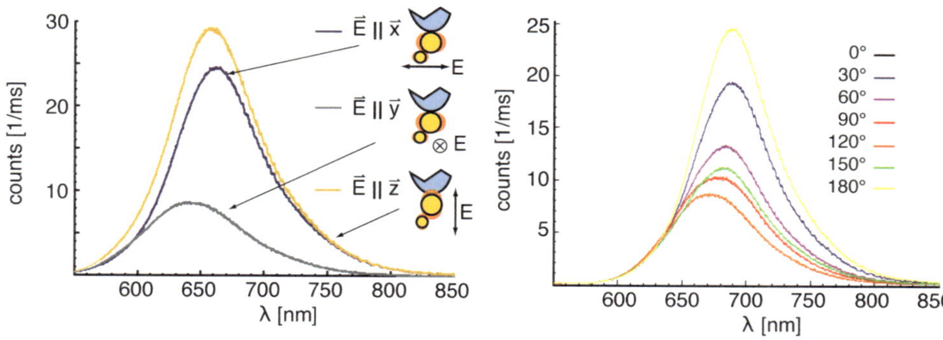


Fig. 12. Orientation-dependent fluorescence in the vicinity of a pair of 80 nm + 30 nm Au spheres. *Left*. The fluorescence intensity for x- and z-oriented excitation polarisation are rather similar, whereas in-plane y-polarisation provides a rather small signal. *Right* A rotational analysis reveals that the peak intensity varies by more than 50%.

check, Fig.12 *right*, shows a significant intensity variation for various relative orientations. This hints that the relative particle orientation is rather close to image VIII. in Fig.8, with the principal axis along x direction. In comparison to the two Au particles of equal size one realises that the principal axis here is not just slightly off the plumb line, but by a significant angle.

What's more: considering the property of the scattering cross section, equation (1), a single 30 nm particle is supposed to be nearly irrelevant in comparison to a 80 nm sphere, as the volume squares $V_{30}^2/V_{80}^2$, and hence their scattering cross sections, $C_{sc30}/C_{sc80}$, have a ratio of less than 0.003. But thanks to near-field-mediated coupling of the two particles' plasmons, the smaller particle has a significant impact on the fluorescence in the vicinity of the particle pair. So by combining particle-decorated probe tips with a straight-forward fluorescence analysis it is possible to learn a lot about the particles and their relative orientation. This is a valuable simplification of a lab workflow: For many applications, it is already sufficient to know the direction of the projected principal axis – the tedious procedure (of mounting the tip, attaching particles, unmounting the tip keeping its orientation in mind, moving it into an electron microscope, get the particle orientation, return the tip to the optical setup while maintaining the orientation according to the SEM coordinates) is dodged. Note that for these experiments, actual *distance-controlled scanning* of the probe tip was used only for the act of mounting the particles – the orientation determination took place well off the cover slide surface.

## 4. Conclusion & outlook

At this point, the principles of scanning particle microscopy were explained. The technical details and "tricks" were given, and the reader should be able to follow up according to her/his own experiments. Not to use the most pointed tips available, and to exploit the properties of immersion liquids may be valuable hints.

The Application section underlined the versatility of particle-decorated probes as mobile hot spots, *i.e.* as optical antennas providing locally enhanced intensities. As an example how well the third dimension can be explored by "pulling the strings", fluorescence in the vicinity of Au nanoparticle pairs was analysed.

The given procedures are not exhaustive, but a dexterous experimentalist, after checking alternatives (Eng et al., 2010), may make use of the information here – the technology and its scientific applications are not maxed out yet, and many interesting experiments can be expected in the future.

## 5. Acknowledgements

The author wishes to thank Thomas Kalkbrenner for long-term inspiration on this topic, and Lukas M. Eng, Marc-Tobias Wenzel, and Thomas Härtling for providing a productive climate in the SNOM lab of the Institute for Applied Photophysics at the TU Dresden, Germany. Helge Weman of the Institute of Electronics and Telecommunications at the NTNU Trondheim, Norway, deserves gratitude for endorsing the writing of this scripture.

## 6. References

E. Abbe (1904). *Gesammelte Abhandlungen. Bd. 1: Theorie des Mikroskops.* G. Fischer, Jena.

M. S. Anderson (2000). Locally enhanced Raman spectroscopy with an atomic force microscope. *Appl. Phys. Lett.*, Vol. 76 no. 21, pp. 3130ff.

L. Aigouy, B. Samson, G. Julié, V. Mathet, N. Lequeux, C. N. Allen, H. Diaf, & B. Dubertret (2006). Scanning near-field optical microscope working with a CdSe/ ZnS quantum dot based optical detector. *Rev. Sci. Instrum.*, Vol. 77, 063702.

E. Bailo & V. Deckert (2008). Tip-enhanced Raman spectroscopy of single RNA strands: Towards a novel direct-sequencing Method. *Angew. Chemie Internat. Ed.*, Vol. 47, pp. 1658-1661.

G. Balasubramanian, I. Y. Chan, R. Kolesov, M. Al-Hmoud, J. Tisler, Chang Shin, Changdong Kim, A. Wojcik, P. R. Hemmer, A. Krueger, T. Hanke, A. Leitenstorfer, R. Bratschitsch, F. Jelezko, & J. Wrachtrup. Nanoscale imaging magnetometry with diamond spins under ambient conditions. *Nature*, Vol. 455, pp. 648-651.

A. Bek, R. Jansen, M. Ringler, S. Mayilo, T. A. Klar, & J. Feldmann (2008). Fluorescence enhancement in hot spots of AFM-designed gold nanoparticle sandwiches. *Nano Lett.* Vol. 8 no. 2, pp. 485-490.

P. Bharadwaj, P. Anger, & L. Novotny (2007). Nanoplasmonic enhancement of single-molecule fluorescence. *Nanotechnology*, Vol. 18, 044017.

D. Boyer, P. Tamarat, A. Maali, B. Lounis, & M. Orrit (2002). Photothermal imaging of nanometer-sized metal particles among scatterers. *Science* Vol. 297, pp. 1160-1163.

M. Brust, D. J. Schiffrin, D. Bethell, & C. J. Kiely (1995). Novel gold-dithiol nano-networks with non-metallic electronic properties. *Adv. Mater.*, Vol. 7 no. 9, pp. 795ff.

M. Danckwerts & L. Novotny. Optical frequency mixing at coupled gold nanoparticles. *Phys. Rev. Lett.*, Vol. 98, 026104.

R. C. Dunn (1999). Near-field scanning microsopy. *Chem. Rev.* Vol. 99 no. 10, pp. 2891-2927.

L. M. Eng, T. Härtling, & P. Olk (2010). Device and method for metallizing scanning probe tips. *Int. patent application*, PCT/DE2010/000579.

M. Fleischer, C. Stanciu, F. Stade, J. Stadler, K. Braun, A. Heeren, M. Häffner, D. P. Kern, & A. J. Meixner. Three-dimensional optical antennas: Nanocones in an apertureless scanning near-field microscope. *Appl. Phys. Lett.*, Vol. 93 no. 11, 111114.

H. G. Frey, S. Witt, K. Felderer, & R. Guckenberger (2004). High-resolution imaging of single fluorescent molecules with the optical near-field of a metal tip. *Phys. Rev. Lett.*, Vol. 93, 200801.

T. Härtling, Y. Alaverdyan, M. T. Wenzel, R. Kullock, M. Käll, & L. M. Eng (2008). Photochemical tuning of plasmon resonances in single gold nanoparticles. *J. Phys. Chem. C*, Vol. 112 no. 13, pp. 4920-4924.

T. Härtling, T. Uhlig, A. Seidenstücker, N. C. Bigall, P. Olk, U. Wiedwald, L. Han, A. Eychmüller, A. Plettl, P. Ziemann, & L. M. Eng (2010). Fabrication of two-dimensional Au@FePt core-shell nanoparticle arrays by photochemical metal deposition. *Appl. Phys. Lett.*, Vol. 96, 183111.

M. J. Howes & D. V. Morgan (1980). *Optical Fiber Communications - Devices, Circuits, and Systems*, John Wiley & Sons Ltd., ISBN 0-471-27611-1.

J. Jiang, K. Bosnick, M. Maillard, & L. Brus (2003) Single molecule Raman spectroscopy at the junctions of large Ag nanocrystals. *J. Phys. Chem. B*, Vol. 107 no. 37, pp. 9964-9972.

J. C. Johnson, H. Yan, R. D. Schaller, L. H. Haber, R. J. Saykally, & P. Yang (2001). Single nanowire lasers. *J. Phys. Chem. B*, Vol. 105 no. 46, pp. 11387-11390.

T. Kalkbrenner, M. Ramstein, J. Mlynek, & V. Sandoghdar (2001). A single gold particle as a probe for apertureless scanning near-field optical microscopy. *J. Microscopy* Vol. 202, pp. 72-76.

T. Kalkbrenner, U. Håkanson, & V. Sandoghdar (2004). Tomographic plasmon spectroscopy of a single gold nanoparticle. *Nano Lett.* Vol. 4 no. 12, pp. 2309-2314.

S. C. Kehr, M. Cebula, O. Mieth, T. Härtling, J. Seidel, S. Grafström, L. M. Eng, S. Winnerl, D. Stehr, & M. Helm. Anisotropy contrast in phonon-enhanced apertureless near-field microscopy using a free-electron laser. *Phys. Rev. Lett.* Vol. 100 no. 25, 256403.

K. Li, M. I. Stockman, & D. J. Bergman (2003). Self-similar chain of metal nanospheres as an efficient nanolens. *Phys. Rev. Lett.*, Vol. 91, 227402.

S. Nie & S. R. Emory (1997). Probing single molecules and single nanoparticles by surface-enhanced Raman scattering. *Science*, Vol. 275 no. 5303, pp. 1102-1106.

P. Olk, J. Renger, T. Härtling, M. T. Wenzel, & L. M. Eng (2007). Two particle enhanced nano Raman microscopy and spectroscopy. *Nano Lett.*, Vol. 7, no. 6, pp. 1736-1740.

P. Olk, J. Renger, M. T. Wenzel, & L. M. Eng (2008). Distance dependent spectral tuning of two coupled metal nanoparticles. *Nano Lett.*, Vol. 8 no. 4, pp. 1174-1178.

P. Olk (2008). *Optical Properties of Individual Nano-Sized Gold Particle Pairs*. Dissertation, TU Dresden, *http://nbn-resolving.de/urn:nbn:de:bsz:14-ds-1218612352686-00553*.

P. Olk, T. Härtling, R. Kullock, & L. M. Eng (2010). Three-dimensional, arbitrary orientation of focal polarization. *Appl. Optics*, Vol. 49 no. 23, pp. 4479-4482.

P. Pramod & K. G. Thomas (2008). Plasmon coupling in dimers of Au nanorods. *Adv. Mater.* Vol. 20 no. 22, pp. 4300-4305.

T. G. Philbin, C.Kuklewicz, S. Robertson, S. Hill, F. König, & U. Leonhardt (2008). Fiber-optical analog of the Event Horizon. *Science*, Vol. 319 no. 5868, pp. 1367-1370.

W. Plieth, H. Dietz, G. Sandmann, A. Meixner, M. Weber, P. Moyer, & J. Schmidt (1999). Nanocrystalline structures of metal deposits studied by locally resolved Raman microscopy. *Electrochimica Acta*, Vol. 44 no. 21-22, pp. 3659-3666.

W. Rechberger, A. Hohenau, A. Leitner, J. R. Krenn, B. Lamprecht, & F. R. Aussenegg (2003). Optical properties of two interacting gold nanoparticles. *Opt. Comm.* Vol. 220 no. 1-3, pp. 137-141.

V. Sandoghdar & J. Mlynek (1999). Prospects of apertureless SNOM with active probes. *J. Opt. A: Pure and Appl. Opt.* Vol. 1 no. 4, pp. 523-530.

C. Sönnichsen (2001). *Plasmons in Metal Nanostructures.* Cuviller, Göttingen.

N. Stephenson (1996). Mother Earth Mother Board. *Wired*, Vol. 12, ISSN 1059-1028.

R. Stöckle, Y. D. Suh, V. Deckert, & R. Zenobi (2000). Nanoscale chemical analysis by tip-enhanced Raman scattering. *Chem. Phys. Lett.* Vol. 318, pp. 131 - 136.

E. Udd (1990). *Fiber Optic Sensors*, Wiley Interscience, ISBN 0-471-83007-0.

B. Uhlig, J.-H. Zollondz, M. Haberjahn, H. Bloeß, & P. Kücher (2007). Nano-Raman: Monitoring nanoscale stress. *AIP Conf. Proc.*, Vol. 931 pp. 84-88.

M. T. Wenzel, T. Härtling, P. Olk, S. C. Kehr, S. Grafström, S. Winnerl, M. Helm, & L. M. Eng (2008). Gold nanoparticle tips for optical field confinement in infrared scattering near-field optical microscopy. *Opt. Express*, Vol. 16, pp. 12302-12312.

J. Wessel (1985). Surface-enhanced optical microscopy. *J. Opt. Soc. Am. B*, Vol .2 no. 9, pp. 1538-1541.

F. Zolla, G. Renversez, A. Nicolet, B. Kuhlmey, S. Guenneau, & D. Felbacq (2005). *Foundations Of Photonic Crystal Fibers*, Imperial College Press, ISBN 1-86094-507-4.